Primer to The Immune Response

2nd Edition

Tak W. Mak

Mary E. Saunders

Bradley D. Jett

Contributors:

Wendy L. Tamminen

Maya R. Chaddah

AMSTERDAM • BOSTON • HEIDELBERG • LONDON
NEW YORK • OXFORD • PARIS • SAN DIEGO
SAN FRANCISCO • SINGAPORE • SYDNEY • TOKYO
AP Cell is an imprint of Elsevier

ELSEVIER

AP Cell is an imprint of Elsevier
30 Corporate Drive, Suite 400, Burlington, MA 01803, USA
525 B Street, Suite 1900, San Diego, California 92101-4495, USA
84 Theobald's Road, London WC1X 8RR, UK

This book is printed on acid-free paper.

Library of Congress Cataloging-in-Publication Data
Mak, Tak W., 1945-
   Primer to the immune response / Tak W. Mak, Mary E. Saunders, Bradley D. Jett;
   contributors, Wendy L. Tamminen, Maya R. Chaddah.
   Includes index.
   ISBN-13: 978-0-12-374163-9 (alk. paper) 1. Immune response. 2. Immunology. I. Mak, Tak W.,
   II. Saunders, Mary E., Ph.D. III. Jett, Bradley D., Immune response. IV. Title.
   [DNLM: 1. Immune System. 2. Immune System Diseases. 3. Immunity. QW 504 M235p 2008]
   QR186.M35 2008
   616.07′9–dc22
2008001748

British Library Cataloguing-in-Publication Data
A catalogue record for this book is available from the British Library.

ISBN: 978-0-12-385245-8

For information on all AP Cell publications, visit
our Web site at www.books.elsevier.com

An Online Study Guide is now available with your textbook, containing case studies, and all of the journal articles.

---

1. To access the Online Study Guide, as well as other online resources for the book, please visit: http://booksite.academicpress.com/Mak/primerAC/

2. For instructor-only materials, please visit: http://textbooks.elsevier.com/web/product_details. aspx?isbn=9780123847430

Preface

In 2008, we published the first edition of *Primer to The Immune Response* (by Drs. Tak W. Mak and Mary E. Saunders). Our goal was to create a compact textbook that would serve as a useful resource for undergraduates in the life sciences or health science professions, or for anyone else who wished to gain a solid grounding in the basic concepts of immunology and its clinical connections. The *Primer* was designed to be a clear and succinct distillation of the immunological essentials that were provided at a more advanced level in our 2005 reference book entitled *The Immune Response*. In 2010, we partnered with Academic Cell to issue the Update Edition of the *Primer to The Immune Response*, which comprised the first edition of the textbook enhanced by an accompanying online study guide. This study guide, authored by Dr. Bradley Jett, featured cases studies in immunology and links to relevant research articles published by Cell Press. Now we are pleased to present a second edition of the *Primer to The Immune Response* textbook and its accompanying online study guide, both of which have been fully updated to include the many exciting advances in immunology over the past few years. Specifically, all chapters now take into account the growing appreciation of the fundamental function of innate immunity as the foundation of all immune responses. As a result, the vital role of chronic inflammation in initiating and perpetuating autoimmune and inflammatory disorders as well as transplant rejection, hypersensitivity and cancer is highlighted. In addition, the critical importance of the body's commensal organism populations to immune protection and maintenance of homeostasis is emphasized, as is the role of local tissue microenvironments in directing immune responses. Lastly, a new chapter is included that draws together current information on immunodeficiencies that are caused by either a genetic abnormality (primary immunodeficiencies) or HIV infection (acquired immunodeficiency syndrome; AIDS).

Our Contributors, educational consultant Wendy Tamminen and illustrator Maya Chaddah, have once again turned their outstanding talents and backgrounds in immunology toward making the *Primer* as useful as possible to readers needing a rapid, accurate and painless introduction to the immune system. We are truly grateful for the sound, logical pedagogy and crystal clear illustrations resulting from their efforts. During its evolution, the *Primer* has also benefitted greatly from the input of numerous experts on a vast array of immunological topics. These experts, many of whom consented to be listed on the Acknowledgements page, gave freely of their valuable time and perceptive insights to improve the quality and accuracy of both the text and the illustrations. Any remaining errors are solely the responsibility of the authors.

As in previous editions, the *Primer to The Immune Response, 2nd Edition* is divided into two major sections: Part I, "Basic Immunology," and Part II, "Clinical Immunology." In both sections, we have attempted to cover the relevant topics in an engaging way that is concise and clear but comprehensive. Part I (Chapters 1–12) describes the cellular and molecular elements of the immune system and immune responses, while Part II (Chapters 13–20) examines how these elements either combine to preserve good health or malfunction to cause disease. Parts I and II are followed by Appendices A–F, which present current information on topics ranging from historical milestones in immunology to comparative immunology to key techniques used in immunology labs. The textbook is completed by the inclusion of an updated and extensive Glossary that defines the key immunological terms shown in bold throughout the text.

With respect to specific textbook features, the most successful of the approaches used in the first edition have been maintained in the second edition, including the use of special topic *Boxes* that provide an extended discussion of a particular point of interest, and the *Take-Home Message* and *Did You Get It? A Self-Test Quiz* at the end of each chapter. Users of our first edition subsequently gave us feedback on additional features that would increase the utility of our book, and we have listened. New features include tips in the page margins that provide small but important pieces of information for the reader, such as a link to a useful website on the topic under discussion or a cross-reference to another relevant part of the textbook or a salient statistic. *Notes* are small boxes that are embedded in the main text between paragraphs and allow a short, crisp expansion of an associated point. As part of the Academic Cell series, our second edition also contains *Focus on Relevant Research* boxes that give the reader a taste of front-line experimental work and introduce the Cell Press journal article used to build the case study in the corresponding chapter of the online study guide. In addition to these text enhancements, the second edition of the *Primer* contains *Full Color Illustrations* that are not only fully updated with respect to content but also use color as a means of identifying cell lineages and their products. Complete *Figure Legends* are now provided for each figure and plate. Our *Tables*, which are helpful in summarizing important points on a topic, have also been updated. Instructors will appreciate our inclusion at the end of each chapter of a new feature entitled *Can You Extrapolate? Some Conceptual Questions*, the answers for which are supplied online only. Also as requested by our audience, we have provided a supplemental reading list for each chapter entitled *Would You Like to Read More?* As always, we welcome any input that will help to make future editions of this book even more useful for its intended audience.

Our hope is that the *Primer to The Immune Response, 2nd Edition* will propel students on a journey of immunological learning that is rewarding and exhilarating. We are confident that students who embark on this journey will be left in no doubt that the immune system is among the most vital and intriguing elements of the human body.

Tak W. Mak, Mary E. Saunders and Bradley D. Jett

The Academic Cell Approach

In attending conferences and speaking with professors across the biological sciences, the editors at Academic Press and Cell Press learned that journal articles were increasingly being incorporated into the undergraduate classroom experience. They were told of the concrete benefits students received from an early introduction to journal content: the ability to view lecture material in a broader context, the acquisition of improved analytical skills, and exposure to the most current and cutting-edge scientific developments in a given field. Instructors also shared their concerns with the editors about how much additional preparation time was required to find relevant articles, obtain images for classroom presentations, and distill the content of the articles into a form suitable for their students. The desire to provide a solution to these difficulties led to a collaborative effort resulting in the birth of the Academic Cell line of textbooks.

The objective of the Academic Cell initiative is to offer instructors and their students the benefits of a traditional textbook combined with access to an online study guide that highlights the use of primary research articles. The textbook serves as a reference for students and a lecture framework for instructors, and the online study guide is divided into chapters that align with those of the textbook. Each study guide chapter contains a brief summary of the textbook chapter material plus a case study based on a relevant research article chosen from a Cell Press journal. Questions are posed that challenge the student to use the textbook information to understand the research article and work through the case study. The textbook and study guide articles are further integrated by Focus on Relevant Research boxes that appear in the textbook. These passages introduce the Cell Press article used for the accompanying case study and provide context that encourages students to delve further into the article. Instructors will be pleased to note that images from the Cell Press articles have been made available in a PowerPoint format that instructors can use freely. Additional materials contained in the online study guide are the answers to the Conceptual Questions posed in the textbook as well as optional test bank questions and flash cards.

Immunology is a complex subject. Over my many years as a research scientist, I have enjoyed the challenge of sharing the mysteries of immunology with students of all ages and educational levels. For younger students, I have tried to explain difficult concepts using simple analogies, such as "The Adventures of Tommy the T Cell," who stands fast against marauders of disease.

For the reader of an immunology textbook—that is, for a true student of immunology—explanations necessarily have to become more sophisticated. In writing the first edition of the *Primer to the Immune Response*, Dr. Mary Saunders and I were extremely happy to have created a textbook that presented complex immunological concepts to the novice student in a clear, interesting and effective way. Reading of the first edition of our book firmly grounded a student in the basics of the subject, but there remained a desire to connect this information with immunology in the "real world." This meant going beyond the facts and figures to give the student a taste of how this body of knowledge was created in the first place, and how it continues to expand on a daily basis in research labs around the world. It also meant giving the student a look at how today's new ideas and observations can translate into medical treatments that improve the health and lives of people everywhere. To make this connection, we partnered with Dr. Bradley Jett to create the Academic Cell *Primer to The Immune Response*, an updated first edition of our book that included an online study guide. Now, in the Academic Cell *Primer to The Immune Response, 2nd Edition*, updated content and new pedagogical features in our textbook have been combined with access to an updated online study guide containing new research articles and case studies.

In each chapter of the online study guide, Dr. Jett has selected a journal article from the current scientific literature that complements the corresponding chapter of the *Primer to The Immune Response, 2nd Edition*. In each case, after offering a concise, easy-to-read summary of the chapter contents, he has provided thought-provoking questions and scenarios that relate to the journal article and challenge the student to think like a scientist or clinician. In this way, the study guide provides a bridge that encourages the reader to cross over from the realm of the "immunology student" into the realm of the "immunologist."

By using the online study guide as part of their immunology course work, students will grasp immunological concepts more fully. They will also face head-on the sometimes frustrating, sometimes invigorating fact that the field of immunology is never static. The science constantly moves forward, driven by the novel, exciting and sometimes controversial work of front-line researchers. It is into this fascinating world that the study guide immerses the budding immunologist.

It is my hope that students and instructors will take advantage of this excellent online study guide and use it as an enriching companion to *Primer to The Immune Response, 2nd Edition*. They will benefit, and I will feel I have taken another step in my evolution as an educator. And to think it all started with "Tommy the T cell."

Tak W. Mak
Toronto, Canada
January 2014

Tak Wah Mak

Tak W. Mak is the Director of the Campbell Family Institute for Breast Cancer Research in the Princess Margaret Hospital, Toronto, Canada, and a University Professor in the Departments of Medical Biophysics and Immunology, University of Toronto. He was trained at the University of Wisconsin in Madison, the University of Alberta, and the Ontario Cancer Institute. He gained worldwide prominence in 1984 as the leader of the team that first cloned the genes of the human T cell antigen receptor. His group went on to create a series of genetically altered mice that have proved critical to understanding intracellular programs governing the development and function of the immune system, and to dissecting signal transduction cascades in various cell survival and apoptotic pathways. His current research remains centered on mechanisms of immune recognition/regulation, malignant cell survival/death, inflammation in autoimmunity and cancer, and metabolic adaptation in tumor cells. Dr. Mak has published over 700 papers and holds many patents. He has been granted honorary doctoral degrees from universities in North America and Europe, is an Officer of the Orders of Canada and Ontario, and has been elected a Foreign Associate of the National Academy of Sciences (U.S.), a Fellow of the Royal Society of London (U.K.), and a Fellow of the AACR Academy. Dr. Mak has won international recognition as the recipient of the Emil von Behring Prize, the King Faisal International Prize for Medicine, the Gairdner Foundation International Award, the Sloan Prize of the General Motors Cancer Foundation, the Novartis Prize in Immunology, the Robert Noble Prize, the Killam Prize, the Stacie Prize, the McLaughlin Medal, and the Paul Ehrlich and Ludwig Darmstaedter Prize.

Mary Evelyn Saunders

Mary E. Saunders holds the position of Scientific Editor for the Campbell Family Institute for Breast Cancer Research, Toronto, Canada. She completed her B.Sc. degree in Genetics at the University of Guelph, Ontario, and received her Ph.D. in Medical Biophysics at the University of Toronto. Dr. Saunders works with Dr. Mak and members of his laboratory on the writing and editing of scientific papers for peer-reviewed journals as well as on various grant applications and book projects. She takes pride and pleasure in producing concise, clear, highly readable text and making complex scientific processes readily understandable.

Bradley Dale Jett

Bradley D. Jett is the James Hurley Professor of Biology at Oklahoma Baptist University (OBU) in Shawnee, Oklahoma, USA. He completed his B.S. degree in Biology from OBU, followed by his Ph.D. in Microbiology and Immunology from the University of Oklahoma College of Medicine. After a postdoctoral fellowship at Washington University in St. Louis, Missouri, he joined the faculty at the University of Oklahoma College of Medicine. His research interests are primarily focused on host–parasite relationships. Much of his published work relates to the virulence factors of Gram-positive bacteria such as *Enterococcus, Staphylococcus* and *Bacillus,* as well as the host immunological responses to these infections. In his current full-time, undergraduate teaching position at his alma mater, he has been awarded Oklahoma Baptist University's Promising Teacher Award and the Distinguished Teaching Award.

Maya Rani Chaddah

Maya R. Chaddah graduated with a B.Sc. in Human Biology and a B.A. in Spanish, followed by an M.Sc. in Immunology at the University of Toronto. In 1996, she started a business focused on the writing and editing of scientific and medical publications. Her expertise has grown to include scientific and medical illustration, and she continues to produce a variety of communications for diverse audiences in the public and private sectors. (www.mayachaddah.com)

Wendy Lynn Tamminen

Wendy L. Tamminen completed her B.Sc. degree in Chemistry and Biochemistry at McMaster University, Hamilton, and received her Ph.D. in Immunology from the University of Toronto. She has taught immunology at the undergraduate level to students in both the biomedical sciences and medicine at the University of Toronto, where her teaching skills have been recognized with an Arts and Science Under-graduate Teaching Award. In her role as writer, editor and lecturer, Dr. Tamminen's main interest is the communication of scientific concepts to both science specialists and non-specialists.

Acknowledgments

The authors are indebted to the following individuals for the reviewing of one or more chapters of one or more editions of this book.

Wiebke Bernhardt
European Patent Office, Biotechnology
Munich, Germany

Bruce Blazar
University of Minnesota
Minneapolis, Minnesota, USA

James R. Carlyle
Sunnybrook Research Institute
Toronto, Ontario, Canada

Radha Chaddah
Neurobiology Research Group, University of Toronto
Toronto, Ontario, Canada

Vijay K. Chaddah
Grey Bruce Regional Health Centre
Owen Sound, Ontario, Canada

Dominique Charron
Institut Universitaire d'Hématologie (IUH), Hôpital Saint Louis
Paris, France

Irvin Y. Chen
David Geffen School of Medicine at UCLA
Los Angeles, California, USA

Dale Godfrey
University of Melbourne
Parkville, Australia

Douglas R. Green
St. Jude Children's Research Hospital
Memphis, Tennessee, USA

Zhenyu Hao
The Campbell Family Institute for Breast Cancer Research
Toronto, Ontario, Canada

William Heath
The Walter and Eliza Hall Institute of Medical Research
Victoria, Australia

Jules Hoffmann
Institut de Biologie Moléculaire et Cellulaire, CNRS
Strasbourg, France

Kristin Ann Hogquist
Center for Immunology, University of Minnesota
Minneapolis, Minnesota, USA

Robert D. Inman
Toronto Western Hospital
Toronto, Ontario, Canada

Robert Lechler
King's College London
London, England, UK

Eddy Liew
University of Glasgow
Glasgow, Scotland, UK

Bernard Malissen
Centre d'Immunologie de Marseille-Luminy
Marseille, France

Ruslan Medzhitov
Howard Hughes Medical Institute, Yale University School of Medicine
New Haven, Connecticut, USA

Mark Minden
Princess Margaret Hospital
Toronto, Ontario, Canada

Thierry Molina
Université Paris-Descartes, Hôtel Dieu
Paris, France

David Nemazee
Scripps Research Institute
La Jolla, California, USA

Pamela Ohashi
The Campbell Family Institute for Breast Cancer Research
Toronto, Ontario, Canada

Marc Pellegrini
The Campbell Family Institute for Breast Cancer Research
Toronto, Ontario, Canada

Noel R. Rose
Johns Hopkins Center for Autoimmune Disease Research
Baltimore, Maryland, USA

Lawrence E. Samelson
Center for Cancer Research, National Cancer Institute
Bethesda, Maryland, USA

Daniel N. Sauder
Robert Wood Johnson Medical School
New Brunswick, New Jersey, USA

Warren Strober
NIAID, National Institutes of Health
Bethesda, Maryland, USA

John Trowsdale
Cambridge Institute for Medical Research
Cambridge, England, UK

Ellen Vitetta
Cancer Immunobiology Center
University of Texas Southwestern Medical Center
Dallas, Texas, USA

Peter A. Ward
University of Michigan Health Systems
Ann Arbor, Michigan, USA

Tania Watts
University of Toronto
Toronto, Ontario, Canada

Hans Wigzell
Microbiology and Tumor Biology Centre, Karolinska Institute
Stockholm, Sweden

David Williams
University of Toronto
Toronto, Ontario, Canada

Gillian E. Wu
York University
Toronto, Ontario, Canada

Juan-Carlos Zúñiga-Pflücker
Sunnybrook Research Institute
Toronto, Ontario, Canada

Contents

CHAPTER 4 The B Cell Receptor: Proteins and Genes 85

CHAPTER 5 B Cell Development, Activation and Effector Functions 111

CHAPTER 16 *Tumor Immunology* 423

CHAPTER 17 *Transplantation* 457

Contents

CHAPTER 20 *Hematopoietic Cancers* **553**

PART I: BASIC IMMUNOLOGY

Introduction to the Immune Response

Books must follow sciences, and not sciences books.

Francis Bacon

A. Historical Orientation

What is immunology? Simply put, **immunology** is the study of the immune system. The **immune system** is a system of cells, tissues and their soluble products that recognizes, attacks and destroys entities that could endanger the health of an individual. The normal functioning of the immune system gives rise to **immunity**, a word derived from the Latin *immunitas,* meaning "to be exempt from." This concept originated in the 1500s, before the causes of disease were understood. Survivors who resisted death during a first exposure to a devastating disease and who did not get sick in a subsequent exposure were said to have become "exempt from" the disease, or "immune."

In 1796, the English physician Edward Jenner carried out experiments that solidified the birth of immunology as an independent science. At that time, smallpox was a disfiguring and often fatal disorder that decimated whole villages (**Plate 1-1**). Jenner observed that dairymaids and farmers lacked the pock-marked complexions of their fellow citizens, and wondered whether those who worked with cattle might be resistant to smallpox because of their close contact with livestock. Cows of that era often suffered from cowpox disease, a disorder similar to smallpox but much less severe. In an experiment that would be prohibited on ethical grounds today, Jenner deliberately exposed an 8-year-old boy to fluid from a cowpox lesion (**Plate 1-2**). Two months later, he inoculated the same boy with infectious material from a smallpox patient. In this first example of successful vaccination, the boy did not develop smallpox. Jenner's approach to smallpox prevention was quickly adopted in countries throughout Europe. The modern story of smallpox vaccination and the global eradication of this scourge is one of the most successful public health endeavors in history.

Jenner's work was the first controlled demonstration of the **immune response**, but little was understood at that time about the cellular and molecular mechanisms underlying the observed immunity. In Jenner's day, the cause of infectious disease was still a mystery, and theories of that time did not envision the transmission of disease-causing germs. In 1884, Robert Koch proposed the "germ theory of disease," which

In 1967, the World Health Organization (WHO) launched a vaccination campaign to eradicate smallpox from the planet, and in 1980 they announced their success. More details about the global eradication of smallpox can be found in Chapter 14.

MAK: Primer to the Immune Response. http://dx.doi.org/10.1016/B978-0-12-385245-8.00001-7

Plate 1-1
Smallpox

Smallpox infection in a young Bangladeshi girl in 1973. Two hundred years earlier, in Edward Jenner's day, this disease was a widespread scourge. *[Reproduced by permission of the Public Health Image Library, CDC]*

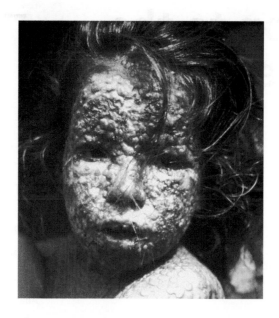

Plate 1-2
The First Smallpox Immunization

A painting by E. Board depicts Edward Jenner vaccinating the young James Phipps against smallpox. *[Reproduced by permission of Wellcome Library, London]*

Selected landmark discoveries in immunology and immunologists who have won Nobel Prizes for their work are featured in Appendices A and B, respectively.

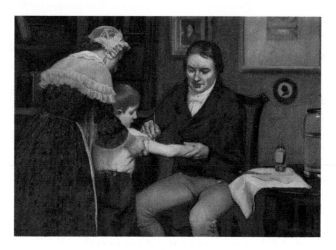

stated that microbes invisible to the naked eye were responsible for specific illnesses, and the first human disease-causing organisms or **pathogens** were identified in the late 1800s. At about the same time, Louis Pasteur applied Jenner's immunization technique for smallpox to the prevention of various animal diseases. Pasteur demonstrated that inoculation with a pathogen that had been weakened in the laboratory could protect against a subsequent exposure to the naturally occurring pathogen. It was Pasteur who coined the term **vaccination** (from the Latin *vaccinus*, meaning "derived from cows") for this procedure, in honor of Jenner's work. The research of Pasteur and other investigators spurred the evolution of immunology as a science distinct from (but related to) the established fields of microbiology, pathology, biochemistry and histology. Today, **immunology** can be defined as *the study of the cells and tissues that mediate immunity and the investigation of the genes and proteins underlying their function.*

B. The Nature of the Immune Response

A normal healthy person's body always strives to maintain **homeostasis**, a natural state of balance of all its organs and the nervous and circulatory systems. When this homeostasis is disturbed by either trauma, pathogens, or the deregulation of body cells (such occurs in cancer), the immune system responds in an attempt to restore balance. At its simplest level, this response involves identifying and clearing damaged and dying cells from the body. More complex responses have evolved to counteract assault on

PATHOGEN ATTACHMENT AND ENTRY INTO HOST

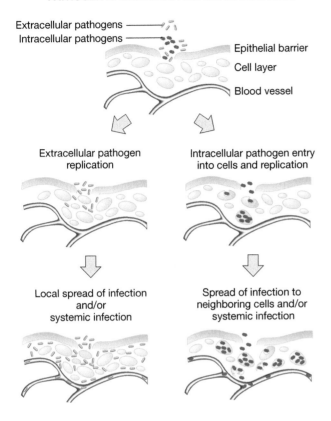

Fig. 1-1
Establishment of Infection by Extracellular and Intracellular Pathogens

Pathogens that are not blocked by epithelial barriers can gain access to underlying cell layers. Extracellular pathogens replicate outside host cells, whereas intracellular pathogens must enter host cells to replicate. In both cases, if infection is not controlled, the pathogen may spread locally in the tissues or enter the bloodstream and spread systemically.

the body by infectious agents, including bacteria, viruses, parasites and fungi. Because these foreign invaders are literally everywhere on Earth and constantly seeking vulnerable hosts, the immune system is constantly occupied with containing attacks from this quarter. Indeed, despite the successful eradication of smallpox and the development of antibiotics, infectious diseases remain among the biggest killers on the planet. Longstanding disorders such as malaria and tuberculosis, as well as "emerging" diseases caused by new pathogens such as West Nile virus, severe acute respiratory syndrome (SARS) virus, and H1N1 influenza virus, present ongoing challenges to the immune system.

Humans are surrounded by other organisms—in the air, in the soil, in the water, on the skin, and on the **mucosae**, the protective layers of epithelial cells that line the gastrointestinal, urogenital and respiratory tracts. While most of these organisms are harmless and some are even beneficial, some are pathogenic. Like all species, pathogens live to reproduce. In order to reproduce, however, many must penetrate a host's body or one of its component cells. **Infection** is defined as the attachment and entry of a pathogen into the host. Once inside the body or cell, the pathogen replicates, generating progeny that spread into the body in a localized or systemic fashion. The manner of this replication determines whether the pathogen is considered extracellular or intracellular.

Extracellular pathogens, such as certain bacteria and parasites, do not need to enter cells to reproduce. After accessing the body, these organisms replicate first in the interstitial fluid bathing the tissues and may then disseminate via the blood (**Fig. 1-1**). **Intracellular pathogens**, such as viruses and other bacteria and parasites, enter a host cell, subvert its metabolic machinery, and cause it to churn out new virus particles, bacteria, or parasites. These pathogens may then also travel systemically by entering the blood. For both intracellular and extracellular pathogens, if the infectious agent overwhelms normal body systems or interferes with cellular functions, the body becomes "sick." This sickness is manifested as a set of characteristic clinical symptoms that we call an "illness" or "disease."

Our bodies are under constant assault by harmful microbes, yet, most of the time, assaults by these organisms are successfully repelled and disease is prevented. This resistance is due to both basic and sophisticated immune responses that combat pathogens. Simply put, the job of the immune response is to "clean up" infections in the interstitial fluid, tissues and blood, and to destroy infected host cells so that neighboring host cells do not share their fate. Because pathogens are constantly evolving mechanisms to evade or block immune defenses, the immune system must constantly adapt to maintain its effectiveness. It is a continual horse race as to which will be the more successful mechanism: the body's immune surveillance or the pathogen's invasion and infection strategy. We note here that the immune response itself may cause limited collateral damage to tissues as part of its larger battle against pathogens, but such **immunopathic** effects are usually short-lived in an otherwise healthy individual.

At the turn of the twentieth century, there were two schools of thought on what mechanisms underlay immune responses. One group of scientists believed that immunity depended primarily on the actions of cells that destroyed or removed unwanted material from the body. This clearance process was referred to as **cell-mediated immunity**. However, another group of researchers was convinced that soluble molecules in the serum of the blood could directly eliminate foreign entities without the need for cellular involvement. In this case, the clearance process was referred to as **humoral immunity**, a term derived from the historical description of body fluids as "humors." Today, we know that both cell-mediated and humoral responses occur simultaneously during an immune response and that both are often required for complete clearance of a threat.

The cells responsible for cell-mediated immunity are collectively called **leukocytes** ("leuko," white; "cyte," small body, i.e., a cell) or white blood cells. However, "blood cell" is a bit of a misnomer, because a majority of leukocytes reside in tissues and specialized organs, and move around the body through both the blood circulation and a system of interconnected vessels called the **lymphatic system**. The soluble molecules responsible for humoral immunity are proteins called **antibodies**, and antibodies are secreted by a particular type of leukocyte. The production of these antibodies and the mounting of cell-mediated immune responses depend on an elaborate signaling system by which leukocytes communicate with each other as well as with other cell types in the body. This signaling is mediated by small secreted proteins called **cytokines**, which are mainly produced by leukocytes.

C. Types of Immune Responses: Innate and Adaptive

The mammalian immune system can mount two types of responses: the **innate response** and the **adaptive response**. Recently, it has become increasingly clear that these two types of responses are less distinct and more interconnected than once thought. Indeed, researchers now believe that vertebrates are capable of a continuum of immune responses that bring more and more precise weapons to bear on a threat as required. More specifically, innate immunity is involved in all levels of immune response, whereas the adaptive response is mounted only when innate mechanisms signal that there is a relatively serious infection. In all cases, the objective is to clear the body of unwanted entities and re-establish homeostasis in the most efficient way possible.

Innate immune responses are triggered by disruptions to homeostasis caused by either non-infectious or infectious means. With respect to the former, innate responses help to repair tissues injured by trauma (such as that caused by a cut, a blow, or surgery) and remove dead cells and cells too damaged or old (senescent) to be of further use to the body. Products synthesized by leukocytes mediating innate responses (innate leukocytes) also help to prevent "bystander injury" of healthy tissues. With respect to invading pathogens, elements of the innate immune system are the first to confront the threat and work to eliminate it from the body. Such elements include pre-existing anatomical and physiological barriers that attempt to block pathogen entry, and other responses that are induced after the pathogen has gained entry. Only if innate responses are insufficient to control the situation and restore homeostasis, as is often the case with infection by a rapidly replicating pathogen, is an adaptive response mounted. Together, the innate and adaptive immune responses allow a seamless escalation of countermeasures that maintain homeostasis in the face of cellular aging, tissue trauma and/or pathogen infection.

What exactly triggers the initial response by the innate system? In a word: recognition. Wherever cells are damaged or dying—either in the presence or absence of infection—particular host macromolecules are released into the extracellular milieu or become accessible on the surface of damaged cells or in cellular debris. These molecules are called **damage-associated molecular patterns (DAMPs)**. When a pathogen attacks, it furnishes common molecular structures on its own surface, or on the surface of cells it has infected, or as part of products it synthesizes. These structures are called **pathogen-associated molecular patterns (PAMPs)**. Cell damage or death in the absence of a pathogen gives rise to DAMPs only, but both DAMPs and PAMPs will be present when a pathogen invades (**Fig. 1-2**). It is the recognition of DAMPs and PAMPs by innate leukocytes that initiates innate responses.

DAMPs and PAMPs are recognized by host proteins called **pattern recognition molecules (PRMs)**, most of which are expressed by innate leukocytes. There is not a large array of different PRMs, meaning that there is a limited repertoire of molecular patterns that can be recognized. However, each PRM recognizes a DAMP or PAMP that is shared by many different damaged cells or pathogens. Thus, PRMs give the innate immune response the property of *broad recognition.* Furthermore, because innate leukocytes expressing PRMs are present in large numbers, these cells don't need to multiply to work effectively: the response is *immediate.* Finally, because innate leukocytes are activated but don't multiply in response to engagement of their PRMs, the speed and strength of their response is exactly the same upon a second exposure to the same pathogen or type of damage. The result is a response that is said to have *no memory.*

Recognition is also the key to initiating the adaptive immune response, but there are important differences. In contrast to the broad recognition of the PRMs mediating the innate response, the receptors expressed by the cells participating in the adaptive response (adaptive leukocytes) are equipped with receptors that recognize unique molecular structures on pathogens. These unique structures are called **antigens**, and the receptors that recognize them are **antigen receptors**. Together as a population, adaptive leukocytes express a vast array of antigen receptors capable of recognizing almost any structure, meaning that they represent a diverse **repertoire** of antigen specificities. However, the antigen receptor of any one adaptive leukocyte binds to an antigen exclusive to a single type of pathogen. Thus, antigen receptors have the property of *specific recognition.* Because adaptive leukocytes are very few in number, these cells must proliferate and expand their ranks after receptor engagement before they can be effective: the response is *delayed.* In addition, after adaptive leukocytes proliferate and act to eliminate the pathogen, some of these cells become long-lived so that the speed and strength of the adaptive response is increased upon a subsequent exposure to the same pathogen. The adaptive response is thus said to have *memory.* The distinguishing features of the innate and adaptive responses are summarized in **Table 1-1**.

Antigens got their names because they were first identified as pathogen components that could bind antibodies; i.e., they induced "<u>anti</u>body <u>gen</u>eration." "Antigen" now refers to structures targeted by humoral or cell-mediated responses.

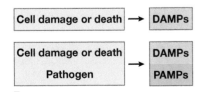

Fig. 1-2
DAMPs vs. PAMPs

A DAMP is a common molecular structure that is produced by host cells when they are damaged or dying (with or without pathogen infection). A PAMP is a common molecular structure shared by a wide variety of pathogens or their components or products. In the case of pathogen attack, both DAMPs and PAMPs will be present.

TABLE 1-1	Comparison of Innate and Adaptive Immune Responses	
Innate	**Adaptive**	
Pre-existing barriers and induced mechanisms	All mechanisms induced	
Innate leukocytes express a small number of pattern recognition molecules (PRMs) of limited diversity that collectively recognize a wide range of threats	Adaptive leukocytes express an almost infinite number of extremely diverse antigen receptors, each of which is specific for a particular pathogen	
Response is triggered by PRM binding to widely shared damage-associated molecular patterns (DAMPs) or pathogen-associated molecular patterns (PAMPs)	Response is triggered by antigen receptor binding to a unique antigen derived from a specific pathogen	
Number of leukocytes responding to a given threat is large, so no proliferation is required and response is fast	Number of leukocytes responding to a given threat is very small, so proliferation is required and response is slower	
Responding cells do not proliferate and are short-lived, so level of defense is similar upon repeated exposure to the same pathogen	Responding cells proliferate and are long-lived, so level of defense is stronger and faster with repeated exposure to the same pathogen	

From the point of view of the body's leukocytes, a complex pathogen represents a collection of many different PAMPs, which evoke an innate response, and antigens, which may evoke an adaptive response if the innate response is not sufficient to eliminate the threat (**Fig. 1-3**). Anatomically, an innate response can be triggered at any point in the body because the majority of innate leukocytes are positioned where microbes most commonly attempt to gain access: just below the skin surface or at the mucosae protecting body tracts such as the gut and respiratory system. Most of the time, the innate response successfully eliminates an invader. Only those entities that succeed in overwhelming the innate defenses trigger adaptive responses that are initiated by adaptive leukocytes positioned in specialized anatomical regions called the **lymphoid tissues**, which include the spleen, tonsils and lymph nodes (see Ch. 2). The general characteristics of innate and adaptive immune responses primarily as they relate to pathogens are described in more detail in the next two sections.

> NOTE: Elements of an innate immune response can be found in all multicellular organisms, whereas the mechanisms of the more recently evolved adaptive immune response are present only in the higher vertebrates (fish and above). Appendix C contains a Table of Comparative Immunology that summarizes the immune system elements present in an evolutionarily diverse range of animals.

I. General Features of Innate Immunity

In this section, we present a general introduction to the central elements of innate immunity, namely *barrier defenses, complement activation, pattern recognition, inflammation, phagocytosis* and *target cell lysis*. More detailed information about the innate immune response can be found in Chapter 3 on innate immunity.

Fig. 1-3
PAMPs vs. Antigens

A given PAMP is expressed by a wide variety of pathogens and is recognized by PRMs whose engagement activates innate leukocytes. An antigen is a molecular structure that is generally unique to a specific pathogen and is recognized by a specific antigen receptor whose engagement activates a particular adaptive leukocyte.

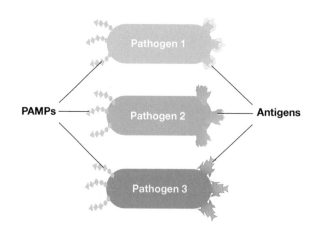

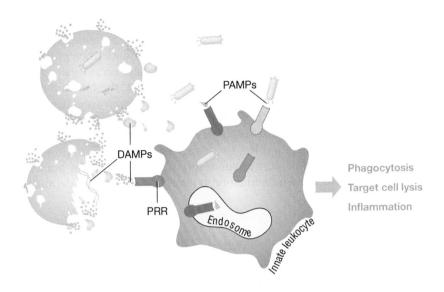

▶ **Fig. 1-4**
PRR Engagement and Effector Functions of Innate Leukocytes

When PAMPs on a pathogen bind to the PRRs of an innate leukocyte, the cell becomes activated and attempts to clear the pathogen through phagocytosis, target cell lysis, and/or the induction of inflammation. The PRR may be expressed on the leukocyte surface, on its endosomal membranes or in its cytoplasm.

i) Barrier Defense

The first level of innate defense is mediated by a pre-existing collection of physical, chemical and molecular barriers that exclude foreign material in a way that is totally non-specific and requires no induction. These elements include **anatomical barriers** and **physiological barriers**. An example of an anatomical barrier is intact skin, while the low pH of stomach acid and the hydrolytic enzymes in body secretions are examples of physiological barriers.

ii) Complement Activation

Should barrier defense prove insufficient, other forms of innate defense are induced. A key player in the induced innate response is **complement**, a complex system of enzymes that circulates in the blood in an inactive state. Once activated, the complement system contributes directly and indirectly to innate and adaptive immune defenses, as described in detail in Chapter 3.

iii) Pattern Recognition

When invaders breach anatomical and physiological barriers, innate leukocytes start to take action as a result of pattern recognition mediated by the binding of PRMs to PAMPs furnished by pathogens and to DAMPs emanating from damaged host cells. As introduced previously, and as illustrated in **Figure 1-4**, PRMs come in several different forms, some of which are membrane-bound and others of which are soluble. PRMs that are expressed by innate leukocytes are called **pattern recognition receptors (PRRs)**. PRRs are located in the plasma membrane of an innate leukocyte, or are soluble molecules free in the leukocyte's cytoplasm, or are fixed in the membranes of intracellular vesicles called **endosomes**. Other PRMs are made by non-leukocytes and are present as soluble molecules free in the extracellular milieu. When these latter PRMs bind to a given PAMP or DAMP, they must then bind to another receptor fixed in a leukocyte membrane in order to trigger the appropriate clearance response. The clearance mechanisms typically induced following PRM or PRR engagement by a DAMP or PAMP are *inflammation, phagocytosis,* and/or *target cell lysis*. A brief description of each of these processes follows, with more detail provided in Chapter 3.

iv) Inflammation

When leukocyte PRRs are engaged by their ligands, intracellular signal transduction leads to the induction of new gene transcription and the synthesis of various "pro-inflammatory" cytokines. These cytokines in turn drive events responsible for the influx of first innate and later (if necessary) adaptive leukocytes into the site of injury or infection. This influx is part of a process called **inflammation** or an **inflammatory response**, and the redness and swelling we commonly associate with inflammation are its outward physical signs. This inflammation is normal and helpful, and, when properly regulated, promotes a localized gathering of the cells and molecules necessary to repair tissue damage and clear pathogens. Once the threat is eliminated, the inflammation resolves naturally with time. However, if the inflammation fails to resolve and becomes chronic, it may become pathological. The powerful molecules secreted by the innate leukocytes participating in the inflammatory response can eventually cause tissue damage and impair immune system function if left to operate unchecked. Thus, properly controlled inflammation is part of a healthy innate response and essential for homeostasis, whereas excessive or prolonged inflammation is immunopathic and undermines homeostasis.

v) Phagocytosis

Innate leukocytes frequently use a sophisticated means of engulfing entities that is called **phagocytosis** ("eating of cells"). Phagocytosis is carried out primarily by three types of PRR-expressing cells: **neutrophils**, **macrophages** and **dendritic cells** (DCs); these cell types are consequently known as **phagocytes**. Neutrophils mainly engulf and destroy pathogens, while macrophages and DCs engulf not only pathogens but also dead host cells, cellular debris and host macromolecules. The phagocytosis of foreign entities by macrophages and DCs allows these cells to "present" unique antigens from this material on their cell surfaces in such a way that the antigens can be recognized by adaptive leukocytes that have been drawn to the site of inflammation. These cells are then activated and mediate antigen-specific protection.

vi) Target Cell Lysis

Cancer cells and cells infected with intracellular pathogens frequently express certain DAMPs and/or PAMPs on their cell surfaces that mark them as "target cells" for destruction by cells of the innate system. When these molecules are recognized and bound by PRRs of innate leukocytes, complex processes are initiated that result in the lysis of the cancer cell or infected cell, preventing it from causing further harm to the body. These types of innate responses are carried out primarily by neutrophils, macrophages and another type of innate leukocyte called a **natural killer cell (NK cell)**.

II. General Features of Adaptive Immunity

In this section, we introduce the main features of adaptive immunity. The adaptive leukocytes referred to in previous sections are called *lymphocytes*. Lymphocytes are categorized as either **B lymphocytes** (B cells) or **T lymphocytes** (T cells). The fundamental functional and developmental characteristics of lymphocytes are responsible for the *specificity, division of labor, memory, diversity* and *tolerance* of the adaptive response. Each of these important concepts is introduced here as a prelude to more detailed discussions of these topics in later chapters.

i) Specificity

The specificity of an adaptive immune response for a particular antigen is determined by the nature of the antigen receptors expressed on the surfaces of T and B lymphocytes. Each lymphocyte expresses thousands of identical copies of a unique antigen receptor protein. Interaction of these antigen receptor molecules with antigen triggers activation of the lymphocyte. In contrast to the broad recognition mediated by PRMs, antigen binding to a lymphocyte antigen receptor usually hinges on the presence of a unique molecular shape unlikely to appear on more than one pathogen.

The antigen receptors on the surface of a B cell are called **B cell receptors** (BCRs), whereas those on the T cell surface are called **T cell receptors** (TCRs). In both cases, the antigen receptor is itself a complex of several proteins. Some of these proteins interact directly and specifically with antigen, while others convey the intracellular

TABLE 1-2	Division of Labor in Adaptive Immune Responses		
Lymphocyte	**Antigen Recognized**	**Pathogens Combatted**	**Mechanism Used by Effector Cells**
B cell	Whole soluble antigen or whole antigen present on a pathogen surface	Extracellular pathogens	Antibody-mediated clearance of pathogen
Tc cell	Antigen peptide + MHC class I	Primarily intracellular pathogens	Lysis of infected host cells to prevent pathogen spread
Th cell	Antigen peptide + MHC class II	Many extracellular and intracellular pathogens	Secretion of cytokines required for activation of B cells and Tc cells, and stimulation of innate leukocytes

signals triggered by antigen binding into the cytoplasm and nucleus of the lymphocyte. These signals are critical for sustaining the activation of the lymphocyte and promoting its proliferation into an army of short-lived daughter **effector cells**. These effector cells then undertake the effector action appropriate for eliminating the pathogen furnishing the antigen.

ii) Division of Labor
"Division of labor" in the adaptive response refers to the different but equally important contributions of B lymphocytes and two types of T lymphocytes. These two major T cell subsets are called **cytotoxic T cells** (Tc) and **helper T cells** (Th). The roles of B, Tc and Th cells in the adaptive response are defined by the distinct mechanisms by which they recognize and respond to antigen (**Table 1-2**).

B cells recognize antigen in a way that is fundamentally different from that used by Tc and Th lymphocytes. The BCR of a B cell binds directly to an intact antigen, which may be either a soluble molecule or a molecule present on the surface of a pathogen (**Fig. 1-5A**). Activation of a B cell in this way causes it to proliferate and produce

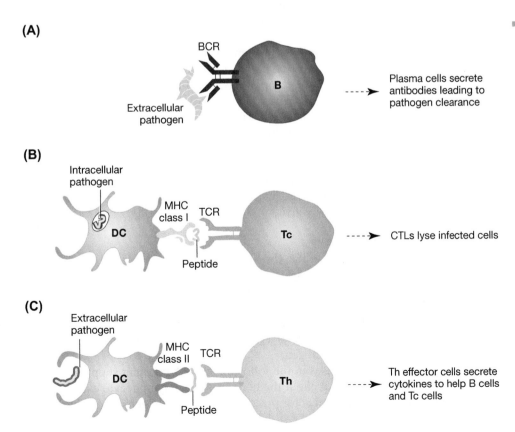

(A)

(B)

(C)

▪► **Fig. 1-5**

Antigen Recognition and Effector Functions of B, Tc and Th Cells

(A) The BCR of a B cell binds directly to an antigen present on an extracellular pathogen, resulting in production of plasma cells (PCs). Antibody secreted by PCs leads to clearance of the pathogen. **(B)** The TCR on a Tc cell recognizes a pMHC complex composed of an antigenic peptide that is derived from an intracellular pathogen and presented on the MHC class I molecule of a DC. The Tc cell becomes activated and produces CTLs that can lyse cells infected with the same pathogen. **(C)** The TCR on a Th cell recognizes a pMHC complex composed of an antigenic peptide that is derived from an extracellular pathogen and presented on the MHC class II molecule of a DC. Th effector cells produced by the activated Th cell secrete cytokines that help activate B and Tc cells.

Antibody production by B cells is described in detail in Chapters 4 and 5.

"MHC" stands for "major histocompatibility complex," a region on human chromosome 6 that contains genes encoding several types of molecules vital for the adaptive immune response. MHC genes and proteins are described in detail in Chapter 6.

identical daughter effector cells called **plasma cells** that secrete vast quantities of specific antibody. An **antibody** is a protein that is a modified, soluble form of the original B cell's membrane-bound antigen receptor. Antibody molecules enter the host's circulation and tissues, bind to the specific antigen, and mark it for clearance from the body, establishing a humoral response. Since antibodies are present in the extracellular milieu, such a response is effective against extracellular pathogens. However, antibodies are unable to penetrate cell membranes and so cannot attack an intracellular pathogen once it has entered a host cell.

Unlike BCRs, TCRs are unable to recognize whole, native antigens. Rather, a TCR recognizes a bipartite structure displayed on the surface of a host cell. This bipartite structure is made up of a host surface protein called an **MHC molecule** bound to a short peptide derived from a protein antigen. This complex is referred to as a **peptide–MHC complex (pMHC)**. Tc and Th cells are activated by the binding of their TCRs to specific pMHCs presented on the surface of a DC. DCs can acquire extracellular pathogen antigens by phagocytosis, and intracellular pathogen antigens either by infection or by phagocytosis of the debris of dead infected cells. Peptides from these antigens are bound to one of two types of MHC molecules, either **MHC class I** or **MHC class II**, and then displayed on the DC surface for inspection by T cells.

Tc cells recognize antigenic peptides that the DC has presented on MHC class I molecules (**Fig. 1-5B**). The activated Tc cell proliferates and differentiates into identical daughter effector cells called, for historical reasons, CTLs (cytotoxic T lymphocytes). CTLs can lyse any host cell displaying the same peptide–MHC class I complex recognized by the original Tc cell. MHC class I molecules are expressed on almost all body cells, making them all potential targets for CTL-mediated lysis. However, unlike DCs, most host cells cannot phagocytose extracellular entities. It is therefore only after infection with an intracellular pathogen that a host cell displays an antigenic pMHC complex that can be recognized by the TCR of first the Tc cell and then its daughter CTLs. The cell-mediated immunity directed against such infected host cells is critical to the clearance of intracellular pathogens.

Th cells recognize antigenic peptides that a DC has presented on MHC class II molecules (**Fig. 1-5C**). The activated Th cell proliferates and differentiates into identical Th effector cells that can recognize the same pMHC structure that activated the original Th cell. However, MHC class II molecules are expressed only by certain cell types that can act as specialized **antigen-presenting cells (APCs)**, including DCs, macrophages and activated B cells. In the course of a pathogen attack, APCs can capture the pathogen and/or its products or components, and then display peptides derived from its antigens on MHC class II. When the TCR of a Th effector cell recognizes this pMHC structure, the cell is stimulated to produce cytokines that assist B cells and Tc cells to become fully activated. In this way, Th cells contribute to both humoral and cell-mediated adaptive responses, and are thus crucial for protection against many extracellular and intracellular pathogens. In addition, cytokines secreted by Th cells stimulate innate leukocytes, thereby reinforcing this arm of immunity. The combined efforts of Tc, Th and B lymphocytes and many types of innate leukocytes may be necessary to eliminate a particularly wily invader.

iii) Immunological Memory

The innate response deals with a given threat in a very similar way each time it enters the body. In contrast, the adaptive system can "remember" that it has seen a particular antigen before. This **immunological memory** means that an enhanced adaptive response is mounted upon a second and subsequent exposures to a given pathogen, so that signs of clinical illness are mitigated or prevented. In other words, immunity is achieved because the body has effectively "adapted" its defenses and acquired the ability to exclude a particular pathogen.

Immunological memory arises in the following way. A constant supply of resting B and T cells is maintained throughout the body, with each lymphocyte expressing its complement of unique antigen receptor proteins. When a pathogen antigen enters the

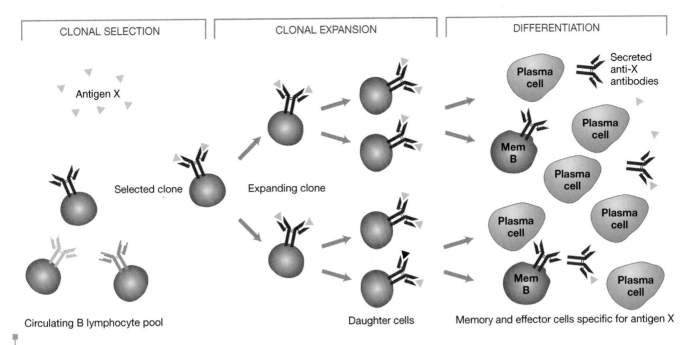

Fig. 1-6
Clonal Selection and Generation of Memory and Effector Lymphocytes

When antigen X enters the body, only those lymphocytes with receptors that specifically recognize X will be selected to proliferate and generate daughter cells that differentiate into effector cells and memory cells. This figure illustrates this principle for B cells, where antigen X triggers the generation of plasma cell effectors and memory B cells (Mem B) specific for antigen X.

body for the first time, a process of **clonal selection** takes place in which only those lymphocytes bearing receptors specific for that antigen are triggered to respond. The selected cells leave the resting state and multiply to generate daughter cells all expressing the same antigen-specific BCR or TCR. The differentiation of these daughter cells gives rise to the short-lived effector cells required to eliminate the pathogen, as well as long-lived **memory cells** that persist in the tissues essentially in a resting state until a subsequent exposure to the same pathogen (**Fig. 1-6**). The attack on the pathogen by this first round of effector cells is called the **primary immune response**. The second (or subsequent) time that the particular pathogen enters the body, it is met by an expanded army of clonally selected, antigen-specific memory cells that undergo much more rapid differentiation into effector cells than occurred during the first antigen encounter. The result is a stronger and faster **secondary immune response** that eliminates the pathogen before it can cause illness. New populations of memory cells are also produced during the secondary response, ensuring that the host maintains long-term or even lifelong immunity to that pathogen.

Clonal selection is explained in detail in Chapter 5 for B cells and in Chapter 9 for T cells.

NOTE: The generation of protective memory lymphocytes is the basis of vaccination. A healthy person is immunized with a vaccine containing pathogen antigens in order to provoke the production of memory cells that will prevent disease from developing if the individual is ever naturally infected by that pathogen. Vaccination is described in detail in Chapter 14.

iv) Diversity

The degree of diversity of antigens recognized also distinguishes the adaptive response from the innate response. Whereas the innate immune system exhibits a genetically fixed and finite capacity for antigen recognition, the recognition capacity of the adaptive immune system is nearly limitless. In fact, our bodies are capable of recognizing totally synthetic antigens that do not occur in nature. This huge diversity arises from the combined actions of several genetic mechanisms that affect the genes encoding the antigen receptors. Some of these mechanisms operate before a lymphocyte encounters antigen, and some after.

The primary source of antigen receptor diversity is a gene rearrangement process that occurs during the development of B cells and T cells *prior* to encounter with antigen. The genes encoding antigen receptors are not individual continuous entities. Rather, the BCR and TCR genes are assembled from a large collection of pre-existing gene segments by a mechanism called **somatic recombination**. A single, random combination of gene segments is thus created in each developing lymphocyte. As a result, a lymphocyte population is generated in which the antigen receptor proteins are vastly diverse in specificity because they are encoded by hundreds of thousands of different DNA sequences. In the case of B cells, additional mutational mechanisms operate that result in further structural diversification of antibody proteins. The sheer numbers of B and T lymphocyte clones generated in a healthy individual guarantee that there will be at least one clone expressing a unique receptor sequence for every antigen encountered during the host's life span. As alluded to previously, these clones, with their array of antigen receptor specificities, are collectively called the individual's *lymphocyte repertoire*.

v) Tolerance

The generation of lymphocyte clones that can theoretically recognize any antigen in the universe raises the question of how the body avoids lymphocyte attacks on molecules present in its tissues. This avoidance is called **tolerance**, the fifth aspect in which the adaptive response exhibits more refinement than the innate response. As stated previously, the specificity of each antigen receptor is randomly determined by somatic recombination during early lymphocyte development. By chance, some of the genetic sequences produced will encode receptors that recognize self molecules (*self antigens*). These lymphocytes must be identified as recognizing self and then either removed from the body entirely, or at least inactivated, to ensure that an individual has an effective lymphocyte repertoire that does not attack healthy tissue. Tolerance is established in two broad stages, each of which involves multiple mechanisms. The first stage, called **central tolerance**, occurs during early lymphocyte development. The mechanisms of central tolerance are designed to eliminate clones that recognize self antigens, thus establishing a lymphocyte repertoire that targets "non-self." In the second stage, called **peripheral tolerance**, any B and T lymphocytes that recognize self but somehow escaped the screening of central tolerance and completed their development are functionally silenced by another set of inactivating mechanisms.

D. Interplay between the Innate and Adaptive Responses

Although we described the innate and adaptive responses separately in the preceding sections, in reality, they occur in a continuum that brings stronger and stronger weaponry to bear when (and only when) it is needed. Our bodies normally deal with a pathogen attack in three broad and overlapping phases, two of which involve innate responses and the last of which requires adaptive immunity (**Fig. 1-7**). An individual pathogen, such as a single influenza virus particle, is dealt with by these phases sequentially, but because the actual assault on the host involves billions of such particles, not every particle is in the same phase at the same time. As a result, in the host as a whole, and for an infecting pathogen population as a whole, all three phases may be happening concurrently.

The first phase is mediated by the innate physical barriers, which offer immediate protection. The components making up these barriers (skin, mucosae, gut enzymes) are non-inducible in that they pre-exist and do not develop or change in response to the presence of a foreign entity. Should these barriers be penetrated, the second, inducible phase of innate defense becomes effective 4–96 hours after the threat enters

---

Somatic recombination, the process that shuffles the DNA of the genes encoding BCRs in B cells and TCRs in T cells, is achieved by specialized genetic mechanisms that are described in Chapter 4 for B cells and Chapter 8 for T cells.

The establishment of central tolerance during lymphocyte development is described in Chapter 5 for B cells and in Chapter 9 for T cells. Processes regulating mature lymphocytes, including mechanisms of peripheral tolerance, are described in Chapter 10.

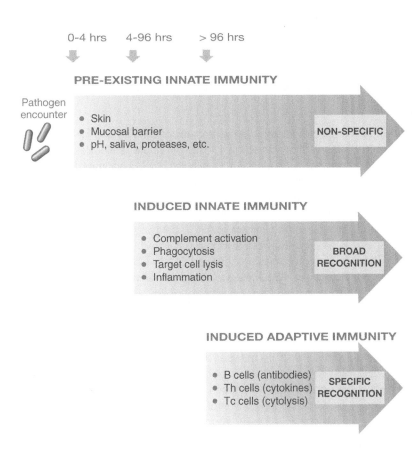

0-4 hrs   4-96 hrs   > 96 hrs

PRE-EXISTING INNATE IMMUNITY

Pathogen encounter

- Skin
- Mucosal barrier
- pH, saliva, proteases, etc.

NON-SPECIFIC

INDUCED INNATE IMMUNITY

- Complement activation
- Phagocytosis
- Target cell lysis
- Inflammation

BROAD RECOGNITION

INDUCED ADAPTIVE IMMUNITY

- B cells (antibodies)
- Th cells (cytokines)
- Tc cells (cytolysis)

SPECIFIC RECOGNITION

Fig. 1-7
The Three Phases of Host Immune Defense

A pathogen attempting to enter the body first encounters pre-existing barriers and enzymes that block access non-specifically. If the pathogen manages to enter the underlying cell layer, mechanisms mediated by complement and innate leukocytes are induced due to relatively broad recognition of PAMPs. If a more targeted, pathogen-specific response becomes necessary, elements of innate immunity then facilitate induction of highly specific adaptive responses initiated by engagement of the antigen receptors of B, Th or Tc lymphocytes.

the body. Innate leukocytes activated by the binding of PRRs to PAMPs provided by the attacking pathogen, or to DAMPs present due to host cell injury or death, work quickly to eliminate the invader using the mechanisms of inflammation, phagocytosis and target cell lysis. Complement activation also contributes to clearance at this stage. In many cases, the incursion is completely controlled by the two phases of innate defense before adaptive immunity is even triggered. However, the cells of the innate response are ultimately limited both in numbers and in recognition capacity. Fortunately, even if the pathogen ultimately manages to thwart complete elimination by innate immunity, some of the invaders and/or their components or products will have been taken up by innate leukocytes and presented as foreign antigens to T lymphocytes. The third and final phase of host defense mediated by adaptive immunity is thus initiated. Th, Tc and B cells are clonally selected and activated, and then proliferate and differentiate into memory cells and large numbers of daughter effector cells that eliminate the pathogen via humoral and/or cell-mediated mechanisms. Because it takes time for the clonally selected lymphocytes to produce effector cells, adaptive responses are not usually observed until at least 96 hours after the start of infection.

The second and third phases of immune defense are more tightly interwoven than it may first appear. Innate and adaptive immunity do not operate in isolation, and each depends on or is enhanced by elements of the other (**Fig. 1-8**). The lymphocytes of the adaptive response require the involvement of cells and cytokines of the innate system to become activated and undergo differentiation into effector and memory lymphocytes. Conversely, activated lymphocytes secrete products that stimulate and improve the effectiveness of innate leukocytes. The interplay between innate and adaptive immunity is sustained by cytokine signaling and through direct intercellular contacts between innate and adaptive leukocytes. By cooperating in this way, innate and adaptive immunity combine to mount an optimal defense against pathogens.

"The Race between Infection and Immunity: How do Pathogens Set the Pace?" by Davenport, M.P., Belz, G. T. and Ribeiro, R.M., (2009) *Trends in Immunology* 30, 61–66.

Focus on Relevant Research

Normal host homeostasis is re-established when the host wins (at least for that moment) the ongoing horse race between the host's immune system and a pathogen attempting to establish infection. Pathogens replicate at widely divergent rates, with some bacteria and viruses able to double their population size within a couple of hours. Fast-replicating pathogens therefore initially get ahead of the host's immune defenses, but then are caught and destroyed as the specialized cells of the immune system become activated. The mystery highlighted by the authors of this article is why slow-replicating pathogens are able to establish infections at all, and why so many of the most damaging chronic infections we experience, like those of the hepatitis B and C viruses and the bacteria causing tuberculosis, are caused by some of the slowest-replicating pathogens. The answer may lie in achieving certain threshold levels of infection that supply adequate amounts of the molecular triggers needed to spark first the innate response and then the adaptive response. If a pathogen replicates rapidly, large amounts of DAMPs and PAMPs are present that immediately activate innate leukocytes, including the APCs needed to initiate T cell responses. These T cells proliferate quickly enough to overwhelm the pathogen and eliminate it. In contrast, a slow-replicating pathogen may generate only low levels of DAMPs and PAMPs that do not activate large numbers of APCs, delaying the T cell response. Even once the T cell response is initiated, without sufficient cytokines produced by activated APCs, T cell proliferation may be slow and insufficient to completely eliminate the pathogen. The pathogen then survives to establish a low level of infection that doesn't kill the host but makes him/her miserable. The memory T cell response is of little help, as it only controls rather than eliminates the slow-replicating pathogen. An uneasy balance is set up between the pathogen and the host immune system that results in chronic infection. The contrast in immune response kinetics between what the authors call "fast" and "slow" pathogens is illustrated in Figure 1 from their article.

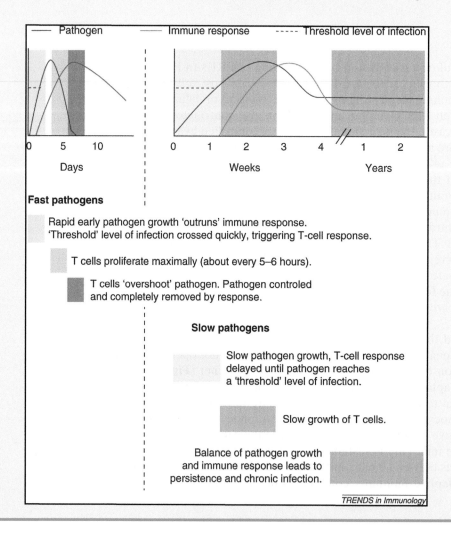

TRENDS in Immunology

Focus on Relevant Research–Cont'd

The pace of pathogen replication may have implications for vaccination strategies. Currently, vaccination often uses a harmless pathogen antigen to induce the host to respond and generate large populations of memory lymphocytes, which stand guard ready to mount a faster and stronger response should the real pathogen attack. For a fast-replicating pathogen, this tactic is successful because these types of pathogens readily generate enough PAMPs, DAMPs and antigen to surpass the thresholds of APC and T cell activation. With the "head start" provided by memory lymphocytes, the pathogen may not get to replicate at all. However, a slow-replicating pathogen may not provide sufficient antigen to reach even the lower threshold required by memory T cells, meaning that the vaccine will not have offered an advantage to the host. Worse, if the large numbers of memory T cells present after vaccination compete for the miserly quantities of antigen present, and so proliferate only slowly, the pathogen may in fact get a better foothold than if the host had NOT been vaccinated. These considerations should encourage us to experimentally test the threshold hypothesis and to refine our approach to vaccination accordingly.

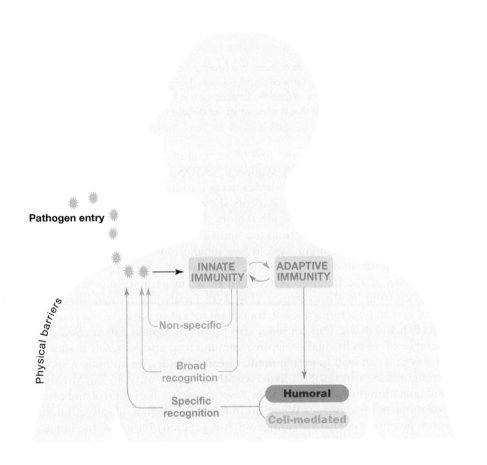

Pathogen entry

Physical barriers

INNATE IMMUNITY

ADAPTIVE IMMUNITY

Non-specific

Broad recognition

Specific recognition

Humoral

Cell-mediated

▪▪ **Fig. 1-8**
Interplay between Innate and Adaptive Responses in Host Defense

Innate and adaptive immune mechanisms work together to allow a full range of responses of appropriate strength and specificity. In addition to their direct and broadly protective role, innate immune responses recruit adaptive leukocytes and contribute to their activation. Once activated, lymphocytes mount humoral and cell-mediated anti-pathogen responses that are highly specific. Cytokines are produced that further stimulate both the innate and adaptive responses. Innate and adaptive immunity thus work synergistically to protect the host from all threats.

E. Clinical Immunology

When the immune system is functioning normally, harmful entities are recognized, and the host is protected from external attack by pathogens and internal attack by cancers. Some localized tissue damage may result from the normal inflammatory response that develops as innate leukocytes work to eliminate the threat, but these immunopathic effects are limited and controlled. As well, the tolerance of the healthy immune system prevents cells of the adaptive response from attacking normal self tissues.

Sometimes a person's immune system is incomplete, either because of **primary immunodeficiency** or **acquired immunodeficiency**. In primary immunodeficiencies, the failure of the immune system is congenital; that is, an affected individual is born with a genetic defect that impairs his/her ability to mount innate and/or adaptive immune responses. In an acquired immunodeficiency, an external factor such as a nutritional

A healthy immune response is crucial for providing an individual with immunity to infectious diseases (Ch. 13) and may be made more robust through vaccination (Ch. 14).

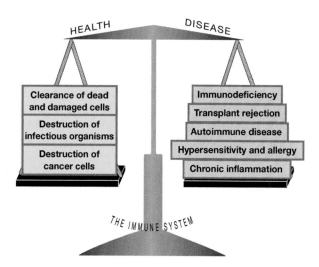

Fig. 1-9
Role of the Immune System in Health and Disease

The way in which the immune system responds can maintain a healthy body or lead to disease. In a healthy individual, the immune system mounts an appropriate response that clears damaged cells and kills infectious organisms or cancerous cells. However, the immune system is a powerful and multi-faceted weapon that requires strict controls to remain in balance. When the immune system fails to respond adequately, or mounts an overzealous or inappropriate attack, or ignores normal control mechanisms, a wide variety of disease states can result.

The pathological consequences of immune system irregularities include immunodeficiency (Ch. 15), cancer (Ch. 16 and 20), transplant rejection (Ch. 17), hypersensitivity (Ch. 18) and autoimmune disease (Ch. 19).

imbalance or a pathogen may cause the loss of an immune system component. For example, in acquired immunodeficiency syndrome (AIDS), the human immunodeficiency virus (HIV) destroys the T lymphocytes needed to fight infection. In both primary and acquired immunodeficiencies, patients show a heightened susceptibility to recurrent infections and sometimes develop tumors. The prognosis for most of these diseases is very poor and treatment options are limited.

What happens when the immune system is complete but malfunctions, or when its normal functioning has undesirable effects? There are several instances in which inappropriate actions of the immune system can result in pathologic consequences that tip the balance from health to disease (**Fig. 1-9**). Firstly, the normal attack of a healthy immune system on a foreign tissue transplant meant to preserve life results in **transplant rejection** and deleterious consequences for the transplant patient. Secondly, when the tolerance of an individual's adaptive immune response fails, self tissues are attacked in a way that can lead to **autoimmune disease**. Thirdly, an immune response that is too strong or long can result in **hypersensitivity** to an entity. In this situation, the normal inflammation that accompanies the response gets out of control and causes significant collateral tissue damage. When an individual's normal adaptive immune system responds inappropriately to an antigen that is generally harmless, the immune response is manifested clinically as a form of hypersensitivity called **allergy**. Lastly, if normal regulation of the inflammatory response is lost for an extended period for any reason, chronic inflammation ensues that can set up conditions allowing the development of tumors, heart disease or autoimmune disorders.

The remainder of this book will take the reader on a logical tour of the immune response. Chapter 2 continues our Basic Immunology section, providing detailed descriptions of the various cells, tissues and cytokines involved in mediating immune responses. Chapters 3–12 cover all aspects of the innate and adaptive immune responses in depth. The chapters in our Clinical Immunology section, Chapters 13–20, address immunological principles not only of practical value to the more medically oriented reader but also of interest to all who seek to understand the pivotal role of the immune system in health and disease. It is our hope that the reader will come away with a solid understanding of the cellular and molecular mechanisms underlying immunity and how these mechanisms can go awry to cause illness. We also provide a glimpse of how researchers seek to manipulate these mechanisms with the goal of ensuring good health for all.

Chapter 1 Take-Home Message

- The immune system is a central player in the maintenance of human health.

- Immune responses depend on the coordinated action of leukocytes that travel throughout the body to recognize and eliminate threats posed by trauma as well as by extracellular and intracellular pathogens, toxins and cancerous cells.

- Cytokines are secreted intercellular messenger proteins that mediate complex interactions among leukocytes.

- Complete protection from and clearance of unwanted entities often involve both innate and adaptive responses.

- Some innate mechanisms require no induction and are completely non-specific, whereas others are inducible and involve broad recognition of a limited number of molecular patterns (PAMPS and DAMPs) by PRMs.

- Adaptive responses are mounted after innate responses have failed to remove an unwanted entity, and are dependent on cytokines produced during the innate response.

- Adaptive immunity involves the selective activation of lymphocytes by the engagement of their antigen receptors by antigens derived from the particular inciting entity.

- The three subsets of lymphocytes are Th, Tc and B cells, which use distinct mechanisms to recognize antigen and carry out effector functions to eliminate specific entities.

- Elements of innate immunity influence adaptive immunity and vice versa.

- Malfunctions of the immune system are related to autoimmune disorders, allergy and other hypersensitivities, transplantation rejection, primary and acquired immunodeficiencies, chronic inflammation, and cancer development.

Did You Get it? A Self-Test Quiz

Section A

1) Can you define these terms? immunology, immune system, immunity, pathogen

2) What was Koch's theory and why was it important?

3) Who coined the term "vaccination" and on what basis?

Section B

1) Can you define these terms? homeostasis, mucosae, leukocyte, cytokine, immunopathic

2) What is the difference between infection and disease?

3) What is the major difference between extracellular and intracellular pathogens?

4) Distinguish between cell-mediated and humoral immunity.

Section C

1) Can you define these terms? senescent, phagocytosis, inflammation, PRM, PRR, MHC, lymphocyte, TCR, BCR, antibody, clonal selection, tolerance, self antigen, immunological memory

2) What are three major differences between innate and adaptive immune responses?

3) Distinguish between DAMPs and PAMPs.

4) Distinguish between PRMs and antigen receptors.

5) Can you give two examples of innate barrier defenses?

6) Describe three types of PRMs.

7) Is inflammation good or bad, and why?

8) How does target cell lysis help protect the body?

9) Describe five important characteristics of the adaptive response.

10) How do Th, Tc and B cells differ in antigen recognition?

11) How does each lymphocyte subset contribute to immunity against extracellular and/or intracellular pathogens?

12) Why is the secondary immune response stronger and faster than the primary response?

13) What are the two stages of tolerance?

Section D

1) Can you describe the three phases of immune defense and the components that mediate them?

2) How do the innate and adaptive responses influence each other and why is this important?

Section E

1) Can you describe four examples of diseases arising from immune system failure?

Can You Extrapolate? Some Conceptual Questions

1) Two patients have experienced the piercing of their skin by a metal object. Patient #1 is having surgery in the course of an elective hip replacement in a modern hospital. Patient #2 is not so lucky and has stepped on a rusty nail in her garden. What sort of DAMPs and/or PAMPs might be present in each case, and what sorts of immune responses might you expect to find in these patients?

2) If a person's immune system was unable to produce memory lymphocytes, how would this affect his/her innate and adaptive responses upon repeated exposure to a given pathogen?

3) Why are B lymphocyte responses usually associated with defense against extracellular rather than intracellular pathogens? How might they also help protect a host against intracellular pathogens?

Would You Like To Read More?

Chaplin, D. D. (2010). Overview of the immune response. *Journal of Allergy & Clinical Immunology, 125*(Issue 2, Suppl 2), S3–23.

Iwasaki, A., & Medzhitov, R. (2010). Regulation of adaptive immunity by the innate immune system. *Science, 327*(5963), 291–295.

Matzinger, P. (2007). Friendly and dangerous signals: Is the tissue in control? *Nature Immunology, 8*(1), 11–13.

Moser, M., & Leo, O. (2010). Key concepts in immunology. *Vaccine, 28*(Suppl 3), 2–13.

Components of the Immune System

WHAT'S IN THIS CHAPTER?

No country can act wisely simultaneously in every part of the globe at every moment of time.

Henry Kissinger

The immune system is not a discrete organ, like a liver or a kidney. It is an integrated partnership, with contributions by the circulatory system, lymphatic system, various lymphoid organs and tissues, and the specialized cells moving among them. In this chapter, we describe the components of the immune system, how communication is carried out among these components, and how leukocyte movement underlies immune responses.

A. Cells of the Immune System

I. Types of Hematopoietic Cells

Mammalian blood is made up of **hematopoietic cells** (red and white blood cells) that are carried along in a fluid phase called **plasma**. As introduced in Chapter 1, the white blood cells are infection-fighting **leukocytes**, whereas the red blood cells are **erythrocytes** that carry oxygen to the tissues. All hematopoietic cells are generated in the bone marrow from a common precursor called the **hematopoietic stem cell (HSC)** in a process called **hematopoiesis** (see later in this chapter). Immunologists generally classify hematopoietic cells not by their red or white color but by the pathway by which they develop from the HSC through various progenitors to generate mature blood cells. Historically, immunologists believed that there were two major developmental pathways that gave rise to what were termed cells of either the **myeloid** or **lymphoid** lineage. Myeloid cell types included **neutrophils, basophils, eosinophils, monocytes** and **macrophages**, and lymphoid cell types included **T** and **B lymphocytes** as well as **natural killer (NK) cells** and a cell type with characteristics of both lymphocytes and NK cells called **natural killer T (NKT) cells**. The physical characteristics and functions of these cell types are summarized in **Figure 2-1**. Although the terms "myeloid" and "lymphoid" are still used today, we now know that hematopoietic cell development is much more complex and that lineage paths cross and overlap. For example, erythrocytes

Fig. 2-1
(Part 1) Characteristics of Mature Hematopoietic Cells *Micrographs of cells reproduced by permission of D.C. Tkachuk, J.V. Hirschmann and J.R. McArthur, Atlas of Clinical Hematology (2002), W.B. Saunders Company.*

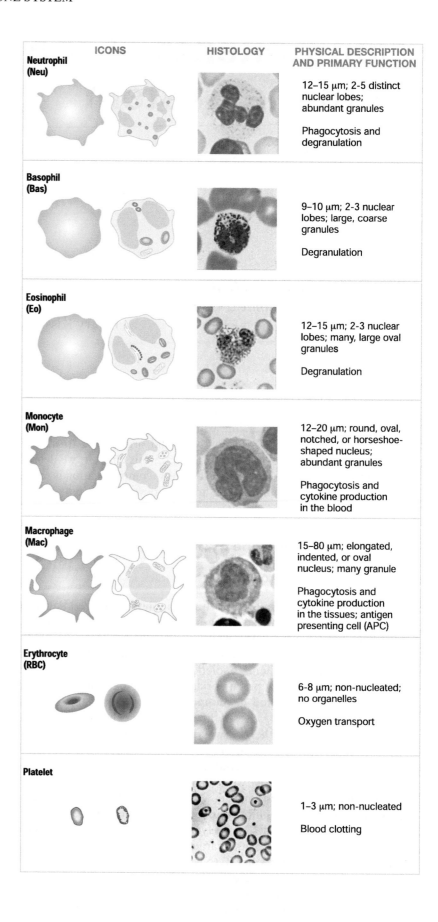

	ICONS	HISTOLOGY	PHYSICAL DESCRIPTION AND PRIMARY FUNCTION
Neutrophil (Neu)			12–15 µm; 2-5 distinct nuclear lobes; abundant granules Phagocytosis and degranulation
Basophil (Bas)			9–10 µm; 2-3 nuclear lobes; large, coarse granules Degranulation
Eosinophil (Eo)			12–15 µm; 2-3 nuclear lobes; many, large oval granules Degranulation
Monocyte (Mon)			12–20 µm; round, oval, notched, or horseshoe-shaped nucleus; abundant granules Phagocytosis and cytokine production in the blood
Macrophage (Mac)			15–80 µm; elongated, indented, or oval nucleus; many granule Phagocytosis and cytokine production in the tissues; antigen presenting cell (APC)
Erythrocyte (RBC)			6-8 µm; non-nucleated; no organelles Oxygen transport
Platelet			1–3 µm; non-nucleated Blood clotting

ICONS	HISTOLOGY	PHYSICAL DESCRIPTION AND PRIMARY FUNCTION
Megakaryocyte (Meg) 		30–160 μm; irregularly shaped nucleus; fine granules Platelet production
Mature B cell **Mature T cell**		9–12 μm; round or slightly indented nucleus; few granules Adaptive immune responses
Plasma cell (PC)		14–18 μm; round or oval nucleus; no granules B effector cell; antibody production
NK cell (NK)		12–16 μm; round nucleus; many granules Cytolysis and cytokine production
NKT cell (NKT)		Phenotypically similar to NK cell Cytokine production
Dendritic cell (DC)	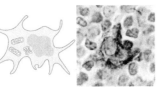	Irregularly shaped cell and nucleus; many cellular projections Phagocytosis and cytokine production; antigen presenting cell (APC)
Mast cell		5–25 μm; non-segmented nucleus; many large granules Degranulation and cytokine production

▪ Fig. 2-1
(Part 2) *Characteristics of Mature Hematopoietic Cells*
Micrographs of cells reproduced by permission of D.C. Tkachuk, J.V. Hirschmann and J.R. McArthur, Atlas of Clinical Hematology (2002), W.B. Saunders Company.

Fig. 2-2
Compartmentalization of Hematopoietic Cells

In the absence of an immune response, different populations of hematopoietic cells tend to be found in the anatomic locations indicated.

BONE MARROW	BLOOD	TISSUES
HSCs	Monocytes	Macrophages
Early hematopoietic progenitors	Neutrophils	Dendritic cells
	Basophils	Mast cells
	Erythrocytes	NK cells
		NKT cells
		Eosinophils
	B lymphocytes	
	T lymphocytes	

and **megakaryocytes** (the source of platelets for blood clotting) were once considered to be mature myeloid cells but are now thought to constitute a separate lineage. [The primary functions of erythrocytes and megakaryocytes are non-immunological and so will not be discussed further in this book.] **Mast cells** appear to arise from HSCs via a unique pathway not shared by any other hematopoietic cell type, whereas DCs are derived from multiple lineages.

Different hematopoietic cell types tend to reside and function in different body compartments. Hematopoietic stem cells and early hematopoietic progenitors are concentrated in the bone marrow. In the absence of any threat to the body (i.e., at steady state), mature lymphocytes move constantly between the blood and the lymphoid organs and tissues, while other mature hematopoietic cell types tend to concentrate in either the blood or the tissues (**Fig. 2-2**). Upon injury or infection, however, mature leukocytes may start to move among these compartments in order to restore homeostasis. In a local site of inflammation, the activated innate leukocytes and the injured or infected cells secrete various cytokines. Some of these cytokines recruit additional leukocytes from the circulation into the affected tissue. Other cytokines called **chemokines** draw these newly arrived leukocytes toward the precise site of injury or attack in a process called **chemotaxis**. As chemokines diffuse into the tissue surrounding the inflammatory site, a concentration gradient is created with the highest chemokine concentration at the site, and steadily decreasing concentrations at increasing distances from this area. Innate and adaptive leukocytes both express chemokine receptors that cause these cells to migrate up the gradient within the tissue toward the source of the chemokine. Once in the inflammatory site, these leukocytes can act to eliminate the threat. Thus, the location of a given leukocyte at any one moment depends both on its state of maturity and whether the host is at steady state.

II. Cells of the Myeloid Lineage

Neutrophils, basophils and eosinophils are all **granulocytes**, which are myeloid cells that harbor intracellular granules containing microbe-killing molecules. Monocytes and macrophages are myeloid cells that do not possess granules and use other means to carry out their functions. All of these cell types make important contributions to the innate immune response.

i) Neutrophils

Neutrophils are the most abundant leukocytes in the circulation and constitute the majority of cells activated during an inflammatory response. These cells were originally called **polymorphonuclear (PMN) leukocytes** because of the appearance under the light microscope of their irregularly shaped, multi-lobed nuclei. At any given moment, there are over 50 billion mature neutrophils in the blood of the average adult human. This huge number represents a substantial army of cells that can be recruited quickly into sites of inflammation anywhere in the body, allowing the prompt elimination of many threats. However, because each neutrophil has a life span of only 1–2 days, the bone marrow must constantly produce new neutrophils at a very high rate.

Focus on Relevant Research

"Neutrophil Kinetics in Health and Disease" by Summers, C., Rankin, S.M., Condliffe, A.M., Singh, N., Peters, A.M., and Chilvers, E.R. (2010) *Trends in Immunology* 8, 318–324.

Much recent immunological research has focused on T cells, B cells and DCs due to their critical and interlocking roles in adaptive immunity. In contrast, relatively little attention has been directed toward the neutrophil, which perhaps has come to be viewed as less glamorous than other leukocytes due to its relatively pedestrian function in the phagocytosis of bacteria and fungi. In this review, however, Summers *et al.* summarize research that corrects this misperception and highlights various aspects of experiments designed to demystify neutrophil production, activation, regulation, and distribution.

Given their staggering production rate of about 50 billion cells per day, neutrophils are undoubtedly the "workhorse" of the immune system and are the most common circulating leukocytes. Neutrophil production in the bone marrow is regulated by complex interactions among various cytokines, cytokine receptors and other molecules expressed by surrounding bone marrow stromal cells (non-hematopoietic support cells). In addition, these interactions determine whether mature neutrophils are retained in the bone marrow or released from it into the blood. The authors also discuss the "biodistribution" of neutrophils and how this process corrals the destructive potential of these cells. Neutrophils released from the marrow do not simply circulate in the bloodstream *en masse*; rather, some are sequestered in the spleen and liver in a process called "margination." Of the neutrophils that do remain in the general circulation, some are drawn into local sites of inflammation in the host's tissues to help fight incipient infections. However, once "primed" by circulation through such inflammatory sites,

any of these highly toxic neutrophils that rejoin the circulation can easily lodge in otherwise healthy capillary beds and inflict considerable damage on uninfected tissues. The authors go on to present controversial data suggesting that primed neutrophils may be diverted from the general circulation into the lungs, where they are "deprimed" in an effort to reduce the chances of collateral damage that might ensue from their random accumulation throughout a host's tissues.

The authors complete their summary of neutrophil kinetics by noting that, 1–2 days after their release from the bone marrow, neutrophils undergo a form of programmed cell death called **apoptosis** (see later in this chapter). Neutrophils present in various organs and tissues undergo apoptosis in these locations and are phagocytosed by resident macrophages. However, aged neutrophils in the blood appear to be drawn back to the bone marrow due to re-expression of the same cytokine receptors that originally held them there transiently after their birth. Once back in the bone marrow, these neutrophils also undergo apoptosis and are consumed by bone marrow-resident macrophages. The continuous death of these neutrophils stimulates hematopoiesis to produce more neutrophils and maintain homeostasis.

Future applications of this increased knowledge of neutrophil behavior are discussed, including the possibility of medically manipulating the production, activation and purification of neutrophils. In short, Summers *et al.* have argued convincingly that the humble neutrophil is deserving of much more respect and research effort.

As early as 30 minutes after an acute injury or the onset of infection, neutrophils enter the site of attack in response to local chemokine gradients. Once in the inflammatory site, the neutrophils are activated by the binding of their PRRs to DAMPs or PAMPs. The activated neutrophils immediately phagocytose any foreign entities and sequester them in intracellular vesicles called **phagosomes**. The neutrophil's cytoplasmic granules then fuse with the phagosomes, triggering the immediate release of granule contents that destroy the engulfed entity. If an entity cannot be phagocytosed by neutrophils, these cells may attempt to destroy it by releasing the destructive contents of their granules extracellularly in a process called **degranulation**. If an inflammatory response is particularly long or intense, extended degranulation may kill bystander healthy host cells and cause some to liquefy. This accumulation of fluid, dead host cells and degraded foreign material forms **pus**, a familiar sign of infection.

Neutrophils are primarily a component of the innate response, since their activation occurs via PRR-mediated recognition of molecular patterns shared by a broad range of pathogens or host stress molecules. However, neutrophils are also indirectly linked to the adaptive response because, in addition to PRRs, neutrophils express surface proteins that are able to bind to the **Fc region** of a soluble antibody molecule. The Fc region of an antibody is that part of the protein which is not involved in antigen binding. The expression of such **Fc receptors** (**FcRs**) on neutrophils allows them to facilitate the clearance of antibody-bound antigens.

The Fc region of the antibody molecule is further described in Chapter 4 and the role of FcRs in clearing antibody-bound antigen is described in detail in Chapter 5.

ii) Basophils and Eosinophils

Basophils and eosinophils are named for the colors of their cytoplasmic granules when stained blood smears are viewed under the microscope. Basophilic granules react with basic dyes such as hematoxylin and stain a dark blue color. Eosinophilic granules stain reddish with acidic dyes such as eosin. Although eosinophils and basophils can capture microbes by phagocytosis, degranulation is the primary means by which these cells defend the body. Eosinophilic granules are filled with highly basic proteins and enzymes that are effective in the killing of large parasites, whereas basophilic granules contain substances that are important for sustaining the inflammatory response. Basophils are present in the body in very low numbers (1% of all leukocytes) and reside primarily in the blood until they move into the tissues during inflammation. In contrast, the vast majority of mature eosinophils (4% of all leukocytes) normally reside in the connective tissues. Eosinophils also play a prominent role in allergy (see Ch. 18).

iii) Monocytes

Among the largest cells resident in the blood circulation are the monocytes. The principal functional features of these phagocytes are their numerous cytoplasmic lysosomes filled with hydrolytic enzymes, and their abundance of the organelles required for the synthesis of secreted and membrane-bound proteins. Monocytes are present in human blood at moderate density, representing approximately 10% of all circulating leukocytes. In a host at steady state, newly produced monocytes patrol the blood for about 1 day, carrying out phagocytic removal of apoptotic cells and toxic macromolecules. These monocytes then enter the tissues, where some remain as monocytes and others are induced by the surrounding microenvironment to differentiate into macrophages (see following subsection). In a host that is experiencing infection, monocytes in the blood produce large quantities of cytokines that regulate or promote inflammation. Monocytes that subsequently enter an inflamed tissue not only also give rise to macrophages but also to inflammatory DCs that can contribute to local innate and adaptive defense.

iv) Macrophages

Macrophages are generally several times larger than the monocytes from which they are derived and display further enhancements of the protein synthesis and secretion machinery. These long-lived (2–4 months) powerful phagocytes reside in all organs and tissues, usually in sites where they are most likely to encounter foreign entities. Subtle differences in the morphology and function of macrophages develop as a result of the influence of a particular microenvironment, giving rise to tissue-specific names for these cells (**Table 2-1**). For example, macrophages in the liver are known as **Kupffer cells**, whereas those in the bone are called **osteoclasts**. Regardless of their location, macrophages constantly explore their tissue of residence and tirelessly engulf and digest not only foreign entities but also spent host cells and cellular debris.

Macrophages play key roles in both the innate and adaptive immune responses. Macrophages express a large battery of PRRs, meaning that these cells are readily activated by direct contact with PAMP-bearing pathogens or their products. DAMPs

TABLE 2-1	Macrophages Named by Tissue
Macrophage Location	**Name**
Liver	Kupffer cells
Kidney	Mesangial phagocytes
Central nervous system	Microglia
Connective tissues	Histiocytes
Bone	Osteoclasts
Lung	Alveolar macrophages
Spleen	Littoral cells
Joints	Synovial A cells

derived from host tissue breakdown products and proteins of the complement or the blood coagulation systems can also activate macrophages during the innate response. As well as carrying out phagocytosis, activated macrophages secrete chemokines that draw neutrophils and other innate (and adaptive) leukocytes to a site of inflammation. Cytokines and growth factors produced by macrophages stimulate innate leukocytes as well as cells involved in wound healing.

With respect to the adaptive response, macrophages produce several cytokines that influence lymphocyte activation, proliferation and effector cell generation. Most importantly, macrophages are one of the few cell types that can function as an APC. As introduced in Chapter 1, APCs take up a protein antigen, digest it into peptides, and combine the peptides with MHC class II molecules. These pMHC complexes are then displayed on the APC surface for inspection by Th cells. Recognition of a pMHC by the TCR of a Th cell triggers the activation of the latter. Unlike DCs, macrophages cannot activate **naïve** T cells (T cells that have never encountered their specific antigen before) but do activate effector and memory T cells. Macrophages also express FcRs, allowing these cells to avidly take up and dispose of antigens that have been coated in soluble antibody (see Ch. 5). Lastly, macrophages participate in cell-mediated immunity because they respond to cytokines secreted by activated T cells by becoming **hyperactivated**. Hyperactivated macrophages gain enhanced antimicrobial and antiparasitic activities and new cytolytic capacity, particularly against tumor cells. The mutual stimulation of T cells and macrophages is vital, since it amplifies both the innate effector mechanisms of macrophage action and the adaptive effector mechanisms of the cell-mediated and humoral responses.

III. Cells of the Lymphoid Lineage

i) T and B Lymphocytes

As introduced earlier, resting mature B and T lymphocytes that have not interacted with specific antigen are said to be **naïve**, **virgin** or **unprimed**. These cells have a modest life span (up to a few weeks) and are programmed to die unless they encounter their specific antigen. As outlined in Chapter 1, T and B cells recognize and bind to antigens using antigen receptors. A mature B lymphocyte carries close to 150,000 identical copies of a particular BCR on its cell surface, whereas a mature T cell bears ~30,000 copies of its TCR. The vast majority of T cells in the body are αβ T cells, so termed because their TCRs are made up of a TCRα protein subunit and a TCRβ protein subunit (TCRαβ; see Ch. 8).

For both T and B lymphocytes, the antigen-binding site of the receptor protein is designed to recognize a particular antigen shape, and only those antigens binding with adequate strength to a sufficient number of receptors will trigger activation of the lymphocyte and initiate an adaptive response. Although it is often said that "one lymphocyte recognizes one antigen," a T or B cell can in fact recognize more than one antigen because different molecules closely related in shape to the "ideal antigen" will fit to some degree into the antigen-binding cleft of the receptor. This very small collection of antigens will therefore be recognized by the lymphocyte with a strength proportional to the quality of the fit. The antigen receptor is said to **cross-react** with such antigens.

Binding of specific antigen activates a lymphocyte and stimulates it to progress through cell division leading to lymphocyte **priming**. The transcription of numerous genes is triggered, causing the progeny cells to undergo morphological and functional changes that result in the production of **lymphoblasts**. This conversion of the resting lymphocyte into daughter lymphoblasts occurs within 18–24 hours of antigen receptor engagement. Lymphoblasts are larger and display more cytoplasmic complexity than resting lymphocytes, and undergo rapid cell division. Lymphoblasts then differentiate into short-lived effector cells and long-lived memory cells. For B cells, the fully differentiated effector cell is the antibody-secreting plasma cell (**Table 2-2**). For Tc cells, the fully differentiated effector cell is the CTL, which is capable of target cell cytolysis and cytotoxic cytokine secretion. For Th cells, the fully differentiated T helper effector cell secretes copious amounts of cytokines supporting both B and Tc responses. The

TABLE 2-2	Effector and Memory Lymphocytes		
Type of Lymphocyte	**Effector Cell**	**Effector Action**	**Memory Cell**
B cell	Plasma cell	Antibody production	Memory B cell
Tc cell	CTL	Target cell cytolysis	Memory Tc cell
Th cell	Effector Th cell	Cytokine production	Memory Th cell

integrity of these effectors and the lymphocyte compartment as a whole is vital for the maintenance of effective adaptive immunity.

NOTE: Primary immunodeficiencies arising from genetic defects that are particularly devastating to lymphocytes are described in Chapter 15. They include ataxia-telangiectasia (AT), Nijmegen breakage syndrome (NBS), Bloom syndrome (BS), and xeroderma pigmentosa (XP).

B and T lymphocytes are difficult to distinguish morphologically. However, immunologists have devised a means of using antibodies to detect differences in the expression of certain cell surface proteins called **CD markers** (see **Box 2-1**). For example, B cells typically express a surface protein called CD19 but not a protein called CD3, while the reverse is true for T cells. T cell subsets are also distinguished by CD marker expression, as Th cells generally express CD4 while Tc cells usually express CD8. Because they bind to invariant regions of the same MHC molecule that presents antigenic peptide to the T cell's TCRs, CD4 and CD8 are known as **coreceptors**.

To find the most complete and updated CD information, the reader should visit the Human Cell Differentiation Molecules (HCDM) website at http://www.hcdm.org.

ii) Natural Killer Cells, γδ T Cells and Natural Killer T Cells

The NK cells and NKT cells mentioned previously, along with a third cell subset called "γδ T cells," are three types of lymphoid cells that make their primary contributions to innate, rather than adaptive, immunity. None of these cell types expresses the highly diverse and specific antigen receptors found on B and T cells; nor do they generate memory cells upon activation. Each cell type is discussed briefly here, with more on their mechanisms of antigen recognition and clearance described in Chapter 11.

NK cells are large leukocytes that morphologically resemble effector T cells and are found in both the blood and lymphoid tissues. Like CTLs, NK cells are capable of cytolytic killing of target cells, but this destruction is achieved without the fine specificity of

Box 2-1 CD Markers

The CD designation system developed as immunologists searched for ways to tell leukocytes (and particularly lymphocytes) apart in the absence of morphological differences. The expression of cell surface proteins differs among cell types and often varies on a given cell type depending on its stage of development or activation. The identification of characteristic surface molecules, or "cell surface markers," on a cell allows it to be defined in a way that does not depend on morphology. To identify these cell surface markers, immunologists originally injected purified human lymphocytes into mice so that proteins on the human lymphocyte surface that were *not* also expressed in mice triggered the production of mouse antibodies able to bind to the human-specific proteins. To their surprise, these researchers discovered that the injected mice often made several different antibodies recognizing the same human protein. Each member of this "cluster" of antibodies was found to bind to a different area on the single human protein in question, but, as a group, the binding of these antibodies demonstrated the presence of that protein on the human lymphocyte surface. This type of antibody cluster was originally called a **cluster of differentiation (CD)** because its binding helped to define the differentiation state of a cell. Today, the cell surface protein recognized by the antibody cluster is referred to as a **CD marker** or **CD molecule**. Each CD marker is given a unique number, such as CD4 or CD19, and a cell expressing CD4 is said to be "CD4+". The function of a CD molecule does not have to be known for it to be useful in identifying specific cell types by either lineage, species, stage of maturation or state of activation. More than 350 CD molecules have been defined at the time of writing. An international organization called Human Cell Differentiation Molecules (HCDM) organizes workshops so that newly identified cell surface molecules can be defined, characterized and assigned a CD number. A description of selected CD markers is contained in Appendix D.

a T lymphocyte. In addition, NK cells carry FcRs that allow them to lyse cells coated in soluble antibody (see Ch. 5). NK cells are particularly important for their ability to recognize and kill many virus-infected and tumor cells. Furthermore, in the presence of high levels of the cytokine **interleukin-2** (IL-2), NK cells undergo additional differentiation and may acquire even greater powers of target cell recognition and cytolysis. Because activated T cells are the principal producers of IL-2, this enhancement of NK cell effector functions is another example of the interplay between the innate and adaptive responses.

γδ T cells are a type of T lymphocyte that supports the innate response due to the nature of its TCR. In contrast to the CD4$^+$ Th and CD8$^+$ Tc cells of the adaptive response, which are αβ T cells expressing αβ TCRs, γδ T cells express γδ TCRs containing two slightly different chains: TCRγ and TCRδ (TCRγδ). γδ T lymphocytes never express αβ TCRs, and αβ T cells never express γδ TCRs. Unlike the enormous diversity of αβ TCRs, γδ TCRs bind to only a limited set of broadly expressed antigens that may or may not be presented by molecules other than MHC class I and II. As well, unlike αβ T cells, γδ T cells often reside in the mucosae where they can respond rapidly to an incipient pathogen attack. Once activated, γδ T cells proliferate and differentiate into effectors that are capable of cytolysis or the secretion of destructive cytokines. Other cytokines produced by γδ T cells may influence the behavior of αβ T cells, establishing another link to the adaptive response.

NKT cells combine features of both αβ T lymphocytes and NK cells. Like an αβ T cell, an NKT cell expresses many copies of a single αβ TCR on its cell surface. However, unlike the diversity of the TCRs expressed by the αβ T cell repertoire, most NKT clones express highly similar TCRαβ molecules, such that the NKT TCR is described as being "semi-invariant." Like the γδ TCR, the semi-invariant NKT TCR recognizes a small collection of antigens. Once activated by engagement of the semi-invariant TCR, NKT cells immediately secrete cytokines that support the activation and differentiation of B and T cells; that is, like NK cells, NKT cells do not have to first differentiate into effector cells like γδ and αβ T cells do. In addition, NKT cells resemble NK cells morphologically and express some of the same surface markers. However, an NKT cell cannot carry out target cell cytolysis like an NK cell can. Interestingly, NKT cells in different anatomical locations function slightly differently. For example, in mice, NKT cells in the liver primarily produce **interferon**-γ (IFNγ), a cytokine with anti-cancer effects. However, NKT cells in murine spleen produce alternative cytokines that have little effect on tumor cells but regulate αβ T cells. The rapidity of the NKT response and the limited range of antigens to which these cells respond mark them as innate leukocytes. However, the influence of NKT-secreted cytokines on B and T lymphocyte functions indicates that NKT cells constitute yet another important link between innate and adaptive immunity.

IV. Dendritic Cells

The first cells identified as DCs were irregularly shaped, short-lived cells that exhibited long, finger-like membrane processes resembling the dendrites of nerve cells. The functions of these cells are crucial to the induction of effective immune responses, and it is these cells we mean when we use the term "DC" in this book. Some immunologists refer to these DCs as "classical" or "conventional" DCs to distinguish them from three other cell types that have similar properties and/or morphologies but distinct functions: plasmacytoid DCs, inflammatory DCs and follicular DCs. All of these cell types are briefly described next, with their specific functions examined in later chapters.

i) Classical or Conventional DCs

Both myeloid and lymphoid precursors present in the bone marrow can give rise to DCs, as can certain precursors present in a lymphoid organ called the **thymus** (see later). With the right stimulus, some DCs can also be derived from more differentiated cell types already resident in the blood or tissues. DCs are important sentinels of both innate and adaptive immunity because these cells use their PRRs and phagocytic powers to sample tissue microenvironments for the presence of DAMPs and PAMPs that indicate "danger," and also present pMHCs in a way that can trigger the activation of antigen-specific T cells.

In the absence of PRR engagement, DCs are considered to be "immature." Some immature DCs are resident in the lymphoid tissues and monitor antigens brought into these structures, while other immature DCs patrol the body's **peripheral tissues** or "periphery." Immature DCs are specialists in internalizing entities from the surrounding milieu, generating peptides from these entities, and combining them with MHC molecules. In an environment without DAMPs/PAMPs, the pMHCs presented on an immature DC's surface cannot activate T cells. However, when a DAMP/PAMP engages a DC's PRRs, the immature DC is stimulated to "mature," a process that involves significant changes to DC gene expression, localization and behavior. Mature DCs undertake two vital functions: (1) secretion of cytokines that induce a local inflammatory response, and (2) presentation of pMHCs containing non-self peptides to naïve T cells under conditions that allow their activation. Indeed, mature DCs are the *only* cell type capable of activating naïve T cells. Mature DCs are influenced both by the microenvironment in which they find themselves and the DAMPs and PAMPs to which they are exposed, and so may show subtle differences in their cytokine production, responses to cytokines, and effects on T cell responses.

More on how DCs mature and become capable of activating naïve T cells can be found in Chapter 7 on antigen presentation.

DCs were first identified and characterized by Ralph Steinman, who steadfastly championed these cells as central players in the immune response. Dr. Steinman was posthumously awarded a share in the 2011 Nobel Prize in Physiology or Medicine for his seminal work.

ii) Plasmacytoid DCs

Plasmacytoid dendritic cells (pDCs) are a relatively recently discovered cell type that was first characterized in the 1990s. The earliest *in vitro* studies found that these cells were developmentally and functionally related to DCs. However, pDCs lack dendrite-like projections and instead have the rounded shape of plasma cells. Their name thus derives from the combination of their DC-like function with their plasma cell-like morphology. pDCs are mainly found in the lymphoid organs and tissues where they remain resident for relatively long periods. These cells express a narrow range of PRRs that are focused on the recognition of viral nucleic acids. In response to engagement of these PRRs by viral RNA or DNA, pDCs secrete large quantities of IFNα and IFNβ, cytokines that have potent antiviral properties (see Ch. 13). The capacity of pDCs for antigen presentation is more limited than that of DCs, and pDCs appear to be less important for the activation of naïve T cells. A comparison of the properties of DCs and pDCs can be found in **Table 2-3**.

TABLE 2-3	**Contrasting Properties of cDCs and pDCs**	
	Conventional DC (cDC)	**Plasmacytoid DC (pDC)**
Morphology	Many dendrites Prominent, finger-like protrusions	No dendrites Rounded plasma cell-like appearance
Main location	All tissues	Sites where lymphocytes reside
Relative life span	Short-lived	Long-lived
PRR expression	Broad range	Narrow range (pDC PRRs recognize viral nucleic acids only)
MHC class II expression	High	Low (until activated)
Main function	Antigen presentation, especially to naïve T cells	Secretion of cytokines effective against viruses
Link to adaptive immunity	Direct T cell activation	Recruitment and activation of many leukocyte subsets

iii) Inflammatory DCs

Detailed examination of inflamed tissues has revealed that some monocytes appear to be selectively recruited to sites of inflammation, and that once in these locations, these monocytes are induced to differentiate into DC-like cells. It is currently unclear whether these "inflammatory DCs" have the capacity to activate naïve T cells, although they can activate effector T cells and may regulate some T cell functions. Inflammatory DCs also produce molecules involved in the direct killing of bacteria.

iv) Follicular Dendritic Cells

Follicular dendritic cells (FDCs) have a dendritic morphology but are completely unrelated to DCs and pDCs in lineage and function. While DCs are critical for T cell responses, FDCs are important for B cell responses and are abundant in the secondary lymphoid organs where B cell activation occurs (see Ch. 5). Within these organs, FDCs excel at trapping antigens on their surfaces, exactly where these entities can be readily bound by the BCRs of B cells. FDCs also help to organize various cell types within the lymphoid organs and promote the ongoing removal of apoptotic cell debris created as a result of programmed cell death during B cell activation and differentiation (see Ch. 5). It has been difficult to isolate pure FDCs because FDC-specific surface markers have yet to be identified; the precise lineage of FDCs therefore remains unresolved. Some evidence suggests that at least some FDCs may arise from stromal cells and thus are of non-hematopoietic origin.

V. Mast Cells

Mast cells are rare, long-lived cells important for defense against worms and other parasites, and are key mediators associated with the form of immune hypersensitivity known as allergy. Mast cells resemble basophils in that they contain basic-staining cytoplasmic granules that harbor an array of pro-inflammatory substances. The degranulation of mast cells is rapidly triggered by tissue invasion or injury, resulting in a flood of cytokines and other molecules that initiate the inflammatory response. However, mast cells differ from basophils in several ways: (1) Mast cell granules are smaller and more numerous than those of basophils. (2) Unlike basophils, mast cells are rarely found in the blood and preferentially reside in the connective tissues and gastrointestinal mucosa. (3) Mast cells are not of the myeloid or lymphoid lineage and take an independent path in their differentiation from HSCs.

The role of mast cells in mediating allergies is described further in Chapter 18 on immune hypersensitivity.

VI. Hematopoiesis

As introduced previously, the generation of all red and white blood cells in the body is called **hematopoiesis**. Hematopoiesis is a continual process that ensures an individual has an adequate supply of erythrocytes and leukocytes throughout life. In humans, all hematopoietic cells arise from HSCs that can first be detected in embryonic structures at 3–4 weeks of gestation (**Fig. 2-3**). The HSCs migrate to the human fetal liver at 5 weeks of gestation, and liver hematopoiesis completely replaces embryonic hematopoiesis by 12 weeks of gestation. At 10–12 weeks of gestation, some HSCs or their descendants commence migration to the spleen and bone marrow. The fetal spleen transiently produces blood cells between the third and seventh months of gestation. The bone marrow of the long bones (tibia, femur) assumes an increasingly important role by about the fourth month and takes over as the major site of hematopoiesis during the second half of gestation. By birth, virtually all hematopoiesis occurs in the marrow of the long bones. After birth, the activity of the long bones steadily declines and is replaced in the adult by the production of hematopoietic cells in the axial skeleton—the sternum, ribs and vertebrae as well as the pelvis and skull. If the bone marrow suffers an injury in an adult, the liver and spleen may resume some level of hematopoiesis.

Hematopoietic stem cells occur at a frequency of just 0.01% of all bone marrow cells. However, HSCs are capable of tremendous proliferation and differentiation in response to an increased demand by the body for hematopoiesis. Hematopoietic stem

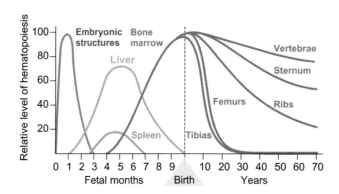

Fig. 2-3
Sites of Hematopoiesis in Humans

The site of blood cell formation shifts during fetal and adult life, as shown by the indicated changes in the relative levels of hematopoiesis in different tissues over time. Note that as an individual ages, hematopoiesis becomes less vigorous. *[Adapted from Klein, J., & Horĕjší V. (1997). Immunology, 2nd ed. Blackwell Science, Osney Mead, Oxford.]*

cells are said to be both *multipotent* and *self-renewing*. Multipotency means that an HSC can differentiate into any type of hematopoietic progenitor, including those eventually giving rise to lymphocytes, granulocytes, macrophages, DCs and mast cells. Self-renewal means that, instead of differentiating into such lineage-committed precursors, an HSC can generate more HSCs. Thus, when stimulated to divide, HSCs in the bone marrow will either self-renew or generate committed progenitors that eventually give rise to mature hematopoietic cells. Immunologists may refer to the production of cells of the myeloid lineage as **myelopoiesis**, and the production of cells of the lymphoid lineage as **lymphopoiesis**. It is important to note that the development of any given mature cell type from the HSC rarely proceeds from start to finish in the bone marrow. Progenitors often leave the bone marrow before maturation is complete, and local immune responses in the tissues often influence the direction of hematopoiesis because particular chemokines released by injured or infected host cells draw progenitor cells with a particular differentiation capacity to the area. Once the progenitors arrive in the local site, cytokines in the microenvironment push the progenitors to proliferate and differentiate into mature cells of the type best suited to deal with the threat.

The development of HSCs into mature blood cells takes place through a continuum in which identifiable progenitors generate successive intermediates that progressively lose their capacity to self-renew and differentiate into multiple cell types. Immunologists differ widely on what these progenitors and intermediates are called, how to identify them, and precisely which mature cell types they generate. **Figure 2-4** shows our simplified model of one scheme by which an HSC can either self-renew, or produce a very early progenitor sometimes called the **multipotent progenitor (MPP)**. The MPP gives rise to (at least) three precursors of more limited potential sometimes called the **common myeloid progenitor (CMP)**, the **common lymphoid progenitor (CLP)**, and the **mast cell progenitor (MCP)**. Although these early progenitors are thought to retain some self-renewal capacity, their primary function is to differentiate into various types of mature hematopoietic cells. As alluded to previously, debate rages as to the exact nature of the cellular intermediates involved in producing any one mature cell type. It is thought that the CMP gives rise to two less multipotential progenitors called the granulocyte/monocyte precursor (GMP) and the megakaryocyte/erythrocyte precursor (MEP). The GMP in turn generates granulocytes (neutrophils, basophils and eosinophils), as well as DCs, pDCs and monocytes. Monocytes can then give rise to macrophages and, under the correct conditions, inflammatory DCs. The CLP is less well established than the CMP as *bona fide* hematopoietic progenitor, but many immunologists believe it gives rise through a series of developmental intermediates to cells of

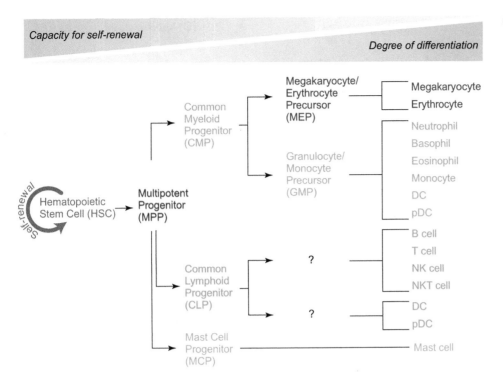

▶ Fig. 2-4

***Simplified Model of Hema-
topoiesis***

A hematopoietic stem cell is influenced by cytokines and other factors
in its microenvironment to either
self-renew or produce multipotent
progenitor cells. These in turn
can give rise to cells such as the
indicated precursor types, which can
respond to factors in their microenvironments and differentiate (via steps
that have yet to be fully defined) into
the indicated mature hematopoietic
cell types. As the degree of differentiation of a hematopoietic cell
progresses, its capacity to self-renew
decreases.

Wiscott–Aldrich Syndrome
(WAS) is a primary
immunodeficiency caused
by mutation of a protein
involved in hematopoietic cell
differentiation and activation.
The mutant WAS protein
(WASP) causes a wide variety
of symptoms described further
in Chapter 15.

the B, T, NK, NKT and DC lineages. With respect to the mast cell lineage, MCPs
are thought to migrate to connective tissues or mucosae where they complete their
differentiation into mature mast cells. For all lineages, once a mature hematopoietic cell is generated, its high level of specialization means that it can be stimulated
to divide and produce effector cells but can no longer give rise to precursors or
even mature cells like itself.

VII. Apoptosis

Apoptosis occurs when it is advantageous to the host for a cell to die, such as during
embryonic development to shape a particular body structure, or in the adult animal
when expended cells must be removed. Apoptosis is also crucial for the homeostasis of
the immune system. The mounting of effective immune responses requires that a huge
number of hematopoietic cells be generated and maintained at all times. Accordingly,
at any given moment, the average healthy adult human possesses an estimated 10^{12}
lymphocytes and an estimated 5×10^{10} circulating neutrophils. However, to ensure
that the required hematopoietic cells are always fresh and fully functional, leukocytes
are programmed to die by apoptosis after only days or months of patrolling the body
for pathogens. Apoptosis removes these aged and spent cells as quickly as the bone
marrow creates new ones, maintaining the immune system in balance. In addition,
the mounting of an immune response in a site of injury or attack results in a temporary need for leukocyte recruitment and perhaps lymphocyte proliferation at that site.
When the threat has been cleared, the apoptosis of "leftover" effector cells ensures a
return to steady state cell numbers in the now healthy tissue.

Apoptotic cell death differs in several important ways from **necrosis**, which is cell
death caused by tissue trauma or pathogen attack. A cell that has been injured in this
way lyses in an uncontrolled manner and becomes necrotic. A necrotic cell loses the
integrity of its nucleus and its plasma membrane, and leaks its intracellular contents
into the surrounding microenvironment. These contents include the deadly hydrolytic
enzymes and oxidative molecules normally confined in lysosomes. Once leaked, these
molecules can precipitate neighboring cell breakdown and an unwanted inflammatory
response. In contrast, apoptosis is a controlled form of cell death that allows the cell
to maintain plasma membrane integrity (**Plate 2-1**). Hallmark morphological changes

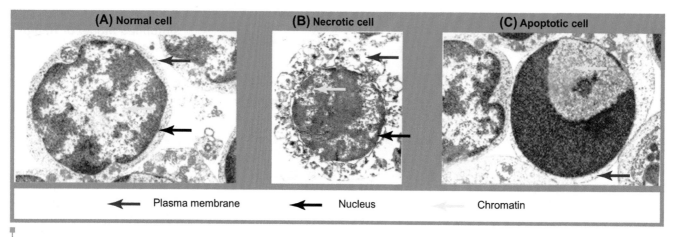

(A) Normal cell   **(B) Necrotic cell**   **(C) Apoptotic cell**

⟵ Plasma membrane   ⟵ Nucleus   ⟵ Chromatin

Plate 2-1
Comparison of Apoptotic and Necrotic Cell Death

(A) A normal cell has a healthy nucleus containing chromatin and an intact plasma membrane. **(B)** A necrotic cell breaks down and releases its nuclear and cytoplasmic contents in a random way. **(C)** An apoptotic cell undergoes programmed death in a controlled way that allows it to reduce its cytoplasm and sequester its fragmented protein and nucleic acid contents from the extracellular milieu. *[Reproduced by permission of Andrea Cossarizza, University of Modena and Reggio Emilia, Italy.]*

occur only to those cells intentionally marked for "suicide." Such a cell undergoes an orderly demise that is dictated by specific biochemical pathways; the cell cleaves its own chromosomal DNA, fragments its nucleus, degrades its proteins, and shrinks its cytoplasmic volume. Eventually, the cell is reduced to an **apoptotic body**, which is a small membrane-bound structure in which the remaining organelles such as lysosomes and mitochondria are held in intact form. The rapid phagocytosis of the apoptotic bodies by macrophages prevents any leakage of lysosomal enzymes or other detrimental contents.

B. How Leukocytes Communicate

I. Intracellular Communication: Signal Transduction

The binding of a ligand to a receptor expressed by a leukocyte is the initiating signal indicating that an immune response is required. Such ligands include DAMPs, PAMPs, antigens, cytokines and molecules present in the membranes of other host cells. Accordingly, the receptors involved include the membrane-bound PRRs of innate leukocytes, the TCRs and BCRs of lymphocytes, and complementary cell surface receptors that bind to molecules involved in phagocytosis, adhesion, or cellular activation or inhibition.

Cell surface receptors usually consist of one or more transmembrane proteins in which the **extracellular domains** bind the ligand or ligand complex, the **transmembrane domains** anchor the receptor chains in the plasma membrane, and the **cytoplasmic domains** transduce intracellular signals into the interior of the cell (**Fig. 2-5**). However, a signal is not actually generated until the binding of ligand induces the aggregation of two or three receptor molecules. This aggregation triggers a change (usually conformational) in the cytoplasmic domains of the component receptor chains. This change in turn sparks the phosphorylation (or dephosphorylation) of other proteins, particularly **protein tyrosine kinases (PTKs)** associated with the cytoplasmic domain of the receptor. PTKs are enzymes that, when activated, phosphorylate the tyrosine residues of substrate proteins. Within seconds of ligand binding and receptor aggregation, the PTKs associated with the receptor "tail" are activated and initiate a complex phosphorylation/dephosphorylation cascade involving other PTKs and additional substrates (**Fig. 2-6**). This cascade of specific enzyme activation eventually induces the release of mediators such as calcium ions from their sequestration in special intracellular storage structures. When the response required by the cell takes the form

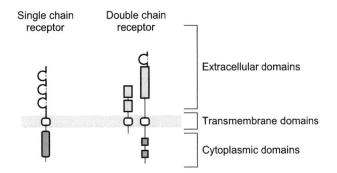

Fig. 2-5
Examples of Cell Surface Receptor Structure

Cell surface receptors are composed of one or more polypeptide chains anchored in the plasma membrane by one or more transmembrane domains. Extracellular domains bind a specific ligand when it is present in the extracellular environment. Upon ligand binding, cytoplasmic domains convey intracellular signals into the interior of the cell.

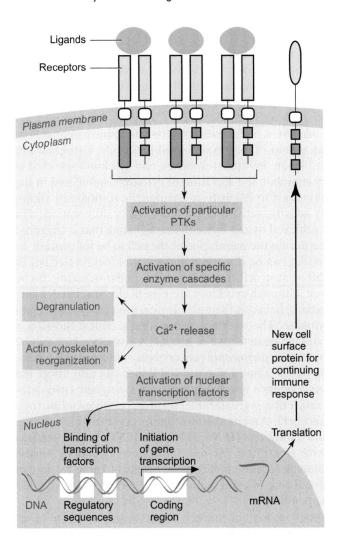

■ **Fig. 2-6**
Simplified Model of an Intracellular Signaling Pathway

Ligand-induced aggregation of receptors anchored in the plasma membrane leads to activation cascades involving protein tyrosine kinases (PTKs) and other intracellular enzymes. The resulting release of intracellular calcium can trigger various cytoplasmic responses and can also activate nuclear transcription factors. These factors translocate into the nucleus, bind to regulatory sequences in gene promoters, and induce the transcription and translation of new cell surface proteins or other molecules needed for the cell's response to the ligand.

of degranulation or actin cytoskeleton reorganization, the end point of the signaling pathway is in the cytoplasm and does not involve the nucleus. However, the proteins needed to carry out an effector action are often not routinely present as part of a cell's "housekeeping" metabolism, so that new synthesis of specialized proteins is needed. New protein synthesis requires new transcription, so that the enzymatic cascade and mediator release continue until one or more **nuclear transcription factors** capable of entering the nucleus are stimulated. The activated nuclear transcription factors induce the transcription of the previously silent genes encoding the proteins responsible for the required effector action.

The **intracellular signaling** pathways in lymphocytes are among the most complex identified. Not only must these cells become activated like innate leukocytes, but they must also proliferate and differentiate into effector cells with vastly different powers from the parental cell. The binding of antigen to TCRs or BCRs therefore involves multiple receptor proteins and the recruitment of numerous PTKs and other molecules. The resulting cascade of PTK activation, calcium ion release and nuclear transcription factor activation leads to new molecules appearing on the lymphocyte surface, including receptors for cytokines that stimulate proliferation and effector cell differentiation. Another facet of the complexity of these signaling pathways is the fine degree of control necessary for regulating an adaptive immune response. Without extensive measures and countermeasures in place to channel the power of effector lymphocytes, the bystander damage to host tissues would be intolerable.

II. Intercellular Communication: Cytokines

i) The Nature of Cytokines

In both innate and adaptive immunity, there are many situations in which leukocytes must communicate with one another in order to achieve a response. As introduced previously, this intercellular communication is most often mediated by cytokines. Cytokines are low molecular weight peptides or glycoproteins of diverse structure and function, and well over 100 of them have been identified. Cytokines are secreted primarily, but not exclusively, by leukocytes, and regulate not only immune responses but also hematopoiesis, wound healing and other physiological processes. Cytokines are induced in both the adaptive and innate immune responses, and cytokines secreted in one type of response frequently stimulate the secretion of cytokines influential in the other type. In this context, cytokines act to intensify or dampen the response by stimulating or inhibiting the secretory activity, activation, proliferation or differentiation of various cells. This regulation is achieved by the intracellular signaling that is triggered when a cytokine binds to its receptor on the membrane of the cell to be influenced.

While several structural families can be discerned among cytokines, molecules of similar function can have very different structures, and those of similar structure can be very different in function. As well, although cytokines are genetically unrelated, many appear to be functionally redundant; that is, the same biological effect may result from the action of more than one cytokine. This overlap ensures that a critical function is preserved should a particular cytokine be defective or absent. A given cytokine can be produced by multiple cell types, may be *pleiotropic* (act on many different cell types), or have several different effects on the same cell type. Cytokines can also affect the action of other cytokines and may behave in a manner that is *synergistic* (two cytokines acting together achieve a result that is greater than additive) or *antagonistic* (one cytokine inhibits the effect of another). Among the most important cytokines are the *interleukins* (IL-1, IL-2, etc.), the *interferons* (IFNα, IFNβ and IFNγ), *tumor necrosis factor (TNF)*, *transforming growth factor beta (TGFβ)*, the *colony-stimulating factors (CSFs)*, and the *chemokines*. These molecules are described briefly in **Table 2-4**, and in more detail in the comprehensive listing of cytokines and their receptors, production and functions that appears in Appendix E.

Cytokines share several properties with hormones and growth factors, the other major communication molecules of the body, but also differ from them in important ways (**Table 2-5**). Like hormones and growth factors, cytokines are soluble proteins present in very low amounts and exert their effects by binding to specific receptors on the cell to be influenced. However, cytokines differ from hormones and growth factors in their sites of production, modes of operation and range of influence. Hormones are induced by specific stimuli and are synthesized in specialized glands. Hormones tend to operate in an *endocrine* fashion (over substantial distances or systemically) and influence only a very limited spectrum of target cells. Growth

TABLE 2-4	Principal Functions of Selected Important Cytokines and Chemokines		
Cytokine (Symbol)	**Principal Functions**	**Cytokine (Symbol)**	**Principal Functions**
Interferons		***TNF-related cytokines***	
Interferon (IFNα, IFNβ)	• Antiviral and antiproliferative • Promote inflammation	Tumor necrosis factor (TNF)	• Mediates inflammation • Has immunoregulatory, cytotoxic, antiviral effects • Can trigger either apoptotic or cell survival signaling
Interferon (IFNγ)	• Antiviral and antiproliferative • Promotes inflammation • Influences T cell subset differentiation and macrophage activation	Lymphotoxin (LT)	• Has TNF-like activities • Promotes formation of secondary lymphoid organs
Interleukin-1 (IL-1)	• Promotes inflammation and induces fever	B cell activating factor (BAFF)	• Essential for certain stages of B cell development
Interleukin-2 (IL-2)	• Stimulates B cell activation and T, B and NK cell proliferation • Promotes tolerance	**Transforming growth factor β (TGFβ)**	• Has immunosuppressive effects • Promotes T cell subset differentiation • Acts as chemoattractant for T cells, monocytes and neutrophils
Interleukin-3 (IL-3)	• Stimulates mast cell growth • Promotes antiparasitic responses of basophils and mast cells	***Colony-stimulating factors***	
Interleukin-4 (IL-4)	• Promotes T cell subset differentiation • Promotes humoral response	Stem cell factor (SCF)	• Promotes HSC survival, self-renewal and differentiation into progenitors
Interleukin-5 (IL-5)	• Promotes survival, differentiation and chemotaxis of eosinophils • Promotes histamine release in mast cells	Granulocyte-monocyte colony-stimulating factor (GM-CSF)	• Promotes generation of GMPs from HSCs
Interleukin-6 (IL-6)	• Promotes hematopoiesis and inflammation • Stimulates B and T cell subset differentiation	Granulocyte colony-stimulating factor (G-CSF)	• Acts on GMPs to generate granulocytes
Interleukin-7 (IL-7)	• Stimulates B and T cell development • Stimulates generation and maintenance of memory T cells	Monocyte colony-stimulating factor (M-CSF)	• Acts on GMPs to generate monocytes
Interleukin-10 (IL-10)	• Damps down inflammation • Has immunosuppressive effects on lymphocytes	***Chemokines***	
		*CCL2	• Chemoattractant for monocytes
Interleukin-12 (IL-12)	• Stimulates T cell subset differentiation • Induces production of IFNγ and other cytokines • Promotes cytotoxicity of CTLs and NK cells	CCL4	• Chemoattractant for monocytes, macrophages, and naïve T and B cells
		CCL5	• Chemoattractant for monocytes, memory T cells, eosinophils, basophils
Interleukin-15 (IL-15)	• Stimulates NK and γδ T cell development and proliferation • Stimulates mast cell proliferation • Promotes T cell activation, differentiation, homing and adhesion	†CXCL8	• Chemoattractant for neutrophils, basophils, T cells
		CXCL6	• Chemoattractant for neutrophils and NK cells
Interleukin-17 (IL-17)	• Promotes inflammation • Implicated in some autoimmune diseases	CXCL12	• Chemoattractant for T cells, DCs

*CCL, chemokine bearing a cysteine–cysteine motif
†CXCL, chemokine bearing a cysteine–X-cysteine motif

factors are usually produced constitutively and by individual cells (both leukocytes and non-leukocytes) rather than by glands. Many (but not all) growth factors can be detected at significant levels in the circulation, and a broad range of cell types is influenced. Taking the middle road, cytokines are synthesized under tight regulatory controls by a sizable variety of leukocyte and non-leukocyte cell types, and generally exert their effects in a fashion that is either *autocrine* (upon themselves) or *paracrine* (upon nearby cells).

TABLE 2-5	Cytokines versus Growth Factors and Hormones		
Property	**Cytokine**	**Growth Factor**	**Hormone**
Solubility	Soluble	Soluble	Soluble
Receptor required?	Yes	Yes	Yes
Site of production	Many cell types with wide tissue distribution	Several cell types with moderate tissue distribution	Specialized cell types in glands
Expression	Induced or upregulated	Constitutive	Induced
Range of effect	Autocrine or paracrine	Paracrine or endocrine*	Endocrine
Cell types influenced	Several	Broad range	Very limited

*In this context, "endocrine" means systemic effects

ii) Production and Control of Cytokines

Most cytokines are synthesized primarily by activated Th cells in response to antigen, or by activated macrophages in response to the presence of microbial or viral products. More modest contributions are made by other leukocytes and some non-leukocyte cell types. Several means are used to regulate cytokine production and effects. The half-lives of cytokines and their mRNAs are generally very short, meaning that new transcription and translation are required when an inducing stimulus is received. The effect is one of a transient flurry of cytokine production followed by resumption of a resting state in the absence of fresh stimulus. The influence of cytokines is also curtailed by controls on receptor expression. Cells lacking expression of the appropriate receptor may be sitting in a pool of cytokine but are unable to respond to it. Lastly, since most cytokines act only over a short distance, only those cells that are in the immediate vicinity of a producing cell (and express the required receptor) will be influenced.

iii) Functions of Cytokines

In general, there are three broad categories of cytokine function: regulation of the innate response, regulation of the adaptive response, and regulation of the growth and differentiation of hematopoietic cells. Many cytokines mediate aspects of all three, resulting in a constellation of effects on the proliferation, differentiation, migration, adhesion and/or function of a range of cell types. Thus, cytokines contribute to pro-. inflammatory and anti-inflammatory responses, antiviral responses, growth and differentiation responses, cell-mediated immune responses, humoral immune responses and chemotaxis.

The intracellular signaling triggered by the binding of a cytokine to its specific receptor results in new gene transcription and changes to cellular activities. The affinity of binding between cytokines and their receptors is high ($K = 10^{10} - 10^{12} \, M^{-1}$), so that picomolar concentrations of cytokine are often sufficient to produce a physiological effect. A frequent result of cytokine action is the induction of expression of another cytokine or its receptor, creating a cascade of receptivity and a coordinated response. It is therefore not unusual for one cytokine to induce the expression of a nuclear transcription factor needed to bind to the promoter of another cytokine gene and trigger production of this second cytokine. Alternatively, one cytokine may repress the expression of another cytokine or its receptor, creating an antagonistic situation.

Because the immune system cannot operate without the signals delivered by cytokines, their activities often overlap. The sharing of protein chains by different cytokine receptors forms the basis for at least some of this observed functional redundancy (**Fig. 2-7**). For example, the interleukins IL-2 and IL-15 show an overlap in function in that both molecules stimulate the proliferation of T lineage cells. The interleukin-2 receptor (IL-2R) contains three chains: the unique IL-2Rα chain that binds to the IL-2 molecule, and the IL-2Rβ and γ_c chains that convey intracellular signals to the cell's nucleus. The interleukin-15 receptor (IL-15R) contains a unique

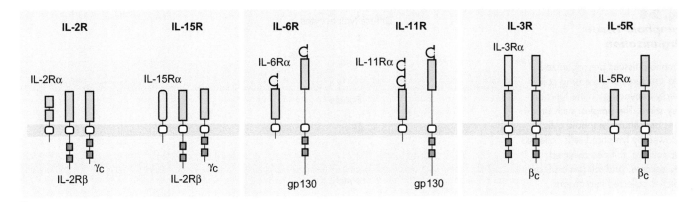

Fig. 2-7
Sharing of Cytokine Receptor Chains

Cytokine receptors are usually composed of two or more polypeptide chains anchored in the plasma membrane (gray shading) by transmembrane domains (white squares). A unique chain is responsible for binding a particular cytokine, while the other chain or chains are responsible for intracellular signaling and may be shared by other cytokine receptors. In the examples shown here, IL-2R and IL-15R share the γc chain and IL-2Rβ chain (left), IL-6R and IL-11R share the gp130 chain (centre), and IL-3R and IL-5R share the βc chain (right). Of interest, the γc chain also appears in several other cytokine receptors, including IL-4R, IL-7R, IL-9R and IL-21R (not shown).

IL-15Rα chain that binds to the IL-15 molecule but also the IL-2Rβ and γ_c chains. In fact, the γ_c chain (the "common gamma chain") is shared quite widely, also being present in the receptors for IL-4, IL-7, IL-9 and IL-21. In addition to multiple cytokines having the same effect, more than one effect of an individual cytokine may impinge on a given cell over a period of time, with some aspects experienced immediately or within minutes, and others delayed by hours or even days. *In vivo*, a cell may be exposed to a complex mixture of cytokines for an extended period, making the cellular outcome difficult to predict.

iv) Rationale for Cytokine Network Complexity

Why should evolution have created this complex web of cytokine interactions? Cytokines are a remarkably flexible means of controlling the immune response, which can be harmful to the host if too vigorous or too long in duration. Positive and negative feedback loops, agonistic (activating) and antagonistic (inhibiting) relationships and redundant functions exist among the cytokines to provide a fine level of control over the powerful cells and effector mechanisms that can be unleashed to contain a pathogen attack. Without such controls, extreme immunopathic damage to host tissues or even autoimmune disease could result. In most cases, such multi-level structuring also ensures that no matter what strategy the pathogen uses to invade the host and avoid immune surveillance, a cytokine can be induced to mobilize an effective response. If this cytokine fails, there are others with overlapping functions that can fill the gap.

C. Lymphoid Tissues

I. Overview

A **lymphoid tissue** is simply a tissue in which lymphocytes are found. Lymphoid tissues range in organization from diffuse arrangements of individual cells to encapsulated organs (**Fig. 2-8**). **Lymphoid follicles** are organized cylindrical clusters of lymphocytes that, when gathered into groups, are called **lymphoid patches**. **Lymphoid organs** are usually groups of follicles that are surrounded, or encapsulated, by specialized supporting tissues and membranes.

Lymphoid tissues are classified as being either **primary** or **secondary** in nature (**Table 2-6**). The **primary lymphoid tissues** in mammals are the bone marrow and the

Fig. 2-8
*Lymphoid Tissue
Organization*

Lymphoid tissues are organized into structures of increasing complexity depending on the function they serve. The simplest such structures are diffuse lymphoid cells, followed by lymphoid cells collected into follicles, follicles collected into patches, and follicles and/or patches collected into organs.

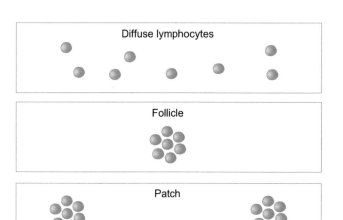

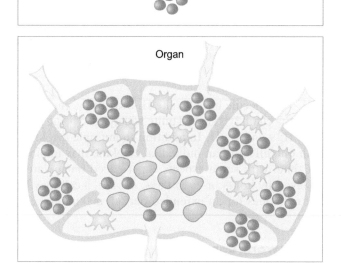

TABLE 2-6	Primary and Secondary Lymphoid Tissues
Primary Lymphoid Tissues	**Secondary Lymphoid Tissues**
Bone marrow	Lymph nodes
Thymus	Spleen
	MALT (mucosa-associated lymphoid tissue)
	• NALT (nasopharynx-associated lymphoid tissue)
	• BALT (bronchi-associated lymphoid tissue)
	• GALT (gut-associated lymphoid tissue)
	SALT (skin associated lymphoid tissue)

thymus. It is in these sites that lymphocytes develop and central tolerance is established; that is, most lymphocytes recognizing self antigens are removed, and only cells recognizing foreign antigens are permitted to mature. All lymphocytes arise from HSCs in the bone marrow and B cells largely mature in this site, whereas newly produced T cell precursors migrate from the bone marrow to the thymus and mature in this location. Once mature, naïve T and B cells then leave the primary lymphoid tissues and migrate through the blood and lymph (see later) to take up residence in the secondary lymphoid tissues.

Secondary lymphoid tissues provide sites for both antigen accumulation and the gathering of the leukocytes (primarily APCs and T and B lymphocytes) that must

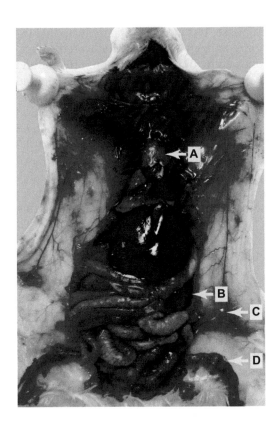

Mouse Primary and Secondary Lymphoid Organs and Tissues

Dissected mouse, showing (A) thymus, (B) spleen, (C) lymph node, and (D) bone marrow. Reprinted with permission.

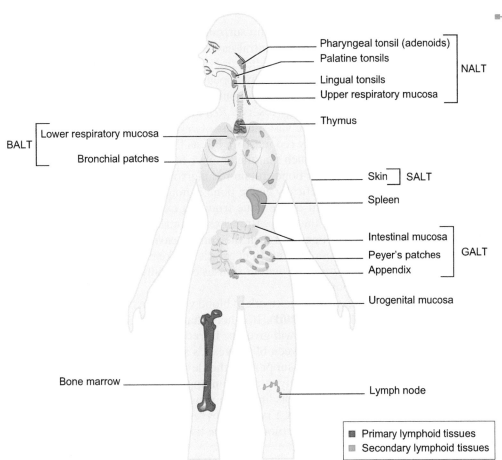

➡ **Fig. 2-9**
Primary and Secondary Lymphoid Tissues

Primary lymphoid organs, where most lymphocytes develop, are shown in purple. Secondary lymphoid organs and tissues, where most lymphocytes become activated by antigen collected in these structures, are shown in blue.

Pharyngeal tonsil (adenoids)
Palatine tonsils
Lingual tonsils
Upper respiratory mucosa
NALT

Thymus

BALT
Lower respiratory mucosa
Bronchial patches

Skin] SALT

Spleen

Intestinal mucosa
Peyer's patches
Appendix
GALT

Urogenital mucosa

Bone marrow

Lymph node

■ Primary lymphoid tissues
■ Secondary lymphoid tissues

cooperate to mount an optimal adaptive response. It is in the secondary lymphoid tissues that mature naïve T and B cells recognize antigen and become activated, undergoing clonal selection and proliferation followed by differentiation into effector and memory cells. Once generated, the effector cells migrate from the secondary lymphoid tissues to infected organs or tissues to wage war on invading pathogens. Since pathogens can attack at any location, the secondary lymphoid tissues are widely distributed throughout the body's periphery. Secondary lymphoid tissues in mammals include the **lymph nodes**, the **spleen**, the **skin-associated lymphoid tissue (SALT)**, and the **mucosa-associated lymphoid tissue (MALT)** present in the bronchi (BALT), nasopharynx (NALT) and gut (GALT). The lymph nodes collect antigen moving through the body in the lymphatic system (see later), the spleen traps antigen circulating in the blood, and the SALT and various types of MALT deal with antigens attempting to penetrate through the skin or mucosae, respectively. A photograph of a dissected mouse showing the positions of the major lymphoid organs and tissues as they sit in the body is shown in **Plate 2-2**, and a diagram of the localization of human lymphoid tissues appears in **Figure 2-9**.

II. Primary Lymphoid Tissues

i) The Bone Marrow

The bone marrow is the primary site of hematopoiesis in the adult human. In a healthy individual, the total bone marrow cell population consists of about 60–70% myeloid lineage cells, 20–30% erythroid lineage cells, and 10% lymphoid lineage cells, with the remainder consisting of mast lineage cells plus various other non-hematopoietic cell types such as stromal cells and adipocytes (fat cells). Many of these other cell types are vital for hematopoiesis because they secrete the cytokines and growth factors that are required for blood cell maturation.

The compact outer matrix of a bone surrounds a *central cavity* (also called the *medullary cavity*) (**Fig. 2-10A**). The central cavity has a honeycomb structure made up of thin strands of connective tissue called *trabeculae*. Within the cavities created by the trabeculae lies the marrow, which may appear red or yellow in color. The *yellow marrow* is usually hematopoietically inactive but contains significant numbers of adipocytes that act as an important energy reserve. *Red marrow* is hematopoietically active tissue that gains its color from the vast numbers of erythrocytes produced. The bones of an infant contain virtually only red marrow, but as the child grows, the demand for hematopoiesis slackens slightly such that the number of bones with active red marrow declines. In the adult, only a few bones retain red marrow, including the sternum, ribs, pelvis and skull. A hematopoietically active bone is nourished by one or more nutrient arteries that enter the shaft of the bone from the exterior, as well as by arteries that penetrate at the ends of the bone. Branches of nutrient arteries thread through the *Haversian canals* of the bone to reach the central cavity. The circulatory loop is completed when the branches of the nutrient arteries make contact with small vascular channels called *venous sinuses*. The venous sinuses eventually feed into nutrient veins that exit the shaft of the bone.

Within a trabecular cavity, the network of blood vessels surrounds groups of developing hematopoietic cells (**Fig. 2-10B**). It is here that the HSCs either self-renew or differentiate into progenitors that will give rise to red and white blood cells of all lineages. Tucked into the spaces between blood vessels are interconnecting stromal cells and fibers that form a framework supporting the developing hematopoietic cells. As the hematopoietic cells of a given type reach an appropriate stage of maturity, they squeeze between the endothelial cells lining the venous sinuses, enter a nutrient vein, and finally join the blood circulation. Erythrocytes remain in the circulation, while leukocytes are distributed between the blood and the tissues and in some cases may complete their maturation in the latter.

(A) Cross-section of Bone

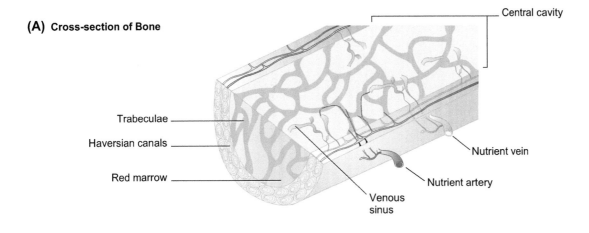

Central cavity

Trabeculae

Haversian canals

Red marrow

Nutrient vein

Nutrient artery

Venous sinus

(B) Red Marrow

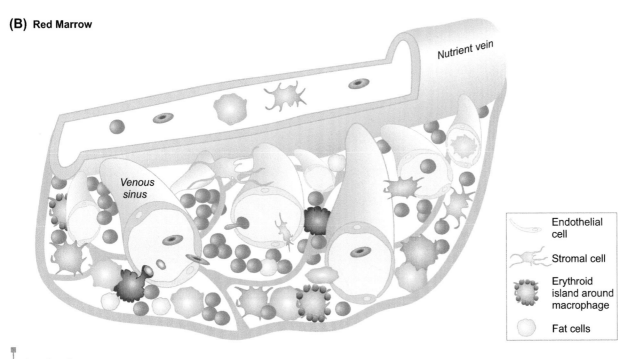

Nutrient vein

Venous sinus

Endothelial cell

Stromal cell

Erythroid island around macrophage

Fat cells

Fig. 2-10
The Bone Marrow

(A) Cross-section showing the central cavity of a long bone (such as a femur). **(B)** Cross-section showing hematopoietic cells developing in the red marrow within the central cavity. At the appropriate developmental stage, these cells enter the venous sinuses, leave the central cavity via the nutrient veins, and travel to a peripheral tissue. Various non-hematopoietic cell types (see Detail) and connective tissue structures provide the microenvironment necessary to nurture hematopoietic cells.

ii) The Thymus

The thymus is a lymphoid organ located above the heart, and it is in the thymus that immature T cells complete their development. Hematopoiesis in the bone marrow generates lymphoid precursors that leave the bone marrow and enter the blood circulation. The epithelial cells of the thymus secrete chemokines that specifically attract these precursors from the blood into the thymus, where many of them become **thymocytes**. During their proliferation and maturation in the thymus, thymocytes undergo a process called **thymic selection** that determines the specificities comprising the mature

Fig. 2-11
The Thymus

(A) Cross-section showing multiple lobules of the thymus, each with its own medulla and cortex. Detail: blood supply of one lobule.
(B) Cross-section showing the various cell types and structures present within the cortex and medulla of a single thymic lobule. *[With information from Klein, J., & Hořejší V. (1997). Immunology, 2nd ed. Blackwell Science, Osney Mead, Oxford.]*

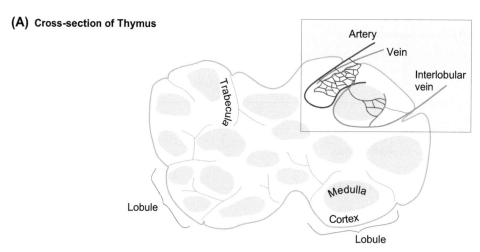

(A) Cross-section of Thymus

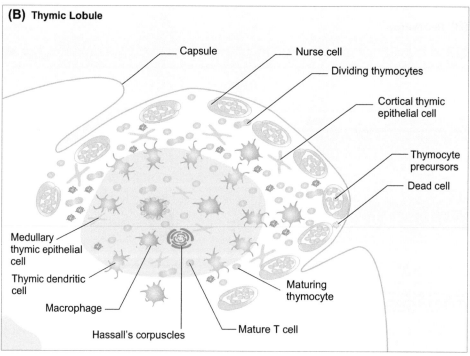

(B) Thymic Lobule

The development of T cells, including thymic selection, is described in detail in Chapter 9.

T cell repertoire. Thymic selection has two parts: **positive selection** and **negative selection**. Positive selection ensures that only thymocytes expressing TCRs with at least some binding affinity for the host's MHC molecules survive. Negative selection ensures that thymocytes with TCRs that recognize pMHC complexes in which the peptide is derived from a self antigen are eliminated. As a result, self-reactive clones are deleted from the T cell repertoire and only the T cells most likely to recognize foreign entities survive. Only about 1% of all thymocytes survive positive and negative thymic selection, making the thymus a site of both tremendous T cell proliferation and T cell apoptosis.

Structurally, the thymus is a bi-lobed organ in which each lobe is encapsulated and composed of multiple lobules that are separated from one another by trabeculae (**Fig. 2-11A**). As thymocytes mature through various stages, they generally move from the densely packed outer **cortex** of a lobule into its sparsely populated inner **medulla**. Along the way, the developing thymocytes interact with and are supported by **cortical thymic epithelial cells (cTECs)**, **medullary thymic epithelial cells (mTECs)**, thymic DCs and macrophages (**Fig. 2-11B**). In the outer edges of the cortex, specialized epithelial cells called *nurse cells* form large multicellular complexes that envelop up to 50 maturing thymocytes within their long processes. At the junction of the cortex and medulla, thymic DCs begin to take on this nurturing role. Epithelial cells that have exhausted their supportive function within the thymus degenerate, forming the **Hassall's corpuscles** observed in the medulla.

After puberty, the thymus starts to regress in a process called **thymic involution**. Most of the lymphoid components of the thymus are eventually replaced with fatty connective tissue, greatly decreasing thymic production of mature T cells. However, certain T cell subsets can mature outside the thymus in parts of the intestine. This **extrathymic development** provides for a limited amount of mature T cell generation in the adult.

III. Secondary Lymphoid Tissues

i) MALT and SALT

The mucosa-associated lymphoid tissues (MALT) and the skin-associated lymphoid tissues (SALT) are the first elements of the immune response encountered by a pathogen that has overwhelmed the body's passive anatomical and physiological barriers. The leukocyte subsets that populate the MALT and SALT are situated at the most common points of antigen entry, behind the mucosae of the respiratory, gastrointestinal and urogenital tracts, and just below the skin (refer to **Fig. 2-9**). DCs, macrophages, γδ T cells, NK cells and NKT cells, as well as mature T and B lymphocytes, are positioned in these locations to counter pathogens as they breach a body surface. Thus, both innate and adaptive leukocytes stand ready in the MALT and SALT to protect the body at all surfaces that interface with the external environment.

The MALT is subdivided by body location and includes the NALT, BALT and GALT. The nasopharynx-associated lymphoid tissue (NALT) is located in the upper respiratory tract. The bronchi-associated lymphoid tissue (BALT) is located in the lower respiratory tract and the lungs. The gut-associated lymphoid tissue (GALT) is located in the digestive tract. Throughout most of the MALT, the leukocytes are dispersed in diffuse masses just under the layer of mucosal epithelial cells that forms the wall of the body tract. In some cases, slightly more organized collections of cells exist, such as the Peyer's patches of the GALT. In other cases, the cells are organized into discrete structures such as the appendix in the GALT and the tonsils of the NALT.

The SALT comprises small populations of leukocytes resident in the epidermis (upper layer) and dermis (lower layer) of the skin. Dendritic cells known as **Langerhans cells** are scattered throughout the epidermis; these cells function as APCs and secrete cytokines drawing lymphocytes to an area under attack. The underlying dermis is dominated by T cells, dermal DCs and macrophages.

How the various elements of the MALT and SALT act to eliminate antigens is discussed in more detail in Chapter 12.

ii) The Lymphatic System

All cells in the body are bathed by nutrient-rich interstitial fluid. This fluid is blood plasma that, under the pressure of the circulation, leaks from the capillaries into spaces between cells (**Fig. 2-12**). Ninety percent of this fluid returns to the circulation via the venules, but 10% filters slowly through the tissues and eventually enters networks of tiny channels known as the *lymphatic capillaries*, where it becomes known as **lymph**. The overlapping structure of the endothelial cells lining the lymphatic capillaries creates specialized pores that allow microbes, leukocytes and large macromolecules to also pass into the lymphatic capillaries, where they are collected in the lymph. Valves in the lymphatic capillaries ensure that the lymph and its contents move only forward as the lymphatic capillaries collect into progressively larger *lymphatic vessels*. These vessels in turn connect with one of two large *lymphatic trunks* called the *right lymphatic duct* and the *thoracic* (or *left lymphatic*) *duct*. The entire network of vessels and ducts that collect and channel the lymph and its contents throughout the body is known as the *lymphatic system* (**Fig. 2-13**). The right lymphatic duct drains the right upper body, while the entire lower body drains into the *cisterna chyli* at the base of the thoracic duct. Lymph from the left upper body also enters the thoracic duct. The right lymphatic duct empties the lymph into the *right subclavian vein* of the blood circulation, while

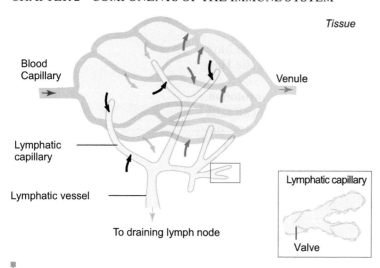

Fig. 2-12
The Collection of Lymph in Peripheral Tissues

Blood enters a tissue's blood capillaries (red arrow), and some plasma leaks into the tissue (light gray arrows). Most of this plasma (dark gray arrows) is reabsorbed by the venules and rejoins the circulation (blue arrow). The 10% of blood plasma remaining in the tissue collects foreign entities as it percolates through the tissue and becomes lymph (black arrows) as it enters the lymphatic capillaries. These small vessels in turn feed into progressively larger lymphatic vessels that eventually connect with the secondary lymphoid organs. Inset: Valves within lymphatic capillaries and lymphatic vessels ensure that the lymph flows in one direction only.

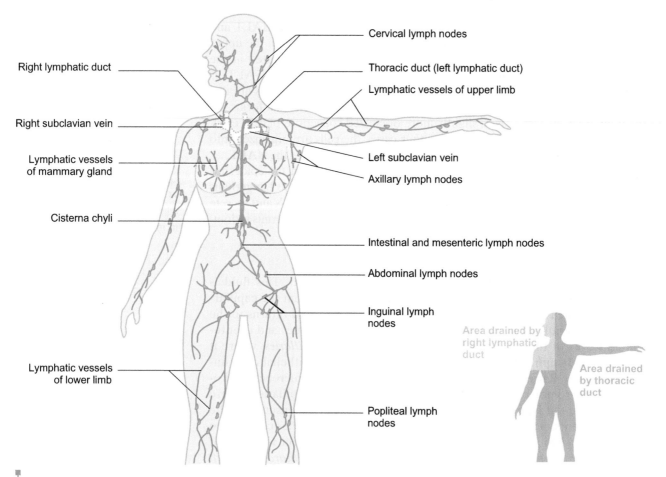

Fig. 2-13
Major Vessels and Nodes of the Lymphatic System

Lymphatic vessels and lymph nodes are named according to their location in the body. Entities collected in the lymph are conveyed through the lymphatic capillaries to lymphatic vessels that ultimately connect with the blood circulation. Detail: All lymphatic vessels draining the upper right quadrant of the body feed into the right lymphatic duct, which empties into the right subclavian vein. All other parts of the body are drained by lymphatic vessels that empty into the thoracic duct, which empties into the left subclavian vein.

the thoracic duct connects with the *left subclavian vein*. As a result, the lymphatic system is connected to the blood circulation, ensuring that no matter how a pathogen accesses the body, it will be trapped and eventually conveyed to a secondary lymphoid organ.

iii) Lymph Nodes

As lymph flows through the lymphatic vessels, it passes through the **lymph nodes**. The lymph node is the major site for the interaction of lymphocytes with antigen during a primary adaptive response. Antigen that finds its way past the body's innate defenses and escapes MALT and SALT is collected in the lymph and brought to the nearest lymph node (referred to as the **draining lymph node**). Lymph nodes occur along the entire length of the lymphatic system but are clustered in a few key regions. For example, the *cervical* lymph nodes drain the head and neck, while the *popliteal* lymph nodes drain the lower legs (refer to **Fig. 2-13**).

Lymph nodes are bean-shaped, encapsulated structures 2–10 mm in diameter that contain large concentrations of lymphocytes, FDCs and APCs (**Fig. 2-14**). Lymph enters a lymph node through several **afferent lymphatic vessels**. It then passes through the **cortex, paracortex** and **medulla** of the node, and exits on the opposite side through a single **efferent lymphatic vessel**. The cortex contains large numbers of resting B cells, FDCs and macrophages arranged in lymphoid follicles, while the paracortex is home to many loosely scattered T cells and DCs. The medulla becomes well stocked with antibody-secreting plasma cells during an adaptive response.

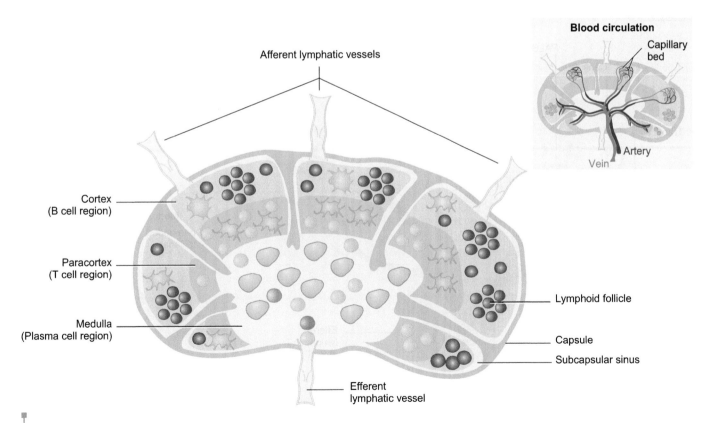

Fig. 2-14
The Lymph Node

Cross-section of a lymph node, showing the capsule and subcapsular sinus that enclose the cortex, paracortex and medulla. Lymph and its molecular and cellular contents enter the node through multiple afferent lymphatic vessels and leave through a single efferent lymphatic vessel. Detail: Blood enters by an artery near the efferent lymphatic, flows through capillary beds in the cortex, and leaves the lymph node via the lymph node vein.

iv) The Spleen

Most antigens escaping the innate immune response, MALT and SALT make their way into the tissues, are collected in the lymphatic system and are channeled into local lymph nodes. However, there are several ways by which an antigen can access the blood circulation. (1) Sometimes an antigen is introduced directly into the blood, as during drug injection or via insect or snake bites. (2) Overwhelming local infection at skin and mucosal sites can result in penetration of underlying blood vessels by the pathogen. (3) Systemic infection that cannot be contained by the lymph nodes may pour into the efferent lymph so that antigens are eventually dumped into the blood. Fortunately, we have the **spleen**, an abdominal organ that traps blood-borne antigens. The entire blood volume of an adult human courses through the spleen four times daily.

The structural framework of the spleen is created by a network of trabeculae that branch out of the *hilus* (base) of the spleen and connect to a thin exterior capsule (**Fig. 2-15**). Antigen carried in the blood circulation enters the spleen by a single artery called the *splenic artery*. The splenic artery branches into a network of smaller arteries that travel through the organ via the hollow spaces created by the trabeculae. These smaller arteries branch into *arterioles* that penetrate the trabeculae. The arterioles in turn branch off into capillaries, some of which connect directly with the venous sinuses. Other capillaries are open-ended and empty into the *splenic cords*, which are collagen-containing lattices through which blood percolates before entering the venous sinuses. Blood collecting in the venous sinuses empties into small veins that connect with the *splenic vein* exiting the spleen.

White and Red Pulp

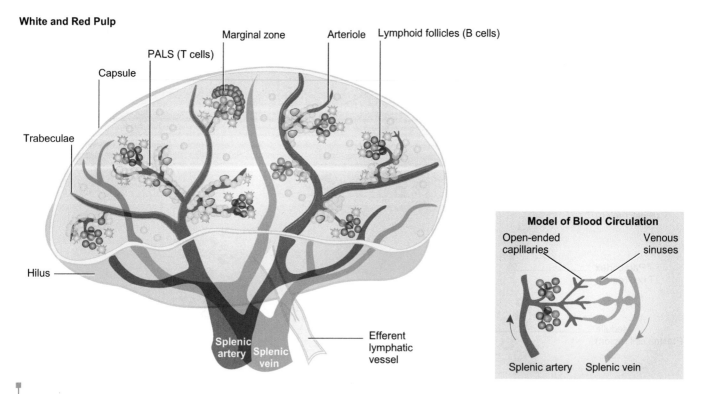

Fig. 2-15
The Spleen

Cross-section of the spleen, showing how trabeculae create a connective tissue network that branches out from the hilus and acts as a system of tunnels for splenic arteries. Smaller arterioles branch outside the trabeculae and are surrounded by the T cell-rich periarteriolar lymphoid sheath (PALS) and the B cell-rich follicles that together make up the white pulp. Detail: Splenic arterioles branch into either closed capillaries or open-ended capillaries that convey blood bearing antigens into the splenic tissue. Blood is reabsorbed by the venous sinuses that eventually rejoin splenic vein.

Each arteriole in the spleen is encased by a **periarteriolar lymphoid sheath (PALS)**, a cylindrical arrangement of cells dominated by mature T cells but also including low numbers of plasma cells, macrophages and cDCs. The splenic tissue surrounding the PALS is filled with lymphoid follicles containing resting B cells and macrophages. Surrounding the follicles is the **marginal zone**, which contains particular B cell subsets. **White pulp** is the name given to that part of the spleen containing the splenic arterioles with their PALS, the follicles and the marginal zone. **Red pulp**, which is named for its abundance of erythrocytes and surrounds the white pulp, consists of the splenic cords and the venous sinuses. The red pulp functions chiefly in the filtering of particulate material from the blood and in the disposal of senescent or defective erythrocytes and leukocytes.

D. Cellular Movement in the Immune System

I. Leukocyte Extravasation

A unique feature of the immune system is that its constituent cells are not fixed in a single organ but instead move around the body in the blood circulation to tissues where they are needed. To exit from the blood into a tissue under attack, leukocytes carry out a migration process called **extravasation** (**Fig. 2-16**). A tissue suffering from trauma or infection releases chemokines and inflammatory molecules that activate endothelial cells in the postcapillary venules nearest the affected site. These activated endothelial cells upregulate their expression of cellular adhesion molecules called *E-selectins*, which can bind to complementary glycoproteins present on the surface of a leukocyte that happens to be circulating through the venule. The leukocyte is thus temporarily "tethered" to the venule wall. However, these bonds are relatively weak and can be reversed in seconds, so that the leukocyte "rolls" across several activated endothelial cells. The sequential binding exerted by these cells increasingly slows the leukocyte, resulting in a phenomenon called "*slow rolling.*" During tethering and rolling, intracellular signaling is delivered that activates the leukocyte and causes it to display new

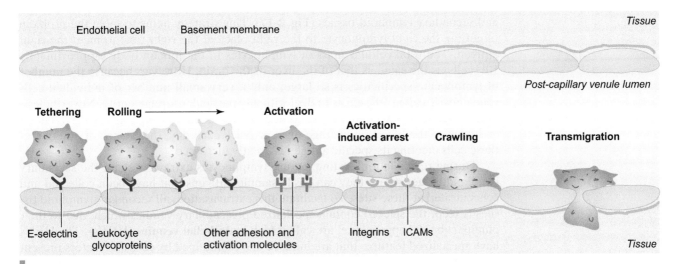

Fig. 2-16
Leukocyte Extravasation

Interactions between complementary adhesion molecules present on leukocytes and activated endothelial cells of a post-capillary venule allow the leukocyte to exit the blood and enter the injured tissue below. The process starts with tethering, in which E-selectins (dark red) expressed by activated endothelial cells bind to glycoprotein receptors (brown) on the resting leukocyte. Rolling over additional endothelial cells induces signaling within the leukocyte that activates it and triggers the expression of new adhesion molecules (pink), which bind to counter-receptors on the endothelial cells (purple). Interaction of newly expressed integrins (green) on the leukocyte with ICAMs (blue) on the endothelial cells causes the leukocyte to undergo activation-induced arrest and express additional new proteins that allow it to crawl over the surface of the endothelial cell. The activated leukocyte eventually transmigrates between two endothelial cells into the tissue.

adhesion and activation molecules on its surface. Among these molecules are the *integrins*, which bind to **intercellular adhesion molecules (ICAMs)** expressed by the endothelial cells. Integrin-mediated binding is strong enough to reduce the movement of the leukocyte and to finally arrest it on the endothelial cell surface. This *activation-induced arrest* of the leukocyte permits it to further strengthen its adhesive contacts with the endothelial cells, and then to initiate new gene transcription that allows the leukocyte to begin *crawling* over the endothelial cell surface. Eventually, the leukocyte gains the capacity to move across the endothelial layer into the affected tissue below in a process known as **transmigration**. Most often, the crawling leukocyte inserts a pseudopod between two endothelial cells of the venule lining at the cell-cell junction, squeezes between them, and secretes enzymes that digest the basement membrane supporting the endothelium. The leukocyte then completes its migration through the venule wall into the tissue. Once the leukocyte has entered the tissue, the binding of additional integrin molecules to components of the extracellular matrix allows it to migrate through the tissue and follow the local chemokine gradient to the specific inflammatory site where it is needed.

> NOTE: The type of transmigration described in the preceding text is sometimes called "paracellular transmigration." There is another rare type of transmigration, called "transcellular transmigration," in which the extravasating leukocyte uses a tight-fitting pore to pass right through an endothelial cell into the tissue below, rather than squeezing between two endothelial cells at their cell-cell junction.

II. Lymphocyte Recirculation

Although all leukocytes can extravasate into inflammatory sites, only lymphocytes can regularly shuttle back and forth between the blood and tissues in the absence of inflammation. This process is called **lymphocyte recirculation** and involves the continuous migration of most resting T cells and some B cells between the blood, lymph and secondary lymphoid tissues (**Fig. 2-17**). This strategy helps to solve the problem of getting the right lymphocyte to the right place at the right time to meet the right antigen. The total cellular mass of lymphocytes in the human body (approximately 10^{12} cells) is the same as that of the liver or the brain. However, because the number of lymphocyte specificities is so large, only a very small number of individual cells (one naïve T cell in 10^5) exists to deal with any particular foreign entity. Nevertheless, because naïve cells recirculate from blood to the secondary lymphoid organs to lymph and back to the blood on average once or twice each day, the chance of any one of these cells meeting its specific antigen is greatly increased.

During recirculation, resting mature lymphocytes extravasate into the secondary lymphoid tissues where they can easily sample antigens that have been collected and concentrated at these sites. To facilitate this extravasation, all secondary lymphoid tissues (except the spleen) contain specialized postcapillary venules which, due to their plump, cuboidal appearance, are called **high endothelial venules (HEVs)**. The HEVs have specialized features that are induced and maintained by cells and factors present in the afferent lymph coming into lymphoid tissues. In particular, the endothelial cells lining the HEVs constitutively express high levels of adhesion molecules that promote lymphocyte extravasation. Indeed, it has been estimated that, throughout the body, a total of 5×10^6 lymphocytes exit the blood through the HEVs into a secondary lymphoid tissue each second. Thus, naïve lymphocytes travel through every lymph node every day, thereby efficiently and thoroughly surveying sites of antigen collection in the host on a continuous basis.

Most lymphocytes that enter a given lymph node fail to encounter their specific antigens and exit directly via the efferent lymphatic vessel without undergoing activation. These cells eventually rejoin the blood circulation via the connections

Naïve B and T cells within a lymph node have been clocked moving at 10 μm/min and 6 μm/min, respectively.

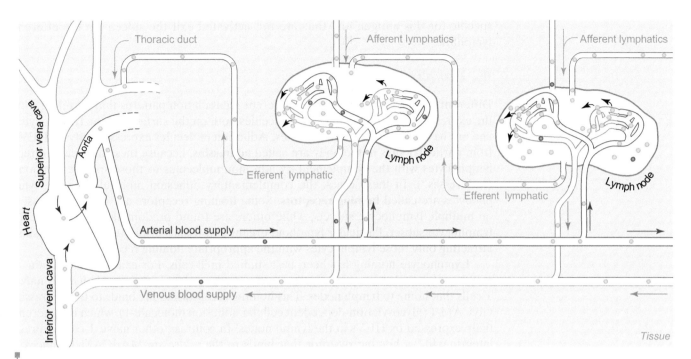

Fig. 2-17
Schematic Representation of Lymphocyte Recirculation

Lymphocytes in the blood circulation enter the heart via the superior and inferior vena cava (purple arrows). The heart pumps these cells out through the aorta into arterial blood (red arrows), which eventually takes them into the capillaries supplying each tissue in the body. In secondary lymphoid tissues, including the lymph nodes illustrated here, the post-capillary venules take the form of specialized high endothelial venules (HEVs) that allow lymphocytes to continuously extravasate from the blood into the interior of the node (black arrows) to search for antigens. Lymphocytes that do not extravasate into the node in this way return to the heart via the venous blood supply (blue arrows). Lymphocytes that extravasate into the node but do not encounter specific antigen (and so are not activated) enter a lymphatic capillary and eventually the efferent lymphatic vessel (gray arrows) of the node. These cells then make their way back to the heart via the lymphatic system. Lymphocytes that do encounter specific antigen while in the node are activated and remain there to generate effector and memory cells, which later exit the node.

between the lymphatic system and the left and right subclavian veins. However, a few lymphocytes may find their specific antigens deposited in the cortex or paracortex of the lymph node. A B cell may recognize antigen held by FDCs in the cortex, whereas a T cell may recognize a pMHC presented by a DC in the paracortex. Engagement of sufficient antigen receptors by antigen will trigger the activation of these B and T lymphocytes. Unlike the original naïve B cell, the effector plasma cells and memory B cells generated following antigen encounter tend to remain in the node and do not recirculate. Instead, the antibodies secreted by the plasma cells diffuse through the medulla of the node, exit it via the efferent lymphatic, and eventually enter the blood to provide systemic protection. In contrast, the effector and memory T lymphocytes generated by naïve T cell activation within a lymph node exit the node via the efferent lymphatic and join the blood circulation. However, as is described later, these cells then migrate to the peripheral tissue that is under attack (i.e., the source of the antigen that was collected in the original lymph node) and do not recirculate further.

In the spleen, the situation is slightly different. Unlike a lymph node, the spleen has no afferent lymphatic or HEVs, so that lymphocytes must enter the spleen via the splenic artery. The cells exit the blood via capillary beds in the marginal zone of the spleen, and then migrate to its follicles and PALS. Antigen in the blood percolating through this area is taken up by resident DCs and used to activate T cells in the PALS. Alternatively, macrophages patrolling the red pulp can capture antigen collected here and convey it to lymphocytes in the PALS. Lymphocytes that are not specific for this antigen and thus are not activated exit the spleen via the efferent lymphatic.

III. Lymphocyte Homing

Different lymphocyte subsets show different recirculation patterns that correlate with the expression of varying adhesion molecules both on the surfaces of the lymphocytes and on the HEVs of various tissue sites. Adhesion molecules expressed only on HEVs from particular sites in the body are called **addressins**, because these molecules direct lymphocytes with the complementary adhesion molecules to those specific locations ("addresses"). In these cases, the complementary adhesion molecules on the lymphocytes are called **homing receptors**. Some homing receptors are widely expressed on multiple lymphocyte subsets, while others are found predominantly on a specific lymphocyte subset. Particular lymphoid organs and tissues display distinct addressins, attracting only those lymphocytes with the appropriate homing receptor.

Lymphocyte homing has been best studied in T cells. For example, L-selectin, a member of the selectin family of adhesion molecules, is highly expressed on many naïve T cells that home to lymph nodes. This homing receptor readily binds to the addressin GlyCAM-1 (glycosylation-dependent cellular adhesion molecule-1), which is preferentially expressed by HEVs in the lymph nodes. In contrast, other naïve T cells express integrin-α4β7, a homing receptor that binds to the addressin MAdCAM-1 (mucosal addressin cellular adhesion molecule-1). MAdCAM-1 is exclusively expressed on HEVs in the GALT, so that T cells expressing integrin-α4β7 home to the gut.

As well as tissue-specific differences in homing receptors, there are homing differences determined by the maturation and differentiation status of a lymphocyte. While naïve T cells express high levels of L-selectin, effector and memory T cells possess very little of this homing receptor and so no longer recirculate through the lymph nodes. Instead, effector and memory T cells express homing receptors such as VLA-4 (very late antigen-4) that mediate their migration to peripheral tissues in which inflammation has been initiated. The endothelial cells of the postcapillary venules in these sites express new vascular addressins like VCAM-1 (vascular cellular adhesion molecule 1) that bind to VLA-4 and permit the extravasation of effector and memory lymphocytes as well as neutrophils and monocytes from the blood into the affected tissue. Because this leukocyte influx includes effector and memory lymphocytes, such sites of inflammation are sometimes called **tertiary lymphoid tissues**. If the inflammation does not resolve and becomes chronic, the postcapillary venules in the affected site take on the characteristics of HEVs, inducing lymphocytes to constantly traffic through the area.

Although effector T cells generally die in the peripheral tissue in which they have been sent to combat an invader, memory T cells generated in a primary response express homing receptors that allow them to return repeatedly (after the inflammation is resolved) to monitor the peripheral tissue in which they first encountered antigen. For example, memory T lymphocytes expressing integrin-α4β7 are drawn to MAdCAM-1-expressing endothelial cells in the intestinal Peyer's patches. Likewise, memory T lymphocytes of the MALT and SALT recirculate in the mucosae and skin, respectively, under the influence of mucosal and cutaneous addressins. This tactic makes evolutionary sense because the triggering antigen is likely to enter the body in the same fashion in a subsequent infection. The restricted homing of memory cells thus increases their chances of being in the right place at the right time to encounter the specific antigen, contributing to the shorter lag time of the secondary adaptive response.

Now that the reader is familiar with the cells and tissues of the immune system, we return in the next chapter to innate immunity, to describe in detail how most foreign entities attacking our bodies are defeated.

Chapter 2 Take-Home Message

- The immune system is a partnership of the blood circulation, the lymphatic system, various lymphoid organs and tissues, and the hematopoietic cells moving among them.

- Hematopoietic cells in the bone marrow give rise to myeloid, lymphoid and mast cell progenitors that differentiate into the mature leukocytes that populate the body. Host immune system homeostasis is achieved by a balance of hematopoiesis and apoptosis.

- The primary lymphoid tissues are the bone marrow and the thymus. The secondary lymphoid tissues are the lymph nodes, the spleen, the MALT and the SALT. Lymphocytes develop in the primary lymphoid tissues and are activated by antigen in the secondary lymphoid tissues. Tertiary lymphoid tissues are inflamed tissues that have been infiltrated by effector lymphocytes.

- The binding of antigen, PAMPs, DAMPs, cytokines or other ligands to the appropriate receptors triggers intracellular signaling that results in particular cytoplasmic or nuclear outcomes with the potential to alter cell behavior.

- The primary function of cytokines is to mediate intercellular signaling between leukocytes. The induced innate and adaptive responses are largely dependent on cytokine-mediated signaling.

- Extravasation of leukocytes is mediated by the binding of complementary adhesion molecules that are upregulated in response to inflammation.

- Lymphocyte recirculation allows naïve lymphocytes to continuously patrol the body's sites of antigen entry and collection, even in the absence of inflammation.

Did You Get it? A Self-Test Quiz

Section A.I–III

1) What are hematopoietic cells? What are leukocytes?

2) Name four types of myeloid cells and three types of lymphoid cells.

3) Can you describe how chemotaxis works?

4) Can you define these terms? granulocyte, phagosome, degranulation, APC, Fc region

5) Describe the functions of four types of myeloid cells.

6) How are neutrophils linked to the adaptive response?

7) How are macrophages linked to the adaptive response?

8) How do macrophages become hyperactivated, and why is this important?

9) Can you define these terms? naïve, cross-reacting antigen, lymphoblast

10) What are CD markers, and why are they useful?

11) Describe the functions of four types of lymphoid cells.

12) Describe how NK cells link innate and adaptive immunity.

13) Why are γδ T cells and NKT cells considered innate leukocytes?

Section A.IV–VII

1) Name four types of DCs.

2) Which DC type is particularly important for primary adaptive responses and why?

3) In what areas of the body are classical DCs generated and from what types of precursors?

4) Distinguish between immature and mature DCs.

5) How did pDCs get their name?

6) From what cell type do inflammatory DCs originate?

7) What is an FDC and what does it do?

8) Describe three ways mast cell differ from basophils.

9) What common affliction is mediated by mast cells?

10) Can you define these terms? hematopoiesis, myelopoiesis, lymphopoiesis, multipotent, self-renewing

11) How does the location of hematopoiesis shift as a human fetus develops into an adult?

12) Name three types of hematopoietic cell progenitors and a mature hematopoietic cell type derived from each.

13) How does an HSC differ from an early progenitor?

14) Why is apoptosis important?

15) What is an apoptotic body?

16) How do apoptosis and necrosis differ?

Section B

1) What are the three principal domains constituting many cell surface receptors?

2) Describe three elements of an intracellular signaling cascade.

3) In the context of cytokines, can you define these terms? endocrine, paracrine, autocrine, pleiotropic, synergistic, antagonistic

4) What are two differences between cytokines, growth factors and hormones?

5) How is the power of cytokines controlled?

6) Name the three principal categories of cytokine function.

7) Why do cytokines overlap in function, and how is this overlap often achieved?

Did You Get it? A Self-Test Quiz (cont'd)

Section C

1) Can you distinguish between lymphoid follicles, lymphoid patches and lymphoid organs?

2) What are the primary and secondary lymphoid tissues?

3) What are two major differences between the primary and secondary lymphoid tissues?

4) Can you define these terms? red marrow, trabeculae, thymocyte, thymic involution, PALS, afferent, efferent, marginal zone

5) What are positive and negative thymic selection, and why are they important?

6) Can you describe MALT, SALT, BALT, NALT and GALT?

7) How does the lymphatic system connect with the blood circulation?

8) Describe the key functions of lymph nodes and the spleen.

Section D

1) Can you define these terms? extravasation, rolling, crawling, transmigration

2) What is an ICAM and why is it important?

3) How does inflammation promote leukocyte extravasation?

4) Where are the HEVs and how do they promote lymphocyte recirculation?

5) How do leukocyte extravasation and lymphocyte recirculation differ?

6) Describe antigen collection in the lymph nodes and the spleen.

7) Can you define these terms? homing receptor, vascular addressin

8) What is the effect of inflammation on vascular addressin expression?

9) How do memory lymphocyte homing patterns contribute to the efficiency of secondary adaptive responses?

Can You Extrapolate? Some Conceptual Questions

1) Hematopoietic cells are very sensitive to the effects of radiation. A patient who undergoes aggressive radiation treatment as part of his/her therapy for a disease is left without leukocytes. To protect this patient from potentially lethal immunodeficiency, he/she can be given a "stem cell transplant"; that is, a relatively low number of normal HSCs from a healthy donor are infused into the patient's blood. These HSCs then find their way from the blood to their "natural home" in the body. Where would you expect the HSCs to take up residence in the patient? How would you account for the ability of the transplant to restore both innate and adaptive immune responses in the patient? How often would the transplant procedure need to be repeated?

2) When a patient visits a doctor complaining of symptoms of an upper respiratory tract infection, the doctor will often palpate (examine by touch) the patient's neck to check for swelling. What would account for such swelling? Would this swelling be more likely to appear within 2–3 hours or 2–3 days after the start of an infection?

3) Leukocytes extravasate from venules into tissues due to interactions between adhesion molecules on both the leukocyte and venule endothelial cells. As outlined in **Fig. 2-16**, the initial step of extravasation is tethering. However, unlike the case for leukocytes, the tethering-related adhesion molecules on the venule endothelial cells are not expressed constitutively. How does the controlled expression of this initial set of adhesion molecules improve the efficiency of the immune response?

Would You Like To Read More?

Aguzzi, A., & Krautler, N. J. (2010). Characterizing follicular dendritic cells: A progress report. *European Journal of Immunology, 40*(8), 2134–2138.

Auffray, C., Sieweke, M. H., & Geissmann, F. (2009). Blood monocytes: Development, heterogeneity, and relationship with dendritic cells. *Annual Review of Immunology, 27*, 669–692.

Deane, J. A., & Hickey, M. J. (2009). Molecular mechanisms of leukocyte trafficking in T-cell-mediated skin inflammation: Insights from intravital imaging. *Expert Reviews in Molecular Medicine, 11*, e25.

Hayasaka, H., Taniguchi, K., Fukai, S., & Miyasaka, M. (2010). Neogenesis and development of the high endothelial venules that mediate lymphocyte trafficking. *Cancer Science, 101*(11), 2302–2308.

Laiosa, C. V., Stadtfeld, M., & Graf, T. (2006). Determinants of lymphoid-myeloid lineage diversification. *Annual Review of Immunology, 24*, 705–738.

Ley, K., Laudanna, C., Cybulsky, M. I., & Nourshargh, S. (2007). Getting to the site of inflammation: The leukocyte adhesion cascade updated. *Nature Reviews Immunology, 7*(9), 678–689.

Reizis, B., Bunin, A., Ghosh, H. S., Lewis, K. L., & Sisirak, V. (2011). Plasmacytoid dendritic cells: Recent progress and open questions. *Annual Review of Immunology, 29*, 163–183.

Ruddle, N. H., & Akirav, E. M. (2009). Secondary lymphoid organs: Responding to genetic and environmental cues in ontogeny and the immune response. *Journal of Immunology, 183*(4), 2205–2212.

So, what exactly is a cytokine? (2009). *Drug & Therapeutics Bulletin, 47*(8), 89–91.

Soehnlein, O. (2009). An elegant defense: How neutrophils shape the immune response. *Trends in Immunology, 30*(11), 511–512.

Yona, S., & Jung, S. (2010). Monocytes: Subsets, origins, fates and functions. *Current Opinion in Hematology, 17*(1), 53–59.

Innate Immunity

A year that's good is evident by its spring.

Persian

As introduced in Chapters 1 and 2, innate immunity is conferred by both non-inducible and inducible mechanisms. With respect to the former, the body has natural anatomical and physiological barriers that act non-specifically to prevent infection. If these barriers are penetrated, pattern recognition that is mediated by either soluble molecules in the extracellular milieu or receptors expressed by innate leukocytes triggers inducible innate mechanisms. Innate defenses do not generally undergo permanent change or establish long-term memory following exposure to a pathogen, and a second encounter provokes a response of a magnitude and character very similar to that induced by the first encounter. The host uses innate immunity to rapidly generate an initial response to almost any pathogen, either eliminating it or at least containing it until the slower but more focused adaptive immune response can be mounted. It is the cytokines produced by innate leukocytes locked in battle with the pathogen that play key roles in recruiting and activating the highly specific lymphocytes of adaptive immunity.

A. Non-Inducible Innate Mechanisms

Several anatomical and physiological barriers substantially decrease the likelihood of infection and reduce its intensity should it occur. Anatomical barriers include structural elements such as the skin and mucosae that physically prevent access through the body surfaces and orifices. Physiological barriers include the actions of body structures (such as sneezing) or substances produced by tissues (such as tears and mucus) that reinforce the anatomical barriers. These elements are illustrated in **Figure 3-1**.

Intact skin provides a tough surface that is relatively difficult for microorganisms or harmful inert entities (e.g., glass shards) to penetrate. The skin is composed of three layers: the outer **epidermis**, the underlying **dermis** and the fatty **hypodermis** (**Fig. 3-2A**). Cells at the exterior surface of the epidermis are dead and filled with keratin, a protein that tends to repel water. Other features of the epidermis that discourage infection are its lack of blood vessels and its rapid

MAK: Primer to the Immune Response. http://dx.doi.org/10.1016/B978-0-12-385245-8.00003-0

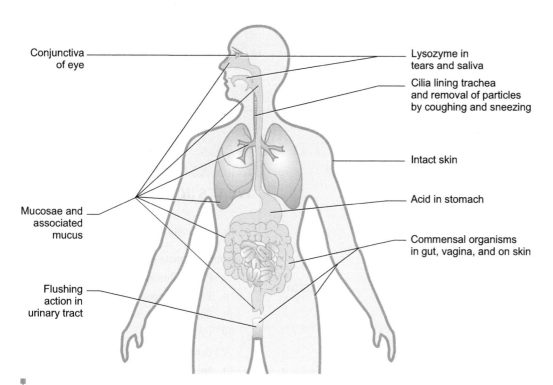

Conjunctiva of eye

Lysozyme in tears and saliva

Cilia lining trachea and removal of particles by coughing and sneezing

Intact skin

Mucosae and associated mucus

Acid in stomach

Commensal organisms in gut, vagina, and on skin

Flushing action in urinary tract

Fig. 3-1
Anatomical and Physiological Barriers

The indicated anatomical and physiological barriers prevent the majority of infectious agents from penetrating the body's external and internal surfaces.

Fig. 3-2
The Skin as an Innate Barrier

(A) The intact skin presents a tough outer layer (epidermis) bathed in acid secretions that fends off most pathogens. **(B)** Damaged skin allows pathogens easy access into the dermis, hypodermis and blood vessels supporting cells in these layers.

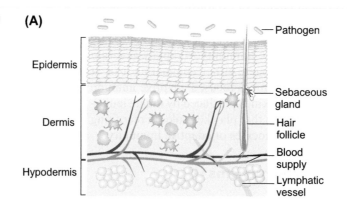

(A)

Pathogen

Epidermis

Sebaceous gland

Dermis

Hair follicle

Blood supply

Hypodermis

Lymphatic vessel

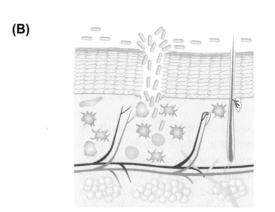

(B)

turnover: complete renewal of the human outer skin layer occurs every 15–30 days. Below the epidermis lies the dermis, which contains all of the blood vessels and other tissues necessary to support the epidermis. Among these are the sebaceous glands. These glands produce sebum, an oily secretion with a pH of 3–5 that inhibits the multiplication of most microbes. (We note here that the dermis also contains populations of leukocytes that can both defend the dermis and migrate into the epidermis; the activation of these cells is considered part of inducible innate immunity.) Below the dermis lies the hypodermis, which provides a barrier of fat that impedes pathogens. Failure of the skin as a defense occurs when it is breached, as in the case of wounds or insect bites (**Fig. 3-2B**).

Other areas of the body that come into contact with the outside world have developed different anatomical barriers. Instead of dry skin, which would not permit the passage of air or food, the surfaces of the gastrointestinal, respiratory and urogenital tracts are covered with a thin layer of mucosal epithelial cells known as a **mucosa**. Similarly, the eye possesses a delicate epithelial layer called the conjunctiva that lines the eyelids and protects the exposed surface of the eye while allowing the passage of light. The conjunctiva of the eye and the mucosae of internal body tracts are defended from penetration by microbes or inert entities by the flushing action of secretions such as mucus, saliva and tears. Mechanical defense in the lower respiratory and gastrointestinal tracts is derived from the sweeping action of tiny oscillating hairs called cilia on the surface of the mucosal epithelial cells. Vomiting, sneezing, coughing and bowel activity also serve to expel mucus-coated microbes and other entities. Failure of these innate measures allows pathogens to survive and penetrate the epithelial layer (**Fig. 3-3**). Infection can also occur when a microorganism possesses a cell surface molecule that permits it to attach to surface proteins of conjunctival or mucosal epithelial cells. This attachment circumvents the barriers offered by tears or mucus and facilitates the entry of the pathogen directly into the epithelial cells.

A microorganism that manages to evade the external and internal barriers of the body may succumb to a hostile physiological environment, such as a body temperature too high or a stomach pH too low to permit pathogen replication. Body fluids often contain antibacterial or antiviral substances. For example, **lysozyme**, an enzyme that hydrolyzes bacterial cell walls, is found in tears and mucus. Another physiological barrier is provided by the complex collections of benign bacteria and fungi that inhabit the mouth, the digestive and respiratory systems, and the skin surface. These beneficial microbes, which can represent over 1000 different species in any one host, are collectively known as **commensal organisms** (or *microbiota* or *microflora*), and their contribution to innate immunity is only now becoming fully appreciated. The gut alone contains 10^{13}–10^{14} commensal bacteria, a number that approximates the total number of eukaryotic cells in a human body! These vast communities of microbes maintain a symbiotic relationship with their host, breaking down food into products the host can

More on the immune defense of the mucosae and the skin appears in Chapter 12.

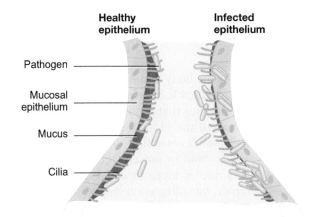

Healthy epithelium   Infected epithelium

Pathogen

Mucosal epithelium

Mucus

Cilia

Fig. 3-3
The Respiratory Mucosal Epithelium as an Innate Barrier

Left side: Invading pathogens are prevented from adhering to epithelial cells in the mucosa of a healthy respiratory tract by mucus that covers the epithelial surface and cilia that sweep pathogens away. Right side: When mucus is decreased and cilia are ineffective, pathogens can easily adhere to and infect respiratory epithelial cells.

absorb while taking what nutrients they need to survive. The microflora also use the host's mucosal surfaces as a base upon which to reproduce and establish colonies. This growth of commensal organisms serves the function of making it difficult for competing pathogens to find available resources and host binding sites; that is, the microbiota prevent the expansion of pathogenic organisms and thus confer "colonization resistance." In this way, commensal organisms play a key non-induced role in maintaining the body's homeostasis.

NOTE: The consequences of our modern society's indiscriminant use of antibiotics extend far beyond the inadvertent selection of the antibiotic-resistant bacteria known as "superbugs." Commensal organisms are also disrupted by antibiotics, leading to loss of homeostasis and defective colonization resistance. Even after antibiotics are stopped and the density of the protective microflora is restored, the species composition of this community may have radically changed. This change may have a deleterious effect on the host because, in addition to their role in non-inducible innate responses, the host microbiota influence the development and activation of specific leukocyte subsets. A loss of such influence could impair the regulation of the host's inflammatory response and adaptive immunity. Thus, the inappropriate use of antibiotics may put the patient at risk of developing not only an infection but also chronic inflammation that can promote conditions such as ulcerative colitis, irritable bowel syndrome, autoimmune disease, and perhaps even colon cancer.

Another important non-inducible innate means of host protection is the sequestration of iron. The cells of the host and any invading pathogen both need iron because this element is essential for many fundamental metabolic reactions. The control of iron stores is therefore a key battleground between a host and an attacking microbe. To maintain careful control over iron uptake and availability, mammalian hosts have evolved the expression of specific iron-binding molecules such as lactoferrin, transferrin and ferritin. These molecules, which are mainly produced by innate leukocytes and epithelial cells, efficiently sequester iron both within host cells and outside of them in the host's blood vessels and interstitial spaces. This sequestration prevents iron acquisition by pathogens, curtailing their growth. Naturally, pathogens have evolved opposing mechanisms intended to increase their access to the host's iron stores. **Siderophores** (Greek: "iron carriers") are a class of small iron-chelating molecules that are produced by microbes attempting to grow under conditions of low iron. In an example of the never-ending horse race between pathogens and hosts, mammals defend their iron supply from siderophores by producing lipocalin, a protein that captures iron-bound bacterial siderophores. In addition, host cells combat some intracellular microbes by adopting the reverse strategy and downregulating their iron uptake while simultaneously increasing their iron export. Interference with the acquisition of iron by pathogens is thought by some researchers to offer a fresh new approach to preventing or treating infections, and so has recently become a field of keen clinical interest.

B. Inducible Innate Mechanisms

If a pathogen or foreign macromolecule penetrates the body's anatomical barriers and escapes destruction by physiological defenses, the inducible innate response is next in line to try to eliminate the invader. "Induction" implies that a recognition event has occurred and has sparked a response. In the case of the inducible innate response, the recognition event is limited to the relatively small collection of broadly shared molecular patterns (PAMPs and DAMPs) able to bind to pattern recognition molecules (PRMs). Engagement of a PRM triggers one or more mechanisms designed to contain the assault and/or mediate tissue repair, including complement activation, inflammation, and/or cellular internalization mechanisms such as phagocytosis. These molecules, as well as the mechanisms associated with innate leukocytes such as DCs, neutrophils and macrophages, are discussed in detail in this chapter. Inducible innate immunity also includes the cytokine secretion and cytolysis of infected cells carried out by activated NK cells, γδ T cells and NKT cells. The role of these innate leukocytes in host defense is introduced later in this chapter and discussed fully in Chapter 11.

I. Pattern Recognition to Detect "Danger"

Over the past couple of decades, immunologists have moved away from viewing the immune response as strictly a means of distinguishing between self and non-self and destroying the latter. Today's concept of immune recognition is that it is designed to identify not only components of non-self entities (like viruses) but also the stress molecules generated by host metabolism when the host is in danger, be that threat from a replicating pathogen or tissue trauma. It is the job of PRMs to immediately recognize "danger" from any quarter and to take at least the initial appropriate actions to restore homeostasis and ensure host survival. We remind the reader here that, in contrast to the genes encoding the TCRs and BCRs of lymphocytes, the genes encoding PRMs are fixed in the germline DNA and do not undergo somatic recombination. Thus, compared to the almost infinite diversity of lymphocyte receptors, the PRMs of innate leukocytes exhibit a fairly limited repertoire of binding specificities.

As introduced in Chapter 1, PRMs have evolved to recognize PAMPs and DAMPs. PAMPs are highly conserved and repetitive structures that are common to a wide variety of microbes or their products but are not usually present in host cells. As illustrated in **Table 3-1**, PAMPs accumulating in sites of infection may include bacterial, viral or parasite products such as **lipopolysaccharide** (LPS) and peptidoglycan in bacterial cell walls or chitin in fungal or parasite cell walls; non-mammalian carbohydrates, such as mannose; CpG dinucleotides in bacterial DNA; and viral RNA genomes. DAMPs are often derived from tissue damage that causes host cells to undergo necrotic death. Once a cell's plasma membrane is breached, harmful molecules and/or cell contents that are not usually extracellular spill out into the surrounding tissue. DAMPs include components of host cell chromatin, such as fragments of genomic DNA and the non-histone DNA-binding protein HMGB1 (high mobility group box-1); mitochondrial DNA; **heat shock proteins** (HSPs), which cells produce as distress signals in response to any kind of stress (not just heat); uric acid crystals, which are released by damaged cells; the products of **complement** activation (see below); and **reactive oxygen**

TABLE 3-1	Examples of PAMPs and DAMPs
Pathogen-Associated Molecular Patterns (PAMPs)	
Source	**Examples**
Bacterial products	LPS, peptidoglycan fragments, CpG, mannose-containing carbohydrates, cell wall glycolipids
Viral products	dsRNA, ssRNA
Fungal products	Chitin, zymosan
Parasite products	Chitin, Plasmodium hemozoin, Schistosome egg antigens
Damage-Associated Molecular Patterns (DAMPs)	
Source	**Examples**
Complement products	C3b, C4b, iC3b
Reactive oxygen intermediates	H_2O_2, OH•, O_2^-
Stress molecules	HMGB1, HSPs, chaperone proteins, S100 proteins, lactoferrin, defensins, hyaluronic acid fragments, fibrinogen, surfactants, heparan sulfate
Metabolic products	K+, ATP, uric acid, cholesterol, saturated fatty acids
Nucleic acids	mRNA, ssRNA, chromatin components
Exogenous substances from inert sources	Alum, silica, asbestos

intermediates (ROIs) and members of the S100 family of calcium-binding proteins, which are released by physically or metabolically stressed cells. When a PRM is engaged, the ensuing response may range from simple wound healing and debris clearance to innate and then adaptive responses to pathogenic threats. It is also important to note that, in the absence of infection (and therefore in the absence of PAMPs), DAMPs can instigate inflammation that contributes to transplant rejection, hypersensitivity, and some types of autoimmune disease.

> NOTE: We mentioned previously that commensal organisms are beneficial to the host and do not induce tissue damage; i.e., no DAMPs are produced in response to their presence. In the absence of such DAMPs, the host's PRMs do not trigger an innate response even if they are engaged by a commensal microbe's molecular structures. This blindness of the innate response to microbial molecular patterns in the absence of DAMPs may have evolved to allow a host's immune system to distinguish between the harmless microflora and invading pathogens.

II. Pattern Recognition Molecules

Some PRMs are totally extracellular (free in the blood, lymph or interstitial spaces), while others are cell-associated. Cell-associated PRMs are commonly called "pattern recognition receptors" (PRRs). PRRs may be fixed in the plasma membrane of an innate leukocyte, positioned in its endosomal membranes, or found as soluble molecules within its cytosol. Upon engagement by a DAMP/PAMP, a PRR initiates intracellular signaling that is transmitted via a pathway which features several different types of enzymes as well as adaptor proteins that act as scaffolding for these enzymes and allow them to carry out their catalytic activities. The end result of this pathway is the activation of nuclear transcription factors and new gene transcription that supports the effector functions of the leukocyte. The engagement of extracellular PRMs has the same result, but after these molecules bind to their DAMPs/PAMPs, they must then engage another receptor on an innate leukocyte's surface to trigger its response. Some examples of the various types of PRMs and the ligands they bind are described next and summarized in **Table 3-2**. A schematic diagram illustrating how some of these molecules might operate in and around a leukocyte such as a DC, neutrophil or macrophage appears in **Figure 3-4**.

> NOTE: Immunologists often use the term "pattern recognition *receptor*" to refer to any "pattern recognition *molecule*" even if it is not associated with a membrane as the term "receptor" would usually imply. Common examples of this usage include the NOD and RLR proteins described later, which are actually free, cytosolic PRMs, but are often referred to as "PRRs."

i) Pattern Recognition Receptors (PRRs)

a) Toll-Like Receptors (TLRs)
The *Toll* gene was originally discovered in Drosophila, where the transmembrane Toll protein plays a dual role in embryonic development and antifungal immunity. In mammals, the **Toll-like receptors (TLRs)** are key PRRs that are structurally similar to Drosophila Toll and mediate antimicrobial innate defense. The TLRs are transmembrane proteins that are fixed either in the plasma membrane of the cell or in its endosomal membranes. TLRs are most highly expressed by DCs, monocytes, macrophages and neutrophils. Some TLRs are also expressed by NK cells, T and B cell subsets and certain non-hematopoietic cells.

Different TLRs recognize different microbial structures. For example, TLR2 recognizes bacterial lipoteichoic acids; TLR3 binds to double-stranded RNA (dsRNA); TLR4 binds to bacterial LPS; and TLR7 binds to viral single-stranded RNA (ssRNA). Once engaged, different TLRs activate similar but not identical signaling pathways that involve a variety of enzymes, adaptor molecules and nuclear transcription factors. Most TLR engagement results in phagocytosis, cellular activation, and/or the production of

TABLE 3-2	**Pattern Recognition Molecules**			
	Location	**Examples of Ligands Recognized**	**Expression Profile**	**Primary Function**
(TLRs)	**Plasma membrane**	Pathogen lipid and protein structures Bacterial peptidoglycan, lipopeptides		
	TLR1	Bacterial peptidoglycan; HSPs;	Monocytes, macrophages,	Activation of phagocytes
	TLR2	yeast, parasite lipoteichoic acids	neutrophils, DCs,	Induction of pro-inflamma-
		Gram-bacterial LPS; viral	NK cells, some T and	tory cytokines
	TLR4	phosphorylcholine; HSPs	B cell subsets, some	
		Bacterial flagellin	non-leukocytes	
	TLR5	Lipopeptides (bacterial, viral, parasite)		
	TLR6	Pathogen nucleic acids		
	Endosomal membrane			
	TLR3	Viral dsRNA, ssRNA		
	TLR7	Viral ssRNA		
	TLR8	Viral ssRNA		
	TLR9	Viral dsRNA, unmethylated viral or bacterial DNA (CpG motifs)		
NLRs	**Cytoplasm**			
	NLRP1	*Bacillus anthracis* toxin	Macrophages, mono-cytes, neutrophils, DCs	Activation of inflamma-somes (e.g. NLRP1, NLRP3)
	NLRP3	Uric acid, ATP, bacterial RNA, viral products	Widely expressed	Production of antimi-
	NLRC4	Flagellin	Macrophages, mono-cytes, neutrophils, DCs	crobial peptides and pro-inflammatory cytokines
	NOD1	Fragment of bacterial peptidoglycan (mostly Gram–)		
	NOD2	Fragment of bacterial peptidoglycan (Gram+ and Gram–)		
RLRs	Cytoplasm	Viral ssRNA, dsRNA	DCs	Pro-inflammatory cytokines
CLRs	Plasma membrane	Microbial carbohydrate structures	DCs, macrophages,	Cytokine secretion
	DEC-205	Mannose or fucose moieties on the	neutrophils	Regulation of Th subsets
	DC-SIGN	surfaces of fungi, bacteria, viruses, parasite egg antigens		Antigen processing and presentation
	Mannose receptor	Fungal mannans		
	Dectin-1	Fungal β-glucans		
Scavenger receptors	Plasma membrane	Modified lipid structures derived from microbial cell wall components or membranes of dying host cells	Macrophages, monocytes, DCs, endothelial cells, other non-leukocytes	Phagocyte activation
RAGE	Plasma membrane	Extracellular DAMPs, HMGB1, S100 proteins Modified proteins, lipids, nucleic acids	Widely expressed by leukocytes, activated endothelial cells	Amplification of signaling leading to new gene transcription
NK receptor	Plasma membrane	Components of infected, stressed or cancerous cells	NK cells	Target cell lysis Pro-inflammatory cytokines
NKT receptor	Plasma membrane	Glycolipid antigens	NKT cells	Cytokine secretion
γδ **TCR**	Plasma membrane	Antigens from infected, stressed or damaged host cells	γδ T cells	Target cell lysis Pro-inflammatory cytokines
Collectins	**Blood plasma**			
	MBL	Microbial polysaccharides	Hepatocytes	Complement activation
	Lung			
	SP-A	Microbial polysaccharides, DNA,	Respiratory epithelial	Opsonized phagocytosis
	SP-D	pollens, apoptotic host cells	cells	Pro-inflammatory cytokines
Acute phase proteins	Blood plasma	Microbial polysaccharides	Hepatocytes	Complement activation Opsonized phagocytosis

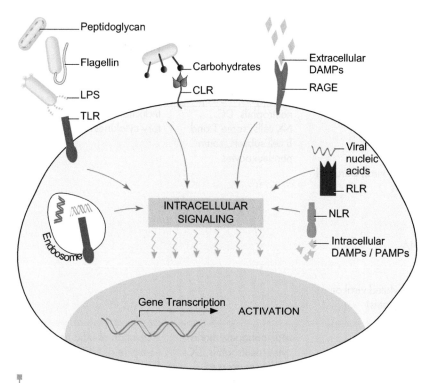

Peptidoglycan

Flagellin

LPS

TLR

Carbohydrates

CLR

Extracellular DAMPs

RAGE

Viral nucleic acids

RLR

NLR

Intracellular DAMPs / PAMPs

Endosome

INTRACELLULAR SIGNALING

Gene Transcription   ACTIVATION

Fig. 3-4
Activation of Innate Leukocytes through PRR Engagement

In this schematic representation, an innate leukocyte (such as a DC, macrophage or neutrophil) has encountered a large collection of DAMPs and PAMPs that engage its cell-associated PRRs. Extracellular PAMPs derived from bacterial flagellin, LPS or peptidoglycan engage TLRs positioned in the leukocyte's plasma membrane, while extracellular bacterial carbohydrates engage its CLRs. Extracellular DAMPs produced by the host in response to an invading pathogen stimulate the leukocyte by engaging its RAGE receptors. The leukocyte depicted here has succeeded in phagocytosing a pathogen, causing pathogen nucleic acids to be present in an endosomal compartment expressing endosomal TLRs. If the pathogen is a virus, cytosolic RLRs may bind to viral nucleic acids during replication. Additional PAMPs and intracellular DAMPs present in the cytosol are captured by NLRs. The end result of the engagement or any or all of these PRRs is intracellular signaling that leads to nuclear transcription factor activation, gene transcription, and the activation of the leukocyte. Note that extracellular PRMs are not depicted in this figure, and not all types of PRRs are shown.

the pro-inflammatory cytokines IL-1, IL-6, IL-18, IL-12, TNF and/or IFNα/β. TLR engagement on macrophages also triggers the synthesis of an enzyme called **inducible nitric oxide synthase (iNOS)** that produces the powerful antimicrobial agent nitric oxide (NO). In addition, TLRs are linked to the adaptive response because TLR engagement on DCs induces their maturation and promotes their migration to the lymph nodes where mature naïve T cells congregate. In the periphery, TLR-stimulated DCs and macrophages act as efficient APCs for effector T cells. Lastly, the particular cytokines produced by DCs in response to the engagement of different TLRs dictate which T helper lymphocyte subsets are generated by the later adaptive response. Signaling initiated by these cytokines induces the differentiation of the Th effectors most appropriate for dealing with the pathogen that triggered the innate leukocyte in the first place.

As well as responding to PAMPs, certain TLRs recognize stress molecules produced by a host when it experiences trauma that does not involve a pathogen. For example, the binding of the DAMPs HSP60 or HSP70 to TLR2 or TLR4 triggers intracellular signaling that leads to pro-inflammatory cytokine secretion. It is thought that the resulting influx of leukocytes into the site of injury may assist in wound healing.

More on how PRM engagement on innate cells influences adaptive responses to pathogens can be found in Chapters 7 and 9.

NOTE: TLRs are part of a powerful, multi-pronged response to incipient infection, meaning that any loss of TLR function can have a detrimental impact on host health. For example, individuals expressing TLR variants associated with altered signaling are more susceptible to malaria, tuberculosis or Legionnaire's disease. Other TLR defects cause primary immunodeficiencies, which are discussed in Chapter 15.

b) NOD Proteins, NOD-like Receptors (NLRs) and Inflammasomes

NOD1 and NOD2 (nucleotide-binding oligomerization domain-containing) were the first free cytoplasmic PRRs identified as detecting PAMPs of intracellular pathogens once these invaders had accessed a host cell's interior. The binding of particular peptides derived from bacterial peptidoglycan to NOD proteins initiates intracellular signaling that causes the cell to produce TNF, IL-1, IL-18 and/or IFNα/β. The NOD proteins are now considered the prototypical molecules of what was subsequently found to be a much larger group of structurally related cytoplasmic proteins called **NLRs** (**NOD-like receptors**). All NLR proteins have three basic domains: a C-terminal domain and a central domain, which are both very similar among all NLR proteins, and an N-terminal domain that varies in structure. Based on differences in these N-terminal domains, the NLR proteins are subdivided into the NLRA, NLRB, NLRC and NLRP families.

When certain NLRs bind to their respective DAMPs/PAMPs, they are induced to both oligomerize and interact with other large cytoplasmic proteins to form multimeric complexes called **inflammasomes**. Accordingly, the engagement of NLRP1, NLRP3 or NLRC4 sparks the formation of the NLRP1, NLRP3 or NLRC4 inflammasomes, respectively. A major function of an inflammasome is to activate *caspases*, which are proteases that are routinely maintained within resting cells in zymogen (inactive) form as pro-caspases. For example, during the formation of the NLRP3 inflammasome, pro-caspase-1 is recruited to the NLRP3 oligomer along with various enzymes and adaptor proteins. Within this complex, the pro-domain of pro-caspase-1 is removed by enzymatic cleavage to yield active caspase-1. The release of active caspase-1 into the cytosol is a critical tipping point in the inflammatory cascade because this enzyme is required for the processing of inactive pro-IL-1 and pro-IL-18 to yield the active forms of these cytokines, and IL-1 and IL-18 are crucial drivers of the inflammatory response. Thus, there is a synergy between TLR and NLR functions because pro-IL-1 and pro-IL-18 synthesized in response to TLR signaling are converted to their active forms by caspase-1 activated when the NLRs of inflammasomes engage DAMPs/PAMPs. **Figure 3-5** gives a hypothetical example of this elegant sequence of events, and **Box 3-1** introduces inflammatory diseases associated with NLR gene defects.

c) Retinoic Acid Inducible Gene-1 (RIG-1)-Like Receptors (RLRs)

In contrast to TLRs that can bind to viral PAMPs while these pathogens are still on the exterior of a cell or within its endosomal compartments, **RLRs** are cytosolic PRMs that specifically recognize intracellular viral RNAs that happen to be outside any membrane-bound compartment. Structurally, RLRs resemble the RNA helicase family of proteins that can detect cytosolic viral RNAs. There is also evidence that some RLRs may be specific for viral DNAs. In a resting cell, RLRs are present as inactive monomers. Engagement of RLRs by viral PAMPs triggers conformational changes to RLR proteins that result in monomer aggregation and intracellular signaling leading to the production of antiviral proteins, including IFNα/β. Some signaling molecules and adaptor proteins are shared between the TLR and RLR cascades, while others are unique to each pathway.

d) C-type Lectin Receptors (CLRs)

CLRs are PRRs that act as **lectins**, which are proteins that bind to carbohydrate moieties such as mannose and fucose. Structurally, all lectins have at least one carbohydrate-binding domain, and the nature of these domains determines which type of carbohydrate is recognized. In the immune system, CLRs are expressed on innate leukocytes, particularly on APCs, and recognize carbohydrate-based DAMPs/PAMPs. Engagement of a CLR by a DAMP/PAMP triggers intracellular signaling that may lead directly to nuclear transcription factor activation and new gene transcription, or which may affect a signaling pathway activated by TLR engagement. In either case, the binding of a CLR by its sugary ligand leads to the induction of a diverse range of innate responses, including cytokine secretion and antigen processing and presentation.

CLRs are classified into two major groups. Group I contains members of the mannose receptor family. These proteins recognize carbohydrates of various bacteria such as those causing tuberculosis and pneumonia, as well as carbohydrates of HIV and dengue

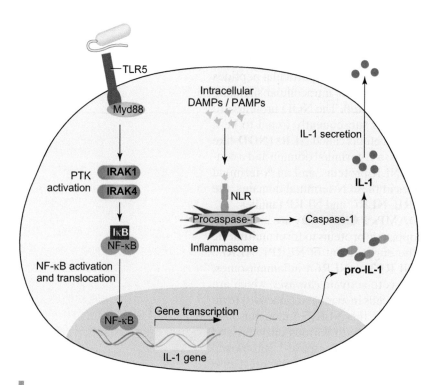

Fig. 3-5
Cooperation between an Innate Leukocyte's TLRs and Inflammasomes to Produce Pro-Inflammatory Cytokines

In this schematic example, the flagellin molecule of a bacterium has engaged TLR5 on the surface of an innate leukocyte. This binding triggers the recruitment of the adaptor protein MyD88 and the activation of the protein tyrosine kinases (PTKs) IRAK1 and IRAK4. A cascade of intracellular signaling events ensues that leads to the degradation of the inhibitor IκB from molecules of the important transcription factor NF-κB; the function of IκB is to hold NF-κB inactive in the cytoplasm until it is needed. Freed from IκB degraded in response to TLR engagement, NF-κB moves into the leukocyte's nucleus and induces transcription of the IL-1 gene. The resulting mRNA is translated in the cytoplasm to produce pro-IL-1 protein. Meanwhile, DAMPs (or PAMPs) that have engaged an NLR protein in the leukocyte's cytoplasm stimulate the formation of an inflammasome containing pro-caspase-1. Enzymes within the inflammasome remove the pro-domain of pro-caspase-1, releasing activated caspase-1 into the cytoplasm. Activated caspase-1 cleaves pro-IL-1 and generates the mature active IL-1 cytokine. IL-1 is secreted from the leukocyte and initiates an inflammatory response.

virus. When CLRs bind these carbohydrate-based PAMPs, the primary outcomes are phagocytosis and antigen processing and presentation. Group II CLRs contain members of the asialoglycoprotein receptor family. These proteins recognize carbohydrates of bacteria such as those causing leprosy or ulcers, as well as carbohydrates on yeast, ticks, the SARS and measles viruses, and the eggs of Schistosome parasites. In addition to phagocytosis and antigen processing and presentation, PAMP engagement by these CLRs results in upregulation of TLR-induced cytokine production and the modulation of T cell responses. Two particularly important Group II CLRs are DEC-205 and DC-SIGN, which are expressed most abundantly on DCs and recognize mannose- and fucose-based PAMPs on a wide range of pathogens. Dectin-1 is a group II CLR that is expressed by both DCs and macrophages, and recognizes β-glucan molecules in fungal cell walls. Engagement of these CLRs by these PAMPs induces signaling that influences TLR-mediated responses initiated by responding leukocytes.

"DC-SIGN" stands for "DC-specific ICAM3-grabbing non-integrin."

"DEC-205" is so named because it is expressed on DCs and epithelial cells and has a molecular weight of 205 kD.

e) Scavenger Receptors (SRs)
Scavenger receptors bind to a wide variety of lipid-related ligands derived either from pathogens or from host cells that are damaged, apoptotic or senescent. Damaged or dying host cells undergo deleterious changes to their membranes such that internal phospholipids that are normally hidden become exposed. The SRs then recognize molecular patterns that are not present on healthy host cells. Binding of a ligand to an SR triggers phagocytosis of the microbe or dying host cell. There are currently eight classes of structurally diverse SRs, the majority of which are expressed on phagocytes,

---

Box 3-1   Inflammatory Diseases Linked to Genetic Defects in NLR Proteins

Recent insights into the function of NLR proteins and inflammasomes has helped clinicians understand the genetic basis of a group of rare but serious diseases known collectively as "cryopyrinopathies." The normal *NLRP3* gene encodes the NLRP3 protein (also known as cryopyrin), which is an essential part of the NLRP3 inflammasome. In patients suffering from cryopyrinopathies, mutations in the *NLRP3* gene cause hyperactivation of the NLRP3 protein, leading to inappropriate and uncontrolled inflammatory responses. These disorders, which are characterized by hyperproduction of IL-1, include (1) familial cold autoinflammatory syndrome; (2) Muckle–Wells syndrome; and (3) neonatal-onset multisystem inflammatory disease (NOMID). Due to their persistent or periodic inflammation, patients experience fever, rashes and joint pain, as well as other debilitating symptoms specific to their particular condition. In some cases, patients have experienced relief after treatment with the targeted anti-inflammatory agent "anakinra," an antagonist of the IL-1 receptor. In addition to the cryopyrinopathies, abnormalities of the NLRP3 inflammasome have been implicated in gout and diabetes, which may be related to NLRP3-mediated recognition of DAMPs such as uric acid and cholesterol crystals, respectively.

Mutations in other NLR family proteins have also been linked to inflammatory diseases. NOD2 mutations are associated with Crohn's disease, a disorder characterized by uncontrolled inflammation in the gastrointestinal tract. Lastly, NLRP1 mutations have been found in patients with Addison's disease (an adrenal gland disorder) or vitiligo (a skin pigment disorder).

---

and most particularly on macrophages. In a healthy host, SRs play an important role in maintaining normal cholesterol and lipoprotein metabolism. However, the normal activities of certain SRs have been implicated in the pathophysiology of atherosclerosis and other cardiovascular diseases as well as in diabetes.

f) RAGE (Receptor for Advanced Glycation End Products)

RAGE is a specialized receptor that is widely expressed on leukocytes as well as on activated endothelial cells and some non-hematopoietic cell types. RAGE binds to a range of extracellular DAMPs, including HMGB1, S100 proteins, and modified lipids, proteins and nucleic acids released from stressed cells. RAGE's main function is to initiate additional intracellular signaling that activates gene transcription and amplifies the innate immune response. In addition, RAGE on the surface of endothelial cells may function as an adhesive receptor and directly interact with integrins expressed by passing leukocytes, enhancing their recruitment. Accordingly, abnormal RAGE function has been associated with various inflammatory diseases as well as with diabetes, kidney dysfunction and cancer.

g) Receptors of NK, NKT and γδ T Cells

As mentioned in Chapter 2, defense mechanisms mediated by NK, NKT and γδ T cells are considered to be part of the inducible innate response because these cells respond much faster than T and B lymphocytes to a threat and carry out ligand recognition that lacks the fine specificity of TCRs and BCRs. For this reason, the receptors by which NK, NKT and γδ T cells carry out their innate functions can be considered PRRs. Information on NK, NKT and γδ T cells is presented briefly later in this chapter and in detail in Chapter 11.

ii) Extracellular PRMs

a) Collectins

Collectins are PRMs that float freely in the blood and other body fluids. Their name is derived from their structure since these proteins are made up of a collagen domain fused to a lectin domain. Like CLRs, different collectins recognize different carbohydrate patterns constituting DAMPs/PAMPs. These patterns are clearly distinct from those present on living eukaryotic cells, so that healthy host cells are ignored. Once bound by ligand, collectins generally mediate pathogen clearance via complement activation, by inducing uptake by phagocytic cells, or by aggregating microbial cells together (**agglutination**). In the case of apoptotic host cells, collectins engaged by a DAMP stimulate leukocytes to phagocytize the dead cell in the absence of significant inflammation. Perhaps the best-known collectin is the **mannose-binding lectin (MBL)** that triggers the lectin pathway of complement activation (see later). The surfactant proteins SP-A and SP-D are two collectins that help to protect the lung.

b) Acute Phase Proteins

Early in a local inflammatory response, macrophages activated by engagement of their TLRs secrete cytokines that trigger hepatocytes to produce **acute phase proteins**. At least some of these soluble proteins, particularly *C-reactive protein (CRP)*, bind to a wide variety of bacteria and fungi through recognition of common cell wall components that are not present on host cells. Once bound to the surface of a microbe, these soluble molecules can activate complement and stimulate phagocytosis by innate leukocytes. Other acute phase proteins, notably serum amyloid A, contribute to innate defense by activating enzymes that degrade the extracellular matrix of tissues. It is then easier for additional leukocytes to follow a chemotactic gradient through a tissue to the precise site of attack. Acute phase proteins also include fibrinogen, which promotes the formation of blood clots that can trap microbes, as well as haptoglobin and ferritin, which bind to hemoglobin and soluble iron, respectively, and thus inhibit microbial iron uptake. Some proteins of the complement system (described next) are also considered acute phase proteins.

III. Complement Activation

The complement system is a vital physiological element that must be induced to carry out its protective functions. Complement was originally named for its ability to assist, or be "complementary" to, antibodies involved in the lysing of bacteria.

i) Nature and Functions

Complement is not a single substance but rather a collection of about 30 serum proteins that make up a complex system of functionally related enzymes. These enzymes are sequentially activated in a tightly regulated cascade to carry out complement's functions. Complement activation has four principal outcomes: (1) lysis of pathogens; (2) opsonization (coating) of foreign entities to enhance phagocytosis; (3) clearance of **immune complexes** (soluble lattices of antigen bound to antibody); and (4) the generation of peptide by-products that are involved in the inflammatory response. All these functions are discussed here and in later chapters.

The complement system can be activated via three biochemical pathways: the **classical pathway**, the **lectin pathway** and the **alternative pathway** (**Fig. 3-6**). All three pathways result in the production of the key complement component C3b and the assembly on the pathogen surface of a structure called the **membrane attack complex (MAC)**. The classical pathway is triggered when antigen present on a pathogen surface binds to antibody, and this antibody in turn is bound by complement component C1. In the lectin pathway, complement activation is initiated by the direct binding of MBL to certain carbohydrates on the surface of a pathogen; no antibody is involved. The alternative pathway is activated when complement component C3 spontaneously hydrolyzes and then interacts with certain enzymatic factors to produce C3b. This C3b can bind to almost any carbohydrate or protein on the surface of a pathogen; again, no antibody is involved. The attachment of C3b to a pathogen surface leads to the formation of the MAC. The MAC then bores a hole in the outer membrane of the pathogen, creating an osmotic imbalance that lyses the invader.

The classical pathway of complement activation was discovered first but probably evolved last. Due to its involvement of antibody, the classical pathway has the advantage of greater specificity than the alternative and lectin pathways but the disadvantage of tardiness due to its dependence on an adaptive response. The lectin and alternative pathways are less specific but very rapid and can activate complement during the lag phase required for B cell activation and antibody synthesis. In addition, because it involves antibody, the classical pathway links the innate and adaptive responses. The alternative and lectin pathways are purely elements of innate immunity.

Many pathogens with membranes can become targets for MAC assembly and thus subject to destruction via MAC-mediated lysis. Most bacteria are susceptible to MAC-mediated killing, although some of these organisms protect their membranes by covering them with thick polysaccharide capsules. Some enveloped virus particles

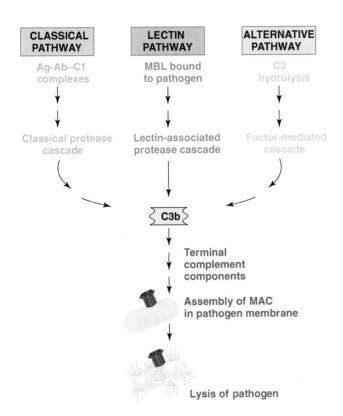

Fig. 3-6
Complement Activation Pathways

The figure shows a summary of the three pathways of complement activation. Please see main text and **Figure 3-7** for details.

and virus-infected host cells that express viral antigens in their membranes (and thus become coated by antiviral antibodies) are also subject to MAC-mediated lysis. Even foreign erythrocytes introduced into a host by a blood transfusion may be destroyed by MAC formation if the erythrocytes bear surface antigens recognized by host antibodies. However, most uninfected nucleated cells resist MAC-mediated lysis because of the action of host regulatory proteins (see later).

Amplification is a key feature of complement activation. The cascade is made up of a large number of sequential activation steps, some of which generate multiple products. Since many of these products are themselves enzymes that trigger subsequent steps, an exponential increase in activated molecules is generated. A huge amplification of an initial signal can be achieved, making it possible to generate a massive response from a single triggering event within a short period of time. However, the complement system is also highly sensitive, since a deficiency of any one enzyme can halt the progression of the cascade completely. With respect to sources of complement components, most are synthesized principally by hepatocytes in the liver but are also secreted by macrophages, monocytes and epithelial cells of the urogenital and gastrointestinal tracts. Cytokines produced during the inflammatory response (such as TNF and IL-1) can also induce the synthesis and secretion of some complement proteins by other types of host cells.

ii) Nomenclature

Biochemists have adopted certain conventions in naming the complement components. Because they were identified first, the enzymes of the classical pathway are designated by the letter "C" and the numbers 1–9 (which reflect their order of discovery, not their order of action). Once a protein is cleaved, the resulting peptide fragments are designated as "a" and "b," where "a" indicates the smaller product and "b" indicates the larger one. Generally speaking, the "b" fragments contribute to the next enzymatic activity in the cascade (e.g., C3b), while the "a" fragments (e.g., C5a) are involved in the inflammatory response (see later). The exception is C2, where C2a (not C2b) is the larger fragment and contains the enzymatic domain. Often several complement components must associate to form an active complex, such as C4b, C2a and C3b coming together to form C4b2a3b. The components of the alternative pathway were discovered later historically and are called "factors." The factors are assigned a single letter (Factor B, Factor D, etc.) and are often abbreviated simply as B, F, D, H and I. The lectin pathway shares most of the components of the classical pathway but differs in the enzymes that initiate the cascade (see later).

iii) Complement Activation Pathways

All three pathways of complement activation achieve the same goal of generating the key molecule C3b. It is C3b that allows the **terminal complement components** (which are the same in all three pathways) to come together and assemble the MAC. The steps of the classical, alternative and lectin complement activation pathways are summarized in **Figure 3-7** and **Table 3-3**.

a) Classical Pathway

Complement component C1 circulates in the plasma as a huge, inactive protein complex containing one C1q subunit, two C1r subunits and two C1s subunits. The C1q subunit can bind to the Fc regions of two antigen-specific antibodies that have bound in close proximity to antigen fixed on the surface of a pathogen. This binding activates the C1r subunits such that they activate the C1s subunits. The activated C1s subunits can then cleave serum C4 into C4a and C4b. C4a diffuses away while C4b attaches to proteins on the surface of the pathogen and then binds serum C2. This interaction causes C2 to become susceptible to cleavage by C1s, generating C2a and C2b. C2a remains bound to C4b, while C2b diffuses away. The C4bC2a structure is known as the *classical C3 convertase*. C3 convertase cleaves serum C3 into C3a and the key molecule C3b. C3b binds to C4bC2a to form another enzyme complex called the *classical C5 convertase*.

b) Lectin Pathway

MBL binds to carbohydrate moieties in pathogen proteins. A protease complex called *MBL-associated serine protease (MASP)* associates with MBL and cleaves serum C4 to generate C4a and C4b. C4b then binds to motifs in proteins on the pathogen surface as described for the classical pathway. C2 binds to C4b and is cleaved by the MASP complex such that a classical C3 convertase is formed. The rest of the pathway is the same as in classical complement activation, such that a classical C5 convertase is formed.

c) Alternative Pathway

Serum C3 spontaneously hydrolyzes an internal thioester bond to form a product called C3i. Factors B and D can act on C3i to generate C3b. This C3b binds to surface macromolecules on pathogens and recruits another molecule of Factor B. The C3b–Factor B complex is cleaved by Factor D to generate C3bBb, the *alternative C3 convertase*. This enzyme can generate additional C3b, amplifying the pathway. A protein called *properdin* (P) then binds to the C3 convertase complex, yielding the stabilized alternative C3 convertase (C3bBbP). The addition of another C3b molecule yields C3bBbPC3b, the *alternative C5 convertase*.

iv) Terminal Steps

Once a C5 convertase (regardless of derivation) is fixed on a pathogen surface, it cleaves serum C5 to yield C5a (which diffuses away) and C5b (which remains bound to the convertase). The binding of C6 then stabilizes the complex, while the binding of C7 exposes hydrophobic regions that facilitate penetration of the complex into the pathogen membrane. The addition of C8 stabilizes the complex in the membrane, and the formation of a pore is initiated. With the addition of at least four molecules of C9, the MAC is completed (**Plate 3-1**). The pore becomes a tunnel, allowing ions and water molecules to pour through the pathogen membrane into the interior. The pathogen lyses due to osmotic imbalance.

v) Other Roles of C3b

C3b has several other important functions in addition to initiating MAC assembly. (1) A pathogen coated in C3b binds to complement receptor 1 (CR1) expressed on the

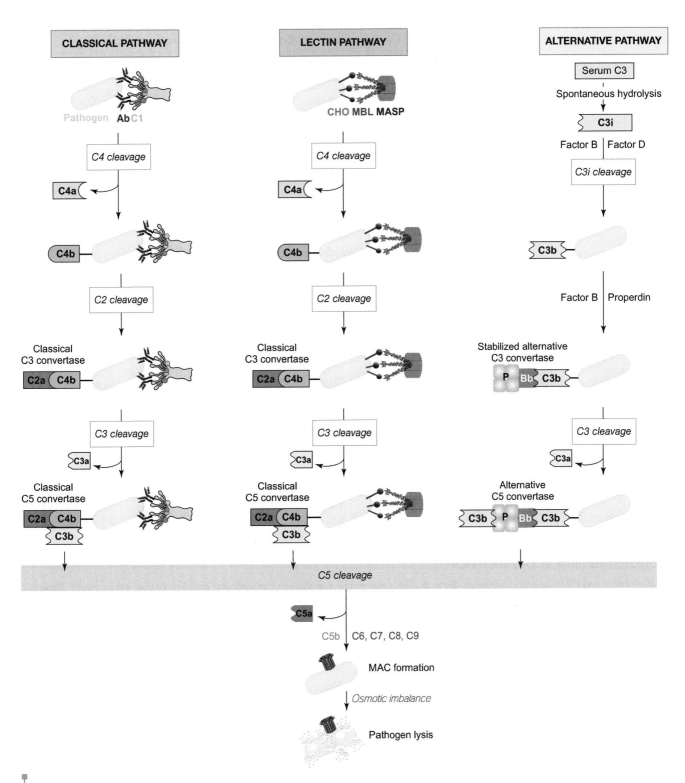

Fig. 3-7
Formation of Convertases and the MAC during Complement Activation

The components of the three pathways of complement activation and the formation of the relevant C3 and C5 convertases in each pathway are illustrated. The generation of the anaphylatoxins (C3a, C4a and C5a) is also shown. All three pathways culminate in C5 cleavage and the production of C5b, which joins with the terminal complement components (C6–C9) to form the MAC. The presence of the MAC in the pathogen membrane inexorably leads to pathogen lysis.

TABLE 3-3	Components of Complement Activation Pathways		
	Classical Pathway	**Lectin Pathway**	**Alternative Pathway**
Initiating components	C1 complex	MBL and MASP complex	C3i
Starting point	Antigen–antibody complex interacts with C1	Microbial carbohydrates interact with MBL	Serum C3 hydrolysis
Early components	C4 C2	C4 C2	Factor B Factor D
Component that attaches to pathogen surface	C4b	C4b	C3b
C3 convertase	C2aC4b	C2aC4b	C3bBbP
C5 convertase	C2aC4bC3b	C2aC4bC3b	C3bBbPC3b
Terminal components	C5–C9	C5–C9	C5–C9

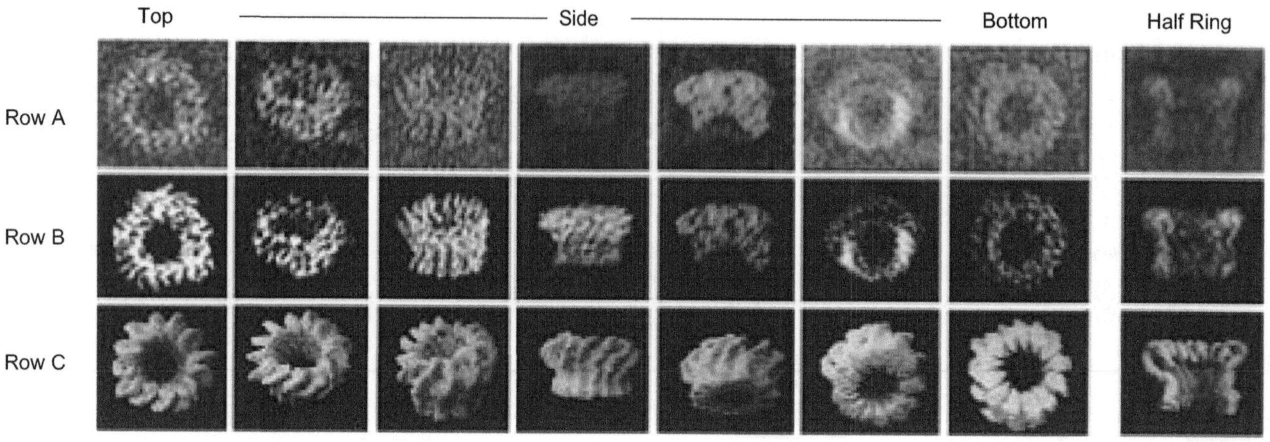

Plate 3-1
Electron Micrographs of Polymerized C9

Row A. Polymerized C9 in several orientations. **Row B.** The same images are shown after computer-assisted contrast enhancement. **Row C.** Photographs of a ceramic model of polymerized C9, positioned in different orientations to match Row A. The central pore of polymerized C9 is approximately 100 angstroms in diameter. [*Reproduced with permission from DiScipio, R. G. & Berlin, C. (1999). The architectural transition of human complement component C9 to poly(C9). Molecular Immunology 36, 575–585.*]

surface of phagocytes (including APCs). These cells then easily engulf and destroy the invader. The C3b is said to be acting as an **opsonin** in this case because it enhances the ability of an entity to be phagocytosed. By encouraging pathogen uptake by APCs in this way, C3b indirectly enhances antigen presentation to T cells and thus the adaptive response. (2) Soluble antigen–antibody complexes can bind to C1 and trigger the classical complement pathway such that C3b is deposited on the complexes themselves. The C3b then blocks the networking between multiple antigen and antibody molecules that results in the formation of large insoluble immune complex lattices (**Fig. 3-8A**). The C3b is said to be "solubilizing" the immune complexes, making them easier to clear from the circulation and preventing the damage they might inflict if they accumulated in the small vessels and channels of the body. Erythrocytes expressing CR1 can bind to C3b-coated antigen–antibody complexes and transport them through the circulation to the liver and spleen. Phagocytic cells in these locations then engulf and destroy the antigen–antibody complexes during the routine disposal of red blood cells. (3) C3b also contributes directly to defense against viruses. When these pathogens become coated in C3b, the C3b blocks the binding of the virus to its

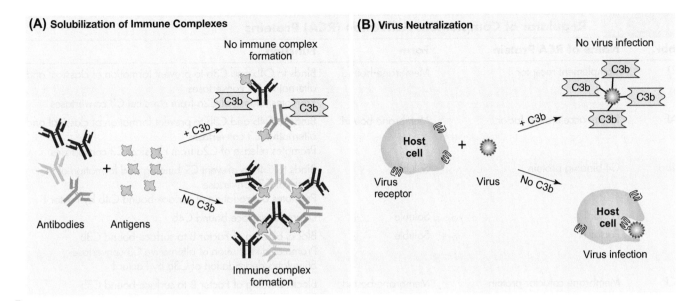

(A) Solubilization of Immune Complexes

No immune complex formation

Antibodies   Antigens

Immune complex formation

(B) Virus Neutralization

No virus infection

Host cell

Virus receptor   Virus

Host cell

Virus infection

Fig. 3-8
Some Other Functions of C3b

(A) Solubilization of immune complexes. Binding of C3b to the Fc region of an antibody bound to antigen prevents the networking necessary to form large insoluble immune complexes. **(B)** Virus neutralization. A coating of C3b on a virus prevents it from binding to receptors on a host cell.

receptors on a host cell and prevents cell entry (**Fig. 3-8B**). The C3b is said to have "neutralized" the virus.

vi) Anaphylatoxins

The small fragments C3a, C4a and C5a that are cleaved from serum C3, C4 and C5 during complement activation are called **anaphylatoxins** (refer to **Fig. 3-7**). Anaphylatoxins are not just inert by-products: C3a and C5a play important roles in inflammation (C4a is not very active). The anaphylatoxins are so named because, at the high systemic concentrations generated in response to a serious bacterial infection, anaphylatoxins can induce dramatic cardiovascular and bronchial effects that resemble **anaphylaxis** (a severe systemic allergic reaction; see Ch. 18). Anaphylatoxins can trigger the degranulation of mast cells and basophils, resulting in the release of potent inflammatory mediators. In addition, C5a is a powerful chemoattractant for neutrophils and stimulates the **respiratory burst** (see later) and degranulation of these cells. C5a and C3a also upregulate adhesion molecule expression on neutrophils and endothelial cells, promoting extravasation. Finally, C5a stimulates macrophages and monocytes to secrete increased amounts of the pro-inflammatory cytokines IL-1 and IL-6. Because these cytokines can also stimulate the proliferation of activated T cells, C5a therefore plays an indirect role in adaptive immunity. All these effects are mediated by the binding of the anaphylatoxins to specific receptors expressed on various types of leukocytes.

vii) Control of Complement Activation

Because of the tremendous number of players involved and their non-specific destructive capacity, the complement cascade is rigidly organized and tightly regulated to minimize damage to host tissues. Firstly, complement enzymes are present in the plasma as zymogens; that is, the protease is inactive until a specific fragment is cleaved off by the enzyme preceding it in the cascade. Secondly, the activated enzymes function for only a short time before being inactivated again. Thirdly, numerous regulatory and inhibitory molecules called **regulator of complement activation (RCA)** proteins ensure that the complement pathways are activated only where and when they should be (**Table 3-4**). Some of these RCA proteins are

	TABLE 3-4	**Regulator of Complement Activation (RCA) Proteins**		

Abbr.	Name of RCA Protein	Form	Function
CR1	Complement receptor 1	Membrane-bound	Binds to C4b and C3b to prevent formation of classical and alternative C3 convertases Promotes release of C2a from classical C3 convertases
DAF	Decay accelerating factor	Membrane-bound	Binds to C4b and C3b to prevent formation of classical and alternative C3 convertases Promotes release of C2a from classical C3 convertases
C4bp	C4 binding protein	Soluble	Binds to C4b to prevent C2 binding and formation of classical C3 convertase Promotes degradation of surface-bound C4b by Factor I
I	Factor I	Soluble	Degrades surface-bound C4b
H	Factor H	Soluble	Blocks binding of Factor B to surface-bound C3b Promotes dissociation of alternative C3 convertase Stimulates degradation of C3b by Factor I
MCP	Membrane cofactor protein	Membrane-bound	Blocks binding of Factor B to surface-bound C3b Promotes dissociation of alternative C3 convertase Stimulates degradation of C3b by Factor I
	Vitronectin	Soluble	Binds and inactivates free terminal complexes
	Clusterin	Soluble	Binds and inactivates free terminal complexes
HRF	Homologous restriction factor	Membrane-bound	Binds to C8 and prevents C9 addition
MIRL	Membrane inhibitor of reactive lysis	Membrane-bound	Binds to C8 and C9 and prevents C9 polymerization and pore formation

Human deficiencies exist for virtually every component of the complement system, including its major enzymes and RCA proteins. More information on some of these diseases appears in Chapter 15.

soluble inhibitors that circulate in the blood, while others are fixed in the plasma membranes of healthy host cells. In general, both types of RCA proteins act to prevent the assembly of C3 convertases and/or the MAC on healthy host cell surfaces. Occasionally, a partially assembled MAC containing the terminal components C5b67 is released as a free terminal complex from the membrane of a pathogen. If not restrained by host regulatory proteins, a free terminal complex could penetrate into the membrane of a nearby healthy host cell and attempt to complete MAC formation.

IV. Inflammation

In the preceding sections, we described some of the molecular events that activate innate leukocytes. The responses of these leukocytes are part of a wider range of non-specific soluble and cellular processes collectively known as **inflammation**. The goal of inflammation is to restore homeostasis to disrupted host tissues. The process of inflammation is designed to promote the elimination of foreign entities, clear cellular debris, and repair damaged tissues. Inflammation can be induced by pathogen attack, inert tissue injury, products of complement activation, or cytokines released by innate or adaptive leukocytes that were activated locally or drawn from elsewhere. Next we describe some of the physiological and clinical features that are considered hallmarks of inflammation.

i) Clinical Signs

We are all familiar with the clinical signs of inflammation: localized heat, redness, swelling, and/or pain at a site of infection or injury. The heat and redness result from **vasodilation** (an expansion in the diameter of local blood vessels), which allows increased blood flow into the affected area. Enhanced leukocyte adhesion to local blood vessel walls and increased permeability of the capillaries in this area encourage an influx of leukocytes into the tissue, causing swelling. The pain results not only from the swelling but also from the stimulation of pain receptors in the skin by peptide mediators. The increased permeability of blood vessels also allows large molecules such as antibodies of the adaptive response and enzymes of the blood clotting system to leak into the affected tissue. The blood clotting enzymes trigger the deposition of fibrin to form a blood clot and commence wound healing, while antibodies bolster infection control and prevent pathogen spread.

ii) Initiators and Mediators

The actual initiation of inflammation is still not well understood. Factors released by injured cells, or pro-inflammatory cytokines produced by activated resident macrophages, are thought to play important roles (**Table 3-5**). Some factors act directly and locally to induce initial changes in blood vessel diameter and permeability, signaling that an inflammatory response is under way. Among these factors are the kinins, a family of small peptides that circulate in the blood in inactive form. Once activated by the blood clotting system or by enzymes released by damaged cells, kinins give rise to potent peptide mediators (including bradykinin) that cause vasodilation, increased vascular permeability, smooth muscle contraction, and pain.

Other inflammatory mediators contribute in an indirect way to the response. Cells infected with viruses are triggered to produce IFNα and IFNβ that induce nearby uninfected cells to adopt an "antiviral state" and resist infection (see Ch. 13). IFNγ produced by stimulated NK and NKT cells activates macrophages. TNF, IL-1 and IL-6 secreted by NK cells, macrophages and activated endothelial cells as well as by damaged tissue cells perpetuate the inflammatory response. Hepatocytes responding to TNF and IL-6 produce acute phase proteins, including CRP. CRP can initiate the complement cascade via C1 activation in a way that does not involve antibody. The C3b generated by complement activation coats microbes, making them vulnerable to opsonized phagocytosis (see later), while C5a and C3a provoke mast cell degranulation and thus the immediate release of pre-formed heparin and vasoactive amines. Heparin is a blood clotting inhibitor that helps maintain the required influx of soluble factors and leukocytes into the affected area. Vasoactive amines, such as histamine, induce vasodilation and increase vascular permeability. Finally, mediators called leukotrienes and prostaglandins increase vasodilation and promote neutrophil chemotaxis during the later stages of inflammation. These molecules are not pre-formed and must be derived from the breakdown of phospholipids in the membranes of activated macrophages, monocytes, neutrophils and mast cells.

TABLE 3-5	**Major Inflammatory Mediators**		
Mediator Class	**Example**	**Principal Source**	**Principal Effect**
Kinins	Bradykinin	Blood	Vasodilation and increased vascular permeability, smooth muscle contraction, pain
Pro-inflammatory cytokines	IFNs	Virus-infected cells	Adoption of antiviral state by uninfected cells, activation of macrophages and NK cells
	TNF, IL-1, IL-6	Activated macrophages and endothelial cells, stimulated NK cells, damaged tissue cells	Induction of acute phase protein synthesis by hepatocytes
Acute phase proteins	CRP	Activated hepatocytes	Activation of complement
Complement products	C5a C3a	Complement activation	Leukocyte chemotaxis, mast cell degranulation, smooth muscle contraction
Granule products	Heparin	Degranulation of mast cells, basophils	Prevention of blood clotting
	Histamine		Increased vascular permeability, smooth muscle contraction, chemotaxis
Leukotrienes Prostaglandins	Leukotriene B4 Prostaglandin 1	Phospholipid products of leukocyte membranes	Increased vascular permeability, neutrophil chemotaxis
Chemokines	CXCL8, CCL3*	Activated macrophages, monocytes, lymphocytes, endothelial cells	Leukocyte chemotaxis

Many YouTube videos of leukocyte extravasation during inflammation are now available. One of the most popular can be viewed at http://www.youtube.com/watch?v=qRqTBgavso.

Watch this classic YouTube video to see a neutrophil chasing down a bacterium in a "cat-and-mouse" game. Visit http://www.youtube.com/watch?v=T5W6VpKPt1Y&feature=related.

iii) Leukocyte Extravasation and Infiltration

The chemical messengers emanating from a site of trauma or pathogen attack prompt the extravasation of leukocytes from the blood into the injured tissue (refer to **Fig. 2-16**). To aid in this process, endothelial cells respond to tissue damage or pro-inflammatory cytokines by becoming activated and upregulating selectins and other adhesion molecules that can bind to oligosaccharides or adhesion proteins on leukocyte surfaces. Tissue damage also increases the production of **platelet-activating factor (PAF)** on the endothelial cell surface. PAF stimulates the integrin-mediated adhesion of neutrophils to the postcapillary venule wall closest to the site of injury.

Once innate leukocytes have extravasated through the blood vessel wall, they employ chemotaxis to follow a gradient of chemotactic factors and migrate to the site of injury or infection (**Fig. 3-9**). Chemotactic factors in an inflammatory site can include the anaphylatoxins, various blood clotting proteins, PAF, leukotriene B4 (LTB4), microbial components and chemokines. Once in the site, activated macrophages and neutrophils vigorously attempt to engulf the foreign entity and may release antimicrobial peptides (such as defensins) that coat and protect the skin and body tract linings. Finally, innate leukocytes in an inflammatory site secrete additional chemokines that specifically attract lymphocytes of the adaptive response.

iv) Good vs. Bad Inflammation: A Mixed Blessing

Although inflammation is essential for the repair of tissue damage and protection of the host from pathogen invasion, this response can cause as many problems as it solves if it gets out of control. **Endotoxic shock** (also known as "sepsis" or "septic shock") is an example of a disorder in which a runaway inflammatory response to PAMPs, rather than the pathogen itself, kills a patient (see **Box 3-2**). In addition, prolonged PRM-mediated recognition of DAMPs in the absence of PAMPs can be detrimental. DAMPs are routinely released by trauma-damaged tissues or cells under metabolic stress. "Good" inflammation is then initiated to effect tissue repair. However, if the

Fig. 3-9
Leukocyte Chemotaxis

Pathogens that have breached the skin and established an infection induce surrounding host cells to send out chemical signals that establish a chemotactic gradient. Leukocytes, mainly neutrophils (Neu) and macrophages (Mac), that extravasate through the wall of the local blood vessel follow the gradient through the tissue to the site of infection. Once in this site, these leukocytes, as well as resident DCs, secrete soluble mediators and attempt to phagocytose the pathogen.

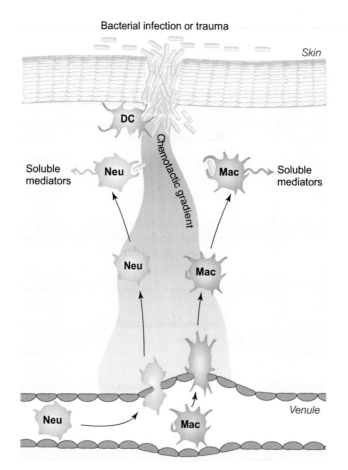

Box 3-2 Bacterial Endotoxic Shock: Too Much of a Good Thing

A little TNF is very good thing, and essential to combat infections. However, when an overwhelming amount of the bacterial cell wall component LPS is present, the **endotoxin** released induces activated macrophages to produce massive quantities of TNF, with potentially catastrophic results for the host. Very high TNF concentrations stimulate the overproduction by macrophages of IL-1, which in turn amplifies LPS-induced production by vascular endothelial cells and macrophages of additional pro-inflammatory cytokines such as IL-6. This systemic tidal wave of pro-inflammatory cytokines, termed septic or **endotoxic shock**, has devastating effects on several key organ systems. Capillaries are blocked both by aggregations of neutrophils and excessive blood clot formation. Clotting factors are subsequently depleted, leading to hemorrhage. Cells in the heart begin to fail due to an accumulation of NO produced by the cytokine-activated enzyme iNOS. Cytokine-induced slowing of the blood circulation contributes to a drop in blood pressure followed by circulatory collapse. Cytokines also stimulate muscles to consume glucose at an increased rate, but then interfere with glucose synthesis by the liver, leading to metabolic failure. The combination of circulatory collapse and metabolic failure results in irreversible, and often lethal, damage to host organs. As a result, death due to endotoxic shock can happen within hours of infection. In North America, about 600,000 people per year experience endotoxic shock, and more than 10% of these cases have a fatal outcome. Sadly, clinical trials evaluating experimental treatments based on antibodies directed against LPS, TNF or other pro-inflammatory cytokines or their receptors have been unsuccessful to date. Increased understanding of PRM engagement and signaling may help researchers to design drugs that can shut down out-of-control inflammation.

TABLE 3-6	**Human Diseases Associated with Persistent Inflammation and Elevated DAMPs**
Disease	**Associated DAMPs**
Atherosclerosis	HSPs, fibrinogen, HMGB1, S100 proteins, heparan sulfate, cholesterol crystals
Cancer	HSPs, fibrinogen, HMGB1, S100 proteins, heparan sulfate
Diabetes	HSPs, HMGB1
Gout	Uric acid crystals
Inflammatory bowel disease	HSPs, S100 proteins, heparan sulfate
Multiple sclerosis	HSPs, fibrinogen, HMGB1
Rheumatoid arthritis	HSPs, fibrinogen, HMGB1, S100 proteins

activation of these PRMs and their consequent downstream signaling do not end when they should, persistent "Bad" inflammation can result that may lead to chronic inflammatory disease or even autoimmune disease or cancer. Some examples of human diseases in which high levels of particular DAMPs have been identified in patient blood or tissues are given in **Table 3-6**.

V. Cellular Degradation of Unwanted Entities

A major activity of many innate leukocytes is the capture and disposal of pathogens or other unwanted entities encountered in the body's blood or tissues. Engulfment followed by degradation within the leukocyte disposes of the vast majority of external invaders. Engulfment is achieved by *macropinocytosis, clathrin-mediated endocytosis*, or *phagocytosis. Autophagy* is a parallel process by which internal entities, which include defunct organelles or intracellular pathogens, are sequestered within the cell and then digested. Both engulfment and autophagy make use of the cell's endosomal compartment, a complex system of connected membrane-bound vesicles. Various forms of these vesicles participate in **endocytic processing** that eventually delivers harmful or spent contents to the degradative enzymes of the lysosomes.

Primary immunodeficiencies resulting from defects in the cells and/or molecules associated with phagocytosis or autophagy are described in Chapter 15.

i) Engulfment of External Entities
As introduced previously, there are three distinct processes for engulfing extracellular entities and importing them into a cell (**Fig. 3-10**). **Macropinocytosis** and **clathrin-mediated**

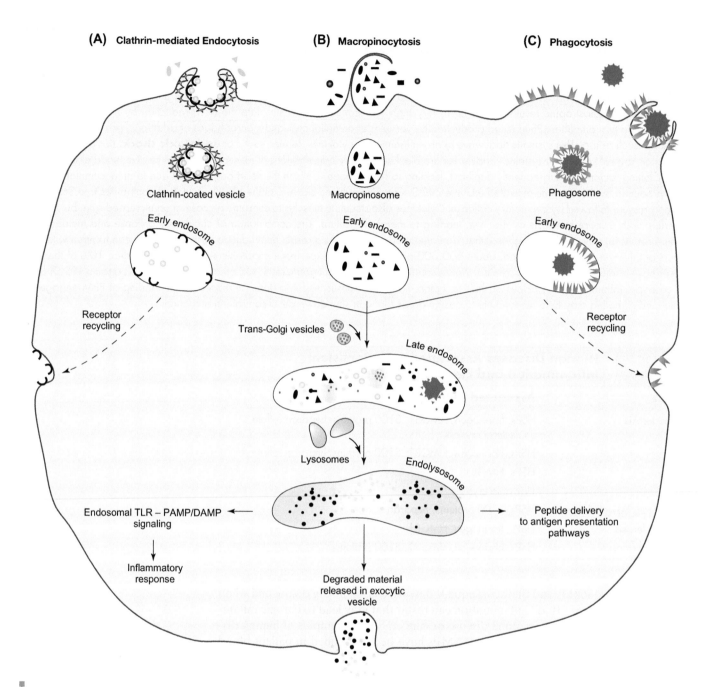

(A) Clathrin-mediated Endocytosis   **(B)** Macropinocytosis   **(C)** Phagocytosis

Clathrin-coated vesicle

Macropinosome

Phagosome

Early endosome

Early endosome

Early endosome

Receptor recycling

Trans-Golgi vesicles

Late endosome

Receptor recycling

Lysosomes

Endolysosome

Endosomal TLR – PAMP/DAMP signaling

Peptide delivery to antigen presentation pathways

Inflammatory response

Degraded material released in exocytic vesicle

Fig. 3-10
Engulfment and Endocytic Processing

In this schematic example, external foreign entities are shown to be engulfed by **(A)** clathrin-mediated endocytosis, **(B)** macropinocytosis or **(C)** phagocytosis. While almost all cell types can carry out clathrin-mediated endocytosis and macropinocytosis, only certain leukocytes, such as neutrophils, DCs and macrophages, can carry out phagocytosis. External entities are initially sequestered in clathrin-coated vesicles, macropinosomes or phagosomes. These membrane-bound vesicles then undergo endocytic processing by fusing with early endosomes. Any receptors used for internalization are recycled to the cell surface. Digestion of the foreign entity commences within early endosomes and advances as these vesicles fuse with various trans-Golgi vesicles to form late endosomes. Digestion is completed after late endosomes fuse with lysosomes to form endolysosomes. Unwanted degradation products are expelled from the cell by exocytosis. In the case of phagocytes, DAMPs/PAMPs resulting from endolysosomal digestion may be shunted to endosomal compartments expressing TLRs, initiating TLR signaling and an inflammatory response. In addition, peptides resulting from endolysosomal digestion may be directed into specialized antigen presentation pathways for the formation of peptide–MHC complexes needed to initiate an adaptive response (see Ch. 7 and 9).

endocytosis involve the internalization of relatively small soluble macromolecules, and are ongoing activities that can be carried out by any cell. **Phagocytosis**, which handles large extracellular particles (including bacteria and viruses), is carried out only by specialized phagocytic leukocytes, predominantly neutrophils, macrophages, monocytes and DCs. Regardless of its method of internalization, the external entity is eventually eliminated by endocytic processing. An elegant feature of this system is that any host receptor proteins originally used to capture the unwanted entity are recycled intact to their original locations.

a) Macropinocytosis

In macropinocytosis, a cell ruffles its plasma membrane to form a small vesicle called a **macropinosome** around a droplet of extracellular fluid. Macropinosomes are highly variable in volume and contain fluid phase solutes, rather than insoluble particles. Macromolecules enter the vesicle with the extracellular fluid on a gradient of passive diffusion. The number of macromolecules caught in the vesicle depends entirely on their concentration. This process provides an efficient but non-specific means of sampling the extracellular environment.

b) Clathrin-Mediated Endocytosis

In clathrin-mediated endocytosis, the uptake of a macromolecule depends on whether it binds to an appropriate cell surface receptor. Such binding triggers the polymerization of **clathrin**, a protein component of the microtubule network located on the cytoplasmic side of the plasma membrane. Invagination of clathrin-coated "pits" internalizes the receptor and its bound ligand into a small clathrin-coated vesicle.

c) Phagocytosis

Phagocytosis is initiated when multiple cell surface receptors on a phagocyte bind sequentially in a "zippering" manner to ligands present on a large particle, whole microbe or apoptotic host cell. The interactions between these receptors and ligands induce the polymerization of the cytoskeletal protein actin at the site of internalization. The plasma membrane then invaginates and forms a large vesicle called a **phagosome** around the entity to be removed. Entire live pathogens engulfed in a phagosome are killed by an oxidation reaction called the **respiratory burst** before entering the endosomal compartment. The end result of the respiratory burst is the generation of toxic reactive oxygen and nitrogen intermediates (ROIs and RNIs) that interfere with pathogen metabolism and replication, eventually causing its destruction. Should the damaging ROIs or RNIs escape from the phagosome into the cytoplasm, the phagocyte protects itself by activating several enzymes capable of neutralizing these reactive molecules. Alternatively, these ROIs and RNIs can be released extracellularly by the phagocyte to combat additional pathogens in the external environment.

Microbes and particles that do not interact with a phagocyte's surface receptors cannot be phagocytosed directly. However, the range of targets is greatly expanded when microbes are coated with **opsonins**. As mentioned previously, an opsonin is a host-derived protein that binds to the exterior of a microbe and facilitates its engulfment by phagocytes expressing the receptors for that opsonin (**Fig. 3-11**). One of the most common opsonins is the complement component C3b, which can bind to particular motifs of pathogen surface proteins and thus acts as an extracellular PRM. When a microbe coated in C3b binds to CR1 on the surface of a phagocyte, this interaction triggers the phagocytosis of the invader. Another common opsonin is antigen-specific antibody, which is available only after the initiation of an adaptive response. A microbe coated in specific antibody can be bound by a phagocyte's FcRs, initiating phagocytosis of the microbe–antibody complex. When a pathogen is coated with more than one type of opsonin, phagocytosis is even more efficient.

> NOTE: Many of the receptors cited earlier in this chapter as being PRRs are also involved in the routine engulfment of entities mediated by clathrin-mediated endocytosis and phagocytosis. They include complement receptors; scavenger receptors like CD36 and CD91; and CLRs like the mannose receptors, DEC-205 and DC-SIGN.

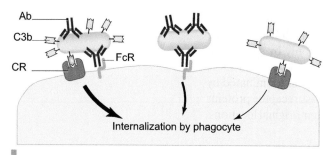

Fig. 3-11
Opsonization

Opsonins are host-derived proteins that coat foreign entities and enhance phagocytosis by leukocytes. Antibodies and complement components (particularly C3b) can act as opsonins, either on their own or in combination, when the appropriate receptors are expressed on the phagocyte. The ability of different opsonins to enhance phagocytosis varies: antibody plus complement component > antibody > complement component. CR, complement receptor; FcR, Fc receptor.

ii) Endocytic Processing

Once a cell has captured and killed (if necessary) entities from the extracellular environment, it must then dispose of its burden via **endocytic processing**. As noted previously, endocytic processing not only is an internal trafficking system for membrane-bound vesicles but also recycles any receptors and membrane components that were used for capture and internalization. Endocytic processing involves four major types of membrane-bound vesicular compartments: early endosomes, late endosomes, lysosomes and endolysosomes (refer to **Fig. 3-10**). When the uptake of entities from the extracellular environment results in the formation of a membrane-bound vesicle (e.g., a macropinosome, clathrin-coated vesicle or phagosome), the vesicle first fuses to an early endosome, which has a mildly acidic environment (about pH 6.5). This increase in pH promotes the dissociation of any receptor proteins that were associated with the captured entity. The receptors are then collected in a tubular extension of the early endosome that buds off and fuses with the plasma membrane, returning the receptors to the cell surface. Early endosomes containing captured entities then fuse with Golgi-derived vesicles to create late endosomes, resulting in a further decrease in internal pH. Fusion of late endosomes with additional trans-Golgi vesicles initiates the digestion of the entity. Complete degradation occurs when the late endosome fuses with a lysosome, forming an **endolysosome**. In the case of engulfment of a whole extracellular pathogen, many immunologists say that the initial phagosome has undergone "maturation" to eventually form a **phagolysosome.** Regardless of its moniker, an endolysosome has an internal pH of 5 and a cargo of hydrolytic enzymes (including lipases, glycosidases, nucleases and proteases) that degrade any macromolecules to their fundamental components (fatty acids, nucleotides, amino acids, sugars, etc.). Useful metabolic building blocks can be re-used by the cell, whereas unwanted products of digestion are collected in an exocytic vesicle that buds off the endolysosome. The exocytic vesicle fuses with the plasma membrane, harmlessly expelling degradation products into the extracellular fluid in a process called **exocytosis.**

Two additional events occur within innate leukocytes that have digested entities within their endosomal processing systems. DAMPs/PAMPs generated by the degradation process may bind to endosomal TLRs, initiating TLR signaling that activates the leukocyte and induces an inflammatory response. In addition, peptides generated as a result of protein degradation can be delivered into specialized late endosomes that are part of the antigen processing pathways in APCs that generate pMHC complexes. These pMHCs are presented on the APC surface for inspection by T cells, allowing the mounting of a T cell–mediated adaptive response against the specific pathogen being encountered by the innate leukocytes.

"Statins Enhance Formation of Phagocyte Extracellular Traps" by Chow, O.A., von Köckritz-Blickwede, M., Bright, A.T., Hensler, M.E., Zinkernagel, A.S., Cogen, A.L., Gallo, R.L., Monestier, M., Wang, Y., Glass, C.K., Nizet, V. (2010) *Cell Host and Microbe* 8, 445–454.

The killing of bacteria by phagocytic cells of the innate immune system, such as neutrophils, has been traditionally ascribed to (1) phagocytosis and oxidative destruction within phagolysosomes, and (2) release of antimicrobial peptides via degranulation. However, recent studies have revealed a third mechanism by which phagocytes corral their prey in **"neutrophil extracellular traps" (NETs)**. This process succeeds in neutralizing the pathogen but unfortunately involves the death of the phagocytic cell. [For an excellent review on NETs, see Papayannopoulous and Zychlinsky, (2009) *Trends in Immunology*, 30, 513–521.)] In certain cases, a neutrophil that interacts with a microbe or microbial product is triggered to undergo a cell death program that causes the membranes of its nucleus and granules to dissolve. The plasma membrane then ruptures and the dying cell expels a molecular network composed of chromatin and its associated histones as well as granule proteases. These substances form the NET structure that not only traps the microbe but also has antimicrobial activity, so that the pathogen is killed in a way analogous to an insect tangled in a spider's web. Importantly, NET formation has little effect on surrounding healthy host tissues, so that the collateral damage sometimes inflicted on bystander tissues by neutrophil degranulation is avoided.

An intriguing connection has been uncovered between NETs and a class of drugs used to prevent heart disease. In this primary research article, Chow *et al.* reveal the molecular mechanism behind a remarkable observation: that patients taking cholesterol-lowering statin drugs are less likely to die of some very serious bacterial infections than are patients not taking statins. As shown in Figure 1, Panel A from this article, when human neutrophils, human macrophages (U937 cells) and mouse macrophages (RAW 264.7 cells) are treated *in vitro* with mevastatin, the survival of *Staphylococcus aureus* in these cell cultures is significantly reduced relative to control cultures not treated with mevastatin (**$p < 0.01$, ***$p < 0.005$). As the reader studies the article, he/she will learn that the bacteriocidal effect of statin drugs is due to their promotion of neutrophil NET formation. The reader will also start to explore the fascinating biochemical pathways enhanced by these drugs.

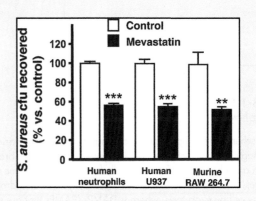

iii) Engulfment of Internal Entities: Autophagy

Autophagy was first identified as a response mounted by cells subjected to environmental or nutritional stress. Under these conditions, a host cell can degrade its own components in an effort to redistribute the resulting amino acids, lipids, and other molecules into pathways that can help the cell to survive until favorable environmental conditions are restored. In this way, autophagy contributes to the maintenance of normal cell numbers and thus host homeostasis. More recently, it has been found that autophagy in leukocytes can clear intracellular bacteria, viruses and parasites, as well as promote additional innate and adaptive responses.

a) Macroautophagy

The process of **macroautophagy** commences with the construction of an "isolation membrane" that partially encloses the cytosolic entity to be engulfed and digested (**Fig. 3-12**). In leukocytes, this entity is often a pathogen. The isolation membrane completes its circularization to form an *autophagosome*. The autophagosome eventually fuses with a lysosome, forming an *autophagolysosome* in which both the inner lipid bilayer of the double membrane and the vesicular cargo are degraded. Under stress or starvation conditions, the breakdown products within the autophagolysosome are redirected for use in the cell's various metabolic pathways. During an

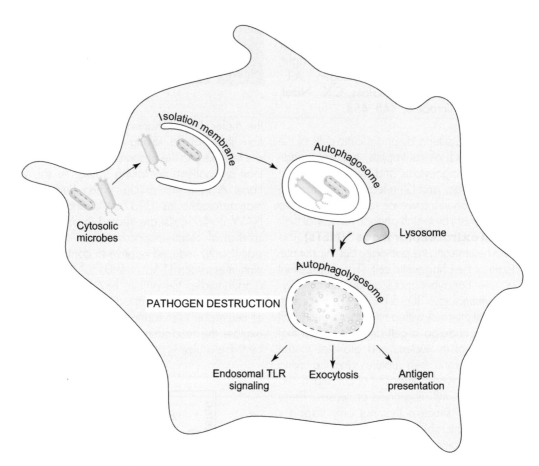

Fig. 3-12
Autophagy

In this schematic example, a leukocyte has experienced invasion by microbes that have attempted to set up shop in its cytosol. The cell initiates autophagy by constructing a double-layered isolation membrane around the microbes. Completion of circularization of this membrane generates an autophagosome that eventually fuses with a lysosome to form an autophagolysosome. The inner membrane of the autophagolysosome dissolves, and the microbes are destroyed by lysosomal hydrolytic enzymes. As was true for phagocytosis, unwanted products are expelled from the cell by exocytosis, whereas DAMPs/PAMPs that are shunted into endosomal compartments bearing TLRs may initiate inflammatory signaling. Autophagy is linked to the adaptive response when peptides generated by protein digestion are shunted into antigen presentation pathways.

More than 30 autophagy-related genes have been identified in yeast, many of which have mammalian counterparts.

infection, some of these breakdown products can be shunted into late endosomes bearing TLRs in their membranes. As described previously, any DAMPs/PAMPs that are present activate endosomal TLR signaling, triggering an inflammatory response. In addition, any peptides present due to protein digestion can be diverted into the specialized late endosomes that support pMHC formation, promoting an adaptive response.

b) Microautophagy and Chaperone-mediated Autophagy
Autophagy can occur by two other mechanisms: *microautophagy* and *chaperone-mediated autophagy*. In microautophagy, a cytoplasmic entity itself buds directly into a lysosome without prior formation of an isolation membrane. This mechanism has been clearly documented only for lower organisms such as yeast. In chaperone-mediated autophagy, proteins that contain a particular signal peptide are bound by cytosolic "chaperone" proteins, which guide the targeted protein to a lysosome and allow it to enter this degradative organelle. Chaperone-mediated autophagy occurs only when a cell is subjected to prolonged metabolic stress.

NOTE: Some pathogens, including *Mycobacterium tuberculosis*, have devised strategies for escaping from phagolysosomes into the cytosol, thereby avoiding immediate destruction. Autophagy may have evolved partly as a way to re-capture these pathogens once they become cytosolic and to return them to the endocytic processing system. Such evolution would be yet another example of the competition that pits pathogen evasion tactics against host defense strategies. More on pathogen evasion mechanisms can be found in Chapter 13.

VI. NK, γδ T and NKT Cell Activities

NK, γδ T and NKT cells are leukocytes that bridge innate and adaptive immunity. Although all three of these cell types are closely related to αβ T cells, their responses to injury or infection are rapid and involve broad ligand recognition, marking them as players in innate immunity. The detailed characterization of NK, γδ T and NKT cells occurred well after the biology of B cells and αβ T cells was established. Early analyses of NK, γδ T and NKT cell receptors showed that the diversity of these molecules was much less than that of TCRs and BCRs, leading some scientists to conclude that NK, γδ T and NKT cells were lymphocytes with minor roles in the adaptive response. However, we now know that the relatively limited diversity in ligand binding exhibited by NK, γδ T and NKT cells is essential to their primary physiological function as sentinels on the front lines of innate defense. When the receptors of NK, γδ T and NKT cells are engaged by the appropriate DAMPs/PAMPs, these cells produce cytokines that stimulate both the innate and adaptive responses. NK cells and γδ T cells also mediate the cytolysis of infected cells and thus contribute directly to innate defense against certain pathogens. Since the biology of NK, γδ T and NKT cells and their receptors has been elucidated and defined relative to that of αβ T cells (described in Chapters 8, 9 and 10), we have reserved a complete discussion of NK, γδ T and NKT cells until Chapter 11.

The innate response deals with foreign entities using mechanisms based on broad recognition and is sufficient to counter most of the threats our bodies encounter every day. However, when the innate response is overwhelmed, the more focused mechanisms of adaptive immunity are required. Many of the cells activated in the course of an innate response then act as bridges to the adaptive response, by interacting directly with T cells or producing cytokines needed to support T and B cell activation. The next two chapters, Chapters 4 and 5, discuss the biology of B cells and their genes and proteins, and the mechanisms of the humoral adaptive response.

Chapter 3 Take-Home Message

- Mechanisms of innate immunity work collectively to inhibit the entry of a pathogen or eliminate it, thereby preventing the establishment of infection. Alternatively, innate immunity holds an infection in check until the slower adaptive immune response can be mounted. Innate immune mechanisms also clear both host cell and microbial debris from the body.

- Some innate immune mechanisms are completely non-specific, whereas others involve broad pattern recognition mediated by soluble extracellular molecules, or cell-associated molecules that are soluble or membrane-fixed.

- Innate immune mechanisms include anatomical and physiological barriers, complement activation, inflammatory responses, cytokine secretion, target cell lysis, and cellular degradation mechanisms such as phagocytosis and autophagy.

- Innate immune responses result in the activation of cells that support the adaptive response, as well as the production of soluble factors (particularly cytokines) that are critical for the recruitment, activation and differentiation of lymphocytes.

- The major types of leukocytes mediating inducible innate immunity are neutrophils, DCs, macrophages, mast cells, NK cells, NKT cells and γδ T cells.

Did You Get it? A Self-Test Quiz

Section A

1) Can you define these terms? hypodermis, lysozyme, commensal organisms, microflora, siderophore

2) Give three examples of non-inducible innate defense and explain how they help prevent infection.

3) What is colonization resistance and why is it important?

4) Name two deleterious effects on the host of excessive antibiotic use.

Section B.I–III

1) Distinguish between PAMPs and DAMPs and give three examples of each.

2) Why is the repertoire diversity of the inducible innate response more limited than that of the adaptive response?

3) Can you define these terms? PRR, PRM, TLR, HSP, SR, CRP, NOD, NLR, CLR, RLR, RAGE

4) Distinguish between cell-associated and extracellular PRMs.

5) Name two functions of TLRs in mammals.

6) Distinguish between NLRs, RLRs and CLRs.

7) What is an inflammasome and what does it do?

8) Name two diseases associated with NLR dysfunction.

9) Name two types of extracellular PRMs and describe how they contribute to inducible innate defense.

10) Can you define these terms? C3 convertase, C5 convertase, Factor B, properdin, MBL, MAC, terminal components, RCA protein

11) Name four outcomes of complement activation.

12) Describe how the three pathways of complement activation differ in their initiation.

13) Which pathway of complement activation is considered part of the adaptive response and why?

14) Name three functions of C3b.

15) What is an anaphylatoxin, and what does it do?

16) Name four RCA proteins and describe how they control the complement cascade.

Section B.IV-VI

1) Can you define these terms? inflammatory response, vasodilation, chemotaxis, defensin

2) Describe three clinical signs of inflammation and how they arise.

3) Can you give four examples of major inflammatory mediators?

4) How does extravasation underpin the inflammatory response?

5) Why is endotoxic shock sometimes fatal?

6) Can you define these terms? macropinosome, clathrin, endosome, phagosome, phagolysosome, exocytosis, ROI, RNI, iNOS, NET, autophagosome

7) Can you describe the three mechanisms of cellular internalization of external entities and how they differ?

8) What is opsonization and how is it helpful to immune defense? Give two examples of opsonins.

9) What is the respiratory burst, and why is it important?

10) Can you describe the endocytic processing pathway?

11) Give two functions of autophagy.

12) Why are NK, γδ T and NKT cells considered to bridge the innate and adaptive responses?

Can You Extrapolate? Some Conceptual Questions

1) A patient has recurrent viral infections but seems to have normal immune responses to bacteria. Preliminary biochemical analyses suggest that the patient may have a PRR deficiency. Which PRR(s) do you suspect might be lacking in function and why?

2) In considering hypothetical complement deficiencies, why might you expect C2 deficiency to have less severe consequences than C3 deficiency?

3) A hypothetical intracellular bacterium replicates in the cytoplasm of an innate leukocyte, activating it to produce inflammatory cytokines. The researcher studying this microbe initially hypothesizes that the leukocyte activation he/she observed might be due to inflammatory signaling through NLRP1, but subsequently finds by experimentation that it is mediated mainly through TLR9. How might the researcher justify his/her initial hypothesis and then explain the actual observation?

Would You Like To Read More?

Blander, J. M., & Medzhitov, R. (2006). On regulation of phagosome maturation and antigen presentation. *Nature Immunology, 7*(10), 1029–1035.

Kawai, T., & Akira, S. (2009). The roles of TLRs, RLRs and NLRs in pathogen recognition. *International Immunology, 21*(4), 317–337.

Litvack, M. L., & Palaniyar, N. (2010). Review: Soluble innate immune pattern-recognition proteins for clearing dying cells and cellular components: Implications on exacerbating or resolving inflammation. *Innate Immunity, 16*(3), 191–200.

McDermott, M. F., & Tschopp, J. (2007). From inflammasomes to fevers, crystals and hypertension: How basic research explains inflammatory diseases. *Trends in Molecular Medicine, 13*(9), 381–388.

Medzhitov, R. (2009). Approaching the asymptote: 20 years later. *Immunity, 30*(6), 766–775.

Munz, C. (2009). Enhancing immunity through autophagy. *Annual Review of Immunology, 27*, 423–449.

Nairz, M., Schroll, A., Sonnweber, T., & Weiss, G. (2010). The struggle for iron—a metal at the host-pathogen interface. *Cellular Microbiology, 12*(12), 1691–1702.

Palaniyar, N. (2010). Antibody equivalent molecules of the innate immune system: Parallels between innate and adaptive immune proteins. *Innate Immunity, 16*(3), 131–137.

Piccinini, A. M., & Midwood, K. S. (2010). DAMPening inflammation by modulating TLR signalling. *Mediators of Inflammation, pii*, 672395. http://dx.doi.org/10.1155/2010/672395 Epub 2010 July 13.

Schroder, K., & Tschopp, J. (2010). The inflammasomes. *Cell, 140*(6), 821–832.

Steinman, R. M., & Banchereau, J. (2007). Taking dendritic cells into medicine. *Nature, 449*(7161), 419–426.

Turvey, S. E., & Broide, D. H. (2010). Innate immunity. *Journal of Allergy & Clinical Immunology, 125*(Issue 2, Suppl 2), S24–32.

Ueno, H., Klechevsky, E., Morita, R., Aspord, C., Cao, T., Matsui, T., et al. (2007). Dendritic cell subsets in health and disease. *Immunological Reviews, 219*, 118–142.

The B Cell Receptor: Proteins and Genes

Great acts are made up of small deeds.

Lao Tzu

Adaptive immunity depends in large part on the production and effector functions of antibodies. As introduced in Chapter 1, the binding of specific antigen to the BCR of a B lymphocyte induces the activation, proliferation and differentiation of that cell into both memory B cells and plasma cells that synthesize antibody able to bind to the antigen. The antibody protein itself is a modified form of the membrane-bound BCR expressed by the original B lymphocyte. Both the BCR and the antibody have the same antigenic specificity, and both are forms of **immunoglobulin** (Ig) proteins. This chapter focuses on the structure and function of Ig proteins as well as the Ig genes encoding them.

A. Immunoglobulin Proteins

I. The Nature of Immunoglobulin Proteins

How did antibodies and immunoglobulins get their names? In the late 1800s, scientists discovered that the serum component of blood could transfer immunity to toxins from an immunized animal to an unimmunized one. The serum preparation from the immunized animal was called the "antiserum," and the active transferable agent in the antiserum that conferred the immunity on the second animal was called an "antitoxin." The label "antitoxin" was changed to "antibody" in the 1930s to account for the fact that transferable agents inducing immunity were also produced in response to substances other than toxins. During the next few decades, scientists made routine use of electrophoresis as a technique for characterizing proteins: a mixture of proteins was introduced into a semi-solid medium (such as a starch, agar or polyacrylamide gel) and separated on the basis of size or charge by application of an electric current. When normal serum was electrophoresed, its proteins separated into a large albumin fraction and three smaller α, β and γ globulin protein fractions. Upon immunization, a dramatic increase in the γ globulin peak was observed and attributed to the production of antibodies, leading to the designation of these proteins first as "gamma globulins" and then as "immunoglobulins" to acknowledge their role in immunity.

MAK: Primer to the Immune Response. http://dx.doi.org/10.1016/B978-0-12-385245-8.00004-2

II. Structure of Immunoglobulin Proteins

i) Basic Structure

In both humans and mice, all Ig molecules are large glycoproteins with the same basic core structure: two heavy (H) chains and two light (L) chains that come together to form a Y-shaped molecule (**Fig. 4-1**). This structure is often abbreviated as "H_2L_2". Genetically, an individual B cell can produce only one kind of H chain and only one kind of L chain, so that the two H chains in a given Ig molecule are identical and the two L chains are identical. In general, the two H chains are joined to each other by two or more covalent disulfide bonds, and each H chain is joined to an L chain by an additional disulfide bond. These disulfide bonds vary in number and position between different types of Ig molecules and among species. The N-termini of the H and L chains come together in H–L pairs to form two identical antigen-binding sites. The much less variable C-terminal end of the Ig molecule contains that part of the antibody that mediates the antibody's effector functions. Brief digestion of an Ig protein with the protease papain results in two identical **Fab** fragments, so called because they retain the antibody's antigen-binding ability (**Fig. 4-2**). The remaining fragment of the Ig protein is called the **Fc** region because it crystallizes at low temperature. The Fc region was introduced in Chapter 2 as the tail of the antibody protein that does not recognize antigen but does bind to Fc receptors (FcRs) expressed by innate leukocytes. It is the interaction of the Fc region of an antigen-bound antibody with the body's antigen clearance mechanisms (including those mediated by FcRs) that is the foundation of the humoral response (see Ch. 5).

ii) Constant and Variable Domains

If the basic structure of all Igs is the same, how do the receptors of different B cells recognize different antigenic structures? Comparison of the complete amino acid sequences of individual Ig molecules reveals a vast diversity in the **variable (V) domains** making up the N-termini of the H and L chains. In contrast, the **constant (C) domains** that make up the rest of the H and L chains are relatively conserved. Each L chain contains one V and one C domain, denoted V_L and C_L, respectively. Similarly, each

Fig. 4-1

Basic H₂L₂ Structure of the Immunoglobulin Molecule

Schematic representation of the basic Ig protein structure. Two heavy (H) chains and two light (L) chains are held together by interchain disulfide bonds, forming two identical antigen-binding sites and one region controlling antibody effector function.

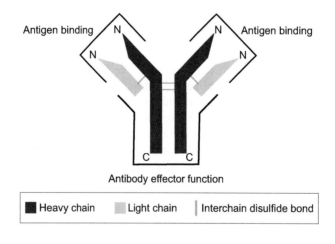

Fig. 4-2

Antibody Fragments Produced by Limited Proteolytic Digestion

Limited digestion of antibody with the enzyme papain yields two Fab fragments, each containing one antigen-binding site, plus one Fc fragment.

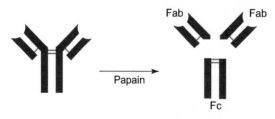

H chain contains one V_H and three or four C_H domains. Within an Ig molecule, the H and L polypeptides are aligned such that the V domains and C domains in the L chain (V_L and C_L) are positioned directly opposite their counterparts in the H chain (V_H and C_H1) (**Fig. 4-3A**). Thus, two **V regions** are formed by the pairing of the V_L domain of an L chain with the V_H domain of the H chain with which it is associated. The **C region** of the Ig comprises all the C_L and C_H domains of the H and L chains.

All V and C domains in H and L chains are based on a common structural unit known as an **Ig domain** (**Fig. 4-3B**). An Ig domain is about 70–110 amino acids in length with cysteine residues at either end that form an intrachain disulfide bond. The amino acid sequences of Ig domains are not identical but are very similar. Within a given Ig domain, the linear amino acid sequence folds back on itself to form an identifiable and characteristic cylindrical structure known as the *immunoglobulin barrel* or *immunoglobulin fold*. Many other proteins involved in intermolecular and intercellular interactions both within the immune system and in other biological processes in the body also contain Ig-like domains with a similar barrel structure. These proteins are all said to be members of the **Ig superfamily.** The presence of the Ig fold in an Ig superfamily member confers enhanced structural stability in hostile environments. Thus, antibodies

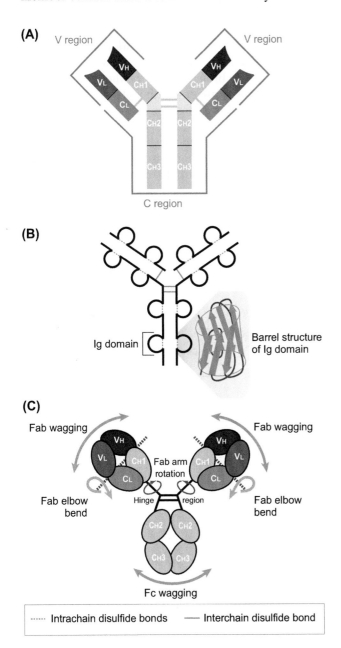

(A)

V region   V region

V_H V_H
V_L C_H1 C_H1 V_L
C_L C_L
C_H2 C_H2
C_H3 C_H3

C region

(B)

Ig domain

Barrel structure of Ig domain

(C)

Fab wagging   Fab wagging

V_H V_H
V_L C_H1 Fab arm C_H1 V_L
C_L rotation C_L

Fab elbow bend   Hinge   region   Fab elbow bend

C_H2 C_H2
C_H3 C_H3

Fc wagging

----- Intrachain disulfide bonds   —— Interchain disulfide bond

Fig. 4-3

Structural Features of the Immunoglobulin Molecule

(A) Schematic representation of the Ig variable (V) and constant (C) regions showing their domain structure. **(B)** Schematic representation of the characteristic Ig domain loop structure, with inset showing the three-dimensional barrel-like structure of the loop. **(C)** The hinge region holds the central area of the Ig molecule rigid, but each Fab arm is able to rotate around the V_H-C_H1 axis (red dotted line). Flexibility outside the hinge region allows both the Fab and Fc portions of the molecule to "wag" back and forth to some degree. The Fab can also bend like an elbow along the interface between its V and C domains. *[Part C: With information from Brekke, O.H. (1996). The structural requirements for complement activation by IgG: Does it hinge on the hinge? Immunology Today 16, 85–90.]*

are remarkably stable under conditions of extreme pH (such as occur in the gut) and are resistant to proteolysis under natural conditions.

As well as stability, Ig molecules are capable of remarkable physical flexibility (**Fig. 4-3C**). In most Ig molecules, the portion of the H chain between the first and second C_H domains (where the Fab region joins the Fc region at the center of the Y) is somewhat extended and is known as the **hinge region**. The hinge region contains many proline residues, which impart rigidity to the central part of the molecule. However, glycine residues in the hinge region create a flexible secondary structure such that the Fab regions and the Fc region of the Ig molecule may "wag," bend or rotate independently of each other around the proline-stabilized anchor point.

III. Structural Variation in the V Region

If one compares the amino acid sequences of the V domains of the L and H chains making up a collection of many different Ig molecules, one finds three short stretches in each chain that exhibit extreme amino acid sequence variability among Ig proteins (**Fig. 4-4A**). These three short sequences of at least 5–7 amino acids each are called the **hypervariable regions**. The hypervariable regions are responsible for the diversity that allows the total repertoire of Igs to recognize almost any molecule in the universe of antigens. The hypervariable regions are separated by four **framework regions (FRs)** of much more restricted variability. Although the hypervariable regions are not contiguous in the amino acid sequence of the Ig, they are brought together when the Ig protein folds into its native conformation. Each of the two antigen-binding sites of the Y-shaped Ig molecule is then formed by three-dimensional juxtaposition of the three hypervariable regions in the V_L domain with those in the associated V_H domain. Because these sequences result in a structure that is basically complementary to the shape of the specific antigen bound by the Ig, the hypervariable regions are also called **complementarity-determining regions (CDRs)**. The CDRs project from the relatively flat surfaces of the framework regions as loops of varying sizes and shapes, and it is these loops that interact with specific antigen (**Fig. 4-4B**). The framework regions are thought to position and stabilize the CDRs in the correct conformation for antigen binding.

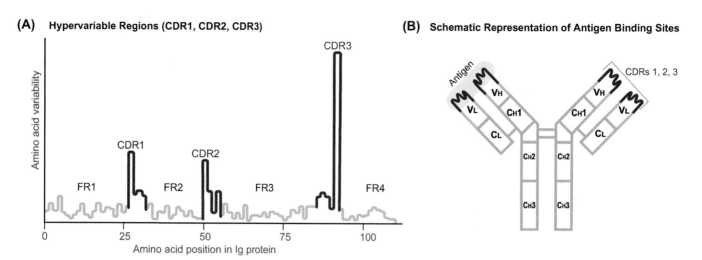

(A) Hypervariable Regions (CDR1, CDR2, CDR3)

(B) Schematic Representation of Antigen Binding Sites

Fig. 4-4
Immunoglobulin Hypervariable Regions

(A) Comparison of the V region amino acid sequences of many purified Ig molecules generates a plot of the amino acid variability observed at each position in the region. There are areas of both extremely high variability (complementarity-determining regions, CDRs) and less variability (framework regions, FRs). **(B)** Schematic representation of the Ig molecule folded into its native conformation such that the CDRs on one arm group together to form one antigen-binding site. [Part A adapted from Wu, T.T., & Kabat E.A. (1970). An analysis of the sequences of the variable regions of Bence Jones proteins and myeloma light chains and their implications for antibody complementarity. Journal of Experimental Medicine 132, 211–250.]

IV. Structural Variation in the C Region

In contrast to a V domain, the amino acid sequence of a C domain shows very little variation among antibodies because it is the C domain that interacts with the invariant molecules responsible for the body's antigen clearance mechanisms. In this way, antibodies are capable of triggering the same effector action in response to a wide variety of antigens. Nevertheless, an Ig of a given antigenic specificity can display two types of variation in its constant region: it can differ in its **isotype** and in its **structural isoform**.

i) Isotypes

Relatively minor differences in the constant domains of H and L chains give rise to Ig **isotypes**, or constant region classes. Differences in isotype can affect the size, charge, solubility and structural features of a particular Ig molecule. These factors in turn influence where that Ig goes in the body and how it interacts with surface receptors and other antigen clearance molecules.

a) Light Chain Isotypes

Humans and mice have two types of L chains: the **kappa** (κ) light chain and the **lambda** (λ) light chain (**Fig. 4-5**). All antibodies containing the κ light chain are of the κ isotype, and all antibodies containing a λ light chain are of the λ isotype. In both humans and mice, there is a single major type of κ chain. However, there may be three different λ chains expressed in a mouse, and four to six different λ chains in a human. Although any κ or λ chain can be combined with essentially any type of H chain in an Ig molecule, any one B cell expresses antibodies containing only the κ chain or one type of λ chain. In an individual human's repertoire of antibody-producing B cells, 60% produce antibodies with κ light chains and 40% produce antibodies with λ light chains. Because there are no known functional differences between κ- and λ-containing antibodies, L chains are considered to contribute to the diversity of antibodies but not to their effector functions.

b) Heavy Chain Isotypes

Although the C_H regions of different Ig molecules do not show the sequence variation seen among V_H regions, five different types of C_H regions are distinguished by subtle amino acid differences and allow Igs to engage different antigen clearance mechanisms. The five H chain isotypes are defined by polypeptides called μ, δ, γ, ε and α that determine whether an Ig is an IgM, IgD, IgG, IgE or IgA molecule, respectively. The structural differences between H chains are such that the μ and ε chains of human IgM and IgE antibodies contain four C_H domains, whereas IgD, IgG and IgA antibodies contain only three C_H domains in their shorter δ, γ and α H chains (**Fig. 4-6**). The amino

Light chain isotype		% Occurrence in adult	
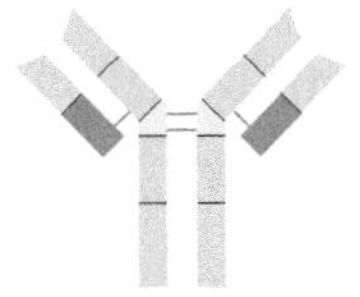 κ		Human	60
		Mouse	5
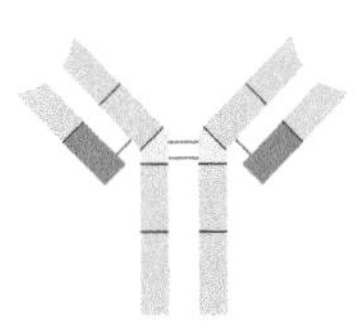 λ		Human	40
		Mouse	95

Fig. 4-5
Light Chain Isotypes

The presence of one of two possible light chain constant regions (κ or λ) determines the light chain isotype of the Ig molecule. *[With information from Gorman, J.R., & Alt, F.W. (1998). Regulation of Ig light chain isotype expression. Advances in Immunology 69, 157.]*

Fig. 4-6
Human Heavy Chain Isotypes

The presence of one of five possible heavy chain constant regions (μ, δ, γ, ε or α) determines the heavy chain isotype of the Ig molecule.

Basic structure	Heavy chain isotypes	Heavy chain domains
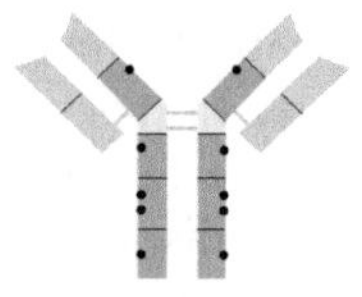 IgM	μ	$C_H 1$-4
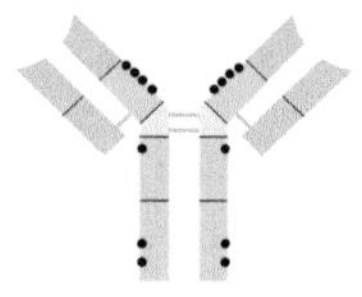 IgD	δ	$C_H 1$-3
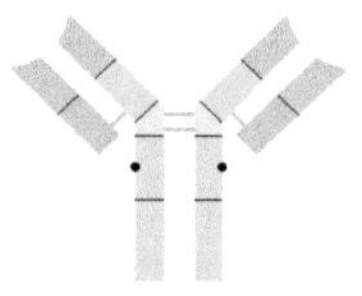 IgG	$\gamma 1$-$\gamma 4$	$C_H 1$-3
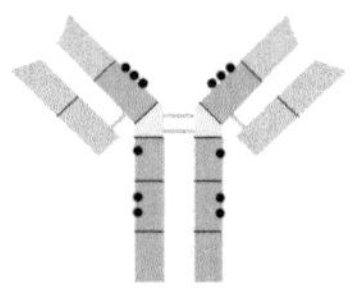 IgE	ε	$C_H 1$-4
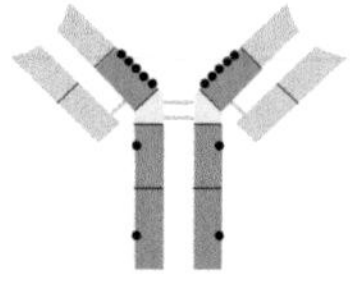 IgA	$\alpha 1$, $\alpha 2$	$C_H 1$-3
— Interchain disulfide bond • Carbohydrate (CHO)		

Ig isotypes are not strictly correlated across species; for example, human IgG3 is equivalent to mouse IgG2b, not mouse IgG3.

acid sequences of the $C_H 1$, $C_H 3$ and $C_H 4$ domains in IgM and IgE correspond to the amino acid sequences of the $C_H 1$, $C_H 2$ and $C_H 3$ domains in IgD, IgG and IgA. Variations among the sequences of the α and γ H chains in humans, and among γ sequences in mice, have given rise to Ig subclasses: IgA1, IgA2, IgG1, IgG2, IgG3 and IgG4 in humans; and IgG1, IgG2a, IgG2b and IgG3 in mice. Thus, a total of nine Ig H chain isotypes can be found in humans and eight in mice. The Ig H chain isotypes also vary in their carbohydrate content. These oligosaccharide side chains, which are generally attached to amino acids in the C_H domains but not in the V_H, V_L or C_L domains, are thought to contribute to the stability of the antibody.

c) Isotype Switching
When a naïve B cell first encounters antigen, its initial set of progeny plasma cells produce only IgM antibodies. However, plasma cells of this clone that are generated later in the response can produce Igs of isotypes other than IgM. These C region changes are due to a genetic mechanism called **isotype switching**, in which the DNA encoding the V region of the Ig H chain is reshuffled to combine with the DNA encoding C_H region sequences other than that of IgM. Thus, toward the end of the primary response, the B cell clone can switch to making IgG, IgE or IgA antibodies of the same antigenic

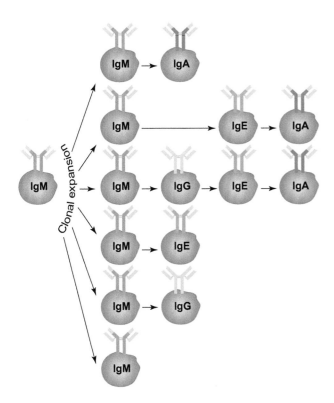

Fig. 4-7
Examples of Isotype Switching

During the primary response of a B cell to antigen, it initially produces and secretes IgM. Later, through the process of isotype switching, different heavy chain isotypes may be expressed by progeny of the original IgM-producing clone. Since such switching changes the C region but not the V region of the Ig protein, the various antibodies produced over time have different isotypes but the same specificity.

specificity (because the V region has not changed) (**Fig. 4-7**). Both memory and plasma cells capable of making the new isotypes are generated. In a subsequent response to the antigen, a memory B cell of this clone can be triggered to proliferate and differentiate into plasma cells that secrete antibodies of the new isotype. Indeed, progeny cells of the memory B cell may switch isotypes again as the cycle of proliferation and differentiation repeats. This capacity of the host to produce antibodies of different isotypes ensures that all effector mechanisms can be brought to bear to eliminate an antigen.

The mechanics of isotype switching, by which a given B cell clone produces IgM, IgG, IgE or IgA, are described in Chapter 5 along with a different genetic mechanism that allows the generation of IgD.

ii) Structural Isoforms of Immunoglobulins

As noted previously, the basic core structure of any Ig is H_2L_2. Depending on the stage of activation of a given B cell and the cytokine cues it receives from its immediate microenvironment, three different isoforms of this basic core structure can be produced (**Fig. 4-8**). These structural isoforms, which vary in amino acid sequence at their C-termini (but not at their N-termini), perform distinct functions in the body. **Membrane-bound** Igs serve as part of the BCR complex. **Secreted** (or serum) Igs serve as circulating antibody in the blood. **Secretory** Igs are secreted antibodies that undergo modifications that enable them to enter and function in the external secretions of the body, such as in tears and mucus. Although their antigenic specificities are identical, a membrane-bound Ig protein does not "turn into" a secreted one, or vice versa. The production of a membrane-bound or secreted Ig is regulated by the B cell at the level of transcription. RNA transcripts of the H chain gene that are slightly different at their 3′ ends are produced, resulting in structural forms of the Ig protein that differ at their C-termini (see later). This control step occurs prior to protein synthesis.

a) Membrane-Bound Immunoglobulins

The membrane-bound Ig protein (often denoted "mIg") of a B cell is very similar in form to the antibody its progeny cells will later secrete. However, compared to secreted antibody, mIg has an extended C-terminal tail and lacks a short "tailpiece" (see later). Membrane-bound Ig is ultimately situated in the plasma membrane such that its extracellular domain containing the Fab regions is displayed to the exterior of the cell, and a transmembrane domain extending beyond the last C_H Ig domain spans the plasma membrane and dangles the cytoplasmic domain of the Ig protein into the cytoplasm (refer to **Fig. 4-8A**). How is the mIg fixed in the membrane? The Ig H chain gene

Fig. 4-8
Structural Isoforms of Immunoglobulins

(A) Membrane-bound Ig is anchored in the B cell membrane by a transmembrane domain and has a cytoplasmic domain that extends inside the cell. **(B)** Secreted Ig lacks the transmembrane and cytoplasmic domains but does have a tailpiece. **(C)** Secretory Ig is formed when monomers of secreted Ig first polymerize around a J chain protein and then acquire a secretory component while passing through a mucosal epithelial cell or a glandular duct cell.

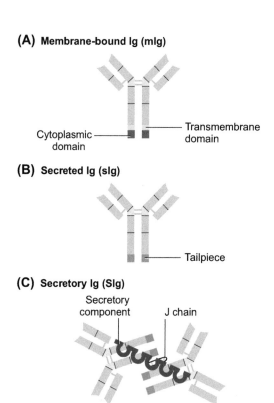

(A) Membrane-bound Ig (mIg)

Cytoplasmic domain — Transmembrane domain

(B) Secreted Ig (sIg)

Tailpiece

(C) Secretory Ig (SIg)

Secretory component   J chain

can be transcribed such that the resulting mRNA includes the sequences that encode the transmembrane domain. The transmembrane domain characteristically contains amino acids with hydrophobic side chains that interact with the lipid bilayer. Thus, when the Ig H chain mRNA is translated and the corresponding protein is synthesized in the endoplasmic reticulum (ER), the hydrophobic residues of the transmembrane domain cause the H chain to become anchored in the ER membrane (**Fig. 4-9**, left side). After the L chains associate with the H chains, the anchored mIg passes through the Golgi apparatus and is fixed in the membrane of a vesicle that transports the mIg to the cell surface. The membrane of the transport vesicle then fuses with the plasma membrane of the B cell, positioning the mIg protein on its external surface.

b) Secreted Immunoglobulins

The mature plasma cell progeny of an activated B cell secrete the shortest and simplest form of an Ig as antibody, often denoted "sIg" (for "secreted"). An sIg antibody has the same antigenic specificity (same N-terminal sequences) as the mIg in the BCR of the original B cell that was activated by specific antigen, but the C-terminal sequences of the Ig end shortly after the last C_H domain. The mRNA encoding an sIg protein lacks the sequences for the transmembrane and cytoplasmic domains and instead specifies a short amino acid sequence called the **tailpiece**, which lies C-terminal to the last C_H domain and facilitates secretion (refer to **Fig. 4-8B**). During Ig synthesis in the ER, Ig chains that are synthesized with only the tailpiece (and not the transmembrane domain) are not fixed in the ER membrane (**Fig. 4-9**, right side). Rather, these Igs are sequestered as free molecules inside secretory vesicles that bud off from the Golgi. These secretory vesicles fuse with the plasma membrane of the cell, releasing the free sIg molecules as antibodies into the extracellular environment.

c) Polymeric Immunoglobulins

The basic H_2L_2 unit of an Ig is sometimes referred to as the Ig "monomer." In most cases, mIg is present in the BCR complex in monomeric form. However, secreted IgM and IgA frequently form soluble polymeric structures because their tailpieces allow interactions between the H chains of several IgM or IgA monomers. Most polymeric forms of IgM and IgA contain a **joining (J) chain**, which is a small acidic polypeptide

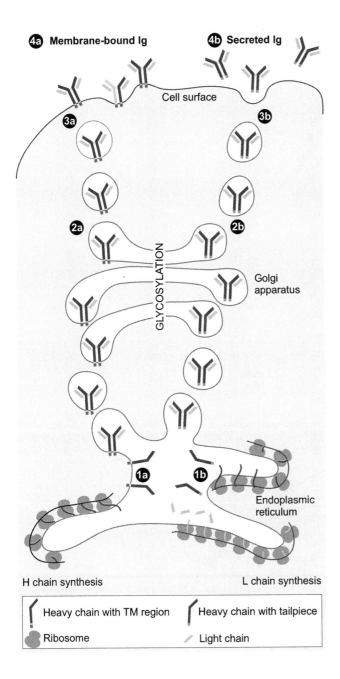

4a Membrane-bound Ig   **4b** Secreted Ig

Cell surface

3a   **3b**

GLYCOSYLATION

2a   **2b**

Golgi apparatus

1a   **1b**

Endoplasmic reticulum

H chain synthesis   L chain synthesis

▌ Heavy chain with TM region	▌ Heavy chain with tailpiece
● Ribosome	╱ Light chain

Fig. 4-9

Synthesis of Membrane-Bound and Secreted Immunoglobulins

Immunoglobulin molecules newly assembled in the endoplasmic reticulum (ER) of a single B cell are targeted differently depending on the amino acid sequences at the C-terminus of the H chain. Left side: Ig chains with transmembrane domains (#1a) are integrated into the membranes of the Golgi apparatus (#2a) and remain in the membranes of the secretory vesicles (#3a) as they are transported to the cell surface. Fusion of the vesicle membrane to the plasma membrane results in the fixing of mIg in the B cell plasma membrane (#4a). Right side: Ig chains with only a tailpiece (#1b) are not integrated into the Golgi membrane (#2b) and instead remain free molecules inside secretory vesicles (#3b). When the vesicle membrane fuses with the plasma membrane, the Ig protein is released as secreted Ig (#4b).

that binds to the tailpiece of a μ or α H chain via a disulfide linkage. The J chain is not structurally related to Igs and is encoded by a separate genetic locus. Although a single J chain appears to be able to stabilize the component monomers of polymeric Igs, it is not crucial for the joining event. Typically, five IgM monomers congregate to form a pentamer, whereas two to three IgA monomers form dimers or trimers, respectively (**Fig. 4-10**).

d) Secretory Immunoglobulins

Most IgA antibodies are found in the external secretions (such as tears, mucus, breast milk and saliva) rather than in the blood. How do they get there? Newly synthesized polymeric sIgA antibodies containing tailpieces and J chains are released into the tissues underlying a gland or mucosal layer by IgA-secreting plasma cells that have homed to this location (see Ch. 12). The sIgA molecules then bind to a receptor called the **poly-Ig receptor (pIgR)** expressed by mucosal epithelial cells that line body tracts or glandular ducts. The pIgR is present on that surface of the mucosal epithelial cells facing the tissues, rather than the surface facing the lumen of the gland or body tract

Fig. 4-10
Polymeric Immunoglobulins

(A) Secreted IgM monomers have tailpieces that allow them to interact with each other and to polymerize around the J chain polypeptide, resulting in the formation of IgM pentamers. **(B)** Two sIgA monomers often polymerize around a J chain to form dimers.

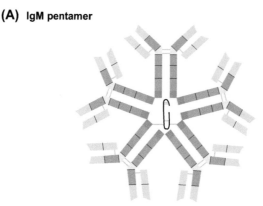

(A) IgM pentamer

(B) IgA dimer

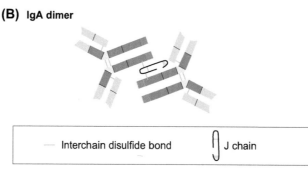

— Interchain disulfide bond	J chain

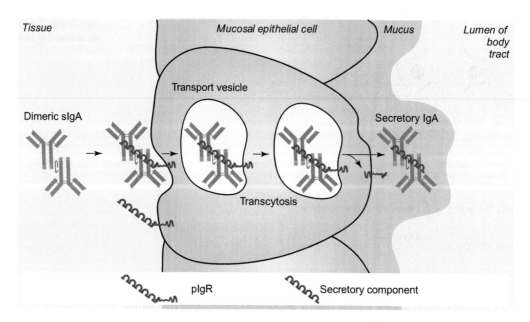

Fig. 4-11
Generation of Secretory IgA

In this example of a mucosal epithelial cell lining a body tract, dimeric sIgA binds via its J chain to a polymeric Ig receptor (pIgR) expressed on the cell surface. The polymeric sIgA molecule is endocytosed by pIgR, transported across the cell (transcytosis), and released into the lumen of the tract. During this release, the pIgR is enzymatically cleaved so that the polymeric IgA and a pIgR fragment (secretory component) remain attached and are released together as secretory IgA.

(**Fig. 4-11**). The pIgR recognizes the J chain present in polymeric sIgA molecules and binds to the C-terminal domain of sIgA, triggering receptor-mediated endocytosis of the pIgR-bound Ig multimer into a transport vesicle. This vesicle conveys the sIgA through the mucosal epithelial cell to its opposite (lumen-facing) side in a process called **transcytosis**. The membrane of the transport vesicle fuses with the plasma membrane of the epithelial cell at its luminal surface and releases the vesicle contents into

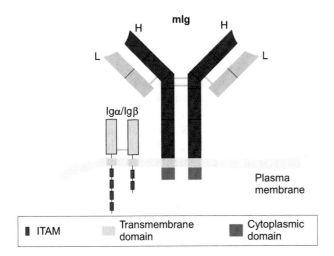

Noncovalent association between membrane-bound Ig and the transmembrane Igα/Igβ heterodimer forms the B cell antigen receptor complex. Upon antigen binding, intracellular signals are transduced through the ITAMs (immunoreceptor tyrosine-based activation motifs) of the Igα/Igβ molecule.

the secretions either coating the body tract or produced by a gland. However, as the IgA molecule is expelled from the transport vesicle, a part of the pIgR molecule is enzymatically released from the vesicle membrane and remains attached to the polymeric IgA molecule as it enters into the secretions. This heavily glycosylated piece is called the **secretory component**, and the antibody then becomes known as secretory IgA (SIgA) (refer to **Fig. 4-8C**).

iii) The BCR Complex

Although an mIg provides the specificity for antigen recognition, the C-terminal cytoplasmic sequences of the Ig protein are so short that other molecules are needed to help convey the intracellular signals indicating that antigen has bound. The tyrosine kinases and phosphatases that carry out the actual enzymatic reactions of intracellular signaling simply cannot bind to the short Ig tails. In addition, the Ig tails contain none of the molecular motifs generally associated with intracellular signaling. Instead, the mIg molecule associates with a disulfide-bonded heterodimer composed of two glycoprotein chains called **Igα** and **Igβ** (**Fig. 4-12**). These chains, which are expressed only in B cells, are co-synthesized with the mIg molecule and co-inserted with it into the ER membrane. An Ig-like domain in the extracellular regions of the Igα and Igβ proteins allows the heterodimer to associate with the C_H domains of any mIg isotype. The long cytoplasmic tails of Igα and Igβ contain tyrosine-rich amino acid motifs called **immunoreceptor tyrosine-based activation motifs (ITAMs)**. ITAMs allow the tail of a protein to recruit kinases and other molecules needed to transduce intracellular signaling and activate nuclear transcription factors. These nuclear transcription factors initiate the new gene transcription needed for the activation and differentiation of the B cell.

B. Immunoglobulin Genes

The antibody repertoire produced over an individual's lifetime includes specificities for essentially any antigen in the universe. How does the plasma cell population produce this plethora of different antibody proteins, each capable of binding to essentially a single specific antigen? The mIgs in the BCRs already exist on the surfaces of B cells *prior* to antigen exposure, meaning that it is not antigen that drives the establishment of the antibody repertoire. (Of course, clonal selection by antigen later determines which B cells are activated to produce and secrete antibody.) As well, the antibody repertoire cannot be directly encoded in the DNA of B cells because the nucleus simply does not have sufficient volume to contain the vast numbers of Ig genes that would be required to individually encode all the Ig proteins of the antibody repertoire. It turns out that our genomes do not contain any complete Ig genes at all, and evolution has come up with an elegant solution in which a relatively small amount of germline DNA can be organized to encode an enormous number of Ig proteins.

I. Structure of the Ig Loci

In the genomic DNA of a given developing B cell, the V domains of the heavy and light chains comprising the variable region of its Ig protein are encoded by **variable (V) exons** that are physically separated from the **constant (C) exon**s encoding the C domains of each chain. Moreover, V exons are assembled from a vast collection of smaller DNA fragments called **gene segments**. There are three types of gene segments: **variable (V)**, **diversity (D)** and **joining (J)**. The V exons of the Ig heavy chains contain V, D and J segments, whereas the V exons of the Ig light chains contain V and J segments but no D segments. Each type of gene segment contributes coding information for a particular set of amino acids in the V region of an Ig heavy or light chain. For any V exon, the participating gene segments are *randomly* brought together by the gene rearrangement process that was introduced in Chapter 1 as somatic recombination and is also known as **V(D)J recombination.** V(D)J recombination mediates a physical and permanent juxtaposition *at the DNA level* that results in a complete V exon. It is this rearrangement in the genomes of individual developing B cells that collectively gives rise to the large, diverse repertoire of Ig proteins. An example of the physical joining of V, D and J segments to form a V exon is shown for an Ig heavy chain in **Figure 4-13**.

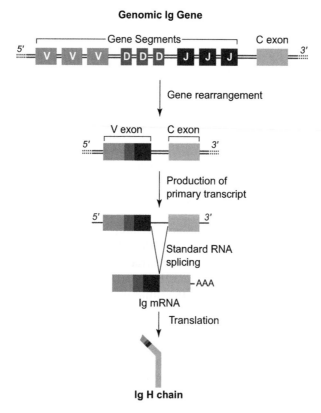

Fig. 4-13
Overview: From Genomic Ig Gene to Heavy Chain Protein

During the development of a B cell, the germline configuration of the Ig loci is rearranged such that separate gene segments are assembled by somatic recombination to form a complete variable exon (V exon). The H chain locus has V, D and J gene segments (as shown), while the L chain loci have only V and J segments (not shown). The rearranged VDJ (or VJ) variable exons remain in close proximity to at least one constant region exon (C exon), allowing the transcription of a complete Ig gene and the production of a primary RNA transcript. This primary transcript is then spliced to form a messenger RNA with contiguous V and C sequences. In this example, translation of the mRNA produces a complete Ig H chain with V and C regions.

The V, D and J gene segments encoding the V domains of Ig proteins are physically resident in three large genetic loci, the *Igh* (H chain), *Igk* (kappa L chain) and *Igl* (lambda L chain) loci, which are located on different chromosomes (**Table 4-1**). The complex structures of these loci in the genomes of humans and mice are shown in **Figures 4-14** and **4-15**, respectively, and the estimated numbers of each type of gene segment in humans and mice are given in **Table 4-2**.

In contrast to the V exon, the C exon of an Ig polypeptide is not formed by the rearrangement of smaller gene segments. Although the *Igk* locus contains a single Cκ exon, the *Igl* locus contains multiple Cλ loci: seven in human and four in mouse. In a given individual, a variable number of these Cλ exons may be pseudogenes.

Each Ig locus contains, among its functional V, D and J gene segments, additional gene segments that are non-functional *pseudogenes*.

TABLE 4-1	Chromosomal Location of Ig Genes		
	Chromosome		
Genetic Locus	**Human**	**Mouse**	
Igh	14	12	
Igk	2	6	
Igl	22	16	

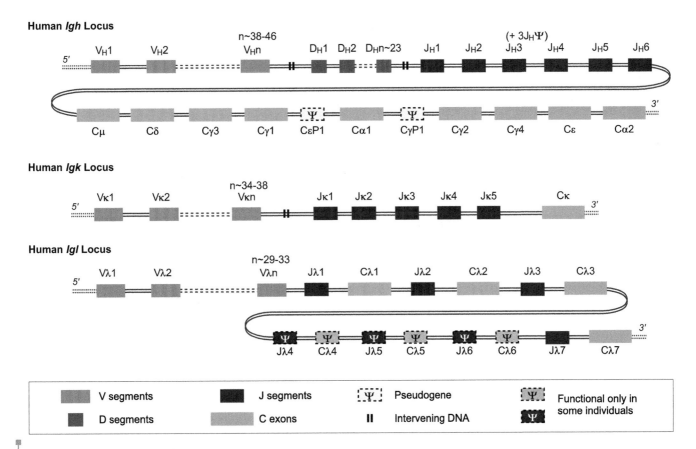

Fig. 4-14
Genetic Organization of the Human Immunoglobulin Loci

The human Ig H chain locus (*Igh*) includes the V_H, D_H and J_H segments that rearrange to form the V_H exon, as well as several different C_H exons that encode amino acids defining the various Ig H chain isotypes. Human Ig L chains are encoded by the κ chain locus (*Igk*) and the λ chain locus (*Igl*), both of which contain V and J segments. Not all segments in a grouping are shown (dotted lines), and groupings of V, J and D segments are separated from each other by long intervening stretches of DNA. Some gene segments function in only some individuals, whereas others are pseudogenes that are completely non-functional. Not all gene segments and pseudogenes are shown.

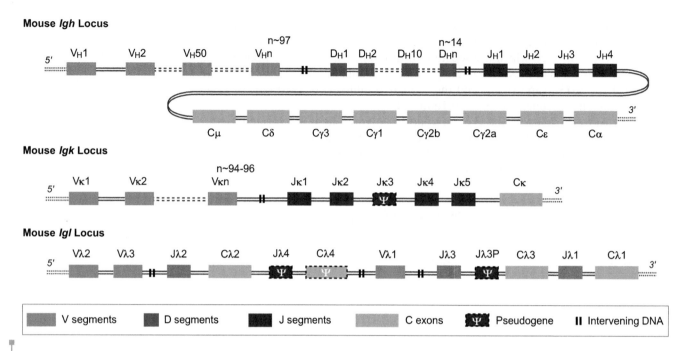

Fig. 4-15
Genetic Organization of the Murine Immunoglobulin Loci

The mouse Ig H chain locus (*Igh*) includes the V_H, D_H and J_H segments that rearrange to form the V_H exon, as well as several different C_H exons that encode amino acids defining the various Ig H chain isotypes. Mouse Ig L chains are encoded by the κ chain locus (*Igk*) and the λ chain locus (*Igl*), both of which contain V and J segments. Not all segments in a grouping are shown (dotted lines), and groupings of V, J and D segments are separated from each other by long intervening stretches of DNA. Some gene segments are pseudogenes that are completely non-functional. Not all gene segments and pseudogenes are shown.

TABLE 4-2	Estimated Numbers of Functional Gene Segments in Mouse and Human Ig Loci		
	Number of Gene Segments in Germline		
		Light Chain	
Gene Segment Type	**Heavy Chain**	**Kappa**	**Lambda**
Mouse			
V	97	94–96*	3 lab mice 8 wild mice
D	14	0	0
J	4	4	3
Human			
V	38–46	34–38	29–33
D	23	0	0
J	6	5	4–5

*Number of functional segments can vary by individual.
http://www.imgt.org/IMGTrepertoire/LocusGenes/genetable/human/geneNumber.html (accessed October 2012).

The *Igh* locus contains multiple C_H exons: nine in human and eight in mouse. As illustrated in **Figures 4-14** and **4-15**, the functional exons in the *Igh* loci are designated Cμ, Cδ, Cγ3, Cγ1, Cα1, Cγ2, Cγ4, Cε1 and Cα2 in humans, and Cμ, Cδ, Cγ3, Cγ1, Cγ2b, Cγ2a, Cε and Cα in mice. Each C_H exon encodes the amino acid sequence defining a particular Ig isotype. For H and L chains, the C exon is joined to the V exon at the RNA level by conventional splicing to produce a translatable Ig mRNA. The Ig mRNA is then translated like any other to generate the H or L protein.

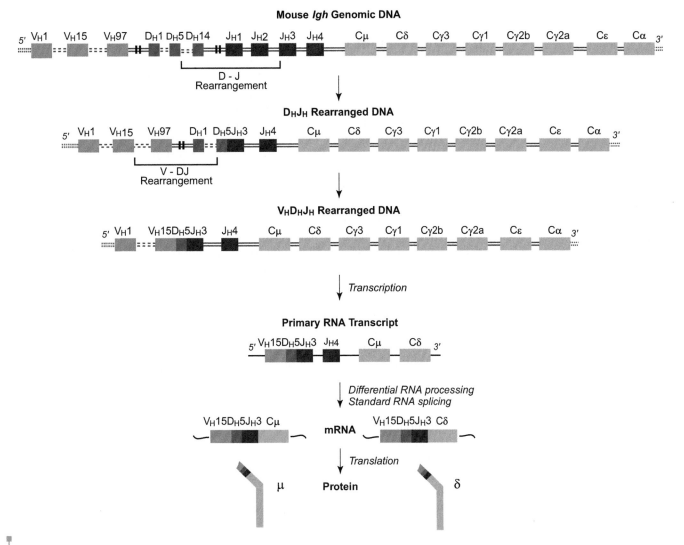

Fig. 4-16
V(D)J Recombination and IgM and IgD H Chain Synthesis

Schematic representation of IgM and IgD H chain synthesis following V(D)J recombination in the mouse *Igh* locus. In this hypothetical example, D_H5 has randomly joined to J_H3, and V_H15 has then randomly joined to D_H5J_H3. After completion of V(D)J recombination, transcription of the rearranged Ig gene can proceed, and primary RNA transcripts are produced. Due to mechanisms described in Chapter 5, only the $C\mu$ and $C\delta$ exons are included in the primary transcripts of a maturing naïve B cell. After differential RNA processing and RNA splicing, mRNAs are produced that encode the newly assembled V region linked to either the $C\mu$ or $C\delta$ sequence. These mRNAs are then translated to produce a mixture of μ and δ H chain proteins whose V regions are identical. Thus, the IGM and IGD antibodies made by the B cell will have identical antigen-binding sites.

A more detailed example of H chain synthesis in a developing B cell is illustrated in **Figure 4-16**. In this hypothetical situation, the D_H5 gene segment has randomly joined to the J_H3 gene segment, and the V_H15 gene segment has randomly joined to D_H5J_H3 to form the V exon. The V exon is then joined at the RNA level to the C exon. During the synthesis of any one H chain, only the most 5′ of the C exons is normally included in the primary transcript of the rearranged Ig gene. An exception occurs in developing B cells, where both the $C\mu$ and $C\delta$ exons are included in the primary transcript (as illustrated in **Fig. 4-16**). Through differential processing and RNA splicing of many copies of this primary transcript (see Ch. 5), two populations of mRNAs are produced: one encoding μ H chains and the other specifying δ H chains. Because they share the same V exon, these H chains give rise to IgM and IgD molecules of identical antigenic specificity. The reader is reminded here that isotype switching is responsible for generating the IgG, IgA and IgE isotypes; this mechanism is also described in Chapter 5.

II. Mechanism of V(D)J Recombination

i) RAG Recombinases

Prior to V(D)J recombination, V, D and J gene segments are distributed over a comparatively large expanse of DNA such that the cell's transcriptional machinery is physically unable to transcribe all the sequences comprising a complete Ig gene. Two specialized recombinase enzymes known as **RAG-1** and **RAG-2** (RAG, recombination activating gene) randomly combine selected V, D and J gene segments in an enzymatic cutting and covalent rejoining of the DNA strand that excludes non-selected V, D and J gene segments and other intervening sequences (refer to **Fig. 4-16**). This rearrangement creates a $V_L J_L$ or $V_H D_H J_H$ exon that, because of its proximity to the C_L or C_H exon, allows the transcription of the complete Ig gene followed by translation of the complete L or H chain. The RAG enzymes that carry out this rearrangement are expressed only in developing B and T lymphocytes. Non-lymphocytes never have the capacity for V(D)J recombination, and mature lymphocytes lose it. Furthermore, through developmental restrictions that are not yet understood, B cells do not rearrange their TCR genes, and T cells do not rearrange their Ig genes. Thus, only precursor B cells in the bone marrow are capable of carrying out V(D)J recombination of germline Ig DNA.

ii) Recombination Signal Sequences

V(D)J recombination depends on the cooperative recognition by RAG-1 and RAG-2 of specific **recombination signal sequences (RSSs)** that flank germline V_H segments on their 3′ sides, J_H segments on their 5′ sides and D_H segments on both sides. Similarly, the V and J segments of both the *Igk* and *Igl* loci are flanked by RSSs on their 3′ sides and 5′ sides, respectively. There are two types of RSSs: the "12-RSS" and the "23-RSS." Both RSSs consist of conserved heptamer and nonamer sequences separated by a non-conserved spacer sequence of either 12 or 23 base pairs of DNA. It is the spacer that identifies the RSS as either a 12-RSS or a 23-RSS. The conserved regions of the 12-RSS and 23-RSS ensure mutually complementary binding. Only when one gene segment is flanked on one side by a 12-RSS and the other gene segment is flanked by a 23-RSS can the pair interact and be recognized by the RAG-1/RAG-2 complex and participate in V(D)J joining. This requirement has been called the **12/23 rule**. An example of this rule in action for an H chain is shown in **Figure 4-17**. The RAG complex first brings together the 3′ 12-RSS of a germline D_H segment with the 5′ 23-RSS of a J_H segment so that the complementary sequences of the RSSs are aligned. The RAG complex then cleaves the DNA between these segments such that the D_H and J_H segments are juxtaposed without any RSS between them, forming a hairpin structure. The DNA repair machinery of the cell then cooperates with RAG to nick the hairpin and ligate the DNA strands back together to form a DJ entity with the 12-RSS on its 5′ side left unchanged (as it did not take part in the recombination). The RSS sequences used for the original alignment are joined together to form a circular DNA product that is subsequently lost. The RAG complex then instigates another round of joining by using the 23-RSS on the 3′ side of a V_H segment to join to the 12-RSS on the 5′ side of the $D_H J_H$ entity to form the complete $V_H D_H J_H$ exon.

NOTE: The enzymes that work with RAG to complete V(D)J recombination also participate in the non-homologous end joining (NHEJ) pathway of DNA repair. These enzymes include Artemis, XRCC4, XRCC5, XRCC6, DNA-dependent protein kinase (DNA-PK) and DNA ligase IV. Mutations of these proteins can give rise to immunodeficiency diseases, some of which are discussed in Chapter 15.

III. Order of Ig Locus Rearrangement

The rearrangement of the Ig loci occurs in a particular order. The *Igh* locus is rearranged first, followed by *Igk* and then *Igl*. After V(D)J recombination is completed in the *Igh* locus on one chromosome 14 in a developing human B cell, transcription

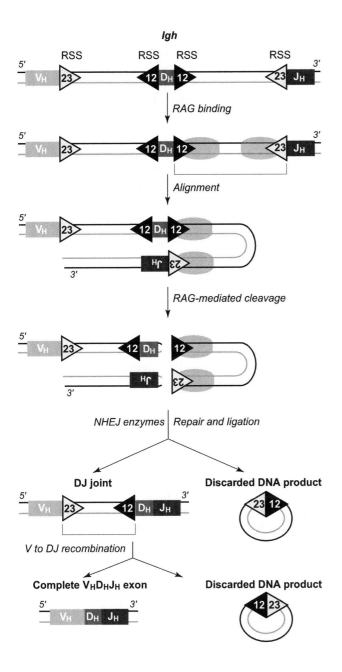

▶ **Fig. 4-17**
Role of Recombination Signal Sequences in V(D)J Recombination

Recombination signal sequences (RSSs) are found on the 5' side, 3' side, or both sides of Ig gene segments. In this example, a D_H segment and a J_H segment undergo recombination mediated by RAG recombinases (gray ovals). RSSs adjacent to each segment are aligned by the RAG recombinases according to the 12/23 rule, whereby a 12-RSS can recombine only with a 23-RSS. RSS alignment results in a looping-out of the DNA between the gene segments to form a hairpin structure. RAG cooperates with enzymes of the non-homologous end joining (NHEJ) pathway to nick the hairpin and prepare the ends of the D_H and J_H segments for joining. DNA repair and ligation are then mediated by additional NHEJ enzymes to form a DJ joint in the Ig gene as well as a circular DNA product that contains an RSS joint; this DNA product is discarded and lost from the genome. This complex recombination process is then repeated to join a V_H segment to the DJ complex, resulting in formation of a complete V exon and another discarded DNA product.

of the Ig H chain gene can proceed, and primary RNA transcripts are produced. The appearance of a newly translated μ chain marks an important checkpoint in Ig synthesis because further gene segment rearrangements in the *Igh* locus that produced the complete chain are blocked. In addition, any V(D)J recombination that started in the *Igh* locus on the other chromosome 14 is halted, preventing the production of a competing H chain in one B cell. The effect is one of **allelic exclusion**, because only one chromosome contributes the H chain gene product. What happens if attempts to rearrange the *Igh* locus on the first chromosome do not result in a functional μ chain? The B cell continues to randomly join gene segments in the *Igh* locus on the other chromosome (allelic exclusion is not invoked). If a functional μ polypeptide still cannot be produced, the B cell dies by apoptosis.

At the same time as the synthesis of a functional μ chain shuts down *Igh* rearrangement, it stimulates the recombination of V and J gene segments first in the *Igk* loci and, if necessary, in the *Igl* loci thereafter. If the recombination of $V_κ$ and $L_κ$ gene segments on one chromosome 2 is successful such that a functional κ light chain is translated, allelic exclusion is again invoked and shuts down any further rearrangements on the originating *Igk* locus and in the *Igk* locus on the other chromosome 2. However, if a functional κ light chain is not produced from either chromosome, rearrangement is stimulated at the *Igl* loci. If a functional L chain is then produced, the viable H and L chains synthesized by the B cell are assembled into complete Ig molecules that are expressed on the B cell surface in the BCRs. If no functional L chain is produced, no further Ig rearrangements can be initiated, and the B cell dies by apoptosis.

NOTE: Allelic exclusion is the reason why an individual B cell produces only one kind of H chain and only one kind of L chain, and why the two H chains in a given Ig molecule are identical and the two L chains are identical. Only one of the paired *Igh* loci, and only one of the four possible light chain loci (two *Igl* and two *Igk*) are functional after successful V(D)J recombination.

IV. Antibody Diversity Generated by Somatic Recombination

Three sources of variability contribute to antibody diversity *before* a given B cell ever encounters its antigen: (1) the existence of multiple V, D and J gene segments in the B cell germline and their combinatorial joining; (2) junctional diversity at the points where the segments are joined; and (3) the randomness of heavy–light chain pairing. These elements, which create a unique variable region sequence for the BCR of each B cell prior to its release into the periphery, are discussed in this chapter. Another process called **somatic hypermutation** contributes to diversity in antigen recognition *after* the naïve B cell is released into the periphery and encounters antigen. This process, which involves point mutations of the rearranged Ig gene, is discussed in Chapter 5.

i) Multiplicity and Combinatorial Joining of Germline Gene Segments

The primary source of diversity in antigen recognition by antibodies is the existence of multiple V, D and J gene segments in the *Igh, Igk* and *Igl* loci (refer to **Table 4-2**). Because single V, D and J segments from these collections are randomly chosen and joined together, thousands of potential combinations can be used to create the V_H and V_L exons in a given B cell. Consider the generation of the V exon of an Ig κ chain in a mouse: each B cell precursor has at its disposal any one of at least 94 functional V_κ gene segments × 4 functional Jκ gene segments = 376 variable exon sequences. The B cell precursor chooses one combination of Vκ and Jκ segments to fuse covalently and irreversibly to form its Vκ exon. Similar combinatorial joining can occur in the *Igh* locus, but with an added multiplier effect attributable to the D_H gene segments. Although the D_H gene segments encode only a small number of amino acids, their presence in the *Igh* locus greatly increases the diversity of heavy chains that can be achieved. Consider the following for a mouse Ig H chain: at least 97 V_H segments × 14 D_H segments × 4 J_H segments = at least 5432 different V_H exons, as compared to the 388 possible without the D_H segment.

ii) Junctional Diversity

Junctional diversity arises when the fusion of V, D and J segments is not exact. Such "imprecise joining" of gene segments can result from the deletion of nucleotides in the joint region and/or the addition of so-called P and/or N nucleotides during V(D)J recombination (**Fig. 4-18**). About 80% of antibodies in adult humans show some type of junctional diversity.

a) Deletion

In the last stages of V(D)J recombination, the DNA strands of the two gene segments to be joined are trimmed by an exonuclease enzyme prior to final linking by DNA ligases. Very often, the exonuclease trims into the coding sequence itself and eliminates a few of the original genomic nucleotides. After DNA ligation is complete, the sequence of the rearranged Ig gene in this region differs slightly from that predicted from the germline sequence. Although each individual change may be subtle, deletion is a common event and so has a significant diversifying effect.

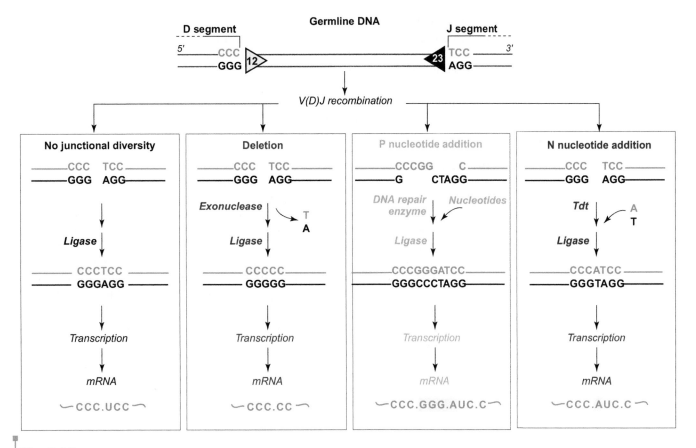

Fig. 4-18
Junctional

If the D_H and J_H segments in this example are joined precisely (far left panel), the nucleotide triplets "CCC" and "TCC" are eventually transcribed into the mRNA codons "CCC" and "UCC" and no junctional diversity occurs. However, prior to DNA ligation, the processes of deletion, P nucleotide addition, and/or N nucleotide addition may result in alterations to the mRNA coding sequence. Such imprecise joining will result in junctional diversity in the Ig H chains made by different B cells that underwent rearrangement of the same D_H and J_H gene segments.

b) P Nucleotide Addition

Sometimes the RAG recombinases carrying out V(D)J recombination nick the DNA in the intervening region between gene segments, rather than at the precise ends of the coding sequences of the gene segments. As a result, one strand on each gene segment will have a recessed end while the other strand will have an overhang. The gaps on both strands are filled in with new nucleotides by DNA repair enzymes, and the gene segments are ligated together. On the coding strand (only), these inserted nucleotides that modify the joint sequence are called **P nucleotides**. Although the insertion of P nucleotides can markedly change the amino acid sequence of the Ig protein, it is a relatively rare event and so has a limited impact on diversity.

c) N Nucleotide Addition

N nucleotide addition occurs almost exclusively in the *Igh* locus. Usually 3–4 but up to 15 extra nucleotides may be found in VD and DJ joints that do not appear in the germline sequence and are not accounted for by P nucleotides. Sometimes DNA strands are nicked such that they have blunt ends, so that an enzyme called *terminal dideoxy transferase (TdT)* can randomly add "non-templated" **N nucleotides** onto the ends of these strands before their final ligation. Light chains do not usually undergo N nucleotide addition because the TdT gene is essentially "turned off" at the stage of B cell development when light chain gene rearrangement occurs.

iii) Heavy–Light Immunoglobulin Chain Pairing

The random pairing of H and L chains also contributes to diversity in antigen recognition prior to antigenic stimulation. A mature Ig protein molecule's antigen-binding or antigen-combining site is composed of the V domains of both the L and H chains. Our previous calculations of VJ and VDJ combinatorial joining in the mouse resulted in a total of at least 376 V_κ exons possible for an L chain gene and at least 5432 V_H exons possible for an H chain gene (not including diversity contributed by deletion and N and P nucleotide addition). Any one B cell synthesizes only one sequence of H chain and one sequence of L chain, but if one assumes that any of the 5432 H chain genes can occur in the same B cell as any of the 376 L chain genes, the number of possible L/H chain gene combinations is at least $376 \times 5432 =$ over 2 million.

iv) Estimates of Total Diversity

Immunologists vigorously debate just how many different antibodies one individual host's repertoire can contain. It is impossible to accurately quantitate actual repertoire diversity because of the variations introduced by junctional diversity and (later on) somatic hypermutation. It is safe to say that the number of different antibody molecules that an individual can produce is considerably greater than the combinatorial diversity contributed by the selection of different gene segments. Taking into account sources of variation both before and after antigen encounter, the total theoretical diversity of an individual's antibody repertoire has been estimated to range from 10^{11} to 10^{14} different antigenic specificities! However, at any one moment, the actual repertoire available to an individual to counter antigens will be more limited than the theoretical, since a certain proportion of B cells will die before ever encountering their specific antigen. Other B cells will be eliminated before joining the mature B cell pool because their randomly generated BCRs have the potential to recognize self antigens. Release of these self-reactive B cells from the bone marrow into the periphery could lead to the development of an autoimmune disease. Developing B cells producing self-reactive BCRs are therefore removed from the bone marrow pool by either B cell negative selection during the establishment of central tolerance (see Ch. 5), or by mechanisms of peripheral tolerance if these cells reach the body's secondary lymphoid tissues (see Ch. 10). As a result, immunologists estimate that, at any given point in time, a human may have only 10^6 different B cell clones able to respond to antigens, and a mouse only 10^5. Nevertheless, this repertoire is sufficiently large to deal with any threat encountered.

Focus on Relevant Research

"Generating Recombinant Antibodies to the Complete Human Proteome" by Dübel, S., Stoevesandt, O., Taussig, M.J., and Hust, M. (2010) *Trends in Biotechnology* 7, 333–339.

One of the most powerful and information-rich scientific achievements of the past century was the complete sequencing of the human genome. However, just knowing where the genes are physically located in this impressively long and complex DNA sequence does not reveal very much about which genes are actually expressed. Ultimately, if one wishes to know what a biological system is doing under certain physiological conditions, one must know which proteins are being synthesized by that system at that time. The collection of all proteins expressed by a cell, tissue or organism at any given moment or under a set of defined conditions is called its *proteome*, and the field of research studying proteomes is known as *proteomics*.

In order to study proteomics effectively, researchers need ways to identify and quantify the proteins that are being expressed by various cells and tissues at various times. In this research article, Dübel *et al.* discuss the merits of antibodies as exceptionally efficient tools for such protein studies. In fact, as they point out, several global organizations have taken on the monumental challenge of generating antibodies against all of the proteins expressed by the human genome. The authors describe ways in which this daunting task can be undertaken *in vitro* using genetic engineering techniques, thereby avoiding the time-consuming and complicated animal immunization strategies usually used for antibody production. As well as helping to define the human proteome, such an antibody library could be used to improve laboratory diagnostics and perhaps facilitate new targeted antibody therapies for infectious diseases and cancers. The relative merits of existing and innovative methods of antibody production and their implications for proteomics are discussed.

C. Antibody–Antigen Interaction

I. Structural Requirements

Whether an Ig protein is membrane-fixed as part of the BCR complex or free as a soluble antibody, it is folded so that the CDRs of both V_L and V_H are grouped at the tip of each arm of the Ig "Y," as illustrated in **Plate 4-1**. In each Fab, this CDR grouping forms loops that project outward to form an antigen-binding "pocket." While small antigens (such as peptides) can fit securely within this pocket, large macromolecular antigens (such as globular proteins on the surfaces of pathogens) not only project into the V_L–V_H pocket but also make specific surface–surface contacts with the framework regions outside the CDR loops.

The small region of a large antigen that binds to the antigen-binding site of the antibody is called the **antigenic determinant** or **epitope**. Studies of many antigen–antibody interfaces have revealed that the contact residues of an antigenic epitope are often discontinuous in sequence but contiguous in space and involve a surface area of about 600–900 $Å^2$. However, since every antibody has a different sequence in its CDRs, the pattern of contact between one antibody and its antigenic epitope is slightly different from that of the next antibody with its epitope. Furthermore, while both the V_L and V_H domains of an Fab are involved in forming the antigen-binding site, not every Fab will use all of its six CDRs to interact with its antigen. In general, antigens form the greatest number of contacts with amino acids in CDR3 of the Ig H chain. CDR3 displays the greatest degree of sequence variability as it is usually encoded by nucleotides at the interfaces between the V, D and J gene segments and is thus the site of junctional diversity.

Although antibody and antigen contact surfaces show a high degree of complementarity prior to binding, there is a degree of flexibility in the antigen-combining site. Upon making contact with the antigenic epitope, some antibodies undergo slight conformational modifications to their CDR3 loops and alter the orientations of certain side chains to form a better bond. Immunologists call this phenomenon **induced fit**, since the structure of the antibody is "induced" to fit by the binding of the antigen.

II. Intermolecular Forces

The antigen–antibody bond is the result of four types of non-covalent intermolecular forces: hydrogen bonds, van der Waals forces, hydrophobic bonds and ionic bonds. None of these forces is itself very strong, but because they are all working simultaneously, they combine to forge a very tight bond. The contribution of each type of force to the overall binding depends on the identity and location of the amino acids or other chemical groups in both the antibody and antigen molecules. The more closely the relevant chemical groups can approach one another, the more efficient the binding will be. Similarly, the more complementary the shapes of the antigenic epitope and the antigen-binding site, the more contact sites will simultaneously be brought into close proximity. The number of non-covalent bonds of all types will be increased, resulting in a stronger overall binding. Mapping of binding sites on antibodies has shown that cavities, grooves or planes are often present that correspond to complementary structural features on the antigen. If an antigen approaches whose shape is less complementary to the conformation of the binding site on the antibody, fewer bonds are formed, and steric hindrance and repulsion by competing electron clouds are more likely to "push" the prospective antigen away. Only those antigens shaped such that they make a sufficient number of contacts of the required strength will succeed in binding to the antigen-combining site of the antibody molecule.

The strength and specificity of the antigen–antibody bond are often used in the laboratory to isolate antigens from mixtures of proteins (see Appendix F). To release the antigen from the antibody, the non-covalent forces that hold them together are disrupted by the application of high salt concentration, detergent or non-physiological pH.

Plate 4-1

CDR Regions in the Antigen Binding Site of a Human Antibody

The CDR loops of a human antibody Fab shown in two different ways. **(A)** Top-down view as a space-filling model. **(B)** Side view of the V domain as a worm structure. CDR1, CDR2 and CDR3 of the H chain and CDR1, CDR2 and CDR3 of the L chain are highlighted in the indicated colors. The numbers represent the positions within the H or L polypeptide chain (counting from the N-terminus) of amino acid residues that are thought to play significant roles in determining antigen–antibody affinity. *[Reproduced by permission of Chingwei V. Lee et al., (2004) High affinity human antibodies from phage-displayed synthetic Fab libraries with a single framework scaffold. Journal of Molecular Biology 340, 1073–1093.]*

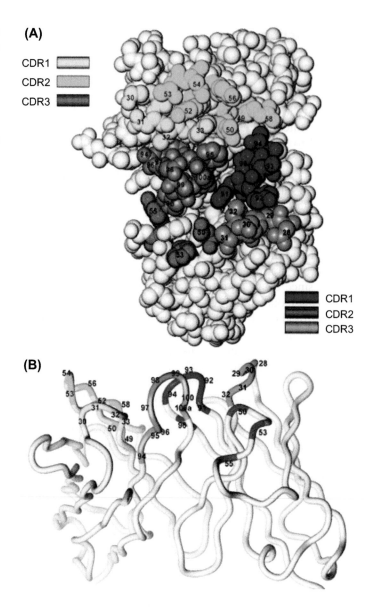

III. The Strength of Antigen–Antibody Binding

i) Affinity

Immunologists define the **affinity** of an antibody for its antigen as the strength of the non-covalent association between *one* antigen-binding site (thus, one Fab arm of an antibody molecule) and *one* antigenic epitope. Affinity is measured in terms of the equilibrium constant (K) of this binding, with a K of 10^9 M^{-1} representing strong affinity and a K of 10^6 M^{-1} representing weak affinity. In a natural immune response to invasion by a complex antigen, many different B cell clones will be activated, leading to a *polyclonal* antibody response. The antibodies generated will be heterogeneous with respect to both specificity and affinity. Not only will antibodies to different epitopes on the antigen be produced, but a single epitope will induce the activation of various B cells whose receptors display a variable accuracy of fit with the epitope. In addition, affinity is not an absolute quality for an antibody: a given antibody may display different affinities for a spectrum of related but slightly different antigens. Importantly, the affinity of antibodies raised during a secondary immune response to a given epitope is increased compared to antibodies induced in the first encounter. This phenomenon, which is called **affinity maturation**, is discussed in Chapter 5.

ii) Avidity

Whereas affinity describes the strength of the bond between one Fab and one antigenic epitope, **avidity** refers to the overall strength of the bond between a multivalent antibody and a multivalent antigen. Microorganisms and eukaryotic cells represent large multivalent antigens when they feature multiple identical protein molecules on their cell surfaces. Other examples of multivalent antigens are macromolecules composed of multiple identical subunits, and macromolecules containing a structural epitope that occurs repeatedly along the length of the molecule. All antibodies are naturally multivalent because they have two or more identical antigen-binding sites. Secreted IgG and IgE molecules have only two Fabs, but polymeric sIgA molecules can exhibit 4 or 6 identical antigen-combining sites, and sIgM molecules are most often found as pentamers with 10 binding sites. Binding at one Fab on a polymeric antibody holds the antigen in place so that other binding sites on the antibody are more likely to bind as well. With each additional binding site engaged by a polymeric antibody, the probability of the simultaneous release of all bonds drops exponentially, resulting in a lower chance of antigen–antibody dissociation and therefore a stronger overall bond. For example, in the case of pentameric IgM, each site in itself may not have a very strong affinity for antigen, but because all 10 sites may be bound simultaneously, IgM antibodies have considerable avidity for their antigens. At any given moment, dissociation of antibody from any one antigen-combining site may occur, but the complete release of antigen from antibody would require that all 10 binding sites be dissociated from the antigen simultaneously. The practical result of high avidity binding by an antibody is that the association between antigen and antibody is formed more rapidly and stably, and the antigen is thus eliminated from the body more efficiently.

IV. Cross-Reactivity

The specificity of an antiserum as a whole is defined by its component antibodies. Any one epitope of an antigen can induce the production of a collection of antibodies with a range of affinities and avidities for that epitope. As well, an antigen usually exhibits more than one epitope, so that antibodies exhibiting a spectrum of epitope specificities will also be included in the antiserum. The antigen used to immunize an animal and induce it to produce an antiserum is known as the **cognate antigen**. A **cross-reaction** is said to have occurred when an antiserum reacts to an antigen other than the cognate antigen. Cross-reactivity results either when one epitope is shared by two antigens, or when two epitopes on separate antigens are similar in structure (**Fig. 4-19**).

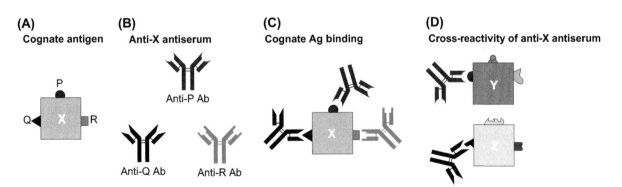

Fig. 4-19
Antibody Cross-reactivity

(A) In this schematic example, cognate antigen X has epitopes P, Q and R. **(B)** When antigen X is used to immunize an individual, the resulting polyclonal anti-X antiserum contains antibodies of three specificities: anti-P Ab, anti-Q Ab and anti-R Ab. **(C)** The antibodies in the anti-X antiserum bind as expected to the corresponding epitopes on cognate antigen X. **(D)** The anti-X antiserum cross-reacts with the non-cognate antigen Y due to the presence of epitope P, which allows the anti-P Ab to bind. In addition, the anti-X antiserum cross-reacts with the non-cognate antigen Z due to the presence of an epitope that is similar to epitope Q and allows the anti-Q Ab to bind. Anti-R antibodies will make no contribution to cross-reactivity in either case because they fail to recognize any epitope on antigens Y or Z.

The differing degrees to which cross-reacting antibodies bind to antigen can be understood in the light of the intermolecular binding forces described earlier. An antigen of a shape slightly different from that of the cognate antigen may not be able to form one or more of the multiple types of bonds required for tight binding to antibody; the antibody is thus more easily dissociated from the non-cognate antigen. The more closely a non-cognate epitope resembles the shape of the cognate epitope, the greater the possible molecular contacts between the non-cognate epitope and the antigen-combining site of the antibody, and the more likely cross-reactive binding will occur.

This chapter has described the structures of the Ig genes and proteins and the mechanisms of gene segment rearrangement and combinatorial joining that contribute to antibody diversity before an encounter with antigen. The next chapter discusses B cell development, the activation of these cells by antigen, the generation of antibody diversity following activation, and the biological expression of Igs. A short section on the sources of antibodies used in the laboratory and the clinic is also included.

Chapter 4 Take-Home Message

- Immunoglobulins (Igs) are antigen-binding proteins produced by B cells. All Igs made by a given B cell have the same antigenic specificity.

- Ig molecules have an H_2L_2 structure in which two identical heavy (H) chains and two identical light (L) chains are held together by disulfide bonds.

- Each Ig chain has an N-terminal variable (V) domain that differs widely in amino acid sequence among B cells. The C-terminal end of each L chain contains a single constant (C) domain of relatively invariant amino acid sequence. Each H chain contains three or four C domains.

- There are two L chain isotypes: κ and λ. There are five H chain isotypes, μ, δ, γ, ε and α, that specify IgM, IgD, IgG, IgE and IgA antibodies.

- There are three isoforms of Ig proteins. Membrane-bound (mIg) molecules have a transmembrane domain and associate with the Igα/Igβ heterodimer to form the BCR complex on the B cell surface. Plasma cells secrete copious amounts of an antibody that is a soluble form (sIg) of the Ig protein. Secretory antibodies (SIg) occur in body fluids.

- The Ig genes are located in the *Igh*, *Igl* and *Igk* loci and contain variable (V) and constant (C) exons.

- V exons are assembled at the DNA level from random combinations of V (variable), D (diversity) and J (joining) gene segments that are flanked by recombination signal sequences (RSSs).

- RAG recombinases initiate V(D)J recombination to join a D segment to a J segment and then DJ to a V segment to form the H chain V exon. V and J segments are joined in the *Igk* and *Igl* loci to form L chain V exons. V exon RNAs are brought together with C exon RNAs by conventional RNA splicing.

- Successful rearrangement of the *Igh* locus on one chromosome such that a functional μ H chain is produced shuts down V(D)J recombination of the other *Igh* locus via allelic exclusion. Rearrangement is then usually stimulated first at the *Igk* locus and subsequently at the *Igl* locus if a functional κ L chain is not produced. Productive rearrangement of at least one H and one L chain must occur in order for the B cell to survive.

- Much of the diversity in the antibody repertoire is derived from multiple germline V, D and J gene segments and their combinatorial joining, junctional diversity and random H–L chain pairing.

- Specificity of antigen–antibody binding is defined by the unique surface created by the combined V_H and V_L CDRs. The strength of antigen–antibody binding depends on the additive effect of several types of non-covalent binding forces.

- The affinity of antibody binding to a given antigen is the strength of the antigen–antibody interaction at one antigen-binding site. Avidity is a measurement of the total strength of binding where multivalent antibody binds multivalent antigen.

- Cross-reactivity of a given antibody may be due to recognition of the same or a similar epitope on different antigens.

Did You Get it? A Self-Test Quiz

Section A

1) Can you define these terms? gamma globulin, Fab, Fc, Ig domain, hinge region

2) What is the basic structure of an Ig molecule, and how do its C and V regions differ?

3) What is the Ig fold? Give two reasons why this feature is useful to a protein.

4) Can you define these terms? CDR, framework region

5) What are the hypervariable regions, and how do they contribute to antibody diversity?

6) What is the function of the framework regions?

7) Can you define these terms? mIg, sIg, SIg, tailpiece, J chain, secretory component, pIgR, transcytosis, ITAM

8) Describe the L chain isotypes. Explain why they don't contribute to antibody effector functions.

9) What are the five major H chain isotypes, and how are they derived?

10) What is isotype switching, and when does it occur?

11) Compare the production and functions of the three structural Ig isoforms.

12) Describe the components of the BCR complex and their functions.

Section B

1) Can you define these terms? gene segment, V exon, C exon, V(D)J recombination

2) Why isn't the antibody repertoire encoded in the germline of B cells?

3) Describe the three Ig loci and their complements of gene segments and C exons.

4) Does RNA splicing contribute to V exon assembly? If not, why not?

5) What C_H exons are first included in the primary transcripts of a rearranged *Igh* locus?

6) How does the production of a functional μ chain affect somatic recombination in the L chain loci? How does it affect cell survival?

7) Can you define these terms? RAG, RSS

8) Why is V(D)J recombination of the Ig genes necessary for their expression?

9) How many hematopoietic cell types can carry out V(D)J recombination of the Ig genes?

10) In what order are V, D and J gene segments joined together, and what does the 12/23 rule have to do with this?

11) Can you define these terms? P nucleotide, N nucleotide

12) Describe four sources of antibody diversity associated with somatic recombination.

13) Describe two types of junctional diversity.

14) Give two reasons why the total diversity of the antibody repertoire is likely to be less than the theoretical estimate of 10^{11}.

Section C

1) Can you define these terms? epitope, determinant, induced fit

2) Which CDR plays the most important role in antigen binding and why?

3) Name four types of intermolecular forces that contribute to the antigen–antibody bond.

4) How can the antigen–antibody bond be reversed in the laboratory?

5) What makes an antibody response polyclonal?

6) Distinguish between antibody affinity and antibody avidity.

7) Can you define these terms? cognate antigen, cross-reaction

8) Describe two ways in which cross-reactivity can arise.

Can You Extrapolate? Some Conceptual Questions

1) IgG is isolated from an antiserum and is digested with papain.

   a) How many different types of fragments are produced?

   b) For each type of fragment produced, how many C_H and V_H domains are present?

   c) If the fragments are treated chemically to break all disulphide bonds, describe the new fragments produced.

2) Serum Ig is separated into IgM, IgA and IgG subfractions, and the molecular weights of each isotype are determined. If the IgG was found to have molecular mass of about 150 kDa, what mass would you expect for the serum IgM and IgA?

3) DNA is isolated from a sample of skin cells and a sample of plasma cells, and the same restriction endonuclease is used to enzymatically digest both samples. When the DNA fragments from each sample are separated by size using gel electrophoresis, and the Ig gene fragments present are visualized, those obtained from the skin cells and plasma cells are different sizes. How would you explain this?

4) A mouse mutant is identified that cannot produce J chain due to a genetic mutation. How would this mutation affect the affinity and avidity of the IgM produced by this mouse strain?

5) In Chapter 1, we described how, in 1796, Edward Jenner was able to vaccinate patients to protect them from smallpox infection. However, for his vaccination procedure, Jenner inoculated patients with the cowpox virus and not the smallpox virus. Why do you think Jenner's approach to developing a smallpox vaccine was successful despite the vaccine being based on a different pathogen?

Would You Like To Read More?

Butler, J. E., Zhao, Y., Sinkora, M., Wertz, N., & Kacskovics, I. (2009). Immunoglobulins, antibody repertoire and B cell development. *Developmental & Comparative Immunology, 33*(3), 321–333.

Chen, K., & Cerutti, A. (2010). New insights into the enigma of immunoglobulin D. *Immunological Reviews, 237*(1), 160–179.

Feeney, A. J. (2011). Epigenetic regulation of antigen receptor gene rearrangement. *Current Opinion in Immunology, 23*(2), 171–177.

Labrijn, A. F., Aalberse, R. C., & Schuurman, J. (2008). When binding is enough: Nonactivating antibody formats. *Current Opinion in Immunology, 20*(4), 479–485.

Raju, T. S. (2008). Terminal sugars of Fc glycans influence antibody effector functions of IgGs. *Current Opinion in Immunology, 20*(4), 471–478.

Schatz, D. G., & Yanhoung, Ji (2011). Recombination centres and the orchestration of V(D)J recombination. *Nature Reviews Immunology, 11*, 251–263.

Schroeder, H. W., Jr, & Cavacini, L. (2010). Structure and function of immunoglobulins. *Journal of Allergy & Clinical Immunology, 125*(2 Suppl 2), S41–52.

Vettermann, C., & Schlissel, M. S. (2010). Allelic exclusion of immunoglobulin genes: Models and mechanisms. *Immunological Reviews, 237*(1), 22–42.

B Cell Development, Activation and Effector Functions

Never give advice unless asked.

German Proverb

B cell development takes place in a series of well-defined stages that can be grouped into two phases: the *maturation* phase and the *differentiation* phase. In the maturation phase, an HSC divides and eventually generates mature naïve B cells through a process that is tightly controlled by cytokines but *independent* of foreign antigen. The major developmental stages of the maturation phase include the HSC, the MPP, the CLP, the pro-B cell (progenitor B cell), the pre-B cell (precursor B cell), the immature naïve B cell, the transitional B cell and the mature naïve B cell. Some stages are subdivided, as in "early" and "late" pro-B cells. The maturation phase begins in the bone marrow and ends with mature naïve B cells taking up residence in the secondary lymphoid tissues in the body's periphery. There are two stages in the differentiation phase: the activation of a mature naïve B cell by its specific antigen, and the generation by that cell of antigen-specific plasma cells and memory B cells. The antibodies produced by the plasma cells then carry out a range of effector functions that work to eliminate the original antigen. Which effector functions are deployed are a function of the isotype of antibody produced and the tissue in which the assault has occurred.

NOTE: The antibody response is crucial to immune defense, as exemplified by the plight of patients with primary immunodeficiencies exclusively affecting B cells. These disorders are caused by mutations in B cells that alter signal transduction pathways, isotype switching, Ig gene expression, somatic hypermutation, and/or the cooperation between B and T cells necessary for a complete adaptive response. Patients harboring such mutations can suffer from abnormally high or low antibody levels, production of inappropriate isotypes, or a complete absence of antibodies. B cell-related immunodeficiencies are explored further in Chapter 15.

WHAT'S IN THIS CHAPTER?

MAK: Primer to the Immune Response. http://dx.doi.org/10.1016/B978-0-12-385245-8.00005-4

A. B Cell Development: Maturation Phase

A general scheme of the maturation phase of B cell development is shown in **Figure 5-1**. For the pro-B and pre-B cell stages in particular, the stromal cells of the bone marrow are absolutely critical for continued development. The stromal cells secrete chemokines and cytokines and establish direct intercellular contacts with both pro-B and pre-B cells that are essential for their survival and progression.

> NOTE: All steps in B cell maturation in the bone marrow are antigen-independent. It is only after the fully mature B cell is released into the body's peripheral tissues that it has the chance to interact with foreign antigen and become activated.

I. Pro-B Cells

Early pro-B cells are the first hematopoietic cells clearly recognizable as being of the B lineage. These cells are identified by their expression of certain B lineage markers and by the fact that all their Ig genes are still in germline configuration (i.e., the *Igh*, *Igk* and *Igl* loci have yet to undergo V(D)J recombination). In late pro-B cells, the first attempts at $D_H J_H$ joining in the *Igh* locus are initiated, such that the $D_H J_H$ sequences on at least one chromosome are clearly distinct from the germline *Igh* sequence.

> NOTE: Since the early 1990s, immunologists have been striving to find ways to differentiate among morphologically very similar stages of B cell development. As well as the V(D)J rearrangement status of the Ig loci, there are several other useful markers, including CD19, CD24, CD43, CD45, CD117 and CD127. Pro-B cells, pre-B cells and immature B cells at slightly different stages may express high, medium or low amounts of these proteins, or fail to express them at all. The overall pattern of expression of these markers can pinpoint a B lineage cell at a precise point in its development.

II. Pre-B Cells

It is in early pre-B cells that complete V(D)J rearrangement of the *Igh* locus can be detected; that is, a V_H gene segment is joined to $D_H J_H$ on one chromosome. Both the *Igk* and *Igl* loci remain in germline configuration and there is no expression yet of Ig L chain transcripts. These early pre-B cells transcribe the newly rearranged *Igh* locus to determine whether it can produce a functional Ig H chain of the μ isotype. Normally, an H chain must associate with an L chain in order to be conveyed to the cell surface, but, at this point, there are no L chains present. Instead, a protein homologous to the N-terminus of the L chain comes together with another protein homologous to the C-terminus of an

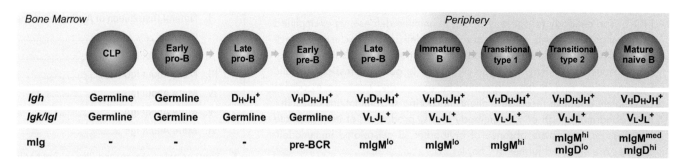

	CLP	Early pro-B	Late pro-B	Early pre-B	Late pre-B	Immature B	Transitional type 1	Transitional type 2	Mature naive B
Igh	Germline	Germline	$D_H J_H^+$	$V_H D_H J_H^+$	$V_H D_H J_H^+$	$V_H D_H J_H^+$	$V_H D_H J_H^+$	$V_H D_H J_H^+$	$V_H D_H J_H^+$
Igk/Igl	Germline	Germline	Germline	Germline	$V_L J_L^+$	$V_L J_L^+$	$V_L J_L^+$	$V_L J_L^+$	$V_L J_L^+$
mIg	-	-	-	pre-BCR	$mIgM^{lo}$	$mIgM^{lo}$	$mIgM^{hi}$	$mIgM^{hi}$ $mIgD^{lo}$	$mIgM^{med}$ $mIgD^{hi}$

Fig. 5-1
Major Stages of B Cell Development

The top panel shows stages of B cell development from CLPs in the bone marrow to mature naïve B cells in the periphery. Panels underneath indicate the status of the heavy and light Ig chain loci at a given stage, and whether and in what form these cells express mIg.

"Sin1-mTORC2 Suppresses *rag* and *il7r* Gene Expression through Akt2 in B Cells" by Lazorchak, A.S., Liu, D., Facchinetti, V., Di Lorenzo, A., Sessa, W.C., Schatz, D.G., and Su, B. (2010) *Molecular Cell* 39, 433–443.

Focus on Relevant Research

Every response of the immune system involves intricate intracellular signal transduction cascades. [A simple example of such a cascade was presented in Fig. 2-6 in Chapter 2.] Accordingly, both B cell development and effector function require numerous signaling pathways that employ specific cell surface receptors and cytoplasmic kinases. The accompanying research article by Lazorchak *et al.* describes experiments aimed at elucidating key signal transduction mediators regulating B cell development.

A central pathway in pre-B cell development is known to depend on the kinases mTORC2 and PI3K and the transcription factor FoxO1, but the precise relationships among these molecules and other pathway components is unclear. Using data acquired by multiple *in vitro* and *in vivo* techniques, Lazorchak *et al.* have proposed a model that defines the elements composing this pathway and clarifies their organization. A key to these researchers' success was their use of a **knockout (KO) mouse**, which is a mutant mouse strain in which a single gene in the DNA of a mouse embryo is deliberately deleted or rendered defective by genetic engineering techniques (see Appendix F). The mutant embryos may then develop into adult mice missing the function of that single gene, making the role of the protein it encodes easier to identify.

In this study, Lazorchak *et al.* had attempted to make a KO mouse for an important component of the mTORC2 complex known as Sin1, but embryos that were homozygous for the inactivating mutation in the *Sin1* gene (*Sin$^{-/-}$* embryos) died after only 10 days of embryonic development. To get around this problem, these scientists harvested HSCs from *Sin$^{-/-}$* embryos before the 10-day mark and transplanted them into wild type (WT; *Sin$^{+/+}$*) adult mice that were previously irradiated to destroy their own hematopoietic systems. The transplanted *Sin$^{-/-}$* HSCs were able to survive with the support of the normal *Sin$^{+/+}$* stromal cells in the bone marrow of these *chimeric* recipients. Using a technique called *flow cytometry* (see Appendix F), these researchers were able to determine the effect of Sin loss on the surface expression of mIg, and thus on B cell development. An example of the data produced by these types of experiments is shown in the accompanying Figure 1, Panel A from the article. The total bone marrow cells of a given mouse are put through a flow cytometer that counts the number of cells expressing CD19 (a B lineage marker) and the number of cells expressing mIgM. Bone marrow cells that express high levels of *both* CD19 and mIgM are B lineage cells maturing from the pro-B to the pre-B stage. In the plots in the figure, each tiny dot represents a single cell counted by the flow cytometer. The y-axis shows the level of CD19 expression, whereas the x-axis shows the level of mIgM expression. The numbers labeling the boxes on the plots represent the relative proportions of each cell type present. When *Sin1$^{-/-}$* HSCs were transplanted into the *Sin1$^{+/+}$* adult recipients (KO → WT), the number of CD19$^+$mIgM$^+$ B cells present in the total bone marrow cell population was significantly reduced compared to the control experiment in which *Sin1$^{+/+}$* embryonic cells were transplanted into *Sin1$^{+/+}$* recipients (WT → WT). Through these results and other experiments examining IL-7R expression and RAG activity, Lazorchak *et al.* clearly demonstrated that Sin1 is very important for B cell development.

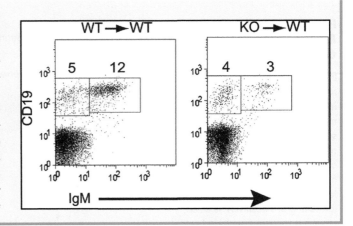

L chain to form the **surrogate light chain (SLC)**. Two SLC molecules associate with two copies of the candidate μ chain plus the Igα/Igβ signaling heterodimer to form a **pre-B cell receptor complex (pre-BCR)**. The pre-BCR is expressed transiently on the pre-B cell membrane to test the functionality of this particular $V_H D_H J_H$ combination. A pre-BCR is not a true Ig and cannot respond to antigen, but it is thought to bind to ligands on bone marrow stromal cells such that an intracellular signal is delivered telling the cell that a functional H chain has been successfully synthesized. If the candidate $V_H D_H J_H$ sequence is not functional, sequential $D_H \rightarrow J_H$ and $V_H \rightarrow D_H J_H$ rearrangements are attempted at the second *Igh* allele, and the resulting candidate μ chain is also tested in pre-BCR form. Failure on both chromosomes occurs in about half of pre-B cells, pre-empting further maturation and leading to the apoptotic death of most of these cells in the bone marrow.

Late pre-B cells develop from those early pre-B cells that do produce a functional pre-BCR. Further *Igh* locus rearrangements are terminated and V(D)J recombination is stimulated at the *Igk* and *Igl* loci as described in Chapter 4. Only late pre-B cells that successfully express a complete Ig molecule containing fully functional H and L chains survive to become immature B cells. The presence of the complete BCR complex (mIg plus Igα/Igβ) on the B cell surface delivers an intracellular signal that terminates all Ig locus rearrangements.

III. Immature B Cells in the Bone Marrow: Receptor Editing

Despite their functional BCRs, immature B cells in the bone marrow do not proliferate or respond to foreign antigen. It is this population of cells that is screened for central tolerance; that is, immature B cells that recognize self antigens are removed. Bone marrow stromal cells express "housekeeping" molecules that are produced by all body cells. A B cell whose BCR binds to one of these molecules with high affinity is potentially autoreactive, and the release of such a B cell from the bone marrow into the periphery could lead to an attack on self tissues. The immature B cell therefore determines *before* it is released from the bone marrow whether its BCR recognizes the housekeeping molecules present on the surrounding stromal cells. If so, the autoreactive B cell receives an intracellular signal to halt development. The cell is then given a brief window of opportunity to try to further rearrange its Ig loci and stave off apoptosis by altering its antigenic specificity. This secondary gene rearrangement is called **receptor editing** (**Fig. 5-2**).

Receptor editing occurs primarily in the L chain because, after $V_H D_H J_H$ recombination in the *Igh* locus is completed, there are rarely any D_H gene segments available to permit further rearrangements satisfying the structural rules of V(D)J recombination. In contrast, the light chain loci usually still have available many upstream V_L gene segments and a few downstream J_L segments after a productive round of *Igk* or *Igl* rearrangement. If receptor editing fails to achieve success, the B cell dies by apoptosis and is said to have been **negatively selected**. However, if receptor editing is successful,

Fig. 5-2
Receptor Editing in B Cell Central Tolerance

Immature B cells whose BCRs bind to broadly-expressed "housekeeping" molecules expressed by bone marrow stromal cells could be self-reactive. These cells undergo receptor editing of their Ig light chain genes in an attempt to avoid negative selection. Cells that are no longer self-reactive after receptor editing are positively selected, receive a survival signal, and enter the peripheral lymphoid tissues. Cells that remain self-reactive despite receptor editing die by apoptosis.

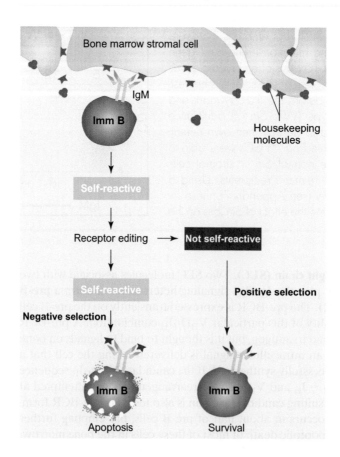

the BCR no longer recognizes self antigen, and the cell appears to receive a **positive selection** signal that sustains survival. On a daily basis, about 2–5% of the entire pool of immature B cells survives the selection processes that establish central B cell tolerance.

NOTE: While receptor editing alters the specificity of the BCR of an autoreactive immature B cell in the bone marrow before it is released to the body's periphery, a process called "receptor revision" may do the same for mature B cells after they start their patrol of the body's lymphoid tissues. Some researchers have found evidence of secondary rearrangements of the Ig genes in peripheral B cells, but whether these rearrangements were undertaken to avoid autoreactivity and how they occur are controversial.

IV. The Transition to Maturity: IgM and IgD Coexpression

Immature B cells remain in the bone marrow for 1–3 days before commencing the expression of new adhesion molecules and homing receptors that allow them to leave the bone marrow and travel in the blood to the secondary lymphoid tissues. At this point, the developing B cell is known as a *transitional type 1 B cell,* or *T1 B* cell. T1 B cells extravasate from the blood first into the red pulp of the spleen and then into its PALS. After about 24 hours in the PALS, T1 B cells become *transitional type 2 B cells,* or *T2 B* cells. Local production of a cytokine called *BAFF* (B lymphocyte activating factor belonging to the TNF family) is essential for this transition and T2 B cell survival. Some T2 B cells start to colonize the B cell-rich areas of the spleen and acquire the ability to emigrate to other secondary lymphoid tissues, particularly the lymph nodes. These cells, which will recirculate throughout the lymphoid system, will become mature *follicular B cells.* Other T2 cells move from the red pulp of the spleen to its marginal zone, and will become *marginal zone B cells.* Both types of T2 B cells commence the surface expression of IgD as well as IgM. Both Igs are derived from the same rearranged H and L chain genes containing the same $V_H D_H J_H$ and $V_L J_L$ exons, respectively, and therefore have the same V domains in their antigen-binding sites and the same antigenic specificity. It is only the C region of the H chain that differs. Coexpression of IgM and IgD occurs because the primary RNA transcripts that are synthesized from the rearranged H chain gene contain the sequences encoded by the $V_H D_H J_H$ exon and *both* the Cμ and Cδ exons. Differential processing of this transcript at alternative polyA addition sites followed by RNA splicing results in the removal of either the Cμ or the Cδ exon from the RNA, leaving an mRNA specifying the production of either a δ or μ heavy chain, respectively, and hence an IgD or IgM antibody (**Fig. 5-3**). Both pathways operate simultaneously to process the population of primary transcripts generated in the B cell, such that a single T2 B cell expresses two different isotypes of mIgs.

V. Mature Naïve B Cells in the Periphery

Once T2 B cells establish themselves in the lymphoid follicles and splenic marginal zone, they are considered to be mature naïve B cells in the periphery. At the molecular level, mature naïve B cells show slightly lower levels of mIgM than do T2 B cells but higher levels of mIgD. Although mIgM is clearly essential for the response of the B cell to antigen, the function of mIgD has been more difficult to ascertain (see later). Mature naïve B cells also lose expression of RAG-1 and RAG-2, so that, in general, no further changes in V(D)J gene segment usage can occur in either the mature B cell itself or in its memory and plasma cell progeny. These B cells are now poised to encounter antigen.

B. B Cell Development: Differentiation Phase

The adult human bone marrow churns out about 10^9 mature naïve B cells every day. However, over the course of their life span, the chances of these B cells being stimulated by specific antigen are extremely limited: only 1 in 10^5 peripheral B cells encounters

Fig. 5-3
IgM and IgD Coexpression

Primary transcripts containing both the Cμ and Cδ exons undergo differential RNA processing and polyadenylation that result in two different mRNAs, one containing Cμ and the other containing Cδ. After RNA splicing brings the common V exon together with either the Cμ or the Cδ exon, these mRNAs are translated into μ or δ Ig heavy chains, respectively. In an individual B cell, both mRNAs are produced and translated simultaneously, allowing co-expression of IgM and IgD antibodies of the same antigenic specificity.

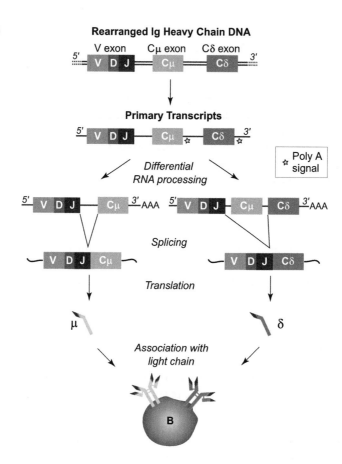

specific antigen and avoids death by apoptosis. It is the binding of specific antigen to a B cell's BCRs that triggers the differentiation phase.

I. The Nature of B Cell Immunogens

As was introduced in Chapter 1, an antigen is any substance that binds specifically to the antigen receptor of a T or B cell. However, some antigens that bind to a TCR or BCR fail to induce the activation of the lymphocyte even under optimal conditions. Immunologists have therefore coined the term **immunogen** to refer to any substance that binds to a BCR or TCR *and* elicits an adaptive response. Thus, all immunogens are antigens, but not all antigens are immunogens. (Despite this very real distinction, the immunological community regularly uses "antigen" when "immunogen" would be more accurate.) Almost every kind of organic molecule (including proteins, carbohydrates, lipids and nucleic acids) can be an antigen, but only macromolecules (primarily proteins and polysaccharides) have the size and properties necessary to be physiological immunogens. In a natural infection, invading bacteria and viruses appear as collections of proteins, polysaccharides and other macromolecules to the host's immune system. Within this collection, there are many immunogens, each inducing its own adaptive immune response.

i) Responses to B Cell Immunogens

Three major classes of B cell antigens (or more properly, immunogens) exist: **T-independent-1 (Ti-1)**, **T-independent-2 (Ti-2)** and **T-dependent (Td)** antigens. Both types of Ti antigens can activate B cells to produce antibodies without interacting directly with T cells. Td antigens can bind to the BCRs of B cells to initiate activation but cannot induce plasma cell differentiation and antibody production unless the B cell interacts directly with a Th effector cell activated by the same antigen. The Th effector cell is said to be supplying **T cell help** to the B cell, which is usually a follicular B cell. T cell help takes the form of cytokines and intercellular contacts that mediate **costimulation**.

TABLE 5-1	**Features of Responses to Td and Ti Antigens**		
Property	**Td Antigen**	**Ti-1 Antigen**	**Ti-2 Antigen**
Requires direct interaction with a T cell for B cell activation	Yes	No	No
Requires T cell cytokines	Yes	No	Yes
Epitope structure	Unique	Mitogen	Repetitive
Protein	Yes	Could be	Could be
Polysaccharide	No	Could be	Could be
Relative response time	Slow	Fast	Fast
Dominant antibody isotypes	IgG, IgE, IgA	IgM IgG (rarely)	IgM IgG (sometimes)
Diversity of antibodies	High	Low	Low
Stimulates immature B cells	No	Yes	No
Polyclonal B cell activator	No	Yes	No
Memory B cells generated	Yes	No	No
Magnitude of response upon a second exposure to antigen	Secondary response level	Primary response level	Primary response level
Examples	Diphtheria toxin Mycobacterium protein	Bacterial LPS	Pneumococcal polysaccharide

Costimulatory molecules are proteins on the surfaces of lymphocytes whose engagement by specific ligand is necessary for complete activation. It is T cell help that allows activated B cells to undergo **somatic hypermutation, isotype switching** and memory B cell production. As a result, responses to Td antigens are dominated by highly diverse IgG, IgE or IgA antibodies and generate memory B cells capable of mounting a faster, stronger secondary response in a subsequent exposure to that antigen.

B cell responses to Ti-1 and Ti-2 antigens do not involve direct interaction with effector Th cells and thus can provide relatively rapid defense against invaders. It is usually B cells in the marginal zone of the spleen that respond to the majority of these antigens. However, the absence of T cell interaction means that B cells responding to Ti antigens do not carry out extensive isotype switching or somatic hypermutation and therefore produce mainly IgM antibodies of limited diversity. In addition, memory B cells are not generated so that the intensity of the response in a subsequent exposure to the antigen is stuck at the primary level rather than progressing to the secondary level. The principal features of responses to Td, Ti-1 and Ti-2 antigens are discussed next and summarized in **Table 5-1**.

ii) Properties of Ti Antigens

Ti-1 antigens contain a molecular region that allows it to act as a **mitogen**. A mitogen is a molecule that non-specifically stimulates cells to initiate mitosis. In the case of a Ti-1 antigen, a portion of the molecule binds to a site on the B cell that is separate from the antigen-binding site of the BCR. This site can be in a different region of the BCR complex, or indeed on a different surface receptor entirely. Once bound, the mitogen sends a strong signal to the nucleus of the B cell to proliferate. Because the mitogen-binding site is not unique to any one B cell clone (unlike the antigen-binding site), many clones of naïve B cells can be activated at once by Ti-1 antigen molecules. *Polyclonal* B cell activation is said to have taken place. In addition, unlike Td and Ti-2 antigens, Ti-1 antigens can activate immature B cells, emphasizing the unconventional nature of the molecular interaction between the mitogen and the B cell and the mode of cellular activation involved.

Ti-2 antigens, which are found in many bacterial and viral structures and products, are generally large polymeric proteins or polysaccharides (and sometimes lipids or nucleic acids) that contain many repetitions of a structural element. This repetitive structure acts as a multivalent antigen that can bind with high avidity to the mIg molecules in neighboring BCR complexes on the surface of a B cell. The BCRs are said to be *cross-linked* because the antigen-binding sites of two (or more) mIg molecules are indirectly linked together by virtue of their binding to the same very large immunogen. This extensive cross-linking triggers intracellular signaling that leads to B cell activation, proliferation and differentiation in a way that bypasses the need for costimulatory contacts with an activated Th cell. However, these antigens cannot activate naïve B cells in the absence of cytokines produced (mainly) by activated T cells.

iii) Properties of Td Antigens

As introduced in earlier chapters, the TCR of a T cell recognizes a complex composed of an MHC molecule bound to a peptide of an antigenic protein; that is, the epitope for a TCR is pMHC. For a Th effector cell to supply T cell help to a B cell that has encountered an antigen, the original naïve Th cell must have been activated by a pMHC derived from the same antigen. Thus, as well as supplying a B cell epitope, a Td antigen must contain protein, since it must supply at least one peptide that can bind to MHC class II and form the T cell epitope. Other physical properties of a protein such as its foreignness, conformation and molecular complexity also affect its ability to be a Td antigen (**Table 5-2**).

a) Foreignness

To be immunogenic, an antigen must be perceived as non-self by the host's immune system. Whether an antigen is perceived as non-self depends on whether there is a lymphocyte in the host's repertoire that can recognize the antigen. In the case of proteins that are similar to self proteins, many of the lymphocyte clones whose antigen receptors might have recognized that antigen should have already been removed during the establishment of central tolerance. Thus, a protein that is widely conserved among different species will not be very immunogenic. In contrast, the more an antigen deviates structurally from self proteins, the more easily it is recognized as non-self. Accordingly, the introduction of such a protein into the host will usually trigger an intense immune response.

b) Complexity

The molecular complexity of a Td antigen encompasses the properties of *molecular size*, *subunit composition*, *conformation*, *charge* and *processing potential*. A protein of large size and/or composed of multiple subunits will likely contain both the T and B epitopes needed for a Td response. In general, Td antigens are at least 4 kDa and

TABLE 5-2	Factors Affecting Td Immunogenicity	
	More Immunogenic	**Less Immunogenic**
Foreignness	Very different from self	Very similar to self
Molecular complexity		
Size	Large	Small
Subunit composition	Many	Few
Conformation	Denatured, particulate	Native, soluble
Charge	Intermediate charge	Highly charged
Processing potential	High	Low
Dose	Intermediate	High or low
Route of entry	Subcutaneous > intraperitoneal > intravenous or gastric	
Host genetics: Allelic variation in MHC and/or Ag processing molecules	Efficient peptide binding and presentation	Inefficient peptide binding and presentation

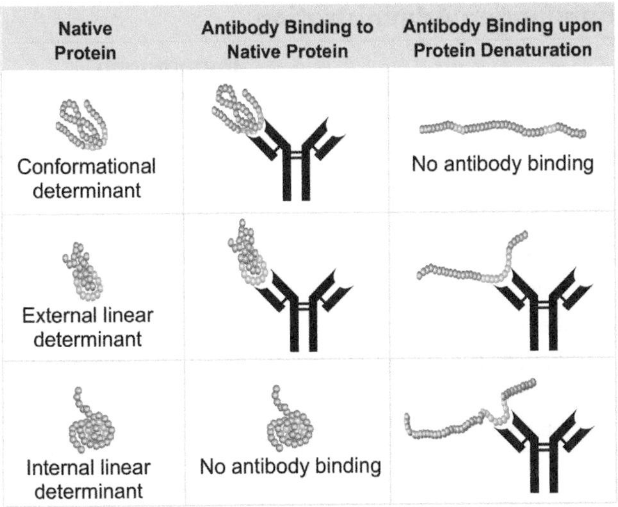

Native Protein	Antibody Binding to Native Protein	Antibody Binding upon Protein Denaturation
Conformational determinant		No antibody binding
External linear determinant		
Internal linear determinant	No antibody binding	

Fig. 5-4
Effect of Protein Conformation on Antibody Binding to Antigenic Determinants

Conformational B cell determinants are lost upon denaturation of the antigenic protein. External linear determinants are accessible to antibody when the protein is in its native conformation or denatured. Internal linear determinants are accessible only upon protein denaturation.

often close to 100 kDa in molecular size. However, large molecules are not good Td antigens if they are too simple in structure. For example, homopolymers (composed of a single type of amino acid) are not effective Td antigens, but copolymers of two different amino acids are because a copolymer provides more possibilities for T and B epitopes. For the same reason, macromolecules with several different subunits are generally more effective Td antigens than are macromolecules containing a single subunit type.

The conformation of a molecule can affect immunogenicity. Most B cell epitopes are structures that project from the external surfaces of macromolecules or pathogens. The majority of these B cell epitopes are **conformational determinants**, a term used by immunologists to refer to epitopes in which the contributing amino acids may be located far apart in their linear sequence but which become juxtaposed when the protein is folded in its natural or *native* shape (**Fig. 5-4**). These epitopes depend on native state folding, so that they disappear when the protein is denatured. Other epitopes are linear, in that they are defined by a particular stretch of consecutive amino acids. Some linear determinants are present on the surface of a protein and are accessible to antibody whether the protein is in native or denatured form (external linear determinants). Other linear determinants are buried deep within a molecule and are accessible to antibody only when the protein is denatured (internal linear determinants). Some determinants are **immunodominant**; that is, even though the antigen contains numerous epitopes, the majority of antibodies are raised to only a few of them. Immunodominant epitopes for B cells tend to be conformational determinants on the surfaces of macromolecules because these sites are more available for binding to BCRs than are internal epitopes. Structural considerations that affect the ease of binding between the epitope and the antigen receptor binding site may make one surface epitope immunodominant over another.

The charge or electronegativity of a molecule is also important for immunogenicity. Since B cell epitopes are most often those exposed on the surface of a macromolecule in the aqueous environment of the tissues, those areas of globular proteins acting as B cell epitopes are often hydrophilic in nature. Proteins containing a high ratio of aromatic to non-aromatic amino acid residues often have numerous hydrophilic epitopes and are very effective Td antigens. However, very highly charged molecules elicit poor antibody production, possibly due to electrostatic interference between the surface of the B cell and the antigen.

c) Dosage

When a pathogen invades the body, it is able to replicate and so delivers substantial doses of a plethora of immunogenic molecules to the host. In the case of experimental immunizations with a single antigen, immunologists have determined that there is a threshold dose below which no adaptive response is detected because insufficient

TABLE 5-3	Modes of Immunogen Administration		
	Abbr.	**Description**	**Immunogen Channeled to**
Oral	p.o.	By mouth	MALT
Parenteral			
Intravenous	i.v.	Into a blood vessel	Spleen
Intraperitoneal	i.p.	Into the peritoneal cavity	Spleen
Intramuscular	i.m.	Into a muscle	Regional lymph node
Intranasal	i.n.	Into the nose	MALT
Subcutaneous	s.c.	Into the fatty hypodermis layer beneath the skin	Regional lymph node
Intradermal	i.d.	Into the dermis layer of the skin	SALT

numbers of lymphocytes are activated. As the dose of antigen increases above the threshold, the production of antibodies rises until a broad plateau is reached that represents the optimal range of doses for immunization with this antigen. At very high antigen doses, antibody production again declines, possibly due to mechanisms responsible for tolerance to self proteins (which are naturally present at high concentrations). Memory lymphocytes generally respond to a lower dose of antigen than naïve cells, but very high or very low doses of antigen can induce memory cells to adopt non-responsive states known as **high zone tolerance** or **low zone tolerance**, respectively (see Ch. 10). Antibody production is minimal at these dosages.

d) Route of Entry or Administration

The immunogenicity of a molecule can depend heavily on its route of administration. The food we eat contains many antigens, but because these antigens are "dead" (cannot replicate) and are introduced via the digestive tract, enzymatic degradation usually destroys them before an immune response is provoked. For this reason, antigens used for immunizations are usually introduced by a **parenteral** route, that is, by a route other than the digestive tract (**Table 5-3**). The route chosen determines how the antigen is channeled to a nearby secondary lymphoid tissue to trigger lymphocyte activation. Interestingly, the dose of a molecule that provokes a strong immune response when injected by one route of administration may induce only a minimal response when introduced by another. More on the administration of immunogenic molecules appears in Chapter 14 on vaccines.

e) Potential for Antigen Processing and Presentation

A protein's potential for uptake and intracellular processing by APCs to generate the peptides for pMHC assembly is also important for Td immunogenicity. Macromolecules that are readily phagocytosed by APCs and proteins that are easily digested by an APC's enzymes tend to be more immunogenic. Subtle sequence differences in the genes encoding MHC molecules or other proteins involved in antigen processing can alter antigen presentation, affecting the type, as well as the intensity, of the immune response to a given immunogen (see Ch. 6).

II. B Cell Activation by Td Immunogens

i) Rationale for T Cell Help

A large proportion of the body's B cells will need T cell help to respond because Ti immunogens are only a small fraction of the antigens that a host encounters. Many of the molecules making up a pathogen are proteins of unique amino acid sequences that lack the large repetitive structures needed to cross-link BCRs and trigger B cell activation directly. Even when a protein is present in thousands of copies on the surface of an intact pathogen, steric hindrance may forestall extensive BCR cross-linking and prevent complete B cell activation. Enter T cell help supplied by a Th effector cell specific for the same antigen. The B cell receives both costimulation and cytokines that deliver signals completing the intracellular signaling pathway leading to complete B cell activation, proliferation and antibody production. Without the signals provided by this "B–T cooperation," activation events such as protein kinase activation and increases in intracellular Ca^{2+} may be initiated in the B cell, but neither cellular proliferation nor antibody production can occur.

Th effector cell development and activation is described in Chapter 9.

ii) The Three Signal Model of B Cell Activation

The activation of a resting mature naïve B cell by a Td antigen occurs in three steps: (1) the binding of antigen molecules to the BCRs; (2) the establishment of intercellular contacts with a CD4+ Th effector cell; and (3) the binding of Th-secreted cytokines to cytokine receptors on the B cell surface. Each of these steps delivers a stimulatory signal to the B cell, but complete activation does not occur unless all three signals are received (**Fig. 5-5**).

a) Signal 1

A resting, mature naïve B cell in the periphery is said to be in a *cognitive state* because it is capable of recognizing and reacting to its specific antigen. However, the binding of one antigen molecule to one antigen-binding site of an mIg is not sufficient to activate

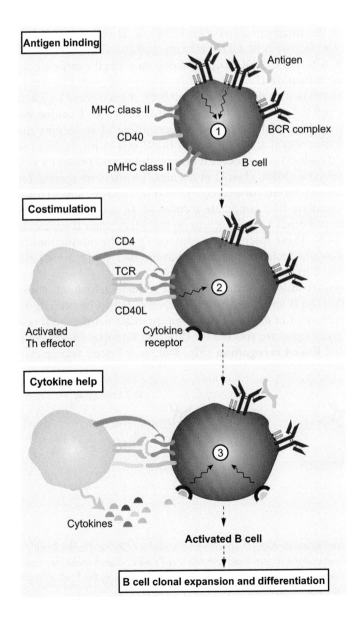

Fig. 5-5
Three Signal Model of B Cell Activation by Td Antigen

A mature B cell in the periphery must receive three distinct signals to achieve full activation, namely antigen binding, costimulation and cytokine help. Signal 1 is delivered when multiple molecules of antigen bind to multiple surface BCR complexes containing mIg plus the Igα/Igβ signaling heterodimer. Antigen bound to the BCR complex is internalized and processed to generate peptide-MHC class II complexes that can engage the TCRs of activated CD4+ Th effector cells. The binding of CD40L on the Th cell to the B cell's CD40 molecules delivers costimulatory Signal 2, which stimulates cytokine receptor expression. Signal 3 is delivered when cytokines produced by the Th effector cell bind to their receptors on the B cell surface. Several pairs of adhesion molecules also help to hold the B-T conjugate together (not shown).

a B cell. Rather, multiple (in the range of 10–12) antigen–BCR pairs are required to reach the response threshold. These antigen–BCR groupings, which are sometimes referred to as *BCR microclusters*, appear to undergo a conformational change that allows the cytoplasmic portion of a BCR to recruit the kinases necessary for signal transduction. Within a minute of reaching the threshold of BCR stimulation, a signal is conveyed down the length of the transmembrane Igα/Igβ proteins to these recruited kinases and induces the activation of nuclear transcription factors. These factors then activate and regulate the gene transcription necessary to prepare the B cell to receive T cell help. Within 12 hours of antigenic stimulation, the B cell expands in size, increases its RNA content, and moves from the resting phase of the cell cycle to the phase preceding cell division. MHC class II molecules are upregulated on the B cell surface, as well as costimulatory molecules that are required for the B cell to receive "signal 2" from the Th cell. The expression of cytokine receptors necessary for the receipt of "signal 3" also commences. This B cell is said to be in a *receptive state*.

> The binding of multiple BCRs by a Td antigen delivers signal 1 in a way that is equivalent to the cross-linking of Fabs imposed by a Ti antigen of repetitive structure.

b) Signal 2

Signal 2 is delivered by the engagement of costimulatory molecules on the B cell surface. If a B cell's BCRs are engaged by a Td antigen in the absence of costimulation, the lymphocyte eventually undergoes apoptosis or becomes **anergic** (unresponsive to its specific antigen). Only after receiving both signal 1 and signal 2 can a B cell accept signal 3 in the form of Th cell-secreted cytokines. The most important costimulatory event for B cell activation is the interaction between CD40 on B cells and CD40L (CD40 ligand) on activated Th effector cells. CD40 is expressed constitutively (all the time) by mature naïve B cells, whereas CD40L is upregulated on a T cell's surface only in response to TCR engagement by pMHC.

In order for a receptive B cell to actually receive T cell help, it must come together with an activated Th effector cell to form a bicellular structure called a **B–T conjugate**. The formation of this conjugate is possible because the recognition of antigen by the BCR does more than just deliver signal 1: it also allows the B cell to internalize the antigen via receptor-mediated endocytosis, process it into peptides, and present T cell *epitopes from the same antigen* on its MHC class II molecules. Th effectors specific for this antigen can then recognize the pMHCs displayed by the B cell and are able to form pMHC-TCR "bridges" that mediate B–T conjugate formation. In addition, the CD4 coreceptor of the Th effector cell binds to a distal site on the MHC class II molecule of the pMHC bound by the TCR. These intercellular contacts (plus the interactions of several ICAM-related adhesion molecule pairs) ensure that the B–T conjugate is held together long enough for signal 2 to be delivered.

The fact that a B–T conjugate must be formed for the complete activation of a B cell by a Td antigen means that the B and T cell epitopes involved must be present *on the same macromolecular structure or in the same pathogen*. In other words, the B cell and Th cell involved in a Td response are specific for different epitopes of the same antigen, a phenomenon called **linked recognition** (**Fig. 5-6**). It is linked recognition

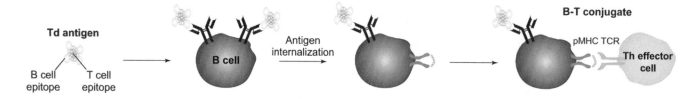

Fig. 5-6
Linked Recognition

Within any Td antigen, the B cell epitope and the T cell epitope must be linked together. Antigen-specific B cells whose BCRs bind to the B cell epitope are triggered to internalize the antigen and process it so that the T cell epitope is combined with the B cell's MHC class II molecules and expressed on the B cell surface as a pMHC complex. A Th effector cell whose TCR is specific for the pMHC complex containing the T cell epitope will bind to it, initiating the formation of the B-T conjugate required for complete B cell activation.

that results in B cells and Th cells mounting an efficient, coordinated response against a foreign entity. This requirement for linked recognition also means that only proteins of a certain size and complexity are effective Td immunogens, as they must contain both B and T cell epitopes. In addition, linked recognition reduces the likelihood that an autoreactive B cell (that has somehow escaped tolerance mechanisms) will be activated, since the Th help required can only come from an antigen-specific Th cell, and there is a relatively small chance that an autoreactive Th cell specific for the same host protein will also have escaped tolerance and be present in the periphery.

c) Signal 3

The binding of antigen to sufficient BCRs on a B cell, coupled with CD40/CD40L costimulation, induces the expression of new cytokine receptors (particularly the receptors for IL-1 and IL-4) on the B cell surface. However, without the binding of the relevant cytokines to these receptors, the B cell usually undergoes only limited proliferation. While some of the pertinent cytokines required for B cell activation can be secreted by a nearby activated macrophage or DC, most are derived from the antigen-activated Th effector cell that has made contact with the B cell. Shortly after adhesive, pMHC-TCR and costimulatory contacts are established between the B cell and the Th effector cell, the Golgi apparatus is reorganized within the Th cell such that this secretory structure is closer to the site of contact. The release of the cytokines is then directed more precisely toward the B cell, increasing the efficiency of the delivery of "signal 3." A B cell that has already received signals 1 and 2 and binds IL-4 starts to proliferate vigorously. Other cytokines, including IL-2, IL-5 and IL-10, support this proliferation. As the progeny B cells continue to divide, IL-2, IL-4, IL-5, IL-6, IL-10, IFNs and TGFβ become important for inducing the differentiation of antibody-secreting plasma cells and memory B cells (see later).

III. Cellular Interactions During B Cell Activation

Naïve B cell activation and the subsequent maturation of progeny cells into plasma cells and memory B cells depends on cellular interactions that occur in various anatomical structures within the secondary lymphoid tissues. These interactions have been best studied in the lymph node.

i) Lymph Node Paracortex and Primary Follicles

One of the most common places for antigen and lymphocytes to meet is in the lymph node. Prior to a B cell's encounter with antigen, the follicles within the cortex of a lymph node are known as **primary follicles** (**Fig. 5-7**). A primary follicle is filled with

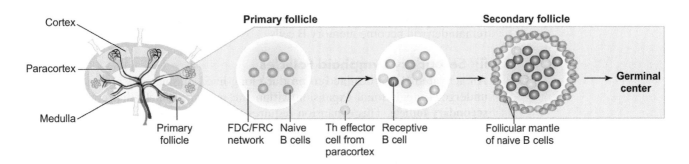

Fig. 5-7
Primary and Secondary Follicle Structure in the Lymph Node

Naïve B cells in primary follicles bind free antigen or antigen trapped in the FDC/FRC network. Simultaneously, Th cells in the paracortex are activated and generate Th effector cells when the same antigen is processed by local APCs. The Th effector cell and receptive B cell meet at the outer edge of the primary follicle, where the B cell receives Th cell help. The B cell becomes fully activated, moves into the center of the follicle and proliferates, pushing naïve B cells to the edge of the follicle to form the follicular mantle characteristic of secondary follicles.

resting naïve B cells, a collection of FDCs that can trap whole antigens on their surfaces, and a network of *fibroblastic reticular cells (FRCs)* possessing extended fibers. These fibers form two types of conduits for small antigens: the *follicular conduit,* which conveys antigen to the area of the primary follicle occupied by B cells, and the *paracortical conduit*, which delivers antigen to areas of the paracortex populated by DCs and T cells. FDCs and FRCs in the follicular region also secrete the chemokine CXCL13, which attracts B cells, whereas the chemokines CCL19 and CCL21 secreted by cells in the paracortical region attract T cells and DCs. These molecules ensure that all the players needed for a humoral response are in the right place at the right time.

Let us suppose that an antigen X has penetrated the skin or mucosae and has been conveyed by the lymphatic system to the cortex and paracortex of the regional lymph node. A resting naïve B cell expressing a BCR specific for an epitope on antigen X may be present in a primary follicle of this node. If molecules of antigen X bind in sufficient numbers to the BCRs of this B cell, activation signal 1 is delivered. The now-receptive B cell displays pMHCs derived from the antigen in anticipation of conjugate formation, and maintains its expression of CD40 in anticipation of signal 2. Meanwhile, the proteins of antigen X are digested and processed by the lymph node's various APCs, including DCs in the paracortex. These cells present pMHC complexes to naïve Th cells located in the paracortex, activating them and causing them to proliferate and generate Th effector cells that express CD40L.

> NOTE: To form the B–T conjugate necessary for naïve B cell activation in a follicle, the Th effector cell must itself be activated. This activation is achieved when the Th cell uses its TCRs to bind to pMHCs presented by the participating B cell. As is described in Chapter 9, there are several different subsets of Th effectors, and the subset that participates in naïve follicular B cell activation is sometimes called a *follicular Th (fTh)* cell.

At this point, the two partners necessary for B–T conjugate formation are activated but localized in different parts of the lymph node. The receptive B cell follows a chemokine gradient emanating from the T cell-rich paracortex to the edge of the primary follicle. On the edge of the follicle, the receptive B cell meets a Th effector cell specific for the same antigen and forms the B–T conjugate necessary for the delivery of signals 2 and 3 to the B cell. Upon delivery of these signals, the B cell achieves full activation.

About 4–6 days after antigen contact, the activated B cell undergoes one of two fates. In some cases, the B cell on the edge of the follicle immediately proliferates and terminally differentiates into a population of **short-lived plasma cells** without carrying out isotype switching or somatic hypermutation. In other cases, the B cell partner of the B–T conjugate drags the Th effector back into the center of the follicle where the Th cell dissociates from the B cell to allow the latter's proliferation (see later). The majority of the progeny of this B cell will become **long-lived plasma cells** while the remainder will become memory B cells.

ii) Secondary Lymphoid Follicles

An activated B cell destined to produce long-lived plasma cells and memory B cells undergoes rapid clonal expansion within the primary follicle that converts it into a **secondary follicle**. This expansion requires the presence of antigen (displayed on the surfaces of FDCs), continued engagement of CD40 by the CD40L of an activated antigen-specific Th effector cell, and stimulatory cytokines secreted by the Th effector. By 6–9 days after antigen contact, the uninvolved naïve B cells that filled the primary follicle are displaced by the proliferating B cell clone and are compressed at the edges of the follicle to form the *follicular mantle*.

iii) Germinal Centers

By 9–12 days after antigen contact, the secondary follicle polarizes into two distinct areas, a **dark zone** and a **light zone**, and becomes a **germinal center (GC)** (**Plate 5-1**). This process of polarization is called the **germinal center reaction.**

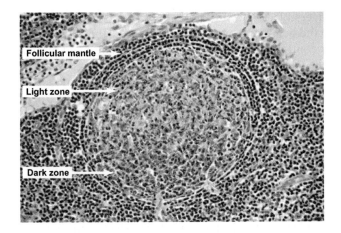

Plate 5-1
Germinal Center

A GC develops from a secondary follicle and contains a distinct dark zone and light zone surrounded by the follicular mantle. *[Reproduced by permission of David Hwang, Department of Pathology, University Health Network, Toronto General Hospital.]*

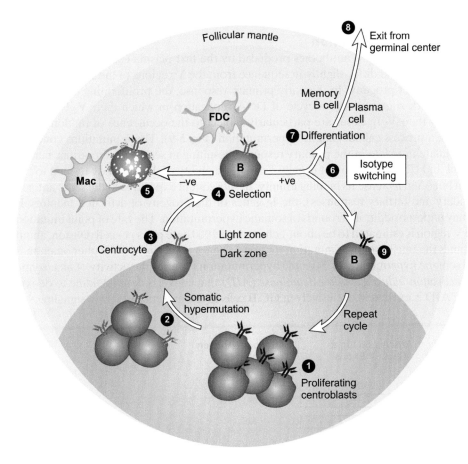

Fig. 5-8
Germinal Center Function

An antigen-activated B cell proliferates and generates centroblasts, converting the secondary follicle into a GC with dark and light zones (1). Somatic hypermutation occurs in the dark zone (2), giving rise to centrocytes that move into the light zone (3) to undergo selection (4). Negatively selected centrocytes die by apoptosis (5), whereas positively selected centrocytes survive and undergo isotype switching (6). Some switched centrocytes immediately differentiate into memory or plasma cells (7) that exit the GC and enter the circulation (8), whereas others re-enter the dark zone to undergo another cycle of somatic hypermutation and selection (9). *[Adapted from McHeyzer-Williams, M.G., & Ahmed R. (1999). B cell memory and the long-lived plasma cell. Current Opinion in Immunology 11, 172–179.]*

Activated B cells are found first in the dark zone, where they continue to proliferate rapidly and become known as **centroblasts** (**Fig. 5-8**). It is within the centroblast population that the antibody repertoire undergoes its final diversification by somatic hypermutation of the V_H and V_L exons. As centroblasts mature and differentiate further, they migrate into the light zone where they become known as **centrocytes.** Centrocytes bearing the newly generated somatic mutations interact with antigen on the FDCs in the light zone and are either negatively or positively selected. In the first step toward establishing **peripheral B cell tolerance**, negative selection in the GC induces B cells that no longer recognize the antigen (and could thus be autoreactive) to undergo apoptosis; the dead B cells are removed by macrophages within the node. Positive selection ensures the survival of B cells that continue to recognize the FDC-displayed antigen with the same or increased affinity (affinity maturation). The light zone is also where the Ig C_H exons undergo isotype switching in progeny

Somatic hypermutation, affinity maturation and isotype switching are all discussed in this chapter. Peripheral B cell tolerance and other forms of lymphocyte regulation are described in Chapter 10.

B cells to increase functional diversity. At the end of all these processes, the surviving centrocytes either return to the beginning of the GC cycle for further expansion, diversification and selection, or continue their differentiation into long-lived plasma cells or memory B cells that exit the GC and enter the circulation and tissues. The tremendous proliferation and differentiation in GCs throughout the lymph node can persist for up to 21 days after antigen encounter, after which the number and size of the GCs decrease unless there is a fresh assault by antigen.

IV. Germinal Center Processes that Diversify Antibodies

In Chapter 4, processes that increased the diversity of the antibody repertoire before encounter with antigen were discussed. Here we describe two processes, **somatic hypermutation** and **affinity maturation**, that take place in the GC and contribute to antibody diversity after the antigen encounter. Isotype switching, which contributes to functional diversity rather than antigen-binding diversity, also occurs in the GC.

i) Somatic Hypermutation

The V regions of the IgM antibodies produced by the first plasma cells generated by an activated B cell clone differ slightly in sequence from the V regions of the IgG antibodies produced by later progeny. In the early primary response, the proliferating centroblasts in the GC dark zone undergo a cycle of DNA replication in which their $V_H D_H J_H$ and occasionally their $V_L J_L$ exons are particularly subject to the occurrence of random point mutations, a process called somatic hypermutation (**Fig. 5-9**). These mutations continue to accumulate until late in the primary response, resulting in sequence alterations mainly in CDR1 and CDR2 that do not usually destroy the ability of the antibody to bind to the antigen but often increase its binding affinity. In subsequent exposures to the same antigen (secondary and tertiary responses), the Ig genes of the progeny of activated memory B cells may undergo additional rounds of somatic hypermutation. The rate of point mutation in the V region is estimated to be about 1 change per 1000 base pairs per cell division, about 1000 times the rate of somatic mutation for non-Ig genes (which is why the phenomenon is called somatic *hyper*mutation). Somatic hypermutation requires the activity of an enzyme called *activation-induced cytidine deaminase (AID)* that converts deoxycytidine to deoxyuridine. AID is expressed exclusively in GC B cells in the presence of CD40 signaling.

Fig. 5-9

Somatic Hypermutation and Affinity Maturation

As an antigen-specific B cell clone progresses from the early primary response through subsequent responses, successive rounds of somatic hypermutation lead to the accumulation of point mutations primarily in the CDR1 and CDR2 regions of the VH and VL exons. Selection occurring after each round of somatic hypermutation favors the survival of GC B cells producing antibody that has an increased affinity for antigen based on complementary fit.

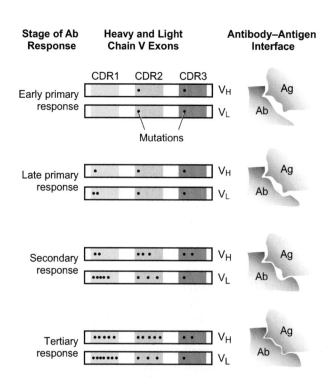

ii) Affinity Maturation

Somatic hypermutation is a random process so that, theoretically, a set of mutations in the antigen-binding site of an antibody could increase, decrease or have no effect on its binding affinity. In reality, the supply of antigen trapped on the FDCs in the GC is limited, so that competition for binding to a particular epitope on an antigen occurs between centrocytes with different somatic mutations in their BCRs. After somatic hypermutation, centrocytes expressing a BCR of very low affinity for the epitope do not succeed in binding to it and thus do not receive a survival signal that rescues them from death by apoptosis. Centrocytes whose somatic mutations confer a BCR with a higher affinity for the epitope are more likely to bind to it and thus receive a survival signal than are centrocytes with BCRs of only moderate affinity for the epitope. In addition, centrocytes with high affinity BCRs are more efficient at internalizing the antigen and presenting pMHC to Th effector cells. This increased antigen presentation means that these B cells also preferentially receive growth stimulatory signals from the Th cells. These factors combine to ensure the survival and proliferation of B cell clones with increased affinity for the epitope. The higher affinity antibodies produced by these successful clones thus predominate later in the response.

Centrocytes with a significant increase in affinity for an epitope after somatic hypermutation preferentially become memory B cells. When these memory B cells are activated in a secondary response, their progeny plasma cells produce antibodies that recognize the same epitope as primary response antibodies but bind to it with increased affinity (hence the name, affinity "maturation"). Those centrocytes with more modest affinities for the epitope tend to become plasma cells.

iii) Isotype Switching

Centrocytes with a high affinity for antigen are also those cells that undergo isotype switching, replacing the original production of IgM antibodies with that of IgG, IgA or IgE antibodies of the same antigenic specificity. Although isotype switching is independent of somatic hypermutation and can occur without it, most B cells that express Igs of new isotypes have already undergone somatic hypermutation.

During isotype switching, the C_H region of the centrocyte undergoes a series of DNA cutting/rejoining events that can bring any of the downstream C_H exons next to the $V_H D_H J_H$ exon previously established by V(D)J recombination. The antigenic specificity of subsequent progeny B cells is the same (because the V exon is unchanged), but this specificity is now linked to a C region that may confer a different effector function. The actual mechanism of isotype switching, called **switch recombination**, is not yet fully understood but requires the same AID enzyme involved in somatic hypermutation. Switch recombination depends on the pairing of highly conserved *switch regions* ($S\mu$, $S\gamma3$, $S\gamma1$, $S\gamma2b$, $S\gamma2a$, $S\varepsilon$ and $S\alpha$ in the mouse) that lie just upstream of each C_H exon (except $C\delta$) (**Fig. 5-10**). Once the signal to switch is received, the DNA is likely looped out such that the selected C_H exon is juxtaposed next to the rearranged $V_H D_H J_H$ exon, the intervening sequences (including unused C_H exons) are deleted, and the DNA is repaired to restore the Ig gene. After switching, the C_H exon closest to the $V_H D_H J_H$ exon is preferentially transcribed, spliced and translated, generating Ig proteins of the new isotype.

Isotype switching cannot occur unless the chromatin of the Ig locus is made more accessible, an event associated with signaling induced by either Td or Ti antigen engagement of the BCR. This increased accessibility allows the initiation of transcription of short stretches of the selected C_H exon *in its germline configuration*. The transcription process exposes the DNA of the two selected switch regions and permits AID to bind to them and deaminate deoxycytidine residues to deoxyuridines. The processing of these deoxyuridines by enzymes of the isotype switching machinery results in double-strand breaks in the switch region DNA. The pairing of the selected switch regions to loop out the intervening DNA is then followed by the ligation of the DNA breaks. As shown in **Table 5-4** for mice, cytokines (particularly TGFβ, IL-4 and IFNγ) can promote or inhibit the initiation of germline transcription of the selected C_H exon and thus the production of Igs of a particular isotype. Isotype switching is therefore heavily influenced by cytokines in the immediate microenvironment of the activated B cell.

Defects in isotype switching cause immunodeficiencies known as Hyper IgM (HIGM) Syndromes. These diseases, which are characterized by normal or elevated IgM but low or absent levels of all other isotypes, are discussed in Chapter 15.

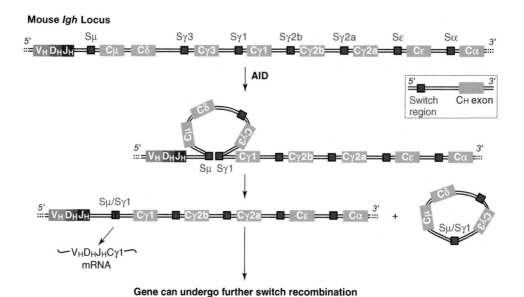

Fig. 5-10
Mechanism of Switch Recombination

In this example of switch recombination from IgM to IgG1 in a mouse B cell, germline transcription (not shown) of short stretches of the Cγ1 exon and the action of AID promote the selection and looping out of the Ig H chain DNA such that the Sμ and Sγ1 switch regions are juxtaposed. The DNA is cut and rejoined at a site within Sμ/Sγ1, excising all C_H exons 5' of Cγ1. The resulting Ig H chain gene lacks the Cμ exon specifying the μ heavy chain but retains the Cγ1 exon specifying the γ1 heavy chain in the correct position for mRNA transcription (3' of the switch site). The $V_H D_H J_H$ exon is spliced to the Cγ1 exon to produce an mRNA that is translated into the γ1 heavy chain, causing the B cell to switch from IgM production to IgG1 production. Note that each daughter of this IgG1-producing B cell may subsequently undergo its own switch, delete one or more of the remaining 3' C_H exons, and start to produce the corresponding IgG2a, IgG2b, IgE or IgA antibodies.

TABLE 5-4	Cytokine Effects on Isotype Switching in Mice		
Cytokine	**Inhibits Isotype Switching to C_H Exon**	**Promotes Isotype Switching to C_H Exon**	**Ig Produced**
IL-4	Cγ2a	Cγ1 Cε	IgG1 IgE
IFNγ	Cγ1, Cε, Cα	Cγ2a Cγ3	IgG2a IgG3-
TGFβ	Cε	Cα Cγ2b	IgA IgG2b

V. Plasma Cell Differentiation

i) Short-Lived Plasma Cells

In a response to a Td antigen, a minority of activated B cells do not experience the GC reaction. When these cells are first activated and positioned on the edge of the primary follicle, they fail to upregulate a transcriptional repressor called Bcl-6 that blocks access to the plasma cell terminal differentiation pathway. As a result, these B cells do not re-enter the follicle and instead immediately differentiate into short-lived plasma cells without undergoing isotype switching or somatic hypermutation. No memory B cells are produced via this pathway. Short-lived plasma cells have a half-life of 3–5 days and secrete only low affinity IgM antibodies. An encounter of a B cell with a Ti antigen also gives rise to short-lived plasma cells because, without the involvement of a Th cell, the activated B cells cannot upregulate Bcl-6. Short-lived plasma cells produced in the spleen are particularly important for the very early stages of the adaptive response against blood-borne Ti antigens.

ii) Long-Lived Plasma Cells

Most of the B cells responding to a Td antigen undergo the GC reaction and make the decision to become either a long-lived plasma cell or a memory cell. A centrocyte in a GC that has been positively selected and undergone isotype switching and somatic hypermutation but later experiences a loss of Bcl-6 function is directed to the plasma cell terminal differentiation path. In the presence of IL-2 and IL-10, these cells first become **plasmablasts** and then long-lived plasma cells that can secrete IgG, IgA and/or IgE antibodies. These mature plasma cells have enlarged ER and Golgi compartments and an increased number of ribosomes. Long-lived plasma cells express little or no CD40, MHC class II or mIg on their cell surfaces; can no longer receive T cell help; and are incapable of cell division. Once generated, these plasma cells migrate from the GCs primarily to the bone marrow but can also take up residence in the medulla of lymph nodes or in the splenic red pulp. In these sites, the plasma cells can produce high affinity antibodies for several months in the absence of any cell division or re-exposure to the original Td antigen. Up to 40% of the total protein synthesized by these mature plasma cells is immunoglobulin, most of which is released into the blood or tissues as secreted antibody.

iii) Mechanism of Antibody Synthesis

Prior to its differentiation, an activated B cell clone can make both the membrane-bound form of its Ig protein to serve in its BCR, or the secreted form of its Ig protein to serve as a circulating antibody. Varying proportions of these forms are produced when progeny of the original activated B cell differentiate into plasmablasts and then mature plasma cells. In the genome of any murine B cell, each C_H exon is followed by three small exons: *Se*, which encodes the Ig tailpiece of a secreted Ig; *M1*, which encodes the transmembrane domain of an mIg; and *M2*, which encodes the cytoplasmic domain of an mIg (**Fig. 5-11**). When the polyadenylation site 3' of M2 is used during transcription, the processed transcripts contain the Se, M1 and M2 sequences. RNA splicing removes the Se sequence but retains M1 and M2, leading to mRNA that is translated into H chains that have transmembrane domains and thus form mIg. When the polyadenylation site 3' of Se is used, only the Se exon is retained at the 3' end of the processed transcript, and the mRNA is translated into an H chain containing a tailpiece; sIg is

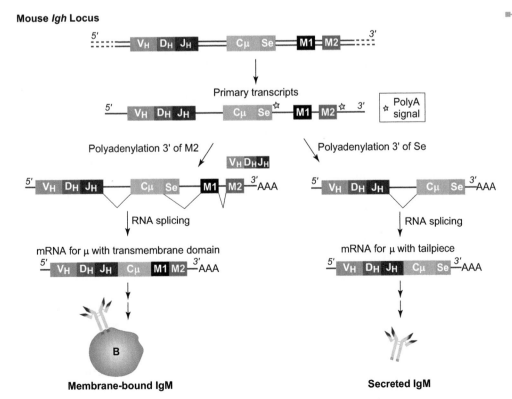

Mouse *Igh* Locus

Primary transcripts

PolyA signal

Polyadenylation 3' of M2

Polyadenylation 3' of Se

RNA splicing

RNA splicing

mRNA for μ with transmembrane domain

mRNA for μ with tailpiece

Membrane-bound IgM

Secreted IgM

■ Fig. 5-11

Membrane versus Secreted Immunoglobulin

In this example, the synthesis of the membrane and secreted forms of an Ig is illustrated for a B cell producing IgM. Like all C_H exons, Cμ is followed by its own Se (tailpiece), M1 (transmembrane domain) and M2 (cytoplasmic domain) exons. Polyadenylation of the primary transcript occurs either 3' of Se or 3' of M2. In the latter case, the M1 sequence is retained after splicing, and the mRNA is translated into the H chain of an mIg; in this case, mIgM. When polyadenylation occurs 3' of Se, RNA splicing produces an mRNA that is translated into the H chain of an sIg; in this case, sIgM.

thus produced. In a resting B cell, the vast majority of H chain transcripts are polyadenylated and processed to produce mIg. However, as the progeny cells of an activated B cell differentiate into plasmablasts, signals are received that cause these cells to preferentially use the other polyadenylation site. The frequency of transcripts containing the transmembrane domain drops precipitously such that the level of mIgM on the cell surface decreases and sIgM is produced in large quantities. Because Se, M1 and M2 exons are associated with each C_H exon, sIgs of all isotypes can be produced after isotype switching. However, the tailpiece sequences that occur in the Cδ exon appear to be seldom used, since IgD is rarely secreted. The same is true in humans, where a very similar mechanism has been identified as producing mIgs and sIgs.

VI. Memory B Cell Differentiation

A GC centrocyte that has been positively selected and undergone isotype switching and somatic hypermutation is directed to the memory cell differentiation path if it receives sustained Bcl-6 signaling. This transcriptional repressor blocks the plasma cell differentiation pathway and forces the centrocyte to become a memory B cell.

i) General Characteristics

Memory B cells resemble naïve B cells in their small size and general morphology but carry different surface markers and have a longer life span. As the primary response terminates, memory B cells often take up residence in a body location where the antigen might next be expected to attack. For example, in response to an antigen first encountered in a lymph node, some of the memory B cells produced remain in the follicular mantle and are ready to react rapidly when a fresh dose of the antigen is conveyed to the lymph node. However, other memory B cells may leave the original lymph node and enter the blood, circulating among the body's chain of lymph nodes and maintaining peripheral surveillance for the antigen. In the case of an antigen first encountered in the spleen, the memory B cells produced during the primary response tend to congregate in the splenic marginal zones, precisely where blood-borne antigens collect. For many different antigens, at least some of the memory B cells produced localize preferentially among the epithelial cells of the skin and mucosae, thereby contributing to SALT and MALT.

ii) Secondary Responses

In a secondary response (**Fig. 5-12**), the activation of memory B cells by antigen occurs in much the same way as in the primary response but is more efficient for several reasons. Firstly, because of their expanded battery of adhesion molecules, memory B cells in the periphery can home to the primary follicles of a lymph node more rapidly than naïve B cells can migrate from the bone marrow. Secondly, due to affinity maturation, the BCR of a memory B cell has an increased affinity for antigen so that the cell is stimulated more easily and efficiently. Thirdly, memory B cells are both present in increased numbers and can act as APCs for memory Th cells, removing the requirement of having to wait for a DC to activate a naïve Th cell. Fourthly, antigen presentation by a memory B cell is associated with faster upregulation of the costimulatory molecules needed for complete activation of Th cells. Fifthly, the progeny of activated memory B cells differentiate into second generation plasma cells that produce antibodies that are already of greater affinity and diversified isotype.

Although no further somatic hypermutation can occur in mature plasma cells, this process can continue in the progeny of the first generation of memory B cell clones in the GCs. Positive selection then favors the survival of second generation memory B cell clones that display even greater affinity for the antigen. As the secondary response succeeds in clearing the antigen, the second generation plasma cells die off, leaving the second generation memory B cells to maintain peripheral surveillance. In a tertiary response, these second generation memory B cells may undergo additional somatic hypermutation and give rise to third generation plasma cells secreting antibodies with even higher affinities,

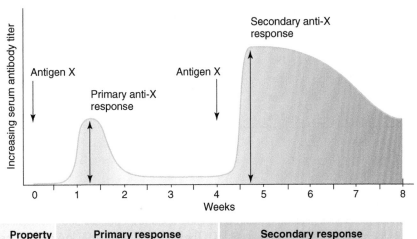

Fig. 5-12
A Comparison of Primary and Secondary Antibody Responses

The graph plots the change in serum antibody concentration after a first and then a second exposure to antigen X. The accompanying chart compares the indicated properties of primary and secondary immune responses.

Property	Primary response	Secondary response
Type of B cell involved	Naive B cell	Memory B cell
Lag time	4–7 days	1–3 days
Time of peak response	7–10 days	3–5 days
Magnitude of peak response	Depends on antigen	100-1000x higher than primary response
Isotype produced	IgM predominates	IgG predominates
Antibody affinity	Lower	Higher

better-suited effector functions, and/or more appropriate physiological localizations. The effectiveness of the immune response will thus be further improved.

iii) Memory Cell Life Span

Some immunologists believe that, after each round of antigenic stimulation, the progeny of memory B cells are more likely to become plasma cells (which die) than new memory cells (which survive). *In vivo*, this could translate into an increased number of effector cells in the secondary and subsequent responses, and a control on the possible overexpansion of one particular memory B cell clone. However, it also means that the host might one day no longer have memory cells of this clone to call upon when the relevant pathogen strikes. In this "decreasing potential hypothesis," immunological memory is ultimately limited. In addition, for reasons that are not yet understood, memory cells specific for different antigens have different life spans. These variations have implications for how frequently a booster shot must be given to ensure complete vaccination against a particular pathogen (see Ch. 14).

C. Effector Functions of Antibodies

The binding of antibody to antigen leads to clearance or destruction of the antigen and protection of the host. Since different antibody isotypes can trigger different effector functions, how an antigen is eliminated often depends on the isotype of the antibodies to which it is bound. Four types of effector functions can be ascribed to antibodies: **neutralization, classical complement activation, opsonization**, and **antibody-dependent cell-mediated cytotoxicity (ADCC)** (**Table 5-5**). Neutralization depends solely on the Fab region of the Ig molecule and so is isotype-independent. Classical complement activation depends on the isotype because complement component C1q binds to the Fc regions of only certain Ig isotypes. For opsonization and ADCC, the Fc region of the antigen-bound antibody must interact with the FcRs present on innate leukocytes

TABLE 5-5	General Characteristics of Antibody Effector Functions	
Effector Function	**Isotype-dependent?**	**FcR-mediated?**
Neutralization	No	No
Complement activation	Yes	No
Opsonization	Yes	Yes
ADCC	Yes	Yes

Box 5-1   Fc Receptors

The physical removal of an antibody–antigen complex often depends on the recognition of the Fc region of the Ig molecule by an FcR. FcRs are expressed on the surfaces of various leukocytes (including neutrophils, macrophages, NK cells and eosinophils) that can act to eliminate the antigen. There are FcRs for all antibody isotypes except IgD. The FcRs binding to IgG, IgA and IgE are the best characterized.

An FcR is named first for the Ig isotype to which it binds, indicated by a Greek letter. For example, the FcγRs bind to the Fc region of IgG antibodies. If major subtypes of an FcR exist, they are denoted by Roman numerals. For example, the three major FcR subtypes that bind to IgG are FcγRI, FcγRII and FcγRIII. These subtypes differ in amino acid sequence, affinity for the various IgG subtypes, cell type distribution and effector functions. Within each FcR subtype, distinct but related receptors are indicated with a Roman capital letter. For example, FcγRIII occurs in two slightly different forms: the FcγRIIIA receptor, which is expressed on NK cells, and the FcγRIIIB receptor, expressed exclusively on neutrophils.

Most FcRs display several Ig-like extracellular domains, a transmembrane domain, and a cytoplasmic domain that often contains ITAMs. Many membrane-bound FcRs are multi-subunit complexes, whereas others are single polypeptide chains. Several FcRs are composed of one subunit that confers Fc region binding specificity plus two other subunits involved in either transport to the cell surface or intracellular signal transduction in response to Ig binding. This intracellular signaling triggers the initiation of phagocytosis or ADCC, and thus destruction of the antigen–antibody complex.

capable of carrying out these effector functions. These interactions are isotype-dependent because different FcR subtypes are expressed only on particular leukocytes and bind only to specific Ig isotypes. FcRs are described further in **Box 5-1**.

I. Neutralization

Neutralization of an antigen is carried out by secreted or secretory antibodies. Certain viruses, bacterial toxins, and the venom of insects or snakes cause disease by binding to proteins on the host cell surface and using them to enter host cells. A neutralizing antibody that can recognize and bind to the virus, toxin or venom can physically prevent it from binding to a host cell protein and triggering internalization, thereby protecting the cell. If preformed neutralizing antibodies exist in a host (due to a previous exposure to a pathogen), the initial spread of the pathogen can be averted. Once an infection is entrenched, however, neutralizing antibodies are no longer sufficient, and the action of CTLs is usually required for successful pathogen elimination (see Ch. 9 and 13).

Antibodies are extracellular and so can protect a cell only from the outside. Once an infectious agent (like a virus) invades a cell, antibodies are no longer effective and cell-mediated mechanisms are required for invader destruction.

II. Classical Complement Activation

A pathogen coated with antibody can initiate the classical pathway of complement activation. In humans, IgM, IgG1, IgG2 and IgG3 are the antibodies best suited for activating complement in this way and so are sometimes called "complement-fixing" antibodies. The binding of antigen to these antibodies opens a site in the Fc region that allows the binding of C1q. However, the C1q molecule must bind simultaneously to *two* C1q-binding sites (and thus two separate Fc regions) for it to activate the cascade (refer to Fig. 3-7). A pathogen must therefore usually be well coated with Ig molecules supplying Fc regions in close proximity before the cascade can commence and a MAC is assembled on the pathogen surface.

III. Opsonization

Opsonization is the process by which an antigen is coated with a host protein (the opsonin) in order to enhance recognition by phagocytic cells such as neutrophils and macrophages (refer to Fig. 3-11). Antibodies are powerful opsonins because phagocytes express FcRs that bind strongly to the Fc regions of particular antibody isotypes. Clathrin-mediated endocytosis of the antigen–antibody complexes is then greatly stimulated. In humans, it is antigen-bound IgG1 and IgG3 that best mediate opsonization.

IV. Antibody-Dependent Cell-Mediated Cytotoxicity (ADCC)

When a pathogen is antibody-coated but too large to be internalized by a phagocyte, ADCC can eliminate the invader. ADCC is carried out by certain leukocytes that express FcRs and have cytolytic capability, such as NK cells, eosinophils, and, to a lesser extent, neutrophils, monocytes and macrophages. Once a target entity (which could be a large bacterium, parasite, virus-infected cell or tumor cell) is coated by antibodies, the Fc regions of these antibodies bind to the FcRs of the lytic cell and trigger its degranulation (**Fig. 5-13**). The release of the hydrolytic contents of the granules in close proximity to the target damages the target's membrane such that its internal salt balance is disrupted and it lyses. NK cells and activated monocytes and macrophages whose FcγRs are engaged also synthesize and secrete TNF and IFNγ, which hasten the demise of the target.

NK cells are the most important mediators of ADCC. In humans, these cells express FcγR molecules that bind to monomeric IgG1 and IgG3 molecules. In general, only targets that are thoroughly coated with IgG will trigger the release of the NK cell's damaging contents. Eosinophils express FcεR and FcαR molecules that can bind to IgE- or IgA-coated parasitic targets, particularly helminth worms. These pathogens are resistant to the cytotoxic mediators released by activated neutrophils and NK cells but are susceptible to the granule contents of eosinophils.

NK cells are discussed in more detail in Chapter 11.

D. Immunoglobulin Isotypes in Biological Context

Each of the five major Ig isotypes has distinct physical properties and biological effector functions that depend on its C-terminal amino acid sequence and carbohydrate content. These isotype-specific properties influence where a given antibody may travel in the body and how it is involved in host defense. The physical and functional properties of human antibody isotypes are described here and summarized in **Table 5-6**. As well as in nature, antibodies are highly useful for many laboratory and clinical applications. Some of these uses are described in Section E at the end of this chapter and in Appendix F.

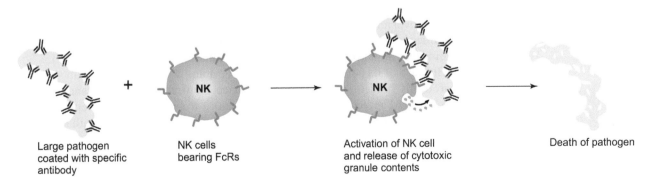

Large pathogen coated with specific antibody

NK cells bearing FcRs

Activation of NK cell and release of cytotoxic granule contents

Death of pathogen

Fig. 5-13
Antibody-Dependent Cell-Mediated Cytotoxicity (ADCC)

Antibodies whose Fab regions are engaged by a large cellular antigen can bind via their Fc regions to the FcRs of a lytic cell, such as the NK cell depicted here. FcR engagement stimulates a lytic cell to release its granule contents in close proximity to the antigen, killing it via membrane destruction. Activated eosinophils, neutrophils, monocytes and macrophages also express abundant FcRs and so can kill pathogens via ADCC.

TABLE 5-6	**Major Physical and Functional Properties of Human Ig Isotypes**			
	IgG	**IgM**	**IgA**	**IgE**
Serum concentration (mg/ml)	3–20	0.1–1.0	1–3	0.0001–0.0011
Half-life in serum (days)	2–4	1	1	<1
Intravascular distribution (%)	45	80	42	50
Secreted form	Monomer	Pentamer	Monomer (IgA1) Monomer, dimer, trimer (IgA2)	Monomer
Classical complement activation	IgG1 ++ IgG2 ++ IgG3 +++ IgG4 –	+++	IgA1 – IgA2 –	–
Placental crossing	+	–	–	–
Involved in allergy	–	–	–	+++

I. Natural Distribution of Antibodies in the Body

The bulk of Ig proteins in the body are present in the form of secretory IgA in the external secretions. These antibodies guard the mucosae where pathogens are likely to attempt entry. Next in relative abundance are the secreted antibodies that circulate throughout the body in the blood. Size considerations dictate that the pentameric sIgM molecule remains primarily in the blood vessels, but that the smaller sIgG, sIgA and sIgE molecules can diffuse freely from the blood into the tissues. No Ig is normally detected in the brain.

II. More About IgM

Monomeric mIgM is always the first Ig produced by naïve B cells that encounter antigen. Early in a primary response, plasma cell progeny of an activated B cell secrete pentameric sIgM exclusively. Because all other antibody isotypes (except IgD) are generated by isotype switching that commences only late in a primary response or not until the secondary response, it is sIgM antibodies that are expressed first in any primary immune response, and those that are synthesized first in a newborn mammal. In an adult human, sIgM antibodies normally comprise only about 5–10% of normal total serum Igs. The detection of increased sIgM levels in an adult indicates exposure to either a novel antigen or a Ti antigen that can activate a B cell to secrete sIgM but does not induce isotype switching. Examples of such Ti antigens are the foreign blood group proteins that might be encountered during a blood transfusion (see Ch. 17).

Because of its pentameric nature, the IgM antibody displays 10 Fab sites that can theoretically bind to a pathogen. In practice, however, steric hindrance usually prevents the IgM molecule from binding to more than five antigenic epitopes at once. Nevertheless, this number is sufficient for the IgM antibody to bind with high avidity to a large antigen or pathogen displaying multiple copies of the same antigenic determinant. IgM is thus able to reduce the infectivity of the pathogen and increase its clearance much more efficiently (using fewer molecules) than a monomeric Ig molecule can. In addition, because the individual binding sites of an IgM molecule are of relatively low affinity, they exhibit correspondingly higher levels of cross-reactivity to related epitopes. This property allows the host to "cast a broad net," maximizing the number of antigens recognized by each IgM-secreting B cell clone.

Pentameric sIgM bound to a pathogen surface is ideally suited for classical complement activation, since multiple Fc regions are already juxtaposed in the pentamer and provide the necessary two C1q-binding sites in close proximity. The classical cascade can thus be triggered by a single molecule of antigen-bound IgM. IgM molecules are also very effective neutralizers and easily prevent pathogens from binding to host receptors on epithelial cells. However, IgM is not an isotype prominent in either opsonization or ADCC because FcRs able to bind to IgM occur only rarely on the surfaces of the appropriate leukocytes.

Although the bulk of IgM is found in the blood, if vascular permeability has been increased during an inflammatory response, sIgM antibodies can exit the blood and enter the tissues to reach sites of infection. In addition, because of the presence of the J chain in the pentameric sIgM molecule, these antibodies can occasionally acquire the secretory component by passage through mucosal or glandular epithelial cells and thus enter the external secretions as SIgM. Although their concentration in the external secretions is very low compared to that of SIgA, SIgM antibodies do make a valuable contribution to mucosal humoral immunity (see Ch. 12).

III. More About IgD

IgD is a monomeric, richly glycosylated Ig that is barely detectable in the blood (0.001% of total serum Ig). It is thought that this sIgD may enhance mucosal immunity (particularly in the respiratory tract) by binding to certain species of pathogenic bacteria and viruses. There is also evidence that sIgD can bind to an unknown receptor on various types of innate leukocytes and stimulate them. Membrane-bound IgD is observed mainly on the surfaces of mature, peripheral B cells that already express mIgM, but a subset of IgM⁻IgD⁺ B cells has been found in the upper respiratory mucosa. The precise function of mIgD is still unknown, but this molecule is capable of sending signals to the B cell nucleus via its associated Igα/Igβ heterodimer. It is thus possible that mIgD either regulates B cell maturation or prolongs the life span of mature B cells in the periphery. Some immunologists speculate that the inherent flexibility of mIgD (due to an extended hinge region) may allow this Ig to bind to antigens featuring epitopes that are widely spaced and cannot be bound by the more rigid mIgM molecule. In any case, mIgD disappears after stimulation of the B cell by antigen.

IV. More About IgG

IgG is the "workhorse" of systemic humoral immunity since it is the isotype most commonly found in the circulation and tissues. In the blood of normal adult humans, 70–75% of serum Ig is monomeric sIgG. The approximate proportions (which vary by individual) of its subclass molecules are IgG1, 67%; IgG2, 22%; IgG3, 7%; IgG4, 4%. IgG is a key opsonin and important for phagocytosis and ADCC exerted by FcγR-bearing phagocytes and lytic cells, respectively. The IgG1 and IgG3 subclasses are particularly good opsonins and mediators of ADCC because FcγRs bind IgG1 and IgG3 antibodies with high affinity. FcγRs generally bind less well to IgG4, and hardly at all to IgG2.

Although not as efficient as sIgM, sIgG is also an important activator of complement. Free IgG molecules display readily accessible Fc regions, but because this Ig is monomeric, it supplies only one C1q-binding site per antibody. Two sIgG molecules must therefore be brought together by mutual binding to antigen in order to furnish two C1q-binding sites in close enough proximity to trigger complement activation. IgG3 is the most efficient complement-fixing IgG subclass, while IgG1 is somewhat less efficient, and IgG2 is even less so. IgG4 is unable to bind C1q and so cannot activate complement at all.

IgG antibodies are unique in their ability to cross the mammalian placenta. The immune system of mammals is not fully developed at birth and is limited in its ability to eliminate microbes. In human infants, although independent IgM synthesis starts at birth, it may take as long as 6–12 months for adequate levels of serum and secretory Igs to be produced. To compensate, evolution has provided protection to fetuses and neonates through **passive immunity**, which is defined as "protection by preformed antibodies transferred to a recipient." In humans, maternal IgG1, IgG3 and IgG4 (but not IgG2) antibodies efficiently cross the placenta and enter the fetal circulation, preparing it for birth when it enters a pathogen-filled environment. Maternal sIgG is detectable in a human infant's blood until about 9 months after birth.

V. More About IgA

More IgA is produced per day in an adult human than all other Ig isotypes combined. In humans, 85–90% of IgA antibodies are produced mainly by plasma cells in the MALT and are found in the secretory form in the body's external secretions. These secretions include tears, saliva, breast milk and prostatic fluid, as well as the mucous secretions of the gastrointestinal, urogenital and respiratory tracts. The remaining 10–15% of IgA antibodies circulate in the blood and are produced by plasma cells in the lymph nodes, bone marrow and spleen. Most sIgA antibodies in the blood are monomeric, but 20% occur as multimers (up to hexamers). Memory B cells located in the diffuse submucosal lymphoid tissues, the Peyer's patches, and the tonsils produce large quantities of IgA antibodies because cytokines secreted by Th cell subsets in these MALT microenvironments promote successive switching to the IgA isotype.

Secretory IgA antibodies are of enormous importance because they facilitate antigen removal right at the mucosal surface, the most common site of initial pathogen attack. Neutralization is the predominant effector mechanism used by SIgA antibodies. Accordingly, these Igs are often said to have "antiviral" activity, since the polymeric nature of this antibody allows it to easily bind repeating epitopes on virus particles, impeding their attachment to mucosal cell surfaces. The expulsion of these particles from the body is then carried out by mechanical means, such as the movement of mucus by the undulating cilia of the respiratory tract (see Ch. 12). Because IgA antibodies are poor complement fixers and opsonins, the delicate mucosae are protected from potential damage caused by the inflammation associated with the activation of phagocytes, lytic cells or complement. Secretory IgA also contributes to the passive immunity that protects mammalian neonates, since this antibody is passed along in breast milk to defend the neonatal gut mucosa. Secreted IgA is useful for the elimination of helminth worms, because sIgA-coated parasites can be dispatched by ADCC carried out by eosinophils bearing FcαR.

VI. More About IgE

Monomeric sIgE is present in the serum at the lowest concentration of all isotypes, a mere 0.000003% of the total Igs in a healthy human. IgE antibodies do not cross the placenta, cannot fix complement and do not function as opsonins. Nevertheless, IgE antibodies have a clinical impact that far outweighs their actual numbers and limited effector functions. Firstly, IgE is essential for combatting large parasitic worms. Th cells responding to these invaders secrete cytokines that influence activated B cells to undergo isotype switching to IgE. The serum concentration of IgE rises dramatically, and antigen-specific IgE molecules coat the surface of the worm. The presence of the IgE permits the recognition of the invader by FcεR-bearing eosinophils, the only lytic cell type competent to destroy these pathogens. Secondly, sIgE antibodies are responsible for the symptoms experienced in allergic reactions such as hay fever, and more severe conditions such as asthma and anaphylactic shock. These disorders are all manifestations of a type of immune reaction called **immediate hypersensitivity**. The symptoms and causes of this and other types of immune hypersensitivity are discussed in Chapter 18.

E. Sources of Laboratory and Clinical Antibodies

Scientists interested in dissecting a biological system often take advantage of the properties of the antigen–antibody bond because this interaction is highly specific. Where standard techniques may not be able to distinguish between very closely related molecules, specific antibodies for distinct epitopes on those molecules can do so with ease. Identification and purification of a single component from a complex mixture becomes a ready possibility because an antibody can detect one antigenic molecule among 10^8 other molecules. In addition, the antigen–antibody interaction is reversible and does not alter the antigen. For these reasons, techniques employing antibodies are used to purify, characterize and quantitate antigens, and to pinpoint their expression in cells or tissues. In the clinic, doctors use antibodies as diagnostic tools and as immunotherapeutic agents. Below we discuss various sources of antibodies that are used for experimental work and clinical applications. Brief descriptions of laboratory techniques employing antibodies and several figures illustrating these approaches are included in Appendix F. Clinical applications of antibodies are discussed in context in Chapters 14–20.

I. Antisera

Serology is the study of antibodies present within a given **antiserum**—the clear liquid serum fraction of clotted blood obtained from an individual who has been immunized or exposed to a foreign substance or infectious agent. Commercial antisera are often produced by immunizing animals such as rabbits or goats with an antigen of interest. An antiserum is first tested for its **titer** (relative concentration of antigen-specific antibodies) by serially diluting samples of the antiserum until binding to specific antigen can no longer be detected. An antiserum that can be diluted extensively and still shows binding activity is said to have a "high titer" of antigen-specific antibodies. Often an antiserum can be used without further purification but, if necessary, non-Ig proteins can be removed from the antiserum by biochemical methods.

Antisera are *polyclonal*, meaning that, when an animal is exposed to an antigen, many B cell clones respond to the antigen's entire collection of epitopes. A plethora of different antibodies specific for different epitopes of the antigen is produced, with each specificity present in a relatively small quantity. This mixture is an advantage to an organism *in vivo* because it offers multiple ways to attack a pathogen. Similarly, a researcher will use an antiserum to identify an antigen as a whole (as opposed to one particular epitope of that antigen) because, even if some epitopes on the antigen have been denatured during handling or altered due to mutation, the mixed population of antibodies in the antiserum will still likely contain at least some antibodies capable of binding to the antigen. However, the heterogeneity of an antiserum is a problem if the objective is to examine antibody binding to one specific epitope. Removal of undesired antibodies to other epitopes is time-consuming, expensive, less than 100% effective (leaves cross-reacting antibodies behind), and can result in a significant decrease in the concentration of the desired antibody. In addition, antisera vary in composition and titer even among inbred animals, and even when the same protocol is followed for the preparation of different batches (see **Box 5-2**). It is for these reasons that researchers developed *monoclonal antibodies*.

II. Monoclonal Antibodies

i) Hybridomas

An enormous technical breakthrough occurred in the 1970s when immunologists discovered how to derive antibodies of a single defined specificity from a clone of antibody-producing cells that could live indefinitely in culture. These cells were called **hybridomas** because they resulted from the hybridization of two cell types: an antibody-secreting B cell and a **myeloma**. Myelomas are B cell cancers that arise from the malignant transformation of a single plasma cell. A myeloma clone secretes antibodies of a single specificity like any B cell clone. However, unlike a normal plasma cell clone that dies after a few days, a myeloma clone has a virtually infinite life span: it is said to be "immortal." This immortality means that unlimited numbers of antibody-producing

Box 5-2   Innovative Production of Antisera Using Chicken Eggs

Since the early 1980s, scientists have known that an unusual Ig isotype called IgY is produced in birds and some reptiles. IgY is functionally analogous to mammalian IgG but differs from it structurally. IgY is the major antibody produced by the B cells of domestic birds, of which the best studied is the chicken. IgY secreted into a chicken's blood eventually makes its way into the yolk of the chicken's eggs, where it accumulates to a very high concentration. This high concentration is a boon to researchers because a chicken can be immunized against a wide variety of pathogens, including those able to infect humans. Large quantities of IgY antibodies raised against an immunogen can then be purified from egg yolks with relative ease. A single hen lays about 325 eggs per year, and the average egg contains 60 milligrams of IgY protein, meaning that about 20 grams per year of IgY can be recovered from one hen. This is a much greater amount of antibody protein than can be recovered from the serum of an immunized rabbit. In addition, chickens produce higher antibody levels after fewer immunizations than do rabbits (meaning the animal endures fewer injections), and the antibody collection procedure is much easier on all concerned (egg collection vs. extensive blood collection). These advantages make it practical and desirable to carry out large-scale production of both therapeutic and diagnostic IgY antibodies in chickens and other bird species. For example, IgY antibodies that can detect and neutralize pandemic H1N1 influenza virus have been produced cost-effectively in ostrich eggs, and it has been suggested these antibodies might be applied as an extra layer of protection to masks and filters designed to prevent inhalation of the virus.

The reader can learn more about the potential value of IgY antibodies for human therapy in a report by Tsukamoto, M. *et al.* (2011) *Molecular Medicine Reports 4*(2), 209–214. To see how researchers immunize chickens and collect IgY, the reader may want to visit the Video Article on IgY production by Pauly *et al.* (2011) in the *Journal of Visualized Experiments*, May 1 (51).

cells can be grown in culture and manipulated in the laboratory. However, the antigenic specificities of myeloma antibodies are for the most part unknown because the B cells involved are not selected with any defined experimental antigen. A hybridoma is created by artificially fusing the plasma membrane of a myeloma cell with the plasma membrane of an isolated B cell of *known* antigenic specificity, such that the cells combine to form a **heterokaryon** (one cell containing two different nuclei) (**Plate 5-2**). The nuclei eventually fuse but then eliminate genetic material during subsequent cell divisions until a normal chromosome complement is restored, stabilizing the nucleus. As a result, a hybridoma has the immortality and production capacity of the myeloma but the known antibody specificity of the B cell. The hybridoma grows to form a colony secreting large quantities of **monoclonal antibody** (mAb); that is, antibody derived from a single, defined B cell clone.

In the laboratory, mAbs are typically employed to identify a specific protein marker on a cell surface or in a tissue or serum sample, or to map individual epitopes on an antigen. Large quantities of mAbs can be used to purify proteins to be used in research studies, or in industrial or clinical applications. Because hybridomas can be clonally expanded and maintained indefinitely, they provide a permanent and uniform source of antibody. However, because mAbs recognize only a single determinant, a virus

Plate 5-2
Heterokaryon Formation

Fluorescence microscopy shows a single orange-staining membrane surrounding the cytoplasm of two fused cells. The separate nuclei are stained in blue. These two nuclei will later fuse into the single nucleus of a hybrid cell. *[Reproduced with permission from Gottfried E. et al. (2002) Characterization of cells prepared by dendritic cell-tumor cell fusion. Cancer Immunity 2, 15.]*

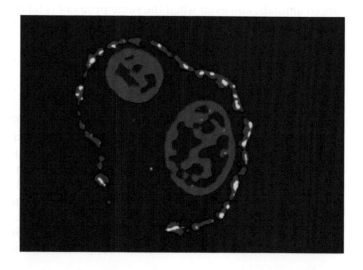

that manages to mutate that precise epitope can escape detection by a mAb where it would not escape detection by a polyclonal antiserum. In addition, mAbs are not ideal for tests based on the detection of large immune complex networks formed between multiple antigen and antibody molecules (see Appendix F). A mAb recognizes only a single epitope, which may be represented only once on an antigen molecule that occurs naturally as a monomer. Thus, networking between multiple antigen and antibody molecules may not occur. It should also be noted that a mAb will be of a single defined isotype that will give it isotype-specific characteristics. These characteristics may make the mAb suitable for use in some applications but not others.

ii) Humanized mAbs

Monoclonal antibodies intended for immunotherapies must usually be **"human-ized"** prior to use. Technically, it is far easier to produce an mAb of the desired specificity in a mouse than in a human. However, a human's immune system will normally mount a response to mouse-specific epitopes on a mouse mAb, decreasing its effectiveness. To avoid this anti-mouse response, a "humanized" mAb is produced by isolating the gene encoding a mouse mAb of the desired specificity and using genetic engineering techniques to combine the sequences encoding the specificity-defining V region of the mouse mAb with the sequences encoding the structural elements of a human Ig. The framework regions of the V region of the Ig may also be "humanized." Such manipulations create a chimeric gene that is translated *in vitro* to yield a mouse/human antibody protein in which most of the mouse Ig sequence has been replaced with human Ig sequence. The patient's immune system therefore does not recognize the humanized mAb as foreign and does not generate large quantities of anti-mAb antibodies that would clear the therapeutic mAb before it could do its job.

iii) mAb Production Using Phage Display

Another approach to deriving mAbs more suitable for use in humans involves artificially constructing soluble Fab fragments. Unlike full-length antibody proteins, Fab fragments penetrate tissues very efficiently and do not have to be processed through the ER. Signal sequences can be added to direct an Fab to an extracellular or intracellular location. Fabs can be used in an unmodified form to neutralize drugs or may be coupled to radioactive markers to visualize tumors. Fabs are usually created by "phage display," which mimics the immune system's selection of antigen-specific antibodies.

A phage display library of Fabs is constructed by joining each of the V exons of a collection of mature B cells to a gene encoding a coat protein of a bacteriophage (a virus that infects bacteria). Each bacteriophage then expresses on its surface a fusion protein consisting of the V domain of the mAb joined to the N-terminus of the coat protein. Selection of the bacteriophage expressing the desired Fab fragment is carried out by "panning" the entire recombinant phage library over the immobilized antigen of interest. Low affinity Fabs that do not bind are simply washed away, and successive rounds of selection ensure that the Fabs that remain are of the highest binding affinity. Bacteria are then infected with the selected phage, which multiplies to yield an abundance of the desired Fab fragment.

A drawback to this approach is that a new phage library must be constructed for every antigen, a time-consuming process. A more clinically significant problem is that Fabs are not full-length antibodies and lack the C region responsible for effector functions. In addition, Fabs produced in bacteria are not glycosylated, leading to a much shorter half-life in a patient's blood. Nevertheless, phage display allows a specific Fab fragment free of mouse C region sequences to be produced completely outside a natural host.

iv) mAb Production Using Plants

It may surprise the reader to learn that plants can also be used to produce mAbs for use in humans. These "plantibodies" have an advantage over phage display Fab fragments in that they are full-length Ig proteins containing the C region necessary for antibody effector functions. Also, because plants are eukaryotic organisms (unlike bacteria), the glycosylation patterns of plantibodies closely resemble those of mAbs produced by hybridomas, allowing them to persist for lengthy periods in a patient's blood. Plantibodies are produced by introducing the entire Ig gene encoding the mAb of interest into the genome of the plant, utilizing techniques routinely used for creating **transgenic** organisms (organisms carrying foreign genes in their DNA). The resulting transgenic plants usually express the human mAb quite stably and without disrupting their own metabolism. As a result, very large quantities of mAb can be produced very cheaply, and the transgenic seeds produced by these plants can be stored almost indefinitely.

This concludes our discussion of B cells and the humoral response. The next two chapters deal with the MHC and antigen processing and presentation. An appreciation of these elements of the T cell adaptive response is necessary for a complete understanding of T cell activation, the lynchpin of both humoral and cell-mediated adaptive immunity.

Chapter 5 Take-Home Message

- B cell maturation proceeds from HSCs through the MPP, CLP, pro-B, pre-B, immature B, transitional B and mature B cell stages. Most of this development occurs in the bone marrow and is independent of antigen.

- Negative selection in the bone marrow removes B cells expressing potentially autoreactive BCRs and establishes central B cell tolerance.

- B cell differentiation takes place in secondary lymphoid tissues and involves the activation of mature B cells by antigen and the generation of memory B cells and antibody-secreting plasma cells.

- Ti-1 antigens contain mitogenic regions and are polyclonal activators. Ti-2 antigens are large polymeric molecules with repetitive structures or subunits capable of cross-linking mIg. Ti antigens activate B cells in the absence of T cell help.

- Td antigens are protein-containing macromolecules that supply both B and T cell epitopes. Complete B cell activation by a Td antigen requires signal 1, antigen binding to BCRs; signal 2, costimulation supplied by an activated Th effector cell specific for the same antigen; and signal 3, Th cell-derived cytokines. Signals 2 and 3 constitute T cell help.

- T cell help is required for somatic hypermutation, affinity maturation, isotype switching and memory B cell differentiation.

- Somatic hypermutation, affinity maturation and isotype switching occur in the GCs and are responsible for antibody diversification after antigen encounter.

- The major effector functions of antibodies are neutralization, classical complement activation, opsonization and ADCC.

- Different Ig isotypes are best suited for different effector functions because an antibody's isotype defines its structure. Antibody structure determines if an Fc region can activate complement via C1q interaction, or engage FcRs expressed by cells mediating opsonized phagocytosis or ADCC.

- Polyclonal antibodies are easy to produce but are of mixed specificity and cross-reactive. Monoclonal antibodies are of single specificity and are often produced *in vitro* from B cell hybridomas. Phage display and plantibody techniques allow large-scale production of customized mAbs. Antibodies are often "humanized" before use as diagnostic tools or as immunotherapeutics.

Did You Get it? A Self-Test Quiz

Section A

1) Can you define these terms? stromal cell, pre-BCR

2) Give two major differences between the maturation and differentiation phases of B cell development.

3) At what B cell stage does V(D)J recombination commence?

4) What is the surrogate light chain, and why is it necessary?

5) What is receptor editing, and why is it useful?

6) Distinguish between negative and positive B cell selection in the bone marrow.

7) Describe at the DNA level how IgM and IgD can be coexpressed by a B cell.

8) To which tissues do immature naïve B cells migrate upon exiting the bone marrow?

Section B.I–III

1) Can you define these terms? T cell help, BCR cross-linking, molecular complexity, conformational determinant, immunodominant, high zone tolerance, parenteral

2) Distinguish between antigens and immunogens.

3) Distinguish between Ti-1 and Ti-2 antigens.

4) How are Ti antigens physically different from Td antigens, and what effect does this have on their immunogenicity?

5) What role do costimulatory molecules play in B cell activation?

6) How do the antibodies produced in response to Ti and Td antigens differ?

7) Describe five properties of a protein that affect its ability to be a Td immunogen.

8) Can you define these terms? B–T cooperation, B–T conjugate, cognitive B cell, receptive B cell, anergic B cell

9) Why does a B cell need a Th cell specific for the same antigen for complete activation?

10) Describe the three signal model of B cell activation.

11) How does linked recognition increase the efficiency of B cell activation?

12) Can you define these terms? centroblast, centrocyte, follicular mantle, GC light zone

13) How does a primary lymphoid follicle become a secondary follicle?

14) What is the germinal center reaction?

Section B.IV–VI

1) Can you define these terms? AID, switch recombination

2) How does somatic hypermutation diversify antibody specificity?

3) Why is affinity maturation useful to the immune response?

4) Distinguish between positive and negative B cell selection in the GC.

5) How does isotype switching diversify antibodies?

6) Give two examples of how cytokines influence isotype switching.

7) Give three differences between short-lived and long-lived plasma cells.

8) Describe at the DNA level how Ig synthesis switches from mIg to sIg production.

9) Why is Bcl-6 important for memory B cell differentiation?

10) Give three reasons why secondary responses are faster and stronger than primary responses.

Section C

1) How do FcRs function to facilitate opsonized phagocytosis and ADCC?

2) How does neutralization by an antibody protect a cell?

3) What human antibodies are best suited for classical complement fixation and why?

4) What human antibodies are best suited for opsonized phagocytosis and why?

5) Describe how ADCC is important for defense against helminth worms.

Section D

1) Which antibody isotype occurs in the largest amount in the body and in what location?

2) What isotypes are the first antibodies produced in a primary response? In a newborn?

3) Why is IgM well suited for classical complement activation?

4) What are some possible functions of IgD?

5) Why is IgG well suited for opsonization and ADCC?

6) What is passive immunity, and how can it protect an infant?

7) Why have IgA antibodies evolved to be poor complement fixers?

8) Which antibody isotype is prominent in allergy?

Section E

1) What is the basis for the laboratory field known as serology?

2) What is a titer?

3) Differentiate between "polyclonal" and "monoclonal" antisera.

4) What is a hybridoma? What is phage display? What is a plantibody?

5) Discuss the advantages and disadvantages of each of the antibody production approaches listed in question 4.

6) Why is it necessary to "humanize" mAbs for therapeutic use?

Can You Extrapolate? Some Conceptual Questions

1) Suppose you have two antibody preparations available that have isotype-specific binding patterns: "Ab1" binds to IgM only, whereas "Ab2" binds to IgD only. Which preparation(s) could you use to distinguish between

   a) early pre-B cells and transitional type 1 B cells?

   b) transitional type 1 and transitional type 2 B cells?

   c) late pro-B cells and mature, naïve B cells?

2) Small protein molecules much less than 4 kDa in size are known in the immunological context as haptens. When injected on their own into an animal, haptens are usually unable to stimulate a Td antibody response even in the presence of substances known as adjuvants (see Ch. 14) that provide the DAMPs necessary for leukocyte activation. However, when the hapten is chemically attached to a much larger carrier protein, injection of the hapten-carrier complex (plus adjuvant) results in a strong antibody response to the hapten (as well as to epitopes on the carrier protein itself).

   a) Why might the hapten–carrier complex elicit a strong anti-hapten response even though the hapten alone is an ineffective immunogen?

   b) If an animal is given a primary immunization with hapten complexed to carrier A, a subsequent secondary response to the hapten is seen only if hapten–carrier A is used for the secondary immunization. Secondary immunization with hapten complexed to carrier B generates only a primary level of antibody response to the hapten. How would you account for this observation?

3) A clone of B cells in an immunized mouse is currently producing antibodies of the IgG2b isotype. Do you expect any daughter cells from this clone to later produce

   a) IgA? Explain.

   b) IgG1? Explain.

Would You Like To Read More?

Harwood, N. E., & Batista, F. D. (2010). Early events in B cell activation. *Annual Review of Immunology*, 28, 185–210.

Herzog, S., Reth, M., & Jumaa, H. (2009). Regulation of B-cell proliferation and differentiation by pre-B-cell receptor signalling. *Nature Reviews Immunology*, 9(3), 195–205.

Klein, U., & Dalla-Favera, R. (2008). Germinal centres: Role in B-cell physiology and malignancy. *Nature Reviews Immunology*, 8(1), 22–33.

Kurosaki, T., Aiba, Y., Kometani, K., Moriyama, S., & Takahashi, Y. (2010). Unique properties of memory B cells of different isotypes. *Immunological Reviews*, 237(1), 104–116.

Lee, S. J., Chinen, J., & Kavanaugh, A. (2010). Immunomodulator therapy: Monoclonal antibodies, fusion proteins, cytokines, and immunoglobulins. *Journal of Allergy & Clinical Immunology*, 125(2 Suppl 2), S314–323.

Liu, X. Y., Pop, L. M., & Vitetta, E. S. (2008). Engineering therapeutic monoclonal antibodies. *Immunological Reviews*, 222, 9–27.

Mandel, E. M., & Grosschedl, R. (2010). Transcription control of early B cell differentiation. *Current Opinion in Immunology*, 22(2), 161–167.

Phan, T. G., Gray, E. E., & Cyster, J. G. (2009). The microanatomy of B cell activation. *Current Opinion in Immunology*, 21(3), 258–265.

Vinuesa, C. G., Linterman, M. A., Goodnow, C. C., & Randall, K. L. (2010). T cells and follicular dendritic cells in germinal center B-cell formation and selection. *Immunological Reviews*, 237(1), 72–89.

Xu, Z., Hong, Z., Pone, E. J., Mai, T., & Casali, P. (2012). Immunoglobulin class-switch recombination: Induction, targeting and beyond. *Nature Reviews Immunology*, 12, 517–531.

The Major Histocompatibility Complex

Near or far, hiddenly,
To each other linked are,
That thou canst not stir a flower
Without troubling a star.

Francis Thompson

A s was introduced in Chapter 2, recognition of antigen by αβ T cells is more complex than antigen recognition by B cells. While the BCR binds directly to a unitary epitope on a pathogen or foreign macromolecule, another host cell must "present" antigens derived from the pathogen to T cells. The epitope recognized by an αβ T cell's TCR is a peptide derived from a protein antigen displayed on a cell surface molecule encoded by one of the genes of the **major histocompatibility complex (MHC)**.

A. Overview of the Major Histocompatibility Complex

Proteins encoded by the MHC were originally discovered in the 1930s during studies of tissue rejection in transplantation experiments. These proteins were therefore named for their association with **histocompatibility** (*histo*, meaning "tissue," and *compatibility*, meaning "getting along"). The genes controlling the histocompatibility of tissue transplantation were localized to a large genetic region containing multiple loci; hence, the term "complex." Moreover, the proteins encoded by these genes were found to have dramatic effects on histocompatibility. To distinguish these proteins from other molecules (encoded elsewhere in the genome) that had relatively minor effects on histocompatibility, these molecules were called the "major" histocompatibility molecules. Thus, the genes encoding these proteins were dubbed the "major histocompatibility complex" (MHC) genes. Soon after, it was discovered that MHC-controlled rejection of transplanted tissue was due to the mounting by the transplant recipient of an immune response against the donated cells (see Ch. 17). Although this finding implied that MHC gene products were directly involved in immune responses, it took several more decades for immunologists to define the normal physiological role of MHC-encoded proteins in presenting antigenic peptides to T cells.

MAK: Primer to the Immune Response. http://dx.doi.org/10.1016/B978-0-12-385245-8.00006-6

Fig. 6-1
Recognition of MHC Class I and II Molecules by T Cells

(A) An MHC class I protein contains a large transmembrane α chain associated with the smaller β2m chain. **(B)** The TCR of a CD8⁺ T cell binds to a peptide-MHC class I complex on a nucleated host cell, while its CD8 coreceptor binds to another site on the same MHC class I molecule. **(C)** An MHC class II protein contains two large transmembrane chains: α and β. **(D)** The TCR of a CD4⁺ T cell binds to a peptide-MHC class II complex on an APC, while its CD4 coreceptor binds to another site on the same MHC class II molecule.

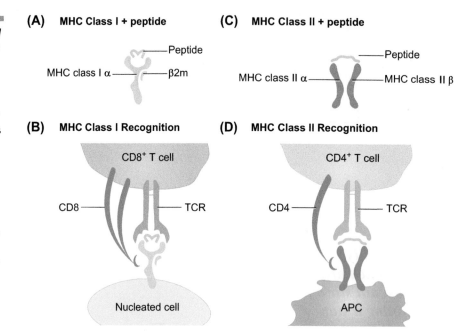

TABLE 6-1	**Chromosomal Location of MHC Class I and II Genes**	
Protein Encoded	**Chromosome**	
	Human	Mouse
MHC class I α chain	6	17
β2-microglobulin	15	2
MHC class II α chain	6	17
MHC class II β chain	6	17

The MHC-encoded proteins that are involved in most instances of antigen recognition by T cells are the **MHC class I** and **MHC class II** molecules. The TCRs of CD8⁺ T cells recognize peptides bound to MHC class I, while the TCRs of CD4⁺ T cells recognize peptides bound to MHC class II (**Fig. 6-1**). As is described in detail in Chapter 8, the CD8 coreceptor of CD8⁺ T cells also binds to MHC class I, while the CD4 coreceptor of CD4⁺ T cells binds to MHC class II. The MHC class I protein is a heterodimer consisting of a large transmembrane α chain non-covalently linked to a small non-transmembrane chain called **β2-microglobulin (β2m)**. The MHC class I α chain is encoded within the MHC, but β2m is not (**Table 6-1**). The MHC class II protein is composed of an α chain and a slightly smaller β chain, both of which are transmembrane proteins encoded by genes in the MHC. Despite this difference in composition, the tertiary structures of MHC class I and class II molecules are highly similar, apart from the peptide-binding groove. While almost all nucleated cells express MHC class I, only the few cell types that function as APCs (including DCs, macrophages and B cells) express MHC class II. Thus, almost any cell can serve as a target cell and present antigen to CTLs derived from CD8⁺ Tc cells, but only APCs can activate CD4⁺ Th cells.

I. HLA Complex

In the human genome, the MHC is called the **HLA complex** (for human leukocyte antigen complex). The HLA complex covers about 3500 kb on chromosome 6 and contains 12 major regions, as shown in **Figure 6-2A**. Each region contains dozens of genes, only some of which are functional and many of which do not appear to be involved in antigen presentation. The HLA-A, HLA-B and HLA-C regions are all MHC class I regions. Each contains a single functional gene encoding a human MHC class I α chain.

(A) Human Leukocyte Antigen (HLA) Complex

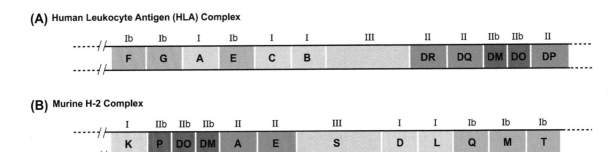

(B) Murine H-2 Complex

Fig. 6-2
General Organization of the MHC in Humans and Mice

Schematic representation of chromosomal regions in which MHC class I, Ib, II, IIb and III genes are found. **(A)** The human leukocyte antigen (HLA) complex on chromosome 6. Note that regions in the HLA complex containing MHC class III genes do not have letter names. **(B)** The murine MHC (H-2) complex on chromosome 17. *[Source: http://imgt.cines.fr/.]*

The DP, DQ and DR regions are all MHC class II regions. Each contains multiple functional genes encoding both MHC class II α and β chains. The single genes within each of the HLA-E, -F and -G regions encode **MHC class Ib** proteins, while several genes in the DM and DO regions encode **MHC class IIb** proteins. MHC class Ib and IIb proteins structurally resemble MHC class I and II proteins, respectively, but are not directly involved in routine antigen presentation to T cells. MHC class Ib and IIb proteins are therefore considered to be "non-classical" MHC molecules (see **Box 6-1**). The **MHC class III** region is not known to encode any peptide-binding presentation molecules but contains many genes relevant to immune responses, including those encoding complement components, HSPs and the cytokines TNF and **lymphotoxin (LT)**.

Box 6-1 Human Non-classical and MHC-like Genes and Proteins

The "non-classical" MHC class Ib and IIb molecules resemble the classical MHC proteins in structure but generally do not present peptides to αβ T cells (see table below). The class Ib genes in humans are located in the HLA-E, -F and -G regions. Some class Ib gene products are secreted (unlike the products of classical MHC class I genes), while others are membrane-bound. Two MHC class Ib proteins called HLA-E and HLA-F may function in antigen presentation to γδ T cells (see Ch. 11). HLA-G is expressed in placental cells during fetal development and may contribute to the prevention of maternal immune responses against the fetus (see Ch. 10). A gene called HFE, which is located in the HLA-E region, encodes an MHC class I-like protein that associates with β2m but does not have a peptide-binding groove. HFE appears to be involved in iron absorption, such that when HFE is defective, excessive iron is deposited in various organs. Two MHC class Ib proteins called **MICA** and **MICB** closely resemble MHC class I molecules in structure but are stress-induced molecules. The binding of MICA to a particular receptor on NK and γδ T cells can stimulate these cells (see Ch. 11). The human MHC class IIb proteins are located in the DO and DM regions. The HLA-DM protein regulates the loading of antigenic peptides onto MHC class II molecules (see Ch. 7), while the HLA-DO protein inhibits HLA-DM activity.

The **CD1** proteins are encoded *outside* the MHC loci but share certain structural similarities and functions with classical MHC molecules. These "MCH-like" CD1 proteins associate with β2m but have binding grooves that are very hydrophobic. This type of groove preferentially binds fragments of lipid and glycolipid antigens. Certain T lineage cells can be activated by this non-peptidic form of antigen presentation (see Ch. 7).

	Class I	Class II	Class Ib	Class IIb	Class III	MHC-like
Gene encoded in MHC region	Yes	Yes	Yes	Yes	Yes	No
Polypeptides	Class I α plus β2m	Class II α plus Class II β	Class I α-like plus β2m	Class II α-like plus Class II β-like	Neither class I nor class II chains	Non-MHC chains plus β2m

	Class I	Class II	Class Ib	Class IIb	Class III	MHC-like
Tissue expression	Almost ubiquitous	APCs	Restricted	APCs	Almost ubiquitous	Restricted
Soluble form?	Very rare	No	Some	Some	Yes	Some
Polymorphism	Extreme	Extreme	Limited	None	None	None
Function	Peptide presentation to CD8+ T cells	Peptide presentation to CD4+ T cells	Stimulation of γδ T or NK cells; fetus protection; iron absorption	Peptide loading of MHC class II	Complement components, inflammatory cytokines, heat shock and stress proteins	Lipid antigen presentation to T lineage cells
Examples	HLA-A HLA-B HLA-C	HLA-DP HLA-DQ HLA-DR	HLA-E HLA-F HLA-G HFE MICA MICB	HLA-DM HLA-DO	C4 TNF HSP70	CD1 isoforms

II. H-2 Complex

In the mouse genome, the MHC is known as the **H-2 complex**. The H-2 complex is spread over 3000 kb on chromosome 17 and contains 12 major regions, as shown in **Figure 6-2B**. The K, D and L regions contain single functional genes that encode mouse MHC class I α chains. The A and E regions each contain a single functional gene encoding an MHC class II α chain, and one or more functional genes encoding an MHC class II β chain. The S region of the H-2 complex contains genes encoding the MHC class III proteins, again including complement proteins, HSPs, TNF and LT. The Q, T and M regions of the H-2 complex contain genes encoding class Ib proteins, whereas class IIb proteins are encoded by genes in the P, DO and DM regions.

B. MHC Class I and Class II Proteins

The MHC class I and class II proteins are heterodimeric molecules composed of an extracellular N-terminal peptide-binding region, an Ig domain-containing extracellular region, a hydrophobic transmembrane region, and a short C-terminal cytoplasmic region. The structure of the peptide-binding region in both MHC class I and class II molecules is such that the affinity of an MHC molecule for peptide is much lower than that of an antibody for its cognate antigen. This relaxed binding is a necessity if a given MHC molecule is to carry out its task of presenting a wide range of peptides for T cell perusal. It should also be noted that a given peptide may be capable of binding to different MHC class I or class II molecules, a phenomenon known as "promiscuous binding."

I. MHC Class I Proteins

Most nucleated cells sport a mixed population of MHC class I proteins. The peptides bound by these MHC class I molecules are generally of **endogenous** origin; that is, they are derived from the degradation of proteins synthesized within the cell. The vast majority of these peptides will be "self" in nature, because most proteins routinely produced within a cell at any one time are of host origin (as opposed to proteins of

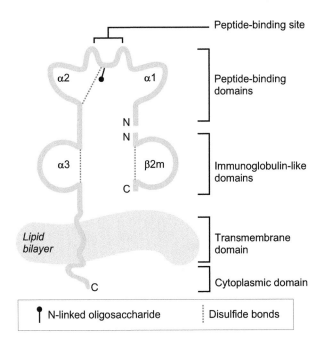

non-self origin, such as those generated during a viral infection). The MHC class I molecule does not discriminate among "self" and "non-self" peptides; that job is left to the TCRs of CD8+ T cells. Self peptide–MHC complexes do not trigger an immune response because T cells with the corresponding specificity are generally absent from the T cell repertoire due to the establishment of central tolerance (see Ch. 9). In contrast, non-self peptides complexed to MHC class I are recognized and trigger CD8+ T cell activation.

How peptides are produced from antigens and loaded into the peptide-binding grooves of MHC class I or II molecules is described in Chapter 7.

i) MHC Class I Component Polypeptides

In both mice and humans, MHC class I α chains are glycoproteins of about 44 kDa and contain three extracellular globular domains (**Fig. 6-3**). Domains α1 and α2 at the N-terminal end of the chain non-covalently pair with each other to form the peptide-binding site, while the Ig-like α3 domain associates non-covalently with the β2m polypeptide. The α chain also supplies the transmembrane domain and the cytoplasmic domain. The α1 domain maintains its shape without disulfide linkage, but the α2 and α3 domains each have an internal disulfide bond. The other partner in the MHC class I molecule, the β2m protein, is a non-transmembrane polypeptide of about 12 kDa. β2m resembles a single Ig-like domain and, through its association with the MHC class I α3 domain, helps to maintain the overall conformation of the MHC class I molecule. Indeed, the binding of β2m to the MHC class I α chain soon after protein synthesis in the ER is essential for the transportation of the complete heterodimer to the cell surface.

ii) MHC Class I Peptide-Binding Site

The groove-like peptide-binding site of the MHC class I molecule is relatively small. As a result, MHC molecules cannot recognize large native antigens. Rather, antigens must be processed into small peptides that can fit into the MHC groove before they can be presented to T cells. It has been estimated that each MHC class I molecule has the ability to bind to several hundred different peptides with moderately high affinity but captures only one peptide at a time.

The MHC class I peptide-binding groove is formed by the juxtaposition and interaction of the α1 and α2 domains of the α chain. The β2m chain contributes by interacting with the amino acids in α1 and α2 that form the floor of the groove. These interactions are strengthened, and the entire MHC class I structure is stabilized when the groove is occupied by a peptide of 8–10 amino acids. The peptide is

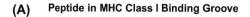

Fig. 6-4
MHC Peptide-Binding Sites

Amino acid positioning within the peptide-binding grooves of MHC class I and II proteins is shown. Blue circles represent individual amino acids of the antigenic peptides.
(A) Peptides binding in an MHC class I groove are usually 8–10 amino acids long. The residues at each end of the peptide are anchored in the groove.
(B) Peptides binding in an MHC class II groove are usually 13–18 amino acids long. The residues at each end of the peptide may overhang the groove and do not serve as anchor residues.

(A) Peptide in MHC Class I Binding Groove

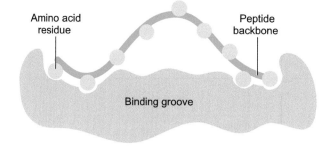

(B) Peptide in MHC Class II Binding Groove

held in place in the groove by interactions between specific amino acids of the α1 and α2 domains and conserved "anchor residues" located in the N- and C-termini of the peptide. The peptide anchor residues point "down" into the groove, while the central peptide residues project "up" toward the TCR (**Fig. 6-4A**). A sufficient degree of conformational flexibility exists such that peptides of widely varying amino acid sequence in the region between the anchor residues can occupy the groove. The ends of the MHC class I groove are closed, which means peptides larger than 8–10 amino acids can fit in only if their central residues can bulge upward out of the groove.

II. MHC Class II Proteins

As mentioned earlier, MHC class II molecules are found almost exclusively on APCs. The peptides bound by MHC class II are of **exogenous** origin; that is, derived from the degradation of proteins that have entered the cell from the exterior via either phagocytosis or receptor-mediated endocytosis. Because APCs also capture and digest spent host proteins, the vast majority of peptides presented on MHC class II molecules are "self" and do not trigger CD4+ T cell activation because these specificities have been removed from the Th cell repertoire by the establishment of central tolerance. A Th response is induced when an APC presents a non-self peptide bound to MHC class II.

i) MHC Class II Component Polypeptides
In both humans and mice, the α and β chains of MHC class II proteins are glycoproteins of similar size and structure (24–32 and 29–31 kDa, respectively). Both chains contain an N-terminal extracellular domain, an extracellular Ig-like domain, a hydrophobic transmembrane domain, and a short cytoplasmic tail (**Fig. 6-5**). The peptide-binding region is made up of the N-terminal α1 and β1 domains of the α and β chains, respectively. The α2 and β2 domains form globular loops that are homologous to the Ig fold but are not involved in peptide binding.

ii) MHC Class II Peptide-Binding Site
The MHC class II peptide-binding groove is similar in overall structure to that of MHC class I molecules (**Fig. 6-4B**). However, the ends of the MHC class II groove are open, permitting the binding of much longer peptides (up to 30 amino acids). Nevertheless, the majority of peptides found in MHC class II grooves are 13–18 amino acids long. The open ends of the MHC class II groove also mean that binding does not depend on

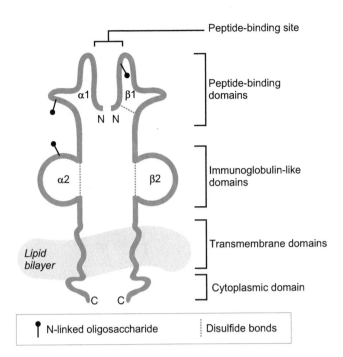

Peptide-binding site

Peptide-binding domains

Immunoglobulin-like domains

Transmembrane domains

Cytoplasmic domain

Lipid bilayer

N N

α1 β1

α2 β2

C C

N-linked oligosaccharide | Disulfide bonds

▶ **Fig. 6-5**
Structure of the MHC Class II Protein

Schematic representation showing component chains and domains of an MHC class II protein and its position in the APC membrane. N, amino-terminus; C, carboxy-terminus.

conserved anchor residues at the ends of the peptides but is instead mediated by hydrogen bonding between the peptide backbone and the sidechains of certain MHC amino acids. Researchers have found that antigenic peptides that are successfully bound to the floor of the MHC class II groove possess a particular conserved secondary structure (resembling a polyproline chain) in the portion of the peptide that aligns with critical acidic MHC residues located in the middle of the groove. As a result of this conformational requirement, MHC class II proteins generally bind a narrower range of proteins than do MHC class I proteins.

III. X-Ray Crystallography of MHC Class I and II Molecules

Much of the information on how MHC class I and II molecules bind to peptides has come from X-ray studies of crystallized pMHC complexes. **Plate 6-1** shows the crystal structures of the carbon backbones of the extracellular regions and peptide-binding grooves of murine MHC class I and MHC class II molecules. The similarity of their tertiary structures can be clearly seen. Analyses of such MHC crystal structures have shown that water plays an important role in peptide–MHC binding. The fit of the peptide in the groove is tightened when water molecules fill any gaps in the complex.

NOTE: As X-ray crystallography techniques become more and more refined, the number of macromolecular 3-D crystal structures available in public databases grows. A searchable database of protein structures is maintained by the U.S. National Center for Biotechnology Information (NCBI) and can be found at the website http://www.ncbi.nlm.nih.gov/sites/structure. At this site, the structures of hundreds of pMHC complexes can be viewed in 3-D.

C. MHC Class I and Class II Genes

I. Polygenicity of MHC Class I and II Genes

Most proteins in our bodies are unique; that is, there is only one functional gene in the genome that encodes a protein carrying out that particular function. The MHC genes are unusual in that, due to gene duplication during evolution, two to three separate, functional genes encoding the same type of MHC class I or II polypeptide exist. This

Plate 6-1
X-Ray Crystal Structures of MHC Class I and II Molecules in the Mouse

Crystal structures showing the carbon backbone of murine MHC class I and II molecules. **(A)** and **(B)** show peptides bound to the extracellular regions of MHC class I or II, respectively. **(C)** and **(D)** show the view looking down at the peptide in the peptide-binding groove of MHC class I or II, respectively. *[Reproduced by permission of Bjorkman, P.J. (1977). MHC restriction in three dimensions: A view of T cell receptor/ligand interactions. Cell 9, 167–170.]*

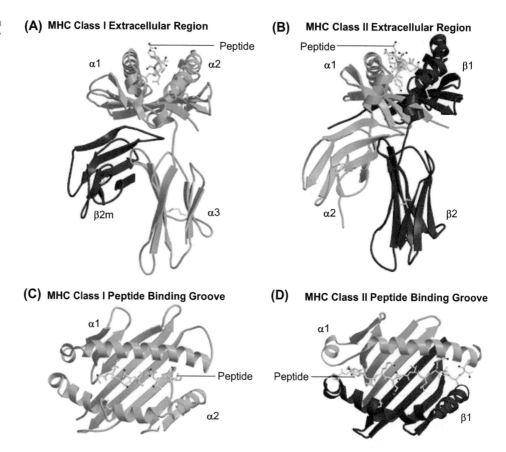

(A) MHC Class I Extracellular Region

(B) MHC Class II Extracellular Region

(C) MHC Class I Peptide Binding Groove

(D) MHC Class II Peptide Binding Groove

phenomenon is called **polygenicity**. These genes are named for their region of location and the chain they specify. For example, the HLA includes three loci, HLA-A, -B and -C, that all encode the same type of polypeptide: an MHC class I α chain. Similarly, genes giving rise to MHC class II α chains can be found in the HLA-DP, -DQ and -DR regions; these genes are called DPA, DQA and DRA genes, respectively. Also in each of the HLA-DP, -DQ and -DR regions are separate genes that encode MHC class II β chains; these are called DPB, DQB and DRB genes, respectively.

The MHC class II loci show further polygenicity in that each of the HLA-DP, -DQ and -DR regions may have more than one α chain gene and more than one β chain gene. For example, within the HLA-DP region, there are two genes that could encode MHC class II α chains, DPA1 and DPA2, and two genes that could encode MHC class II β chains, DPB1 and DPB2. However, only the DPA1 and DPB1 genes are functional. Similarly, the HLA-DQ region contains the DQA1 and DQA2 genes that could encode MHC class II α chains, and the DQB1, DQB2 and DQB3 genes that could encode MHC class II β chains. However, only DQA1 and DQB1 are functional. In the HLA-DR region, a single gene designated DRA encodes the DR α chain and is functional in all humans. In contrast, not every individual carries the same number of DRB loci on his/her chromosomes. Nine different DRB genes have been identified, designated DRB1 to DRB9. While every individual has the DRB1 and DRB9 loci, different individuals may also have one or more DRB loci selected from among the DRB2 to DRB8 genes. However, only DRB1, DRB3, DRB4 and DRB5 are functional and encode DR β chains.

For unknown reasons, an α chain derived from a DP region gene almost always combines with a β chain derived from the DP region (and not from the DQ or DR regions) to form a complete MHC class II molecule. Similarly, a DQ α chain combines with a DQ β chain and a DR α chain with a DR β chain. Only very rarely do mixed MHC class II molecules such as HLA-DRA/DQB occur. **Figure 6-6A** illustrates how the products of the HLA loci can come together to form complete human MHC molecules.

"The 2.5 Å Structure of CD1c in Complex with a Mycobacterial Lipid Reveals an Open Groove Ideally Suited for Diverse Antigen Presentation." by Scharf, L., Li, N.S., Hawk, A.J., Garzón, D., Zhang, T., Fox, L.M., Kazen, A.R., Shah, S., Haddadian, E.J., Gumperz, J.E., Saghatelian, A., Faraldo-Gómez, J.D., Meredith, S.C., Piccirilli, J.A., and Adams, E.J. (2010) *Immunity 33*, 853–862.

Focus on Relevant Research

Nearly one-third of the world's population is infected with the intracellular bacterium *Mycobacterium tuberculosis*, the causative agent of tuberculosis (TB), and nearly 2 million people die worldwide each year from this infection. It is therefore no surprise that much research has been aimed at understanding the immune responses associated with TB. One hallmark feature of this bacterium is its cell wall, which contains unusual and abundant lipids. Scientists have known for some time that T cell-mediated responses are important for protection against TB but that these responses are not triggered by conventional pMHC complexes presented by MHC class I and class II molecules. Instead, anti-TB T cell responses are mounted against bacterial lipids presented by CD1 molecules, which are "MHC-like" proteins encoded by a gene outside the MHC (see **Box 6-1**). Structurally, CD1 molecules have evolved to possess deep, narrow binding grooves that excel at accommodating and anchoring the hydrophobic alkyl chains of lipid molecules.

In this article, Scharf *et al.* report on the precise way in which CD1c, a particularly interesting CD1 isoform, displays lipid antigens to T cells. The authors use the powerful technique of X-ray diffraction crystallography to construct a three-dimensional (3-D) model showing how lipids derived from *M. tuberculosis* are displayed in the open groove of the CD1c molecule. They then use this information to explain why CD1c is uniquely suited to displaying a wide variety of bacterial lipids. Moreover, they determine the lipid-CD1c interaction to a resolution of 2.5 angstroms (2.5 Å), arguably the finest level of detail of macromolecular structure that can be determined experimentally. In Figure 3 from the article (shown here), we see 3-D representations of the space occupied by the antigen-binding cavities of CD1c (magenta), CD1a (yellow), CD1b (cyan) and CD1d (light purple), with the

red, blue and yellow areas representing the contributions of oxygen, nitrogen and sulfur atoms, respectively, to each cavity surface. The volume of each cavity is given in cubic angstroms ($Å^3$), and two major pockets can be identified (A' and F'). The small red and yellow stick models represent lipid antigens bound inside the cavities. In the case of CD1c, we also see amino acid residues that form conserved cavity features (L66, L162, T166) and an important inter-residue contact (V12/F70). Using these types of models to better understand how APCs present lipid antigens to T cells via CD1 isoforms may enhance our ability to create more effective vaccines against important pathogens like *M. tuberculosis*.

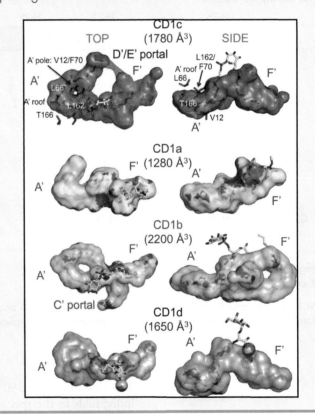

Polygenicity also occurs in the mouse H-2 complex. Two loci, H-2K and H-2D, contain single genes encoding MHC class I α chains. MHC class II α chains are encoded by one functional gene called Aa (or Aα) within the A region of the mouse H-2, as well as by the Ea (or Eα) gene in the E region. Similarly, MHC class II β chains are encoded by the Ab (or Aβ) gene in the H-2A region and the Eb (or Eβ) gene in the H-2E region. **Figure 6-6B** shows examples of how products of the H-2 complex give rise to complete murine MHC class I and II molecules. Again, an α chain derived from an A region gene almost always combines with a β chain derived from the A region, and not with an E region β chain (and vice versa).

(A) Human Leukocyte Antigen (HLA) Complex

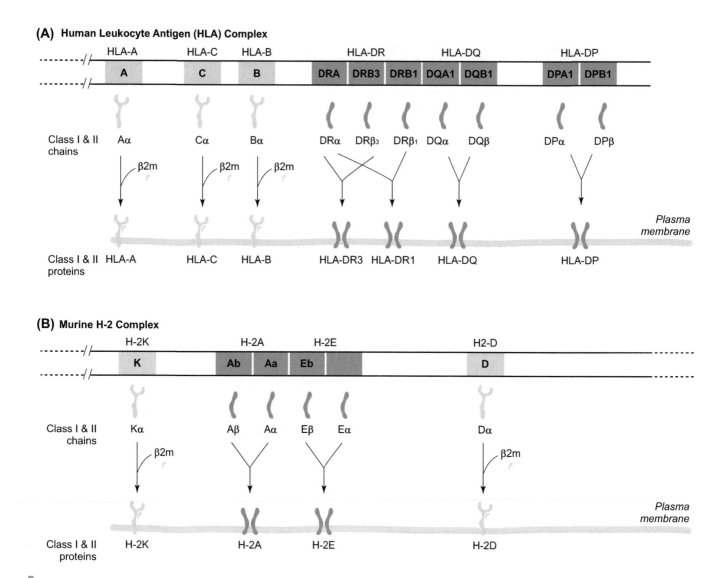

Fig. 6-6
Examples of Polygenicity in the MHC Loci

Multiple genes encode MHC class I and II proteins in the **(A)** HLA and **(B)** H-2 complexes. Examples of how chains derived from each locus can combine to form complete MHC heterodimers are shown. For simplicity, the MHC class Ib, IIb and III loci are not shown.

II. Polymorphism of MHC Class I and II Genes

The vast majority (>90%) of vertebrate genes are **monomorphic**; that is, almost all individuals in the species share the same nucleotide sequence at that locus. In contrast, the MHC loci exhibit extreme polymorphism. **Polymorphism** is the existence in a species of several different alleles at one genetic locus. **Alleles** are slightly different nucleotide sequences of a gene; the protein products of alleles have the same function. For example, close to 1700 alleles have been identified for the HLA-A gene, over 2000 for HLA-B, and more than 1200 for HLA-C (**Table 6-2**). A functional MHC class I molecule can consist of the protein product of any one of these HLA-A, -B or -C alleles associated with the invariant β2m chain. Multiple alleles also exist for the MHC class II genes, so that the product of any DPA allele can combine with the product of any DPB allele (and DQA with DQB, and DRA with DRB) to form a functional MHC class II protein. The degree of sequence variation among MHC alleles can be astonishing: differences of as many as 56 amino acids have been identified between individual alleles. Not surprisingly, this amino acid variation is concentrated in the peptide-binding site of

TABLE 6-2	Numbers of HLA Alleles	
		Number of Alleles*
MHC Class I Genes		
HLA-A		1698
HLA-B		2271
HLA-C		1213
MHC Class II Genes		
HLA-DPA1		32
HLA-DPB1		149
HLA-DQA1		44
HLA-DQB1		158
HLA-DRA		7
HLA-DRB		1074

*Data are from the Immunogenetics HLA (IMGT HLA) database maintained by the European Bioinformatics Institute at EMBL (http://www.ebi.ac.uk/imgt/hla/) and represent alleles reported as of September 2011.

the MHC protein. In MHC class I α chains, most of the polymorphism is localized in the α1 and α2 domains. The α3 domain is less polymorphic and more Ig-like, and the transmembrane and cytoplasmic domains are more conserved than any of the α domains. The β2m protein exhibits almost no polymorphism within a species or variation among species. In the case of MHC class II molecules, polymorphic variation among alleles is found in the α1 and β1 domains that constitute the peptide-binding site. Again, the transmembrane and cytoplasmic domains are highly conserved.

Due to the high level of polymorphism in the HLA, humans (which are outbred) are generally *heterozygous* at their MHC loci (have different alleles on the maternal and paternal chromosomes). In contrast, experimental mice (which have been repeatedly inbred to create pure strains) are *homozygous* for any given MHC gene (have the same allele on both the maternal and paternal chromosomes). In addition, in outbred populations, two individuals are very likely to have different nucleotide sequences at each HLA locus. These two individuals are said to be **allogeneic** to each other at their MHC loci (*allo*, meaning "other"). In an inbred population, not only is each individual homozygous at each MHC locus, but all individuals in the population express the same MHC allele at a given locus. Such inbred animals express exactly the same spectrum of MHC molecules and are said to be **syngeneic** at their MHC loci (*syn*, meaning "same").

III. Codominance of MHC Expression

The polygenicity and polymorphism of the MHC genes underlie the vast diversity of MHC molecules expressed within an outbred population. In an individual, the breadth of MHC diversity is increased by the fact that, at *each* MHC locus, the gene on both chromosomes is expressed independently, or **codominantly**. In other words, when a given MHC locus is expressed in an individual, the genes on *both* the maternal and paternal chromosomes produce the corresponding proteins. For example, in an individual heterozygous at the HLA-A locus, there are two MHC class I α chains produced (maternal and paternal) that can combine with β2m to form two different HLA-A proteins. Similarly, for an MHC class II molecule such as HLA-DP, two different α chains and two different β chains are produced that can combine to form four different HLA-DP proteins. For a locus such as HLA-DR, which comprises more than one DRB gene in most individuals, the number of possible HLA-DR heterodimers is much higher. **Figure 6-7** illustrates the net effects of MHC polygenicity, polymorphism and codominant expression in an outbred individual. A nucleated host cell features a wide spectrum of MHC class I molecules on its surface, whereas a typical APC expresses

New MHC alleles are discovered every day. Up-to-date numbers of MHC alleles can be found in the IMGT/HLA database maintained by the European Bioinformatics Institute at EMBL.

Fig. 6-7
Spectrum of MHC Class I and II Expression in an Outbred Individual

The existence of multiple MHC class I and class II loci, combined with high levels of polymorphism at each locus as well as codominant expression of maternal and paternal alleles at each locus, results in a wide spectrum of MHC molecules being present on the cells of any given individual. All nucleated cells express MHC class I proteins, but only APCs express both MHC class I and II molecules.

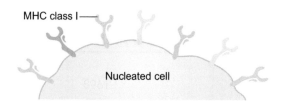

MHC class I —

Nucleated cell

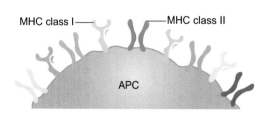

MHC class I — | — MHC class II

APC

multiple types of MHC class II molecules as well as multiple types of MHC class I molecules.

IV. MHC Haplotypes

The MHC loci are closely linked, meaning that the specific set of alleles for all MHC loci on a single chromosome is usually passed on to the next generation as an intact block of DNA. This set of alleles is called a **haplotype**, and any individual inherits two haplotypes from his/her parents: the MHC block on the paternal chromosome and the MHC block on the maternal chromosome. In an outbred population, the high degree of polymorphism of the MHC can lead to great variation in haplotypes among unrelated individuals. Interestingly, within the human population, researchers have identified over 30 *ancestral haplotypes* that are shared not only within a single family but also among a large number of families. These ancestral haplotypes are thought to have originated in "founder" populations that settled in diverse geographic regions, so that a given ancestral haplotype is often associated with a particular ethnic background. For example, one ancestral haplotype is found predominantly among Basques and Sardinians, while another is specific to Eastern European Jews, and a third is exclusive to Southeast Asians.

In inbred mice, both parents have the same allele at each MHC locus, the maternal and paternal haplotypes are the same, and all offspring inherit the same single haplotype on both chromosomes. Immunologists frequently use a single term "short form" to indicate the haplotype of a particular strain. For example, the haplotype of the C57BL/6 mouse strain is denoted "H-2^b", where the "b" in H-2^b means that allele number 12 is present at the K locus, allele number 74 is at the Aβ locus, allele number 3 is at the Aα locus, allele number 18 is at the Eβ locus, and so on. In contrast, the CBA mouse strain has a haplotype of "H-2^k", where allele number 3 is present at the K locus, allele number 22 at Aβ, and so on. More detailed short forms can be used to indicate specific alleles in a haplotype. For example, the term "H-2D^b" means that the D allele being discussed is that which occurs in a mouse strain of the "b" haplotype. Researchers might also identify this allele as D^b and verbalize it as "D of b."

V. Expression of MHC Genes

The expression of MHC genes is tightly and differentially regulated, such that MHC class I is expressed on almost all healthy host cells, but MHC class II expression is limited to APCs. As well, MHC protein expression may be upregulated or induced by cytokines and other stimuli released in a host cell's vicinity. Depending on the type of host cell and the tissue in which it resides, these stimuli may be either constitutively produced or induced during an immune response to injury, pathogens or tumors. For example, molecules in the walls of invading bacteria stimulate macrophages to produce TNF and LT, and viral infection induces the infected cells to synthesize IFNs. The interaction of these cytokines with specific receptors on a host cell triggers intracellular signaling pathways that activate transcription factors. The activated transcription factors enter the host cell nucleus and bind to 5′ regulatory motifs in the DNA upstream of the MHC genes, altering their expression. An increase in MHC expression facilitates the amplification of an adaptive response by enhancing antigen presentation to T cells.

D. Physiology of the MHC

I. Polymorphism and MHC Restriction

How did the MHC loci come to be so polymorphic? In an ancient, antigenically simple world, a primeval MHC molecule that displayed endogenous and exogenous protein fragments to T cells likely existed but was of very limited (or non-existent) variability. As the world became more antigenically complicated, it was individuals with multiple duplications of this primordial MHC gene that likely survived because they possessed more than one gene dedicated to presenting protein fragments. Perhaps concurrently, different alleles of each gene also evolved, each with a different sequence in the peptide-binding groove. A broader range of peptide binding and presentation molecules would have been generated. Today, the resulting polymorphism at multiple MHC loci ensures that each member of an outbred species is heterozygous at most if not all MHC loci, and thus has a very good chance of possessing at least one MHC allele capable of binding to any given antigenic peptide. For the species as a whole, MHC polymorphism means a large catalog of MHC alleles is spread over the entire population. In the case of a devastating pathogen attack, a significant fraction of the population (but not all individuals) will be able to respond to the pathogen and survive to perpetuate the species. However, the multiplicity of MHC molecules does not allow for a free-for-all in terms of presentation to T cells. A phenomenon called **MHC restriction** exists, which dictates that the epitope seen by a given TCR is a combination of a specific peptide with a specific MHC molecule. The discovery of MHC restriction is outlined in **Box 6-2**.

Box 6-2   The Discovery of MHC Restriction

In the early 1970s, neither the TCR nor the structure it recognized on host cells had been defined. However, in 1974, immunologists Rolf Zinkernagel and Peter Doherty published remarkable results from their studies of T cell responses to lymphocytic choriomeningitis virus (LCMV), a pathogen that is deadly in mice. Their experiments showed that CTLs from virus-infected mice would only kill infected cells derived from inbred strains of mice that had the same MHC haplotype. These researchers hypothesized that CTLs must simultaneously recognize both the antigen from the virus *and* the MHC molecule displaying that antigen on the surface of the target cell. This meant that the daughter effector cells of a naïve Tc cell that initially recognized a given antigen in conjunction with a particular MHC allele would only recognize the same antigen if it was presented in association with exactly the same MHC allele. This phenomenon came to be called "MHC restriction," and finally provided a explanation for why T cells could not bind directly to soluble antigens like B cells can. A great deal of work by many other researchers eventually revealed the nature of the structure that engaged the TCRs of T cells: a small antigen-derived peptide associated with an MHC molecule positioned on the surface of a host cell. In order for a CTL to kill an infected body cell, the TCR of that CTL must have engaged an MHC class I molecule that was displaying peptide from a pathogen antigen and was situated on the surface of the infected cell. Similarly, in order to produce the cytokines needed to support responses by B and Tc cells, the TCR of a Th cell must have engaged an MHC class II molecule that was displaying peptide from a pathogen antigen and was situated on the surface of an APC such as a DC. The fundamental roles of the MHC class I and II molecules were therefore found to be remarkably alike, despite the fact that these molecules were recognized by different T cell subsets. Thus, Zinkernagel and Doherty's revolutionary work helped to establish the very basis of modern cellular immunology. Logical extensions of their work have directly influenced modern vaccine development, transplantation medicine, and the exploration of potential new treatments for autoimmune diseases and cancer. In 1996, these two scientists were justly rewarded for their efforts with the Nobel Prize in Physiology and Medicine.

II. MHC and Immune Responsiveness

Immunologists have long observed that some foreign proteins that provoke strong immune responses in some individuals fail to do so in others. Those individuals failing to mount a response were originally called "non-responders," while those that did react were called "responders." Among responders, there were subtle differences in the level of the response, leading to the description of individuals as either low or high responders. Immunologists soon mapped the genes behind immune responsiveness to the MHC and showed that mice of different MHC haplotypes sometimes respond differently to a given peptide (**Table 6-3**). These variations in response levels to a given antigen can be interpreted as differences in the ability of particular MHC alleles to effectively present peptides from that antigen that can be recognized by T cells. In an inbred population, there is a greater possibility that an individual will be a non-responder; that is, an individual will lack an MHC allele that can lead to specific T cell activation during a challenge with a particular antigen. Two hypotheses, which may not be mutually exclusive, have been proposed to account for non-responsiveness: the *determinant selection* model and the *hole in the T cell repertoire* model.

i) Determinant Selection Model

For a T cell response to be mounted against a foreign protein, a host must possess at least one MHC allele with a groove that accommodates a peptide derived from that protein. Responsiveness then depends on the strength of binding between a given MHC allele and a given determinant (peptide), which in turn depends on structural compatibility. In other words, the MHC proteins in an individual "select" which determinants will be immunogenic in that individual as well as the extent of the response. Since a foreign protein is usually processed into three to four strongly immunogenic peptides, an outbred individual is very likely to possess an MHC allele capable of binding to at least one of these peptides and provoking an immune response. Such an individual is then a responder to this particular antigen, and his/her status as a high or low responder correlates with strong or weak binding of the peptide to MHC, respectively. On the other hand, if none of the individual's MHC molecules can bind to any of the peptides generated from the protein, the individual is a non-responder to this antigen.

The determinant selection model has been supported by experiments in which a given peptide is immunogenic only when it is bound to a particular MHC allele. For example, in inbred mice of the H-2^k haplotype, a certain peptide of an influenza virus protein readily provokes an immune response. More specifically, the determinant is recognized when presented to T cells on the MHC class I H-2K^k molecule. However, this same peptide fails to stimulate T cells in mice of the H-2^b haplotype. Instead, a peptide from a different part of the same influenza virus protein triggers an antiviral response in H-2^b mice when presented on the MHC class I H-2D^b molecule.

TABLE 6-3	MHC Haplotype Correlated with Immune Responsiveness	
Mouse Strain	**H-2 Haplotype**	**Response to TGAL* Peptide**
C3H	k	Low
C3H.SW	b	High
A	a	Low
A.BY	b	High
B10	b	High
B10BR	k	Low

*TGAL, synthetic peptide containing lysine, alanine, tyrosine and glutamic acid residues.

ii) Hole in the T Cell Repertoire Model

Immune non-responsiveness may also result from tolerance mechanisms. It may be that, in non-responders, a particular foreign peptide–MHC combination very closely resembles the structure of a self peptide–MHC combination, such that any T cell clones capable of recognizing the foreign peptide–MHC combination were eliminated as autoreactive during the establishment of central tolerance. In a non-responder, this would result in a missing T cell specificity or a "hole" in the T cell repertoire relative to the repertoire of a responder.

III. MHC and Disease Predisposition

An individual's MHC haplotype determines his/her responsiveness to immunogens. If an individual cannot mount an appropriate immune response to an immunogen associated with infection or cancer, the individual will likely suffer disease. If an immune response is mounted when it is inappropriate, disease in the form of autoimmunity or hypersensitivity (including allergy) can result. The direct link between immune responsiveness and particular MHC alleles means that certain MHC haplotypes may predispose individuals to specific susceptibilities or disorders.

NOTE: The Human Genome Project (HGP), which identified the ~25,000 genes in human DNA and sequenced the 3 billion base pairs of the human genome, was completed in 2003. The HGP has made it possible to determine the genomic locations and DNA sequences of HLA alleles with unprecedented precision. By comparing the sequences of HLA alleles derived from diverse populations around the world, researchers have examined the links between certain HLA alleles and disease susceptibility among people of many different nationalities. In addition, relationships between numerous non-HLA genes and disease incidence have been discovered. Studies that utilize the vast information of the HGP to map the location of genes linked to particular diseases have been termed "Genome-Wide Association Studies" (GWAS).

In humans, many of the disorders linked to the possession of specific MHC alleles manifest as **autoimmune disease**. Autoimmune disease results when self-reactive T cell clones escape the tolerance mechanisms that would normally prevent these cells from entering or acting in the periphery. The individual may then possess T cells that can recognize self components and may attack tissues expressing these components. For example, type 1 (insulin-dependent) diabetes mellitus is thought to arise from an autoimmune attack on antigens expressed by the insulin-producing β cells of the pancreatic islets. Immune destruction of the β islet cells results in insulin deficiency and thus diabetes. For unknown reasons, the HLA-DQ8 allele is eight times more prevalent in groups of humans suffering from type 1 diabetes than it is in healthy populations. Similarly, 90% of Caucasian patients suffering from a degenerative disease of the spine called ankylosing spondylitis carry the HLA-B27 allele, whereas only 9% of healthy Caucasians do. Additional autoimmune diseases and their association with particular HLA alleles are included in **Table 6-4**. Note, however, that mere possession of a

TABLE 6-4	Examples of HLA-Associated Disorders in Humans
Disease	**Examples of Associated HLA Alleles**
Ankylosing spondylitis	B27
Birdshot retinopathy	A29
Celiac disease	DR3, DR5, DR7
Graves' disease	DR3
Narcolepsy	DR2
Multiple sclerosis	DR2
Rheumatoid arthritis	DR4
Type 1 diabetes mellitus	DQ8, DQ2, DR3, DR4

predisposing HLA allele is not usually sufficient to cause disease; other genetic and environmental factors are thought to be involved. A complete discussion of autoimmune disease is presented in Chapter 19.

This concludes our discussion of the structure and physiology of the MHC. In the next chapter, titled "Antigen Processing and Presentation," we describe the derivation of antigenic peptides and how MHC molecules associate with these peptides and present them to T cells.

Chapter 6   Take-Home Message

- The MHC class I and II genes in the MHC encode cell surface proteins that present peptides to T cells.

- Non-classical MHC class I and class II genes as well as MHC class III genes also constitute part of the MHC.

- MHC class I and class II proteins are heterodimeric molecules with highly variant N-terminal domains that form a peptide-binding site.

- MHC class I is expressed on almost all cells in the body and generally presents peptides of endogenous origin.

- MHC class II is expressed by APCs and generally presents peptides of exogenous origin.

- MHC class I interacts with the CD8 coreceptor found on Tc cells and CTLs, while MHC class II interacts with the CD4 coreceptor on naïve and effector Th cells.

- The MHC genes are characterized by polygenicity, extreme polymorphism and codominant expression.

- Outbred populations are heterozygous at the MHC loci, while inbred mice are homozygous. Syngeneic individuals have the same MHC genotype, while allogeneic individuals have different MHC genotypes.

- Differences in MHC alleles are largely responsible for transplant rejection and variations in immune responsiveness to a given pathogen.

- Expression of particular MHC alleles is linked to autoimmune disease predisposition.

Did You Get It? A Self-Test Quiz

Section A

1) Describe the derivation of the term "major histocompatibility complex."

2) What polypeptide chains make up the MHC class I molecule? MHC class II?

3) What cell types express MHC class I? MHC class II?

4) To what type of cells does MHC class I present peptide? MHC class II?

5) Name five regions of the HLA complex and describe the nature of their gene products.

6) Name five regions of the H-2 complex and describe the nature of their gene products.

7) Name three non-classical MHC genes and describe the nature of their gene products.

8) What MHC-like proteins present non-peptidic antigens to T lineage cells?

Section B

1) Why do MHC proteins have relatively low binding affinity for peptides?

2) What is an anchor residue?

3) Do self peptides bound to MHC usually provoke immune responses? If not, why not?

4) What domains form the peptide-binding grooves of MHC class I molecules? Of MHC class II molecules? How do these grooves differ in structure?

5) Describe two ways in which the peptides binding to MHC class I differ from those binding to MHC class II.

Section C

1) Can you define these terms? polygenicity, monomorphic, polymorphic, allele, codominance

2) Which MHC molecule is more common in a human: DQA/DRB or DQA/DQB?

3) Where is amino acid variation concentrated in the MHC protein and why?

4) Distinguish between the terms "heterozygous" and "allogeneic."

5) How many different types of HLA-DQ proteins are likely present in an outbred individual?

6) Can you define these terms? haplotype, ancestral haplotype

7) What does the term "H-2K^b" represent? How would you verbalize this term?

8) What effect does inflammation have on MHC expression?

Section D

1) Why does inbreeding sometimes put a species at risk for decimation by a pathogen?

2) What is MHC restriction, and how was it discovered?

3) Outline two theories accounting for variation in immune responsiveness to an antigen.

4) What is a GWAS, and how might it be helpful for studying HLA alleles?

1) The HLA-A2 MHC class I molecule is found to present an 8 amino acid peptide "X" to a T cell clone. In the lab, researchers produce two mutated versions of X, which they call "Xm1" and "Xm2." Xm1 is mutated at amino acid #4 of the peptide sequence, and Xm2 is mutated at amino acid #7. Each peptide is tested to see if its presentation by HLA-A2 will stimulate the T cell clone. How would you explain the following observations?

   a) The T cell activation response to Xm1 is less than the T cell activation response to X.

   b) The response to Xm1 is greater than the response to X.

   c) The response to Xm2 is less than the response to X.

   d) The response to Xm2 is greater than the response to X.

2) MHC molecules and Ig molecules can both be said to show diversity. Briefly explain how the origin of diversity in each of these cases is fundamentally different.

3) You are a researcher studying the amino acid sequences of MHC class I and class II alleles. In which domains of these molecules would you expect to find the most allelic variations? Explain briefly.

Apostolopoulos, V., Yuriev, E., Lazoura, E., Yu, M., & Ramsland, P. A. (2008). MHC and MHC-like molecules: Structural perspectives on the design of molecular vaccines. *Human Vaccines, 4*(6), 400–409.

Corse, E., Gottschalk, R. A., & Allison, J. P. (2011). Strength of TCR-peptide/MHC interactions and *in vivo* T cell responses. *Journal of Immunology, 186*(9), 5039–5045.

Hofstetter, A. R., Sullivan, L. C., Lukacher, A. E., & Brooks, A. G. (2011). Diverse roles of non-diverse molecules: MHC class Ib molecules in host defense and control of autoimmunity. *Current Opinion in Immunology, 23*(1), 104–110.

Liao, W. W., & Arthur, J. W. (2011). Predicting peptide binding to major histocompatibility complex molecules. *Autoimmunity Reviews, 10*(8), 469–473.

Marrack, P., Scott-Browne, J. P., Dai, S., Gapin, L., & Kappler, J. W. (2008). Evolutionarily conserved amino acids that control TCR-MHC interaction. *Annual Review of Immunology, 26*, 171–203.

Menier, C., Rouas-Freiss, N., Favier, B., LeMaoult, J., Moreau, P., & Carosella, E. D. (2010). Recent advances on the non-classical major histocompatibility complex class I HLA-G molecule. *Tissue Antigens, 75*(3), 201–206.

Antigen Processing and Presentation

Real generosity toward the future lies in giving all to the present.

Albert Camus

A. Overview of Antigen Processing and Presentation

Antigen processing is the complex process by which antigens are produced from macromolecules. Most often, antigen processing refers to the generation of antigenic peptides from proteins. **Antigen presentation** refers to the binding of these peptides to MHC molecules and the positioning of the resulting pMHC complexes on a host cell surface so that they can be inspected by T cells. The processing and presentation of antigenic peptides, as well as some non-peptidic antigens, are discussed in this chapter, whereas the recognition of pMHCs by the TCRs of T cells is discussed in Chapter 8.

Antigen processing provides the host with a means of scanning the molecules constantly being produced and turned over in the body. At any one time, almost every cell in the body displays several hundred thousand pMHCs on its surface. This population represents hundreds of distinct peptides, the vast majority of which are "self" in origin and elicit no T cell response in a healthy individual. In the case of an infection, a substantial proportion (up to 10%) of the peptides may be pathogen-derived, a number more than sufficient to trigger the activation of T cells specific for these pMHCs.

With respect to protein antigens, there are four major pathways of antigen processing, two of which are well defined and two of which remain to be completely elucidated (**Fig. 7-1**). The **exogenous processing pathway** acquires proteins from *outside* the host cell (extracellular proteins) and degrades them to peptides within endocytic compartments. In contrast, the **endogenous processing pathway** acquires proteins that are synthesized inside the host cell (intracellular proteins) and degrades them to peptides in the cytoplasm; these peptides are then delivered into the ER. The two more recently discovered pathways allow a protein antigen to be transferred from one of the preceding pathways into

Fig. 7-1

Overview of Antigen Processing Pathways

Antigens processed via the exogenous pathway are presented on MHC class II and activate CD4⁺ Th cells. Antigens processed via the endogenous pathway are presented on MHC class I and activate CD8⁺ Tc cells. Peptides escaping from the exogenous pathway may be displayed on MHC class I via cross-presentation. Peptides from intracellular entities that are captured by autophagy can be diverted into the endocytic compartment and presented on MHC class II.

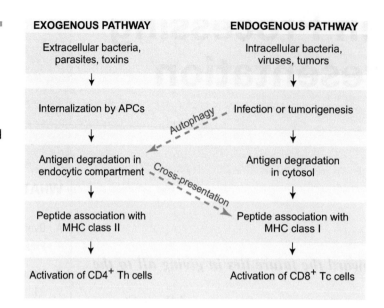

the other. The **cross-presentation** pathway transfers peptides from the exogenous pathway into the endogenous pathway, and the **autophagic** pathway captures cytoplasmic entities and macromolecules and delivers them into the exogenous pathway. These four processing pathways allow antigenic peptides to be presented efficiently to both CD4⁺ and CD8⁺ T cells, so that both subsets will respond to extracellular and intracellular entities as required.

The outcome of antigen presentation to T cells is determined in large part by the MHC molecules involved. MHC class I and class II molecules follow different intracellular trafficking routes after synthesis, show critical differences in cell type expression, and bind to different coreceptors on T cells. With respect to MHC class I molecules, after their synthesis in the ER, these proteins remain in this organelle and bind to peptides delivered into the ER by endogenous processing or cross-presentation. The pMHC complexes generated by this binding are then expressed on the host cell surface, where the CD8 coreceptor binding site of the MHC class I protein ensures that CD8⁺ CTL effectors recognize the host cell as a target. Because MHC class I molecules are expressed on all nucleated cells, CD8⁺ CTLs can effectively target and kill any cell that has fallen prey to the internal afflictions threatening all host cells, namely infection and cancerous transformation. In contrast, only APCs express MHC class II molecules. Within an APC, newly synthesized MHC class II molecules traffic from the ER to specialized endosomal compartments where they receive peptides generated through exogenous processing or autophagy. The resulting pMHC complexes are expressed on the APC surface where the CD4 coreceptor binding site of the MHC class II protein ensures that CD4⁺ Th cells are able to survey the proffered peptides. This limitation of MHC class II expression to APCs ensures that the attention of Th cells is efficiently focused on the only cells capable of Th cell activation. Thus, the various antigen presentation systems at work in different host cell types ensure that any peptide associated with "danger," whether from an extracellular or intracellular source, will activate the Th cells that are the lynchpin of adaptive immunity. The products of these Th cells then support the activation of Tc cells that may be needed to combat intracellular threats, B cells that may be required to produce antibody against extracellular threats, and other Th subsets that may be required to fine-tune and regulate the immune response.

B. Nature of Cells That Can Activate T Cells

To become activated, naïve CD4⁺ and CD8⁺ T cells must interact with host cells that display pMHCs on their cell surfaces. With respect to CD4⁺ Th cells, activation is carried out by so-called professional APCs, which are those few cell types that constitutively or inducibly express high levels of MHC class II and costimulatory molecules. Professional APCs include mature DCs (which can activate naïve Th cells as well as effector and memory Th cells) and macrophages and B cells (which can activate effector and memory Th cells but not naïve T cells). "Non-professional" APCs are cell types (such as fibroblasts and epithelial cells) that can transiently express low levels of MHC class II if exposed to IFNγ during an inflammatory response. With respect to CD8$^+$ Tc cells, although all nucleated host cells express the MHC class I required to present peptides to this lymphocyte subset, again, only mature DCs are able to activate naïve Tc cells. In contrast, CD8$^+$ CTL effectors and memory CD8$^+$ Tc cells can be activated by any cell type expressing the specific peptide-MHC class I complex recognized by the TCR of that effector or memory CD8$^+$ T cell. Thus, any cell that has become aberrant due to cancer or intracellular infection becomes a target cell for CTL-mediated cytolysis.

Mechanisms of T cell activation are discussed in Chapter 9.

NOTE: Intracellular proteins that are released into the extracellular milieu by infected or transformed cells can be internalized by an APC and directed into its exogenous processing pathway. Then, thanks to cross-presentation, the antigen or its fragments can be transferred into the endogenous processing pathway, allowing the APC to display the resulting peptides on MHC class I. Thus, although not under internal attack itself, the APC can activate the Tc cells required to respond to an intracellular threat.

I. Dendritic Cells as APCs

As described in Chapter 2, multiple types of DCs in mammals can be distinguished by their locations in the body as well as by the surface markers and PRRs they express. By virtue of these differences, these various types of DCs play a key role in inducing the T cell response appropriate for dealing with the specific pathogenic threat encountered. We also introduced in Chapter 2 the concept that microenvironmental conditions can directly influence DC differentiation, in that the rare group of DCs called *inflammatory DCs* appears to be derived from monocytes that extravasated into inflamed tissue and subsequently became activated. However, under the steady-state conditions prevailing in healthy individuals, the main types of DCs present in the blood and tissues are either the comparatively rare *plasmacytoid DCs (pDCs)* or the much more numerous *conventional DCs* (which we refer to simply as DCs). In the absence of a threat, most of these pDCs and DCs remain in the immature state; they are induced to undergo maturation only when the healthy host becomes infected or sustains tissue damage. The pDCs specialize in sensing viral RNA and DNA through endosomal PRRs and respond with vigorous production of IFNα and IFNβ, cytokines that have direct antiviral effects. Thus, pDCs make their most important contribution to the innate, rather than the adaptive, immune response. In contrast, conventional DCs are the major drivers of T cell activation and are therefore the subset we focus on in this chapter when discussing the processing and presentation of antigen to T cells.

See Table 2-3 in Chapter 2 for a comparison of the properties of pDCs and DCs.

i) Migratory vs. Lymphoid-Resident DCs

The conventional DCs involved in naïve T cell activation can be further categorized as *migratory* DCs or *lymphoid-resident* DCs (often shortened to "resident DCs"). After their release from the bone marrow into the blood, migratory DCs first access a peripheral tissue rather than a lymphoid site, and collect self and foreign antigens in this location. Antigen-laden migratory DCs can then enter a lymphatic vessel and travel to the nearest secondary lymphoid tissue (most often a lymph node), where they either interact directly with T cells in the node, or act as an antigen source for resident DCs in that node. In contrast, when first released from the bone marrow into the blood, resident DCs move directly into one lymphoid site and do not travel the body in the lymphatic system. These cells remain in their lymphoid tissue of residence, collecting and presenting self and foreign antigens that are either dumped into the tissue via a lymphatic vessel, or are conveyed there by migratory DCs traveling in the lymphatic

TABLE 7-1	Migratory vs. Lymphoid-Resident DCs	
DC Subset	**Function**	**Examples**
Migratory	Front-line defense in peripheral tissues Acquire antigens in peripheral tissues and migrate through lymphatics to lymph nodes Deliver antigens to lymphoid-resident DCs or directly initiate T cell responses in local lymph nodes	Langerhans cells in epidermis Dermal DCs in dermis Mucosal DCs in mucosae of body tracts Interstitial DCs in non-lymphoid tissues
Lymphoid-resident	Front-line defense in lymphoid tissues Do not migrate but acquire antigens from migratory DCs or antigens that have accumulated in lymphoid tissues Initiate T cell responses in lymph nodes Initiate T cell responses to blood-borne antigens in the spleen Participate in central tolerance	Resident DCs in lymph nodes Splenic DCs in spleen Thymic DCs in thymus

system. The properties of migratory and lymphoid-resident DCs at steady-state are summarized in **Table 7-1**.

Migratory DCs are abundant at junctures between the body and the outside world, including just under the skin and the mucosae lining the respiratory and gastrointestinal tracts. Several subsets of migratory DCs have been identified that differ slightly in their surface markers, PRR repertoire, tissue distribution and cytokines secreted upon maturation. Consequently, each DC subset may have a different effect on T cell activation, allowing a tailored response to the specific threat encountered. Examples of migratory DCs are **Langerhans cells (LCs)**, which are relatively long-lived DCs present in the epidermis of the skin; *dermal DCs* present in the dermis of the skin; *mucosal DCs* in the mucosae lining the gastrointestinal, respiratory and urogenital tracts; and *interstitial DCs* present in almost all other non-lymphoid peripheral tissues.

Lymphoid-resident DCs include those present in the thymus and spleen and about half of those present in lymph nodes. *Thymic DCs* remain in this organ throughout their short life span and most likely participate in the establishment of central tolerance, presenting peptides from self antigens to T cells developing within the thymus. Immature T cells that strongly recognize these pMHCs (and thus are self-reactive T cells) are then eliminated. *Splenic DCs* reside in the spleen and monitor blood-borne antigens. At least three subtly different subsets of splenic DCs have been identified in mouse spleen based on differential surface marker expression.

Mechanisms of central T cell tolerance induction in the thymus are discussed in Chapter 9. Mechanisms of peripheral tolerance are discussed in Chapter 10.

ii) Immature vs. Mature DCs

In Chapter 2, we introduced the terms "immature DC" and "mature DC." The distinction is an important one because DCs become capable of activating naïve T cells only after they have undergone the maturation process. Until they receive specific maturation signals associated with infection or tissue damage, migratory and resident DCs remain in the immature state, which is specialized for rapid sampling of the surrounding microenvironment and the monitoring of tissue health. Immature DCs continually form and retract their long finger-like processes to capture entities in the surrounding tissue, and can also extend these processes harmlessly through the "tight junctions" that hold epithelial cells together to sample macromolecules in the external environment. Such sampling is mediated by the general engulfment processes of macropinocytosis, clathrin-mediated endocytosis, phagocytosis and autophagy described in Chapter 3, and the receptors mediating this uptake include scavenger receptors (like CD91 and CD36), complement receptors (like CR1 and CR3), and CLRs (like the mannose receptors, DEC-205 and DC-SIGN). Immature DCs also express high levels of FcγRII, which is a low affinity IgG receptor that can facilitate the uptake of protein antigens complexed to IgG (through opsonization) (**Fig. 7-2**).

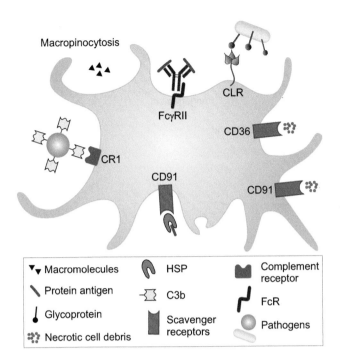

Macropinocytosis

CLR

FcγRII

CD36

CR1

CD91

CD91

▼▼ Macromolecules	⌒ HSP	Complement receptor
╲ Protein antigen	⬡ C3b	FcR
ⵊ Glycoprotein	Scavenger receptors	Pathogens
⁛ Necrotic cell debris		

➤ Fig. 7-2
Examples of Antigen Capture by an Immature DC

Immature DCs sample entities in the surrounding tissue microenvironment using macropinocytosis, receptor-mediated endocytosis and phagocytosis. Major antigen uptake receptors on these cells include scavenger receptors, complement receptors, CLRs and FcRs.

NOTE: Immunologists often call the engulfment of macromolecules from the external milieu "antigen uptake," and the receptors on DCs and other APCs that facilitate this process "antigen uptake receptors." These terms are widely used despite the fact that both foreign and host-derived entities are internalized in the same way, and not all captured entities will act as "antigens" (or, more properly, immunogens) with respect to lymphocyte activation.

Antigen uptake constantly supplies immature DCs with myriad proteins from which they derive peptides for display on MHC class I and II. The regular turnover of these pMHCs permits DCs to carry out ongoing surveillance of the protein environment of the host. Immature DCs do not initiate T cell responses because: (1) the supply of MHC class II on an immature DC is low and the pMHCs created by sampling are turned over very quickly, limiting antigen display and thus the DC's capacity to interact with and activate T cells; (2) under steady-state conditions, the pMHCs displayed do not contain foreign or aberrant antigen and so are not recognized by TCRs; and (3) no DAMPS/PAMPs are present to engage the DC's PRRs in such a way that the cell becomes activated.

In a host experiencing inflammation or infection, phagocytes (including immature DCs) efficiently capture whole pathogens by phagocytosis and use their antigen uptake receptors to capture macromolecular carbohydrate and protein antigens released by necrotic host cells. Large quantities of antigenic peptides are displayed on the immature DC's MHC molecules. Because the situation involves infection or trauma, DAMPs/PAMPs are present and engage the TLRs, NLRs, RAGE, and/or RLRs of immature migratory and resident DCs. The maturation of these cells is thus triggered and inflammatory cytokines are released, as described in Chapter 3. A mature migratory DC uses its large array of cytokine and chemokine receptors (particularly CCR1 and CCR7) as well as adhesion molecules to migrate rapidly and efficiently through the lymphatic system to the nearest lymph node (**Fig. 7-3**). Within the node, resident DCs that acquire antigen from either the afferent lymph or migratory DCs are also triggered to mature. Thus, the node becomes filled with mature migratory and resident DCs that are laden with foreign antigen and present pMHCs derived from the antigen directly to naïve Th and Tc cells also present in the node. Any naïve Th and Tc cells that are activated by these mature DCs proliferate and differentiate into resting Th effectors or CTLs that then leave the node and use their homing receptors to find the tissue under

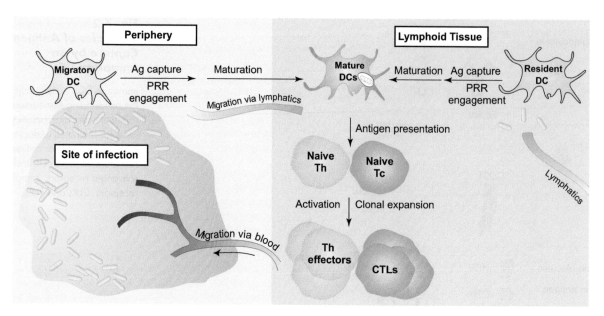

Fig. 7-3
Antigen Presentation by DCs to Naïve T Cells

Immature migratory DCs in peripheral tissues and immature resident DCs in lymphoid tissues continuously capture proteins from their surrounding environments. The maturation of these DCs is triggered only if they acquire antigen and have their PRRs engaged by DAMPs/PAMPs. Migratory DCs that have captured antigen directly from a site of infection initiate maturation as they move via the lymphatics to a lymphoid tissue such as a lymph node. Resident DCs that have either captured antigen conveyed into the lymphoid tissue via a lymphatic, or acquired antigen from an incoming migratory DC, are also triggered to mature. Within the lymphoid tissue, mature DCs of both types present pMHC complexes to naïve Th and Tc cells. Activated antigen-specific Th and Tc cells proliferate and generate Th effectors or CTLs, respectively, that migrate via the circulation to inflamed peripheral tissues. Immature and mature DCs are shown in light and dark brown, respectively.

TABLE 7-2	**Immature vs. Mature DCs**	
	Immature DCs	**Mature DCs**
Location	Peripheral tissues Secondary lymphoid tissues	Secondary lymphoid tissues
Surface MHC class II	Low	High
Antigen internalization capacity	High	Low
Costimulatory molecules	Low	High
Antigen presentation to T cells	Inefficient	Very efficient
Chemokine receptors	High CCR1, low CCR7	Low CCR1, high CCR7
Arrays of actin filaments	Present	Absent

attack. Once in this tissue, the effector Th cells are activated by pMHCs presented by any APC (including mature DCs, macrophages and activated B cells) that have congregated in response to inflammatory signals, and the CTLs are activated by any host cell (including APCs) presenting antigenic peptides on MHC class I. At the conclusion of the primary response, the mature DCs do not "de-differentiate" and resume immature status but instead die by apoptosis, helping to dampen the T cell response after it is no longer needed. The properties of immature and mature DCs are summarized in **Table 7-2**.

iii) Mechanism of DC Maturation

The precise mechanism underlying DC maturation has yet to be completely elucidated. Complex changes to gene transcription programs are initiated by PRR engagement and the engagement of a DC's cytokine receptors by cytokines produced by other

TABLE 7-3	Comparison of Professional APCs		
	Mature DCs	**Macrophages**	**B Cells**
Level of MHC class II	Very high	High	High
Level of constitutive costimulatory molecule expression	High	Moderate	Low
Capable of cross-presentation	+++	++	+/−
Activates naïve T cells	Yes	No	No
Activates effector and memory T cells	Yes	Yes	Yes

activated innate leukocytes (such as macrophages) in the immediate area. The maturing DC's actin cytoskeleton is reorganized, and the efficiency of its endocytic compartment is increased. The receptors used by immature DCs to internalize antigen are downregulated, and the turnover of pMHCs on the DC surface is dramatically slowed, freezing the antigens displayed into a "peptide snapshot" of the tissue under attack. Surface expression of MHC class II increases by 5- to 20-fold, allowing a mature DC to rapidly present many copies of different antigenic pMHCs to Th cells. If the TCR expressed by a naïve Th cell recognizes one of the pMHCs displayed by the mature DC, costimulatory molecules such as CD28 and CD40L are upregulated on the Th cell surface (see Ch. 9). The binding of CD40L on the Th cell to CD40 on the DC greatly upregulates DC expression of the B7 costimulatory molecules (also called CD80/86) that bind to CD28. The high levels of these molecules, which greatly exceed those expressed by other types of APCs, are necessary to supplement the signal delivered by TCR binding to pMHC. The DC is able to push the naïve Th cell over the activation threshold in a way that other APCs cannot. The properties of mature DCs and other professional APCs are compared in **Table 7-3**.

As well as activating naïve Th cells by presenting peptide on MHC class II, mature DCs can present peptides on MHC class I to activate naïve CD8$^+$ Tc cells. Such presentation readily occurs if the DC is infected with a pathogen that replicates intracellularly. In addition, when a DC phagocytoses a whole pathogen or its components, some of the captured proteins may be diverted from the exogenous antigen processing pathway into the endogenous pathway via cross-presentation (see later). As a result, antigenic peptides from an intracellular pathogen may appear on MHC class I even if the pathogen has not infected the DC. Upon TCR engagement by these pMHCs, the same sequence of costimulatory events is triggered such that the naïve Tc cell is pushed over the activation threshold by the mature DC.

Excellent images of T cells interacting with immature and mature DCs can be seen at http://www.springerimages. com/Images/Medicine AndPublicHealth/1-10.1007 _s00277-006-0117-1-1.

iv) Influence of DCs on T Cell Response Type

DC subsets are a key means by which the immune system tailors its response to a particular pathogen. Different DC subsets are present in various sites throughout the body. These subsets express different collections of PRRs, and different pathogens engage different subsets of these PRRs. Depending on which PRRs are engaged, a mature DC not only participates in innate responses by producing the cytokines, chemokines and other molecules best suited to eliminating the attacking pathogen directly, but also develops the properties needed to drive the differentiation of any naïve Th cell it activates down one of several parallel paths. For example, a DC subset that preferentially produces IL-12 and the IFNs in response to engagement of its PRRs causes the naïve Th cell it has activated to differentiate into **Th1 effector cells.** Th1 cells secrete cytokines that assist in adaptive responses against intracellular threats. In contrast, a DC subset that preferentially produces IL-13 helps to influence the naïve Th cell it has activated to differentiate into **Th2 effector cells.** Th2 cells secrete cytokines that assist in adaptive responses against extracellular threats. Finally, a DC subset that preferentially produces IL-6 helps to influence the naïve Th cell it has activated to differentiate into **Th17 effector cells.** Th17 cells both participate in adaptive responses against extracellular threats and play a role in autoimmunity.

Th subsets are discussed in more detail in Chapter 9.

II. Macrophages as APCs

Macrophages excel at ingesting whole bacteria or parasites or other large native antigens present in peripheral tissues. These phagocytes quickly digest these bulky entities, producing a spectrum of antigenic peptides that are combined with MHC class II and can be presented to Th cells. However, unlike mature DCs, macrophages express only moderate levels of costimulatory molecules and so cannot activate naïve Th cells. Instead, macrophages make their contribution as APCs by activating memory or effector T cells that have homed to an inflamed tissue. The IFNγ secreted by a Th effector interacting with a macrophage hyperactivates the macrophage, increasing antigen presentation and thus pathogen clearance. Activated macrophages also produce cytokines that promote Th effector cell differentiation and upregulate MHC class II on other APCs (including DCs) in the immediate vicinity. Macrophages are thus important amplifiers of the adaptive response.

The differing activation requirements of naïve, effector and memory T cells are discussed in Chapter 9.

III. B Cells as APCs

B cells are considered professional APCs because they constitutively express MHC class II. B cells use their aggregated BCRs to internalize protein antigens by receptor-mediated endocytosis. The internalized antigen enters the exogenous processing pathway of the B cell such that peptide–MHC class II complexes appear on its surface. A B cell acting as an APC is one of the most efficient antigen presenters in the body because the BCR binds specific antigen with high affinity and can thus capture antigens present at very low concentrations. However, B cells do not generally serve as APCs in the primary response to a Td antigen because antigen-specific B cells and the antigen-specific Th cells required to help them are very rare in an unimmunized individual. Thus, the chance that a naïve B cell recognizing antigen X and able to act as an APC is in close proximity to an equally rare naïve anti-X Th cell is exceedingly small. In addition, resting mature naïve B cells express only low levels of the costimulatory molecules required for full Th activation. However, once activated, B cells quickly upregulate B7 expression and become effective APCs. Moreover, in a secondary response, anti-X memory B and Th cells are present in significantly greater numbers. Memory B cells thus frequently serve as APCs in the secondary response and become increasingly prominent in this function upon each subsequent encounter with antigen X.

C. Major Antigen Processing and Presentation Pathways

As introduced previously, there are four major pathways of protein antigen processing and presentation, which we now discuss in detail. We conclude this chapter with a discussion of other antigen processing and presentation pathways that are less well defined but still important for immune defense.

NOTE: Several medically important viruses, including HIV, cytomegalovirus (CMV) and human papillomavirus (HPV), escape immune system destruction by interfering with an infected host cell's antigen processing and presentation pathways. These mechanisms are discussed in Chapter 13.

I. Exogenous Antigen Processing Pathway

i) Generation of Peptides via the Exogenous Pathway

An APC that has internalized a whole microbe or macromolecule encloses the entity in a membrane-bound transport vesicle. As introduced in Chapter 3, the transport vesicle enters the endocytic processing system and is successively fused to a series of protease-containing endosomes of increasingly acidic pH. Within the endolysosome, the proteins of the microbe or macromolecule are degraded into peptides of 10–30 amino acids. The peptides are then conveyed in a transport vesicle to specialized late

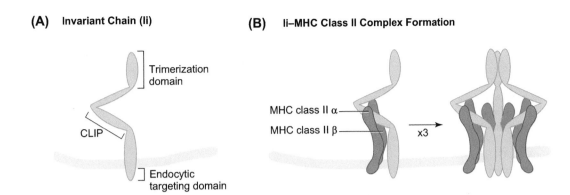

Fig. 7-4
Interaction of MHC Class II and Invariant Chain

(A) Diagram of an Ii monomer showing the trimerization, CLIP, and endocytic targeting domains. **(B)** Within the ER, an MHC class II αβ heterodimer interacts with an Ii monomer such that the CLIP domain is positioned in the peptide-binding groove of the MHC class II protein. Three MHC class II molecules bind to three Ii chains to form a nonameric complex in which each MHC class II binding groove is blocked by a CLIP. [*Adapted from Pieters J. (1997). MHC class II restricted presentation.* Current Opinion in Immunology 9, *89–96.*]

endosomal compartments known as **MIICs** (MHC class II compartments; see later). It is in the MIICs where the peptides bind to newly synthesized MHC class II molecules. Peptides and MHC class II molecules must arrive in the MIICs in a synchronized fashion: if a peptide is not immediately bound by an MHC class II molecule, the peptide is rapidly degraded by lysosomal enzymes contained within the MIIC.

ii) MHC Class II Molecules in the ER and Endosomes
During their synthesis on membrane-bound ribosomes, the α and β chains of the MHC class II molecule are cotranslationally inserted into the ER membrane. A third protein called the **invariant chain** (Ii) is coordinately expressed with the MHC class II chains and is also cotranslationally inserted into the ER membrane. The function of Ii is to bind to newly assembled MHC class II heterodimers and protect the binding groove from being occupied by endogenous peptides present in the ER (see later). The MHC class II molecules are thus "preserved" for the exogenous peptides generated in the endocytic compartment. The part of the Ii molecule that sits in the MHC class II binding groove is called **CLIP** (class II associated invariant chain peptide) (**Fig. 7-4A**). Ii also contains a trimerization domain that allows the formation of a nonameric complex consisting of three MHC class II heterodimers (three αβ pairs) and one Ii trimer (three Ii polypeptides) (**Fig. 7-4B**). Once complexed to Ii, the MHC class II molecules enter the Golgi complex but are then deflected away from the cell's secretory pathway and into its endocytic system by localization sequences in the Ii protein (**Fig. 7-5**). In an endolysosome, most of the Ii protein is degraded by the sequential action of a family of cathepsin enzymes, leaving CLIP stuck in the grooves of the now monomeric MHC class II molecules. The MHC-CLIP complexes then enter the MIICs.

iii) Peptide Loading onto MHC Class II
The next step in exogenous antigen processing is the exchange of the CLIP peptide in the MHC class II binding groove for an exogenous peptide. A non-classical MHC molecule called HLA-DM in humans is essential for this process, but the mechanism remains unclear. (The equivalent molecule in mice is called H-2DM.) Although HLA-DM closely resembles a conventional MHC class II molecule, it does not bind peptides. Instead, the association of HLA-DM with MHC-CLIP likely induces a conformational change that promotes the release of CLIP. HLA-DM then stabilizes the empty MHC class II heterodimer until the exogenous peptide is loaded (refer to **Fig. 7-5**). The actual mechanism of peptide loading has yet to be completely defined, but some studies have indicated that HLA-DM plays an active role in determining which peptides can bind in a given MHC class II antigen-binding site. In any case, once an exogenous peptide lodges stably in the MHC class II groove, the conformation of the MHC

Fig. 7-5
MHC Class II Antigen Presentation Pathway

Within the ER, nonameric complexes are formed in which the peptide-binding groove of each MHC class II heterodimer is blocked by the CLIP domain of an Ii chain (1). These complexes exit the ER, pass through the Golgi (2), and enter an endolysosome (3). If the APC also internalizes an antigen (4), the antigen is degraded by endolysosomal proteases to generate exogenous peptides (5) that are transferred into an MIIC (6). Meanwhile, back in the original endolysosome, the Ii is partially degraded, leaving the CLIP peptide in the MHC class II binding groove (7). The MHC-CLIP complexes are transferred to the MIIC, where the exchange of CLIP for exogenous peptide is mediated by HLA-DM (8). A transport vesicle (9) takes the peptide-loaded MHC class II molecule to the APC surface where it is inserted in the plasma membrane (10).

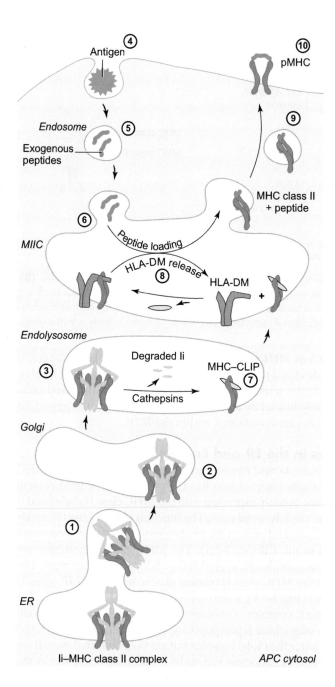

class II molecule alters again to force dissociation of HLA-DM. The pMHC is then transported out of the MIICs in a vesicle, inserted into the APC membrane by reverse vesicle fusion, and displayed to CD4+ Th cells (refer to **Fig. 7-5**).

II. Endogenous Antigen Processing Pathway

Endogenous antigen processing differs from exogenous processing in three ways: (1) Unlike the limited number of cell types that express MHC class II and so can function as professional APCs, almost any nucleated body cell expresses MHC class I and so can present peptides derived from intracellular antigens to CD8+ CTL effectors (but not naïve Tc cells). Thus, almost any body cell that has become aberrant due to cancer or intracellular infection becomes a target cell for CTL-mediated cytolysis. (2) The processing of the protein antigen takes place in the cytosol rather than in the endocytic system. (3) Peptides generated in the cytosol meet newly synthesized MHC class I molecules in the ER rather than in the MIICs.

i) Generation of Peptides via the Endogenous Pathway

Viruses or intracellular bacteria that have taken over a host cell force it to use its protein synthesis machinery to make viral or bacterial proteins. Antigenic proteins are thus derived from the translation of viral or bacterial mRNA on host cytosolic ribosomes. Similarly, abnormal proteins synthesized in the cytoplasm of tumor cells can give rise to peptides that appear to be "non-self" to the host's immune system and thus warrant an immune response. The process used to generate peptides from such foreign proteins is basically the same as that used to deal with misfolded or damaged host proteins. In the cytosol of every host cell are large numbers of huge, multi-subunit protease complexes called **proteasomes**. The function of proteasomes is to degrade proteins into peptides. There are two major types of proteasomes: the *standard proteasome* and the *immunoproteasome* **(Fig.7-6)**. Both types of proteasome contain a basic structure called the 20S core proteasome, a hollow cylinder of four stacked polypeptide rings made up of α and β subunits. The α subunits maintain the conformation of the proteasome core while the β subunits are its catalytically active components. The standard proteasome and immunoproteasome differ slightly in the catalytic β subunits included in their 20S cores. In addition to the 20S core, the standard proteasome contains two copies of a structure called the 19S regulatory complex. In addition to its slightly modified 20S core, the immunoproteasome contains two copies of the 19S regulatory complex, and two copies of a proteasome activator (PA) regulatory complex called PA28.

Standard proteasomes are present in all host cells and carry out a housekeeping function by routinely degrading spent and unwanted self proteins. The peptides produced by a standard proteasome are usually 8–18 residues in length. About 20% are the size (8–10 amino acids) that fits neatly into an MHC class I binding groove. Additional trimming of peptides that are initially too large can be carried out by peptidases in the cytosol. The degradation of host proteins in this way and subsequent loading of these peptides onto MHC class I allow constant scanning of self components and the monitoring of the cell's internal health. In contrast to standard proteasomes, immunoproteasomes are not present in most resting host cells, the exceptions being a narrow range of cell types that includes DCs. However, immunoproteasome formation is induced in most host cells by exposure to pro-inflammatory cytokines (such as IFNγ and TNF) that are present at high concentrations during a pathogen attack. Accordingly, immunoproteasomes most often produce peptides from foreign rather than self proteins and are responsible for the majority of the antigen processing associated with immune responses. Most peptides produced by the immunoproteasome are the optimal 8–10 amino acids in length.

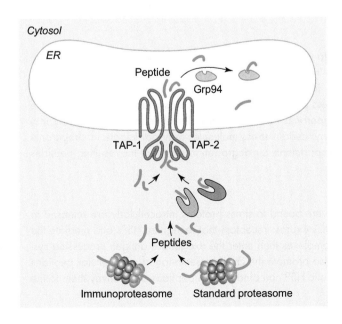

Fig. 7-6
Proteasomes and the TAP Transporter

The TAP transporter is composed of the TAP-1 and TAP-2 subunits and is positioned in the ER membrane. Peptides produced by the immunoproteasome or the standard proteasome in the vicinity of the TAP can bind directly to its peptide-binding site. Alternatively, peptides bound to chaperone proteins (such as the HSP molecules depicted here) are conveyed to the TAP, where they are released. The TAP then transports the peptides into the ER interior, where they bind to ER-resident chaperones (such as the Grp94 molecule depicted here). Peptides of the correct size are then loaded onto MHC class I molecules (not shown).

ii) Transport of Peptides into the Endoplasmic Reticulum

To induce an immune response, the peptides generated by the proteasomes in a host cell's cytosol must access the binding site of an MHC class I molecule. However, the peptide and the peptide-binding site of the MHC class I molecule are on topologically opposite sides of the cell's membrane system. Unlike the case for MHC class II molecules and their peptides, no vesicle fusion event brings MHC class I molecules and their peptides together. Instead, peptides generated in the cytosol are transported directly into the ER where MHC class I molecules are synthesized.

In the membrane of the ER are positioned transporter structures known as **TAP** (transporter associated with antigen processing) (refer to **Fig. 7-6**). TAP is a heterodimeric molecule composed of two subunits, TAP-1 and TAP-2, which are encoded by genes in the MHC. Structurally, TAP-1 and TAP-2 contain domains that project into the ER lumen, hydrophobic domains that span the ER membrane, and domains that extend into the cytosol and combine to form a single peptide-binding site. Peptides produced by the action of proteasomes are normally subject to very rapid degradation in the cytoplasm but can be rescued from this outcome by binding directly to TAP. Alternatively, the peptides can be bound by "chaperone" proteins, including HSPs and other stress molecules (**see Box 7-1**) that protect the peptides from degradation and escort them to TAP for transfer into the ER. It has been estimated that TAP molecules can translocate 20,000 peptides/min/cell, more than enough to ensure a steady supply for loading onto nascent MHC class I molecules generated in the ER at a rate of 10–100/min. TAP preferentially imports peptides of 8–12 residues in length, although longer peptides can be transported with lower efficiency. Once in the ER lumen, the peptides meet one of four fates. Some peptides are bound immediately to MHC class I molecules, whereas others are temporarily taken up by chaperone proteins resident in the ER (such as Grp94; glucose-regulated protein 94) and protected from further degradation prior to loading onto MHC class I. Still other peptides are trimmed by ER-resident peptidases to achieve the correct length and C-terminus necessary for fitting into the MHC class I groove. Lastly, some peptides are rapidly re-translocated back

---

Box 7-1 Heat Shock/Stress Proteins as Peptide Chaperones for Antigen Processing

Heat shock proteins (HSPs) are highly conserved members of a larger group of proteins called *stress proteins*. Stress protein expression is sharply increased in cells subjected to environmental assaults such as a sudden temperature increase, cancerous transformation or inflammation. Stress proteins are important for immunity because they bind to proteins and peptides and facilitate constant immune system surveillance of both the intracellular and extracellular protein environment.

Intracellular Surveillance

Some stress proteins act as quality control monitors in the ER, binding to misfolded proteins and preventing them from leaving the ER. Other stress proteins are cytosolic and have a "chaperone" function in that they facilitate polypeptide folding and protect newly synthesized proteins from intracellular degradation. Several HSPs act as intracellular disposal tags, binding to an unwanted protein and conveying it to the proteasome for destruction. HSPs and other stress proteins can also function as chaperones protecting the peptide products of proteasomal degradation in the cytosol. These chaperones facilitate the transfer of endogenous peptides into the ER and the loading of these peptides onto MHC class I. Conversely, certain stress molecules may participate in chaperone-mediated autophagy, diverting intracellular proteins into endocytic compartments for degradation such that the resulting peptides are presented on MHC class II.

Extracellular Surveillance

When a cell dies of necrosis, complexes of peptides or proteins that were bound to stress proteins intracellularly are released to the extracellular environment. Activated macrophages and immature DCs express receptors that recognize HSPs and mediate the uptake of HSP–peptide complexes by these APCs. The HSP–peptide complexes then enter the exogenous antigen processing system, and the peptides emerge displayed on MHC class II. HSPs can also promote the cross-presentation of extracellular peptides on MHC class I if these peptides happen to access the cytosol. A cytosolic HSP can bind to such peptides and convey them to the ER for loading on MHC class I.

into the cytosol where they are either degraded or re-imported back into the ER via TAP and subjected to further trimming.

iii) MHC Class I Molecules in the ER

The α chain of an MHC class I molecule is synthesized on a membrane-bound ribosome and is cotranslationally inserted into the membrane of the ER. As it enters the ER membrane, the MHC class I α chain associates with a transmembrane chaperone protein called calnexin and a non-transmembrane enzyme called ERp57. Calnexin facilitates proper polypeptide folding and association of the α chain with the coordinately expressed β2m chain. In humans, calnexin is then replaced by a soluble chaperone protein called calreticulin. (In mice, either the original calnexin molecule or an incoming calreticulin molecule associates with the heterodimer after association with β2m.) ERp57 binds to both calnexin and calreticulin and works with these molecules to catalyze the formation of disulfide bonds in MHC class I α chains. ERp57 also promotes the loading of peptide into the MHC class I binding groove.

iv) Peptide Loading onto MHC Class I

To bring a newly synthesized MHC class I heterodimer (with its chaperones) into the vicinity of TAP and the peptides, the MHC class I molecule transiently interacts with a protein called **tapasin** that binds to ERp57, MHC class I and TAP (**Fig. 7-7**). Tapasin helps to stabilize the empty MHC class I heterodimer in a conformation suitable for peptide loading. Exactly how a peptide, either free or bound to a chaperone, accesses the MHC class I binding groove has yet to be determined, but tapasin is important for this process. Tapasin also works with calreticulin to prevent improperly loaded peptide–MHC class I complexes from leaving the ER. Peptide loading is a crucial step in antigen presentation because an MHC class I heterodimer that is transported to the cell surface without a peptide in its groove is unstable and rapidly lost. With the insertion of a pMHC into the membrane of a host cell, it is ready for inspection by CD8+ T cells.

Focus on Relevant Research

"Immunogenetics of *Toxoplasma gondii* Informs Vaccine Design" by Henriquez, F.L., Woods, S., Cong, H., McLeod, R., and Roberts, C.W. (2010) *Trends in Parasitology 26*, 550–555.

The parasite *Toxoplasma gondii* is an important pathogen worldwide and among the leading causes of death in AIDS patients. Antiparasitic drugs are only minimally effective against this scourge, making the development of a prophylactic *T. gondii* vaccine a key objective for researchers in this field. In this article, Henriquez *et al.* review cutting-edge research that (1) points to particular MHC class I alleles as being critical for effective protection against *T. gondii* and (2) tentatively identifies important parasite proteins and short peptides whose processing and presentation may promote CD8+ T cell-mediated immunity against the parasite. Techniques reviewed by the authors include the generation and analysis of mutant mice missing key MHC class I alleles, "caged" MHC tetramer technology, and the use of predictive bioinformatics algorithms.

Figures in the article illustrate (1) the natural mechanisms by which *T. gondii* infects a cell and presumably elicits an immune response and (2) the means by which a vaccination strategy might "mimic" the natural infection such that the *T. gondii* proteins are properly processed and displayed. Epitopes for both CD4+ and CD8+ T cells must be provided by the vaccine immunogen, and a PAMP or "TLR ligand" must be present that can engage the appropriate PRRs of immature DCs and induce their maturation. Ideally, these mature DCs would then display the vaccine epitopes on their surfaces in a suitable context for activating pathogen-specific naïve T cells. In vaccine formulations, components that play the role of TLR ligands in the absence of actual infection are known as "adjuvants." Adjuvants are described further in Chapter 14.

Fig. 7-7

MHC Class I Antigen Presentation Pathway

An MHC class I protein assembled in the ER and bound to its chaperones calreticulin and ERp57 first interacts with tapasin to stabilize the empty molecule and bring it close to the TAP transporter (1). Meanwhile, antigenic peptides generated by proteasomal degradation, which may be bound to HSP chaperones, arrive on the cytosolic side of the TAP (2). The TAP imports the peptides into the ER (3), where they are bound by ER-resident chaperones. The peptides are then transferred into the peptide-binding groove of the MHC class I molecule (4), and the loaded MHC molecule passes through the Golgi (5) and into a transport vesicle (6) prior to insertion in the host cell membrane (7).

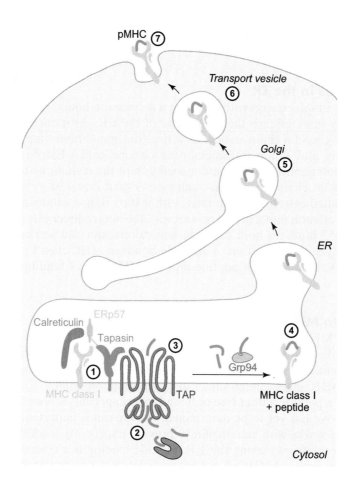

III. Cross-Presentation on MHC Class I

Cross-presentation refers to the display on MHC class I of peptides from *extracellularly acquired* antigens. Although this process is thought to be a major means by which DCs can activate naïve Tc cells, the underlying mechanisms are still not well understood. Cross-presentation was discovered when researchers found that CD8+ CTL responses could be mounted to certain antigens that were known to be extracellular, and that peptides from viral antigens could be presented on MHC class I even when the endogenous processing pathway was blocked. The phenomenon was called "cross-presentation" because the viral antigen appeared to physically "cross over" from an infected host cell to an uninfected APC that presented peptides from the antigen on MHC class I as if the antigen had originated in the interior of the APC. Although various APCs appear to be capable of cross-presentation *in vitro*, DCs are the main cell type responsible for antigen cross-presentation *in vivo*. DC subsets vary in their ability to carry out cross-presentation, with resident DCs located in the lungs and blood vessel walls being particularly proficient. In contrast, migratory DCs are less able to cross-present and concentrate on displaying peptides on MHC class II.

How can exogenous peptides be loaded onto MHC class I molecules? APCs (particularly DCs) most often acquire viral antigens by internalizing debris from infected cells that have undergone necrotic or apoptotic death. In rare cases, DCs may also acquire portions of the membranes of live infected cells (by an ill-defined process sometimes called "nibbling"). In all these situations, because the viral protein has entered the APC from the extracellular environment, it is initially directed to the endocytic system in the usual way. Thus, viral peptides appear on the APC surface associated with MHC class II, and a Th response to the antigen can be mounted. However, during the initial processing of the viral proteins in the early endosomes, a fraction of the resulting polypeptides may be actively transported from the endosomes into the cytosol by endosomal membrane

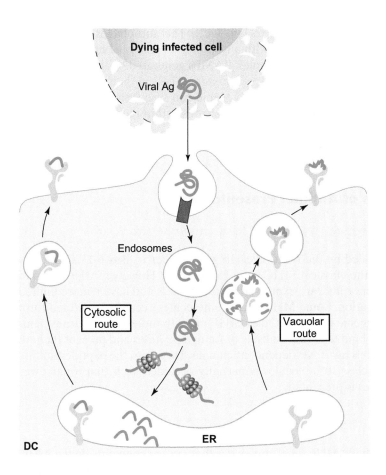

Dying infected cell

Viral Ag

Endosomes

Cytosolic
route

Vacuolar
route

ER

DC

▶ **Fig. 7-8**
***Models of Cross-Presen-
tation on MHC Class I***

When an APC takes up antigen from
its extracellular environment, such as
the viral protein illustrated here, the
antigen enters the cell by the usual
exogenous processing pathway.
However, the antigen can then
cross into the endogenous processing pathway by either a cytosolic
route involving proteasomes (left), or
a vacuolar route involving diver-
sion of MHC class I molecules into
endosomes (right). In both cases,
peptides derived from the antigen
are combined with MHC class I
molecules and are presented on the
APC surface.

proteins whose role in this translocation process is not yet clear. In the cytosol, the polypeptides are taken up by proteasomes and degraded to peptides (**Fig. 7-8**). The viral
peptides are then transported via TAP into the ER and loaded onto MHC class I just as
if the viral protein had originated within the APC itself. This pathway has been dubbed
the "cytosolic" route of cross-presentation. Alternatively, there is some evidence for
cross-presentation via a "vacuolar" route that does not involve proteasomes. According
to this model, some MHC class I molecules may be diverted from the ER into an early
endosome where an extracellularly acquired viral antigen is undergoing degradation by
lysosomal enzymes. The combination of these diverted MHC class I molecules with the
viral peptides within the endosome creates pMHC complexes that subsequently can be
transported to the cell surface for antigen presentation. Lastly, a very few extracellular
proteins appear to be able to cross the plasma membrane directly, bypass the endocytic
system entirely and enter a proteasome. Peptides generated via this latter mechanism
could presumably associate with MHC class I via the cytosolic route. Regardless of their
pathway of cross-presentation, the end result is the display of viral peptide-MHC class I
complexes on the APC surface that can be inspected by antiviral CD8+ T cells.

IV. Autophagic Presentation on MHC Class II

Just as cross-presentation enables the display of exogenous peptides on MHC class
I molecules, autophagy allows the display of endogenous peptides on MHC class II.
Recent analyses of peptides bound to MHC class II molecules of mouse and human
APCs have shown that 20–30% of these ligands originate from cytosolic or nuclear
proteins. As described in Chapter 3, unwanted cytosolic entities are routinely dealt
with by a form of autophagy known as macroautophagy. Macroautophagy is generally
reserved for intracellular entities that are too large for direct proteasomal degradation and/or cannot be kept soluble by chaperone proteins. The reader will recall that,
during macroautophagy, a double-layered isolation membrane circularizes around the

Antigens from intracellular
bacteria and parasites can
also be processed and their
peptides displayed on MHC
class I via cross-presentation.

entity to form an autophagosome (refer to Fig. 3-12). The autophagosome can then participate in innate responses by fusing with a lysosome to generate an autophagolysome in which the entire structure is degraded. PAMPs generated by this degradation can then be shunted into endosomal vesicles expressing TLRs in their membranes. To participate in adaptive responses, an autophagosome can fuse instead with an MIIC containing MHC class II molecules. Lysosomal enzymes within the MIIC degrade the autophagosome and its contents, releasing peptides that can be loaded onto MHC class II via HLA-DM. These endogenous peptide-MHC class II complexes are then displayed on the APC surface for inspection by CD4+ T cells.

D. Other Methods of Antigen Presentation

I. Antigen Presentation by MHC Class Ib Molecules

The glycoproteins encoded by the MHC class Ib genes (refer to **Box 6-1**) are closely related in structure to the classical MHC class I molecules. However, MHC class Ib molecules are less polymorphic, are expressed at lower levels and have a more limited pattern of tissue distribution. Some MHC class Ib molecules occur in a secreted form and do not bind to antigenic peptides. Other MHC class Ib molecules are transmembrane proteins that can bind to certain subsets of foreign peptides and present them to subsets of αβ and γδ T cells in a TAP-dependent manner. However, the peptide-binding groove in these MHC class Ib molecules is partially occluded such that a narrower range of shorter peptides is presented.

Antigen presentation to γδ T cells and NKT cells is discussed in depth in Chapter 11.

II. Non-Peptide Antigen Presentation by CD1 Molecules

In Chapter 6, we described "MHC-like" molecules that are encoded outside the MHC but feature an MHC-like fold in their structures. Five MHC-like **CD1 molecules** have been identified: CD1a, CD1b and CD1c (which make up "Group 1"); CD1d (the sole "Group 2" member); and CD1e (which shows only low homology to the other CD1 molecules and so is not considered a member of either group). The CD1 proteins are of particular interest with respect to antigen presentation because these non-polymorphic proteins can present lipid-based (rather than peptide) antigens to certain T cell subsets. The function of CD1 molecules is therefore thought to be the monitoring of the health of the host's cellular lipids, as well as the detection of pathogen lipids. Human APCs can express all five CD1 isoforms, whereas mouse APCs express only CD1d. CD1d is constitutively expressed on APCs of both species, whereas the expression of the Group 1 CD1 molecules on human APCs is upregulated by TLR signaling. Indeed, the expression of a particular CD1 isoform by human APCs may be dictated by cytokines or other signals present in the immediate inflammatory microenvironment.

The antigen-binding groove in CD1 molecules is much more hydrophobic than that of classical MHC molecules, suiting it to lipid binding. In humans, CD1a, CD1b and CD1c molecules present lipid antigens derived from diacylglycerol, phospholipid, lipopeptide, polyketide, glycolipid, and glycosphingolipid molecules to subsets of αβ Th and Tc cells. CD1e has a large lipid-binding pocket but is not expressed on the APC surface and so does not participate directly in antigen presentation. In both mice and humans, CD1d molecules present a very restricted collection of ceramide-based antigens to NKT cells and some T cell subsets. It is thought that the TCRs of these cells "see" a combined epitope composed of amino acids of the CD1 molecule plus a small portion of the carbohydrate head group of the lipid. NKT and T cells can be very discriminating in their recognition of CD1-presented antigens, failing to respond if the orientation of even a single hydroxyl group of the antigen is changed. The consequences of CD1-mediated antigen presentation are virtually identical to those of

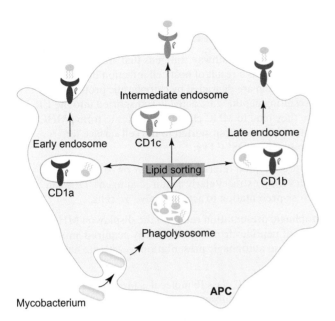

Fig. 7-9
Presentation of Mycobacterial Lipid Antigens by Human CD1 Molecules

A mycobacterium that has been phagocytosed by an APC is digested in a phagolysosome. The bacterial lipids are sorted by structure into the indicated endosome where they are loaded onto the indicated CD1 family members. The CD1-lipid complexes are then displayed on the APC surface for presentation to particular subsets of αβ and γδ T cells and NKT cells.

MHC-mediated peptide presentation; that is, activated NKT cells and Th effectors secrete cytokines, while Tc cells generate CTLs that kill target cells by cytolysis. The activities of these CD1-restricted effectors have been shown to influence the outcome of infections with various viruses and species of bacteria, fungi and protozoa.

Antigen processing and presentation by CD1 molecules appears to utilize elements of both the exogenous and endogenous pathways. Like MHC class I, CD1 chains must be associated with β2m to be transported to the cell surface, but, unlike MHC class I, antigen loading of CD1 molecules does not take place in the ER. Instead, like MHC class II, CD1 molecules are targeted to the endocytic system where they are loaded with antigen, but no association with either Ii or HLA-DM is required. Because of the presence or absence of particular amino acid motifs, different human CD1 molecules accumulate in different endosomal compartments. For example, CD1a molecules are found in early endosomes, while CD1c molecules tend to collect in intermediate endosomes, and CD1b molecules are directed to late endosomes. CD1e appears to translocate from the Golgi to the lysosomes in response to infection and then facilitates the selection of exogenous and endogenous lipids to be displayed by other CD1 family members. In addition to this endosomal sorting of CD1 molecules, different lipid antigens also tend to accumulate in different endosomal compartments. Depending on the structure of the lipid, it is sorted and confined to either an early endosome, an intermediate endosome or a late endosome. At the molecular level, this sorting seems to be determined at least partially by the length of the lipid's alkyl chains, with the earliest endosomes containing lipids with the longest alkyl chains. These measures allow the APC to ensure that the right lipid is loaded onto the right CD1 molecule. Once loaded with lipid, a CD1 molecule is then transported to the plasma membrane for display on the APC surface. In addition, extracellular lipids with short alkyl chains can be loaded directly into the binding grooves of CD1 molecules that have already made it to the APC surface. The actual kinetics and mechanisms involved in the antigen loading of CD1 molecules and their subsequent presentation on the APC surface remain unclear. A schematic representation of CD1-mediated antigen presentation appears in **Figure 7-9**.

MHC molecules presenting foreign peptides are the body's signposts to the immune system that a T cell–mediated adaptive response is required. The next chapter discusses the genes and proteins of TCRs, the antigen receptor molecules that carry out pMHC recognition.

Chapter 7 Take-Home Message

- The pMHCs recognized by the TCRs of T cells are assembled by four major antigen processing and presentation pathways: exogenous, endogenous, cross-presentation and autophagy.

- Professional APCs include mature DCs, macrophages and B cells. These cell types take up antigen efficiently and express MHC class II and costimulatory molecules inducibly or constitutively. However, only mature DCs can activate naïve Tc and Th cells.

- Immature DCs may be migratory or lymphoid-resident. DC maturation is triggered by PRR engagement by DAMPs/PAMPs plus pro-inflammatory cytokine signaling. Macrophages and B cells are efficient APCs for effector and memory T cells.

- In the exogenous pathway, extracellular antigens are internalized by APCs and degraded within endosomal compartments. The resulting peptides bind to MHC class II molecules to form pMHCs that are transported from the endosomes to the APC surface for recognition by CD4+ Th cells.

- In the endogenous pathway, antigens that are produced intracellularly as a result of host cell infection or transformation are degraded by cytoplasmic proteasomes. The resulting peptides are actively transported into the ER where they bind to MHC class I molecules to form pMHC complexes that are transported to the cell surface for recognition by CD8+ CTLs.

- Cross-presentation refers to the display on MHC class I of peptides from extracellularly acquired antigens. DCs can use cross-presentation to activate naïve Tc cells.

- Autophagic presentation refers to the display on MHC class II of peptides from intracellularly acquired antigens. DCs can use autophagic presentation to activate naïve Th cells.

- Non-classical MHC class Ib molecules present very short peptide antigens to subsets of αβ and γδ T cells.

- The CD1 proteins are MHC-like molecules that present lipid-based antigens to subsets of αβ and γδ T cells and NKT cells.

Did You Get it? a Self-Test Quiz

Section A

1) Distinguish between antigen processing and antigen presentation.

2) Distinguish between the exogenous and endogenous antigen processing and presentation pathways.

3) How did cross-presentation get its name?

4) Which major antigen processing pathway would be used to initiate an immune response against a liver cancer and why?

5) Why is it CD4+ rather than CD8+ T cells that respond to extracellular pathogens?

Section B

1) Can you define these terms? Langerhans cell, inflammatory DC, interstitial DC

2) What cell types can function as professional APCs and why?

3) Distinguish between the two main types of DCs present in a host at steady state.

4) Distinguish between migratory and lymphoid-resident DCs.

5) What is the main function of pDCs? thymic DCs? splenic DCs?

6) Why can't immature DCs activate T cells?

7) Does apoptotic cell death induce DC maturation? If not, why not?

8) Give three ways in which immature DCs differ from mature DCs.

9) Briefly outline how DCs influence Th cell differentiation.

10) Why are macrophages considered amplifiers of the adaptive response?

11) Why are B cells efficient APCs for the secondary response?

Section C

1) Can you define these terms? MIICs, CLIP, Ii

2) Where in the cell does the final degradation of extracellularly derived proteins most often occur?

3) Describe the structure and function of the invariant chain.

4) Where in the cell do exogenous peptides and MHC class II meet?

5) What is the role of HLA-DM during the loading of peptides onto MHC class II?

6) How can stress molecules contribute to antigen presentation?

7) Distinguish between standard proteasomes and immunoproteasomes in terms of structure and function.

8) Why are endogenous peptides transported into the ER?

Did You Get it? a Self-Test Quiz—Continued

9) Describe the structure and function of TAP.

10) How does tapasin facilitate pMHC formation?

11) What is a chaperone protein? Give three examples of such proteins and their functions.

12) How is cross-presentation thought to allow DCs to activate naïve Tc cells?

13) Describe two ways in which an uninfected APC might acquire antigens from intracellular pathogens.

14) How do the cytosolic and vacuolar routes of cross-presentation differ?

15) How does autophagy promote the display of intracellular peptides on MHC class II?

Section D

1) How does antigen presentation by MHC class I and Ib molecules differ?

2) What types of antigens are presented by CD1 molecules?

3) Describe how elements of both the standard exogenous and endogenous antigen processing pathways are used in antigen processing and presentation by CD1 molecules.

Can You Extrapolate? Some Conceptual Questions

1) Why might poor lymphatic drainage result in less effective immune responses?

2) In the case of the following mutations, each of which renders the affected component non-functional, would you expect the greatest effect to be on the activation of $CD4^+$ T cells, $CD8^+$ T cells, or both?

Functional mutation of

a) HLA-DM

b) TAP

c) Standard proteasome

d) Immunoproteasome

e) CLIP

f) β2m

3) The HA (hemagglutinin) molecule of influenza virus is a membrane-bound glycoprotein found on the surfaces of both virions and influenza-infected cells. The amino acid sequence of HA includes a leader (L) sequence that directs the protein into transport vesicles after it is translated in the ER. The HA protein is then displayed on the surface of the infected host cell after the transport vesicle fuses with the host cell's plasma membrane. If a new strain of influenza virus (HA-L⁻) is engineered in which the HA gene has no leader sequence, how might the CTL-mediated response to the HA-L⁻ virus compare with the CTL-mediated response to the wild type virus? You may assume that the wild type and HA-L⁻ viruses infect cells equally well.

Would You Like To Read More?

Davis, M. M., Altman, J. D., & Newell, E. W. (2011). Interrogating the repertoire: broadening the scope of peptide-MHC multimer analysis. *Nature Reviews Immunology, 11*(8), 551–558.

Haig, N. A., Guan, Z., Li, D., McMichael, A., Raetz, C. R., & Xu, X. N. (2011). Identification of self-lipids presented by CD1c and CD1d proteins. *Journal of Biological Chemistry, 286*(43), 37692–37701.

Kyewski, B., & Haskins, K. (2012). The classical dichotomy between presentation of endogenous antigens via the MHC class I pathway and exogenous antigens via the MHC class-II pathway. *Current Opinion in Immunology., 24*(1), 67–70.

Neefjes, J., Jongsma, M. L., Paul, P., & Bakke, O. (2011). Towards a systems understanding of MHC class I and MHC class II antigen presentation. *Nature Reviews Immunology, 11*(12), 823–836.

Purcell, A. W., & Elliott, T. (2008). Molecular machinations of the MHC-I peptide loading complex. *Current Opinion in Immunology, 20*(1), 75–81.

Ramachandra, L., Simmons, D., & Harding, C. V. (2009). MHC molecules and microbial antigen processing in phagosomes. *Current Opinion in Immunology, 21*(1), 98–104.

Sadegh-Nasseri, S., Natarajan, S., Chou, C. L., Hartman, I. Z., Narayan, K., & Kim, A. (2010). Conformational heterogeneity of MHC class II induced upon binding to different peptides is a key regulator in antigen presentation and epitope selection. *Immunologic Research, 47*(1–3), 56–64.

Saunders, P. M., & van Endert, P. (2011). Running the gauntlet: From peptide generation to antigen presentation by MHC class I. *Tissue Antigens, 78*(3), 161–170.

Watts, C., West, M. A., & Zaru, R. (2010). TLR signalling regulated antigen presentation in dendritic cells. *Current Opinion in Immunology, 22*(1), 124–130.

The T Cell Receptor: Proteins and Genes

The best way to have a good idea is to have lots of ideas.

Linus Pauling

A. TCR Proteins and Associated Molecules

As introduced in earlier chapters, the T cell receptor (TCR) is responsible for antigen recognition by T cells. TCRs are expressed by all T cells except their earliest precursors, so that most thymocytes and all mature T cells bear TCRs. Like B cells, the antigenic specificities of T cells are clonal in nature, meaning that (with rare exceptions) all members of a given T cell clone carry 10,000–30,000 identical copies of a receptor protein with a unique binding site. Like BCRs, TCRs possess V and C regions, and, like the Ig genes, the TCR genes undergo RAG-mediated recombination of V, D and J gene segments to produce a repertoire of receptors with considerable sequence diversity in the V region. However, TCRs differ from BCRs in two fundamental ways. Firstly, while the BCR repertoire can recognize and bind to virtually any structure, the spectrum of antigens recognized by TCRs is much more restricted. The vast majority of T cells bind to antigenic peptides that must be complexed to MHC molecules displayed on the surfaces of APCs or target cells. Only a small percentage of T cells recognize lipids or unprocessed antigens that may or may not be associated with MHC-related molecules. Secondly, while B cells secrete a form of their BCRs as antibody, T cells do not secrete their TCRs.

As outlined in Chapter 2, there are two types of TCRs defined by their component chains: TCRαβ and TCRγδ (**Fig. 8-1**). Mutually exclusive expression of these TCRs characterizes two distinct T cell subsets that develop independently: αβ T cells and γδ T cells. In humans, most mature T cells are αβ T cells, with only 5–10% being γδ T cells. While αβ T cells are concentrated in the secondary lymphoid tissues and function in adaptive responses, most γδ T cells are **intraepithelial** in location (tucked between the mucosal epithelial cells lining the body tracts) and participate in innate responses. Rather than pMHCs, γδ TCRs recognize a broad range of cell surface molecules that may be encountered in their natural, unprocessed forms. The biology of γδ T cells is discussed in more detail in Chapter 11.

MAK: Primer to the Immune Response. http://dx.doi.org/10.1016/B978-0-12-385245-8.00008-X

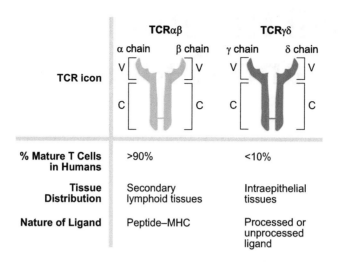

Fig. 8-1

Basic Characteristics of TCRαβ and TCRγδ

Two TCRs exist as defined by their component chains. Expression of these TCRs delineates two distinct T cell subsets: αβ T cells and γδ T cells. The proportions of these subsets in the total mature T lymphocyte population in humans are indicated, as are their general tissue distribution and type of ligand recognized.

	TCRαβ	TCRγδ
% Mature T Cells in Humans	>90%	<10%
Tissue Distribution	Secondary lymphoid tissues	Intraepithelial tissues
Nature of Ligand	Peptide–MHC	Processed or unprocessed ligand

I. Basic TCR Structure

Unlike mIg, which is made up of two light chains and two heavy chains and has two identical antigen-binding sites, a TCR is a heterodimeric glycoprotein with a single antigen-binding site. TCRαβ is composed of a TCRα chain (49 kDa) linked via a disulfide bond to a TCRβ chain (43 kDa), whereas TCRγδ consists of a TCRγ chain (40–55 kDa) linked via a disulfide bond to a TCRδ chain (45 kDa); TCRαδ or γβ structures have not been found in nature. Each TCR chain contains an Ig-like V domain, an Ig-like C domain, a cysteine-containing connecting sequence, a charged transmembrane portion, and a short cytoplasmic tail (**Fig. 8-2**). The V and C domains are arranged in Ig fold structures that are stabilized by intrachain disulfide bonds. The TCR V region is composed of the N-terminal ends of both TCR polypeptides and contains the antigen-binding site. Amino acids in the binding site establish contacts with both the antigenic peptide and the MHC molecule to which it is bound. Unlike the case for Igs (which undergo isotype switching), a TCR's C region is fixed for the life of a given T cell clone. The short connecting sequence located between the TCR C domain and the transmembrane domain is analogous to the Ig hinge.

As was true for the H and L chains of Igs, the V domain of each TCR polypeptide contains sites of increased amino acid variability. There are four such complementarity-determining or hypervariable (HV) regions in a TCR chain: CDR1, CDR2, CDR3 and HV4 (**Plate 8-1**). In TCRαβ, various CDR regions of both the TCRα and TCRβ chains are involved in pMHC recognition, depending on the particular TCR and pMHC involved. In some cases, the CDR1, CDR2 and HV4 regions interact with residues on the MHC class I or

Fig. 8-2

Schematic Representations of TCRαβ and TCRγδ Proteins

The Ig-like domains present in each of the TCRα, TCRβ, TCRγ and TCRδ chains are indicated, as are glycosylation sites and disulfide bonds. Note that the two types of TCRs have identical domain structures. N, N-terminus; C, C-terminus.
[With information from Klein, J., & Horejsí V. (1997). Immunology, 2nd ed., Blackwell Science, Oxford.]

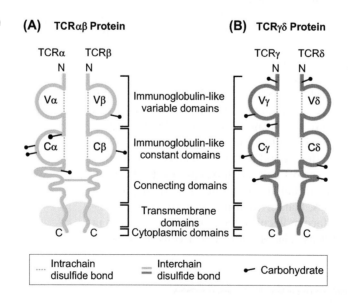

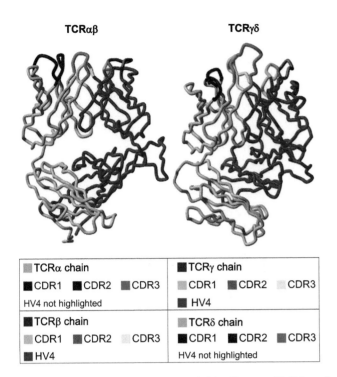

TCRαβ   **TCRγδ**

▨ TCRα chain	■ TCRγ chain
■ CDR1 ■ CDR2 ▨ CDR3	▨ CDR1 ■ CDR2 ▨ CDR3
HV4 not highlighted	■ HV4
■ TCRβ chain	▨ TCRδ chain
▨ CDR1 ■ CDR2 ▨ CDR3	■ CDR1 ■ CDR2 ■ CDR3
■ HV4	HV4 not highlighted

▸ **Plate 8-1**
X-Ray Crystal Structures of TCRαβ and TCRγδ

X-ray crystal structures showing the carbon backbones of TCRαβ and TCRγδ, with the hypervariable regions highlighted in the indicated colors. Additional such structures can be viewed using the NCBI structural database at http://www.ncbi.nlm.nih.gov/sites/structure. *[Reproduced by permission of Ruldph, M. G., & Wilson, I. A. (2002). The specificity of TCR/pMHC interaction. Current Opinion in Immunology 14, 52–65.]*

class II protein itself, while the highly diverse CDR3 regions preferentially make contact with the antigenic peptide nestled in the MHC groove. In other cases, the CDR1 and/or CDR2 regions may bind to part of the peptide, while the CDR3 regions contact the MHC molecule. The V domains of the TCRγ and δ chains also contain CDR1, CDR2, CDR3 and HV4 regions. However, because TCRγδ ligands are often non-peptides, the precise roles of the hypervariability regions in the binding of antigen to these receptors are thought to be slightly different. Antigen recognition by γδ TCRs is discussed in Chapter 11.

While HV4 is a region of amino acid hypervariability, it does not contact the peptide within the pMHC complex directly and so does not "determine complementarity" in the same way as CDRs 1–3.

II. The CD3 Complex

i) Structure

The short cytoplasmic tails of TCR chains are too short for signal transduction. This type of problem is solved in the BCR complex by the association of mIg with the Igα/Igβ heterodimer. The ITAMs present in the cytoplasmic tails of Igα/Igβ intracellularly transduce the signal triggered by antigen binding to mIg. In T cells, while the extracellular V domains of TCRαβ or TCRγδ recognize antigen, the TCR heterodimer as a whole must associate non-covalently with a collection of invariant transmembrane proteins known as the **CD3 complex** to transduce the signal. The CD3 complex contains three heterodimeric proteins made up of variable combinations of five ITAM-containing polypeptides designated CD3γ, CD3δ, CD3ε, CD3ζ (ζ, zeta), and CD3η (η, eta). In general, the CD3 complex that clusters around a human or mouse TCRαβ molecule is composed of a CD3εδ heterodimer, a CD3εγ heterodimer and a CD3ζζ homodimer (**Fig. 8-3**). Sometimes a CD3ζη heterodimer may replace the CD3ζζ homodimer. The CD3 complex in human TCRγδ contains CD3εδ, CD3εγ and CD3ζζ whereas the mouse TCRγδ contains two CD3εγ heterodimers plus CD3ζζ. A TCR-CD3 assembly is often called a **TCR complex**.

ii) Functions

The CD3 complex has two major functions. Firstly, as mentioned in the preceding section, the CD3 chains are required for intracellular signaling. Upon engagement of the TCR by pMHC, tyrosine residues in the CD3 ITAMs are phosphorylated by an intracellular signaling kinase called *Lck*. Additional signaling kinases can then be recruited to the receptor complex to propagate the signaling cascade. Secondly, the CD3 complex is required for TCR surface expression. In the ER, the TCR heterodimer physically associates with the CD3 complex before moving to the Golgi for glycosylation and finally transport to the T cell surface. The invariant, Ig-like extracellular domains present in the

Fig. 8-3
Structure of a TCR–CD3 Complex

The most usual form of the complete TCRαβ–CD3 complex in humans and mice is shown. The human TCRγδ–CD3 structure involves the TCRγ and TCRδ chains plus the same array of CD3 dimers. In mouse TCRγδ–CD3 complexes, the CD3εδ heterodimer is replaced with another CD3εγ heterodimer. In all cases, signaling initiated by pMHC binding is conveyed to the interior of the T cell via the ITAMs of the CD3 molecules. Inset: TCR-CD3 complex icon used in this book.

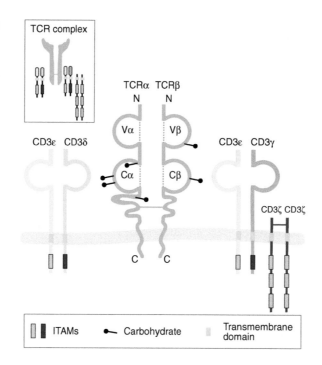

CD3γ, CD3δ and CD3ε chains interact with the Ig-like extracellular domains in the TCR chains to help to keep the TCR and the CD3 complex together throughout transport and on the cell surface. In the absence of CD3 expression, the TCR remains stalled in the ER. In fact, it is the synthesis and incorporation of the CD3ζ chain into the CD3 complex that controls the assembly and transport of the entire TCR–CD3 assembly.

"T Cell Receptor Gene Therapy: Strategies for Optimizing Transgenic TCR Pairing" by Govers, C., Sebestyén, Z., Coccoris, M., Willemsen, R.A., and Debets, R. (2010) *Trends in Molecular Medicine* 16, 77–87

Focus on Relevant Research

Immunosurveillance by T cells is the primary means by which the immune system defends against cancer. However, one of the main challenges presented by cancer cells is their similarity to normal self cells; that is, the lack of "foreign" antigens to engage a T cell's TCRs. T cell-mediated responses against tumor cells are therefore often slow and less efficient than responses against pathogen antigens. Numerous research groups are currently attempting to stem the modern tide of cancer deaths by genetically engineering T cells to be more specific and efficient killers of cancer cells. In this article, Govers *et al.* review these efforts, which are generally focused on (1) selection of TCR genes that recognize specific tumor cell antigens, (2) alteration of TCR subunit pairing, (3) enhancement of TCR cell surface expression, and (4) elimination of the need for CD3-mediated signaling. As an example of the latter case, the figure shown here (Figure 2, panel b from the article) is a representation of an experimental single-chain TCR containing its own intracellular CD3 signal transduction motif. It is hoped that T cells that are genetically engineered to express such optimized TCRs can be infused into patients to enhance the destruction of cancer cells. This approach is currently being studied in patients with various forms of cancer such as melanoma and lymphoma.

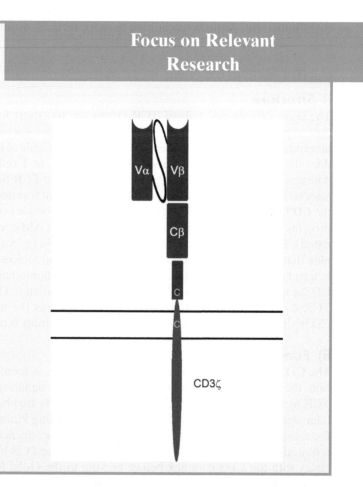

III. The CD4 and CD8 Coreceptors

i) Nature

As introduced in Chapters 2 and 6, mature αβ T cells patrolling the body's periphery bear either the CD4 or CD8 coreceptor. In humans, about two-thirds of mature αβ T cells are CD4⁺ cells, whereas one-third are CD8⁺ cells. Most mature γδ T cells express neither CD4 nor CD8, although some γδ T cells in the gut are CD8⁺ (see Ch. 11). CD4 and CD8 are called "coreceptors" because a molecule of either one of these proteins colocalizes with a TCR on the T cell surface and then binds to the *same* MHC molecule on the APC or target cell that is engaged by that particular TCR. CD4 binds to MHC class II molecules, whereas CD8 binds to MHC class I molecules. However, because CD4 and CD8 bind to sites on their respective MHC molecules that are in invariant regions *outside* the peptide-binding groove, coreceptor binding does not depend on the identity of the antigenic peptide.

The binding by the coreceptors to MHC molecules stabilizes the interaction so that the TCR can determine whether the peptide of the pMHC fits into the TCR's binding site. It is still not known why the expression of CD4 and recognition of peptide–MHC class II are almost exclusively associated with T helper cell functions, while CD8 expression and peptide–MHC class I recognition are features of T cells with cytotoxic powers.

ii) Structure

Despite their ostensibly equivalent functions, the CD4 and CD8 proteins show little similarity in either structure or amino acid sequence. In both mice and humans, CD4 is a transmembrane glycoprotein that is expressed as a single polypeptide on the cell surface. The CD4 protein contains four extracellular Ig-like domains that interact with the α2 and β2 domains of MHC class II; a transmembrane domain; and a cytoplasmic tail with sites that promote relatively strong association with Lck kinase (**Fig. 8-4A**). In

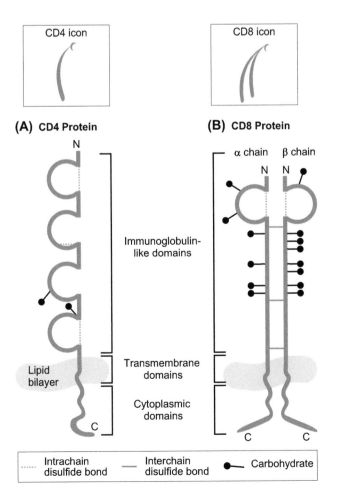

(A) CD4 Protein

(B) CD8 Protein

| Intrachain disulfide bond | Interchain disulfide bond | Carbohydrate |

▶ **Fig. 8-4**

Structures of the CD4 and CD8 Coreceptors

Schematic representation of the structures of the **(A)** CD4 and **(B)** CD8 coreceptors, including Ig-like domains and glycosylation sites. A CD8αβ heterodimer is shown, but CDαα homodimers also exist. The number of carbohydrate sites in the CD8β chain varies with the developmental stage of a given thymocyte. Insets: CD4 and CD8 coreceptor icons used in this book. [*With information from Klein, J., & Horejsí V. (1997). Immunology, 2nd ed., Blackwell Science, Oxford.*]

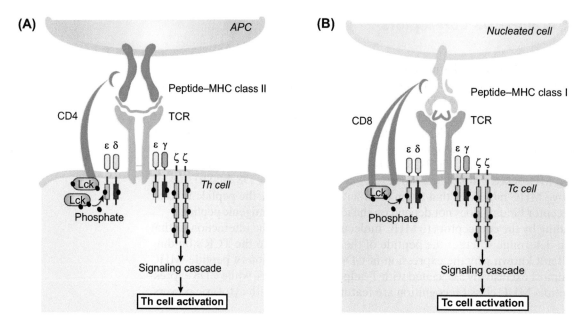

Fig. 8-5
Role of Coreceptors in TCR Signaling

Coreceptors promote T cell binding to an APC or nucleated target cell displaying the appropriate pMHC. Lck associated with the coreceptor tail phosphorylates CD3 ITAMs, triggering a signaling cascade that delivers a stimulatory message to the T cell nucleus. **(A)** CD4 interacts with a non-polymorphic region of MHC class II. Substantial amounts of Lck are associated with the CD4 cytoplasmic tail. **(B)** CD8 interacts with a non-polymorphic region of MHC class I. Modest amounts of Lck are associated with the CD8 cytoplasmic tail.

both humans and mice, the majority of CD8 molecules expressed on a T cell surface are CD8αβ heterodimers made up of CD8α and CD8β chains (**Fig. 8-4B**). Some intestinal intraepithelial T cells express a CD8αα homodimer. In both cases, the complete CD8 protein contains one Ig-like extracellular domain that binds to the α3 domain of MHC class I; a transmembrane domain; and a cytoplasmic tail that associates relatively weakly with Lck kinase.

iii) Functions

CD4 and CD8 have two major functions: (1) stabilization of TCR–pMHC binding by the interaction of CD4 with MHC class II and CD8 with MHC class I, and (2) recruitment of Lck to the TCR–CD3 complex. Although neither CD4 nor CD8 is absolutely required for the initial engagement of TCRαβ by pMHC, the adhesive contacts these molecules establish with the MHC molecule greatly enhance TCR–pMHC binding. With respect to recruitment, because of the positioning of the coreceptors near the TCR in the membrane, Lck that is physically associated with the cytoplasmic tail of CD4 or CD8 is brought into close proximity with the tails of the CD3 chains (**Fig. 8-5**). Lck then can phosphorylate the ITAMs in the CD3 tails, propagating the intracellular signaling cascade that leads to T cell activation (see Ch. 9).

B. TCR Genes

I. Structure of the TCR Loci

The TCRα, β, γ and δ polypeptide chains are encoded by the *TCRA*, *TCRB*, *TCRG* and *TCRD* loci, respectively. The chromosomal locations of these loci in humans and mice are given in **Table 8-1**, and their exon/intron structures are shown in **Figures 8-6** and **8-7**. Like the genes encoding the Ig proteins, the genes encoding

TABLE 8-1	Chromosomal Localization of TCR Loci	
	Chromosome	
Genetic Locus	Human	Mouse
TCRA	14	14
TCRB	7	6
TCRG	7	13
TCRD	14	14

the TCR proteins are composed of V and C exons, with the V exon being made up of small V, D and J gene segments that are assembled at the DNA level by V(D)J recombination.

The *TCRA* and *TCRG* loci contain V and J gene segments but no D segments, making these loci analogous to the Ig light chain loci. Whereas there is one Cα exon in *TCRA* in mice and humans, there are two Cγ exons in human *TCRG* and three functional Cγ exons in mouse *TCRG*. The *TCRB* locus is like *Igh* in that it contains multiple V, D and J gene segments. Two Cβ exons are present in both species, but it is unlikely that these exons are functionally different because, in T cells, there is no mechanism analogous to the isotype switching that occurs in B cells. The sole function of the TCR C domains appears to be association with the CD3 complex, and Cβ1 and Cβ2 have equivalent roles in this respect. Although *TCRD* also has similarities to *Igh*, there are some startling differences. Firstly, in both mice and humans, *TCRD* is nested *within* the

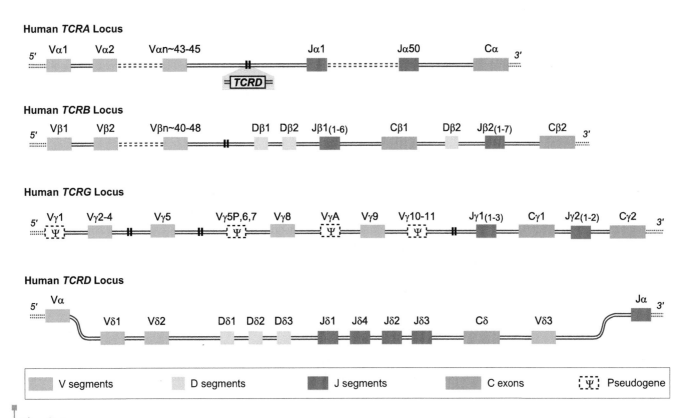

Fig. 8-6
Genomic Organization of the Human TCR Loci

Genomic organization schemes for the human *TCRA*, *TCRB*, *TCRG* and *TCRD* loci are shown. For each locus, a few of the functional V segments are shown, with additional gene segments present in the areas represented by the dashed lines. With respect to the V gene segments of the TCRδ chain, three Vδ segments are in the *TCRD* locus (as shown), with another four to five Vδ segments in the *TCRA* locus (not shown). *TCRA* and *TCRD* have no D segments. [With information from http://imgt.cines.fr/.]

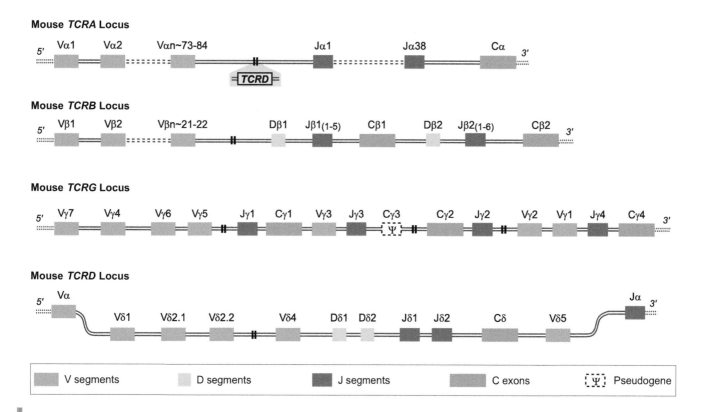

Fig. 8-7
Genomic Organization of the Mouse TCR Loci

Genomic organization schemes for the mouse *TCRA, TCRB, TCRG* and *TCRD* loci are shown. For each locus, a few of the functional V segments are shown, with additional gene segments present in the areas represented by the dashed lines. With respect to the V gene segments of the TCRδ chain, 4 Vδ segments are in the *TCRD* locus (as shown), with another 10–11 Vδ segments in the *TCRA* locus (not shown). Also, a second Cδ exon is found outside *TCRD* just 3′ of Jα (not shown). *TCRA* and *TCRD* have no D segments. *[With information from http://imgt.cines.fr/.]*

TCRA locus. *TCRD* contains its own Vδ, Dδ and Jδ gene segments and Cδ exon but also shares the use of some Vα gene segments. However, Vδ gene segments recombine only with Jδ and Cδ sequences and never with Jα or Cα sequences. The unconventional location of *TCRD* prevents the expression of both *TCRD* and *TCRA* on the same T cell, since the recombination of the Vα and Jα gene segments deletes the entire *TCRD* locus. Secondly, in addition to the use of one Dδ segment to create a VDJ exon, multiple Dδ gene segments can be used in tandem to create a VDDJ exon in mice or a VDDJ or even a VDDDJ exon in humans. The additional D–D joints present in these exons dramatically enhance the junctional diversity found in TCRδ chains.

II. Order of Rearrangement

When a T cell progenitor leaves the bone marrow and enters the thymus to become an immature thymocyte, its TCR genes are in the germline configuration. In the thymus, the immature thymocyte rearranges its TCR genes and eventually becomes either an αβ T cell or a γδ T cell. Most immunologists believe that a given thymocyte is influenced to become either an αβ or γδ T cell by signals it receives both through its antigen receptor proteins and from the local microenvironment. V(D)J recombination of the TCR loci is intimately tied to T cell development and is discussed in more detail in this context in Chapter 9.

i) *TCRA* and *TCRB* Rearrangement
Irrevocable commitment of a thymocyte to the TCRαβ lineage and the continued maturation of the clone depend on V(D)J recombination resulting in a functional TCRβ

gene. The *TCRB* locus rearranges prior to the *TCRA* locus. Simultaneously in *TCRB* on the maternal and paternal chromosomes, V(D)J recombination first joins a Dβ gene segment to a Jβ segment, and then a Vβ segment to DβJβ. When a gene is completed, it is then tested for functionality via formation of the **pre-TCR**, a signaling complex composed of a newly produced candidate TCRβ chain combined with a surrogate TCRα chain called the **pre-T alpha chain** (plus the CD3 chains). Successful intracellular signaling initiated by the pre-TCR indicates that the candidate TCRβ protein is functional and thus that the rearrangement of the *TCRB* gene has been successful. Because signaling through the pre-TCR governs the continued differentiation of thymocytes, this molecule is discussed in more detail in Chapter 9.

The assembly of a functional TCRβ gene on one chromosome signals to the cell to suppress V(D)J rearrangement of *TCRB* on the other chromosome; i.e., allelic exclusion is invoked. In a thymocyte in which *TCRB* rearrangement has been unsuccessful on both chromosomes, the cell neither attempts to rearrange *TCRA* nor becomes a γδ T cell; instead, it dies by apoptosis. If a functional TCRβ gene is produced, V(D)J recombination of *TCRA* commences on both chromosomes. If *TCRA* rearranges productively on either chromosome, it can generate a TCRα chain that can combine with the newly synthesized TCRβ chain and appear on the cell surface as a functional TCRαβ. The *TCRD* locus is deleted by successful *TCRA* rearrangement. If *TCRA* rearrangement fails on both chromosomes, the cell dies by apoptosis.

ii) *TCRG* and *TCRD* Rearrangement

In thymocytes that eventually become γδ T cells, rearrangement commences simultaneously but independently in the *TCRG* and *TCRD* loci on both chromosomes. Despite the fact that the *TCRA* locus physically surrounds *TCRD*, *TCRA* does not undergo rearrangement. VJ joining of TCRγ gene segments occurs in the usual way, but the TCRδ D gene segments can be combined with each other to form tandem D–D or D–D–D units. The D, D–D or D–D–D entities are in turn joined to Jδ and then finally to Vδ to complete the V exon. More on gene rearrangement during the development of γδ T cells appears in Chapter 11.

III. V(D)J Recombination

The same RAG recombinases and DNA repair enzymes that execute V(D)J recombination in the Ig loci in developing B cells act on the TCR loci in thymocytes to produce functional TCR genes. Accordingly, the TCR gene segments are flanked by the same 12-RSS and 23-RSS sequences discussed in Chapter 4. Moreover, as shown for *TCRB* in **Figure 8-8**, the RAG recombinases follow the same 12/23 rule to juxtapose only those RSSs that are not of the same type. Importantly, D segments in the *TCRB* and *TCRD* loci are flanked on the 5′ side by a 12-RSS and on the 3′ side by a 23-RSS. In the *TCRD* locus, where the Dδ segments are clustered together, this arrangement of the RSSs facilitates the tandem joining of Dδ segments prior to the addition of the Jδ segment.

Despite the apparent duplication of the V(D)J recombination apparatus in B cells and T cells, the Ig genes are not rearranged in developing T cells, and the TCR genes are not rearranged in developing B cells. The mechanisms that underlie this stricture are still not clear. There is some evidence indicating that regulatory elements in the RAG genes that influence their transcription are involved, as well as post-translational modifications of the RAG proteins.

IV. TCR Gene Transcription and Protein Assembly

Following V(D)J recombination to generate the V exon, a rearranged TCR gene undergoes conventional transcription from the promoter associated with the participating V gene segment. A single primary transcript that includes V and C exons is generated (**Fig. 8-9**). The primary transcript undergoes RNA splicing to bring the V

Fig. 8-8
RSS-Mediated V(D)J
Recombination of TCRB

Somatic recombination of V, D and J gene segments in the murine *TCRB* locus is illustrated. Each gene segment is flanked by a 23-RSS or a 12-RSS, which are aligned according to the 12/23 rule. The D and J segments are brought together first, followed by the joining of the V segment. Gene products excised by the RAG recombinase complex (gray ovals) are not shown.

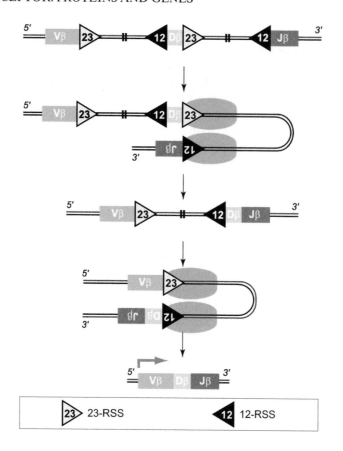

Fig. 8-9
V(D)J Recombination and
TCRβ Chain Synthesis

An example of V(D)J recombination to generate a rearranged mouse TCRβ gene is shown. Gene segment Dβ1 has been arbitrarily selected to join with the fourth gene segment of the Jβ1 cluster [Jβ1$_{(4)}$] and then to Vβ2. Transcription results in a primary transcript that contains the V exon [Vβ2:Dβ1:Jβ1$_{(4)}$] and the Cβ1 exon. After conventional RNA splicing, an mRNA containing the indicated sequences is produced that is translated into a TCRβ chain.

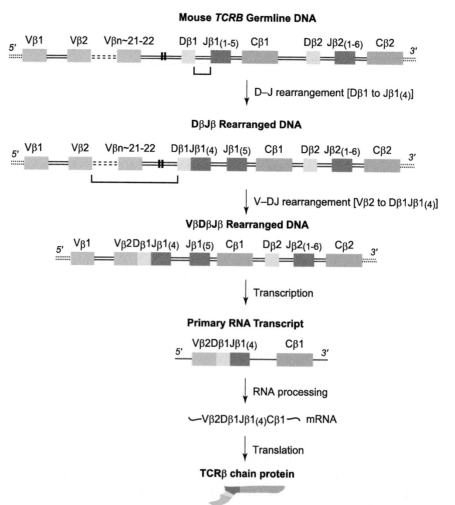

and C exons together, thereby generating mature mRNAs that are translated into TCR polypeptides. In contrast to *Igh* genes, which contain separate exons specifying mIg or sIg, the TCR genes have only an exon encoding a transmembrane domain. Thus, there is no production of alternative mRNA transcripts specifying membrane-bound versus secreted TCR proteins. After translation of the TCR transcripts in the ER, disulfide bonding links the TCRα and β chains, or the TCRγ and δ chains, to form membrane-bound TCRαβ or TCRγδ molecules, respectively. These heterodimers associate with the CD3 complex and are then transported to the plasma membrane where they appear on the cell surface as complete antigen receptor complexes.

V. TCR Diversity

The mechanisms of isotype switching and somatic hypermutation that create diversity in antigen-activated B cells do not operate in T cells, so that the diversity in the T cell repertoire is established entirely by mechanisms that function prior to antigenic stimulation. These are multiplicity of germline segments, combinatorial diversity, junctional diversity, and αβ (or γδ) chain pairing.

i) Multiplicity and Combinatorial Joining of Germline Gene Segments

In mice and humans, the numbers of different V and D gene segments available for recombination in the TCR loci are much lower than the number of corresponding segments in the Ig loci, but the number of *TCRA* J segments is greater than the number of Ig J segments (**Table 8-2**). Overall, the contribution of this source of diversity to the maximum theoretical TCR repertoire is less than for the Ig repertoire. The random juxtaposition of TCR V, D and J segments during V(D)J recombination then contributes diversity that can be calculated just as for the Ig genes. For example, for the mouse TCRα chain, the number of possible combinations (considering functional segments only) is theoretically 84 Vα × 38 Jα × 1 Cα = 3192, whereas that for mouse TCRβ is 22 Vβ × 2 Dβ × 11 Jβ × 2 Cβ = 968. Using this methodology, one might also conclude that there are 7 Vγ × 4 Jγ × 4 Cγ = 112 possible combinations for the mouse TCRγ chain, and 15 Vδ × 2 Dδ × 2 Jδ × 1 Cδ = 60 combinations for the mouse TCRδ chain. However, these theoretical calculations do not take into account certain joining preferences that occur in the TCR loci. For example, Cβ1 is found only in conjunction with Dβ1 and Jβ1, and Vγ segments tend to rearrange only with the closest DJγ. In addition, the gene segments that make up γδ TCRs are not chosen entirely at random. Different γδ TCRs appear to contain specific Vγ and Vδ gene segments depending on the cellular subset

TABLE 8-2	Estimated Numbers of Functional Gene Segments in Mouse and Human TCR Loci

	TCRA	TCRB	TCRG	TCRD
Number of Gene Segments in Germline*				
Mouse Gene Segments				
V	73–84	21–22	7	14–15
D	0	2	0	2
J	38	11	4	2
Human Gene Segments				
V	43–45	40–48	4–6	7–8
D	0	2	0	3
J	50	12–13	5	4

http://www.imgt.org/IMGTrepertoire/LocusGenes/genetable/human/geneNumber.html (accessed December 2012).

*Number of functional segments can vary by individual.

or anatomic location in which they are found. For example, in mouse skin, the γδ T cells present almost exclusively express TCRs containing a TCRγ chain made up of the Vγ3, Jγ1 and Cγ1 segments, coupled to a TCRδ chain made up of Vδ 1, Dδ2, Jδ2 and Cδ. In contrast, the vast majority of γδ T cells in murine tongue express a TCR containing Vγ3, Jγ1 and Cγ1 in the TCRγ chain, and Vδ1, Dδ2, Jδ2 and Cδ in the TCRδ chain. Thus, the actual diversity derived from combinatorial sources is more limited than the theoretical diversity.

Fortunately, what is lost in combinatorial diversity is compensated for by variable D segment inclusion. Although the Ig loci contain higher numbers of D gene segments, only one D segment can join to an Ig J segment during a given rearrangement. Diversity in the γδ TCR repertoire is increased because TCR Dδ segments may join to each other as well as to Jδ segments to form VDJ, VDDJ or VDDDJ variable exons.

ii) Junctional Diversity

The mechanisms of generating junctional diversity that were discussed in the context of B cells in Chapter 4 also apply to T cells. Both P nucleotides and N nucleotides can be added to VD and DJ joints in TCR chains and give rise to amino acids that are not encoded in the germline. Because more than one Dδ segment may be included in tandem in a TCRδ chain, many more opportunities for P and N nucleotide addition occur at each D–D or D–J joint. It has been estimated that junctional diversity contributes billions of possible TCRδ chains to the TCR repertoire.

iii) Chain Pairing

The random pairing of TCRα and β chains (or TCRγ and δ chains) within a given αβ (or γδ) T cell also contributes to TCR repertoire diversity. In the case of an αβ T cell, the TCR's antigen-binding site is composed of the V domains of the one TCRα chain and the one TCRβ chain synthesized in that cell. However, since any one of the vast number of possible sequences for a TCRα chain gene can occur in the same T cell as any one of the even more numerous possibilities for a TCRβ chain, the total number of possible αβ heterodimers approaches 10^{18} in humans and 10^{15} in mice. These numbers compare very favorably to the 10^{11} specificities estimated for the Ig repertoire. Again, however, due to the death of T cell clones before they ever meet their antigens, as well as the processes of central and peripheral tolerance, it has been estimated that a human has a repertoire of 2×10^7 functional αβ T cell clones, whereas a mouse can draw on 2×10^6 such clones.

C. TCR–Antigen Interaction

The interaction between a TCRαβ protein and its pMHC epitope underlies fundamental aspects of the cell-mediated adaptive immune response. Firstly, the strength of binding between a thymocyte's TCR and various pMHCs encountered in the thymus determines whether the thymocyte is negatively selected and dies, dies of "neglect," or is positively selected and survives to become a mature T cell (see Ch. 9). Secondly, the strength of binding between a mature αβ T cell's TCR and pMHC presented by an APC in the periphery determines whether the T cell will be activated to proliferate and differentiate into effector cells, or will become anergic (non-responsive). Immunologists still do not fully understand the molecular pathways governing these cell fate decisions. The structural aspects of TCR binding to pMHC are presented here, whereas issues associated with T cell activation/differentiation and peripheral T cell tolerance are discussed in Chapters 9 and 10, respectively.

Studies of TCR X-ray crystal structures have shown that the V domains of the TCRαβ heterodimer resemble the V domains of the Ig molecule, but that the interdomain pairing of the Cα and Cβ regions differs from that in the Ig C regions. In addition, in contrast to the relative independence of the Ig V and C domains, TCR Vβ and TCR Cβ are closely associated within the crystals. This association may confer a degree of inflexibility to that region of the TCR that is analogous to the Ig Fab region. **Plate 8-2A** depicts the V domains of a human αβ TCR interacting with peptide bound to the extracellular

(A) Human TCR-pMHC Class I Interaction   **(B)** Mouse TCR-pMHC Class I Interaction

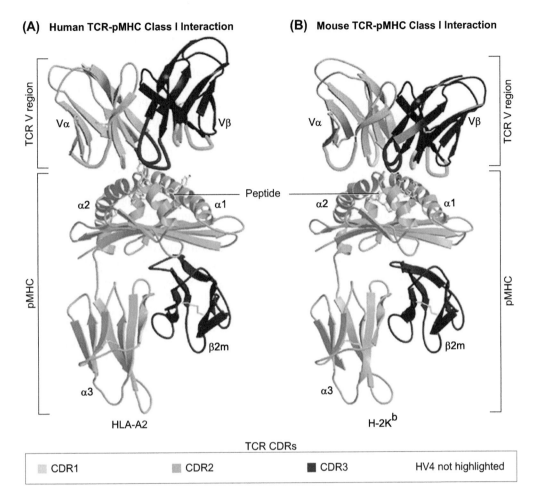

Plate 8-2
X-Ray Crystal Structures of Human and Mouse TCR–pMHC Interaction

X-ray crystal structures showing the carbon backbone of a human **(A)** and a mouse **(B)** TCR interacting with peptide-MHC class I. *[Reproduced by permission of Bjorkman, P. J. (1997). MHC restriction in three dimensions: a view of T cell receptor/ligand interactions. Cell 89, 167–170.]*

region of MHC class I, while **Plate 8-2B** shows the corresponding murine molecules. Comparable analyses of TCRγδ crystal structures have shown that TCRγδ differs physically from TCRαβ. In particular, the structure of the TCR Vδ domain looks more like Ig V$_H$ than TCR Vα or Vβ. This finding is consistent with the results of functional studies showing that γδ T cells recognize antigenic structures other than pMHCs (see Ch. 11).

NOTE: The reader is referred once again to the NCBI database at http://www.ncbi.nlm.nih. gov/sites/structure where hundreds of pMHC complexes can be viewed in 3-D.

In many TCRs, the TCR peptide-binding site itself is relatively flat except for a deep hydrophilic cavity between the TCRα CDR3 and TCRβ CDR3. When bound, the TCR is oriented in a diagonal position over the pMHC such that the flat region can interact with the peptide. Both the TCRα and β chains are usually involved in binding to both the MHC molecule and the peptide, and this binding occurs virtually simultaneously. In general, the highly variable CDR3 regions of the TCRα and β chains bind to the middle of a peptide lodged in the MHC binding groove as well as to points on the MHC protein backbone. The less variable CDR1 and CDR2 regions tend to bind to the ends of the peptide and to conserved sites on the MHC backbone. The sequence of binding events is variable: sometimes CDR3 initiates interaction with peptide first, and sometimes CDR1 or CDR2 binding to pMHC is established first.

The area of contact between the TCR and the pMHC is relatively small such that only a few of the residues in the peptide generally make contact with a TCR chain. This limited opportunity for intermolecular bonding means that the binding affinity of a TCR for pMHC ($K = \sim 5 \times 10^5$ M^{-1}) is significantly lower than that of an antibody for its antigen ($K = 10^7$ –10^{11} M^{-1}). This relatively modest affinity of TCR binding has two implications. Firstly, the initial contact between T cells and APCs or target cells is established not by TCR–pMHC interaction but rather by the binding of complementary pairs of adhesion molecules. Specific TCR–pMHC contacts are made only after the cells are held in close enough proximity by the adhesion molecules to permit the T cell to scan the pMHCs in the APC or target cell membrane. At this point, contacts between CD4 or CD8 and the MHC class II or I molecule, respectively, also become important in holding the cells together. Secondly, because of their modest affinity for their cognate ligands, TCRs can bind (with varying strength) to a surprisingly broad range of pMHCs. Such promiscuity facilitates thymic selection because one peptide can positively select several thymocyte clones, amplifying the T cell repertoire. Thymic selection is discussed in more detail in Chapter 9.

The ability of a TCR to bind to several different pMHCs is largely due to two properties unique to the CDR3 regions of its TCRα and TCRβ chains. Firstly, whereas the CDR1, CDR2 and HV4 hypervariable regions are encoded by the V gene segment of a variable exon, the DNA sequences encoding the CDR3 regions span the VJ joint in the rearranged TCRα gene and the VD and DJ joints in the rearranged TCRβ gene. Junctional diversity at these joints then imparts extreme variability to the amino acid sequence, length and conformation of the CDR3 region. Secondly, comparisons of the conformations of TCRs that have not bound to pMHC versus TCRs bound to pMHC have demonstrated that the CDR3 regions are capable of undergoing an enormous conformational shift in order to achieve the diagonal orientation favored for binding to pMHC. The adoption of this "induced fit" affects only the CDR3 regions and does not alter the conformation of either the rest of the TCR molecule or the MHC molecule. In any case, once a TCR finalizes its contacts with a given pMHC, the entire complex is stabilized, and the flexibility of both the TCR and pMHC binding surfaces is lost.

This marks the end of our discussion of the TCR proteins and genes. In Chapter 9, we examine the development of T cells and the crucial role that TCRs play in the positive and negative selection processes that shape the T cell repertoire. Chapter 9 also describes T cell activation by antigen and the differentiation and functions of effector and memory T cells.

Chapter 8 Take-Home Message

- There are two types of TCRs, TCRαβ and TCRγδ, which are expressed by αβ T cells and γδ T cells, respectively.

- TCRαβ molecules recognize peptides bound to either MHC class I or class II, whereas γδ TCRs can recognize antigens in their natural, unprocessed forms.

- TCR chains are incapable of signal transduction. This function is carried out by the ITAM-containing CD3 complex that associates with TCRαβ or TCRγδ.

- Each chain of the TCRαβ molecule has four hypervariable regions that promote binding to a small collection of highly similar pMHCs.

- The *TCRA*, *TCRB*, *TCRG* and *TCRD* loci contain multiple V, D and J gene segments and one or two C exons. *TCRD* is nested within *TCRA*. V(D)J recombination assembles functional TCR genes in a strict order tied to T cell development.

- Although isotype switching and somatic hypermutation do not occur in T cells, the overall diversity of the T cell repertoire is greater than that of the Ig repertoire because of increased junctional diversity.

- αβ T cells express either the CD4 coreceptor that binds to a non-polymorphic region of MHC class II or the CD8 coreceptor that binds to a non-polymorphic region of MHC class I.

- Coreceptor binding to MHC increases the adhesion between T cells and APCs or target cells, and facilitates Lck recruitment.

- The promiscuity of the TCR binding site is largely due to the presence of four hypervariable regions in each TCR chain. CDR3 is particularly important for peptide binding by a TCR.

Did You Get It? A Self-Test Quiz

Section A

1) What are intraepithelial cells?

2) Give two differences between TCRs and BCRs.

3) Give three differences between αβ T cells and γδ T cells.

4) What protein chains come together to form TCRs?

5) How do the hypervariability sites in the TCR chains differ from those in the Ig chains?

6) Describe the composition of the CD3 complex.

7) Why does the TCR need the CD3 complex?

8) What is Lck, what does it do, and why is this important?

9) Why are CD4 and CD8 called "coreceptors"?

10) How do CD4 and CD8 differ in structure? Does this difference affect their function?

11) Give three functions of the coreceptors.

Section B

1) What loci encode the TCR chains? Is there anything unusual about the structure of these loci?

2) T cells do not undergo isotype switching. How is this reflected in their gene structure?

3) How do the D segments in the TCR loci differ from those in the Ig loci?

4) Which TCR locus is the first to rearrange in a thymocyte that will become an αβ T cell?

5) Which TCR locus is the first to rearrange in a thymocyte that will become a γδ T cell?

6) Why aren't the Ig genes expressed in developing T cells?

7) Why aren't TCR proteins secreted?

8) Give two reasons why the actual diversity in the TCR repertoire is less than the theoretical diversity.

9) Why does diversity in the TCR repertoire exceed that in the Ig repertoire?

Section C

1) Describe two cellular events governed by the affinity of binding between a TCRαβ and its pMHC epitope.

2) How does the structure of the TCRαβ V domain differ from that of the TCRγδ V domain, and what effect does this have on antigen recognition?

3) Describe how the hypervariable regions of the TCRαβ chain interact with peptide in the MHC binding groove.

4) How does the binding affinity of TCRαβ for pMHC differ from that of Ig for its antigen, and what are two implications of this difference?

5) Why is the promiscuity of TCRαβ binding largely due to the CDR3 region?

Can You Extrapolate? Some Conceptual Questions

1) Consider the V region domains of the TCRα and TCRβ chains and the genetic loci that encode them.

   a) Which TCR chain is more genetically analogous to the Ig heavy chain?

   b) Which TCR chain is more genetically analogous to Ig light chain? Why?

   c) How is a complete TCRαβ molecule similar to an Ig Fab fragment? (Consider V and C domains.)

2) A genetic probe is developed that binds to the human Dδ2 gene segment in the *TCRD* locus. Give a possible explanation for each of the following:

   a) The probe binds to DNA isolated from some γδ T cells but not others.

   b) The probe does not bind to DNA isolated from any αβ T cell.

3) A number of monoclonal antibodies (mAbs) are produced that bind to epitopes on a TCRαβ protein derived from T cell clone "X." Whereas mAb1 binds to an epitope in the Cα domain, mAb2 binds to an epitope in the Vβ domain. Explain which mAb you would choose to attempt to block the activity of T cell clone X.

Would You Like To Read More?

Garcia, K. C., Adams, J. J., Feng, D., & Ely, L. K. (2009). The molecular basis of TCR germline bias for MHC is surprisingly simple. *Nature Immunology*, *10*(2), 143–147.

Godfrey, D. I., Rossjohn, J., & McCluskey, J. (2008). The fidelity, occasional promiscuity, and versatility of T cell receptor recognition. *Immunity*, *28*(3), 304–314.

Krangel, M. S. (2009). Mechanics of T cell receptor gene rearrangement. *Current Opinion in Immunology*, *21*(2), 133–139.

Marrack, P., Scott-Browne, J. P., Dai, S., Gapin, L., & Kappler, J. W. (2008). Evolutionarily conserved amino acids that control TCR-MHC interaction. *Annual Review of Immunology*, *26*, 171–203.

Morris, G. P., & Allen, P. M. (2012). How the TCR balances sensitivity and specificity for the recognition of self and pathogens. *Nature Immunology*, *13*(2), 121–128.

van der Merwe, P. A., & Dushek, O. (2011). Mechanisms for T cell receptor triggering. *Nature Reviews Immunology*, *11*(1), 47–55.

T Cell Development, Activation and Effector Functions

Chapter

9

Don't fall before you're pushed.

English Proverb

A. T Cell Development

I. Comparison of B and T Cell Development

B and T cells are both lymphocytes and are derived from the same very early hematopoietic progenitors, but their development differs in several important ways:

(1) The thymus is required for the generation of the vast majority of mature peripheral T cells but not for that of mature B cells. T cells developing in thymus are called **thymocytes**.

(2) Naïve B cells are freshly produced at a virtually constant rate for the life of the individual. In contrast, once the involution of the thymus commences around puberty, new naïve T cell production is sharply reduced, and the maturing adult becomes increasingly dependent on the existing repertoire of T cells.

(3) MHC molecules are involved in the establishment of central tolerance of T cells but not B cells. The TCR on a thymocyte must not only be functional (be derived from productive rearrangements of the TCR genes) but must also recognize the host's MHC molecules (so that it can "see" pMHC). This requirement imposes an additional layer of selection on T cells that developing B cells do not experience.

(4) The TCR expressed on a T cell's surface is fixed for the life of the clone and cannot undergo the somatic hypermutation in response to activation that occurs following B cell activation.

(5) Whereas the vast majority of functional B cells result from a single developmental program, functional T cells can result from several different paths. A developing thymocyte may give rise to γδ T cells or αβ T cells, and among these, Th or Tc cells. Once activated by antigen, an αβ T cell clone can further differentiate into subsets that differ slightly in their effector functions. In addition, certain thymocytes have the capacity to differentiate into **regulatory T cells** that can control the responses of activated Th and Tc cells.

Regulatory T cells contribute to peripheral tolerance and are discussed in Chapter 10.

MAK: Primer to the Immune Response. http://dx.doi.org/10.1016/B978-0-12-385245-8.00009-1

> NOTE: The complex pathways mediating lymphocyte development, activation and differentiation involve many genes that can become targets for mutations leading to primary immunodeficiencies (PIDs). PIDs are discussed in detail in Chapter 15.

II. Colonization of the Thymus

During the early embryonic development of a mammalian fetus, the thymus is empty of hematopoietic progenitors. The fetal thymus must be "colonized" or seeded with hematopoietic progenitors that subsequently proliferate and mature in the thymus into functional naïve T cells. As introduced in Chapter 2, T cells (like all hematopoietic cells) are derived from HSCs. Although HSCs are located in the bone marrow in an adolescent or adult individual, they are generated in the liver in a fetus. In either location, a proliferating HSC can differentiate into early progenitors, including the postulated MPPs and CLPs of the simplified model of hematopoiesis shown in Figure 2-4. These MPPs and CLPs leave the bone marrow or liver and enter the blood circulation. Many immunologists believe that circulating CLPs eventually give rise to a slightly more differentiated progenitor called the **NK/T precursor** that can generate NK cells and T cells but not B cells. When circulating NK/T precursors that express high levels of the chemokine receptors CCR9 and CCR7 exit the blood, they are drawn to the thymic endothelium by the presence of the ligands for these receptors, the chemokines CCL25 and CCL19 (respectively). Once a given NK/T precursor enters the thymus, it may eventually differentiate into αβ or γδ T cells, lymphoid DCs, or NK or NKT cells, depending on the cytokines and stromal cell ligands it encounters.

Development of γδ T, NK and NKT cells from NK/T precursors is described in Chapter 11.

The fetal thymus is colonized by NK/T precursors in a limited number of distinct waves that occur both before and after birth. The earliest prenatal waves migrate from the fetal liver, enter the fetal thymus and give rise only to γδ thymocytes. Subsequent waves of NK/T precursors entering the thymus just before birth and shortly thereafter give rise to both αβ and γδ thymocytes. However, after birth, those NK/T precursors destined to become T cells are more and more biased toward the αβ T cell lineage such that γδ T cells become a minor population. In addition, the bone marrow becomes the dominant site of generation of the NK/T precursors needed to replenish the thymus.

Age-dependent changes in the dominant site of hematopoiesis were illustrated in Figure 2-3.

Study of the developmental path of HSCs to mature T cells in mice has shown that the pattern is generally the same in fetal, neonatal, adolescent and adult animals but displays slower kinetics in adolescents and adults (**Table 9-1**). Importantly, the TdT enzyme responsible for much of the junctional diversity generated during TCR gene rearrangement is not fully active until shortly after birth. Thus, the repertoire of T cell specificities available in the neonate is significantly less diverse than in older individuals.

> NOTE: Identifying the factors driving thymic colonization by NK/T precursors is of more than just academic interest. A patient that has undergone a **bone marrow transplant** (BMT) or a **hematopoietic stem cell transplant** (HSCT) in an attempt to treat disease depends on the ability of donor-derived CLPs (or another early progenitor) to generate NK/T precursors able to repopulate his/her thymus and generate mature T cells. On its own, this process can be quite slow, putting the patient at high risk for infection. Clinicians therefore use laboratory methods to manipulate the cells to be transplanted in an effort to accelerate thymic repopulation and T cell generation (see Ch. 17).

III. Thymocyte Maturation in the Thymus

Thymocytes at different developmental stages are morphologically very similar and so are usually distinguished by either their patterns of surface marker expression or by their TCR gene rearrangement status. These parameters have been used to divide thymocyte maturation into three broad phases: the **double negative** phase (DN, stages

TABLE 9-1	Comparison of Murine Thymocyte Development at Different Life Stages		
Property	**Fetus**	**Neonate**	**Post-Neonate***
Origin of NK/T precursors	Fetal liver	Fetal liver	Bone marrow
TCRs in the periphery	Only TCRγδ No TCRαβ	Majority TCRγδ Minority TCRαβ	Majority TCRαβ Minority TCRγδ
Kinetics of progression from NK/T precursors to mature T cells	Fast	Fast	Slow
TdT expression	None	Initiated	Fully active
T cell repertoire diversity	Limited	Limited	Fully diversified
Generation of thymocytes	Continuous	Continuous	Minimal after thymic involution

*Post-neonate includes adolescent and adult mice

DN1–DN4), in which thymocytes express neither CD4 nor CD8; the **double positive** (DP) phase, in which thymocytes express both CD4 and CD8; and the **single positive** (SP) phase, in which thymocytes express either CD4 or CD8 but not both (**Fig. 9-1**). Several selection processes occur during these transitions that remove non-functional and potentially autoreactive thymocytes (see later). Once SP thymocytes emerge from the thymus and enter the circulation and secondary lymphoid tissues, they are considered to be mature naïve CD4+ or CD8+ peripheral T cells.

> DiGeorge syndrome is a PID in which thymus development is impaired (see Ch. 15).

i) The Thymic Environment

The development of thymocytes through the DN, DP and SP phases is totally dependent on the stromal cells that make up the thymic architecture (**Fig. 9-2**). As normal thymocytes mature, they pass through a succession of thymic microenvironments characterized by different mixes of stromal cell types. Among the most important of these stromal cells are *cortical thymic epithelial cells (cTECs), medullary thymic epithelial cells (mTECs), thymic DCs* and *thymic fibroblasts*. As discussed later, thymic DCs, cTECs and mTECs are vital for the establishment of T cell central tolerance during the DP phase. cTECs and mTECs also express cell surface ligands for *Notch1*, a cell fate protein expressed on the surface of thymocytes. Once Notch1 has bound to its ligand, the cytoplasmic domain of Notch1 interacts with transcription factors to promote T cell development while suppressing B cell development. Continued Notch1 signaling

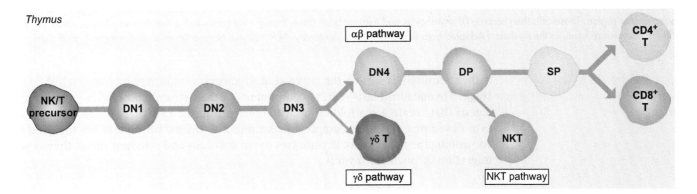

Fig. 9-1
Model of Thymocyte Maturation in the Thymus

This simplified pathway shows the generation of mature naïve αβ and γδ T cells and NKT cells from NK/T precursors through the double negative (DN), double positive (DP) and single positive (SP) stages of thymocyte development. Most DN3 thymocytes become αβ T cells, but some generate γδ T cells. Most DP thymocytes that survive various thymic selection processes become CD4+ or CD8+ SP T cells, but some generate NKT cells. The development of γδ T cells and NKT cells is discussed further in Chapter 11.

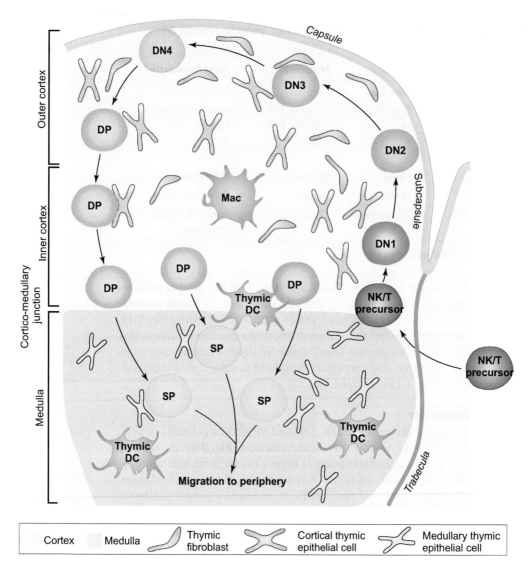

Fig. 9-2
Thymic Microenvironment and Location of Developing Thymocytes

NK/T precursors enter the thymus at its cortico-medullary junction. DN1 thymocytes are found in the inner cortex but soon migrate toward the outer cortex and become DN2 cells. DN3 thymocytes predominate in the outer cortex and make the transition to the DN4 stage in the subcapsular region. These cells then become DP thymocytes and migrate back down through the cortex to the cortico-medullary junction. Only SP thymocytes are found in the medulla. [*Adapted from Blackburn, C.C., & Manley, N.R. (2004). Nature Reviews Immunology 4, 278–289.*]

is then required to sustain the survival of thymocytes until they pass through the DN stage. Thymic fibroblasts secrete components of the extracellular matrix (such as collagen) that create a scaffolding used to concentrate the cytokines crucial for thymocyte development. Other components secreted by thymic fibroblasts are involved in controlling the adhesion of thymocytes to stromal cells and thus may direct thymocyte migration through the thymus.

NOTE: In humans, a mutation of the transcription factor FoxN1 blocks TEC development. Without cTECs and mTECs, mature T cells cannot develop properly from thymocytes, rendering these individuals immunodeficient (see Ch. 15). Similarly, the *nude* mutant mouse strain has a defect in the development of thymic stromal cells and so lacks all mature T cells. This immunodeficient animal is used in many laboratory experiments in which immune system reconstitution is required.

TABLE 9-2	**Comparison of Murine and Human Thymocyte Markers**	
Stage	**Murine**	**Human**
DN1	ckit⁺ CD44⁺CD25⁻	CD34⁺CD38⁻CD1α⁻
DN2	ckit⁺ CD44⁺CD25⁺	CD34⁺CD38⁺CD1α⁻
DN3	ckit⁻ CD44⁻CD25⁺	CD34⁺CD38⁺CD1α⁺
DN4	ckit⁻ CD44⁻CD25ˡᵒ	CD34⁻CD38⁻CD1α⁺CD4⁺
DP-TCRβ	CD4⁺CD8⁺	CD4⁺CD8⁺
DP-TCRαβ	CD4⁺CD8⁺	CD4⁺CD8⁺
SP-CD4	CD4⁺CD8⁻	CD4⁺CD8⁻
SP-CD8	CD4⁻CD8⁺	CD4⁻CD8⁺

ii) DN Phase

As mentioned above, the earliest thymocytes are said to be in the double negative or *DN phase* because they express neither CD4 nor CD8. These cells are also negative for TCR expression, cannot bind pMHC and do not carry out effector functions. Within the DN phase are four subsets of thymocytes labeled DN1–4. In mice, these subsets are distinguished from each other by their expression of the surface markers c-kit (a cytokine receptor), CD44 (an adhesion protein) and CD25 (the α chain of the IL-2 receptor). Note that IL-2 is not required for thymocyte development, and the function of CD25 in thymocytes is unknown. In humans, expression patterns of the markers CD34 (a putative adhesion protein), CD38 (an adhesion protein) and CD1a (an MHC-like protein) distinguish the DN1–4 thymocyte subsets. Major markers expressed by developing human and murine T cells are listed in **Table 9-2**, and additional molecules expressed during murine thymocyte development appear in **Figure 9-3**. The following sections discuss the better-studied developmental path of murine thymocytes.

a) DN1 Subset

Murine DN1 thymocytes express both c-kit and CD44 (but not CD25) and reside in the thymic cortex near its junction with the medulla (refer to **Fig. 9-2**). The TCR

Subset Marker	DN Phase				DP Phase		SP Phase	
	DN1	**DN2**	**DN3**	**DN4**	**TCRβ**	**TCRαβ**	**CD4⁺**	**CD8⁺**
c-kit	+	+	−	−	−	−	−	−
CD44	+	+	−	−	−	−	−	−
CD25	−	+	+	Low	−	−	+	+
CD3	−	+	+	+	+	+	+	+
Rearranging TCR genes	−	−	TCRB TCRG TCRD	TCRB	TCRA	−	−	−
pTα	−	−	+	+	Low	−	−	−
RAG	−	−	+	Low	+	+	−	−
TdT	−	−	+	Low	+	+	−	−
TCR	−	−	−	−	−	+	+	+
CD4	−	−	−	Low	Med	+	+	−
CD8	−	−	−	Low	Med	+	−	+

Figure 9-3
Markers Characterizing the Phases of αβ T Cell Development in the Mouse

The expression status of various molecules can be used to monitor the phases of T cell development as a given thymocyte moves through each phase sequentially. pTα = pre-T alpha chain (see main text).

genes remain in germline configuration. DN1 thymocytes are small and closely packed together among the cTECs. cTECs supply *stem cell factor (SCF)* that binds to c-kit on DN1 cells and delivers a survival signal. Without c-kit signaling, the maturation process ceases and the DN1 cells die. The transcription factor GATA-3 is also vital for the generation of DN1 thymocytes from thymic NK/T precursors.

b) DN2 Subset

Murine DN2 thymocytes express CD25 as well as c-kit and CD44 and are sometimes known as *pro-T cells* (progenitor T cells). These thymocytes start to migrate toward the subcapsule of the thymus and thus are present primarily in the outer cortex. The TCR genes remain in germline configuration. DN2 thymocytes commence expression of the CD3 chains, but it is unclear whether these proteins have a signaling function at this stage. Under the influence of IL-7 and SCF, DN2 thymocytes start to proliferate rapidly.

c) DN3 Subset

Murine DN3 thymocytes lose their expression of c-kit and CD44 but continue to express CD25. These cells stop proliferating and remain in the outer cortex. The DN3 stage is critical in T cell development because five key events occur: (1) DN3 thymocytes become restricted to the T lineage and eventually generate mature αβ and γδ T cells. (2) The *TCRG, TCRD* and *TCRB* loci commence V(D)J recombination with concomitant upregulation of RAG and TdT. (3) DN3 thymocytes that eventually generate mature αβ T cells express a functional **pre-TCR** complex that allows them to determine if a functional TCRβ chain has been produced. (4) Successful rearrangement at the *TCRB* locus induces the cessation of further rearrangements at the *TCRG* and *TCRD* loci in these cells. (5) These DN3 thymocytes become *early pre-T cells* that are fully committed to the αβ T cell lineage and express a diverse repertoire of TCRβ chains.

In DN3 thymocytes that eventually generate αβ T cells, the *TCRB* locus is the first to undergo V(D)J recombination. Some rearrangement of the *TCRG* loci may also occur, but functional chains are not produced. As introduced in Chapter 8, the productivity of a given rearranged TCRβ gene in a DN3 thymocyte is tested by the synthesis of the candidate TCRβ chain and the formation of a pre-TCR analogous to the pre-BCR structure in pre-B cells. The pre-TCR counterpart of the surrogate light chain in the pre-BCR is the **pre-T alpha chain** (pTα). pTα is an invariant protein first expressed in DN3 thymocytes that develop into αβ T cells; there is no equivalent in DN3 thymocytes that generate γδ T cells. pTα functions as a "surrogate TCRα chain" and brings the newly translated TCRβ chain (and the CD3 signaling chains) to the thymocyte membrane to form a pre-TCR complex. The pre-TCR complex acts as a sensor so that if the TCRβ chain is functional, the cell receives a survival/differentiation signal that involves Notch1 signaling. The cell also starts to proliferate vigorously, generating a clone of DN4 thymocytes that will eventually become αβ T cells. The process of testing newly produced TCRβ chains is called **β-selection**, and cells that survive β-selection are said to have passed the **pre-TCR checkpoint (Fig. 9-4)**. The successful rearrangement of the *TCRB* gene on one chromosome signals to the cell to suppress V(D)J rearrangement of the *TCRB* locus on the other chromosome and rearrangement of both *TCRG* loci. If *TCRB* rearrangement on both chromosomes has been unsuccessful, the cell neither attempts to rearrange its *TCRA* genes nor becomes a γδ T cell; instead, it dies by apoptosis. Indeed, only 10% of DN3 thymocytes successfully rearrange their TCRβ genes, are β-selected and enter the cell cycle. β-selection is thus directly linked to the proliferation of thymocytes that can proceed further in maturation.

d) DN4 Subset

Murine DN4 thymocytes, also called *late pre-T cells*, are slightly larger in size than DN3 cells and are concentrated in the subcapsular region of the thymic cortex (refer to **Fig. 9-2**). DN4 cells contain a functionally rearranged TCRβ gene; downregulate their expression of CD25, RAG and TdT; and start to express very low levels of CD4 and CD8.

Signals that commit DN3 thymocytes to the γδ T cell lineage are discussed in Chapter 11.

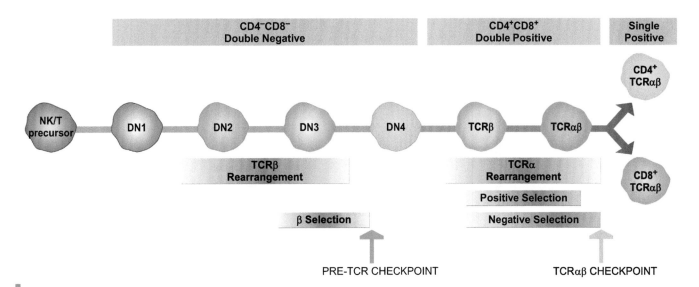

Fig. 9-4
Checkpoints of T Cell Development

Under the influence of particular cytokines and intercellular contacts, NK/T precursors pass through the DN1–DN4 stages of the double negative phase (CD4⁻CD8⁻). The pre-TCR checkpoint marks the end of this phase and entry into the double positive (CD4⁺CD8⁺) phase. After passing through the TCRαβ checkpoint, the developing thymocytes enter the single positive phase and become mature CD4⁺CD8⁻ or CD4⁻CD8⁺ T cells. *[Adapted from Yeung, R.S., Ohashi, P., & Mak, T.W. (2007). T cell development. In Ochs, H.D., Smith, C.I., & Puck, J.M., Eds.,* Primary Immunodeficiency Diseases. A Molecular and Genetic Approach, *2nd ed. Oxford University Press Inc., New York.]*

> NOTE: A group of PIDs characterized by **severe combined immunodeficiency** (SCID) result from a failure in the V(D)J recombination required in both B cells and T cells. *Alymphocytosis* results from null mutations of either the RAG-1 or RAG-2 genes, while *Omenn syndrome* is caused by amino acid substitution mutations in the RAG genes that reduce their activity to 1–25% of normal. *Artemis SCID* is caused by mutations of the Artemis gene, which is important for both V(D)J recombination and DNA repair (see Ch. 15).

iii) The DP Phase

In both humans and mice, the DP phase of αβ T cell development is dominated by the thymic selection processes that shape the mature αβ T cell repertoire. CD4 and CD8 expression levels are steadily upregulated, and these coreceptors play increasingly important roles in directing thymocyte development. As they mature, DP thymocytes move from the subcapsular region through the outer cortex and back through the inner cortex toward the medulla (refer to **Fig. 9-2**).

a) TCRαβ Pool Expansion and TCRA Locus Rearrangement

Early DP thymocytes receive signals through their pre-TCRs that drive their rapid proliferation. These signals appear to depend upon the assembly of pre-TCR itself and not upon interaction with a specific ligand. RAG and TdT expression resume, and V(D)J recombination in both *TCRA* loci commences, resulting in the deletion of the *TCRD* loci (refer to Ch. 8). With the production of the first TCRα chains, pTα expression is gradually downregulated. Newly synthesized (but untested) TCRα chains combine with the proven TCRβ chains to form complete TCRαβ heterodimers (which may or may not be functional). *TCRA* rearrangement continues on both chromosomes until *positive selection* (see below) delivers a survival signal to those thymocytes with functional TCRs that recognize the host's MHC molecules with moderate affinity.

b) Thymic Selection and the Establishment of Central T Cell Tolerance

The establishment of central T cell tolerance requires that thymocytes with TCRs that recognize self antigen be eliminated before they leave the thymus. Central T cell

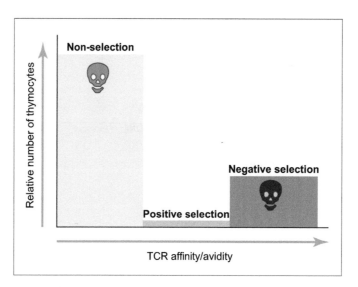

Fig. 9-5
Affinity/Avidity Model of Thymic Selection

The overall affinity/avidity of the interaction between a developing thymocyte's TCR and self pMHC results in intracellular signaling whose nature and level determine the thymocyte's fate. Most thymocytes bear TCRs that fail to bind to self pMHC or do so very weakly; these cells are non-selected and die by apoptosis. Thymocytes with TCRs that bind strongly to self pMHC are negatively selected and also die. Thymocytes with TCRs interacting with self pMHC with moderate affinity/avidity are positively selected and survive. *[Adapted from Yeung, R.S., Ohashi, P., & Mak, T.W. (2007). T cell development. In Ochs, H.D., Smith, C.I., & Puck, J.M., Eds.,* Primary Immunodeficiency Diseases. A Molecular and Genetic Approach, *2nd ed. Oxford University Press Inc., New York.]*

tolerance is established by the processes of **non-selection** (or "neglect"), **positive selection** and **negative selection**. Mature T cells must recognize both MHC and peptide simultaneously to mount an immune response, meaning that the TCRs of a DP thymocyte must bind to the host's MHC molecules (self MHC) with at least moderate affinity. T cells that fail to produce functional TCRs, or produce TCRs that have no affinity for self MHC, cannot be activated in the periphery and thus are useless with respect to defending the host. These cells, which can number as high as 80% of developing thymocytes, are therefore "non-selected" and undergo apoptosis (**Fig. 9-5**). Negative selection removes from the T cell repertoire DP cells whose TCRs bind strongly to pMHCs involving self peptides bound to self MHC. The activation of such T cells in the periphery could spark a damaging autoimmune response in the host. Strong binding of a TCR to self pMHC triggers signaling in the thymocyte that induces its apoptosis. It is estimated that almost 20% of developing thymocytes are "deleted" from the thymus in this way. Positive selection preserves the 1–2% of developing thymocytes whose TCRs recognize self pMHCs neither too strongly nor too weakly. It is a very fine line that separates ligands that are bound to a given TCR too avidly from those that are bound just tightly enough, but only in the second case is a survival signal delivered that allows positively selected thymocytes to proliferate and mature further. Once released to the periphery, it is this small population of new T cells that is most likely to bear TCRs recognizing non-self peptide on self MHC, which is exactly what an APC or target cell will present when an immune response is required. The few remaining potentially autoreactive clones that escape thymic deletion, and clones that recognize self antigens that emerge only later in life (after thymic involution), are neutralized by the mechanisms of peripheral tolerance (see Ch. 10).

Immunologists continue to debate exactly when and where in the thymus each type of selection occurs. Some maintain that positive selection takes place primarily in the cortex and is mediated mainly by cTECs, whereas negative selection occurs later when the DP cells approach the medulla and is mediated by mTECs. Others believe that both positive and negative selection can occur in either the cortex or the medulla, and that these processes are temporally independent. What is now clear is that, regardless of timing, mTECs have a specialized role to play during negative selection that establishes central tolerance to *tissue-specific antigens*. Tissue-specific antigens are self antigens that are normally expressed only in particular host tissues, such as insulin in the pancreas and saliva proteins in the salivary glands. mTECs express a transcription factor called AIRE (*autoimmune regulator*) that allows the expression of many of these tissue-specific antigens in the thymus, so that thymocytes bearing TCRs that recognize these antigens can be deleted. Once transcribed and translated within an mTEC, a given tissue-specific antigen may be processed through the endogenous pathway and presented on MHC class I, or processed via autophagy and presented on MHC class II. In either case, a DP thymocyte bearing a TCR specific for this antigen is negatively selected.

NOTE: Both humans and mice with a genetic deficiency of AIRE show symptoms of autoimmune disease because, without this transcription factor, tissue-specific antigens are not expressed in the thymus. Thymocytes capable of responding to these tissue-specific antigens are therefore not deleted and escape to the periphery where they can mount autoreactive responses against host tissues expressing these antigens (see Ch.19). Inhibition of autophagy in the thymus has also been associated with autoimmunity in mouse models.

c) Nature of Signaling during Thymic Selection

The intracellular signals received by thymocytes during non-selection, positive selection and negative selection depend on the overall affinity/avidity of the interaction between the TCRs of a DP thymocyte and the pMHCs presented by the thymic APC it encounters. The level of intracellular signaling triggered by this interaction is also influenced by the level of aggregation of the TCRs and coreceptor molecules, by the type of thymic APC presenting the pMHC (thymic DC, mTEC or cTEC), and by the costimulatory molecules expressed by the thymic APC. As noted earlier, the majority of DP thymocytes are non-selected because their TCRs cannot interact at all with the pMHCs presented by thymic APCs. Some of these thymocytes have out-of-frame rearrangements of the Vα and Jα TCR gene segments such that no TCRα protein can be produced. Other thymocytes have undergone successful V(D)J recombination, but the TCR produced has a conformation that simply cannot bind to self MHC with any level of affinity. In both cases, TCR signaling is not triggered, and these DP cells proceed down a default path of apoptosis and die in the cortex.

When pMHC co-ligates a TCR and its coreceptor, molecules of Lck kinase associated with the cytoplasmic tail of the coreceptor begin to phosphorylate the ITAMs in the CD3 chains. A negatively selecting pMHC presented by a thymic APC (often an mTEC) binds to the TCR and coreceptor for a longer period of time than a positively selecting pMHC (often presented by a cTEC), enabling complete CD3 ITAM phosphorylation. These phosphorylated ITAMs in turn recruit molecules of ZAP70 kinase, which participates in signaling driving new gene expression. The amount of ZAP70 that accumulates around the base of the TCR complex appears to dictate the direction of selection: the binding of negatively selecting pMHCs results in the recruitment of three times more ZAP70 molecules to the TCR than does the binding of a positively selecting pMHC. The sum total of all the signaling downstream of a TCR bound to a negatively selecting pMHC results in the expression of pro-apoptotic genes that induce cell death. In contrast, the signaling triggered by TCR engagement by a positively selecting pMHC is sufficient to induce the transcription of anti-apoptotic genes that rescue the thymocyte from death. While several molecular events have been identified as contributing to positive selection signaling, how they all fit together remains unclear. Ca^{2+} flux and the activity of Erk kinase are extremely important for positive selection, but it is not understood how they are regulated by the weak binding of a DP thymocyte's TCR to the self pMHC of a thymic APC. A relatively recently discovered molecule important for positive selection is Themis ("thymocyte expressed molecule involved in selection"). Themis is expressed in DN4 thymocytes and early DP thymocytes and is required for passage of a thymocyte through the *TCRαβ checkpoint* (see next section). Themis is rapidly phosphorylated after TCR engagement and is downregulated after positive selection, but its precise molecular function is unknown.

d) The TCRαβ Checkpoint

DP thymocytes that have survived the gauntlet of thymic selection have passed the second developmental checkpoint, the **TCRαβ checkpoint** (refer to **Fig. 9-4**). These positively selected DP cells express a fully functional TCRαβ and both CD4 and CD8, and can thus interact with both MHC class I and class II. It is these thymocytes that proceed to the final phase of αβ T cell development. As described in Chapter 11, it is also these cells that will eventually give rise to NKT cells.

iv) The SP Phase

The SP phase is entered when the class of MHC recognized by a positively selected DP thymocyte becomes fixed by the loss of expression of either CD4 or CD8. This cell and its immediate progeny will thus be either CD4+ or CD8+ SP thymocytes. Which coreceptor is lost and which is retained on the thymocyte surface is determined by complex intracellular signaling pathways and the class of MHC molecule that has participated in the positive selection of the DP thymocyte. As described in Chapter 8, CD4 and CD8 molecules bind to specific regions (outside the TCR binding site) on MHC class II and MHC class I molecules, respectively. If the TCR of the DP thymocyte has bound to a peptide-MHC class I complex presented by the selecting thymic APC, there is an interaction between the MHC molecule and the CD8 coreceptor that causes CD8 expression to be retained and CD4 expression to be lost. Conversely, if the DP thymocyte's TCR has interacted with a peptide-MHC class II complex, an interaction occurs between the MHC molecule and CD4 coreceptor such that CD4 expression is retained while CD8 expression is lost. Thus, the descendants of SP CD4+ thymocytes will bind to pMHCs containing MHC class II, and descendants of SP CD8+ thymocytes will bind pMHCs containing MHC class I. We stress that this discrimination is not due to differences in the type of TCR expressed by CD4+ and CD8+ T cells; all TCRs expressed by all αβ T cells are derived from the same pool of TCRα genes and TCRβ genes.

In addition to TCR-pMHC interaction, there are likely many more downstream intracellular signaling events involved in determining CD4/CD8 lineage commitment. For example, there is some evidence that positively selected DP thymocytes expressing higher levels of ZAP70 become CD8+ SP thymocytes, whereas those with lower ZAP70 levels become CD4+ SP thymocytes. In addition, activation of a transcription factor called *cKrox* may promote CD4+ SP development at the expense of CD8+ SP cells. Other studies have suggested that cytokines within the thymus, particularly IL-7, may nudge DP thymocytes to commit to the CD8+ SP lineage.

Once committed, both CD4+ and CD8+ SP thymocytes loiter in the medulla of the thymus for a short time (2–3 days in the mouse) before they receive a final proliferative signal and expand their numbers. These progeny exit the thymus into the blood and travel to the secondary lymphoid organs, taking up residence as fully functional mature CD4+ or CD8+ T cells. They survive in these locations (in the apparent absence of significant antigenic stimulation) for at least 5–7 weeks, and are poised to react upon meeting their specific antigens. For reasons that are not yet understood, the eventual effector function acquired by a T cell clone is largely dependent on which coreceptor it expresses, such that the vast majority of CD4+ SP thymocytes differentiate into mature naïve Th cells and CD8+ SP thymocytes usually differentiate into mature naïve Tc cells.

B. T Cell Activation

Like B cells, the complete activation of naïve T cells generally requires three signals: (1) the engagement of the antigen receptor by antigen, (2) costimulation, and (3) the receipt of cytokines. However, these signals differ slightly between B and T cell activation, and between naïve Th and Tc cell activation. Additional differences in the activation of effector and memory T cells also exist and are addressed in Section C of this chapter.

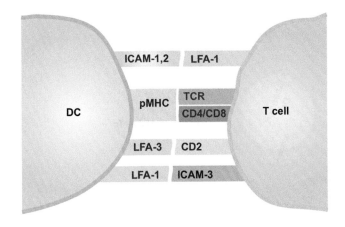

■-**Fig. 9-6**
**Important Adhesion
Contacts between Human
T Cells and DCs**

In addition to the interaction
between the pMHC displayed by
a mature DC and the TCR complex
plus coreceptor molecule present
on a naïve T cell surface, binding
between the indicated pairs of
adhesion receptors and counter
receptors allows the T cell enough
time to determine whether its TCR
recognizes the pMHC.

I. Meeting of Naïve T Cells and DCs

The activation of most naïve T cells takes place in the paracortex of the lymph nodes, the secondary lymphoid organs where antigen-loaded mature DCs congregate and through which naïve T cells recirculate. As described in Chapter 7, immature migratory and lymphoid-resident DCs are experts at capturing entities from the surrounding microenvironment. If the host is experiencing infection or inflammation, pro-inflammatory cytokines and DAMPs/PAMPs that engage the PRRs of these DCs will be present such that these cells are induced to mature. Different pathogens will influence the panel of cytokines the maturing DC will eventually produce, and thus the direction of T cell differentiation (see later). A maturing migratory DC enters a lymph node via an afferent lymphatic vessel and settles in the paracortical region surrounding the HEVs within the node. Maturing lymphoid-resident DCs are already in this location. In both cases, the mature DCs process their captured antigens and display antigenic peptides derived from them on MHC class II via exogenous processing and on MHC class I via cross-presentation. If a DC has become infected by the pathogen, intracellular antigens may also be processed via the endogenous pathway and presented on MHC class I or displayed on MHC class II via autophagy.

Meanwhile, as described in Chapter 2, naïve Th and Tc cells are recirculating in the blood and throughout the secondary lymphoid tissues. In most cases, a naïve T cell enters the node via its HEVs and inspects the pMHCs displayed by the mature DCs in the immediate vicinity of these vessels. The T cell "crawls" slowly over the surface of a DC in a process facilitated by several adhesion molecule pairs (**Fig. 9-6**). These molecules loosely hold the T cell and DC together so that the fit of a particular pMHC in the TCR's binding site can be evaluated. pMHCs that are bound with sufficient affinity/avidity by the TCR have the potential to activate the T cell.

Look back at Figures 7-3
and 2-17 for illustrations of
DC and lymphocyte migration,
respectively.

> NOTE: The crawling of a T cell over a DC's surface has been visualized in a lymph node using live tissue imaging based on two-photon laser microscopy. To see an example, you can visit the following website: https://www.youtube.com/watch?v=PsOZkfj-DTA&feature.

II. Signal 1

Signal 1 is delivered when specific pMHCs displayed on a DC surface bind to multiple copies of a TCR expressed on a naïve Th or Tc cell surface (**Fig. 9-7**). The engagement of a TCR by pMHC likely leads to a conformational change of its associated CD3 chains that allows the phosphorylation of the CD3 ITAMs by molecules of Lck kinase associated with the CD4 and CD8 cytoplasmic domains. This aggregation of pMHC-bound TCRs coupled with the conformational shift in CD3 chains may deflect large molecules

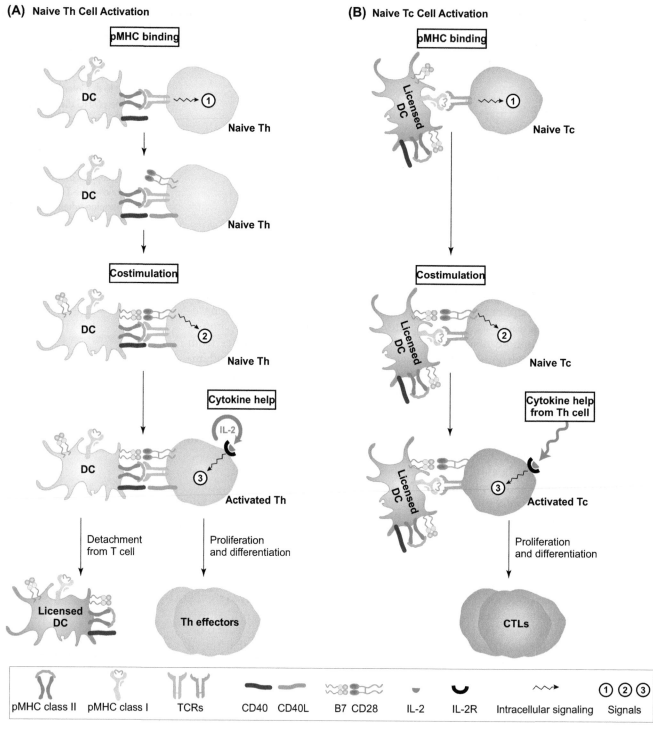

(A) Naive Th Cell Activation

(B) Naive Tc Cell Activation

pMHC class II   pMHC class I   TCRs   CD40   CD40L   B7 CD28   IL-2   IL-2R   Intracellular signaling   Signals ① ② ③

Fig. 9-7
Three Signal Model of Naïve Th and Tc Cell Activation

(A) For naïve Th cells, the binding of the TCR to antigenic peptide displayed on MHC class II by a mature DC delivers signal 1 to the Th nucleus and also upregulates CD40L on the Th cell surface. The binding of this CD40L to CD40 on the DC upregulates the latter's expression of B7 molecules. Signal 2 to the Th nucleus is delivered when these B7 molecules interact with costimulatory CD28 molecules on the Th cell. Expression of cytokine receptors is upregulated, and the Th cell commences production of IL-2. Signal 3 is delivered when the Th cell's IL-2 receptors are engaged by IL-2. The DC, which is now considered to be "licensed" due to its upregulated B7 expression, detaches from the Th cell. The activated Th cell proliferates and generates daughter cells that differentiate into Th effector cells. **(B)** A naïve Tc cell receives signal 1 when its TCR binds to peptide presented on MHC class I expressed by a licensed DC. Costimulation is mediated even in the absence of CD40/CD40L binding because the B7 already expressed on the licensed DC binds to the upregulated CD28 on the Tc cell. Signal 2 is thus delivered to the Tc cell nucleus and results in upregulated IL-2R expression. Signal 3 is delivered when IL-2 produced by a Th cell engages the Tc cell's IL-2Rs. The Tc cell detaches from the licensed DC, proliferates and generates daughter cells that differentiate into effector CTLs. For simplicity, the CD4/CD8 coreceptors and the aggregation of TCRs necessary for T cell activation are not shown.

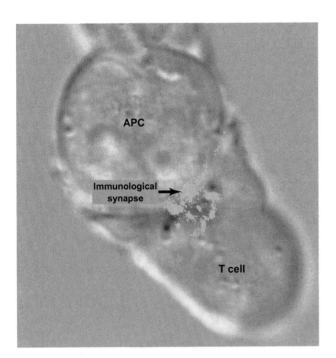

Plate 9-1
The Immunological Synapse

Green dots indicate points of contact at the interface between an APC and a T cell. In the case of naïve T cell activation, the APC is a DC. [*Reproduced by permission of Vincent Das and Andres Alcover, Institut Pasteur.*]

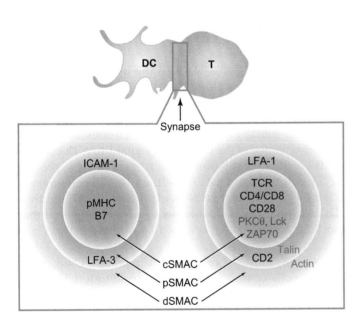

Fig. 9-8
The Supramolecular Activation Complex (SMAC) in the Immunological Synapse

The expansion box shows a cross-sectional view of the three layers of the SMAC (central, peripheral and distal) at the interface between a DC and a naïve T cell. Cell surface molecules are shown in black, whereas intracellular molecules are shown in gray.

(such as CD45 phosphatase) that normally inhibit TCR signaling in the absence of specific pMHC. Additional intracellular signaling enzymes are then recruited to the cytoplasmic tails of the CD4 or CD8 coreceptor and the CD3 chains. Together, these enzymes mediate a cascade of chemical reactions that leads to the activation of several other enzymes. When this activation cascade occurs for multiple TCRs, the T cell receives signal 1.

Because the affinity of binding between a given pMHC and a TCR is relatively low (has a high "off" rate), a single pMHC does not engage a single TCR long enough to achieve complete activation of a naïve T cell. Neither is a transient interaction between a few pMHC–TCR pairs sufficient. Sustained interaction between the naïve T cell and DC for several hours is needed to properly trigger the intracellular signaling pathways within the T cell that lead to the activation of the nuclear transcription factors necessary for new gene transcription. The TCRs and pMHCs required for sustained signaling are gathered together by the formation of an **immunological synapse** at the interface between the T cell and DC (**Plate 9-1**).

Both cells undergo rearrangements of their actin cytoskeletons that are induced by the initial TCR–pMHC binding. The T cell also undergoes polarization, in that its *microtubule organizing center (MTOC)* and Golgi apparatus move toward the site of contact with the DC. A parallel reorganization of these cytoplasmic organelles may also be occurring in the DC. In the T cell at least, these alterations result in the formation of three concentric rings, each containing various signaling, adhesion and cytoskeletal molecules that cluster around the TCR–pMHC pairs (**Fig. 9-8**). The inner ring is called the **central supramolecular activation cluster (cSMAC)** and is composed mainly of the aggregated TCRs and costimulatory molecules. The middle ring is called the **peripheral supramolecular activation cluster (pSMAC)** and contains the signaling adaptor talin as well as large numbers of integrins and other adhesion molecules. The outer ring, called the **distal supramolecular activation cluster**, mainly contains actin-based cytoskeletal structures as well as large proteins excluded from the inner area of TCR aggregation. On the side of the T cell opposite from the immunological synapse is the *distal pole complex*, which is thought to actively sequester negative signaling regulators away from the site of TCR aggregation.

III. Signal 2

In most cases, the engagement of TCRs by pMHCs is not sufficient to fully activate a naïve Th or Tc cell, and signal 2 in the form of costimulatory signaling is required. Occasionally, a Tc cell will encounter a pMHC (usually derived from a virus) that delivers such a strong signal 1 that costimulation is not required; this response is then independent of both costimulation and Th cell help.

In the case of Th cells, the receipt of signal 1 leads to the upregulation of the important costimulatory molecule *CD28* on the Th cell's surface (refer to **Fig. 9-7A**). However, in order for CD28 to convey signal 2 to the Th cell nucleus, it must bind to its ligand **B7** on the surface of the DC presenting the activating pMHC. When the immunological synapse first starts to form, the mature DC involved does not express optimal levels of B7. A critical consequence of the delivery of signal 1 to a Th cell and the initial binding of CD28 to B7 is the upregulation of the transmembrane protein *CD40 ligand* (CD40L) on the Th cell surface. Once CD40L on the Th cell engages CD40 expressed by the DC, the DC greatly increases its expression of B7 and thus its binding to CD28 on the Th cell. As a result, a vigorous signal 2 is delivered that enhances the activatory intracellular signaling that is occurring within the T cell as a result of signal 1.

Although they upregulate CD28, most Tc cells do not express CD40L even after receiving signal 1. Thus, Tc cells cannot induce a DC to initiate CD40 signaling and upregulate B7 expression. Instead, Tc cells rely on CD28 engagement resulting from interaction with a DC that *already* expresses B7 due to a previous interaction with an antigen-activated Th cell. Some immunologists say that these DCs have been "licensed" for Tc activation (refer to **Fig. 9-7B**). This licensing of DCs by Th cells is one component of the T cell help provided by Th cells for Tc responses.

For both Tc and Th cells, CD28 signaling lowers the T cell activation threshold necessary to activate new gene transcription in the T cell and push it to proliferate and differentiate. In the absence of CD28 costimulation, naïve T cells are anergized instead of activated and fail to respond to pMHC (see Ch. 10). Costimulation via CD28 has several important molecular effects: (1) IL-2R expression is induced on the T cell surface, allowing the cell to receive signal 3. (2) Th cells start to secrete large quantities of IL-2, as well as other important cytokines and chemokines. (3) The expression or upregulation of additional costimulatory and regulatory molecules is induced in both Th and Tc cells. (4) Intracellular signaling supporting T cell survival, proliferation, and metabolism is promoted.

B7 actually refers to two closely related costimulatory proteins: B7-1 (CD80) and B7-2 (CD86).

NOTE: The potentially destructive power of T cells must be tightly controlled to ensure it is applied only where and when it is necessary. The TCR and costimulatory signaling pathways are therefore subject to negative regulation at multiple steps. The two most important negative regulators of T cell activation are *PD-1* (programmed death-1) and *CTLA-4* (cytotoxic T lymphocyte associated molecule 4). While PD-1 is expressed by T cells, B cells and some DCs, CTLA-4 expression is exclusive to T cells. PD-1 expression on a T cell is induced within hours of its activation, while expression of the PD-1 ligands PD-L1 and PD-L2 on the DC surface is induced by inflammatory cytokines. When PD-1 and the TCR are both engaged, PD-1 transmits an inhibitory signal that shuts down early steps of the TCR signaling pathway. In contrast to PD-1, CTLA-4 is not expressed on the T cell surface until 1–2 days after activation is initiated by TCR/pMHC interaction, giving the adaptive response time to eliminate the threat before T cell activation is damped down. CTLA-4 competes with CD28 for binding to the B7 costimulatory ligands. Because CTLA-4 has a much higher affinity for B7 proteins than does CD28, CTLA-4 displaces CD28 and recruits inhibitory molecules to the TCR complex.

IV. Signal 3

A naïve Th or Tc cell that has received signals 1 and 2 upregulates the receptors (particularly IL-2R) which permit it to receive signal 3 in the form of cytokines, chemokines and growth factors (refer to **Fig. 9-7**). Activated Th cells produce many of the cytokines that can bind to these newly expressed receptors, the chief among them being IL-2. IL-2 is believed to deliver the most important signal for the proliferation of newly activated naïve T cells. Although a Th cell on its own can make sufficient IL-2 to meet its requirements (i.e., carries out *autocrine* IL-2 production), a Tc cell usually cannot. Thus, another component of T cell help provided to Tc cells by Th cells is the production of IL-2 (and possibly other cytokines) necessary for Tc proliferation.

A naïve T cell that achieves activation proliferates and generates daughter T cells that differentiate into effector T cells. Effector T cells differ from naïve T cells in several important ways besides function, including tissue of residence, preferred APC, costimulatory requirements, duration of TCR signaling needed for activation, dominant metabolic pathways, rate of cell division, sensitivity to cell death mechanisms, and life span. The next two sections address the properties of Th effector cells and CTLs.

C. Th Cell Differentiation and Effector Function

I. Overview

Once a naïve Th cell is fully activated, it starts to produce copious amounts of IL-2 and proliferates vigorously. The progeny generated are called *Th0* cells. About 48–72 hours after the original antigenic stimulation of the naïve Th cell, these Th0 cells terminally differentiate into various subsets of resting effector Th cells. Of these subsets, *Th1* cells, *Th2* cells and *Th17* cells are arguably the most important. Other Th effector subsets include *Th9* cells, *Th22* cells and follicular Th (*fTh*) cells. Th0 cells can also generate *induced regulatory T cells,* which function in peripheral tolerance and so are introduced here and discussed further in Chapter 10.

The type of effector Th subset generated from a proliferating Th0 cell is determined by (1) the cytokines and other factors present in the immediate microenvironment and (2) the nature of the DC by which the original naïve Th cell was activated. Different pathogens supply PAMPs that bind to different PRRs, causing the DCs expressing these receptors to mature into subtly different subsets. These DCs then secrete different panels of cytokines and deliver various intercellular signals that help to direct Th differentiation such that the Th effectors best-suited for eliminating the pathogen are produced. Some Th effector subsets secrete cytokines that facilitate effector functions specialized for the disposal of intracellular invaders, whereas the cytokines produced by other Th effector subsets promote effector functions designed to counter extracellular threats. In any case, following their generation and differentiation in a secondary lymphoid tissue, most resting Th effectors migrate back to the site of inflammation containing the antigen that sparked the activation of the original naïve Th cell. In this site, presentation of the same antigenic pMHCs by an APC (which can be a DC, macrophage or B cell at this stage) activates these Th effectors and causes them to take action in the form of secreting their subset-specific panels of cytokines.

NOTE: After newly produced naïve T cells migrate from the bone marrow and enter the secondary lymphoid tissues, they adopt a resting state in which the metabolic demands of perusing APCs for foreign pMHCs are met by ATP generated via mitochondrial oxidative phosphorylation of glucose and glucosamine. However, when the naïve T cell is activated and undergoes the proliferation and differentiation needed to generate effector T cells, an entirely new set of metabolic requirements arises that is met by a shift from oxidative phosphorylation to glycolysis. Glycolysis is much more efficient at providing the energy needed for the synthesis of new proteins, sugars, lipids and nucleotides, and for the execution of T cell effector functions. At the conclusion of the primary response, surviving effector T cells and memory T cells revert to quiescence, decrease glycolysis, and resume energy generation via oxidative phosphorylation.

II. Differentiation of Th Effector Cell Subsets

Th cell differentiation has been best studied in mice. Many of the cytokines and transcription factors involved in human Th cell differentiation overlap.

The differentiation paths of the most important Th effector cell subsets identified to date are summarized in **Figure 9-9**, including relevant cytokines and transcription factors. It should be noted, however, that accumulating evidence suggests that these differentiation paths are not fixed for the life of a particular T cell clone, and that one type of effector can become another type if circumstances change and its transcriptional programs shift in response.

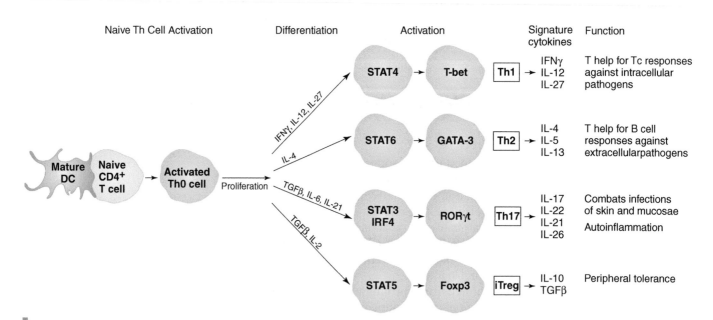

Fig. 9-9
Th Effector Cell Differentiation and Functions

A naïve Th cell activated by a mature DC proliferates and generates daughter cells that are induced to differentiate into the indicated Th effector subsets depending on the cytokines in the immediate microenvironment. For a given Th subset, the indicated cytokines activate the indicated transcription factors within a Th0 cell to direct differentiation down a specific path. Once differentiated, each Th subset produces a signature cytokine profile that mediates its effector function. Not all Th subsets are shown.

i) Th1 Cells

Intracellular pathogens such as viruses and intracellular bacteria trigger macrophages and DCs to produce IFNγ, IL-12 and IL-27. A Th0 cell that has its TCR engaged by specific pMHC and encounters these cytokines experiences activation of the transcription factor STAT4. STAT4 drives a gene expression program that causes the Th0 cell to commit to the Th1 subset 5–7 days after antigen stimulation. IL-18 then supports the survival and proliferation of the newly produced Th1 effector cells. Once in the site of inflammation where their effector function is required, Th1 cells are stimulated by antigen in this site and activate the transcription factor T-bet. T-bet drives IFNγ production and opposes intracellular signaling promoting Th2 differentiation.

ii) Th2 Cells

Most extracellular pathogens do not induce IL-12 production by macrophages and DCs. Instead, these invaders stimulate an unknown cell type (which might be a mast cell or NKT cell) to secrete IL-4. In the absence of IL-12 and IFNγ but in the presence of IL-4, a Th0 cell experiences activation of the transcription factor STAT6. STAT6 drives a gene expression program that causes the Th0 cells to generate Th2 effectors. IL-4 then sustains the survival and proliferation of the newly produced Th2 cells. Once in the inflammatory site where they are needed, Th2 cells stimulated by specific antigen activate the transcription factor GATA-3. GATA-3 both drives the production of the Th2 signature cytokines IL-4, IL-5 and IL-13 and reciprocally opposes intracellular signaling promoting Th1 differentiation.

iii) Th17 Cells

Th17 effector cells counter infections of the skin and mucosae (particularly in the lung and intestine) that are initiated by certain species of extracellular bacteria and fungi. Th0 cells exposed to a combination of the immunosuppressive cytokine TGFβ plus the pro-inflammatory cytokines IL-6 and/or IL-21 produced by local innate leukocytes experience activation of the transcription factors STAT3 and IRF4, which implement a gene expression program resulting in Th17 effector generation. IL-23 appears to be required for the continued survival and terminal differentiation of Th17 cells into fully functioning effectors. In an inflammatory site, antigen-stimulated Th17 effectors activate the transcription factor RORγt that drives a gene expression program resulting in the production of IL-17, IL-21, IL-22 and IL-26. Both IFNγ (produced by Th1 cells) and IL-4 (produced by Th2 cells) suppress Th17 differentiation.

Th17 cells were first identified through studies of autoimmune disease in mice. Although these disorders had been previously blamed on the actions of dysregulated Th1 cells, it has turned out that the autoinflammatory lesions in these animals contain large numbers of Th17 cells. Indeed, knockout mice lacking IL-17R are resistant to the induction of experimental autoimmune diseases. In humans, elevated levels of IL-17 have been detected in the blood and tissues of patients with autoimmune diseases such as multiple sclerosis, rheumatoid arthritis, psoriasis, and inflammatory bowel disease (see Ch. 19). Moreover, T cells isolated from the affected tissues of these patients show the phenotype of Th17 cells.

iv) Other Th Effector Subsets

Th9 cells help to combat infections by helminth worms. Th0 cells exposed to TGFβ and IL-4 differentiate into Th9 effectors that secrete IL-9 upon antigen stimulation. *Th22 cells* help to protect the skin and mucosae, and also promote wound healing after trauma. Th0 cells exposed to IL-6 plus TNF differentiate into Th22 cells that secrete IL-22. IL-22 combines with other cytokines to induce skin cells to produce antimicrobial peptides. *Follicular Th (fTh) cells* is the name some immunologists have given to the subset of Th cells that provide help in the form of cytokine secretion and intercellular contacts to antigen-activated B cells within germinal centers (see Ch. 5). Within the T cell zone of a lymph node, a Th0 cell that is exposed to IL-6 plus IL-27 and/or IL-12 switches expression of its chemokine receptors and makes its way into the GC of a

B cell follicle. In this location, the T cell receives additional differentiation signals from antigen-presenting GC B cells and becomes an fTh effector cell.

In addition to these Th effector subsets, Th0 cells can differentiate into *induced regulatory T (iTreg) cells* if exposed to TGFβ plus IL-2. As mentioned earlier, regulatory T cells are T cells that can shut down the functions of effector T cell subsets, suppressing the adaptive response. Once differentiated, iTreg cells activate the Foxp3 transcription factor, which induces the secretion of IL-10 and TGFβ. IL-10 and TGFβ then act on effector T cells to curtail their responses. Interestingly, IL-6 and IL-21 are potent repressors of TGFβ–driven Foxp3 expression, so that Th0 cells exposed to TGFβ are induced to become Th17 cells rather than iTreg cells if IL-6 and/or IL-21 is also present. Thus, the balance between the generation of Th17 cells and iTreg cells is fine-tuned by the surrounding microenvironment.

NOTE: The plasticity of Th subsets is an issue of growing interest. It has become clear that Th cell fate decisions are influenced by two general mechanisms of gene expression regulation: *microRNA* expression and *epigenetic modification*. MicroRNAs (miRNAs) are short, single-stranded pieces of RNA that do not encode a protein themselves but repress the expression of particular mRNAs by promoting their degradation and inhibiting their translation. Epigenetic modifications involve the methylation of cytosine and guanine residues in genomic DNA in the promoter and enhancer regions of genes. The presence of these methylated CpG motifs blocks gene transcription. At least *in vitro*, different Th effector subsets show different patterns of miRNA expression and CpG methylation, and these patterns often change when the Th cell clone switches effector fates.

III. Activation of Th Effector Cells

i) Localization

Once differentiated, Th effector cells may remain in the lymph node to supply T cell help to naïve Tc cells in the paracortex and naïve B cells in the primary follicles. Alternatively, Th effectors may leave the node and bolster defense of the diffuse lymphoid tissues under the skin and mucosae (SALT and MALT), or follow chemokine gradients to sites of inflammation. All Th effector subsets initially express CCR7, which acts within a lymph node to permit migration of the effector T cells from the paracortex to the edges of the primary lymphoid follicles (where naïve B cells are located). However, as the response progresses, fully differentiated Th1, Th2 and Th17 cells express different panels of chemokine receptors and thus exhibit differential trafficking patterns. Later members of Th1 effector clones express CCR1, CCR5 and CXCR3 that draw the Th1 cells to sites of inflammation in the peripheral tissues where defense against intracellular pathogens is usually required. In contrast, Th2 effectors begin to preferentially express CCR3, CCR4 and CCR8. These receptors direct Th2 cells to sites such as the mucosae where responses against extracellular pathogens and toxins are needed. CCR6 expressed by Th17 cells also promotes migration to the SALT and MALT as well as to inflamed tissues.

ii) Interaction with APCs

Effector Th cells encountering pMHCs presented by APCs either in the lymph node or in the site of attack are activated essentially in the same way as naïve Th cells but with some important differences. Compared to naïve Th cells, effector Th cells express higher levels of adhesion molecules that stabilize the immunological synapse more rapidly, facilitating TCR triggering. While an estimated 20–30 hours of sustained TCR signaling is required for naïve T cell activation, only 1 hour is required for effector Th cell activation. Effector Th cells are thus activated by significantly lower quantities of pMHC. In addition, far less costimulation by the APC is required. As a result, effector Th cells respond efficiently to pMHC presented by DCs, macrophages or B cells or sometimes even by non-hematopoietic cells such as gut or skin epithelial cells. In general, B cells are the principal APCs presenting antigen to Th2 cells, whereas macrophages predominate as APCs in interactions with Th1 and Th17 cells.

iii) Differential Costimulatory Requirements

While CD28-B7 interaction is the major costimulatory mechanism for naïve T cell activation, effector T cells appear to require only low levels of CD28-B7 costimulation for activation. Nevertheless, CD28 engagement on effectors is critical because it reduces the time required to achieve activation and avoids prolonged stimulation. In addition, CD28 signaling downregulates the expression of chemokine receptors, preventing the effector cell from migrating away from the site where antigen has been encountered. Two supplementary costimulatory pairs that are important for effector T cell activation are OX40-OX40L and ICOS-ICOSL. In Th1 responses, *OX40* expressed on a Th1 cell surface binds to *OX40 ligand* (OX40L) expressed on APCs. Similarly, the *inducible costimulatory* (ICOS) molecule, which is upregulated on Th2 and Th17 cells only after activation, binds to *ICOS ligand* (ICOSL) expressed on APCs. ICOS is rarely expressed by Th1 effectors.

IV. Functions of Th Effector Cells

A brief comparison of the properties of Th1, Th2 and Th17 effectors is given **Table 9-3**.

i) Th1 Effector Functions

Th1 effectors supply T cell help to Tc and B cells providing cell-mediated and humoral defense against intracellular pathogens. Th1 cells secrete a panel of cytokines dominated by IL-2, IFNγ and lymphotoxin (LT) (sometimes called *Th1 cytokines*). IL-2 drives T and B cell proliferation and enhances ROI production by macrophages. IFNγ and LT hyperactivate macrophages and spur them to secrete additional cytokines, undertake vigorous phagocytosis and upregulate NO production. IFNγ also increases NK cell and macrophage expression of high affinity FcγR molecules that promote ADCC and influences B cells to switch to the production of the Ig isotypes most effective against intracellular pathogens (IgG1 and IgG3 in humans). IgG1 and IgG3 are the antibodies best suited for opsonization, phagocytosis and complement activation, and also bind with high affinity to FcR on NK cells, macrophages and other phagocytes, further increasing ADCC. In addition, Th1 cytokines increase the antigen-presenting potential of macrophages by upregulating MHC class II and TAP. Th1 cells support the activation of Tc cells by producing IL-2 and by providing CD40/CD40L contacts for **DC licensing**.

TABLE 9-3	Comparison of the Properties of Th Effector Cells		
Property	**Th1 Effectors**	**Th2 Effectors**	**Th17 Effectors**
Cytokines important for differentiation	IL-2, IL-12, IFNγ, IL-27	IL-4	TGFβ, IL-6, IL-21
Transcription factors important for differentiation	STAT4	STAT6	STAT3, IRF4
Transcription factors important for effector function	T-bet	GATA-3	RORγt
Distinguishing surface markers	IL-12R	IFNγR	IL-23R
Chemokine receptors	CXCR3, CCR5, CCR1	CCR3, CCR4, CCR8	CCR4, CCR5, CCR6, CXCR6
Preferred APCs	Macrophages	B cells	Macrophages
Costimulation	CD28/B7 (low) OX40/OX40L	CD28/B7 (very low) ICOS/ICOSL	CD28/B7 (low) ICOS/ICOSL
Cytokines secreted	IFNγ, IL-2, LT	IL-4, IL-5, IL-13, IL-10, IL-6, IL-3, IL-1	IL-17, IL-21, IL-22, IL-26
Type of immune response promoted	Humoral, cell-mediated	Humoral	Inflammatory
Pathogens combatted	Intracellular	Extracellular	Bacteria and fungi not eliminated by Th1/Th2
Associated with	Transplant rejection	Allergy	Autoimmune disease

ii) Th2 Effector Functions

Th2 differentiation is usually induced upon invasion by extracellular pathogens. Th2 cells tend to promote humoral responses because these cells secrete IL-3, IL-4, IL-5, IL-6, IL-10 and IL-13 (sometimes called *Th2 cytokines*). A major function of Th2 cells is to establish CD40–CD40L contacts with B cells and to secrete IL-4 and IL-5, cytokines that induce switching to the Ig isotypes most effective against extracellular pathogens. Such isotypes, which include IgG4 in humans, are those best suited for neutralization. IgG4 is not very proficient at complement activation or ADCC, which is an advantage in combatting pathogens in mucosal sites where the inflammation induced by these effector functions could be damaging (see Ch. 12). IL-4 and IL-13 inhibit pro-inflammatory cytokine production, downregulate NO production, and decrease FcγR expression on macrophages, blocking ADCC. However, IL-4 upregulates MHC class II expression on APCs such as macrophages, DCs and B cells, and thereby contributes to Th cell stimulation. IL-4 and IL-13 also enhance the humoral response by stimulating B cell proliferation. IL-5 promotes the growth, differentiation and activation of eosinophils crucial for the elimination of large parasites such as helminth worms. IL-3, IL-4 and IL-10 combine to promote the activation and proliferation of mast cells, also effective against large parasites. In general, however, IL-10 acts as a brake on immune responses and balances the stimulation exerted by other cytokines. For example, IL-10 inhibits the pro-inflammatory functions of macrophages and abrogates their production of IL-12 and MHC class II. IL-10 also downregulates B7 expression on macrophages and DCs.

iii) Th17 Effector Functions

Until recently, Th17 cells were most often thought of in the context of promoting autoimmune disease. Obviously, this is unlikely to be their physiological function! Rather, the massive inflammatory responses mounted by these cells, which are dominated by IL-17, IL-21, IL-22 and IL-26, are designed to protect mucosal surfaces against pathogens that can resist assault by Th1 and Th2 cells. Pathogens triggering strong Th17 cell-mediated responses include several bacterial species, such as *Borrelia burgdorferi* and *Klebsiella pneumoniae*, as well as the AIDS-associated organism *Pneumocystis carinii* and the fungus *Candida albicans* (among others). IL-17 produced by Th17 cells induces nearby non-hematopoietic cells to produce destructive pro-inflammatory cytokines such as TNF, IL-1 and IL-6.

iv) Th Effector Cell Cross-Regulation and Amplification

Because of the cytokines they produce, various Th cell subsets can cross-regulate each other's differentiation and activities, either positively or negatively. For example, Th1 cells produce large amounts of IL-2 that can promote the proliferation of both Th1 and Th2 cells. However, the IFNγ produced by Th1 cells has a direct antiproliferative effect on Th2 cells and inhibits further Th2 differentiation. On the other hand, IFNγ stimulates macrophages to produce IL-12, which promotes Th1 differentiation. Th2 cells do not make substantial amounts of IL-2 or IFNγ and instead secrete IL-4, IL-13 and IL-10. These cytokines suppress IFNγ and IL-2 secretion by Th1 cells, inhibit further Th1 differentiation and downregulate macrophage production of IL-12. In an example of positive feedback, the IL-4 produced by Th2 cells promotes the continued differentiation of this subset. A similar effect is seen for Th17 cells, in that the IL-21 (but not IL-17) produced by activated Th17 cells supports continued Th17 cell differentiation. IL-21 also helps to repress the expression of Foxp3 that drives Treg differentiation.

V. Nature of Th Responses

Among immunologists, it is said that an immune response has either a Th1, Th2 or Th17 character or phenotype, depending on the predominant Th subset and cytokines observed in the host during that response. An attack on a host by intracellular

T cell responses to specific pathogens are discussed in Chapter 13.

pathogens stimulates the production by DCs and macrophages of cytokines favoring Th1 development, leading to the mounting of a *Th1 response*. Conversely, invasion by extracellular pathogens most often promotes the development of a *Th2 response*, or, depending on the specific invader, a *Th17 response*. Immunological disease states also tend to have a Th1, Th2 or Th17 phenotype. For example, allergies are associated with a prevalence of Th2 cells, whereas Th1 cells dominate in transplant rejection, and Th17 cells are associated with many autoimmune disorders. Despite these generalizations, however, the overall phenotype of an immune response to a given pathogen can change with time. For example, mice infected with the parasite causing malaria first develop a Th1 response designed to deal with the intracellular stage of the infection. Th1 effectors are generated that secrete IFNγ, which activates macrophages to secrete cytotoxic cytokines and produce large quantities of NO. IFNγ also induces B cells to switch to the production of parasite-specific IgG2a antibodies. However, 10 days after the initial attack, the parasite adopts an extracellular phase that triggers a Th2 response characterized by high serum levels of IL-4, IL-10 and parasite-specific IgG1 antibodies. Both types of Th responses are needed to keep the pathogen in check.

D. Tc Cell Differentiation and Effector Function

I. Overview

Cytotoxic T cell responses can be thought of as occurring in five stages:

(1) Activation of the naïve Tc cell by a licensed DC in a secondary lymphoid tissue;
(2) Proliferation and differentiation of the activated Tc cell into daughter cells called *pre-CTLs*;
(3) Differentiation of a pre-CTL in an inflammatory site into an *"armed" CTL*;
(4) Activation of the armed CTL by encounter with specific non-self peptide presented by MHC class I on a target cell; and
(5) CTL-mediated destruction of the target cell as well as other cells displaying the identical pMHC.

Target cells of CTLs include cells infected with intracellularly replicating pathogens, tumor cells and foreign cells entering the body as part of a tissue transplant. We emphasize that an activated Tc cell has no lytic powers at all: only its mature CTL progeny develop cytotoxicity.

> NOTE: Until relatively recently, it was thought that activated naïve CD8⁺ Tc cells developed only into CTLs with ability to lyse target cells. There is now *in vitro* and *in vivo* evidence that naïve Tc cells exposed to specific antigen in the presence of cytokines (TGFβ, IL-6, IL-21) that influence Th17 effector cell differentiation can result in the development of so-called *Tc17 cells*. Tc17 cells show greatly repressed cytotoxicity functions and instead secrete copious amounts of IL-17. Tc17 cells have been found along with Th17 cells in the lesions of mice with autoimmune diseases and in the lungs of mice challenged with a lethal dose of influenza virus.

II. Generation and Activation of CTLs

i) Differentiation of CTLs

In the presence of IL-2 secreted by an activated Th cell, an activated naïve Tc cell proliferates and generates pre-CTL precursor cells (**Fig. 9-10**). These pre-CTLs leave the lymph node and travel to the site of pathogen attack. In the presence of IL-12, IFNγ and IL-6 produced by activated macrophages and DCs, the pre-CTLs differentiate into mature CTLs whose cytoplasm contains cytotoxic granules. These mature CTLs (that have yet to encounter antigen) are said to be "armed" and do not need to carry out any additional protein synthesis to be effective killers. This effector generation process is completed within 24–48 hours of TCR stimulation of the original Tc cell. Importantly, because pre-CTL differentiation into armed CTLs requires inflammatory cytokines, the development of cytotoxicity and the power of the CTL response are reserved for situations in which a threat is actually present.

ii) Activation of Armed CTLs and Conjugate Formation

Within an inflammatory site, an armed CTL binds weakly to one host cell after another in search of its specific pMHC. The armed CTL detaches without incident if the affinity of TCR–pMHC binding is too low. However, should the armed CTL encounter a pMHC for which it is specific (the TCR binds with sufficient affinity), the host cell becomes a

Fig. 9-10
CTL Generation and Cytotoxicity

A naïve Tc cell encountering a licensed DC presenting cognate pMHC in the presence of IL-2 is activated, proliferates and generates pre-CTLs. In the presence of the indicated cytokines, the pre-CTLs differentiate into armed CTLs containing cytoplasmic granules and expressing FasL. When the TCR of an armed CTL is engaged by specific pMHC displayed by a target cell, the apoptotic death of this cell is induced via cytotoxic granule release, Fas ligation or the secretion of cytotoxic cytokines. The ultimate death of the target cell occurs after the CTL has detached.

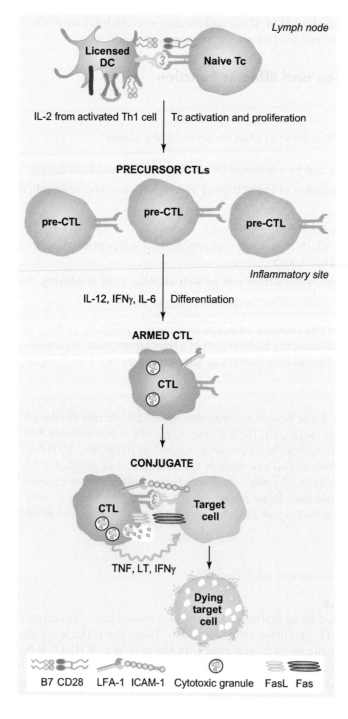

target. Stimulation of the TCR of an armed CTL rapidly increases the binding affinity of adhesion molecule pairs between the CTL and the target cell, forming a bicellular conjugate (refer to **Fig. 9-10**). The CTL then delivers a "lethal hit" of chemical mediators that rapidly causes target cell death. This speed is important to ensure the killing of an infected cell before too many of the progeny of the replicating pathogen can escape to new cells. Much lower concentrations of specific pMHC and the engagement of far fewer TCRs are required to activate armed CTLs compared to naïve Tc cells: only the engagement of a single TCR by a single specific pMHC is needed, and no costimulation is required.

III. Mechanisms of Target Cell Destruction

Target cell destruction by CTLs can occur via the *granule exocytosis pathway,* the *Fas pathway*, and/or the release of *cytotoxic cytokines* such as TNF and LT (**Fig. 9-11**). The pathway used depends on the nature of the attacking intracellular pathogen, but granule exocytosis accounts for the majority of target cell killing by CTLs.

i) Granule Exocytosis
The granule exocytosis pathway refers to the release of the contents of the CTL's cytotoxic granules. Soon after conjugate formation, the cytoskeleton of the CTL reorganizes

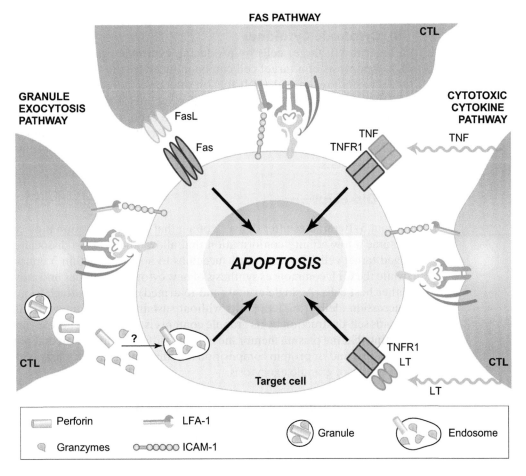

Fig. 9-11
Mechanisms of CTL Cytotoxicity
The three major ways a CTL can kill a target cell are illustrated. In the granule exocytosis pathway, granules are exocytosed by a CTL in close proximity to the target cell membrane. The granzymes and perforin released from the granule are endocytosed by the target cell and captured in endosomes. The perforin forms pores in the endosomal membrane, causing the granzymes to be released into the cytoplasm. In the Fas pathway, FasL expressed by the armed CTL binds to Fas expressed by the target cell. In the cytotoxic cytokine pathway, TNF or LT secreted by the CTL binds to TNFR1 on the target cell. Each of these three pathways triggers apoptosis that may or may not involve caspases.

so that its cytotoxic granules are brought to the site of CTL–target cell contact. The granules fuse with the CTL membrane, and the cytotoxic contents of the granules are directionally exocytosed toward the target cell membrane. Among these granule contents are *perforin* and *granzymes*. Perforin is a pore-forming protein, and the granzymes are a family of serine proteases. It is not clear how these proteins actually enter the target cell, but after they do, they are immediately confined to its endocytic system. Perforin then facilitates the release of the granzymes from the endolysosomal vesicles into the cytoplasm of the target cell. Granzyme A initiates a caspase-independent pathway of DNA damage, while granzyme B triggers classical caspase-mediated apoptosis. Upon the degradation of its DNA and other important intracellular substrates, the target cell dies. This form of death is called **perforin/granzyme-mediated cytotoxicity**.

ii) Fas Pathway

Fas is a transmembrane "death receptor" that is widely expressed on mammalian cells. Engagement of Fas on a target cell by *Fas ligand* (FasL) expressed by an armed CTL results in the death of the target cell. Naïve Tc cells do not express FasL, but after activation by encounter with antigen, FasL is synthesized and stored in specialized transport vesicles in differentiating CTLs. Upon conjugate formation, the FasL-containing transport vesicles fuse with the CTL plasma membrane and anchor FasL on the CTL surface. The FasL engages Fas on the target cell and induces its apoptosis.

iii) Cytotoxic Cytokines

CTLs can kill target cells by producing **cytotoxic cytokines**, particularly TNF and LT. Apoptosis of a target cell can be induced by the binding of TNF produced by a CTL to TNF receptor 1 (TNFR1) on the target cell surface. LT, which also binds to TNFR1, has a similar effect. CTLs also secrete IFNγ, whose action in this context is more indirect. IFNγ stimulates B cells to produce antibodies that facilitate killing via ADCC or complement activation. As well, IFNγ upregulates MHC class I on nearby host cells, enhancing antigen display and making target cells more visible to scanning CTLs.

IV. Dissociation

About 5–10 minutes after delivery of a lethal hit, the adhesion molecules on the CTL resume a low affinity conformation that allows the CTL to dissociate from the damaged target cell. The target cell succumbs to apoptosis within 3 hours of dissociation, while the CTL commences synthesis of new cytotoxic granules and moves off to inspect other host cells. A single armed (and re-armed) CTL can attach to many host cells in succession, delivering lethal hits without sustaining any damage itself. How the CTL avoids self-destruction by its granule contents is a mystery. There is some evidence suggesting that the plasma membrane of CTLs differs slightly from that of most target cells in its lipid and/or protein composition and that this difference may confer resistance to the effects of granule exocytosis.

E. Termination of Effector T Cell Responses

The Th effector cells and CTLs that are actively proliferating and employed in eliminating a pathogen that sparked a primary immune response are sustained by signals delivered by inflammatory cytokines (such as IL-12) and transcription factors (such as Id2 and Bcl-3). However, after the effectors have removed the threat, there is no further need for their presence. Continued exposure to the inflammatory environment in the absence of antigen causes the effectors to downregulate IL-7R and IL-15R, reducing their ability to receive survival signals. Three mechanisms then act in concert to further bias the balance of pro-apoptotic/anti-apoptotic gene expression and induce effector cell death: **activation-induced cell death (AICD), cytokine "withdrawal,"** and **T cell clonal exhaustion.**

AICD is a form of apoptosis induced when the intracellular signaling triggered by TCR engagement by antigen becomes prolonged. This extended signaling is thought to induce the transcription of pro-apoptotic genes such as FasL and TNFR1 within the effector T cell, and to decrease its expression of anti-apoptotic molecules. These changes set the effector T cell up to be killed by contact with Fas or TNF expressed by a neighboring cell. In the early stages of a primary response, effector T cells are protected from AICD by a constant low level of CD28–B7 signaling. This minimal costimulation initially ensures that pro-apoptotic genes are not transcribed to a level that overwhelms the effects of anti-apoptotic genes.

Cytokine withdrawal is a mechanism that was first observed for IL-2 *in vitro*. Clones of T cells die if all IL-2 is removed from the culture medium after activation. It seems that this phenomenon also happens *in vivo*, when antigen is being mopped up and less of it is around to stimulate DCs and other APCs to secrete cytokines. In the absence of antigen and these cytokines, T cell activation cannot be sustained, IL-2 production falls, and effector T cells die. In this case, the induction of T cell apoptosis does not involve Fas or TNF. Rather, the lack of cytokines prevents cytokine receptor engagement and blocks the delivery of a vital survival signal. In the absence of this survival signal, pro-apoptotic gene expression dominates and kills the cell.

Although most effector T cells eventually succumb to AICD or cytokine withdrawal, entire activated T cell clones are sometimes eliminated by "clonal exhaustion." In this case, continuous exposure to antigen causes the T cells to divide so relentlessly into effectors that they burn out metabolically without generating memory cells. Effector cells, as well as the memory cells that normally would have dealt with a subsequent assault, are completely absent from the host.

As noted previously, effector T cell responses can also be suppressed by regulatory T cells in a manner that does not involve killing the effector T cell (see Ch. 20).

F. Memory T Cells

I. Types of Memory T Cells

For both CD4+ and CD8+ T cells, about 5–10% of the antigen-specific progeny T cells generated in a primary response survive AICD or IL-2 withdrawal. These cells are, or give rise to, long-lived memory T cells. Memory T cells recognize the same pMHC as naïve and effector T cells but have properties intermediate between them (**Table 9-4**). Memory T cells are usually found in a resting state but occasionally undergo self-renewal to ensure their long-term survival. Upon a second assault by the same pathogen, memory T cells mount a secondary response that is faster and stronger than the primary response. These differences are attributable to the localization, increased numbers and enhanced capacities of memory T cells, and the significantly faster rate at which they differentiate into effector cells when activated by antigen.

Research over the past decade has revealed that there are at least two major classes of memory T cells: **effector memory T (Tem) cells** and **central memory T (Tcm)**

TABLE 9-4	Comparison of Properties of Naïve, Effector and Memory T Cells		
Property	**Naïve T Cell**	**Effector T Cell**	**Memory T Cell**
Preferred tissue	Secondary lymphoid tissues	Peripheral tissues, inflammatory sites, secondary lymphoid tissues	Peripheral tissues, secondary lymphoid tissues, diffuse lymphoid tissues, bone marrow, inflammatory sites
Preferred APCs	DCs	Macrophages, B cells, DCs	DCs, B cells, macrophages
Required costimulation	CD28/B7 (high)	CD28/B7 (low to none)	Minimal to none
Duration of TCR signaling for activation	20–30 hours	<1 hour	<1 hour
Cell division	Slow	Rapid	Moderate
Sensitivity to AICD	Low	High	High
Life span	5–7 weeks	2–3 days	Up to 50 years

cells. These subsets differ in some important properties. Firstly, in the absence of specific antigen, Tem cells appear to have a shorter life span than Tcm cells. Secondly, after a primary response, differences in homing receptor expression between these populations lead to differences in their distribution throughout the body. In particular, Tcm cells express high levels of the lymph node homing molecules CD62L and CCR7. Accordingly, these cells tend to migrate through the lymph nodes and other secondary lymphoid tissues, thereby maintaining a long-term, central reservoir of memory cells. In contrast, Tem cells express only low levels of CD62L and CCR7, so that they circulate mainly through non-lymphoid tissues where pathogens are likely to attack a second time. Such tissues include the lung, intestines, reproductive tract, liver and fat. Tem cells also express homing and chemokine receptors that allow their active recruitment into sites of inflammation. As a result, Tem cells constantly patrol the peripheral tissues and are able to migrate quickly into sites of infection. Together, Tem and Tcm cells ensure the ability of the host to mount a strong and immediate response upon subsequent exposure to a previously encountered pathogen. The cytokine milieu in which the original naïve T cell was activated appears to influence whether its progeny will have a greater Tem or Tcm character. For example, exposure to high levels of inflammatory cytokines during the primary response can suppress CD62L and CCR7 expression, thus favoring the generation of Tem cells. Researchers continue to debate the precise origin of Tem and Tcm cells because the transcription factors, cytokines and possible intercellular interactions directing the naïve to memory cell transition are as yet unclear.

II. Memory T Cell Activation and Differentiation

Naïve Th and Tc cells are activated exclusively by pMHCs presented by DCs, and this most often takes place in lymph nodes or other secondary lymphoid tissues, In contrast, memory T cells have less stringent requirements. Memory Th cells not only are dispersed in a broader range of anatomical sites than are naïve Th cells, but can also respond to pMHC presented by DCs, B cells or macrophages. Similarly, memory Tc cells can respond to infected host cells located almost anywhere in the body provided that these cells display the appropriate pMHC on their surfaces. In terms of signaling, the activation of both memory Th and memory Tc cells more closely resembles that of an effector T cell than a naïve T cell. Activation can occur at very low concentrations of antigen with only minimal costimulation (if any), and the duration of TCR signaling required is much shorter. Although naïve Tc cells usually require T cell help from antigen-specific Th cells to become activated, memory Tc cells often do not. Once activated, most memory Th and memory Tc cells proliferate more readily and for longer periods than their naïve counterparts.

> Note: While memory T cells in general proliferate more readily than naïve T cells, it appears that Tcm cells proliferate to a much greater extent than Tem cells.

With respect to differentiation, most activated memory Th and Tc cells follow much the same pathways as naïve Th and Tc cells but complete them more quickly (within 24 hours as opposed to 4–5 days). Some immunologists maintain that many memory T cells are not really resting but instead are maintained in a type of "pre-activation" state (which may correlate with their intermediate marker phenotype). This theory holds that pre-activation may make it easier for the cells to immediately differentiate into new effectors capable of quickly combatting an aggressive pathogen. Thus, for example, Tem cells of the CD4[+] Th phenotype that are recruited to a site of attack give rise to effector cells that rapidly produce cytokines such as IFNγ , IL-4 and IL-5. Their Tcm counterparts in the secondary lymphoid organs undergo rapid differentiation and produce copious amounts of IL-2. Similarly, CD8[+] Tem cells in peripheral tissues exhibit high levels of granzyme B expression and, once activated, quickly generate CTLs that can immediately engage in cytolytic activity.

III. Memory T Cell Life Span

Most memory T cells persist in the host for at least several months and often years, greatly exceeding the longevity of both naïve and effector T cells. The maintenance of the cellular pool depends on IL-7 because this cytokine drives the expression of anti-apoptotic molecules that protect against AICD. IL-2 and IL-15 also support the survival of memory T cells over the long-term, allowing them to proliferate as needed to maintain homeostasis. Studies of CD4$^+$ memory T cells have shown that Tem cells express lower levels of IL-2R, IL-7R and IL-15R on their surfaces than Tcm cells, likely accounting for the shorter life spans and less prolific proliferation of Tem cells.

The length of the life span of a memory T cell clone varies with the nature of the antigen that provoked the primary response. We see evidence of this variability in the immunization schedules of different vaccines (see Ch. 14). Just one dose of some vaccines (e.g., against the polio virus) provides immunity for life, whereas "booster" doses of other vaccines (e.g., against the bacterium causing tetanus) must be given every few years to maintain protection. While the cytokines needed to maintain memory T cell pools are now quite well defined, immunologists are still divided over whether the persistence of memory lymphocytes also requires a periodic low level of stimulation by tiny amounts of residual antigen. Such stimulation might help to induce the expression of base levels of anti-apoptotic molecules that would permit the memory T cell to survive. It may be that the requirement for stimulation varies with the antigen: significant numbers of memory T cells can be detected 15 years after infection with at least some viruses, with no evidence of re-infection or persistence of viral antigen.

Can memory T cells protect a host forever? Studies of the aging of the immune system indicate that memory T cells arising from a given clone can be stimulated only so many times before they fail to proliferate in response to antigen. As well, the production of new naïve T cells by the thymus declines precipitously after involution, also curtailing the generation of new memory T cells. Thus, as an individual ages, the numbers of both naïve and memory T cells that can be activated to generate effector T cells is ultimately limited, and the host becomes increasingly susceptible to pathogens toward the end of his/her life. In general, the longevity of T cell memory in humans pales in comparison to that of B cell memory, where antibodies raised during a natural infection by certain pathogens have been documented in some individuals as persisting for over 50 years.

We have now described all the cellular components of an adaptive immune response and have discussed how that response removes non-self entities. In the next chapter, we examine peripheral tolerance, a collection of mechanisms that control those mature naïve lymphocytes in the periphery which escaped the establishment of central tolerance and whose antigen receptors are directed against self antigens.

Chapter 9 Take-Home Message

- HSCs give rise to NK/T precursors that colonize the thymus and generate thymocytes. Thymocytes mature through the DN, DP and SP phases defined by CD4/CD8 expression.

- Development of $\alpha\beta$ T cells is controlled by two checkpoints: the pre-TCR checkpoint and the TCR$\alpha\beta$ checkpoint.

- Thymic selection establishing central T cell tolerance involves the non-selection, positive selection or negative selection of DP thymocytes. Selection is determined by the affinity/avidity of TCR binding to pMHCs presented by TECs.

- Thymocytes that survive selection lose expression of one coreceptor to become SP thymocytes, eventually exiting to the periphery as mature CD4$^+$ Th and CD8$^+$ Tc cells.

- Naïve Th and Tc cells are activated by engagement of multiple copies of their TCRs by pMHCs presented on mature DCs in the lymph node.

- Th cell activation involves the formation of the immunological synapse between the naïve Th cell and the DC. The synapse allows the sustained triggering of multiple TCRs that deliver "signal 1."

- "Signal 2" is delivered by costimulatory contacts such as CD28–B7, whereas "signal 3" is delivered by the binding of cytokines such as IL-2.

- Signal 1 induces the phosphorylation of the ITAMs in the CD3 chains, and signals 2 and 3 trigger subsequent intracellular signaling that leads to the new gene transcription necessary to support proliferation and effector cell differentiation.

- Naïve Tc cells are activated by interaction with DCs that have been "licensed" by Th cells to express B7.

- An activated naïve T cell proliferates and differentiates into effector T cells that eliminate antigens, and into memory T cells that mediate secondary immune responses.

- Effector and memory T cells require lower levels of TCR engagement and costimulation than do naïve cells, and express different adhesion molecules and chemokine receptors.

- Depending on the cytokines in the microenvironment, activated Th cells generate Th effector subsets that secrete different panels of cytokines. These cytokines either act against pathogens or support B cell and Tc cell activation. Regulatory T cells may also be produced.

- Activated Tc cells generate armed CTL effectors that kill tumor cells and infected target cells by perforin- and granzyme-mediated cytotoxicity, Fas ligation, and/or secretion of cytotoxic cytokines.

- The duration of Th effector and CTL responses is controlled by AICD, cytokine withdrawal, and T cell exhaustion. Regulatory T cells can suppress effector T cell responses.

- Memory T cells (both CD4$^+$ and CD8$^+$) remain in the host after the majority of effector T cells responding to an antigen have died off. Tem cells patrol peripheral tissues and sites of inflammation, whereas Tcm cells recirculate among the secondary lymphoid organs. Upon a second exposure to an antigen, Tem and Tcm cells rapidly generate new effector T cells that can act in the tissue where the original naïve T cell was activated.

Did You Get It? A Self-Test Quiz

Section A.I–II

1) Give four ways in which B cell and T cell development differ.

2) What cell types arise from NK/T precursors?

3) Which waves of NK/T precursors entering the thymus give rise to $\alpha\beta$ T cells?

4) Why is T cell diversity in the neonatal repertoire less than in the adult repertoire?

Section A.III

1) What are cTECs and mTECs, and why are they important?

2) Why is Notch1 described as a "cell fate protein"?

3) How do thymic fibroblasts contribute to T cell development?

4) What are the four stages of the DN phase of T cell development? How are these phases similar in surface marker expression? How do they differ?

5) What is SCF, and what is its function in the thymus?

6) What five key events occur during the DN3 stage?

7) Describe the composition and function of the pre-TCR complex.

8) What is β-selection, and why is it important?

9) Name the three components of thymic selection and describe how affinity/avidity of TCR engagement defines each of them.

10) Why is negative selection essential for the establishment of central T cell tolerance?

11) What is AIRE, and why is it important for thymic selection?

12) What factors influence the intracellular signaling triggered by TCR engagement during thymic selection?

13) Give two reasons why a thymocyte might be non-selected.

14) What are the effects on thymic selection of the expression of anti- and pro-apoptotic genes?

15) What is the TCRαβ checkpoint?

16) Describe how coreceptor expression in SP thymocytes determines which class of MHC the T lineage clone will later respond to?

17) Name two molecules (other than the coreceptors) that influence SP thymocyte lineage determination.

18) Do newly produced CD4+ and CD8+ mature T cells need antigenic stimulation to survive?

19) How does SP thymocyte coreceptor expression determine the mature T cell's effector function?

Section B

1) What are the three signals of naïve T cell activation?

2) Briefly outline how naïve T cells and antigen-bearing DCs meet in the lymph node.

3) Name two adhesion molecule pairs that help hold T cells and DCs together.

4) Why is sustained TCR triggering necessary to activate naïve T cells?

5) What is the immunological synapse?

6) Distinguish between the cSMAC, pSMAC and dSMAC.

7) What is the most important costimulatory interaction for naïve Th cells, and why is it necessary?

8) What is DC licensing, and why is it necessary?

9) Give three effects of CD28-mediated costimulation.

10) What is the function of CTLA-4? PD-1? How does the timing of these functions differ?

11) What are the two most important components of Th cell help for naïve Tc cells?

Section C

1) Name four types of effector T cells derived from Th0 cells.

2) Which cytokines drive Th1 differentiation, and what is their source? Th2 differentiation? Th17 differentiation?

3) Name three transcription factors important for Th cell differentiation.

4) What pathogens are countered by Th1 cells? Th2 cells? Th9 cells?

5) What class of diseases is associated with the normal function of Th17 cells?

6) What are fTh cells, and what do they do?

7) Are iTreg cells effector cells? If not, why not?

8) How does the localization of Th1, Th2 and Th17 cells differ, and what is the basis for these differences?

9) Give two ways in which the activation of Th effectors differs from that of naïve cells.

10) What molecules provide supplementary costimulation for Th1 cells? Th2 cells?

11) Give two ways in which Th1 cells support cell-mediated immunity and two ways in which Th2 cells support humoral immunity.

12) Why is IL-10 considered to act as a "brake" on immune responses?

13) Give two examples of Th effector cells that can cross-regulate each other.

14) Name two factors influencing the nature of a Th response.

Section D

1) Can you define these terms? pre-CTL, armed CTL, Tc17 cell, granule exocytosis, dissociation

2) Describe the five stages of cytotoxic T cell responses.

3) How does the activation of the armed CTL differ from that of a naïve Tc cell?

4) What are perforin and granzymes, and what do they do?

5) Describe how CTLs use the Fas pathway to kill target cells.

6) Name three cytotoxic cytokines and describe how they contribute to target cell elimination.

Section E

1) Name two molecules contributing to the survival of proliferating effector T cells.

2) Distinguish between AICD and T cell exhaustion.

3) What happens if IL-2 is withdrawn from a culture of activated T cells?

4) What T cell subset can control the responses of effector T cells?

Section F

1) Give three reasons why memory T cells react faster and stronger than naïve T cells.

2) How is the phenotype of memory T cells intermediate between naïve and effector T cell phenotypes?

3) Compare the properties of Tem and Tcm cells.

4) Name three receptors important for memory T cell survival.

5) Give two reasons why older individuals have less effective immune systems.

Can You Extrapolate? Some Conceptual Questions

1) Where in the thymus (outer cortex vs. inner cortex vs. medulla) would you expect to find thymocytes with the following characteristics? (Hint: Figures 9-2 and 9-3 may be helpful.)

   a) Active rearrangement of the *TCRB*, *TCRG* and *TCRD* genes.

   b) Expression of CD3 and TCRαβ, but no expression of Tdt or RAG.

   c) Expression of TDt and RAG but no expression of CD25, plus interaction with thymic DCs.

2) Before newly developed medicines are approved for public use, they are routinely tested for both "efficacy" (will they work adequately?) and for "safety" (will they cause unwarranted harm?). If you think of the process of thymic selection in terms of "approving" the release of mature T cells to the body, how would you relate non-selection and negative selection to efficacy and safety?

3) At a site of infection, an immature migratory DC captures antigen and binds DAMPs and PAMPs to its PRRs, triggering maturation of the DC. Once in the local lymph node, the mature DC activates a naïve CD4⁺ Th cell, triggering it to proliferate and give rise to activated Th0 cells. What type of Th effectors would you expect to develop from these Th0 cells if

   a) The local environment has high levels of TGFβ and IL-6.

   b) The transcription factors STAT4 and T-bet are expressed during T cell activation.

   c) The local environment has high levels of IL-4.

Would You Like To Read More?

Anderson, M. S., & Su, M. A. (2011). Aire and T cell development. *Current Opinion in Immunology, 23*(2), 198–206.

Bannard, O., Kraman, M., & Fearon, D. (2009). Pathways of memory CD8+ T-cell development. *European Journal of Immunology, 39*(8), 2083–2087.

Derbinski, J., & Kyewski, B. (2010). How thymic antigen presenting cells sample the body's self-antigens. *Current Opinion in Immunology, 22*(5), 592–600.

Dustin, M. L., & Depoil, D. (2011). New insights into the T cell synapse from single molecule techniques. *Nature Reviews Immunology, 11*(10), 672–684.

Klein, L., Hinterberger, M., Wirnsberger, G., & Kyewski, B. (2009). Antigen presentation in the thymus for positive selection and central tolerance induction. *Nature Reviews Immunology, 9*(12), 833–844.

Schlenner, S. M., & Rodewald, H. R. (2010). Early T cell development and the pitfalls of potential. *Trends in Immunology, 31*(8), 303–310.

Simpson, T. R., Quezada, S. A., & Allison, J. P. (2010). Regulation of CD4 T cell activation and effector function by inducible costimulator (ICOS). *Current Opinion in Immunology, 22*(3), 326–332.

Smith-Garvin, J. E., Koretzky, G. A., & Jordan, M. S. (2009). T cell activation. *Annual Review of Immunology, 27*, 591–619.

van Leeuwen, E. M., Sprent, J., & Surh, C. D. (2009). Generation and maintenance of memory CD4(+) T cells. *Current Opinion in Immunology, 21*(2), 167–172.

Zhang, N., & Bevan, M. J. (2011). CD8(+) T cells: Foot soldiers of the immune system. *Immunity, 35*(2), 161–168.

Zhu, J., & Paul, W. E. (2010). Peripheral CD4+ T-cell differentiation regulated by networks of cytokines and transcription factors. *Immunological Reviews, 238*(1), 247–262.

Zlotoff, D. A., & Bhandoola, A. (2011). Hematopoietic progenitor migration to the adult thymus. *Annals of the New York Academy of Sciences, 1217*, 122–138.

Regulation of Immune Responses in the Periphery

If you wish to be brothers, drop your weapons.

Pope John Paul II

T he adaptive immune responses that act in the peripheral tissues of the body are very powerful and, in a healthy individual, tightly controlled. This control is exerted in two ways: **tolerance** and **immune regulation**. Tolerance mechanisms prevent lymphocyte activation, whereas immune regulatory mechanisms rein in the actions of effector cells.

At the cellular level, **peripheral tolerance** is manifested when the interaction between a mature peripheral lymphocyte and its cognate antigen does *not* result in activation of that lymphocyte. The lymphocyte is either induced to undergo apoptosis or is functionally inactivated; immunologists say that the lymphocyte has been *tolerized*. A major role of such lymphocyte inhibition is the maintenance of tolerance to self tissues. This peripheral "self tolerance" is vital because it prevents the activation of **autoreactive** lymphocytes that have escaped central tolerance mechanisms and persist in a repertoire of lymphocytes that have been otherwise shaped to recognize non-self. Peripheral tolerance to innocuous non-self antigens also exists and helps to prevent inflammatory responses that would otherwise inflict unnecessary tissue damage. Peripheral tolerance to harmful non-self entities is not physiological but can be induced experimentally. The same laboratory techniques have been used in attempts to induce experimental tolerance to self antigens in the hopes of mitigating symptoms of autoimmune disease. Studies of these efforts to induce experimental tolerance have helped immunologists to understand much about tolerance mechanisms in general.

Whether a lymphocyte recognizes a non-self or self antigen, if it becomes activated, its response ultimately needs to be damped down to prevent or minimize collateral damage to surrounding healthy tissues. Cellular and biochemical regulatory mechanisms eventually return the body to a steady state. When an autoreactive lymphocyte becomes activated despite attempts to tolerize it, these regulatory measures must act to forestall sustained injury to the healthy self tissues that are the target of the response. In addition, regulation of any responses made to innocuous non-self entities, such as the commensal microbes in the gut, or the proteins in the food we eat or the air we breathe, is essential for normal health.

WHAT'S IN THIS CHAPTER?

Many autoimmune diseases are caused by failures in central and/or peripheral tolerance mechanisms. These diseases, their pathophysiology and treatment are discussed in detail in Chapter 19.

MAK: Primer to the Immune Response. http://dx.doi.org/10.1016/B978-0-12-385245-8.00010-8

Success in implementing both tolerance and regulatory mechanisms ensures that the host focuses the power of the immune response on harmful non-self antigens for the appropriate duration; failure paves the way for uncontrolled tissue damage and the potential development of autoimmune disease. In this chapter, we describe the processes that mediate tolerance and immune response regulation in the periphery. We also discuss two special peripheral tolerance situations: **maternal–fetal tolerance** and **neonatal tolerance**. Finally, we examine **experimental tolerance**, the situation in which an antigen is introduced artificially into the body under circumstances that suppress the induction of an immune response to that antigen.

A. Lymphocyte Tolerance in the Periphery

Because the antigen receptors of B and T cells are randomly generated, a certain proportion of developing lymphocytes will inevitably bear receptors directed against self antigens. Most B cells with such potentially autoreactive BCRs undergo receptor editing in the bone marrow during the establishment of B cell central tolerance such that these B cells no longer recognize self antigen (refer to Ch. 5). Most T cells with potentially autoreactive TCRs are eliminated by deletion during the establishment of central T cell tolerance in the thymus (refer to Ch. 9). However, if an autoreactive lymphocyte is released into the periphery because receptor editing fails, or a relevant self antigen is not expressed at sufficient levels in the bone marrow or thymus to induce negative selection, mechanisms of peripheral tolerance will attempt to ensure that the autoreactive cell cannot be activated to attack self tissues.

> NOTE: Peripheral tolerance mechanisms are discussed next in the context of tolerance to self antigens. However, the same mechanisms prevent the inappropriate and potentially harmful activation of lymphocytes specific for non-self antigens that pose no threat to the host.

I. T Cell Tolerance

i) DC-Mediated Tolerization

All professional APCs (DCs, macrophages, B cells) have the ability to acquire proteins, produce peptides and present them to T cells in the form of pMHCs with the potential to bind to TCRs. In other words, all APCs play a role in making a given protein *antigenic*, that is, able to bind to lymphocyte antigen receptors. However, only DCs have the unique capacity to determine whether the interaction of a given pMHC with the TCR of a naïve T cell will be *immunogenic* (trigger a response by the lymphocyte) or *tolerogenic* (suppress a response by the lymphocyte). How the DC shapes the outcome of the interaction between a lymphocyte's TCR and the DC's pMHC is dictated both by the inherent nature of the DC subtype involved and the external influences acting on the DC in a given tissue microenvironment.

As described in Chapter 7, immature DCs are broadly distributed in the peripheral tissues, both lymphoid and non-lymphoid, and constantly take up antigens from their surroundings. In the absence of pathogen attack or injury, the only antigens handled by these DCs are self or innocuous non-self antigens, such as those shed by healthy tissues, acquired by the engulfment of apoptotic cells during normal turnover, or routinely encountered in the host's environment (e.g., in the air and food). DAMPs/PAMPs are not present, and the DCs are not induced to mature. However, if one of these immature DCs encounters a naïve T cell expressing a TCR specific for an innocuous antigen, the DC exhibits tolerogenic properties and inactivates the T cell rather than activating it (**Fig. 10-1**). The tolerogenicity of immature DCs thus helps to preserve peripheral tolerance to harmless antigens of both self and non-self origin.

For reasons that are not yet fully understood, tolerogenic DCs are incapable of delivering signal 2 to T cells, even though these DCs sometimes still express costimulatory molecules and may resemble mature DCs in their surface marker phenotype.

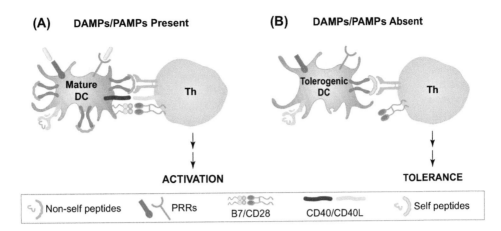

(A) DAMPs/PAMPs Present   **(B) DAMPs/PAMPs Absent**

Mature DC — Th → **ACTIVATION**

Tolerogenic DC — Th → **TOLERANCE**

| Non-self peptides | PRRs | B7/CD28 | CD40/CD40L | Self peptides |

Fig. 10-1
Comparison of Th Cell Activation and Tolerization

(A) Upon injury or infection, DAMPs/PAMPs are present that engage the PRRs of immature DCs, triggering their maturation so that they can activate any T cell whose TCR binds to the appropriate pMHC presented by the DC. **(B)** In the absence of injury or infection, no DAMPs/PAMPs are present, the DC's PRRs are not engaged, and the DC presents pMHCs to T cells in the absence of signal 2. The DC is "tolerogenic," and the T cell is "tolerized" rather than activated.

Without signal 2, signal 3 cannot be delivered either, and the T cell is not activated despite receiving signal 1. For example, let us suppose that a naïve autoreactive Th cell has escaped negative selection during the establishment of central tolerance and has found its way to a lymph node. No DAMPs/PAMPs are present, so if the TCR of this Th cell recognizes the self peptide-MHC complex presented by an immature DC, the DC assumes a tolerogenic function and inactivates the Th cell. In many cases, these events also result in a lack of response by autoreactive Tc cells because a Tc cell is only rarely fully activated in the absence of a licensed DC and the signal 3 cytokines produced by an activated Th cell. If, for some reason, the naïve autoreactive Th cell escapes tolerization by DCs in the lymph node, becomes activated and generates effector Th cells, migratory DCs in the peripheral tissues presenting the same pMHC can take on a tolerogenic function and prevent the activation of these effectors in these sites.

> DCs involved in the induction of regulatory T cells that can suppress the responses of conventional T cells are also deemed to be tolerogenic (see later).

There are two main processes by which tolerogenic DCs inactivate naïve T cells: **clonal deletion** and **anergization**.

a) Clonal Deletion
The most important mechanism by which peripheral Th cell tolerance is maintained is the clonal deletion of autoreactive Th cells. Naïve autoreactive Th cells that interact with pMHC presented by a tolerogenic DC are usually induced to undergo apoptosis. The precise mechanism has yet to be fully elucidated but appears to be independent of the apoptosis triggered by Fas or TNFR engagement.

b) Anergization
Those autoreactive Th cells that receive signal 1 alone from a tolerogenic DC but do not undergo apoptosis are **anergized**. An anergized Th cell survives but is inactivated and cannot produce effector cells (**Fig. 10-2A**). Moreover, if the antigenic pMHC is encountered a second time, the anergic Th cell still cannot respond even if the pMHC is presented by a fully mature DC expressing the appropriate costimulatory molecules and cytokines (**Fig. 10-2B**). While the intracellular signaling pathways that induce anergy have yet to be fully defined, it is clear that they involve the activation of molecules that drive a characteristic gene expression program. Ubiquitin ligases that promote the degradation of the TCR/CD3 complex are upregulated, as are factors that actively suppress IL-2 transcription. Changes to T cell metabolism that inhibit activation also occur. An anergic Th cell can maintain its unresponsive state for up to several months.

Fig. 10-2
Anergization of Th Cells

(A) A naïve autoreactive Th cell that encounters its cognate pMHC (signal 1) presented by a tolerogenic DC does not receive signals 2 and 3. The Th cell is anergized rather than activated and does not generate effector Th cells. **(B)** If this anergized Th cell later meets a mature DC presenting cognate antigen and capable of delivering signals 2 and 3, the anergized Th cell is still unable to respond.

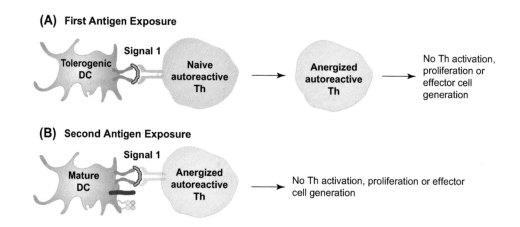

> NOTE: Another means of controlling immune responses is to keep a tight rein on DC numbers. The programmed cell death of DCs is a normal process by which routine DC turnover is maintained. However, DCs can also be killed by antigen-activated CTLs via perforin/granzyme-mediated cytotoxicity, a feedback mechanism that helps to limit the duration of the adaptive response. Defects in this mechanism or in DC apoptosis have been linked to elevated DC numbers, excessive T cell activation, loss of immune response regulation and ultimately autoimmune disease.

ii) Clonal Exhaustion

As well as by DC-mediated mechanisms, peripheral tolerance can be invoked by the elimination of an entire T cell clone due to clonal exhaustion (refer to Ch. 9). In this situation, continuous exposure to an antigen forces the responding T cells to proliferate and generate effectors so rapidly that they burn out without generating memory T cells. Some immunologists believe that tolerance to many self antigens that are present in the body in high abundance may be established this way very early in life. That is, during embryogenesis, the presence of large amounts of self antigen causes the exhaustion of autoreactive clones that escaped central T cell tolerance, ensuring peripheral tolerance to these self elements.

II. B Cell Tolerance

Autoreactive B cells that escape central B cell tolerance are controlled by peripheral tolerance mechanisms that differ slightly from those discussed previously for autoreactive T cells. If an autoreactive B cell encounters self antigen in the periphery, it receives signal 1 but still depends on an antigen-specific Th effector cell to deliver signals 2 and 3. Thus, even if DAMPs/PAMPs are present, if the required Th cell has already been deleted, anergized or exhausted by central or peripheral tolerance mechanisms, the B cell cannot be activated. Instead, the B cell is anergized and subsequently succumbs to apoptosis within 3–4 days (**Fig. 10-3A**). This reliance of the B cell on the Th cell for activation allows the host to benefit, without undue risk of increased autoreactivity, from the somatic hypermutation that occurs in the Ig genes. Although somatic hypermutation might produce a mutation that causes a B cell to become autoreactive, this B cell is unlikely to encounter a Th cell able to supply signal 2 because most autoreactive T cells are eliminated during the establishment of central T cell tolerance. Moreover, the TCR genes do not undergo somatic hypermutation, so that a non-autoreactive T cell cannot suddenly become autoreactive.

   Sometimes an appropriate autoreactive Th cell happens to be present in the peripheral repertoire so that the potential for activation of an autoreactive B cell exists. However, in these cases, in the absence of DAMPs/PAMPs, DCs displaying the appropriate self peptide-MHC are tolerogenic and delete or anergize the autoreactive Th cell rather than activate it. Thus, even though an autoreactive B cell may have had

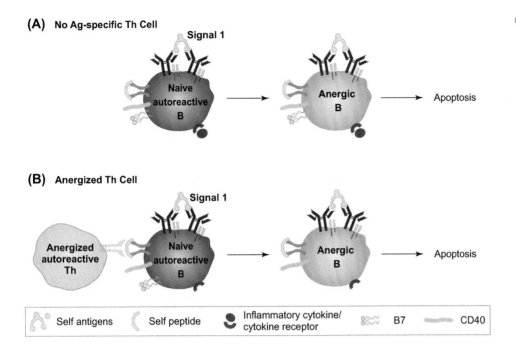

Fig. 10-3
Anergization of B Cells

(A) A naïve autoreactive B cell that binds its cognate antigen to its BCR (signal 1) but does not receive any Th cell help (signals 2 and 3) is anergized and eventually dies. **(B)** A naïve autoreactive B cell that binds its cognate antigen to its BCR is also anergized and dies if it interacts with a Th cell that has already been anergized.

its BCRs engaged by self antigen such that it receives signal 1, in the absence of an activated Th cell responding to the same self antigen, there is no delivery of signal 2 to the B cell. Again, the B cell is anergized and forced into apoptotic death (**Fig. 10-3B**).

B. Regulation of Lymphocyte Responses in the Periphery

Once lymphocytes have been activated, various mechanisms in the periphery control the quality, intensity and duration of the resulting adaptive response. These measures most often involve *regulatory T cells, immunosuppressive molecules, immune deviation,* and/or *immune privilege.* Recent studies of peripheral tolerance in the gut have also implicated the *intestinal microflora* in controlling immune responses in this tissue.

I. Regulatory T Cells

Certain subpopulations of T cells, collectively known as **regulatory T cells**, can control the responses of activated conventional T cells (including autoreactive effector T cells) *regardless of the antigenic specificity of the conventional T cell.* Regulatory T cells therefore help to balance the host's requirement for protective anti-pathogen and anti-tumor responses with its need to avoid excessive inflammatory and autoimmune responses. For example, aberrant activation of Th1 cells and Th17 cells is associated with chronic inflammation and autoimmune diseases, whereas uncontrolled or inappropriate Th2 responses are linked to allergic inflammation and other hypersensitivities. In a healthy host, various subsets of regulatory T cells interact with conventional Th1, Th2 and Th17 (and other) effectors to limit their activities if they threaten to upset immune homeostasis. Multiple types of both CD4+ and CD8+ regulatory T cells exist, but the former are better characterized.

Hypersensitivities are inflammatory diseases caused by harmful secondary immune responses to generally innocuous antigens (see Ch. 18).

i) CD4+ Regulatory T Cells

There are four major subsets of CD4+ regulatory T cells, one of which (**nTreg cells**) is derived directly from thymic precursors. Three others (**iTreg, Tr1** and **Th3** cells) are induced to differentiate from conventional T cells in secondary lymphoid tissues. As described next and summarized in **Table 10-1**, these subsets differ with respect to their derivation, phenotypic markers, suppressive mechanisms and regulatory effects.

TABLE 10-1	Types of CD4+ Regulatory T Cells			
	nTreg	**iTreg**	**Tr1**	**Th3**
Characteristic markers	CD4+CD25hi CTLA-4hi Foxp3+	CD4+CD25hi CTLA-4hi Foxp3+	CD4+CD25lo CTLA-4lo Foxp3−	CD4+CD25lo CTLA-4med Foxp3−
Derivation	Thymic precursor	Th0 cell + mature DC + IL-2 +TGFβ	Naïve Th cell + tolerogenic DC + IL-10	Naïve Th cell + tolerogenic DC + TGFβ
Suppressive mechanism	Intercellular contact	IL-10, TGFβ secretion	IL-10, TGFβ secretion	TGFβ secretion
Regulatory effects	Suppresses activated T cells May induce Th0 cells to produce Tr1 and Th3 cells	Suppresses activated T cells	Suppresses activated T cells	Suppresses activated T cells

a) Thymus-Derived Regulatory T Cells

The best-documented CD4+ regulatory T cells are commonly called "natural T_{reg}" or nTreg cells. The nTreg population comprises about 6–10% of all peripheral CD4+ T cells in a healthy adult mouse or human and displays a TCR repertoire that is biased toward self antigens. nTreg cells are characterized by the expression of the transcription factor Foxp3 as well as high surface levels of CD4 and CD25 (the IL-2Rα chain).

nTreg cells cells develop from precursors in the thymus, like conventional T cells, and can first be identified at the DP stage. Epithelial cells comprising the Hassall's corpuscles in the thymus secrete the cytokine *thymic stromal lymphopoietin (TSLP)*, which induces immature thymic DCs to upregulate costimulatory molecules. It is these TSLP-stimulated DCs that appear to play a prominent role in activating Foxp3 expression in the thymocytes with which they interact, generating nTreg cells. Upon their release from the thymus, newly produced nTreg cells express a huge array of chemokine receptors that allows them to recirculate through the secondary lymphoid tissues, migrate to the GALT, and take up residence in almost any peripheral tissue (**Table 10-2**). Additional chemokine receptors are expressed if the host experiences stress. Indeed, nTreg cells can follow chemotactic stimuli that do not affect conventional T cell populations.

"GALT" means "gut-associated lymphoid tissue" (see Ch. 12).

TABLE 10-2	Homing Receptors of Human CD4+ Regulatory T Cells
Receptor Expressed	**Destination**
Resting conditions	
CD62L	Lymph nodes
αEβ7 integrin	Epithelia
A4β7 integrin	GALT
CCR7	Spleen, lymph nodes
CCR8	Skin
CCR9	Intestine
CCR10	Mucosae, skin
CXCR4	Bone marrow, Peyer's patches
CXCR5	B cell follicles, GCs
CXCR6	Liver
Stress conditions	
CCR2, CCR5	Inflamed tissues
CCR6	Th17 inflammatory site
CCR8	Th2 inflammatory site
CXCR3	Th1 inflammatory site
CXCR4	Tumor sites

Like all regulatory T cells, an nTreg cell has the ability to block the proliferation and IL-2 production of conventional CD4+ T cells of *any* antigenic specificity. These suppressive effects are *not* cytokine-mediated but rather require direct intercellular contact that is independent of the TCRs of both T cells. In some cases, the apoptosis of the conventional T cell is induced by Fas-FasL interaction. In other cases, the death of the conventional T cell may be mediated by the high concentrations of the negative regulators CTLA-4 and PD-1 on the nTreg surface. Particularly in a tumor microenvironment, nTreg cells (despite their Th-like CD4+CD8- surface phenotype) can induce the perforin/granzyme-mediated cytolysis of effector Th cells, CTLs, APCs and/or NK cells. However, in a highly inflammatory environment, nTreg cells may lose the Foxp3 expression necessary to maintain their regulatory properties and convert to an effector T cell, a dangerous proposition for the host since the TCRs of nTreg cells tend to recognize self antigens.

Recent work has suggested that, as well as their effects on conventional T cells, nTreg cells may have direct effects on APCs that cause these cells to become tolerogenic, thereby further contributing to the control of T cell responses. For example, human or murine DCs that interact with CTLA-4-expressing nTreg cells downregulate their B7 costimulatory molecules. Examination of T cell-APC interactions by laser scanning microscopy has shown that Th effector cells stay in contact with APCs for longer when nTreg cells are absent. Conversely, when nTreg cells are present, they bind to DCs for long periods and may obstruct DC maturation and/or interaction with conventional naïve T cells.

> NOTE: TSLP (thymic stromal lymphopoietin) is an IL-7–like cytokine that was first isolated from mouse thymic stromal cells and named for its growth-promoting effects on lymphocytes and DCs. TSLP is now known to have much broader effects on a wide variety of hematopoietic cell types. The receptor for TSLP is a heterodimer composed of the IL-7Rα chain plus the TSLPR chain, and is most highly expressed by DCs. TSLP encourages the DCs to which it binds to generate nTreg cells. In the secondary lymphoid tissues, the cytokine promotes Th2 cell differentiation at the expense of Th1 or Th17 cell differentiation. TSLP is therefore associated with immune responses characterized by prominent B cell activation and only modest production of pro-inflammatory cytokines. However, this proclivity for Th2 responses means that TSLP is also linked to allergy and other hypersensitivities (see Ch. 18). In addition, TSLP has effects on other cell types, promoting cytokine production by mast cells and NKT cells, and stimulating the recruitment and activation of basophils and eosinophils. More on TSLP's role in the gut appears in Chapter 12.

b) Induced Regulatory T Cells

As introduced in Chapter 9, while the interaction of a mature DC with a naïve T cell most often results in the generation of effector T cells capable of dispatching a threat, if the right cytokines are present, regulatory cells may be induced to differentiate. The interaction of naïve T cells with tolerogenic DCs can also result in induced regulatory T cell differentiation, depending on the cytokines in the immediate microenvironment. The influence of DC and T cell status on the development of these cell types and the outcomes of their interaction are summarized in **Figure 10-4**.

- *iTreg cells.* As described in Chapter 9, Th0 cells that interact with a *mature* DC in a lymphoid microenvironment rich in TGFβ and IL-2 can give rise to "induced T$_{reg}$" (iTreg) cells. Like nTreg cells, iTreg cells express CD4, CD25 and Foxp3. However, iTreg cells suppress effector T cells via secretion of the immunosuppressive cytokines IL-10 and TGFβ rather than by intercellular contacts. These cytokines can also attenuate DC functions, allowing iTreg cells to inhibit naïve T cell activation.

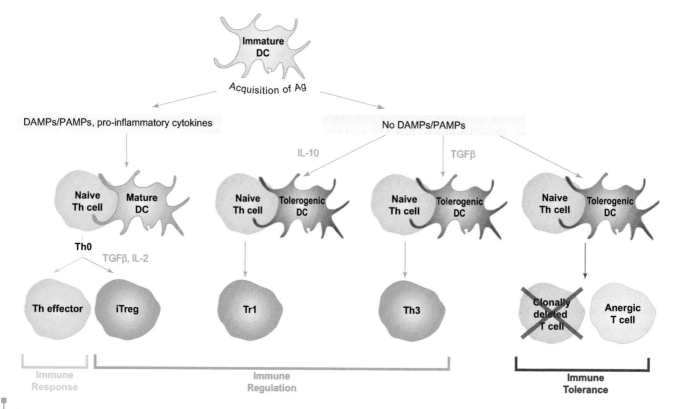

Fig. 10-4
Influence of DC Status on T Cell Activation, Regulation and Tolerance

An immature DC that has acquired antigen in the presence of DAMPs/PAMPs and pro-inflammatory cytokines is triggered to mature. Most often, this mature DC interacts with a naïve T cell and activates it such that it generates Th0 cells that differentiate into conventional Th effector cells. An *immune response* is mounted by the T cells. However, if the immediate microenvironment also contains high levels of TGFβ and IL-2, the mature DC can induce some Th0 cells to become iTregs capable of shutting down Th effectors. In this case, *immune regulation* occurs. Immune regulation is also invoked if an immature DC acquires antigen in a microenvironment that lacks DAMPs/PAMPs but contains immunosuppressive cytokines. The DC does not mature but instead becomes tolerogenic and can induce any naïve T cell with which it interacts to generate Tr1 or Th3 cells. If an immature DC acquires antigen in the absence of DAMPs/PAMPs and the absence of cytokines, it is also deemed to be tolerogenic and induces the clonal deletion or anergization of any naïve T cell it with which interacts, invoking *immune tolerance*.

- *Tr1 and Th3 cells.* Most Tr1 and Th3 cells are thought to arise from *naïve* Th cells (not Th0 cells) that interact with tolerogenic DCs in a lymphoid environment that is devoid of DAMPs/PAMPs but rich in IL-10 and/or TGFβ. Under these conditions, the naïve cell is not anergized or deleted but instead proliferates and differentiates into Tr1 cells (if IL-10 is dominant) or Th3 cells (if TGFβ is dominant). Other Tr1 and Th3 cells may arise from a naïve T cell that establishes intercellular contacts with an nTreg cell. In still other cases, effector T cells exposed to high levels of IL-10 can convert into Tr1 cells. Like iTreg cells, Tr1 and Th3 cells shut down activated effector T cells in an antigen non-specific manner by secreting immunosuppressive cytokines. Tr1 cells secrete IL-10 plus low amounts of TGFβ, whereas Th3 cells preferentially secrete TGFβ. In terms of markers, Tr1 cells express only low levels of both CD25 and CTLA-4, whereas Th3 cells express low levels of CD25 and moderate levels of CTLA-4. Unlike nTreg and iTreg cells, neither Tr1 nor Th3 cells express Foxp3.

NOTE: Foxp3 was discovered by immunologists studying the *scurfy* strain of mutant mice, which die of a "lymphoproliferative" disorder characterized by the uncontrolled cell division of conventional T cells. Researchers discovered that this lethal deregulation was caused by a natural mutation in the *scurfin* gene, which was subsequently found to encode Foxp3. The *scurfin* mutation inhibits the expression of a translatable *Foxp3* mRNA, meaning that *scurfy* mice lack both nTreg and iTreg cells. Conventional T cell responses run riot in these animals and lead to excessive lymphocyte proliferation, but treatment with normal mouse nTreg cells suppresses their autoimmune disease symptoms. In humans, mutations in the *FOXP3* gene cause a rare, X-linked PID called IPEX (immune dysregulation, polyendocrinopathy, enteropathy X-linked syndrome)." Within a few years of birth, 70% of IPEX patients show inflammation of the thyroid gland, and 90% die of complications of type I diabetes. Interestingly, despite the fact that Tr1 and Th3 cells do not require Foxp3 expression for development, the presumed presence of these regulatory T cell subsets in *scurfy* mice and IPEX patients apparently cannot compensate for their loss of nTreg and iTreg cells. These observations suggest that the suppressive functions of Tr1 and Th3 cells do not overlap with those of nTreg and iTreg cells, although the distinctions among the roles played by these subsets in autoimmune disease prevention are not clear.

c) Control of CD4⁺ Regulatory T Cells

Weapons as powerful as regulatory T cells cannot be left uncontrolled, and many immunologists are engaged in defining the mechanisms that regulate the regulators. For example, nTreg cells have been found to express TLRs, and engagement of these TLRs by the appropriate PAMPs early in a pathogen infection stimulates the proliferation of nTreg cells while diminishing their suppressive activity. The anti-pathogen response thus proceeds full force to eliminate the invader. Toward the end of the infection, however, the expanded nTreg cell population regains its suppressive potency and shuts down the now-unnecessary anti-pathogen T cell response.

Another mechanism of regulatory T cell control ties in directly with the inflammatory cytokines driving Th effector subset differentiation. For example, under resting conditions in which IL-6 is not produced, TGFβ induces Th0 cells to differentiate into iTreg cells. However, in the presence of IL-6 plus TGFβ, iTreg differentiation is suppressed and Th17 differentiation is stimulated. Moreover, in sites where pro-inflammatory cytokines persist, some iTreg and even nTreg cells may lose their Foxp3 expression and re-differentiate into conventional Th effector cells. Recent work has linked these effects of pro-inflammatory cytokines to specific transcription factors important for regulatory T cell control. While IFN and IL-12 promote T-bet activation in Th0 cells and thus Th1 effector differentiation, IFN and IL-12 can also promote T-bet activation in regulatory T cells and enhance their migration and suppressive capacity. The means to mount and restrain the Th1 response are thus driven by the same cytokines and same transcription factor. Studies of genetically engineered mice have demonstrated that the lack of a specific transcription factor can have a profound effect on the functionality of regulatory T cells. For example, in mice deficient for the gene encoding STAT3, which is a transcription factor essential for Th17 cells, the regulatory T cells in these animals cannot suppress the responses of conventional Th17 cells. Similarly, regulatory T cells in mice lacking IRF4, a transcription factor important for Th2 responses, can control conventional Th1 and Th17 cells but not Th2 cells.

NOTE: Regulatory T cells are critical for preventing immune responses that would disrupt normal homeostasis and are particularly important for protecting the gut microflora from host attack (see Ch. 12). However, a consequence of this immune response suppression is the potential for the increased survival of pathogens. In the human disease trypanosomiasis, it has been found that IL-10-producing regulatory T cells at the site of infection sometimes prevent effector T cells from eliminating the parasite. Increased regulatory T cell numbers and/or activity have also been cited as supporting the persistence of the pathogens responsible for tuberculosis, malaria, ulcers and various viral diseases. Treatments with agents that target molecules mediating regulatory T cell functions, including CTLA-4, IL-10 and TGFβ, have therefore proven effective in ameliorating some chronic infections.

ii) CD8+ Regulatory T Cells

Natural and induced subsets of CD8+ regulatory T cells also exist in humans and mice. To date, many of these subsets have been found to express both Foxp3 and CD25, but among these cells are a very large number of subtypes distinguished by their expression patterns of other surface markers (**Table 10-3**). Some CD8+ regulatory T cells block the proliferation of conventional naïve and effector T cells via direct intercellular contacts, whereas other CD8+ regulatory T cells secrete IL-10 and/or TGFβ. Still other CD8+ regulatory T cell subsets exert their primary effects on APCs, in that an immature DC that interacts with such a CD8+ regulatory T cell becomes tolerogenic. Interestingly, despite these obvious general overlaps in function with CD4+ regulatory T cells, some CD8+ regulatory T cells appear to play a unique role under certain circumstances. For example, in mice, there is a subset of CD8+ nTreg cells that expresses CD122, which is the IL-2Rβ chain. Mice deficient for CD122, and so lacking these CD8+CD122+ nTreg cells, have a lethal autoimmune disease even though these mutants still possess the powerful CD4+CD25+ nTreg cells. Disease can be prevented if these mutants are administered CD8+CD122+ nTreg cells from healthy mice, but not if they are given donated CD4+CD25+ nTreg cells. These findings imply that regulatory T cells may have coreceptor- and subset-specific functions in preventing autoimmune disease, an issue that remains under investigation. In humans, reduced numbers of CD8+ regulatory T cells have been documented in patients with various autoimmune disorders (see Ch. 19).

> NOTE: In addition to regulatory T cells, some recent experimental work suggests that other leukocytes, specifically subsets of regulatory B cells and neutrophils, may play roles in adaptive response regulation. In mice, there is a small population of B cells that secretes IL-10, leading to their designation as "B10 cells." In mouse models of autoimmune disease, animals lacking B10 cells show more severe symptoms, implicating these cells in contributing to the control of autoreactive T cells. IL-10-secreting B cells have also been found in humans and are depleted in at least some autoimmune disease patients, but much more investigation is needed to define their precise involvement. In the case of neutrophils, preliminary studies suggest that, after these cells have been recruited to sites of inflammation and have carried out their well-defined effector actions to kill and clear pathogens, they may participate directly in damping down T cell responses. Whereas the bulk of neutrophils die in the site of inflammation after giving their all, some appear to migrate to local lymph nodes, where they compete with DCs and macrophages for antigen and thus may help to suppress T cell activation. Again, this intriguing possibility remains under scrutiny.

TABLE 10-3	**Marker Expression Patterns of Some Human and Mouse CD8+ Regulatory T Cell Subsets**	
Treg Type	**Subtype**	**Mechanism of Action**
Human		
nTreg	CD8+CD25+Foxp3+CTLA4+	Intercellular contact via CTLA4
iTreg	CD8+CD28-Foxp3+	Intercellular contact via ILT3, ILT4*
iTreg	CD8+CD28-Foxp3-	Soluble factors, including IL-10
iTreg	CD8+CD25+Foxp3+CD28+	Intercellular contact (unknown)
iTreg	CD8+CD28-CD56+†	Intercellular contact (unknown)
Mouse		
nTreg	CD8+CD122+	Soluble factor (IL-10)
nTreg	CD8+CD25+	Intercellular contact (unknown)
iTreg	CD8+CD28-	Soluble factors (IL-10, TGFβ)
iTreg	CD8+CD28-Foxp3+	Undetermined
iTreg	CD8+Foxp3+PD-1+TGFβ+	Soluble factor (TGFβ)

*ILT= immunoglobulin-like transcript. ILT3 and ILT4 are inhibitory receptors similar to those described for NK cells in Chapter 11.

†CD56 = adhesion molecule expressed by NK cells and some cytolytic T cells.

II. Immunosuppressive Molecules

As mentioned previously, certain cytokines, particularly IL-10 and TGFβ, have immunosuppressive effects and act as brakes on innate and adaptive immune responses, including those initiated against self antigens. Although most often produced by innate leukocytes activated during a pathogen invasion, immunosuppressive cytokines are also synthesized in significant amounts by CD4$^+$ iTreg, CD8$^+$ iTreg, Th3, and Tr1 cells. IL-10 downregulates TCR-induced intracellular signaling in a responding T cell; inhibits macrophage activation and inflammatory cytokine secretion; blocks APC function; prevents the proliferation of Th cells; and destabilizes the mRNAs of many cytokines, including IL-2. The ensuing lack of IL-2 compromises signal 3 delivery, thus promoting T cell anergization. TGFβ inhibits macrophage and NK cell activation, blocks the proliferation and IL-2 production of activated T cells, downregulates Ig synthesis by B cells, and interferes with the stimulatory effects of IL-2 on T and B cells.

Other non-cytokine molecules also help to control the inflammation associated with innate and adaptive responses. In general, inflammation is resolved by phagocyte-mediated clearance of apoptotic cells, promotion of the exit from the site of innate and adaptive leukocytes, and prevention of the recruitment of any more leukocytes to the site. The immunosuppressive cytokines described earlier contribute to these processes, but lipid-based molecules such as *lipoxins, resolvins, protectins* and *prostaglandins* are also important. Neutrophils are important sources of these lipid-based mediators.

NOTE: An interesting link exists between excessive inflammation and cardiovascular disease. When certain lipoproteins are retained in an arterial wall, they trigger a chronic inflammatory response. Failure to resolve this inflammation leads to the accumulation of inflammatory cells and necrotic material that can promote collagen deposition and the formation of arterial plaques. If the plaques reach a dangerous size, they are vulnerable to structural disruption, which can precipitate an acute cardiac event. Clinical researchers are therefore exploring therapeutic strategies to enhance the body's ability to resolve inflammatory responses as a means of preventing cardiac diseases.

III. Immune Deviation

Immune deviation describes a phenomenon in which an adaptive response that has the potential to cause direct or indirect damage to a tissue appears to be "converted" to a less harmful response. This mitigation of damage was originally viewed as a form of tolerance. Immune deviation is caused by a bias toward the differentiation of a specific type of effector Th cell following the activation of a Th0 cell. Depending on the tissue involved, the differentiation of one particular subset can avoid the potential damage that might have been mediated by the differentiation of another subset. For example, the tissues of the eye are very sensitive to Th1 responses, which result in IFNγ production and macrophage hyperactivation. Factors within the microenvironment of the eye, including particular cytokines and/or presentation of pMHC by specific DC subsets, therefore direct the differentiation of an activated Th0 cell toward Th2 effectors. Th2 cells promote the mounting of relatively mild humoral responses rather than tissue-damaging cell-mediated responses. Thus, the immune response is still present, but its effects have been blunted by "deviating" effector generation toward the Th2 phenotype.

IV. Immune Privilege

Sites with **immune privilege** are anatomical regions that are naturally less subject to immune responses than most other areas of the body. Immune-privileged sites include the central nervous system and brain, the eyes and the testes. Even foreign antigens accessing these tissues do not generally trigger immune responses. Immune privilege is thought to exist because the collateral damage accompanying typical immune responses would irreparably damage these highly sensitive tissues, threatening the very existence of the species. Originally, it was thought that self tolerance in immune-privileged sites stemmed from physical barriers that blocked lymphocyte access. However, pioneering research into the reasons why retinal cell and corneal transplants are rejected much more slowly than transplants of other tissues or organs implicated the Fas death receptor. Fas is widely expressed on non-lymphoid cells and is induced on T cells after activation, whereas FasL is expressed on only a very few cell types. Significantly, these cell types are often found in immune-privileged sites. For example, the cells lining the inner chambers of the eye as well as Sertoli cells in the testis constitutively express high levels of FasL. Interaction of Fas on an activated T cell with FasL on a non-lymphoid cell induces the activated T cell to undergo apoptosis before it has a chance to differentiate into effector cells that could attack the eye or testicular tissue, or secrete damaging inflammatory cytokines. Lastly, it is now clear that immunosuppressive cytokines, immune deviation and the actions of regulatory T cells can also contribute to immune privilege.

V. Intestinal Microflora

Researchers have long known that innate and adaptive leukocytes in a healthy gut do not mount aggressive inflammatory responses against the intestinal microbiota. However, it is now clear that more than mere tolerance to gut microbes is involved, and that commensal bacteria are in fact critical for maintaining overall immune homeostasis. These beneficial microbes actively help to support a gut environment that must be ready to respond to acute infection by pathogens but must also maintain ongoing **oral tolerance** to innocuous antigens ingested as food.

The mechanisms underlying oral tolerance were discovered by experimental induction and are discussed in Section D of this chapter.

Immune responses in the gut are discussed in Chapter 12.

The gut microbiota have particular effects on both DCs and regulatory T cells. DCs that interact with certain species of commensal bacteria are rendered tolerogenic and steer the differentiation of Th0 cells toward regulatory T cells rather than effector T cells. Other commensal organisms have direct effects on these regulatory T cells, promoting their accumulation and influencing their cytokine production profiles. Thus, an anti-inflammatory microenvironment dominated by regulatory T cells prevails in the healthy gut. In addition, cells specialized for sampling antigens in the gut, as well as intestinal mucosal DCs and intestinal epithelial cells, have effects on each other that modulate immune responses in this tissue. For example, gut epithelial cells release TSLP, which inhibits IL-12 production by DCs and polarizes Th0 cell differentiation to the less inflammatory Th2 effector subtype. Gut epithelial cells also secrete TGFβ, which influences DCs to become tolerogenic and thus further promotes the generation of regulatory T cell subsets. The importance of commensal bacteria in maintaining peripheral tolerance in the gut is highlighted by the harmful effects of altering the normal microbiota, an event that can trigger autoimmune intestinal disorders such as colitis and Crohn's disease (see Ch. 19). The nature and number of antigens and receptors interacting to induce the tolerogenic state of the gut are the subject of much current research.

C. Special Tolerance Situations

I. Maternal–Fetal Tolerance

A mammalian mother does not normally reject her fetus, despite the fact that half of the histocompatibility molecules the fetus expresses are derived from the father and are therefore frequently "foreign" to the mother. Maternal tolerance to paternal histocompatibility molecules is maintained during fetal development but the potential for reactivity to these antigens eventually returns after birth. Maternal–fetal tolerance is not primarily due to physical barriers, since the fetal and maternal tissues that come together in the placenta are not separated by basement membranes. Furthermore, although the maternal and fetal circulatory systems are physically distinct, the barrier is not absolute, and fetal antigens and cells do enter the maternal circulation and reach the mother's secondary lymphoid organs.

Several mechanisms are thought to contribute to maternal–fetal tolerance.

(1) A critical factor is the presence of CD4$^+$ nTreg cells, whose numbers double or triple during pregnancy in response to changes in maternal hormones. Early during a healthy pregnancy, these nTreg cells can comprise up to 15% of all peripheral CD4$^+$ T cells. Under the influence of the uterine chemokine CCL4, nTreg cells home to the uterus and are thought to carry out their immunosuppressive functions through direct intercellular contacts with effector T cells and DCs in this location. In humans, spontaneous abortion, recurrent miscarriage and infertility have all been associated with abnormally low levels of nTreg cells in the uterus. As a pregnancy advances, however, the contribution of nTreg cells to immunosuppression becomes less important. In mice, depletion of CD25$^+$ cells early during pregnancy results in implantation failure but has no effect in later pregnancy.

(2) The hormonal changes associated with reproduction may prepare the uterus to accept the "fetal graft" so that the fetus is not seen as a foreign and injurious entity. That is, once pregnancy is established, the unique hormonal and cytokine microenvironment of the placenta favors immunosuppression and immune deviation to Th2 responses. Accordingly, Th1 responses are aberrant and associated with fetal loss. IL-6 production is normally low in the uterus, and uterine DCs tend to be tolerogenic and produce copious IL-10. In addition, certain placental cells produce high levels of progesterone, which dampens intracellular signaling such that the actions of effector T cells are generally suppressed.

(3) The tissues forming the maternal–fetal interface are populated with non-professional APCs that lack costimulatory molecules. Moreover, pregnancy does not usually generate the DAMPs that would induce the production of inflammatory cytokines and stimulate professional APCs.

(4) If a professional APC somehow becomes activated, local Tr1 and Th3 cells at the maternal–fetal interface usually suppress any incipient immune response. Both Tr1 and Th3 cells have been implicated in the prevention of miscarriage in animal models.

(5) The placenta and tissues of the developing fetus are almost devoid of MHC class I and II molecules. In fact, the only MHC genes that an early human fetus is known to express are the non-polymorphic HLA-E and HLA-G genes (described in Ch. 6). The products of the HLA-E and HLA-G genes inhibit maternal uterine NK cells that would otherwise attack fetal tissues (see Ch. 11).

(6) Some placental cells express FasL, so that Fas-expressing maternal T cells activated by an encounter with fetal antigens can be killed before fetal tissues are attacked. Fetal cells also express the RCA proteins DAF and MCP that can interfere with maternal complement activation (refer to Ch. 3).

NOTE: Pre-eclampsia is a condition in which a pregnant woman has very high blood pressure and shows increased amounts of protein in her urine. These abnormalities pose a significant danger to the health of both the mother and fetus. Dysregulation of regulatory T cells may be involved in this disease, since the cytokines secreted by these cells (particularly IL-10 and TGFβ) influence blood flow and the homeostasis of the circulatory system.

II. Neonatal Tolerance

When a mammalian fetus emerges from the sterile environment of the uterus into the outside world, it becomes a neonate and is exposed through the oral and respiratory routes to both harmful entities (i.e., pathogens) as well as a myriad of harmless entities (i.e., food, environmental and commensal antigens). If this neonate

The transfer of antibodies to provide antigen-specific protection to a recipient whose own immune system has not been activated by that antigen is a form of **passive immunization** (see Ch. 14).

immediately mounted vigorous immune responses to all these new non-self entities, tissue-damaging inflammation might be triggered that, in the case of the harmless antigens, is totally unnecessary. Evolution has therefore ensured that there is a time shortly after birth when the neonatal immune system is skewed toward inducing tolerance of new antigens rather than immune responsiveness against them. During this period, neonates are protected against pathogen infection by IgG and IgA antibodies transferred from the mother first via the placenta and subsequently in breast milk.

A mammalian mother's milk is also the source of a newborn's first dietary, environmental and microbial antigens. During the processing of these antigens needed to allow them to enter the milk, the antigens may undergo changes that increase their tolerogenicity. In addition, although the neonatal gut itself produces only low levels of TGFβ, the maternal milk contains high levels of IL-10 and TGFβ. Once consumed, these cytokines suppress any neonatal immune response to harmless antigens and also promote regulatory T cell differentiation.

In addition to the presence of immunosuppressive cytokines in maternal milk, there are several other mechanisms that promote neonatal tolerance. Mouse studies have shown that neonates generally have very few T cells, such that absolute numbers of mature naïve T cells are 10,000-fold lower in newborns than in adult animals. Moreover, the few neonatal T cells that are present appear to secrete immunosuppressive cytokines, much more so than the T cells in adult mice. Lymphocyte trafficking patterns are also different in neonates and adults, in that naïve T cells can access non-lymphoid peripheral tissues much more easily in the former and tend to accumulate in lung and skin rather than in the spleen and lymph nodes where antigen and mature DCs are concentrated. Furthermore, APCs in neonates often show reduced antigen presentation capacity. On the B cell side, FDCs in neonates are inefficient at trapping antigen, and B cells in neonates are highly susceptible to tolerance induction. When antigen exposure does result in a response, neonatal B cells do not differentiate into plasma cells as quickly as do adult B cells. Studies in mice have demonstrated that, while neonatal antibody responses to most (but not all) Ti antigens are strong, responses to Td antigens are weak and often delayed, a state that can persist for 2–8 weeks after birth. Lastly, the neonatal murine gut contains a higher proportion of IL-10–secreting B10 cells than does the adult gut.

D. Experimental Tolerance

Much of what is known about natural peripheral tolerance has been derived from experimental models in which a lack of responsiveness to a *foreign* antigen is induced. That is, various experimental contrivances are used to convince the host animal's body that a foreign antigen is a harmless, self-like entity so that an immune response is not mounted against it. Such an animal is said to be "tolerized" to the antigen. A **tolerogen** is an experimental foreign antigen that binds to the antigen receptors of lymphocytes but suppresses, rather than induces, activation.

I. Characteristics of Experimental Tolerance

Experimental tolerance is most easily induced in an animal with an immune system that is not at mature full strength, such as in a neonatal animal or an animal whose lymphocytes have been damaged by treatment with drugs or irradiation. Under these conditions, the animals fail to mount an immune response to a normally immunogenic molecule. Even after the immunocompromising treatment is discontinued and the immune system is allowed to recover, a subsequent exposure to the same immunogen elicits no response. However, the immune system is not globally suppressed because responses to other, unrelated immunogens remain intact. In other words, experimental tolerance is antigen-specific. However, experimental tolerance is not usually permanent, and its maintenance depends on continued exposure to the tolerogen in either a persistent or intermittent fashion. Loss of the

"Immunotherapy of Type I Diabetes: Where Are We and Where Should We Be Going?" by Luo, X., Herold, K.C., and Miller, S.D. (2010) *Immunity* 32, 488–499.

An estimated 350 million people worldwide live with diabetes, a disease that has a global economic impact of nearly half a trillion U.S. dollars! Accordingly, a great deal of research is currently devoted to understanding, preventing and treating diabetes. As is discussed in detail in Chapter 19, there are two types of diabetes: type I and type II. The majority of patients have type II diabetes (insulin-independent) in which external factors such as improper diet and lack of exercise lead to the disease. However, about 20% of patients have type I (insulin-dependent) diabetes, an autoimmune disease in which the individual's immune system launches an attack on the insulin-producing β-islet cells in the pancreas. In particular, the enzyme glutamate decarboxylase-65, which is expressed by β-islet cells, is a target for autoreactive antibodies. In this review by Luo *et al.*, recent advances in molecular and genetic research on type I diabe-

tes are summarized, with particular attention paid to experimental therapies and their relative success in clinical trials. The therapies described originated as strategies to modulate autoreactive responses in a mutant strain of mice called NOD (non-obese diabetic), which serves as a model for human diabetes. Examples of therapies being explored for their safety and efficacy in humans include several that involve experimental manipulation or regulation of peripheral tolerance, including mAb-mediated depletion of autoreactive T cells or B cells that have escaped central tolerance; induction of experimental tolerance via oral insulin administration; induction of tolerance to glutamate decarboxylase-65; and infusion of tolerogenic APCs bearing insulin peptide. It is hoped that such research will lead to informed strategies aimed at preventing and/or treating type I diabetes.

tolerogen for a prolonged period slowly restores normal responsiveness to the foreign antigen. This reversal can occur if the previously tolerized cells recover their immune reactivity or if newly generated lymphocytes with normal reactivity to the antigen emerge. In general, experimental B cell tolerance usually lasts only a few weeks in the absence of antigen, while experimental T cell tolerance can persist for several months.

II. Characteristics of Tolerogens

The physical and behavioral characteristics of tolerogens are compared to those of immunogens in **Table 10-4.**

i) Nature of Tolerogenic Molecules

For reasons that are still unclear, some molecules are naturally more tolerogenic than others. For example, while L-amino acid polymers are immunogenic at almost any dose, D-amino acid polymers are tolerogenic at the same doses. For other molecules, a slight chemical modification is enough to turn an immunogen into a tolerogen for the same lymphocyte clone. Also important is the density of the antigenic epitope on the molecule. Having a moderate number of epitopes per molecule promotes immunogenicity, but a large number of epitopes per molecule favors tolerance induction.

TABLE 10-4	Immunogens vs. Tolerogens	
	Immunogen	**Tolerogen**
Number of antigenic epitopes per molecule of antigen	Moderate	High
Size	Large, aggregated, polyvalent	Small, disaggregated
Solubility	Insoluble	Soluble
Dose	Moderate	Very high or very low
Schedule of Ag administration	Low number of moderate doses	Many small doses
Route of Ag administration	Subcutaneous	Intravenous, oral

ii) Molecular Size

Small, soluble molecules do not often make good immunogens, but they can make good tolerogens. Immunologists speculate that small molecules may induce tolerance because they do not provide PAMPs nor induce the production of DAMPs necessary to trigger the maturation of the DCs that process them. A T cell whose TCR recognizes pMHCs derived from the small molecule therefore receives signal 1 but not signal 2, and is consequently anergized.

For B cells, a tolerogen must be capable of stimulating the BCRs such that signal 1 is delivered. However, the stimulation must not be so complete that the B cell is activated even in the absence of signal 2 (CD40L) delivered by an activated Th cell, as happens with many Ti antigens (refer to Ch. 5). Very small soluble molecules cannot cross-link BCRs at all and therefore do not tolerize (or activate) B cells.

iii) Dose

Although moderate doses of an antigen induce an immune response, very high and very low doses of the same antigen may induce tolerance. These types of tolerance are called **high zone tolerance** and **low zone tolerance**, respectively.

a) High Zone Tolerance

To illustrate high zone tolerance, consider the fate of mice that have been inoculated with either a very high dose or a moderate dose of a purified bacterial antigen. When these animals are later injected with live bacteria, the mice that received the moderate dose mount a humoral response that rescues them, but those that received the high dose do not make anti-bacterial antibodies and die. In some cases, this high zone tolerance is caused by a **B cell receptor blockade**, in which very large amounts of an antigen persistently occupy the BCRs without cross-linking them. This blockade affects signal 1 delivery because the ITAMs on the Igα and Igβ chains in the BCR complex are incompletely phosphorylated. An abnormal and possibly inhibitory signal is transduced that alters downstream signaling events such that B7 expression is not upregulated. Extensive downregulation of mIgM also occurs on these anergic B cells, further precluding signal 1 delivery. However, the failed antibody response could also be due to T cell tolerization, since B cells require T cell help to respond to Td antigens. High zone tolerance for T cells occurs at much lower antigen doses than those required for the same effect on B cells. In addition, whereas B cell tolerization is not observed until about 2 days after antigen administration, high zone tolerance for T cells is evident within just hours of exposure. T cell high zone tolerance often results from clonal exhaustion.

b) Low Zone Tolerance

Tolerization can also occur if an animal is exposed to a very low dose of antigen (too low to provoke an immune response) over a long period of time. Studies in mice have shown that repeated administration of very low doses of an antigen suppresses antigen-specific antibody production. Low zone tolerance has been demonstrated for a wide range of antigens in neonatal animals and immunocompromised adults, but for only a few antigens in healthy adults. The precise mechanism underlying low zone tolerance is still undefined, but some recent evidence implicates T cells with antigen-specific suppressive activity.

iv) Route of Administration

The route by which an antigen accesses the body determines how it is processed and affects whether it is immunogenic or tolerogenic. The outcome is likely related to the frequencies and subtypes of DCs that take up the antigen and the microenvironment in which they do so.

a) Non-Oral Routes

If one *wants* to induce an immune response to an antigen, the most immunogenic mode of delivery is the subcutaneous route. Langerhans cells, which are DCs resident in the skin, are very immunogenic, meaning that antigen delivered in this way often triggers a robust immune response. In contrast, antigen administered intravenously often results in tolerance. Intravenous antigen is conveyed to the spleen, where it is presented primarily by naïve splenic B cells and non-professional APCs. These cells can process and present antigen but express only low levels of costimulatory molecules (if any) and cannot deliver the robust signal 2 needed to activate naïve T cells. Any T cell specific for a pMHC presented by these APCs is therefore usually anergized, and the immune response to the antigen is negligible.

b) Oral Route

Oral administration is a very effective way of inducing peripheral tolerance to immunogens, at least under laboratory conditions. An individual is administered an antigen in his/her food or drinking water. Most such dietary antigens are degraded in the stomach before ever reaching the small intestine, but some partially digested or even intact molecules do get absorbed into the circulation and thus are distributed systemically. In animal models, tolerance to such a systemic antigen can be observed within 5–7 days after the antigen is consumed. T cell tolerance after a single feeding can last for up to 18 months, whereas B cell oral tolerance lasts for 3–6 months.

Several mechanisms are thought to make a contribution to oral tolerance.

(1) Since food is usually harmless, there are no DAMPs/PAMPs associated with the incoming antigen that could trigger DC maturation. The presenting DCs are therefore tolerogenic, and the T cells interacting with them are anergized or deleted (if the antigen is present at a high level), or induced to differentiate into iTreg cells (low antigen level). Any iTreg cells present secrete immunosuppressive cytokines that block the actions of effector T cells in the gut.

(2) As noted previously, the gut microflora and gut epithelial cells secrete cytokines and other factors that help to render mucosal DCs tolerogenic.

(3) Presentation of gut antigens by non-professional APCs lacking costimulatory capacity can anergize antigen-specific T cells.

(4) NKT cells in the GALT are activated by oral antigen administration. Once activated, these cells secrete a panel of cytokines that include IL-10.

The ability to induce antigen-specific tolerance by the oral route suggests a clinical approach to controlling diseases that are caused by damaging immune responses. Indeed, using mouse models of immune hypersensitivity or autoimmune disease, scientists have shown that feeding animals with large quantities of a disease-inducing antigen reduces the severity of disease symptoms. However, despite much effort and some positive results in early clinical trials, the application of orally induced tolerance to human disease mitigation has not been as successful as once hoped. Many immunologists now believe that the manipulation of regulatory T cells may hold the key to success in tolerance induction strategies, but much more needs to be learned about how to induce and channel the activities of these cells before this approach will translate into concrete clinical benefits for patients. To accelerate progress in therapeutic tolerance induction, the Immune Tolerance Network (see **Box 10-1**) has been established to promote the global sharing of experience with all forms of tolerance induction used in clinical applications.

Nasal tolerance is analogous to oral tolerance and involves mucosal tolerance mechanisms in the respiratory tract.

As we reach the end of this chapter, we have completed our study of two important cellular components of the adaptive immune response: B cells and αβ T cells. In the next chapter, we examine cells that are considered to bridge adaptive and innate immunity: NK cells, γδ T cells and NKT cells.

Box 10-1   The Immune Tolerance Network

The Immune Tolerance Network (ITN) is an international organization that is led by the U.S. National Institutes of Health and involves basic and clinical immunologists, clinical trial centers, and biotechnology and pharmaceutical companies. The ITN has the mandate to increase our scientific understanding of immune tolerance and to advance the use of this knowledge for clinical trials of immunotherapy. **Immunotherapy** is the manipulation of the immune system to prevent, mitigate or cure disease. Strategies inducing tolerance should be applicable to a wide range of disorders, including allergies, autoimmune diseases, chronic inflammatory ailments and transplant rejection. On the flip side, because many tumors have the capacity to induce tolerance to themselves and thus shut down host anti-tumor immune responses (see Ch. 16), understanding how to "break" this tolerance could permit the use of a patient's own immune system to fight his/her cancer. Of course, manipulation of the immune response is not without inherent risks, since inducing global tolerance would suppress anti-pathogen or anti-tumor responses as well, increasing the vulnerability of the patient to infection or cancer.

The ITN is currently focused on three main areas: prevention of rejection of solid organ transplants (see Ch. 17), prevention or mitigation of allergic reactions (see Ch. 18), and prevention or mitigation of autoimmune diseases (see Ch. 19). In each of these cases, the antigen-specific effector lymphocytes that are rejecting the transplanted tissue, causing an inappropriate immune response or attacking self tissues must be deleted, inactivated or converted to a less harmful effector subset. The overall objective is to restore normal immune homeostasis by blocking inflammatory pathways, redirecting innate responses, and reactivating or enhancing regulatory signaling. Much research has gone into finding ways to control disease-causing lymphocytes, and modest success has been achieved with some protocols in some settings. However, it is becoming increasingly clear that cytokines and other factors in the tissue microenvironment play key, and sometimes unpredictable, roles in determining whether a particular strategy will block or enhance an unwanted immune response. The ITN continues to delve into both the basic and clinical aspects of therapeutic tolerance manipulation.

More information on the ITN can be found at its website (http://www.immunetolerance.org/).

Chapter 10 Take-Home Message

- Adaptive immune responses are tightly controlled by mechanisms of tolerance and immune regulation. Tolerance mechanisms prevent lymphocyte activation, whereas regulatory mechanisms rein in the actions of activated lymphocytes.

- The major physiological role of peripheral self tolerance is to ensure that any autoreactive lymphocytes that escaped negative selection during the establishment of central tolerance cannot initiate a response that damages host tissues.

- In the absence of DAMPs/PAMPs, immature DCs can have tolerogenic effects on antigen-specific naïve T cells they encounter. Immunosuppressive cytokines may also influence an immature DC to become tolerogenic.

- If a naïve autoreactive T cell encounters its cognate pMHC presented by a tolerogenic DC, the T cell most often dies by apoptosis due to clonal deletion or is inactivated due to anergization.

- An anergized T cell does not generate effector or memory cells and is unable to respond to cognate antigen the next time it meets the antigen under optimal conditions.

- T cell clonal exhaustion may establish tolerance to highly abundant self antigens.

- Anergization of a B cell may occur if its BCRs are engaged by a Td antigen but the appropriate Th cell is not available, or if the Th cell fails to upregulate costimulatory molecules.

- Subsets of CD4$^+$ regulatory T cells include nTreg, iTreg, Tr1 and Th3 cells. Subsets of regulatory CD8$^+$ nTreg and iTreg cells, as well as regulatory B cells, also exist.

- Regulatory T cells may inactivate effector T cells or render DCs tolerogenic via direct intercellular contacts or via secretion of immunosuppressive cytokines.

- Immune deviation refers to the mounting of a less harmful response due to skewing of the direction of Th cell differentiation toward a particular subset.

- Immune-privileged sites show a decreased frequency of immune reactivity due the presence of immunosuppressive cytokines, immune deviation, death receptor upregulation, and/or the actions of regulatory lymphocytes.

- Intestinal microflora contribute to the anti-inflammatory character of the gut microenvironment, which influences DCs to be become tolerogenic and promotes regulatory T cell differentiation.

Chapter 10 Take-Home Message—Continued

- Maternal–fetal tolerance is mediated by regulatory T cells, uterine hormonal status, lack of classical MHC class I expression, expression of non-classical MHC molecules, immunosuppressive cytokines and the presence of non-professional APCs.

- Neonatal tolerance refers to the fact that adaptive responses in neonates are weak and delayed compared to responses in adults due to differences in B and T lymphocytes, FDCs and APCs.

- Experimental tolerance is tolerance artificially induced to a self or foreign antigen and is often mediated by introducing the antigen via a non-standard route or in a tolerizing dose.

- Oral tolerance is mediated mainly by the activities of tolerogenic gut DCs that induce T cell anergization or clonal deletion, or regulatory T cell induction.

Did You Get it? A Self-Test Quiz

Introduction and Section A

1) Distinguish between natural peripheral tolerance and experimental tolerance.

2) Distinguish between tolerance mechanisms and regulatory mechanisms.

3) Under what circumstances is an immature DC tolerogenic?

4) A DC expressing costimulatory molecules always induces an immune response. True or false?

5) Distinguish between clonal deletion and anergy.

6) What is the response of an anergized T cell to a second encounter with antigen presented by a mature DC?

7) What is clonal exhaustion, and how might it contribute to self tolerance?

8) Describe two mechanisms by which a B cell can be tolerized.

Section B

1) Distinguish among the four main types of CD4$^+$ regulatory T cells (T$_{reg}$, iTregs, Th3 and Tr1) in terms of derivation, surface markers, transcription factors and mechanisms of action.

2) Give two possible outcomes of an encounter between a regulatory T cell and a DC.

3) Why is the expression of TLRs on nTreg cells significant?

4) Describe two ways in which CD8$^+$ regulatory T cells might suppress conventional T cell responses.

5) Can CD4$^+$ nTreg cells always substitute for CD8$^+$ nTreg cells? Give an example to support your answer.

6) Why are IL-10 and TGFβ considered to be immunosuppressive? Give three examples for each.

7) What is immune deviation, and why was it originally considered a tolerance mechanism?

8) Describe three mechanisms that can suppress lymphocyte responses in an immune-privileged site.

9) Describe how a change in the species composition of the gut microflora might result in an inflammatory intestinal disorder.

Section C

1) Why is maternal–fetal tolerance required?

2) Describe four mechanisms that help to maintain maternal–fetal tolerance.

3) What substances in a mammalian mother's milk help to induce immune tolerance in her newborn offspring?

4) Give four other reasons why neonates are naturally more tolerant to antigens than are adults.

Section D

1) Describe two situations in which experimental tolerance is easily induced.

2) How can experimental tolerance to an antigen be maintained?

3) Give three characteristics of effective tolerogens.

4) Define "high zone" and "low zone" tolerance.

5) Why is the intravenous administration of a molecule usually tolerogenic?

6) Give two mechanisms thought to contribute to oral tolerance.

7) What is the mandate of the ITN?

Can You Extrapolate? Some Conceptual Questions

1) The mature T cells in a newly developed strain of mice are analyzed, and a very unusual observation is made: the TCR genes are found to undergo somatic hypermutation like the Ig genes in mature B cells. With B cell peripheral tolerance mechanisms in mind, would you expect this strain of mice to develop antibody-mediated autoimmune disease at a higher, lower, or equal rate relative to wild type mice? What is your rationale?

2) An immature DC acquires antigen in an environment in which there are no DAMPs or PAMPs, but there are high levels of IL-10. If this DC presents pMHC to naïve Th cell, what is the likely differentiation path of this Th cell?

3) For each of the following mechanisms/situations that are thought to contribute to maternal–fetal tolerance, state whether you would consider them to be an example of T cell tolerization, control of responses by regulatory lymphocytes, immune deviation or immune privilege:

a) Placental cells express FasL and can kill Fas-expressing maternal T cells.

b) When effector T cells are generated, Th2 responses dominate.

c) Tissues at the maternal–fetal interface are populated with non-professional APCs that lack costimulatory molecules.

d) Tr1 and Th3 cells are found at the maternal–fetal interface.

4) As is described in more detail in Chapter 14, a vaccine against an infectious pathogen contains an antigen derived from it that will be immunogenic (but not harmful) when administered to an individual. You are an immunologist tasked with developing an effective vaccine against such an infectious pathogen. Why might it be important to test different doses and routes of administration of your vaccine antigen?

Would You Like To Read More?

Belkaid, Y., & Chen, W. (2010). Regulatory ripples. *Nature Immunology*, 11(12), 1077–1078.

Campbell, D. J., & Koch, M. A. (2011). Phenotypical and functional specialization of FOXP3+ regulatory T cells. *Nature Reviews Immunology*, 11(2), 119–130.

Chen, S. J., Liu, Y. L., & Sytwu, H. K. (2012). Immunologic regulation in pregnancy: From mechanism to therapeutic strategy for immunomodulation. *Clinical & Developmental Immunology*, 2012, 258391.

Chen, M., & Wang, J. (2010). Programmed cell death of dendritic cells in immune regulation. *Immunological Reviews*, 236, 11–27.

Filaci, G., Fenoglio, D., & Indiveri, F. (2011). CD8+ regulatory/suppressor cells and their relationships with autoreactivity and autoimmunity. *Autoimmunity*, 44(1), 51–57.

Joffre, O., Nolte, M. A., Sporri, R., & Reis e Sousa, C. (2009). Inflammatory signals in dendritic cell activation and the induction of adaptive immunity. *Immunological Reviews*, 227, 234–247.

Mantovani, A., Cassatella, M. A., Costantini, C., & Jaillon, S. (2011). Neutrophils in the activation and regulation of innate and adaptive immunity. *Nature Reviews Immunology*, 11(8), 519–531.

Mauri, C. (2010). Regulation of immunity and autoimmunity by B cells. *Current Opinion in Immunology*, 22(6), 761–767.

Mueller, D. L. (2010). Mechanisms maintaining peripheral tolerance. *Nature Immunology*, 11(1), 21–27.

Nepom, G. T., St. Clair, E. W., & Turka, L. A. (2011). Challenges in the pursuit of immune tolerance. *Immunological Reviews*, 241(1), 49–62.

Reis e Sousa, C. (2006). Dendritic cells in a mature age. *Nature Reviews Immunology*, 6, 476–483.

Sakaguchi, S., Miyara, M., Costantino, C. M., & Hafler, D. A. (2010). FOXP3+ regulatory T cells in the human immune system. *Nature Reviews Immunology*, 10(7), 490–500.

Tabas, I. (2010). Macrophage death and defective inflammation resolution in atherosclerosis. *Nature Reviews Immunology*, 10(1), 36–46.

Weiner, H. L., da Cunha, A. P., Quintana, F., & Wu, H. (2011). Oral tolerance. *Immunological Reviews*, 241(1), 241–259.

Wing, K., & Sakaguchi, S. (2010). Regulatory T cells exert checks and balances on self tolerance and autoimmunity. *Nature Immunology*, 11(1), 7–13.

NK, γδ T and NKT Cells

WHAT'S IN THIS CHAPTER?

If it's natural to kill, why do men have to go into training to learn how?

Joan Baez

In Chapter 3, we introduced NK cells, γδ T cells and NKT cells as cell types that bridge innate and adaptive immunity in both form and function. On one hand, the responses of NK, γδ T and NKT cells to infection or injury can be considered part of the induced innate response because they are rapid and involve broad recognition of antigen that is independent of classical pMHC complexes. On the other hand, NK cells, γδ T cells and NKT cells can be considered part of adaptive immunity because these cells are related to the T cell lineage, and γδ T cells and NKT cells express TCRs derived from V(D)J recombination. In addition, NK, γδ T and NKT cells can directly influence αβ T cells and B cells and their effector actions. In this chapter, we discuss the distribution, activation, function and development of NK cells, γδ T cells and NKT cells. A schematic representation of the major cell surface receptors distinguishing these cell types from αβ T cells appears in **Figure 11-1**.

A. Natural Killer (NK) Cells

I. Overview

NK cells are large, non-phagocytic lymphoid cells that possess cytoplasmic granules containing perforin and granzymes. At the molecular level, NK cells are distinguished from NKT, B and T cells by their lack of expression of TCRs or BCRs and the germline configuration of their TCR and BCR genes. NK cells also often show surface expression of (neural cell adhesion molecule (NCAM; CD56) and/or the low affinity IgG receptor FcγRIIIA (CD16). Resting NK cells are found at their highest frequency in spleen, liver, uterus and peripheral blood, with more moderate numbers in the bone marrow, lymph nodes and peritoneum. In situations of infection or inflammation, however, NK cells can be rapidly recruited to almost any tissue in the body. The life span of a mature NK cell (in the absence of activation) is about 7–10 days.

Although they function primarily in the tissues, NK cells make up 10–15% of human peripheral blood cells.

MAK: Primer to the Immune Response. http://dx.doi.org/10.1016/B978-0-12-385245-8.00011-X

Fig. 11-1
Characteristic Surface Receptors of NK, γδ T, NKT and αβ T Cells

The dominant receptors characteristic of the indicated lymphoid cell types are shown. For simplicity, CD3 complexes have been omitted. AR, NK activatory receptor; IR, NK inhibitory receptor.

The primary functions of NK cells are to induce the cytolysis of tumor cells or virus-infected cells and to secrete cytokines. Cytolysis may be induced by **natural cytotoxicity** (see later), ADCC or cytotoxic cytokines. Many cytokines secreted by NK cells directly regulate T and B cell functions and differentiation. In addition, NK cells may interact with DCs, establishing intercellular contacts and secreting inflammatory cytokines that promote DC maturation and efficient induction of Th and CTL responses. Importantly, unlike activated T and B cells, activated NK cells do not need to proliferate and differentiate into separate effector cells in order to function. Thus, the peak NK cell response can be detected within hours of infection. Interestingly, NK cells that survive after a threat has been cleared take on characteristics associated with memory lymphocytes. These "experienced" NK cells increase their longevity (for up to 2 months), and if the threat appears a second time, these NK cells proliferate and exhibit enhanced cytolytic and cytokine secretion capacities.

> As described in Chapter 5 and illustrated in Fig. 5-13, ADCC means "antibody-dependent cell-mediated cytotoxicity."

II. Ligand Recognition, Activation and Effector Functions

i) Natural Cytotoxicity

a) "Missing Self" Model of NK-Mediated Natural Cytotoxicity
Just like the granules of CTLs, the granules of NK cells contain perforin and granzymes that induce target cells to undergo apoptosis. However, these granules are preformed in an NK cell and do not have to be synthesized in response to activation, as occurs in CD8+ Tc cells. The triggering of NK-mediated natural cytotoxicity depends on a balance between competing signals initiated by two sets of surface receptors of broad binding specificity: the **NK activatory receptors** and the **NK inhibitory receptors**. The NK activatory receptors are triggered by ligands that may be constitutively expressed on healthy cells, or by ligands that may be induced or upregulated in response to viral infection, malignant transformation or other cellular stresses. Some non-classical MHC molecules have also been identified as activatory ligands. In contrast, most NK inhibitory receptors bind only to classical MHC class I (Ia or Ib) molecules expressed by the host. When an NK cell uses its inhibitory and activatory receptors to scan the surface of a host cell, the overall amount of activatory signaling received by the NK cell must be counterbalanced by adequate inhibitory signaling, or the NK cell is activated and releases the cytotoxic contents of its granules. The intensity of the NK cell response is thus determined by a combination of two factors: how many inhibitory versus activatory receptors are engaged and the affinity of those receptors for their ligands.

In the case of normal host cells (almost all of which express MHC class I), enough NK inhibitory receptors are engaged to send an inhibitory signal that dominates normal activatory signaling (**Fig. 11-2A**). The NK cell is not activated, and the normal cell is spared. In contrast, infected and cancerous cells frequently downregulate their MHC class I expression. In these cases, an NK cell's inhibitory receptors are not engaged in sufficient numbers to prevent the activatory receptors from completing their intracellular signaling (**Fig. 11-2B**). In short, the abnormal host cell is "missing self" (i.e., self MHC) relative to the level of MHC class I present on healthy host cells and so cannot

(A) Normal

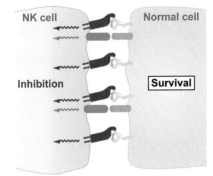

(B) Missing self

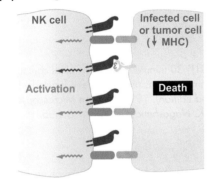

(C) Induced self

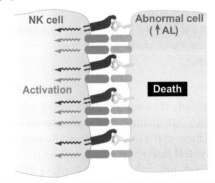

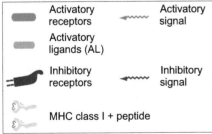

	Activatory receptors		Activatory signal
Activatory ligands (AL)			
Inhibitory receptors			Inhibitory signal
MHC class I + peptide			

▶ Fig. 11-2
Outcomes of NK Activatory/Inhibitory Receptor Signaling

(A) When an NK cell encounters a healthy host cell, normal levels of MHC class I plus peptide (which may be self or non-self in origin) engage the inhibitory receptors of the NK cell. Due to the relatively low expression of activatory ligands associated with stress, tumorigenesis or pathogen attack, the signals received through the NK cell's activatory receptors cannot overpower those received through the inhibitory receptors. The inhibitory signal dominates and the host cell survives. **(B)** When a host cell is infected or has become cancerous, it downregulates its expression of MHC class I and also expresses ligands that engage the NK cell's activatory receptors. Activatory signaling dominates, and the NK cell kills the target. **(C)** When a host cell expresses abnormally high levels of activatory ligands, these molecules engage large numbers of activatory receptors on an NK cell. Any inhibitory signaling is overwhelmed, and the host cell is destroyed.

stop the activation of the NK cell, which then kills the target (**Plate 11-1**). Killing by natural cytotoxicity will also result if the host cell expresses normal levels of self MHC class I but abnormally high levels of activatory ligands. Such a host cell is said to display "induced self" because the activatory ligands are self molecules and the cell has been induced to express them in elevated amounts. The signaling generated when the NK cell's activatory receptors are engaged by this excess of activatory ligands overwhelms the signaling generated by MHC class I engagement of the NK cell's inhibitory receptors, and the abnormal host cell is killed (**Fig. 11-2C**).

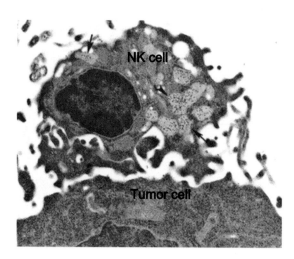

Plate 11-1
Interaction of an NK Cell with a Tumor Cell

Electron micrograph of an NK cell (clearly showing its nucleus) and the tumor cell it has targeted (T) 30 minutes after their initial contact. The black arrows within the NK cell cytoplasm point to destructive granules containing pore-forming proteins that have been labeled with colloidal gold particles. *Reproduced by permission from the* Encyclopedia of Immunology *(Second Edition), 1998, Electron Microscopy: Immunological Applications by Matthew A. Gonda. Micrograph was contributed by J. Ortaldo, K. Nagashima, and M. A. Gonda.*

> NOTE: The term "missing self" arose because the original studies in this field involved situations in which self MHC class I was completely absent from the target cells used in the experiment. It has since become clear that a cell displaying some degree of "reduced self" (if the "self" is MHC) or "induced self" (if the "self" consists of activatory ligands) may also become a target for NK cell killing if activatory signaling exceeds inhibitory signaling. Nevertheless, when referring to models of NK-mediated natural cytotoxicity, immunologists have generally stuck to the historical term "missing self" for convenience.

Although no effector differentiation is required, resting NK cells don't usually acquire significant cytolytic competence until they are primed through exposure to cytokines such as IFNα/β, IL-2, IL-12 or IL-15. These cytokines are often present in a site of infection due to the actions of nearby activated innate leukocytes. The priming cytokines induce upregulation of multiple activatory and inhibitory receptors on NK cells as well as adhesion molecules that stabilize the binding of an NK cell to a potential target cell.

b) Activatory and Inhibitory Receptors

Activatory and inhibitory receptors on NK cells are generally transmembrane proteins. The extracellular domains of both types of receptors are responsible for ligand recognition and often share molecular features. On this basis, they can be categorized into structural classes. The *natural cytotoxicity receptor (NCR)* class contains only activatory receptors, whereas the *natural killer group 2 (NKG2)* and *killer Ig-like receptor (KIR)* classes include both activatory and inhibitory members. The NKG2 receptors are C-type lectin-like receptors, meaning that they possess an extracellular domain that can bind to carbohydrate. The KIR receptors are named according to the number of Ig domains present as well as the length of the cytoplasmic tail. For example, KIR2DS means "KIR, two Ig-like domains, short tail," and KIR3DL means "KIR, three Ig-like domains, long tail."

The opposing functions of activatory and inhibitory receptors can be attributed to differences in their intracellular domains. The NK activatory receptor proteins possess positively charged transmembrane residues and short cytoplasmic tails that contain few intracellular signaling domains. These chains are not expressed on

As described in Chapter 3, C-type lectins expressed by APCs often act as PRRs.

TABLE 11-1	Examples of Human NK Activatory and Inhibitory Receptors		
Receptor Name	**Receptor Class**	**Associated Signaling Chain**	**Example of Ligand**
NK Activatory Receptors			
NKp46	NCR	CD3ζ	Viral hemagglutinin
NKp44	NCR	DAP12	Viral hemagglutinin
NKp30	NCR	CD3ζ	B7-H6
CD94/NKG2C	NKG2	DAP12	HLA-E
NKG2D	NKG2	DAP10	MICA, MICB
KIR2DS1	KIR	DAP12	HLA-C
NK Inhibitory Receptors			
CD94/NKG2A	NKG2	None	HLA-E
CD94/NKG2B	NKG2	None	HLA-E
KIR3DL2	KIR	None	HLA-A
KIR3DL1	KIR	None	HLA-B
KIR2DL1	KIR	None	HLA-C
KIR2DL4	KIR	None	HLA-G

their own and do not mediate any signal transduction in isolation. Instead, activatory receptor proteins associate with a homodimer composed of an accessory signaling molecule such as CD3ζ, the γc chain, or one of two adaptor proteins called DAP10 and DAP12. All of these molecules possess negatively charged transmembrane domains and all (except DAP10, which transduces signals in a slightly different way) contain ITAMs that facilitate signal transduction. Upon the binding of an activatory ligand to an activatory receptor complex, the ITAMs in the associated chain are phosphorylated, and a signal that promotes natural cytotoxicity is conveyed to the interior of the NK cell. In contrast, inhibitory receptor proteins usually have long cytoplasmic tails containing **immunoreceptor tyrosine-based inhibition motifs (ITIMs)**. When an inhibitory receptor is stimulated by the binding of MHC class I, kinases and phosphatases are recruited to the receptor complex. The ITIMs are phosphorylated such that signal transduction within the NK cell is inhibited. It is the balance of these inhibitory and activatory signals that determines whether the NK cell is activated.

As described in Chapter 4, ITAM stands for "immunoreceptor tyrosine-based activation motif."

Some well-studied human NK activatory receptors are listed in **Table 11-1** (upper half) and illustrated in **Figure 11-3** (left panels). NKp46, NKp44 and NKp30 are members of the NCR class of activatory receptors and are expressed exclusively on NK cells. These receptors are largely responsible for direct NK cell killing of virus-infected and tumor cells. An example of an activatory receptor of the NKG2 class is the CD94/NKG2C receptor, composed of the NKG2C protein bound to the CD94 protein, plus the ITAM-containing DAP12 protein. A ligand for CD94/NKG2C is HLA-E, a non-classical MHC class Ib molecule. Perhaps the most important NKG2 activatory receptor is NKG2D, in which a NKG2D homodimer associates with DAP10. (In mice, NKG2D associates with both DAP10 and DAP12.) Human NKG2D binds to a number of MHC class I-related molecules, including the MICA and MICB stress molecules introduced in Chapter 6. NKG2D ligands tend to be transmembrane proteins that are induced on many types of epithelial cells in response to heat shock or other cellular stresses. NKG2D ligands are also frequently upregulated on virus-infected cells and cancer cells, making these cells targets for NK-mediated cytotoxicity. Some KIR receptors are also activatory. For example, KIR2DS1 binds to the conventional MHC class I molecule HLA-C and promotes NK cell activation.

Fig. 11-3
Examples of NK Activatory and Inhibitory Receptors

Schematic diagrams of the structures of selected NK receptors from the indicated classes are shown. Among members of the same class, the presence of ITAMs or ITIMs determines whether the molecule has an activatory or inhibitory effect, respectively.

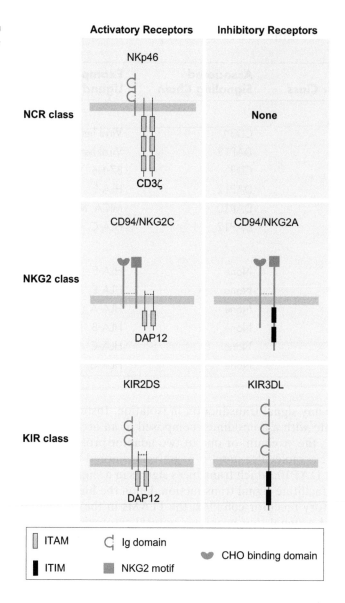

Examples of some important NK inhibitory receptors are listed in the lower half of **Table 11-1** and illustrated in **Figure 11-3** (right panels). Inhibitory members of the NKG2 class include CD94/NKG2A and CD94/NKG2B. CD94/NKG2A binds specifically to HLA-E molecules, providing an inhibitory counterpart to the CD94/NKG2C activatory receptor. CD94/NKG2A function is particularly important for blocking maternal uterine NK cell function during pregnancy and protecting the trophoblast and fetus (whose cells express paternal "non-self" MHC molecules; refer to Ch. 10). KIR receptors that block NK cell activation include KIR3DL2, KIR3DL1 and KIR2DL1, which bind to HLA-A, B or C, respectively. Interestingly,

NOTE: NK cell recognition of the activatory ligand MICA may offer a novel avenue of immunotherapy for brain cancers. The transcription factor STAT3 is overexpressed by many cancers, including human brain tumors. STAT3 activation downregulates MICA transcription in cancer cells, making these cell less visible to the NK cells tasked with killing them. In mouse models of brain cancers, treatment with a STAT3 inhibitor increases NK-mediated lysis of tumor cells. Through the use of special viral vectors, it may be possible to deliver a STAT3 inhibitor directly into the brain tumors of patients, upregulating MICA expression on the malignant cells and enhancing their death by NK-mediated cytotoxicity. Pre-clinical investigation of this possibility is under way.

engagement of KIR2DL4 on maternal NK cells by HLA-G expressed on the placenta does not trigger cytotoxicity but instead induces maternal uterine NK cells to secrete cytokines that promote the formation of new blood vessels, helping to ensure a successful pregnancy.

ii) NK-Mediated ADCC

As noted previously, NK cells express large amounts of FcγRIIIA (CD16). This FcR can trigger ADCC by binding to IgG molecules that have engaged epitopes on tumor cells or virus-infected cells. The engagement of FcγRIIIA activates the NK cell and causes it to release the contents of its cytotoxic granules.

iii) Cytokine Secretion

In response to infection, phagocytes and other innate leukocytes produce IFNα/β, IL-12, IL-15, TNF and additional cytokines that first prime NK cells and then induce them to synthesize large quantities of IFNγ. Subsequent activation of primed NK cells by overwhelming activatory receptor engagement or FcγRIIIA stimulation leads to the production of a whole battery of chemokines, growth factors and cytokines, including IFNγ, TNF, IL-1, IL-3 and IL-6. These molecules have various effects on cells of both the innate and adaptive immune responses, as detailed in Appendix E.

"CD94 Is Essential for NK Cell-Mediated Resistance to a Lethal Viral Disease." by Fang, M., Orr, M.T., Spee, P., Egebjerg, T., Lanier, L.L., and Sigal, L.J. (2011) *Immunity* 34, 579–589.

Focus on Relevant Research

Natural killer (NK) cells have the ability to detect and destroy some, but not all, virus-infected cells. It is suspected that NK cells may respond to infection by different viruses by employing specific strategies, but the nature of this discrimination and the precise mechanisms used are largely unknown. In this research article, Fang *et al.* demonstrate that an NK activatory receptor in which CD94 combines with NKG2E plays a key role in NK cell-mediated protection of mouse cells against ectromelia virus (ECTV), a member of the *Orthopoxvirus* family. The authors infected mice that were deficient for expression of CD94, as well as CD94-expressing control mice, with ECTV. As illustrated in the left panel below (Figure 4, panel A from the article), the spleen and liver of the CD94-deficient mice (white columns) harbored

significantly more virus after infection compared to infected normal mice (gray columns). The researchers then blocked the function of the CD94/NKG2E activatory receptors in the control mice by treating them with a monoclonal antibody called "20D5." As illustrated in the right panel (Figure 3, panel E from the article), when mouse survival was evaluated on various days post-infection (dpi), it was found that significantly more of the 20D5-treated animals succumbed to ECTV infection (gray squares) compared to control mice that were infected with ECTV but treated with an IgG antibody of irrelevant specificity (black circles). Because the CD94/NKG2E activatory receptor is so highly conserved between mice and humans, the findings of this study may lead to improved therapies and/or vaccines against medically important viruses.

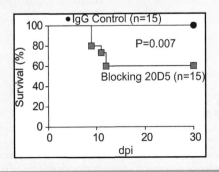

III. Development

i) Developmental Pathway

NK cells originate from the same bone marrow–derived hematopoietic precursors that give rise to T cells. **Figure 11-4** outlines the later stages of NK cell development in mouse bone marrow, which has been better studied than the human situation. CLPs that remain in the bone marrow can generate NK/T precursors in this tissue. These NK/T precursors become subject to signaling delivered by IL-15 as well as a molecule crucial for NK development called *Flt3 ligand* (*Flt3-L*). Bone marrow stromal contacts are also essential for NK development at this stage. Under the continued influence of IL-15, *pro-NK cells* are produced that eventually develop into immature NK cells expressing only a few NK inhibitory receptors. Mature NK cells then emerge that express the full complement of activatory and inhibitory receptors. A secondary source of NK cells arises from NK/T precursors that have migrated from the bone marrow to the thymus. As described in Chapter 9, these precursors give rise mainly to T cells but may also generate a subset of thymic NK cells.

ii) NK Inhibitory Receptor Repertoire

Within an individual, different NK cells express different combinations of inhibitory receptors on their cell surfaces. How is each NK cell's "repertoire" of inhibitory receptors assembled? A sequential expression model has been proposed in which developing NK cells gradually accumulate new types of self MHC-specific inhibitory receptors. It may be that certain activatory receptors, triggered by interaction with their ligands on normal host cells, build up a cascade of intracellular signaling inside the developing NK cell. At some point, the inhibitory receptors start to be expressed, and different classes accumulate on the cell surface until the signaling generated by the triggering of all the inhibitory receptors exceeds the level of signaling generated by all the activatory receptors. It is thought that the cell then receives a maturation signal that halts the expression of new inhibitory receptors.

iii) NK Cell Tolerance

Mature NK cells are usually described as "self-tolerant" because they do not attack normal self cells. According to the "missing self" model, an NK cell with the potential to attack a self cell would be an effector whose collection of inhibitory receptors failed to adequately recognize self MHC class I. As a result, the inhibitory receptors of this NK cell would not be sufficiently engaged to prevent the activatory ligands on normal self cells from triggering NK-mediated cytotoxicity. Thus, during NK cell development, an immature NK cell that does not express sufficient inhibitory receptors recognizing self MHC should undergo some kind of tolerance process to either delete the cell or render it anergic. Some immunologists refer to the establishment of both functional capacity and self tolerance in these cells as "NK cell education." However, some aspects of this

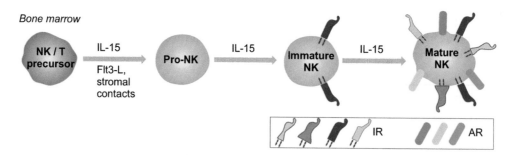

Fig. 11-4
Model of NK Cell Development

An NK/T precursor in the bone marrow that encounters IL-15, Flt-3 ligand (Flt3-L) and certain stromal cell contacts becomes a pro-NK cell. In the continued presence of IL-15, the pro-NK cell starts to express inhibitory receptors, marking it as an immature NK cell. Still in the presence of IL-15, activatory receptor expression commences such that the mature functional NK cell expresses a balance of activatory and inhibitory receptors.

"education" are difficult to explain based solely on a balance between inhibitory and activatory receptor engagement in developing NK cells. For example, in both humans and mice, individuals lacking MHC class I expression do not have a problem with NK cell "autoreactivity" in the periphery. In addition, developing NK cells do not inflict tissue damage even though there appears to be a short period during which the maturing NK cells can execute effector functions but still lack expression of inhibitory receptors. To explain these findings, it has been proposed that NK cells can acquire the capacity to be activated only if there is some degree of inhibitory receptor engagement by self MHC class I. Passage through this MHC-dependent checkpoint has been dubbed "NK cell licensing."

Two models have been proposed to explain MHC-dependent NK cell licensing: *arming* and *disarming*. In the "arming" model, the default status of any activatory receptor expressed by an immature NK cell is assumed to be non-responsive. It is only after signaling conveyed by inhibitory receptor engagement is received that the activatory receptors become competent. This scenario would ensure that the NK cell cannot be activated if it does not recognize self MHC. In other words, before becoming functionally competent, the NK cell must first develop the means of controlling its cytotoxic power. In contrast, in the "disarming" model, the default status of any activatory receptor is assumed to be responsive. If the NK cell fails to express an inhibitory receptor recognizing self MHC so that the cell can be shut down, its activatory receptors would be constantly engaged by self activatory ligands. The resulting continuous signaling should cause the NK cell to become anergic. The disarming model can explain the observation that NK cells lacking expression of any known self-specific inhibitory receptors can exist in a host's periphery but are hypofunctional. Both of these models remain under investigation.

MHC-independent NK cell education also appears to exist, since NK inhibitory receptors can be engaged by ligands other than MHC class I. For example, NK-mediated cytotoxicity is inhibited when the calcium-dependent adhesion molecule E-cadherin binds to KLRG1, an inhibitory receptor belonging to the NKG2 class. Invasive tumors often downregulate their expression of E-cadherin, which frees them to migrate more easily but also makes them more vulnerable to killing by NK cells expressing KLRG1.

Interestingly, studies of mature NK cells indicate that they can be educated, "un-educated," and "re-educated," depending on the host's status. The cells of a host under attack by a persistent virus or growing tumor will display large amounts of activatory ligands and may downregulate MHC class I such that NK cells become overstimulated. These overstimulated NK cells may then become hyporesponsive and reduce the functionality of activatory receptors specific for the viral or tumor ligands; these NK cells have effectively "un-educated" themselves with respect to these threats and have become inappropriately tolerant. "Re-education" of these NK cells such that responsiveness to these ligands is restored (recovery of activatory receptor functionality) can be induced by culturing the NK cells in the absence of the virus-infected or tumor cells. Alternatively, stimulation of these NK cells in the presence of pro-inflammatory cytokines can restore activatory receptor functionality. A better understanding of the effects on NK cells of chronic activatory signaling, how viruses and tumors might exploit this avenue to escape NK killing, and how NK cells can be re-educated to kill again may lead to improved clinical treatments for both chronic infections and cancers.

> NK cell licensing is unrelated to the DC licensing necessary for Tc cell activation (see Ch. 9).

> NK cell education may be skewed by the cytokines in a particular microenvironment because these proteins can influence the ability of NK activatory receptors to become fully functional.

NOTE: An example of the "un-education" of NK cells has been observed in patients with acute myeloid leukemia, a form of hematopoietic cell cancer described further in Chapter 20. NK cells in these patients lose expression of activatory receptors in the face of chronic exposure to activatory ligands expressed on the surface of the leukemic cells. Without their activatory receptors, the NK cells can no longer kill the malignant cells they encounter. In other cases, constant assault by activatory ligands causes actual loss of the NK cells themselves in a situation similar to clonal T cell exhaustion.

B. γδ T Cells

I. Overview

Refer to Figure 8-3 for a schematic drawing of a TCR complex.

In Chapters 8 and 9, we described how two distinct types of T lymphocytes can be distinguished based on the polypeptide chains making up their TCRs. In both humans and mice, the majority of T cells in the body are αβ T cells bearing TCRs containing the TCRα and TCRβ chains. Mature αβ T cells carry either monomeric CD4 coreceptors or heterodimeric CD8αβ coreceptors. In contrast, γδ T cells express TCRs composed of the TCRγ and TCRδ chains plus the CD3 complex and carry either homodimeric CD8αα coreceptors or no coreceptor at all.

Whereas cells bearing TCRαβ recognize only pMHC complexes, γδ TCRs interact with ligands in a way that is similar to ligand recognition by PRRs. That is, γδ T cells can respond to antigens that are derived from a broad range of pathogens or from abnormal or stressed host cells. In so doing, γδ T cells do not require the involvement of conventional MHC nor the processing and presentation of peptide antigens by professional APCs. Intact proteins or peptides from pathogens or stressed host cells, and non-protein antigens such as lipids and phosphorylated nucleotides can serve as TCRγδ ligands. Furthermore, these ligands may be either soluble or bound to a cell surface. Importantly, some stress molecules that bind to the TCRs of γδ T cells are not recognized by myeloid defenders such as neutrophils and macrophages, so that γδ T cells fill critical gaps in host protection.

Once activated, γδ T cells respond by proliferating and differentiating into γδ Th and γδ CTL effectors much like αβ T cells, but do so much more rapidly, in smaller numbers, and often in the apparent absence of conventional costimulation. Some immunologists therefore picture γδ T cells as being in a "resting but preactivated" state ideal for contributing to innate defense. Indeed, the release of cytokines by γδ T cells can precede the activation of αβ T cells by several days. A comparison of the properties of γδ and αβ T cells is given in **Table 11-2**.

TABLE 11-2	Comparison of the Properties of γδ and αβ T Cells	
	γδ T Cells	**αβ T Cells**
Type of immunity	Induced innate	Adaptive
Frequency among T cells	0.5–5%	95–99%
Anatomical distribution	Epithelial layers in the skin and mucosae	Primary and secondary lymphoid organs and tissues
TCR	TCRγδ	TCRαβ
Repertoire diversity	Limited	Almost unlimited
Receptor specificity	Promiscuous	Specific
Distribution of epitope recognized by a given TCR	Broadly expressed on range of pathogens or stressed cells	Expressed on one pathogen
Ligand recognized	Intact pathogen proteins and host stress proteins (e.g., HSPs) Phospholipids, glycolipids, phosphonucleotides, pyrophosphates Antigen-CD1c complexes	Peptide–MHC complexes
APCs required for antigen presentation	No	Yes
Mature T cell coreceptor expression	CD4-CD8- or CD4-CD8αα+	CD4+CD8- or CD4-CD8αβ+
CD28- or CD40-mediated costimulation required for activation	Not necessarily but other types of receptors may contribute	Yes
Days to effector cell generation	1–2 days	7 days
Types of effector cells generated	γδ Th1, Th2, Th17, CTL, regulatory T cells	αβ Th1, Th2, Th17, CTL, regulatory T cells
Capable of long-lived memory response	Not in general	Yes

II. Anatomical Distribution

The anatomical distribution of γδ T cells is strikingly different from that of αβ T cells. Just 5% of all peripheral blood lymphocytes are γδ T cells, and only very low numbers of γδ T cells are found in the secondary lymphoid tissues and thymus of mice and humans. Instead, many γδ T cells are interspersed among the epithelial cells of the skin (SALT) and in the mucosae of various body tracts (MALT). This localization of γδ T cells in the top layers of the skin and mucosae allows them to be among the first defenders to confront invading pathogens or injurious substances.

In addition, unlike αβ T cells, which utilize highly random combinations of V gene segments for their TCRs regardless of their anatomical location, the γδ T cells resident in a particular tissue express a dominant or "canonical" TCR containing specific V gene segments. For example, γδ T cells in mouse skin predominantly express TCRs containing Vγ3Vδ1 or Vγ5Vδ1, whereas the genital epithelium and the tongue feature an abundance of Vγ6Vδ1-expressing cells. Vγ7 appears more often than not in the TCRs of γδ T cells in murine gut epithelium, whereas Vγ2Vδ2 cells dominate in the tonsils, spleen and peripheral blood. In humans, Vγ1Vδ2 TCRs are prevalent on intestinal γδ T cells, whereas Vγ9Vδ2 TCRs occur on γδ T cells in the skin and peripheral blood.

III. Antigen Recognition and Activation

Unlike αβ TCRs, the TCRs of many γδ T cells can bind directly to low molecular weight non-peptide antigens, without the need for presentation by another molecule or cell (**Fig. 11-5A**). However, the TCRs of other γδ T cell subsets depend on non-peptide antigen presentation by non-classical MHC class Ib molecules. Still other γδ T cell subsets express TCRs that recognize non-peptide antigens displayed by members of the non-polymorphic CD1 family of MHC-like molecules (**Fig. 11-5B**). In general, γδ TCRs lack the fine antigenic specificity of αβ TCRs and are often broadly cross-reactive.

Non-peptide antigen presentation to αβ T cells was described in Chapter 7.

γδ T cells have been shown to respond to a wide variety of bacterial, protozoan and viral molecules and products, including peptides, proteins, pyrophosphates, phospholipids, lipoproteins, phosphorylated oligonucleotides and alkyl amines. Some γδ T cell subsets (as defined by their V gene segment usage) appear to be specific for certain types of determinants and thus may counter a whole group of pathogens. The non-random anatomical distribution of such subsets may increase the chance that γδ T cells of the appropriate specificity will be in local abundance to defend against a given invader (**Table 11-3**). For example, the Vγ9Vδ2 subset found in human skin and peripheral blood expresses a TCRγδ that recognizes antigens from pathogens normally found in these locations, including Epstein–Barr virus and *Mycobacteria*.

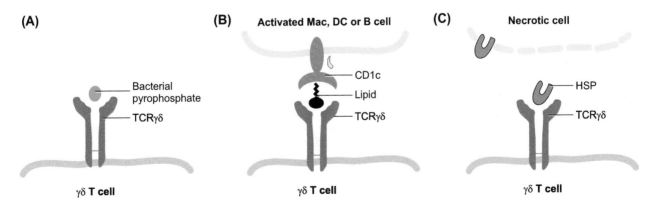

Fig. 11-5
Examples of Antigen Binding to γδ TCRs

The TCRγδ molecules of different γδ T cell clones may recognize: **(A)** Small antigens released by bacteria, such as pyrophosphate molecules; **(B)** Lipid antigens presented by CD1c on the surface of an APC; or **(C)** Stress proteins, such as heat shock proteins (HSPs), leaked by necrotic cells. In all three cases, intracellular signaling activating the γδ T cell is conveyed by the CD3 complex associated with TCRγδ (not shown).

TABLE 11-3		Examples of V Gene Usage and γδ T Cell Antigens Recognized	
Species	**γδ Subset**	**Anatomical Location**	**Antigen Recognized**
Human	Vγ2Vδ2	Skin, peripheral blood	Small pyrophosphate epitopes derived from Epstein–Barr virus, *Mycobacterium, Plasmodium, Leishmania* or *Salmonella*
	Vγ9Vδ2	Skin	HSP 58
Murine	Vγ3	Skin	Stress antigen expressed by skin cells
	Vγ6	Uterus, tongue	Stress antigen expressed by lung epithelial cells

Other γδ T cells are specific for stress molecules that are expressed only by host cells suffering injury, infection or cancerous transformation and do not appear on the surfaces of healthy host cells. The expression of a single stress antigen in response to a wide variety of different infections or injuries allows a γδ T cell population with a limited antigen receptor repertoire to monitor a wide range of assaults to the host epithelium. Whereas some stress molecules recognized by γδ T cells are small pyrophosphate-like molecules, others are peptides or whole proteins, such as the HSPs released in the debris of necrotic cells (**Fig. 11-5C**).

> NOTE: The activation of some γδ T cell subsets may be regulated more like that of NK cells than αβ T cells, since some γδ T cells express NK inhibitory receptors that recognize MHC class I. Indeed, host cells that are deficient in MHC class I expression (like tumor cells and virus-infected cells) have been shown to activate these γδ T cell subsets.

While small phosphorylated metabolites are able to activate some γδ T cells without the need for CD28- or CD40-mediated costimulation (at least *in vitro*), other γδ T cell subsets appear to experience some kind of costimulation mediated by receptors whose identity may be dictated by anatomical location. The γδ T cells in peripheral tissues tend to undergo conventional CD28-mediated costimulation, whereas epidermal and intestinal γδ T cells are costimulated when other cell surface receptors are engaged by stress ligands. For example, many γδ T cells in the epidermis and intestinal epithelium express the NKG2D NK activatory receptor that binds to the stress antigens MICA and MICB. Accordingly, MICA and MICB have been found to stimulate the responses of certain γδ T cell subsets. Another costimulatory molecule that is expressed on the surface of epidermal γδ T cells and is important for their activation is JAML ("junctional adhesion molecule-like" protein). JAML binds to a ligand called CAR ("coxsackie and adenovirus receptor") expressed by damaged skin cells. It remains unclear how signaling through TCRγδ plus NKG2D and/or JAML might cooperate, and whether all these signals are absolutely required for the activation of these γδ T cells. Finally, there is evidence that, in sites of inflammation, APCs may help to activate γδ T cells by supplying stimulatory cytokines or (unknown) intercellular contacts.

IV. Effector Functions

Once activated by antigen, γδ T cells generate effectors in a manner similar to αβ T cells, although the signaling pathways linking TCR stimulation to new gene transcription appear to be slightly different. Some γδ T cells resemble αβ Tc cells and generate γδ CTLs that eliminate infected cells and tumor cells via Fas-mediated killing, perforin/granzyme-mediated cytotoxicity, or secretion of cytotoxic cytokines (**Fig. 11-6**). Other γδ T cells resemble αβ Th cells and generate γδ Th effectors that secrete cytokines and growth factors. Indeed, Th1, Th2 and Th17 subtypes of γδ Th effectors have been identified based on their secretion of Th1 cytokines (IL-2, TNF and IFNγ), Th2 cytokines (IL-4, IL-5, IL-10 and IL-13), or Th17 cytokines (IL-17, IL-21 and IL-22), respectively. Cytokines produced by these γδ Th effectors then activate NK cells and macrophages; support the differentiation of activated αβ Th0 cells into conventional αβ Th1, Th2 or Th17 effectors; and influence isotype switching in B cells. Interestingly, the IL-17

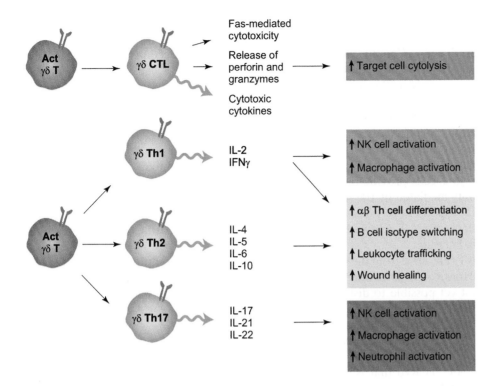

Fig. 11-6
γδ *T Cell Effector Functions*

Activated γδ T cells can generate γδ CTL or γδ Th1, Th2 or Th17 effectors, depending on the circumstances. γδ CTLs kill target cells using the same mechanisms as αβ CTLs, whereas the cytokines secreted by γδ Th effectors influence both innate and adaptive leukocytes as indicated.

produced specifically by γδ Th17 cells in the lung appears to be very important for activating neutrophils in this tissue during the early stages of innate responses to bacterial attack. On the flip side, like αβTh17 cells, γδ Th17 cells have been implicated as contributors to inflammatory autoimmune diseases.

In addition to cytokines, γδ T cells secrete molecules that influence leukocyte trafficking and wound healing. For example, chemokines produced by activated γδ T cells promote the migration of neutrophils and macrophages toward damaged epithelium. Activated γδ T cells can also induce neighboring epithelial cells to produce the antimicrobial compound NO. In the skin, activated γδ T cells secrete *keratinocyte growth factor (KGF)*, a molecule that stimulates the growth and differentiation of skin epithelial cells necessary to close wounds.

Immunological memory does not appear to be a feature of γδ T cell responses, although two phases of γδ T cell proliferation are observed in response to some infections. The first phase is a rapid response occurring immediately after infection and prior to the primary αβ T cell response. The second phase is a proliferative burst that occurs 2–3 days after the primary αβ T cell response concludes. However, the γδ T cells in this second phase do not appear to be fully functional memory cells and have only a restricted capacity to develop into second generation effectors. Thus, defense by γδ T cells appears to be stuck at the primary response level during a secondary challenge with antigen. Investigation of the capacity of γδ T cells to mount true secondary responses is ongoing.

NOTE: Recent intriguing results suggest that human γδ T cells may sometimes act as APCs for αβ T cells. In addition to their γδ TCRs, many γδ T cells express PRRs that allow them to readily engulf antigens. In one study, scientists incubated γδ T cells with experimental antigens and found that the antigens were taken up by phagocytosis and processed, and that antigenic peptides were presented on MHC class II or cross-presented on MHC class I. Importantly, unlike most leukocytes but like professional APCs, these γδ T cells were found to contain immunoproteasomes rather than standard proteasomes. If this finding withstands further scrutiny, it would represent a direct link between the fast-acting γδ T cells of the innate response and the slower-acting αβ T cells of the adaptive response.

V. Development

As introduced in Chapter 9, the first waves of T cells produced in a human or murine embryo are γδ T cells. Rearrangement of the TCRγδ genes can be detected in thymocytes as early as 8 weeks in the human fetus and by day 12.5 of gestation in the mouse. Under the influence of fetal thymic stromal cells, distinct waves of γδ T cells fan out to populate specific regions of the body. In cells of the very first wave exiting the murine thymus, the rearranged TCRγδ invariably contains Vγ5, and these γδ T cells take up residence in the skin. Subsequent waves include Vγ6⁺ cells heading for the mucosae of the urogenital tract and then Vγ7⁺ cells destined to reside in the intestine. These γδ T cells provide immune defense in the fetus and neonate before adaptive immunity mediated by the more powerful αβ T cells is fully established. In particular, αβ Th1 responses are weak in very young animals. Perhaps not coincidentally, higher frequencies of γδ T cells with a Th1 phenotype are present during the neonatal period than in later life.

γδ T cells arise from the same NK/T precursor as that generating αβ T cells and NK cells. However, despite intensive study of NK/T precursors and early DN thymocytes, it is not yet possible to distinguish which among them is destined to become a γδ T cell and which is destined to become an αβ T cell. In the late DN2–early DN3 stage of thymocyte development, the *TCRB*, *TCRG* and *TCRD* loci begin to rearrange in all thymocytes. At some unknown point prior to *TCRA* rearrangement, the commitment to the αβ or γδ T cell lineage is made. As a result, some developing DN3 cells express a complete TCRγδ that is functional and initiates intracellular signaling precluding further *TCRB* rearrangement. The β-selection process characteristic of αβ T cell development does not occur, and access to the αβ T cell developmental path is blocked. These cells eventually leave the thymus as mature γδ T cells and migrate mainly to the epithelial layers of the skin, gut, and urogenital and respiratory tracts.

The differentiation pathway that occurs in the thymus and gives rise to αβ T cells, γδ T cells and NKT cells was shown in Figure 9-1.

NOTE: Some immunologists believe that the strength of intracellular signaling delivered through TCRγδ is the final determinant of whether a DN3 cell becomes a mature γδ T cell or a mature αβ T cell. A DN3 thymocyte that expresses a pre-TCR (made up of the pTα chain and a candidate TCRβ chain; see Ch. 9) experiences relatively weak signaling that nonetheless commits it to the TCRαβ lineage if the candidate β chain is functional and there is no opposing signaling delivered by a TCRγδ. However, if this DN3 thymocyte also expresses a newly assembled TCRγδ that is functional, the signaling delivered through this TCRγδ may outweigh any signaling conveyed through the pre-TCR, committing the cell to the γδ lineage. *In vitro*, late DN3 thymocytes transgenically expressing a genetically defined TCRγδ generated mature γδ T cells if the TCR was engaged by a ligand inducing strong signaling by downstream kinases, but generated mature αβ T cells if the TCR was engaged by a ligand inducing very weak downstream signaling. Researchers speculated that, in the latter case, the signal delivered through the TCRγδ was unable to overcome the signaling mediated by the pre-TCR. Two intracellular molecules important for this putative αβ/γδ T lineage commitment pathway appear to be ERK kinase and the negative transcriptional regulator ID3. The search is also on for a transcription factor uniquely linked to γδ T cell development in the same manner as Foxp3 expression defines nTreg and iTreg cells. A candidate of current interest is Sox13.

Positive and negative thymic selection appear to occur for γδ T cells maturing in the thymus in a fashion resembling the selection that shapes the αβ T cell repertoire, although the molecular details are not understood. At least some γδ T cell clones seem to be positively selected by host cell stress molecules and MHC class Ib molecules expressed on cTECs. Autoreactive γδ T cells are assumed to be negatively selected or anergized in the thymus, but the mechanism remains undefined. Similarly, γδ T cell peripheral tolerance appears to exist, but the underlying mechanisms and ligands are unclear. More complete identification and study of the physiological ligands of γδ T cells should clarify how and when selection and tolerance operate for these cells.

C. NKT Cells

I. Overview

Natural killer T (NKT) cells are T lineage cells that share morphological and functional characteristics with both T cells and NK cells. Low numbers of NKT cells are found virtually everywhere T and NK cells are found: in peripheral blood, spleen, liver, thymus, bone marrow and lymph nodes. Following their activation, NKT cells can immediately commence cytokine secretion without first having to differentiate into effector cells. The rapidity of their response makes NKT cells important players in the very first lines of innate defense against some types of bacterial and viral infections. In addition, many of the cytokines secreted by NKT cells have powerful effects on $\alpha\beta$ T cell differentiation and functions, linking NKT cells to adaptive defense.

> NOTE: Recent work has defined two classes of human and mouse NKT cells, type 1 and type 2, but very little is known about the latter. In this book, we focus on type 1 NKT cells, which some immunologists refer to as "classical" or "invariant" NKT (iNKT) cells. For simplicity, we call these cells "NKT cells."

II. Antigen Recognition and Activation

Rather than interacting with the polymorphic peptide-binding MHC molecules recognized by conventional $\alpha\beta$ T cells, the TCRs of NKT cells recognize glycolipid, glycosphingolipid, or lipid structures presented on the non-polymorphic CD1d molecules expressed by professional and non-professional APCs (including hepatocytes and endothelial cells). As described in Chapter 7, CD1d structurally resembles MHC class I but traffics through the endosomes of the exogenous antigen presentation pathway. The binding groove of the CD1d molecule tethers the lipid tail of a glycolipid antigen, while the carbohydrate head group of the antigen projects out of the groove for recognition by the TCR of the NKT cell.

i) NKT Cell Receptors

In contrast to the vast diversity of TCR$\alpha\beta$ sequences expressed by the $\alpha\beta$ T cell population, NKT cells carry a "semi-invariant" TCR$\alpha\beta$. "Semi-invariant" refers to the fact that the TCRα chain is essentially invariant among the NKT cells in a species, whereas the TCRβ chain can be diversified. For example, in humans, all NKT cells express a TCR in which the TCRα chain contains Vα24 plus Jα18, and the TCRβ chain usually contains Vβ11. In mice, all NKT cells express a TCR in which the TCRα chain contains Vα14 plus Jα18, and the TCRβ chain contains Vβ2, 7 or 8. As is true for all TCR$\alpha\beta$ receptor complexes, intracellular signaling is conveyed by an associated CD3 complex. With respect to coreceptors, human NKT cells are CD4$^+$CD8$^-$, CD4$^-$CD8$^+$ or CD4$^-$CD8$^-$, whereas mouse NKT cells are either CD4$^+$CD8$^-$ or CD4$^-$CD8$^-$.

For both mice and humans, researchers have traditionally identified NKT cells in the laboratory by their ability to recognize the prototypic antigen α-galactosylceramide (α-GalCer). This glycosphingolipid is derived from a marine sponge and is therefore an antigen highly unlikely to be encountered by terrestrial mammals. Nevertheless, in the long absence of the isolation of any mammalian antigen able to activate NKT cells in the laboratory, αGalCer has been routinely used to study NKT cell functions. *In vivo*, experimental administration of α-GalCer to a mouse potently activates its NKT cells, which then help to promote tumor rejection or protect the animal against infection with various pathogens.

Type 2 NKT cells express a more diverse TCR repertoire than type 1 NKT cells, but none of these TCRs recognizes α-GalCer.

ii) NKT Cell Antigens

After considerable struggle, several pathogen- and host-derived glycolipids that act as NKT antigens have now been defined (**Table 11-4**). Gluco- and galactoceramides from *Novosphingobium* and other bacterial species can activate NKT cells directly, as can various bacterial phospholipids. Other pathogens that do not contain glycolipid antigens may still be able to activate NKT cells by inducing the host to produce glycolipid stress molecules that can bind to NKT cell TCRs. The best known host-derived NKT

TABLE 11-4	Examples of Antigens Recognized by Type 1 NKT Cells
Antigen	**Source**
Pathogen-derived	
α-Glucuronosylceramide	*Novosphingobium (Sphingomonas)*
α-Galacturonosylceramide	species
α-Galactosyldiacylglycerol	*Borrelia burgdorferi*
Phosphatidylinositol-mannoside	*Mycobacterium bovis*
α-Glucosyldiacyglycerol	*Streptococcus pneumoniae*
Cholesteryl α-glucoside	*Helicobacter pylori*
Host-derived	
β-Galactosylceramide	
β-Glucosylceramide	
Isoglobotrihexosylceramide (iGb3)	
Disialoganglioside	
Phosphatidylethanolamine	
Phosphatidylcholine	
Phosphatidylinositol	

stress antigen is a lysosomal glycolipid called *iGb3*, which has been shown to stimulate NKT cells *in vitro*. The confirmation of the relevance of such host stress antigens and their production *in vivo* is pending.

iii) NKT Cell Activation

Microbes that contain CD1d-restricted glycolipid antigens which engage the TCR of an NKT cell can activate it directly without the need for costimulation or cytokine receptor engagement (**Fig. 11-7A**). Microbes that do not contain such glycolipid antigens can still activate NKT cells via indirect routes that have been suggested by studies in mice. For example, let us suppose that a mouse has been infected with a bacterium whose cell wall contains LPS. LPS does not bind to the NKT TCR, but the presence of the pathogen may induce host cells to produce a glycolipid stress antigen that can be presented by an APC on CD1d. However, despite engagement of its TCR by the host glycolipid-CD1d complex, the NKT cell may not be activated unless its IL-12 and/or IL-18 receptors are also engaged. Fortunately, the APC presenting the antigen-CD1d complex will frequently

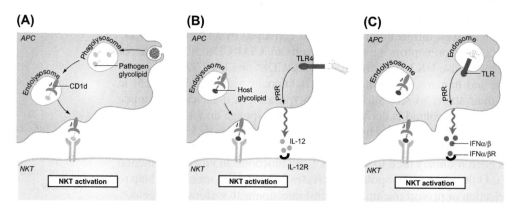

Fig. 11-7
Examples of NKT Cell Activation

(A) Microbes can activate NKT cells via CD1d presentation when microbial glycolipid antigen (blue) is processed and presented to the NKT TCR on CD1d. **(B)** and **(C)** Alternatively, an NKT cell can be activated when self glycolipid antigen (red) presented on CD1d engages the NKT TCR at the same time that microbial PAMPS initiate proinflammatory cytokine signaling by binding to either surface **(B)** or endosomal **(C)** TLRs. *Adapted from Matsuda et al., (2008)* Current Opinion in Immunology 20:358–368.

secrete IL-12 and IL-18 if its surface TLR4 molecules are bound by LPS (**Fig. 11-7B**). Similarly, an intracellular pathogen that attacks an APC may induce the production and CD1d-mediated presentation of a host glycolipid stress antigen. Again, the NKT cell may not be activated by engagement of its TCR alone. However, the attacking pathogen may also supply a PAMP that can engage the endosomal TLR7 or TLR9 molecules of the APC. The APC is triggered to secrete IFNα/β that can bind to the IFN receptors of the NKT cell, providing the additional stimulation needed to push the NKT cell over the activation threshold (**Fig. 11-7C**). In all these scenarios, once activated, the NKT cells then alert many other leukocytes, both innate and adaptive, as to the need for a response.

> NOTE: Like NK cells, NKT cells express inhibitory and activatory NK receptors, including NKG2D and CD94/NKG2A in mice and humans, and certain KIRs in humans. Thus, the activation of NKT cells, like that of NK cells, may be regulated by a balance of activatory and inhibitory signaling. However, NK receptor expression by NKT cells varies with the developmental stage of an NKT cell, its activation status, and the genetic background of the host. It thus remains unclear exactly what NK receptors do for NKT cells. Similarly, although NKT cells express CD40L, ICOS and PD-1, whether these molecules perform the same costimulatory/coinhibitory functions for NKT cells as they do for αβ T cells is unknown.

III. Effector Functions

The characteristics of NKT cell activation are consistent with those of a cell type prominent in induced innate immunity. Mature NKT cells activated in the spleen, liver or bone marrow are stimulated to undergo rapid clonal expansion such that their numbers peak within 3 days of antigen encounter. However, as mentioned previously, these activated NKT cells can immediately carry out their effector functions without the need for differentiation. Thus, as was true for γδ T cells, this accelerated response of NKT cells has prompted many immunologists to consider them to be in a "preactivated" state. Indeed, NKT cells supply timely and effective defense during the interval needed by conventional αβ T cells for proliferation and differentiation into the effectors of the more finely tailored adaptive response. By 9–12 days after first encountering the antigen, NKT cell numbers return to their resting levels, apparently without generating identifiable memory cells.

The diverse effector functions of NKT cells and the broad range of cell types they influence have caused some witty immunologists to dub them the "Swiss Army knife" of the immune system.

The most important *in vivo* effector function of activated NKT cells is cytokine and chemokine secretion (**Fig. 11-8A**). NKT cells carry preformed mRNAs for IL-4 and IFNγ so that massive amounts of these cytokines are produced within 1–2 hours of NKT cell activation. NKT cells can also rapidly synthesize IL-2, IL-10, and IL-17 (among other interleukins), as well as TGFβ and TNF, and a large array of chemokines. This simultaneous production of batteries of pro- and anti-inflammatory cytokines and chemokines means that NKT cells may promote or suppress immune responses in a manner that is hard to predict. The early burst of IL-4 produced by activated NKT cells may contribute to the initiation of Th2 differentiation by nearby activated αβ Th0 cells. Similarly, the IL-2 and IFNγ produced by activated NKT cells may help to prime some NK cells and supply help for αβ Tc activation where αβ Th help might be suboptimal. In contrast, the IL-10 secreted by NKT cells can promote regulatory T cell differentiation, and thus helps to damp down immune reactivity. Conversely, IL-2 produced by NKT cells may modulate regulatory T cell functions.

In addition to influencing αβ T cells, regulatory T cells and NK cells, NKT cells have important effects on γδ T cells, B cells, DCs, macrophages, neutrophils, and eosinophils (**Fig. 11-8B**). For example, upon interaction of its TCR with antigen-CD1d presented by a DC, an NKT cell upregulates its expression of CD40L. Interaction of this CD40L with CD40 on the DC activates the latter, both licensing it for αβ Tc cell activation and spurring it to produce IL-12 to promote αβ Th1 cell differentiation. Exposure to IL-12 also causes the activated NKT cell to selectively increase its secretion of IFNγ. This IFNγ helps to drive the differentiation of αβ Th1 cells and amplifies the NK cell response.

It remains controversial whether, in addition to their cytokine/chemokine secretion, NKT cells can kill target cells by perforin/granzyme-mediated cytotoxicity

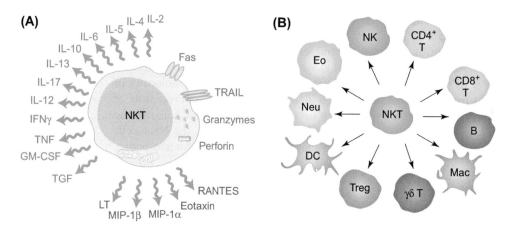

Fig. 11-8
Effector Functions of NKT Cells

Activated NKT cells release numerous cytokines and chemokines **(A)**, which stimulate the effector functions of the indicated innate and adaptive leukocytes **(B)**. TRAIL, "tumor necrosis factor-related apoptosis-inducing ligand," is a death receptor. RANTES, "regulated on activation, normal T cell expressed and secreted," is a chemokine for T cells, eosinophils and basophils; also known as CCL5. Eotaxin is a chemokine for eosinophils; also known as CCL11. MIP-1α, "macrophage inflammatory protein-1α," is a chemokine for neutrophils; also known as CCL3. MIP-1β, "macrophage inflammatory protein-1β," is a chemokine for NK cells and monocytes; also known as CCL4. *Adapted from Matsuda et al., (2008) Current Opinion in Immunology 20:358–368.*

in vivo. In vitro, activated NKT cells can definitely use death receptors or perforin/granzymes to kill the same types of targets as NK cells (infected cells and tumor cells). However, although studies of mouse cancer models have suggested that NKT cells are important for tumor surveillance and rejection, it is unclear whether the large amounts of IFNγ produced by these cells or their cytotoxicity capacity is involved. Some researchers speculate that the ability of NKT cells to promote NK cell activation through cytokine production is more relevant for these *in vivo* effects. Neither has it been determined whether one or a few subsets of mature NKT cells dominate in the periphery and can carry out any NKT effector function as dictated by the surrounding microenvironment, or whether subtly different subtypes of NKT cells exist in specific tissue niches to perform only certain effector functions. For example, liver NKT cells are much better at rejecting tumors in mouse models than are spleen NKT cells. In any case, it is clear that various NKT subpopulations differing in their expression of CD4 and NK activatory and inhibitory receptors exist, and that cytokines in the microenvironment, as well as the nature of the antigen and the APC, can shape the NKT response.

> NOTE: A potential therapeutic application of NKT cells is *ex vivo costimulation*. In this procedure, NKT cells are harvested from patients and cultured in the laboratory in the presence of a glycolipid antigen such as α-GalCer. Once the NKT cells have been sufficiently activated and expanded in numbers, they are infused back into the patient to promote the clearance of tumors or infectious pathogens. Most current trials of *ex vivo* NKT cell costimulation are experimental in nature and focused on the treatment of solid tumors such as head and neck cancers, or viral diseases such as hepatitis B or hepatitis C infection.

As well as their antitumor and antipathogen effects, NKT cells may also play a role in preventing autoimmune disease. Mice that are prone to the development of type 1 diabetes show abnormally low levels of NKT cells, and the NKT cells that are present have functional defects. Interestingly, if diabetes-prone mice are treated in advance with normal NKT cells, the experimental onset of diabetes can be prevented. Whether NKT cell defects occur in humans with type 1 diabetes is currently unclear.

NOTE: Despite their impressive versatility, NKT cells are not always beneficial and can sometimes have deleterious effects on a host. For example, the TCRs of some NKT cells recognize the oxidized lipid antigens deposited in the arteries of patients with *atherosclerosis* ("hardening of the arteries"). The activation of these NKT cells and the ensuing inflammatory response can exacerbate the disease.

IV. Development

NKT cells develop in the thymus from the same NK/T precursors that give rise to $\alpha\beta$ T and $\gamma\delta$ T cells. Cells destined to become NKT cells follow the $\alpha\beta$ T cell developmental pathway until they diverge from them during the DP stage. It is estimated that, in mice, one in a million developing thymocytes becomes committed to the NKT lineage. Because of structural constraints in the *TCRA* locus, the Vα14 gene segment contributing to the TCRα chain of the murine NKT TCR can only start to recombine with the Jα18 gene segment shortly before birth. Developing cells clearly identifiable as being of the NKT lineage therefore appear in the thymus slightly later than developing $\alpha\beta$ T cells. In humans, immature NKT cells are present in the fetal and neonatal thymus and cord blood but may not yet be fully competent to secrete cytokines. Indeed, many NKT cells do not appear to complete their maturation until they reach the periphery at about 3–5 days after birth. This situation stands in contrast to that of NK cells, which are fully mature and functional in fetal life. Interestingly, mature NKT cells appear to accumulate in the peripheral blood of humans as they age.

The reader will recall from Chapter 9 that the positive selection of conventional $\alpha\beta$ T cells in the thymus involves the interaction of TCR$\alpha\beta$ molecules on DP thymocytes with pMHCs on mTECs and cTECs. In contrast, the positive selection of NKT cells involves the interaction of semi-invariant TCR$\alpha\beta$ molecules on certain DP thymocytes with CD1d on neighboring DP thymocytes. It is not known which self glycolipid ligands are presented by CD1d to facilitate positive selection. Negative selection of NKT cells can be experimentally induced by engineering an encounter between developing NKT cells and CD1d bound to an experimental agonist ligand, or even by exposing the NKT cells to high levels of CD1d in the absence of ligand. These findings have yet to be confirmed *in vivo* but, given that NKT cells do have some degree of diversity in their repertoire of semi-invariant TCRs, it is highly likely that NKT cells must undergo negative selection to avoid overt autoreactivity. In any case, once immature NKT cells are positively selected, IL-15 produced by a non-hematopoietic cell type in the thymus drives continued NKT maturation.

We have now come to the end of our study of the basic cell types of the immune system, the modes of antigen recognition of these cells, and their effector functions. These various cell types, some of which are elements of innate immunity, some of which strictly mediate adaptive immunity, and others that bridge the innate and the adaptive responses, combine to distinguish self from non-self, and danger from benign circumstance. In Chapter 12, we explore SALT and MALT, where many of these cell types combine to provide frontline defense at the most common ports of pathogen entry.

Chapter 11 Take-Home Message

- NK, γδ T and NKT cells are considered to bridge the innate and adaptive responses because these cells are closely related to the T lymphocyte lineage, recognize antigens in a more general way than do αβ T cells, have direct and rapid effects on pathogens, and influence B cells and αβ T cells by the cytokines they secrete.

- A balance of signaling by activatory and inhibitory receptors expressed on the NK cell surface controls the activation of NK cells. NK activation occurs when the NK cell encounters a target cell exhibiting "missing self," a relative deficit of self MHC class I.

- NK cells kill virus-infected cells and tumor cells by ADCC or perforin/granzyme-mediated natural cytotoxicity.

- γδ T cells are prominent in the mucosal and cutaneous epithelial layers of the body.

- γδ TCRs bind to non-peptide antigens derived from pathogen proteins and host stress molecules that are either recognized directly, or presented on CD1c or non-classical MHC class Ib molecules.

- Activated γδ T cells generate CTLs that kill infected target cells by perforin/granzyme-mediated cytotoxicity, as well as Th effectors that secrete cytokines influencing αβ T cells and other leukocytes.

- NKT cells bear a semi-invariant TCRαβ that recognizes glycolipid and lipid ligands presented on CD1d by APCs. NKT cells do not have to differentiate before they acquire functional competence.

- Activated NKT cells rapidly secrete large amounts of chemokines and pro-inflammatory and anti-inflammatory cytokines that allow these cells to enhance some immune responses (e.g., antipathogen, antitumor) while suppressing others (e.g., autoreactivity).

- NKT cells have broad effects on other leukocytes and promote Th1, Th2 and Th17 differentiation as well as NK cell functions.

- *In vitro*, activated NKT cells can kill target cells by perforin/granzyme-mediated cytotoxicity. It is unclear whether this capacity is relevant *in vivo*.

Did You Get It? A Self-Test Quiz

Introduction and Section A

1) Why are NK, γδ T and NKT cells considered to bridge the innate and adaptive responses?

2) How is an NK cell similar to a CTL? How is it different?

3) What are the primary functions of NK cells?

4) Name two ways by which NK cells carry out cytolysis of target cells.

5) Do activated NK cells generate separate effector cells? If not, why is this significant?

6) What is the "missing self" model, and why is this term not always accurate? What is "induced self"?

7) Describe the structure and ligand of one member of each of three classes of NK activatory receptors.

8) Why do NK activatory receptors have to associate with a signaling chain?

9) Describe the structures and ligands of one member of each of two classes of NK inhibitory receptors.

10) Describe how a balance of NK inhibitory and activatory receptors controls NK cell activation.

11) What is an ITIM and why is it relevant to NK cell activation?

12) Why are NK cells experts at ADCC?

13) What cytokines are secreted by NK cells?

14) What two molecules are essential for the development of NK cells from NK/T precursors?

15) Why are NK cells considered "self-tolerant"?

16) Distinguish between the terms "education," "un-education," and "re-education" of NK cells.

17) Briefly outline the "arming" and "disarming" models of NK cell tolerance.

Section B

1) Give three ways in which γδ T cells differ from αβ T cells.

2) What kinds of molecules can serve as ligands for γδ TCRs?

3) How does the anatomical distribution of γδ T cells differ from that of αβ T cells?

4) Give two examples of canonical γδ TCRs.

5) Describe three modes of antigen recognition by TCRγδ molecules.

6) Give an example of a stress antigen recognized by γδ T cells.

7) Name two molecules other than CD28 and CD40 that may costimulate γδ T cells.

8) What kinds of effector cells are generated by activated γδ T cells?

9) Describe two ways each in which γδ T cells regulate the innate response and the adaptive response.

10) How do γδ T cells contribute to wound healing?

11) Is memory a feature of γδ T cell responses? If not, what impact does this have?

Did You Get It? A Self-Test Quiz—Continued

12) Describe how the TCR loci rearrange in a thymocyte destined to generate mature γδ T cells.

13) What markers distinguish an NK/T or DN thymocyte destined to become a γδ T cell from an NK/T or DN thymocyte destined to become a αβ T cell?

14) Name two factors thought to influence αβ/γδ T lineage commitment.

15) Why is our knowledge of γδ T cell selection and tolerance incomplete?

Section C

1) What is a semi-invariant TCR?

2) What ligands are recognized by type 1 NKT cells, and how are they presented?

3) Do activated NKT cells need to generate separate effector cells in order to function?

4) Give two ways in which type 1 and type 2 NKT cells differ.

5) Give two examples each of pathogen and host molecules that can serve as NKT cell antigens.

6) How do TLR signaling and host stress NKT antigens work together in the NKT response?

7) What are the primary effector functions of NKT cells *in vivo* and *in vitro*?

8) Give two effects each of NKT-secreted cytokines on the innate and adaptive responses.

9) True or false: NKT cells promote autoimmune disease.

10) What evidence exists for the generation of memory NKT cells?

11) At what stage of thymocyte development do NKT cells arise?

12) Do NKT cells mature in the thymus or the periphery?

13) What is the contribution of NKT cells to neonatal protection?

14) How are positive and negative selection of NKT cells thought to occur?

Can You Extrapolate? Some Conceptual Questions

1) In discussing immune regulatory mechanisms in Chapter 10, we noted that the HLA-A, -B and -C genes are not expressed by the developing fetus. This omission prevents fetal tissues from activating maternal lymphocytes, which might otherwise destroy the growing embryo. Instead, the early human fetus expresses the non-polymorphic HLA-E and HLA-G genes, which encode MHC class I molecules that (unlike HLA-A, -B and -C molecules) are not recognized by maternal lymphocytes but are recognized by maternal NK cells. Referring to the "Missing Self" hypothesis of NK cell target recognition, explain the importance of HLA-E and HLA-G expression in preventing maternal NK reactivity against the fetus.

2) Much early research on T cell populations involved studying mature lymphocytes isolated from the spleen and lymph nodes of mice. How might this have affected progress in our understanding of the biology of γδ T cells?

3) We learned in Chapter 6 that an individual could experience non-responsiveness to a particular pathogen due to an inability of his/her MHC molecules to present any pathogen-derived peptides that could be recognized by any of his/her αβ T cells. How might γδ T cells and NKT cells help to compensate for such a gap in a host's defense?

Would You Like To Read More?

Born, W. K., Zhang, L., Nakayama, M., Jin, N., Chain, J. L., Huang, Y., et al. (2011). Peptide antigens for gamma/delta T cells. *Cellular & Molecular Life Sciences*, 68(14), 2335–2343.

Ciofani, M., & Zuniga-Pflucker, J. C. (2010). Determining gammadelta versus alpha T cell development. *Nature Reviews Immunology*, 10(9), 657–663.

Cruz-Munoz, M. E., & Veillette, A. (2010). Do NK cells always need a license to kill? *Nature Immunology*, 11(4), 279–280.

Darmoise, A., Teneberg, S., Bouzonville, L., Brady, R. O., Beck, M., Kaufmann, S. H., et al. (2010). Lysosomal alpha-galactosidase controls the generation of self lipid antigens for natural killer cells. *Immunity*, 33(2), 216–228.

Elliott, J. M., & Yokoyama, W. M. (2011). Unifying concepts of MHC-dependent natural killer cell education. *Trends in Immunology*, 32(8), 364–372.

Godfrey, D. I., & Rossjohn, J. (2011). New ways to turn on NKT cells. *Journal of Experimental Medicine*, 208(6), 1121–1125.

Haig, N. A., Guan, Z., Li, D., McMichael, A., Raetz, C. R., & Xu, X. N. (2011). Identification of self-lipids presented by CD1c and CD1d proteins. *Journal of Biological Chemistry*, 286(43), 37692–37701.

Held, W., Kijima, M., Angelov, G., & Bessoles, S. (2011). The function of natural killer cells: Education, reminders and some good memories. *Current Opinion in Immunology*, 23(2), 228–233.

Would You Like To Read More? —Continued

Hu, T., Gimferrer, I., & Alberola-Ila, J. (2011). Control of early stages in invariant natural killer T-cell development. *Immunology*, *134*(1), 1–7.

Matsuda, J. L., Mallevaey, T., Scott-Browne, J., & Gapin, L. (2008). CD1d-restricted iNKT cells, the 'Swiss-army knife' of the immune system. *Current Opinion in Immunology*, *20*(3), 358–368.

Meraviglia, S., El Daker, S., Dieli, F., Martini, F., & Martino, A. (2011). Gammadelta T cells cross-link innate and adaptive immunity in mycobacterium tuberculosis infection. *Clinical & Developmental Immunology*, *2011*, 587315.

Moretta, L., Locatelli, F., Pende, D., Sivori, S., Falco, M., Bottino, C., et al. (2011). Human NK receptors: From the molecules to the therapy of high risk leukemias. *FEBS Letters*, *585*(11), 1563–1567.

Orr, M. T., & Lanier, L. L. (2010). Natural killer cell education and tolerance. *Cell*, *142*(6), 847–856.

Salio, M., Silk, J. D., & Cerundolo, V. (2010). Recent advances in processing and presentation of CD1 bound lipid antigens. *Current Opinion in Immunology*, *22*(1), 81–88.

van den Heuvel, M. J., Garg, N., Van Kaer, L., & Haeryfar, S. M. (2011). NKT cell costimulation: Experimental progress and therapeutic promise. *Trends in Molecular Medicine*, *17*(2), 65–77.

Witherden, D. A., & Havran, W. L. (2011). Molecular aspects of epithelial gammadelta T cell regulation. *Trends in Immunology*, *32*(6), 265–271.

Zajonc, D. M., & Kronenberg, M. (2009). Carbohydrate specificity of the recognition of diverse glycolipids by natural killer T cells. *Immunological Reviews*, *230*(1), 188–200.

Zhu, Y., Yao, S., & Chen, L. (2011). Cell surface signaling molecules in the control of immune responses: A tide model. *Immunity*, *34*(4), 466–478.

Mucosal and Cutaneous Immunity

True realism consists in revealing the surprising things which habit keeps covered and prevents us from seeing.

Jean Cocteau

Most human pathogens gain access to the body by penetrating its skin or mucosae. In Chapter 2, we introduced the concepts of the MALT (mucosa-associated lymphoid tissue) and SALT (skin-associated lymphoid tissue). The MALT and SALT are made up of collections of APCs and lymphocytes that are located just under the mucosae or skin, respectively. These collections function as independent arms of the immune system that are responsible for **mucosal immune responses** and **cutaneous immune responses**. Mucosal and cutaneous responses are distinct from the systemic immune responses discussed so far in this book, i.e., those responses most often initiated by lymphocyte–APC interaction in a draining lymph node or the spleen. The majority of mucosal and cutaneous immune responses are mounted locally where antigen is first encountered—that is, without being conveyed to a draining lymph node via its afferent lymphatics. The effector cells that are generated may enter efferent lymphatics or the circulation but then home specifically to mucosal and cutaneous sites to exert their effector functions. However, an accompanying systemic immune response in which effector cells attack the antigen in other locations throughout the body can be induced when antigen-bearing APCs of the MALT or SALT migrate to the local draining lymph node and activate naïve T and B cells in this location.

A. Mucosal Immunity

I. Overview

As introduced in earlier chapters, the mucosae are relatively thin layers of epithelial cells that line a body passage such as the gut, respiratory tract or urogenital tract. In adult humans, the area of all these mucosae exceeds 400 m²! Histologists have identified two categories of mucosae that differ in their epithelial structures. *Type I mucosae*, such as occur in the gut and lungs, are composed of a single layer of columnar epithelial cells. *Type II mucosae* are found where tougher tissue is required, such

MAK: Primer to the Immune Response. http://dx.doi.org/10.1016/B978-0-12-385245-8.00012-1

TABLE 12-1	The "ALTs"		
System Name	**Subsystem**	**Definition**	**Tissue Defended**
MALT		Mucosa-associated lymphoid tissue	Mucosae of body tracts
	GALT	Gut-associated lymphoid tissue	Mucosae of small and large intestines
	NALT	Nasopharynx-associated lymphoid tissue	Mucosae of nose, tonsils, throat
	BALT	Bronchi-associated lymphoid tissue	Mucosae of bronchi and bronchioles in the lungs
SALT		Skin-associated lymphoid tissue	Skin on body surface

as in the mouth, nose and vagina, and are characterized by a top layer of flat, scale-like *squamous epithelial cells*. Both types of mucosae get their name from their capacity to produce **mucus**, a very viscous solution of polysaccharides in water that covers the *apical* (lumen-facing) membrane of an epithelial cell. Mucus contains various antibodies, predominantly SIgA in type I mucosae and IgG in type II mucosae, as well as antimicrobial molecules that help protect the mucosae from pathogen invasion. These antimicrobial molecules include *lysozyme,* which breaks down cell wall components, and *lactoferrin,* which sequesters iron needed for bacterial growth.

As well as by mucus, the mucosae are constantly defended against pathogen assaults by immune responses mounted in the MALT. The MALT includes several subsystems of lymphoid elements associated with each of the body tracts. These subsystems, which differ in structure and cellular composition, are called the GALT, NALT and BALT based on their locations in the gut, nasopharynx and bronchi, respectively (**Table 12-1**).

Scientists working in the MALT field speak of *inductive sites* and *effector sites* (**Fig. 12-1**). An **inductive site** is a specific area in a mucosa where an antigen is encountered and a primary adaptive response is initiated. Inductive sites in the GALT include organized lymphoid structures in the small intestine called **Peyer's patches (PPs)**, the appendix, and diffuse collections of lymphocytes and APCs scattered within and just under the gut epithelium. In the BALT and NALT, the main inductive sites are the bronchial epithelium and the collection of tonsils in the nasopharynx, respectively. A mucosal **effector site** is a specific area in a mucosa to which effector lymphocytes are dispatched. Among the most important mucosal effector sites are the *exocrine glands*, such as the salivary and lacrimal glands. These glands produce protective external secretions (like saliva and tears) that contain antimicrobial molecules and secretory antibodies. In general, the regions of MALT underlying the inductive and effector sites associated with type I mucosae are more densely populated with DCs and lymphocytes than are regions of MALT underlying type II mucosae.

Recent investigations have revealed that, in addition to constitutive inductive sites such as PPs, additional inductive sites can arise in the gut and lungs in response to infection or injury. These new sites are considered to be patches of *induced MALT*, and are a form of tertiary lymphoid tissue (refer to Ch. 2). The appearance of induced MALT (iMALT), sometimes specified as iBALT or iGALT, depends on many of the same molecules involved in generating secondary lymphoid tissues during mammalian embryogenesis. In particular, lymphotoxin (LT) and IL-17, whose local production is induced by inflammation, drive the ectopic development of lymphoid follicles and lymphatic vessels. These cytokines also promote the recruitment of DCs and T and B lymphocytes to the iMALT. The cell types producing this essential LT and IL-17 include Th17 cells and a rare leukocyte subset called *lymphoid tissue inducer cells (LTi cells),* which are also of the CD4+ lymphoid lineage. A good example of MALT induction is the appearance of areas of iBALT in the lungs of mice following the delivery of a vaccine by the intranasal route. Cells within the iBALT are able to mount and sustain immune responses effective against not only the pathogen targeted by the vaccine but also against other microbes reaching the lower respiratory tract. MALT induction may not always be desirable, however, because studies in humans have linked the presence of iBALT to chronic inflammatory diseases of the airway.

After mucosal T and B cells are activated in a given inductive site, they migrate through the lymphatics and blood to various effector sites and complete their

"Ectopic" means "present outside of the normal anatomical location."

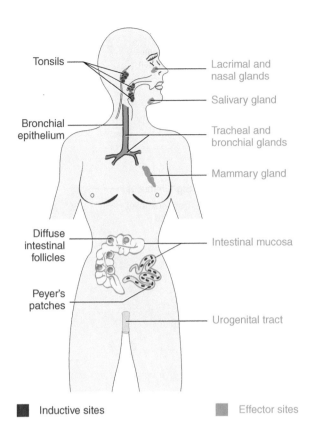

Tonsils

Lacrimal and
nasal glands

Salivary gland

Bronchial
epithelium

Tracheal and
bronchial glands

Mammary gland

Diffuse
intestinal
follicles

Intestinal mucosa

Peyer's
patches

Urogenital tract

■ Inductive sites   ■ Effector sites

Fig. 12-1
*Examples of Mucosal
Inductive and Effector Sites*

An antigen accessing an inductive
site provokes a detectable immune
response in one or more effector
sites, some of which can be quite
remote.

differentiation into Th effectors, CTLs and plasma cells in these locations. Thus, defense at multiple and widely separated mucosal effector sites may occur in response to lymphocyte activation in one inductive site. For example, when an antigen is captured in a PP (inductive site), an antibody response may be detected not only in the intestinal mucosa but also in the urogenital tract and in tissues as remote as the mammary glands (effector sites). This concept is referred to as the **common mucosal immune system** and is discussed later in this chapter.

II. Components of the Gut-Associated Lymphoid Tissue (GALT)

The integration of various components of the GALT provides effective immune defense against ingested pathogens and toxins.

i) Basic Structure

The two major sections of the GALT are the *gut epithelium,* which faces the gut lumen and forms a type I mucosa, and the underlying *lamina propria,* which is composed of loose connective tissue between the *basolateral* (tissue-facing) surface of the epithelium and the muscle layer. In general, the gut epithelium is not flat but rather folded into repeated *crypt* ("cave") and *villus* ("hill") structures (**Fig. 12-2A**). The exceptions are small flat areas of the gut epithelium called *follicle-associated epithelium (FAE),* which contain *M cells* and overlie *intestinal follicles, interfollicular regions,* and *dome regions* (see below).

a) Elements of the Gut Epithelium

The gut is designed to both absorb nutrients from food and repel pathogens and noxious substances. These functions depend on the combined efforts of several different types of gut epithelial cells. *Enterocytes* comprise the majority of the single layer of epithelial cells that line the gut and form the crypts and villi. Enterocytes have a primary function in nutrient absorption, but research in the past few years has highlighted

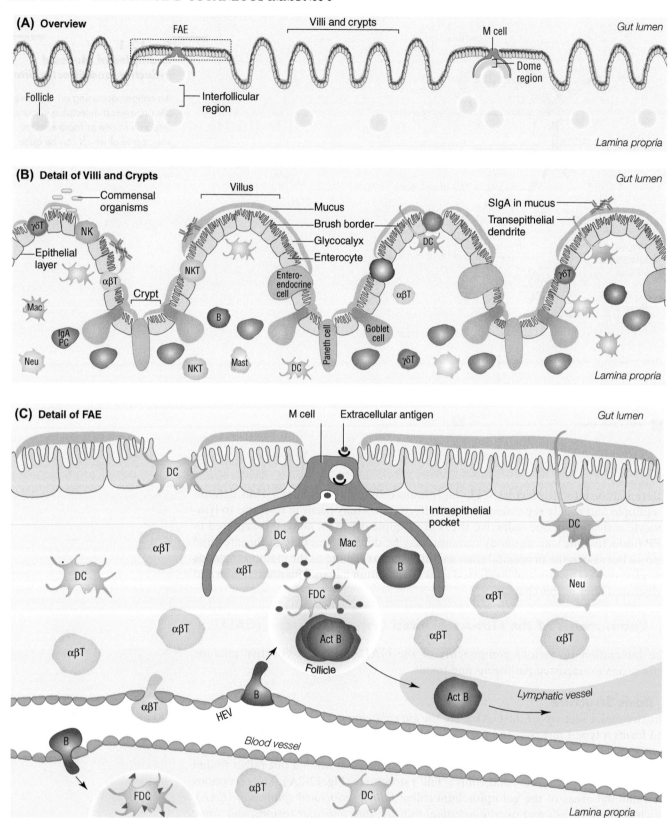

Fig. 12-2
GALT Components

(A) An overview of the lymphoid tissue structures in the gut. Intestinal villi, crypts and regions of follicle-associated epithelium (FAE) are illustrated. **(B)** The detail of villi and crypts shows the various barrier structures and cell types associated with the epithelium and lamina propria of the gut. **(C)** The detail of the FAE illustrates the uptake of antigen by M cells and the subsequent activation of antigen-specific lymphocytes. Once activated, mucosal T and B cells can travel via the lymphatics to multiple mucosal effector sites. *[Adapted from Fagarasan, S. & Honjo, T. (2002). Intestinal IgA synthesis: regulation of front-line defences.* Nature Reviews Immunology *3, 63–72.]*

their importance in integrating myriad internal and external signals and coordinating immune responses in the GALT. Several other epithelial cell types are interspersed among the enterocytes at various sites in the gut mucosa. *Enteroendocrine cells* tend to be localized in the villi, while *Paneth cells* and *goblet cells* tend to be located at the bottom or on the sides of intestinal crypts, respectively. Enteroendocrine cells produce mucus and secrete hormone-like molecules with stimulatory effects on surrounding cells, whereas Paneth cells produce antimicrobial peptides, and goblet cells produce both mucus and antimicrobial molecules.

The apical surfaces of almost all gut epithelial cells are protected against pathogen penetration by several non-induced innate barriers (**Fig. 12-2B**). First of all, any invader has to compete for nutrients and living space with the trillions of **commensal organisms** (also called **microbiota** or **microflora**) that normally live in the intestinal tract. In addition to providing colonization resistance, some of these beneficial microbes secrete toxins that can seriously impede a pathogen. The microbiota also influence the underlying mucus layer, because the more commensal bacteria are present at a given point in the gut, the thicker the mucus in that area.

> NOTE: Many commensal organisms cannot be cultured in the laboratory, making their characterization difficult. Previously, this limitation also obscured their enormous diversity. However, recent advances in gene sequencing techniques have allowed researchers to genetically fingerprint these myriad species without having to culture them first. Our knowledge of the identities of commensal organisms is thus expanding rapidly.

A pathogen that manages to elbow its way past the microbiota and avoids becoming trapped or degraded in the mucus next encounters the *brush border*. The brush border is the name given to the structure formed by the folding of the exterior surfaces of gut epithelial cells into dense *microvilli*. The brush border is coated with the *glycocalyx*, a thick layer of glue-like molecules anchored in the apical membrane. The glycocalyx bears a negative charge that repels many pathogens, and also contains several types of hydrolytic enzymes that degrade microbes and macromolecules.

A pathogen that succeeds in penetrating the non-induced innate barriers protecting the gut mucosa then encounters its cellular defenders. Enterocytes interact with both the microbiota collected on their apical surfaces, and with the innate and adaptive leukocytes positioned below their basolateral surfaces. Enterocytes express TLRs, NLRs and RLRs (among other PRRs; refer to Ch. 3) that can be engaged by components of commensal organisms under steady state conditions, or by PAMPs/DAMPs under stress conditions. The response of the enterocytes to PRR engagement is usually cytokine secretion, and which cytokines are produced depends on the nature of the PRR ligand and factors in the surrounding microenvironment. Additional innate defense is provided by γδ T cells lurking within or near the gut epithelial layer. These cells produce antimicrobial peptides such as *defensins* and *cathelicidins* that kill invading pathogens. Other antimicrobial peptides bar the entry of microbes into any gash in the gut epithelium, and help to shape the species composition of the microbiota. NK cells and NKT cells scattered among the enterocytes also mediate first-line defense using the effector functions described in Chapter 11.

b) Elements of the Lamina Propria

Beneath the basolateral surface of the gut epithelium is the loose connective tissue comprising the **lamina propria** (refer to **Fig. 12-2B**). The lamina propria is home to numerous macrophages and neutrophils as well as low numbers of NKT cells, mast cells and immature DCs. Only small numbers of γδ T cells are present and almost no NK cells. However, memory αβ T cells (both CD4+ and CD8+) and memory B cells are abundant, as are Th17 effector cells. Many of these lymphocytes are diffusely distributed in the lamina propria, whereas others are organized into **intestinal follicles**. Intestinal follicles may occur singly, as can be found scattered along the entire length of the intestine, or in small groups, or in larger groups of 30–40, as occur in inductive

Plate 12-1
Germinal Center in the GALT

Diagram **(A)** and histological slide **(B)** of a germinal center in human ileum. As described in Chapter 5 and illustrated in Figure 5-8, GCs are the site of B cell terminal differentiation into plasma cells and memory B cells. *[Source: Lewis, D. E., Harriman, G. R., & Blutt, S E., Clinical Immunology: Principles and Practice, 3rd edition, Mosby, 2008.]*

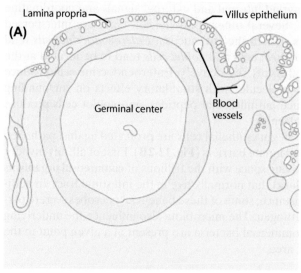

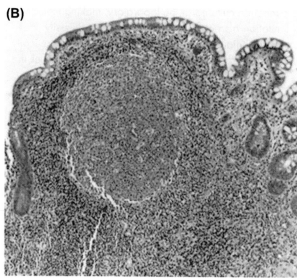

sites such as the PPs and appendix. During a response to antigen in the lamina propria, the activation of B cells can lead to formation of a germinal center (**Plate 12-1**). The constant attack by pathogens on the gut mucosa means that, at any one time, about 10–15% of lamina propria B cells have differentiated into plasma cells, the vast majority of which synthesize IgA. The total secretion of antibody by these IgA-producing plasma cells outstrips the combined output of antibodies synthesized by plasma cells in the spleen, bone marrow and lymph nodes.

ii) Antigen Sampling

a) Follicle-Associated Epithelium (FAE)

Where intestinal follicles are grouped, they lie directly under the small flat sections of gut epithelium that are called **follicle-associated epithelium (FAE)** and are specialized for the capture of gut antigens (**Fig. 12-2C**). The vast majority of cells in the FAE are enterocytes, but 10–20% of FAE cells are large, odd-shaped cells called **M cells** ("membranous" or "microfold" cells) that are experts at antigen transcytosis. The region between the M cell and the underlying intestinal follicles is called the **dome region**. Within the dome are populations of APCs that take delivery of antigen transcytosed by the M cell. The follicles beneath the FAE are separated from one another by **interfollicular regions** that contain high concentrations of mature αβ T cells surrounding an HEV.

The apical surface of an M cell lacks the glycocalyx and thick brush border present on enterocytes, allowing the M cell to easily internalize antigens either by macropinocytosis, clathrin-mediated endocytosis or phagocytosis. M cells possess both FcαRs, which allow them to sample IgA-coated microbes (both commensal and pathogenic), and *glycoprotein-2*, which binds to FimH, a component of the pili often present on bacterial outer membranes. Once the antigens are transcytosed across the cytoplasm of the M cell, they are released into an **intraepithelial pocket** created by the invagination of the M cell's basolateral membrane. The antigens can then be taken up by APCs in the dome. As well as DCs and macrophages, the dome contains CD4+ and CD8+ αβ T cells, resting B cells, and various regulatory T cell subsets. The intestinal follicle itself is made up of a GC containing activated B cells and FDCs.

> NOTE: Unlike the follicles in most secondary lymphoid tissues, the follicles in the "ALTs" lack afferent lymphatics. Thus, the DCs and lymphocytes in the ALT follicles must acquire their antigens by sampling directly from mucosal surfaces. Accordingly, these surfaces have evolved refinements to facilitate this purpose—most notably the M cells in the FAE, which lack epithelial barrier components and actively internalize antigens.

b) GALT DCs

Several DC subsets have been defined in the murine GALT, and different subsets appear to have slightly different functions. Among these cells are two DC subtypes distinguished by their expression of two particular chemokine receptors. DCs that express CX3CR1, which binds to the chemokine fractalkine (CX3CL1), are found throughout the gut, including in the villi, FAE regions, and domes. As well as receive antigen by M cell-mediated transcytosis, CX3CR1+ DCs can participate directly in gut antigen sampling by extending cellular processes, called *transepithelial dendrites,* between epithelial cells into the gut lumen. These dendrites routinely capture soluble food antigens as well as commensal organisms and any pathogens present. In contrast, DCs that express CCR6, a chemokine receptor that binds to the chemokine CCL20, are found only in the domes of the PPs. These DCs acquire antigen solely by M cell–mediated transcytosis.

Under steady state conditions, most GALT DCs function to tolerize naïve T cells with the potential to respond to commensal organisms or food antigens, either by inducing the anergy or death of these cells, or by promoting their differentiation into regulatory T cells. However, when significant inflammation is present due to injury or infection, the PRRs of GALT DCs in the site of attack are engaged by PAMPs/DAMPs under conditions that induce them to mature. Mucosal T cells recognizing the relevant antigens can then be activated. *In vitro*, these mature GALT DCs preferentially direct the differentiation of the activated T cells down the Th17 path. Maturing CCR6+ DCs (but not CX3CR1+ DCs) may also migrate from domes in the GALT to the draining mesenteric lymph node, where the antigens they bear may be used to activate naïve T and B cells that can mount a systemic response.

As well as conventional DCs, the PPs contain plasmacytoid DCs (pDCs; refer to Ch. 7) that contribute to gut defense by secreting IFNα/β.

III. Components of Nasopharynx- and Bronchi-Associated Lymphoid Tissues (NALT and BALT)

The importance of mucosal immune responses in the NALT and BALT is highlighted by the fact that bacterial pneumonia is the third leading cause of mortality worldwide.

i) Basic Structure

The NALT and BALT provide immune defense against dubious substances and pathogens in inhaled air (**Fig. 12-3**). The NALT comprises the nasal submucosal glands, the **tonsils**, and the epithelial layers lining the nasopharynx (upper airway). The components of the BALT include the bronchial submucosal glands; the epithelial layers lining the trachea, the bronchi and the lungs; and the follicles and diffuse collections of lymphocytes underlying the epithelium of the lower airway. Like the gut, the bronchi and

· **Fig. 12-3** ————————■
NALT/BALT Components

Immune defense in the
nasopharyngeal-associated and
bronchi-associated lymphoid tissues
is mediated by various elements
of innate and adaptive immunity,
including non-induced barrier
defenses, commensal organisms,
and the innate and adaptive
leukocytes found below the FAE
regions and within the tonsils.

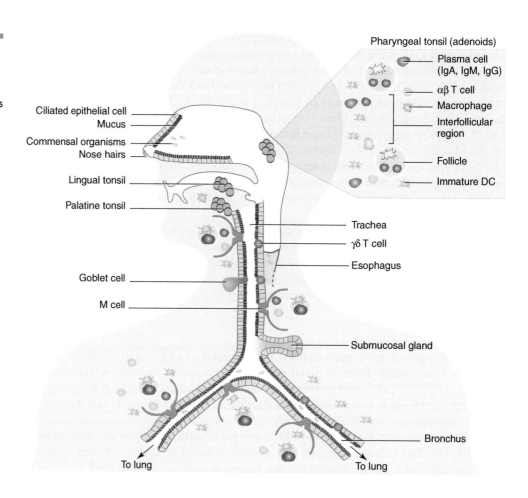

Alterations to the microbiota
or epithelial layer in the airway
can promote the development
of asthma, an allergic
hypersensitivity described
further in Chapter 18.

lungs are covered in a single layer of columnar epithelial cells typical of type I mucosae. In contrast, a type II mucosa protects much of the nasopharynx.

The hairs in the nose constitute the first non-induced innate barrier against harmful entities attempting to breach the NALT. As well, the sweeping movement of cilia on epithelial cells in the upper respiratory tract and coughing in the throat can help to clear invaders attempting to gain a foothold. Nasopharyngeal epithelial cells are not only coated with mucus containing SIgA, lactoferrin, lysozyme and other antimicrobial molecules, but also support a flourishing microbiota equivalent to that in the gut. Under stress conditions, nasopharyngeal epithelial cells themselves can secrete inflammatory cytokines, growth factors and chemokines that summon innate leukocytes to the site of assault. A similar defense strategy prevails in the BALT, where colonization resistance by commensal organisms and sweeping by cilia on epithelial cells in the lower respiratory tract push back invaders. Bronchial and bronchiolar epithelial cells are important sources of IL-5, IL-6, IL-10 and TGFβ, which are cytokines that promote B cell isotype switching to IgA and plasma cell differentiation. In addition, bronchial and bronchiolar epithelial cells secrete the surfactant proteins SP-A and SP-D, which are collectins that can opsonize pathogenic bacteria and promote their phagocytic clearance. The binding of SP-A and SP-D to TLRs of innate leukocytes also stimulates inflammatory cytokine production.

In the human NALT, the tonsils are the most important inductive sites. A tonsil is composed of a network of reticular cells that support lymphoid follicles and interfollicular tissue. The follicles contain prominent GCs populated by FDCs and numerous B cells. The interfollicular regions house large numbers of αβ T cells and professional APCs. Interestingly, the tonsils also appear to serve as effector sites in the NALT, because high concentrations of fully mature plasma cells that express IgG as well as IgM and IgA can be found in these structures. In the BALT, many of the inductive sites are structurally similar to those in the gut. Groups of mucosal follicles that resemble PPs and underlie M cells can be found in the lamina propria underlying the bronchial epithelium.

ii) Antigen Sampling

Antigen uptake in the airway is thought to be easier than that in the GALT because the respiratory tract lacks the harsh degradative enzymes and low pH of the gut. In areas of the respiratory tract lined by a single epithelial layer (type I mucosa), antigen sampling can be carried out by M cells. In the tonsils and regions of the airway covered by the stratified epithelium of a type II mucosa, antigens are conveyed to inductive sites by DCs that can extend their processes between epithelial cells into the tract lumen and acquire inhaled antigens. These DCs commence maturation and return to the underlying diffuse lymphoid tissues in the lamina propria to initiate a mucosal response, or migrate to the more distant draining lymph nodes to initiate a systemic response.

IV. Immune Responses in the GALT, NALT and BALT

Much of our knowledge of mucosal immune responses comes from studies of the GALT. Similar principles are believed to guide most immune reactivity in the NALT and BALT.

i) Contribution of the Microbiota

The unexpectedly prominent role of the microbiota in mucosal immunity has been revealed by investigations of their functions in the GALT. As noted previously, enterocytes are on the frontlines of host interaction with both commensal organisms and pathogens, and so express large batteries of PRRs that bind to molecular patterns expressed by these microbes. The results of this PRR engagement depend on whether the host is at rest or under attack. Under steady state conditions, the engagement of enterocyte TLRs by components of commensal organisms tends to induce the production of molecules that damp down inflammation; adaptive responses to innocuous antigens entering the gut are thus forestalled. For example, engagement of enterocyte TLR2 molecules by microbiota results in the synthesis of TGFβ and TSLP, which are cytokines that influence immature gut DCs to become tolerogenic. These DCs in turn induce the differentiation of iTreg cells producing IL-10, which has a beneficial anti-inflammatory effect that helps to maintain oral tolerance (refer to Ch. 10). Enterocyte TLR2 engagement by commensal organisms in a host at rest is also essential for preserving the physical integrity of the intestinal mucosa. Innocuous encapsulated bacteria whose capsules contain a molecule called *polysaccharide A (PSA)* are able to bind to TLR2, triggering signaling that maintains the tight junctions between enterocytes. In the absence of this steady state TLR2 signaling, these junctions are loosened and the epithelial barrier is compromised.

NOTE: Enterocyte production of TSLP is also essential for normal gut immune homeostasis because this cytokine inhibits IL-12 production by mature DCs. Any T cell that is activated is thus driven toward Th1 differentiation rather than Th1 or Th17 differentiation, resulting in immune responses that are not highly inflammatory. It is likely not a coincidence that gut epithelial cells from some patients with intestinal inflammation show decreased TSLP expression (see Ch. 19).

TLR5, TLR9 and NOD2 are three other PRRs whose roles in normal gut homeostasis have recently been uncovered. TLR5 binds to flagellin, which is expressed by many of the microbiota as well as by pathogens. The engagement of enterocyte TLR5 by the flagellin of a commensal organism leads to the upregulation of anti-apoptotic gene expression in the enterocytes, bolstering their viability. TLR9, which binds to CpG, has been implicated in mucosal barrier homeostasis in a different way from its role as an endosomal PRR of innate leukocytes responding to a pathogen attack. TLR9 is expressed on the basolateral surfaces of enterocytes and Paneth cells. Under steady state conditions, engagement of TLR9 by the CpG of a commensal microbe slipping through the epithelial barrier damps down any incipient inflammation and decreases the chance of release of antimicrobial peptides by the Paneth cell. NOD2 is expressed by Paneth cells and is required not only for their production of antimicrobial peptides

TABLE 12-2	Abnormalities in Intestinal Immunity in Mice with Deficient PRR Signaling
Deficiency For	**Results In**
TLR2	Impaired epithelial cell tight junctions and loss of barrier integrity
TLR5	Spontaneous gut inflammation
TLR9	Increased susceptibility to chemically-induced gut inflammation
NOD2	Abnormal PP development/function; increased penetration of mucosal barrier by microbiota
MyD88	Abnormal PP development; spontaneous intestinal tumors
TRIF*	Increased penetration of mucosal barrier by microflora
STAT3	Spontaneous gut inflammation
SOCS1[†]	Increased susceptibility to chemically-induced gut inflammation

*TRIF (TIR domain-containing adapter-inducing interferon-β) is an adaptor protein for TLR signaling.
[†]SOCS1 (suppressor of cytokine signaling-1) is a protein that inhibits STAT signaling.

but also for normal PP development and regulation of M cell numbers. **Table 12-2** summarizes abnormalities in mice caused by the loss of the normal functions of several PRR signaling molecules involved in normal gut homeostasis.

Recent mouse studies have demonstrated the dramatic effect that the microbiota can have on the building of the GALT itself. Mice raised under germ-free conditions have no microbiota in the gut (**Plate 12-2**) and exhibit decreased GALT, abnormally high numbers of Th2 cells, and abnormally low numbers of Th1 and Th17 cells. The enterocytes in these germ-free animals show decreased expression of the TLRs needed for innate defense, and their intestinal DCs cannot generate transepithelial dendrites. Researchers have used germ-free mice as useful tools for examining the impact on the gut of a controlled encounter with a single commensal organism. For example, scientists exposed germ-free mice to *segmented filamentous bacteria (SFB)*, which are gram-positive, anaerobic bacteria possessing an extended filamentous morphology. These bacteria were able to penetrate the gut mucus of germ-free mice and make contact with enterocytes and other cell types in the gut epithelial layer. After these otherwise germ-free animals were "mono-colonized" with SFB, Th17 cells appeared in the gut, and the animals exhibited increased expression of genes involved in inflammation and antimicrobial defense. When later exposed

Plate 12-2
Segmented Filamentous Bacteria in Mouse Intestine

Scanning electron micrographs of the small intestines of mice that are of the same strain but raised under either germ-free or standard conditions. Top row: The intestinal villi of mice raised under standard conditions (Tac) show normal colonization by commensal segmented filamentous bacteria (SFB), whereas the intestinal villi of mice raised under germ-free conditions (Jax) lack these organisms. Bottom row: Low magnification (left) and high magnification (right) views of an intestinal villus of a Tac mouse, clearly showing the presence of SFB. [*Source: Ivaylo, I., et al. (2009) Cell 139:485–498*].

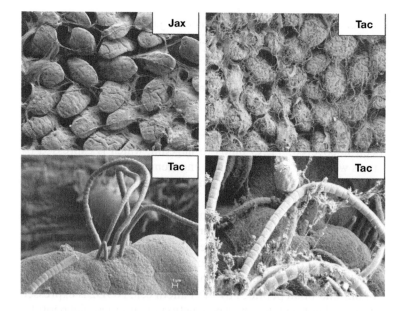

"The Intestinal Epithelial Barrier in the Control of Homeostasis and Immunity" by Rescigno, M. (2011) *Trends in Immunology* 32, 256–264.

Focus on Relevant Research

The intestinal mucosa is perhaps the most complex immunological milieu in the human body. Intestinal epithelial cells and resident innate and adaptive leukocytes are in a constant state of interaction with trillions of gut microbes. In this review article, Maria Rescigno summarizes the host-microbe interactions that are thought to promote healthy intestinal immune homeostasis. Although the gut microbiota comprises hundreds of species of microbes, evidence is presented that points to the key bacteria involved in maintaining the balance between inflammatory and non-inflammatory states. The author reviews bacterial structural components (e.g., LPS, flagellin), molecules released by necrotic host epithelial cells (e.g., ATP), as well as host cell and leukocyte receptors and secreted products (e.g., TLRs, NLRs, cytokines), that are thought to contribute to immune regulation in the gut. The signaling pathways activated by the engagement of these receptors or the actions of these cytokines are also examined. The collected results suggest that overall gut health stems from highly complex and dynamic processes that must, on one hand, limit damaging intestinal inflammation and, on the other hand, be ready to respond with appropriate inflammation and adaptive immunity in the case of assault by a dangerous pathogen. Figure 2 from the article is shown here and illustrates the delicate balance in the gut between the effect of inflammatory triggers (left side) and non-inflammatory triggers (right side); EC, epithelial cell; SAA, serum amyloid A; RA, retinoic acid).

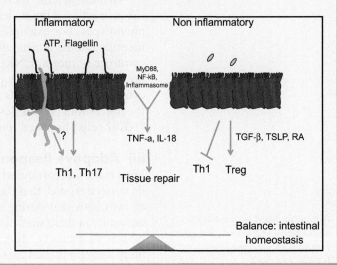

to the mouse intestinal pathogen *Citrobacter rodentium*, the SFB-colonized mice showed enhanced resistance to the invader compared to non-colonized germ-free mice.

ii) Innate Responses to Pathogens

If a host's steady state is lost due to injury or an assault by an aggressive pathogen or toxin, the engagement of enterocyte surface-bound PRRs by PAMPs/DAMPs, as well as that of soluble PRMs such as various collectins and CRP coating the gut epithelial cell surface, activates these cells to produce pro-inflammatory cytokines and chemokines. The lectin pathway of complement activation may also be triggered. Nitric oxide produced by activated phagocytes is particularly effective in inhibiting viral replication and can alter the pH of nearby tissues. Antimicrobial peptide secretion induced by the binding of PAMPs/DAMPs to TLR9 and/or NOD2 of Paneth cells in the gut is also effective against enteric viruses and bacteria. These peptides, which include lactoferrin and the defensins, are directly toxic to many gut viruses and bacteria and block the cell entry of others. In response to the inflammatory distress call, neutrophils and mast cells are recruited from the circulation to the site of infection, and αβ T cells and γδ T cells within or near the gut epithelial layer are activated. The low numbers of NK and NKT cells tucked into the gut epithelium are also stimulated by locally produced cytokines. The engagement of the PRRs and/or antigen receptors of these leukocytes by an attacking pathogen activates them, leading to all the effector actions described in Chapters 3 and 11. In addition, along with their degranulation and phagocytic functions, neutrophils produce IL-18 that combines with DC-produced IL-12 to stimulate NK cells to release IFNγ. IFNγ activates DCs and macrophages and causes them to increase their production of inflammatory cytokines. Neutrophils also capture extracellular bacteria in *neutrophil extracellular traps (NETs)*, which are composed of a chromatin framework incorporating antimicrobial granule proteins.

Interesting research on NETs was described in the Focus on Relevant Research box in Chapter 3.

NOTE: Studies of induced innate responses in the BALT have shown that the large numbers of neutrophils recruited to sites of bacterial attack in the lung can be just as damaging as the pathogens they are trying to clear. Although these innate leukocytes are vital for defense against species such as *Klebsiella pneumoniae* and *Streptococcus pneumoniae*, the accumulation of too many neutrophils can lead to *acute lung injury* or *acute respiratory distress syndrome*. Excessive IL-18 released by activated neutrophils can stimulate other innate leukocytes to release inflammatory cytokines that can damage the delicate lung mucosa. In addition, the destruction of extracellular bacteria by neutrophil NETs may cause the inadvertent release of harmful microbial substances.

In mice, at least, there exist lymphoid lineage cells in the MALT that have many of the properties of innate leukocytes in the peripheral tissues but display distinct surface phenotypes. For example, a subset of NK-like cells dubbed "NCR22 cells" has been identified in the murine intestine. NCR22 cells produce IL-22 and express the natural cytotoxicity receptor NKp46 but do not express any other NK cell markers. Neither do NCR22 cells possess perforin or granzymes, and so cannot kill infected cells by cytotoxicity. NCR22 cells are greatly reduced in number in germ-free mice, implicating the microbiota in their development. A parallel subset of NK-like cells currently called "NK22 cells" has been found in human intestines.

iii) Adaptive Responses to Pathogens

Most mucosae, particularly the type I mucosae in the gut and lungs, are inherently fragile structures, such that they can easily be injured by the products (e.g., TNF and IFNγ) of cells activated during vigorous inflammatory responses. For this reason, mucosal immunity in these sites depends to a large extent on adaptive defense by secretory antibodies, particularly SIgA. These antibodies are key components of mucus and other body secretions such as saliva and tears, and protect against attack by pathogens or toxins without inducing severe inflammation.

a) Rationale for Mucosal SIgA

Secretory IgA has several features and functions that make it ideal for mucosal defense. First, SIgA is constitutively localized in mucus, ensuring that this antibody is ready to neutralize almost any pathogen or toxin trying to make contact with epithelial cells. Second, independent of antigenic specificity, the carbohydrate moieties of SIgA molecules can bind to adhesion molecules expressed by many pathogens, trapping the invaders on the luminal surface. Third, at least in the gut, about half of all SIgA antibodies are unusually cross-reactive, meaning that a broader range of threats can be countered with fewer antibodies. Fourth, SIgA is not an efficient activator of complement, so there is less chance of triggering the cascade and initiating damaging inflammation. Fifth, SIgA is highly resistant to a wide variety of host and microbial proteases, including those in the mammalian gut.

b) Production of Mucosal SIgA

The body's production of SIgA far exceeds that of any other isotype, with about 2–3 grams of SIgA synthesized in the average adult human gut every day. SIgA-producing plasma cells are present at mucosal surfaces even under steady state conditions due to the influence of the microbiota on mucosal B cells. Although the microbiota are mainly confined to the mucosal surface, some do manage to access the follicles in the lamina propria where they promote the ongoing activation of B cells. The constant production of SIgA directed against these commensals keeps their numbers under control, protecting the mucosae and maintaining homeostasis under steady state conditions.

Enterocytes play a key role in SIgA production. When enterocyte PRRs (especially TLR2) are engaged by commensal bacteria, the enterocytes release BAFF, APRIL and IL-10, which are cytokines that cause 80% of activated B cells in the lamina propria to undergo isotype switching to IgA production. Enterocytes also secrete TSLP that causes DCs to secrete more of these cytokines. DCs whose TLR5 molecules

Although SIgM is present in some body secretions, the vast majority of secretory antibodies are SIgA.

Mice unable to produce SIgA due to disruption of the Cα exon or defects in J chain synthesis show increased susceptibility to infection and undesirable changes to the composition of the microbiota.

Humans with PIDs involving impaired IgA production suffer from recurrent respiratory and gastrointestinal infections (see Ch. 15).

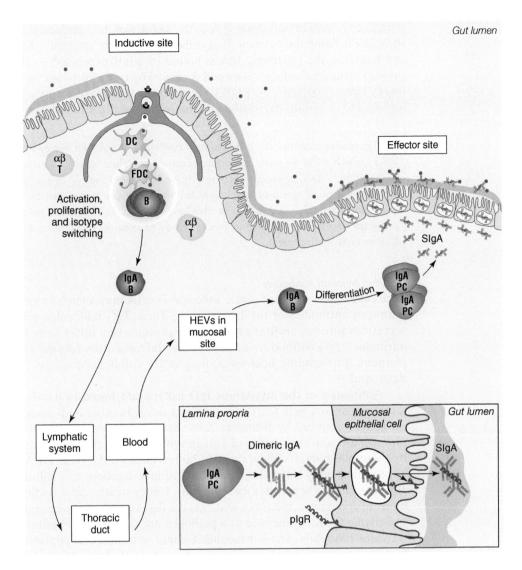

Fig. 12-4
From Antigen Uptake to Secretory IgA Production

Antigen acquired at a gut inductive site activates B cells in the underlying follicle, leading to the generation of B cells that have undergone isotype switching to IgA production. These B cells travel through the lymphatics and blood to various gut effector sites, where they mature into plasma cells that produce secretory IgA. This antibody reaches the mucus covering the apical surface of gut cells through the process of transcytosis (see inset).

are engaged by flagellin-expressing bacteria start to express an enzyme that converts vitamin A into retinoic acid, and retinoic acid promotes the differentiation of mucosal B cells into IgA-producing plasma cells. Other *in vitro* studies have identified a particular subset of DCs that reside in the dome region of the murine PPs (inductive site) and secrete high levels of IL-10. Researchers believe that naïve mucosal Th0 cells that interact with this dome population of DCs are induced to generate three T cell subsets: mucosal fTh cells (refer to Ch. 9) that activate mucosal B cells and produce IL-21; regulatory T cells that secrete TGFβ and IL-10; and mucosal Th2 effectors that synthesize IL-4, IL-5 and IL-10. Th1 responses and IFNγ production are concomitantly suppressed. TGFβ and IL-10 direct antigen-activated B cells in the intestinal follicles to undergo isotype switching to IgA, whereas IL-21 supports TGFβ activity and promotes the migration of the antigen-activated B cells out of the follicles. As illustrated in **Figure 12-4**, these migrating B cells circulate in the lymph and blood, home to mucosal effector sites, and extravasate through local HEVs. In these locations, the activated B cells complete their differentiation into mature IgA-producing plasma cells. The plasma cells secrete polymeric IgA in the vicinity of the basolateral surface of the epithelial cells lining the gut as well as cells making up the exocrine glands. As described in Chapter 4, the polymeric IgA is bound by pIgR expressed on the epithelial cells present in mucosal effector sites and is carried by transcytosis across the cell (**Fig. 12-4, inset**). Upon exocytosis, the pIgR is cleaved such that the secretory component remains attached to the antibody, resulting in the release of SIgA into the mucus.

NOTE: Mouse plasma cells produce only one class of IgA, but humans produce IgA1 and IgA2, which differ in some important properties. While IgA1 production is a feature of both systemic and mucosal responses, IgA2 is present mainly at mucosal surfaces (particularly the distal gut and the urogenital tract) that are populated by large numbers of commensal organisms. Due to a shorter hinge region, IgA2 is more resistant than IgA1 to protease degradation, allowing this antibody to provide sustained protection in a harsh biochemical environment.

c) Other Mucosal Antibodies

If circumstances warrant, some mucosal B cells may switch to production of anti-pathogen antibodies of the IgG isotype. These IgG molecules gain access to body secretions through antibody transporter proteins that differ from pIgR. Unlike IgA antibodies, IgG antibodies induce a robust inflammatory response and activate complement. Surrounding host tissue may suffer collateral damage, but the invader is destroyed.

Antibodies of the mysterious IgD isotype are found in nasal, salivary and lung secretions as well as in tears, but only very rarely in other body secretions. This mucosal IgD is produced by B lineage cells that have the surface phenotype of IgD⁺IgM⁻ plasmablasts and are capable of J chain synthesis. The chemokine receptors expressed by these plasmablasts allow them to be recruited to the NALT/BALT but generally not to other body tracts. Despite intensive investigation, it is still unknown how this antibody enters the respiratory secretions. Under steady state conditions, mucosal IgD binds to commensal microbes and ensures that they do not penetrate too far into the underlying tissue. In the case of a pathogen attack, IgD both binds to the invader and activates basophils. These basophils release antimicrobial peptides and pro-inflammatory cytokines and chemokines, and produce IL-4 and IL-13 that promote B cell switching to IgG.

d) Th1 and Th17 Responses

When a particularly aggressive pathogen attacks a mucosal surface, a Th1 and/or Th17 response may be needed to help clear the invader. With respect to Th1 responses in the murine gut, a distinct DC subset found in the interfollicular areas of the PPs comes into play. Interfollicular DCs that acquire antigen in the presence of PAMPs/DAMPs preferentially produce IL-12, which induces locally activated mucosal T cells to differentiate into Th1 effectors. Some of the interfollicular DCs may also migrate to the draining lymph node and activate naïve T cells in this location, inducing a systemic Th1 response against the pathogen.

As described previously, the presence of the microbiota induces a constitutively high number of Th17 cells in the gut lamina propria. These cells have proven to be critical for mucosal defense against bacterial and fungal infections. Intestinal Th17 cells respond to engagement of their TCRs with the production of IL-17 and IL-21, which cause nearby non-leukocytes to release destructive inflammatory molecules. As well as in the gut, IL-17 signaling is very important in the lung for defense against bacteria. The predominant sources of this cytokine in the NALT/BALT are pulmonary Th17 cells and γδ T cells that have been stimulated by IL-23 produced by local APCs. Th17 cells in the GALT and NALT/BALT also produce IL-22, which combines with IL-17 to induce pro-inflammatory cytokine secretion by intestinal and bronchial epithelial cells.

NOTE: The microbiota not only promote the differentiation of Th17 cells in the GALT but can also modulate their responses. Intestinal epithelial cells that recognize certain commensal microbes are induced to secrete IL-25, a cytokine that inhibits the production of IL-23 by cells in the lamina propria. Without sufficient IL-23, the proliferation of Th17 cells is blocked. This regulatory loop may have evolved to protect the gut from uncontrolled Th17 responses, which can lead to *colitis* (severe intestinal inflammation). More on inflammatory diseases associated with Th17 responses appears in Chapter 19.

e) CTL Responses

If a pathogen avoids being trapped by mucus or bound by secretory antibodies and succeeds in penetrating the mucosa, it may be captured by phagocytic APCs and its antigens processed to activate naïve Tc cells that are resident in mucosal inductive sites. In PPs, Tc cells are found in the interfollicular areas surrounding the intestinal follicles. Nearby mucosal DCs in the dome of the FAE can acquire pathogen antigens from M cells, or from infected epithelial cells that have become necrotic or apoptotic, and then cross-present pMHCs that activate antigen-specific Tc cells. The activated Tc cells then generate antigen-specific CTLs that can migrate to multiple effector sites.

iv) A Common Mucosal Immune System

It has been observed that pathogen invasion at one location in the intestine can lead to the appearance of SIgA not only in the entire gut but also in the respiratory tract, salivary glands, lacrimal glands, ocular tissue, middle ear, and even in the lactating mammary glands. Similarly, antigen introduction intranasally can result in detectable antigen-specific SIgA in the saliva, tonsils, trachea, lung and gut. This disseminated protection has given rise to a concept called the **common mucosal immune system (CMIS)**, in which the migration of mucosal T and B cells from an inductive site through the blood and lymphatics to several effector sites is governed by shared expression of mucosal homing receptors. These receptors, which differ from those expressed by conventional T and B cells activated in the lymph nodes, bind to addressin proteins expressed exclusively in mucosal effector sites. For example, a conventional B cell bearing the α4β1 homing receptor circulates systemically and binds to the addressin VCAM-1 expressed by activated endothelial cells in sites of inflammation. In contrast, a mucosal B cell expresses the α4β7 integrin and ignores sites in peripheral tissues, homing instead to mucosal effector sites where its α4β7 integrin molecules can bind to MAdCAM-1 expressed by endothelial cells in the mucosae. In the case of mucosal T cells, the expression of α4β7 integrin, as well as the chemokine receptor CCR9, is induced specifically by interaction with mucosal DCs in inductive sites. CCR9 binds to the chemokine TECK secreted by mucosal epithelial cells in effector sites.

The immune responses induced at different mucosal effector sites are not of uniform strength, being strongest at those sites closest to the inductive site or in tissues sharing lymph drainage. For example, if the PPs in the GALT are the inductive site, a strong antibody response will be detected in effector sites in the nearby small intestinal mucosa (GALT), but only a weak response will be observed in the more distant tonsils (NALT). Conversely, responses initiated in the NALT induce strong antigen-specific immunity in the respiratory tract but a much weaker response in the gut. Interestingly, there is growing evidence that a response to one pathogen in one inductive site can influence the response to a completely different pathogen in a different inductive site. For example, in individuals infected with *Mycobacterium tuberculosis*, the pathogenic bacterium that causes TB, the immune response in the BALT mounted against this invader appears to be influenced by responses in the GALT against other pathogens. In one study of patients with *M. tuberculosis* infections, researchers found that an individual who also suffered from infection by intestinal helminth worms (which elicit mainly a Th2 response; see Ch. 13) had a weak response against *M. tuberculosis*. However, an individual who was infected with *M. tuberculosis* and co-infected with the intestinal bacterium *Helicobacter pylori* (which elicits mainly a Th1 response) had a much stronger response against *M. tuberculosis*. Variations in co-infection patterns and in the efficiency of the CMIS in different individuals may account for the fact that, globally, over 2 billion people are infected with *Mycobacterium tuberculosis*, but only 20 million suffer from active TB disease.

> NOTE: The concept of a CMIS has led some clinical immunologists to attempt to suppress the immune responses in the airway associated with asthma and other allergies (see Ch. 18) by feeding individuals supplements containing commensal bacterial species (such as *Lactobacillus*). These supplements are sometimes referred to as "probiotics." In animal models, the feeding of such probiotics (and thus the induction of a response in the GALT) appears to attenuate airway inflammation (suppression of a response in the BALT). However, translation of this approach to the clinic and human allergy sufferers has yet to be successful. Research is ongoing into how to use the CMIS to manipulate responses to beneficial microbes and turn them into tools to improve human health.

V. Immune Responses in Other MALT

i) MALT in the Urogenital Tract

Different regions of the female and male urogenital tracts feature different types of mucosae. For example, the vagina and inner foreskin of the penis are protected by tougher type II mucosae, whereas the uterus and the urethra each have a more delicate type I mucosa. In general, the composition of the microbiota associated with urogenital mucosae is quite different from the commensal populations protecting the gut and respiratory mucosae.

The presence of a type II mucosa in the vagina means that it generally lacks the organized lymphoid structures typically present in MALT inductive sites. Intraepithelial DCs and macrophages occur in the cervical and vaginal epithelium, but M cells and lymphoid follicles are absent. Thus, introduction of an antigen into the vagina promotes only weak mucosal responses and no systemic responses. This lack of responsiveness is evolutionarily desirable because an immune response to incoming sperm could block reproduction and thus species survival. In addition, the mucus protecting the cervix is much less acidic than in other locations, allowing sperm to penetrate. The composition of the microbiota in the vagina is also unique, involving species of *Lactobacillus* that maintain an appropriate microenvironment and provide protection against a wide variety of pathogens. Secretory IgA can be found in some upper vaginal secretions, confirming that at least some regions of the vagina are mucosal effector sites.

The penile urethra is both an inductive and effector site. Epithelial cells lining the penile urethra express pIgR, and the lamina propria underlying the urethral mucosa contains many IgM- and IgA-secreting plasma cells. As a result, high concentrations of SIgA and SIgM can be found in the secretions coating the urethral mucosa. IgG-producing plasma cells may also be present in the urethral lamina propria. Mucosal DCs reside among urethral epithelial cells, whereas macrophages and large populations of CD4+ and CD8+ memory T cells are found in the lamina propria.

> NOTE: In mice, delivery of an immunogen via the intranasal route results in measurable effector T and B cell responses (in the form of SIgA) in the genital mucosae, a demonstration of the protection offered by the CMIS. Recent studies have indicated that such a link between nasal inductive sites and genital effector sites may also exist in humans, at least for B cell responses. Thus, it may be possible in the near future to prevent certain sexually transmitted diseases by delivering a vaccine intranasally.

ii) MALT in the Ear

The middle ear cavity is lined with a thin covering of mucus that overlies a type II mucosa made up of several types of secretory, ciliated and non-ciliated epithelial cells. The mucus is constantly conveyed toward the Eustachian tube and nasopharynx by the beating of the cilia on the ciliated epithelial cells. This action helps to keep the middle ear cavity sterile because the tide of organisms trying to access the middle ear from the nasopharynx is continually swept backward. The antimicrobial molecules present in the mucus also take their toll on potential invaders. Comparatively few pathogens access the middle ear from the exterior through the auditory canal, and those that attempt it are usually thwarted by the tough keratinized layer of squamous epithelium covering the exterior side of the tympanic membrane.

There are very few organized lymphoid structures or cells in a healthy middle ear cavity, meaning that it is not an inductive site. However, when infection occurs, the cavity becomes a mucosal effector site, complete with local production of SIgA. Antigen-specific SIgA, antibodies of other Ig isotypes, and pro-inflammatory cytokines such as IL-1, IL-6 and TNF can be detected in the middle ear fluid of infected individuals.

iii) MALT in the Eye

The conjunctiva and anterior ocular surface of the eye are particularly delicate tissues. Thus, as discussed in Chapter 10, the eye is an immune-privileged site in which immune responses and inflammation are generally discouraged. Cells and macromolecules cannot readily pass through the walls of the blood vessels supplying the eye, and the eye is not connected to a draining lymph node. Antigens that do access the eye are captured by intraocular APCs (including specialized subsets of DCs) that have been influenced to promote Th2 development by the high concentrations of TGFβ present in the aqueous humor of the eye. The intraocular APCs migrate from the eye into the blood and thence to the spleen, where lymphocyte activation occurs. Effector Th2 cells home back to the eye where they support non-inflammatory humoral responses.

B. Cutaneous Immunity

The skin protects the body from excessive loss of both water and heat. Cutaneous immunity defends the skin against damage caused by infection or injury. The immune system elements that underlie this defense are collectively known as the **SALT (skin-associated lymphoid tissue)**.

I. Components of the Skin-Associated Lymphoid Tissue (SALT)

As introduced in Chapter 2, the skin is composed of the **epidermis**, the **dermis** and **hypodermis** (**Plate 12-3** and **Fig. 12-5**). The epidermis is not vascularized and is separated from the underlying dermis by the **basement membrane**. The dermis contains both lymphatic and blood vessels. Beneath the dermis is the hypodermis, a fatty layer that provides passive barrier defense and support for the lymphatics and blood vessels. However, the hypodermis functions chiefly as an energy source and so will not be discussed further here.

i) Epidermis
a) Keratin Layer
The tough outer layer of the skin that resists penetration by inert stimuli as well as by microbes is made up of filaments of a resilient, fibrous protein called *keratin*. Keratin is produced by specialized squamous epithelial cells called **keratinocytes**, which comprise over 90% of the cells in the epidermis. The epidermis is divided into several stratified layers, with the outermost representing the oldest keratinocytes. New keratinocytes are constantly being produced from beneath in the lower layers of the epidermis such that the skin eternally renews itself from the inside out. Keratinocytes are generated in organized waves, with the cells in each wave being physically connected by specialized intracellular junctions known as **desmosomes**. The desmosomes ensure the formation of regimented horizontal layers of keratinocytes that divide and migrate upward as a unit. As they age and are pushed up to the skin surface by younger cells beneath them, the older keratinocytes increase their production of keratin fibrils. As the keratinocytes approach the skin surface, their nuclei disintegrate and their lysosomes burst, releasing contents that both

Plate 12-3
The Skin

Histological cross-section of human skin. Leukocytes are stained blue. *[Reproduced by permission of Danny Ghazarian, Princess Margaret Hospital, University Health Network, Toronto.]*

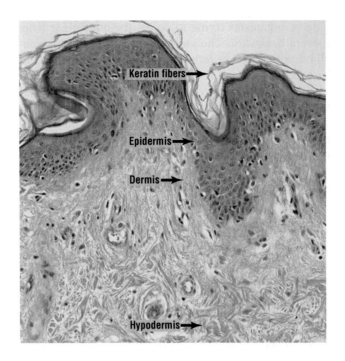

Fig. 12-5
SALT Components

The skin is composed of the epidermis, the basement membrane, the dermis, and the hypodermis. In the epidermis, innate and adaptive leukocytes are interspersed among the various strata of keratinocytes. Langerhans cells (LCs) are DCs that are found only in the epidermis. The dermis contains the skin's blood supply and lymphatics as well as additional concentrations of innate and adaptive leukocytes, dermal DCs, and other cell types such as fibroblasts. In contrast, the hypodermis mainly contains fat cells. Note that B cells are rare in the SALT.

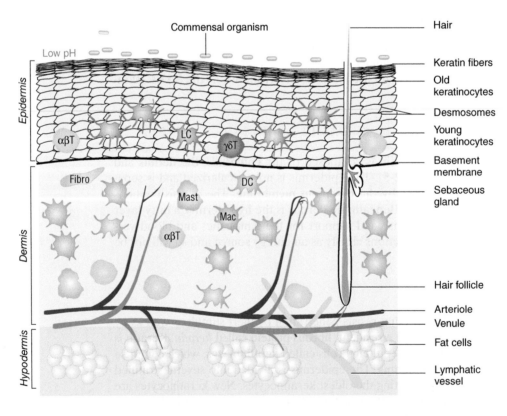

kill the cell and polymerize the keratin into a thick, inanimate layer. The most exterior layers of keratinized shells are eventually lost as flakes of dead skin. This constant turnover of the keratinocytes prevents microbes from becoming entrenched.

In addition to the physical barrier thrown up by the keratin layer, trillions of commensal organisms accumulate on the skin surface. As in the GALT and NALT/BALT, these organisms compete with pathogens for space and nutrients. The skin microbiota also secrete antimicrobial substances to which they themselves are resistant. Among these substances are lipases that break down fats in the skin into free fatty acids, thereby reducing the pH of the skin surface and discouraging pathogen replication. The acidity of the skin is also maintained by sebum produced by *sebaceous glands* originating in the dermis.

b) Lower Epidermis
Below the keratin layer lie the differentiating strata of living keratinocytes. Below the keratinocytes, just above the basement membrane, are relatively small numbers of αβ and γδ epidermal T cells and immature skin DCs known as **Langerhans cells (LCs)**. (B cells are not generally found in skin.) The LCs acquire antigen by infiltrating their long, slender processes between keratinocytes to capture antigens that have penetrated the keratin layer. The survival and activation of the LCs and epidermal T cells depend on low levels of growth factors and cytokines (including TSLP) routinely secreted by keratinocytes. Keratinocytes also constitutively express several TLRs and rapidly release inflammatory cytokines and chemokines in response to TLR engagement. Innate leukocytes responding to these cytokines produce IFNγ, which promotes LC maturation. Because LCs express CD1 proteins as well as MHC class I and II, they are ideal presenters of peptide antigens to αβ epidermal T cells, and glycolipid antigens to γδ epidermal T cells.

It has been estimated that there are ~20 billion T cells in normal human skin, twice as many as in the blood circulation.

ii) Basement Membrane
The youngest keratinocyte layer of the epidermis is separated from the underlying dermis by the basement membrane. The basement membrane is composed of collagen and other molecules produced by epidermal keratinocytes in combination with fibronectin produced by dermal fibroblasts. Because there are no blood vessels in the epidermis, the nutrients required to sustain the keratinocytes must exit the circulation in the dermal blood vessels and diffuse across the basement membrane. Leukocytes that access the epidermis, including the T cell and LC populations, also extravasate from the dermal blood vessels. These migrating cells secrete enzymes that dissolve small regions of the basement membrane, allowing passage of the leukocytes into the epidermis.

iii) Dermis
Compared to the tightly packed cells of the epidermis, the dermis is a much roomier mixture of structural fibers, nerve fibers, blood vessels, lymphatics, hair follicles, and low numbers of cells. Non-leukocytes in the dermis include neurons whose dendrites penetrate the basement membrane, and dermal fibroblasts that synthesize collagen, elastin and hyaluronic acid. Collagen fibers provide structural support for the skin, whereas elastin gives skin its resilience. Hyaluronic acid traps water molecules and keeps the skin taut and moist. The most abundant leukocytes in the dermis are macrophages, mast cells, dermal DCs (which are distinct from LCs), and αβ memory T cells. Leukocytes access the dermis by extravasating through the endothelial cell layer lining the dermal post-capillary venules. While macrophages are scattered throughout the dermis, dermal T cells and mast cells cluster around the arterioles and venules penetrating the dermis. When a pathogen attacks, the endothelial cells of the dermal post-capillary venules secrete chemokines and express vascular addressins that promote the extravasation of additional leukocytes into the dermis.

II. Immune Responses in the SALT

i) Innate Responses to Pathogens
A pathogen that breaches the outer keratinized layer of the skin and penetrates into the epidermis not only provides PAMPs but also causes damage to living keratinocytes,

generating DAMPs. γδ T cells in the lower epidermis immediately recognize these skin stress antigens and are activated upon engagement of their TCRs. Damage to a keratinocyte's membrane also triggers that cell to release IL-1 and TNF. These inflammatory cytokines induce other keratinocytes to produce chemokines, growth factors and additional cytokines within the epidermis. A diffusion gradient is established that penetrates through the basement membrane into the dermis. Dermal fibroblasts and macrophages respond to these molecules with the synthesis of additional inflammatory cytokines and chemokines. Some of these proteins reach the endothelial cells of the dermal blood vessels, promoting local vasodilation and selectin expression. The extravasation into the dermis of additional leukocytes, particularly neutrophils and other granulocytes, is thus facilitated. Once in the dermis, neutrophils produce hydrolases that degrade the basement membrane, allowing leukocytes of all types to penetrate into the epidermis. As neutrophils and macrophages enter the site of assault, they are activated by the presence of pro-inflammatory cytokines and upregulate their PRRs (particularly TLRs) and phagocytic receptors. These cells then internalize any pathogens present and often deploy the respiratory burst to kill them. Phagocytosis may be enhanced if the pathogens are opsonized by complement components that have diffused from local dermal blood vessels into the site of attack. These complement components may also trigger the degranulation of resident mast cells, which release substances that increase local blood vessel dilation and sustain inflammation. NK cells are not normally resident in the skin but can be recruited to sites of injury or infection. Plasmacytoid DCs are present in normal skin in very low numbers but have a significant protective role. Upon injury or infection, these cells are rapidly recruited to the affected site and secrete large amounts of IFNα/β. These cytokines have strong antiviral activity and also promote wound healing.

NOTE: The importance of the cutaneous innate response to defense of the skin has recently been highlighted by the development of a new class of drugs that bind to certain TLRs. Upon engagement by one of these drugs, the TLR initiates signaling that enhances the activation or functions of innate leukocytes. Indeed, members of this class of drugs that stimulate TLR7, TLR3 or TLR9 have shown promise in enhancing host immune responses against skin cancers.

ii) Adaptive Responses to Pathogens

The adaptive response in the SALT is initiated when antigens released from dying keratinocytes and attacking microbes are taken up by LCs. If the cytokine milieu is rich enough, the LCs mature within the epidermis and present peptides derived from stress or pathogen antigens to epidermal αβ Th and Tc cells. Because these αβ T cells are primarily memory cells that have homed to the skin and have taken up residence in the epidermis, the response is almost as rapid as that of the γδ T cells. Within 24 hours, the memory αβ T cells commence differentiation into CTLs and Th effectors. Cells that have internalized the pathogen and thus display its antigenic peptides on MHC class I are destroyed by antigen-specific CTLs. If there are high local concentrations of IL-12 in the site of assault, the differentiating Th effectors are biased toward Th1 and/or Th17 development. In contrast to the gut, Th1 and Th17 responses are well tolerated by the skin due to its inherent toughness. IL-17 produced by Th17 cells combines with IFNγ produced by Th1 cells to upregulate adhesion molecules on keratinocytes and stimulate their production of TNF and other pro-inflammatory cytokines and chemokines. IL-17 also promotes CTL-mediated cytotoxicity against infected keratinocytes.

The interaction of Th0 cells with LCs sometimes results in the differentiation of Th22 cells. As introduced in Chapter 9, Th22 cells are preferentially found in the epidermis and produce (not surprisingly) large amounts of IL-22, as well as fibroblast growth factors. In normal skin, IL-22 promotes keratinocyte proliferation and injury repair, and blocks keratinocyte terminal differentiation. In individuals with inflammatory skin disorders, Th22 cells appear to secrete both IL-22 and TNF. Unlike Th17 cells in the gut and bronchi, which secrete IL-22 in addition to IL-17, Th22 cells in the skin do not secrete IL-17.

Th22 cells have also been found in the lung, where they induce lung epithelial cells to secrete antimicrobial substances.

Following the activation of epidermal T cells, IFNγ and bacterial products that diffuse from the epidermis into the dermis stimulate dermal macrophages and DCs. Activated dermal macrophages produce enzymes that degrade the basement membrane, making it easier for later waves of leukocytes to access the epidermis. The dermal macrophages themselves may cross into the epidermis under the influence of chemokines secreted by LCs, and undertake vigorous phagocytosis in this location. If the response becomes prolonged, the IFNγ secreted by epidermal Th1 cells will drive the macrophages to become hyperactivated and gain enhanced microbicidal powers.

The traffic across the basement membrane can also go the other way. LCs bearing antigen may enter the dermis from the epidermis and access lymphatic channels leading to the local draining lymph node. Naïve T cells in the node may then be activated and generate Th effector cells and CTLs that express the homing receptor *cutaneous lymphocyte antigen (CLA)*. Once these effector cells are released into the blood, their expression of CLA directs them back to inflammatory sites in the dermis.

Should the CTL, Th1 and macrophage responses in a local skin site not prove sufficient to contain a pathogen attack, a switch may be made to Th2 conditions that promote a humoral response. Hyperactivated macrophages that fail to dispose of an invader start to produce more IL-10 than IL-12, and dermal mast cells contribute large amounts of IL-4. In the continuing presence of antigen, Th2 effectors that are generated from epidermal T cells migrate to the draining lymph node and interact with antigen-stimulated B cells in this location. Plasma cells are produced that secrete antibodies into the blood. The blood circulation eventually carries these antibodies back to the dermis and the site of attack.

NOTE: Immune responses in the skin are regulated primarily by nTregs and iTregs (refer to Ch. 9), which are known to express CLA and constitute 5–10% of all skin-resident T cells. In a host at steady state, these regulatory T cells recirculate quietly between the skin and local lymph nodes. During an immune response, these lymphocytes work to control inflammation and limit leukocyte infiltration. For reasons that are not yet clear, numbers of nTregs and iTregs increase in the skin with age, a factor that contributes to the decreased effectiveness of cutaneous immune responses in older individuals. Perhaps not coincidentally, older individuals show a higher incidence of skin cancers (such as melanomas) that correlates with the increased regulatory T cell population.

We have now covered all the basic elements of the immune system and have described their roles in innate and adaptive immune responses. In Chapter 13, we present a discussion of the major classes of pathogens and how each is dealt with by the mammalian immune system.

Chapter 12 Take-Home Message

- The MALT and SALT arms of immunity mount mucosal and cutaneous responses, respectively, to protect the linings of the body tracts and the skin.

- Most mucosal and cutaneous immune responses are initiated locally rather than in a draining lymph node, and the effector cells produced home to mucosal or cutaneous sites.

- Passive anatomical barriers, SIgA-containing mucus, epithelial cells, and innate leukocytes provide initial defense in the MALT.

- M cells in the FAE of inductive sites capture pathogens and convey them to mucosal APCs and T cells residing in the dome covering the B cell-containing lymphoid follicles. DCs may also capture antigen by extending transepithelial dendrites through the epithelium into a tract lumen.

- Effector lymphocytes migrate to multiple mucosal effector sites, including the exocrine glands, and provide coordinated protection at a broad range of mucosae. This phenomenon is called the "common mucosal immune system."

- Immune responses in the MALT are generally biased toward SIgA production to reduce inflammatory damage to fragile mucosae. Aggressive pathogens requiring inflammation for control trigger Th1/Th17 responses.

- The SALT consists of diffuse collections of APCs and T cells in the epidermis and dermis. B cells are not prominent in the SALT.

- Keratinocytes provide a physical barrier and secrete pro-inflammatory cytokines and chemokines that mobilize and activate phagocytes and LCs in the lower epidermis.

- Pathogen antigens captured by LCs either activate epidermal memory T cells or are conveyed to naïve T cells in the local lymph node. Effector T cells home back to the skin site under attack.

- The skin is tough enough to support Th1/Th17 responses, at least initially. However, infection with a persistent pathogen may lead to damage caused by hyperactivated macrophages.

Did You Get It? A Self-Test Quiz

Introduction and Section A.I

1) Can you define these terms? MALT, SALT, GALT, BALT, NALT, apical, ectopic

2) How do mucosal and cutaneous immune responses differ from systemic immune responses? Are they mutually exclusive?

3) Distinguish between type I and type II mucosae.

4) What is mucus?

5) Distinguish between mucosal inductive sites and effector sites. Give two examples of each.

6) Under what conditions might iBALT arise?

Section A.II

1) Can you define these terms? villus, crypt, brush border, glycocalyx, basolateral, FAE, SFB, colitis

2) What are the functions of enterocytes? Paneth cells? Enteroendocrine cells? Goblet cells?

3) Name three non-induced innate barriers and two antimicrobial molecules that protect the mucosae.

4) Describe the structure and cellular components of the lamina propria.

5) What is a Peyer's patch?

6) Describe the structure and function of M cells.

7) Why is an FAE region like an afferent lymphatic?

8) What is a transepithelial dendrite, and what does it do?

9) What is the principal function of GALT DCs in a host at steady state?

Section A.III–IV

1) What body structures are major components of the NALT? The BALT?

2) Describe four elements of non-induced innate defense in the NALT.

3) What is a tonsil, and what does it do?

4) Describe two methods of antigen sampling in the airway.

5) What type of disease might arise from alterations to the normal airway microbiota or epithelial layer?

6) Give two examples of how the microflora contribute to non-induced innate defense in the MALT.

7) Name three immune deficits found in germ-free mice.

8) Describe two ways in which the microbiota can have a tolerogenic influence on immune responses in the GALT.

9) Name four types of leukocytes mediating induced innate defense in the MALT.

10) What are NCR22 cells, and what do they do?

11) Give four reasons why SIgA production is the major means of humoral defense in the MALT.

12) Describe how antigen uptake leads to the appearance of SIgA in the body's secretions.

13) How do the microbiota contribute to SIgA production?

14) Why is IgA2 more resistant than IgA1 to proteases, and why is this important?

15) When and how does IgG contribute to immune defense in the MALT?

16) Give three functions of mucosal IgD.

17) Why is IL-23 important for mucosal immunity? IL-25?

18) In PPs, where are Tc cells usually found?

19) What is the common mucosal immune system, and why might clinicians want to make use of it?

Section A.V

1) The introduction of an antigen into the vagina does not usually promote a systemic immune response. True or false, and why?

2) Give three differences between the vaginal and gut mucosae.

3) Why is the vagina considered a mucosal effector site?

4) Is the penile urethra an inductive site or an effector site?

5) Describe three elements of mucosal immunity in the middle ear.

6) How is the spleen involved in immune defense of the eye?

Section B

1) Can you define these terms? epidermis, dermis, hypodermis, keratin, desmosome, basement membrane.

2) How do dead keratinocytes contribute to immune defense of the skin?

3) Besides keratin, what are two other elements involved in non-induced innate defense of the skin?

4) Describe the localization and function of Langerhans cells.

5) How do the structure and cellular composition of the dermis differ from that of the epidermis?

6) Describe the roles of living keratinocytes and dermal fibroblasts during immune responses in the SALT.

7) Describe the localization and functions of γδ T cells, LCs and pDCs during immune responses in the SALT.

8) Are αβ T cells in the skin of the naïve or memory phenotype?

9) What is the function of Th22 cells in the skin?

10) Give two functions of dermal macrophages during immune responses in the SALT.

11) What is CLA, and why is it important?

12) When and how would a Th1 response in the skin be switched to a Th2 response?

1) How does MALT compare to lymph nodes with respect to the following?

   a) Encapsulation

   b) Afferent lymphatics

   c) Efferent lymphatics

   d) Isotype switching in activated B cells

   e) Ability to serve as tertiary lymphoid tissue

2) Adenovirus is a pathogen that often causes lower respiratory tract disease. An adenovirus vaccine was recently developed that could be given orally in the form of a capsule that would degrade in the small intestine, allowing limited replication of the virus in the gut. How might clinical testing of this vaccine provide evidence for a common mucosal immune system?

3) Adaptive immune responses in the skin are biased toward the Th1/Th17 type and away from the Th2 type. The reverse is true in MALT. Can you briefly account for this difference?

Would You Like To Read More?

Atarashi, K., Umesaki, Y., & Honda, K. (2011). Microbiotal influence on T cell subset development. *Seminars in Immunology, 23*(2), 146–153.

Cerutti, A., Chen, K., & Chorny, A. (2011). Immunoglobulin responses at the mucosal interface. *Annual Review of Immunology, 29,* 273–293.

Di Santo, J. P., Vosshenrich, C. A., & Satoh-Takayama, N. (2010). A 'natural' way to provide innate mucosal immunity. *Current Opinion in Immunology, 22*(4), 435–441.

Forsythe, P. (2011). Probiotics and lung diseases. *Chest, 139*(4), 901–908.

Halle, S., Dujardin, H. C., Bakocevic, N., Fleige, H., Danzer, H., Willenzon, S., et al. (2009). Induced bronchus-associated lymphoid tissue serves as a general priming site for T cells and is maintained by dendritic cells. *Journal of Experimental Medicine, 206*(12), 2593–2601.

Heath, W. R., & Carbone, F. R. (2009). Dendritic cell subsets in primary and secondary T cell responses at body surfaces. *Nature Immunology, 10*(12), 1237–1244.

Hill, D. A., & Artis, D. (2010). Intestinal bacteria and the regulation of immune cell homeostasis. *Annual Review of Immunology, 28,* 623–667.

Iwasaki, A. (2010). Antiviral immune responses in the genital tract: Clues for vaccines. *Nature Reviews Immunology, 10*(10), 699–711.

Jarchum, I., & Pamer, E. G. (2011). Regulation of innate and adaptive immunity by the commensal microbiota. *Current Opinion in Immunology, 23*(3), 353–360.

Kau, A. L., Ahern, P. P., Griffin, N. W., Goodman, A. L., & Gordon, J. I. (2011). Human nutrition, the gut microbiome and the immune system. *Nature, 474*(7351), 327–336.

Littman, D. R., & Rudensky, A. Y. (2010). Th17 and regulatory T cells in mediating and restraining inflammation. *Cell, 140*(6), 845–858.

Maloy, K. J., & Powrie, F. (2011). Intestinal homeostasis and its breakdown in inflammatory bowel disease. *Nature, 474*(7351), 298–306.

Perry, S., Hussain, R., & Parsonnet, J. (2011). The impact of mucosal infections on acquisition and progression of tuberculosis. *Mucosal Immunology, 4*(3), 246–251.

Rangel-Moreno, J., Carragher, D. M., de la Luz Garcia-Hernandez, M., Hwang, J. Y., Kusser, K., Hartson, L., et al. (2011). The development of inducible bronchus-associated lymphoid tissue depends on IL-17. *Nature Immunology, 12*(7), 639–646.

Spits, H., & Di Santo, J. P. (2011). The expanding family of innate lymphoid cells: Regulators and effectors of immunity and tissue remodeling. *Nature Immunology, 12*(1), 21–27.

Strober, W. (2009). The multifaceted influence of the mucosal microflora on mucosal dendritic cell responses. *Immunity, 31*(3), 377–388.

Tezuka, H., & Ohteki, T. (2010). Regulation of intestinal homeostasis by dendritic cells. *Immunological Reviews, 234*(1), 247–258.

Vukmanovic-Stejic, M., Rustin, M. H., Nikolich-Zugich, J., & Akbar, A. N. (2011). Immune responses in the skin in old age. *Current Opinion in Immunology, 23*(4), 525–531.

Weiner, H. L., da Cunha, A. P., Quintana, F., & Wu, H. (2011). Oral tolerance. *Immunological Reviews, 241*(1), 241–259.

Wright, P. F. (2011). Inductive/effector mechanisms for humoral immunity at mucosal sites. *American Journal of Reproductive Immunology, 65*(3), 248–252.

Ziegler, S. F., & Artis, D. (2010). Sensing the outside world: TSLP regulates barrier immunity. *Nature Immunology, 11*(4), 289–293.

PART II:
CLINICAL IMMUNOLOGY

Immunity to Infection

Chapter 13

If one way be better than another, that you may be sure is Nature's way.

Aristotle

Infectious diseases lead to about 14 million human deaths annually. These maladies are caused by six types of pathogens: **extracellular bacteria, intracellular bacteria, viruses, parasites, fungi** and **prions**. Bacteria are microscopic, single-celled, prokaryotic organisms. Extracellular bacteria do not have to enter host cells to reproduce, whereas intracellular bacteria do. Viruses are submicroscopic, acellular particles that consist of a protein coat surrounding an RNA or DNA genome. To propagate, a virus must enter a host cell and exploit its protein synthesis machinery. Parasites are eukaryotic organisms that take advantage of a host for habitat and nutrition at some point in their life cycles. Parasites often damage a host but kill it only slowly. Parasites may be tiny, single-celled *protozoans;* large, multicellular *helminth worms;* or arthropod *ectoparasites*. Fungi are eukaryotic organisms that can exist comfortably outside a host but will invade and colonize that host if given the opportunity. Fungi may be single-celled or multicellular. Prions are infectious proteins that cause neurological disease by altering normal proteins in the brain of the infected host.

Infection occurs when an organism successfully avoids innate defense and colonizes a niche in the body. What follows is a biological "horse race" in which the pathogen tries to replicate and expand its niche, while the immune system tries to eliminate the pathogen (or at least confine it). Only if the replication of the pathogen results in detectable clinical damage does the host experience "disease." Microbial **toxins** released by a pathogen can cause disease even in the absence of widespread colonization. **Immunopathic damage** may occur if host tissues are unintentionally injured by the immune response as it strives to destroy a pathogen. As detailed in Sections A–G that follow, the innate and adaptive effector mechanisms best suited to countering a particular pathogen are determined by the invader's lifestyle and mode of replication.

MAK: Primer to the Immune Response. http://dx.doi.org/10.1016/B978-0-12-385245-8.00013-3

NOTE: Although patients go to hospitals to be cured, about 5% of them will acquire an infection after admission, and about 5% of these individuals will die of these infections. Indeed, in the USA and Europe, *nosocomial* (hospital-acquired) infections are the sixth leading cause of death. In both jurisdictions, billions are spent every year to deal with this problem, even though an estimated one-third of these infections are preventable. Gram-negative bacteria are often the culprits, and pneumonia is the most common life-threatening clinical consequence. Infections of the bloodstream, urinary tract, and surgical sites are also frequent. Individuals who are immunosuppressed are particularly vulnerable to hospital-acquired infections and may succumb to organisms that would otherwise be successfully repelled. Such individuals include cancer patients treated with chemotherapy or radiation, and transplant patients taking medications designed to suppress their immune systems and prevent transplant rejection.

A. General Features of Host–Pathogen Encounters

Most of the mechanisms of innate defense described in detail in Chapter 3 can help the host combat any type of pathogen. The first obstacles encountered by an invader are the intact skin and mucosae. Pathogens are prevented from gaining a firm foothold on the skin by the toughness and routine shedding of the keratin layers protecting the epidermis, and also by having to compete with commensal microorganisms. Pathogens ingested into the gut or inhaled into the respiratory tract are trapped by mucus or succumb to microbicidal molecules in the body secretions or to the low pH and hydrolases of the gut. However, a breach of the skin or mucosae may allow a pathogen access to subepithelial tissues. Barrier penetration may also occur in individuals whose immune systems have been compromised by either disease or therapeutic immunosuppression. These lapses in immune defense may allow **opportunistic pathogens**, which are normally harmless to a healthy individual, to cause disease. In contrast, **invasive pathogens** can enter the body even when surface defenses are intact. Invasive organisms assaulting the mucosae frequently gain access via the M cells of the FAE or by binding to host cell surface molecules that initiate receptor-mediated internalization.

A pathogen that penetrates the skin or mucosae triggers the flooding of the site with acute phase proteins, pro-inflammatory cytokines such as IL-1 and TNF, and complement components. Coating of the pathogen by C3b or MBL facilitates its elimination by the alternative or lectin complement cascades, respectively. At a cellular level, general innate defense is mediated by the PRRs of resident DCs, neutrophils and other granulocytes, macrophages, NK cells, γδ T cells and NKT cells. These PRRs include TLRs, NLRs, RLRs, CLRs, scavenger receptors, and cell-bound collectins, as well as the antigen recognition receptors of NK, NKT and γδ T cells. In addition, soluble collectins in the extracellular matrix that have bound to pathogens or their products may activate complement or stimulate phagocytosis.

NOTE: Recent research has revealed a prominent antipathogen role for the inflammasomes generated following NLR engagement. As described in Chapter 3 and illustrated in Figure 3-5, the engagement of the NLRs NLRP1, NLRP3 or NLRC4 triggers the formation of the NLRP1, NLRP3 or NLRC4 inflammasome, respectively. For example, the NLRP3 inflammasome is activated in response to DAMPs such as host-derived uric acid or cholesterol crystals, or PAMPs derived from extracellular bacteria such as *Streptococcus pneumoniae* and *Yersinia enterocolitica*, or from intracellular bacteria such as *Listeria monocytogenes*, *Bordatella pertussis* and *Legionella pneumophila*. Viral PAMPs (such as those derived from influenza virus), parasite PAMPs (such as those derived from *Schistosoma mansoni* or *Plasmodia falciparum*), or fungal PAMPs (such as those derived from *Candida albicans*) may also induce NLRP3 formation. The PAMPs in these cases include bacterial toxins, viral ssRNA or dsRNA, fungal cell wall components, or parasite egg antigens. NLRC4 inflammasomes also respond to PAMPs from *Salmonella*, *Legionella* or *Pseudomonas* species. NLRP1 inflammasomes are activated by a toxin of *Bacillus anthracis* and have been implicated in combatting some herpesvirus infections.

Recall that FAE is a region of follicle-associated epithelium in a body tract mucosa as described in Chapter 12 and illustrated in Figure 12-2.

Recall that several classes of PRRs expressed by innate leukocytes were illustrated in Figure 3-4 and their features summarized in Table 3-2.

Recall that inflammasome assembly results in the processing and activation of the key pro-inflammatory cytokines IL-1 and IL-18 (see Ch. 3).

In a site of pathogen attack, local leukocytes activated by PRR engagement attempt to eliminate the pathogen or infected cells by clathrin-mediated endocytosis or phagocytosis, secretion of cytotoxic cytokines, or perforin/granzyme-mediated cytotoxicity. These cells also contribute toxic NO and ROIs to the extracellular milieu. Chemokines produced in the ensuing inflammatory response draw neutrophils and other leukocytes from the circulation into the area of infection to assist in the fight. If a pathogen enters the blood, innate defense falls to monocytes and neutrophils in the circulation. Organisms that reach the liver or the spleen are confronted by resident macrophages.

As the innate response proceeds, local DCs that have matured due to exposure to pathogen components become competent to present pathogen-derived pMHCs to naïve T cells, triggering the adaptive response. In many cases, this T cell activation and subsequent B cell activation will take place in inductive sites in the MALT or in the SALT, and the effector cells generated will migrate to effector sites at the body's portals to fight the pathogen. A systemic immune response will soon follow if mature DCs bearing pathogen antigens migrate to lymphoid follicles in the draining lymph node or spleen and activate naïve T and B cells in these locations.

B. Immunity to Extracellular Bacteria

I. Disease Mechanisms

Extracellular bacteria attempting to establish an infection tend to accumulate in interstitial regions in connective tissues; in the lumens of the respiratory, urogenital and gastrointestinal tracts; and in the blood. These organisms often secrete proteins that penetrate or enzymatically cleave components of the mucosal epithelium, allowing access to underlying tissues **(Plate 13-1)**. A wide variety of extracellular bacteria enter the M cells in the FAE, whereas others exploit surface receptors on other host cell types. Examples of diseases caused by infections with extracellular bacteria are given in **Table 13-1**.

Many disease symptoms caused by extracellular bacteria can be attributed to their toxins. **Exotoxins** are toxic proteins actively secreted by either *Gram-positive* or *Gram-negative* bacteria. **Gram-positive bacteria** have cell walls containing a thick layer of peptidoglycan that is colored purple after Gram staining. **Gram-negative bacteria** have cell walls containing a thin layer of peptidoglycan plus LPS that is colored red after Gram staining. **Endotoxins** are the lipid portions of the LPS molecules embedded in the walls of Gram-negative bacteria. Endotoxins are not

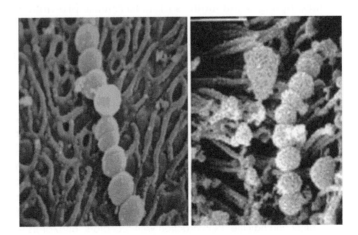

Plate 13-1
Attack by Extracellular Bacteria

Scanning electron micrograph showing *Streptococcus* cells (left panel) attached to the epithelial cells forming the surface of the lingual tonsil (magnification: 10,000x); and (right panel) trapped among the cilia of the nasopharyngeal tonsil (magnification: 7,000x). *[Source: Timoney, J. F., Kumar, P. & Muthupalani, S. (2006) Interaction of Streptococcus equi with the equine nasopharynx. International Congress Series. 1289:267–270.]*

TABLE 13-1	Examples of Extracellular Bacteria and the Diseases They Cause
Pathogen	**Disease**
Bacillus anthracis	Anthrax
Borrelia burgdorferi	Lyme disease
Clostridium botulinum	Botulism
Clostridium tetani	Tetanus
Corynebacterium diphtheriae	Diphtheria
Escherichia coli O157:H7	Hemorrhagic colitis
Helicobacter pylori	Ulcers
Haemophilus influenzae	Bacterial meningitis
Neisseria meningitides	Bacterial meningitis
Neisseria gonorrhoeae	Gonorrhea
Staphylococcus aureus	Food poisoning, toxic shock
Streptococcus pyogenes	Strep throat, flesh-eating disease
Streptococcus pneumoniae	Pneumonia, otitis media
Treponema pallidum	Syphilis
Vibrio cholerae	Cholera
Yersinia enterocolitica	Severe diarrhea
Yersinia pestis	Bubonic plague

secreted but rather are released only when the cell walls of Gram-negative bacteria are damaged. A given Gram-negative bacterial species may supply both exotoxins and endotoxins.

Different exotoxins and endotoxins cause disease by different means and in different locations. For example, infection with *Vibrio cholerae* results in the local release of an exotoxin that binds to gut epithelial cells and induces the severe diarrhea that characterizes cholera. *Clostridium botulinum* produces a neuro-exotoxin that blocks the transmission of nerve impulses to the muscles, resulting in the paralysis characteristic of botulism. In contrast, damage to a host caused by an endotoxin is always immunopathic. The LPS of Gram-negative bacteria activates macrophages and induces them to release pro-inflammatory cytokines, particularly TNF and IL-1. As described in Box 3-2 in Chapter 3, although a little TNF and IL-1 is a good thing, the very high concentrations of these cytokines that are secreted in response to a significant Gram-negative bacterial infection can induce high fever and endotoxic (septic) shock.

NOTE: The ability of an individual to fight off infection can be influenced by the particular allele of a given PRR gene he/she expresses. As defined in Chapter 6, the varying nucleotide sequences of alleles of the same gene are known as *polymorphisms*. **Single nucleotide polymorphisms (SNPs)** are alleles that differ from the cognate gene by one nucleotide. It is estimated that there are ~10 million SNPs in the human genome. An SNP may affect the rate of transcription or translation of the resulting protein product, its amino acid sequence, its stability and half-life, its interaction with receptors, and/or its function. For example, a particular TLR4 SNP is associated with an increased risk of endotoxic shock following infection by Gram-negative bacteria, while a certain TLR2 SNP renders individuals highly susceptible to endotoxic shock following infection by Gram-positive bacteria. A database of defined human SNPs is maintained by the U.S. National Institutes of Health at www.ncbi.nlm.nih.gov/projects/SNP so that medical scientists can easily access this growing resource.

II. Immune Effector Mechanisms

i) Humoral Defense

Because extracellular bacteria cannot routinely "hide" within host cells, antibodies are generally highly effective against these species. Polysaccharides present in bacterial cell walls make perfect Ti antigens for B cell activation (**Fig. 13-1, #1**), while other bacterial components supplying Td antigens induce primarily a Th2 response that provides T help for antibacterial B cells (**#2**). Neutralizing IgM antibodies dominate in the vascular system, while smaller IgG antibodies protect the tissues. These antibodies neutralize bacteria by physically preventing them from attaching to host cell surfaces (**#3**). Even though they do not need to enter host cells for replication, most extracellular bacteria try to adhere to host cells to avoid being swept off or out of the host by skin sloughing or movement of the intestinal contents. Antibodies can also serve as opsonins, coating the bacterium such that it is engulfed by phagocytic leukocytes expressing FcRs (**#4**). Once captured inside the phagocyte, extracellular bacteria are usually very vulnerable to killing via pH changes, defensins, and the ROIs and RNIs associated with the phagosomal respiratory burst. Antibodies made against bacterial exotoxins are called **antitoxins**. Antitoxins neutralize a toxin by preventing it from binding to the cells it would otherwise damage (**#5**). If the toxin is the sole element causing disease in the host, the production of the antitoxin alone will be enough to restore health. For example, human resistance to tetanus or diphtheria relies solely on antitoxins directed against the *Clostridium tetani* exotoxin or *Corynebacterium diphtheriae* exotoxin, respectively.

ii) Complement

All three pathways of complement activation can be brought to bear on extracellular bacteria (**Fig. 13-1, #6**). Antibacterial antibodies of the appropriate isotype (particularly IgM) will bind to complement component C1q to trigger the classical cascade. The alternative pathway can be activated by the binding of C3b to peptidoglycan in Gram-positive bacterial cell walls or LPS in Gram-negative bacterial cell walls. The lectin pathway is activated by the binding of MBL to distinctive sugars arrayed on bacterial cell surfaces. Almost all types of extracellular bacteria can be eliminated by phagocytosis facilitated by the binding of opsonins such as C3b that are produced during complement activation. In addition, bacteria possessing a membrane can be dispatched by MAC-mediated lysis. Complement is particularly crucial for defense against the *Neisseria* group of Gram-negative bacteria.

Mechanisms of complement activation were illustrated in detail in Figure 3-7.

NOTE: There is growing evidence that Th17 responses linked to the apoptosis of infected host cells are important for defense against extracellular bacteria. Infection of a host with a pathogen that preferentially colonizes the mucosae, such as *S. pneumoniae* or *Helicobacter pylori*, often results in the generation of Th17 effector cells that play a key role in bacterial clearance. Some immunologists believe that an infected mucosal cell which undergoes apoptosis in an inflammatory environment furnishes a combination of PAMPs and DAMPs that induce DCs to produce TGFβ and IL-6. Any Th0 cell interacting with such a DC is then directed to undergo Th17 cell differentiation. In contrast, a host cell that undergoes routine apoptosis in the absence of infection produces only DAMPs that induce the DC to secrete TGFβ alone, a situation that favors iTreg cell generation (refer to Ch. 10).

III. Evasion Strategies

Strategies used by extracellular bacteria to evade immune responses are summarized in **Table 13-2**.

i) Interfere with Host PRRs

Some extracellular bacteria are able to manipulate the host's induced innate response by avoiding or modifying the outcome of PRR engagement. For example, *H. pylori* contains modified forms of LPS and flagellin that bind abnormally to TLR4 or TLR5, respectively, and fail to induce proper TLR signaling. *Y. enterocolitica* produces

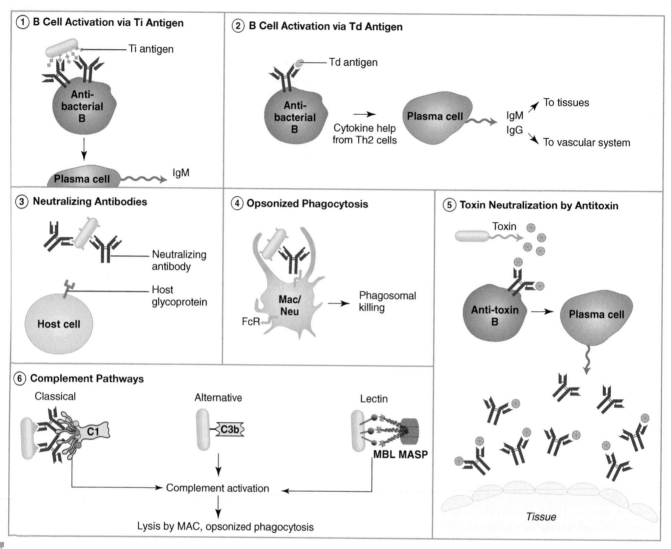

Fig. 13-1
Major Mechanisms of Immune Defense against Extracellular Bacteria

(**1**) Bacterial polysaccharides acting as Ti antigens activate B cells that generate plasma cells producing antibacterial IgM antibodies.
(**2**) Bacterial Td antigens activate additional antibacterial B cells during Th2 responses. (**3**) Neutralizing antibodies recognizing bacterial components block bacterial access to host cell glycoprotein receptors. (**4**) Antibody-bound bacteria are recognized by FcRs on macrophages and neutrophils, which engulf and kill the pathogen. (**5**) Neutralizing antitoxin antibodies bind to bacterial toxin molecules and prevent them from damaging cell surfaces. (**6**) Bacteria bound by antibody plus C1, or C3b, or MBL activate complement.

a protein called the V antigen that binds to TLR2 and stimulates production of the immunosuppressive cytokine IL-10. This IL-10 then inhibits host cell secretion of IFNγ and TNF. Mice lacking TLR2 are thus actually less susceptible than wild type animals to *Y. enterocolitica* infection because their immune systems cannot be co-opted in this way and continue to produce IFNγ and TNF.

ii) Avoid Antibodies

Some extracellular bacteria, such as the *Gonococci*, ensure their adhesion to host tissues by routinely and spontaneously changing the amino acid sequence of the bacterial proteins used to stick to the host cell surface. Neutralizing antibodies directed against the original bacterial protein may not "see" the new version, allowing the bacteria to establish an infection. Other bacteria secrete proteases that cleave antibody proteins and render them non-functional. For example, *Haemophilus influenzae* expresses IgA-specific proteases that degrade sIgA in the blood and SIgA in the mucus.

TABLE 13-2	Evasion of the Immune System by Extracellular Bacteria
Immune System Element Thwarted	**Bacterial Mechanism**
Host PRRs	Produce modified PAMPs Alter PRR signaling and produce IL-10
Antibodies	Alter expression of surface molecules Secrete anti-Ig proteases
Neutrophil recruitment	Secrete a toxin that blocks host cell production of neutrophil chemokines
Phagocytosis	Block binding of phagocyte receptors to bacterial capsule Hide temporarily in non-phagocytes Inject bacterial protein that disrupts phagocyte function
Complement	Prevent C3b binding by lack of suitable surface protein, steric hindrance by surface proteins, C3b degradation Inactivate various steps of complement cascade Capture host RCA proteins Induce host production of antibody isotypes that are poor complement-fixers

iii) Avoid Neutrophils

As we saw in Chapter 3, the chemotaxis and extravasation of neutrophils into a site of pathogen attack are among the first elements of induced innate defense. Studies of *Streptococcus pyogenes* have shown that this bacterium produces a toxin that blocks the production by host cells of the chemokines needed to draw neutrophils into an infected site. While *S. pyogenes* infection of the topmost layer of a tissue causes relatively mild disease (like strep throat), deeper infections can cause *necrotizing fasciitis* (flesh-eating disease), which can be lethal. Histological examination has revealed that this lethality correlates with a deficit in neutrophils in the affected tissue.

iv) Avoid Phagocytosis

The polysaccharide coating of encapsulated bacteria protects them from phagocytosis by conferring a charge on the bacterial surface that inhibits binding to phagocyte receptors. In addition, although C3b may still attach to the bacterial surface, the capsule sterically interferes with the binding of phagocyte receptors to the C3b so that opsonized phagocytosis of the bacterium is much less efficient. Some non-encapsulated extracellular bacteria avoid capture by phagocytes by temporarily entering non-phagocytes such as epithelial cells and fibroblasts. To gain access to these cells, the pathogens may inject bacterial proteins into the host cell that promote either macropinocytosis or cytoskeletal rearrangements facilitating bacterial uptake. Extracellular bacteria may also inject bacterial proteins that have direct antiphagocyte activity. For example, *Y. enterocolitica* injects into macrophages a bacterial phosphatase that binds to certain tyrosine-phosphorylated host proteins required for intracellular signaling and actin reorganization. When the bacterial phosphatase dephosphorylates these host proteins, phagocytosis of the bacterium is blocked.

v) Avoid Complement

Some extracellular bacteria can avoid complement by virtue of their basic structure. For example, *Treponema pallidum*, the organism that causes syphilis, has an outer membrane devoid of transmembrane proteins and so offers almost no place suitable for C3b deposition. Other bacteria have cell wall lipopolysaccharides that contain long, outwardly projecting chains that prevent the MAC from assembling on the bacterial surface. Many extracellular bacteria synthesize substances that inactivate various steps of the complement cascade. For example, group B *Streptococci* contain sialic acid in their cell walls that degrades C3b and blocks alternative complement activation. Other *Streptococci* produce proteins that bind to the normally fluid phase RCA protein Factor H and fix it onto the bacterial surface. In its hijacked site, the recruited Factor H makes any C3b that has attached susceptible to degradation. Certain *Salmonella*

RCA proteins are "regulators of complement activation" that are expressed on host cell surfaces and protect them from complement-mediated destruction (refer to Ch. 3).

species express proteins that interfere with the terminal steps of complement activation, while *Gonococci* and *Meningococci* induce the host to preferentially produce antibody isotypes (such as IgA) that are poor at fixing complement. These "blocking antibodies" compete with complement-fixing antibodies for binding to the bacterial surface, reducing MAC formation. Steric hindrance by blocking antibodies also interferes with C3b deposition.

C. Immunity to Intracellular Bacteria

I. Disease Mechanisms

Like extracellular bacteria, most intracellular bacteria access the host via breaches in the mucosae and skin, but some are introduced directly into the bloodstream by the bites of *vectors* such as ticks, mosquitoes and mites. Once inside the host, intracellular bacteria elude phagocytes, complement and antibodies by moving right inside host cells to reproduce. Epithelial and endothelial cells, hepatocytes and macrophages are popular targets. Because macrophages are mobile, bacteria that infect these cells are quickly disseminated all over the body.

A *vector* is an intermediary organism that introduces the pathogen into the ultimate host.

Intracellular bacteria generally enter host cells by clathrin-mediated endocytosis and are thus first confined to a clathrin-coated vesicle. Some species remain in the vesicle, whereas others escape and take up residence in the cytoplasm. Because of their desire to replicate within a host cell and keep it alive for this purpose, intracellular bacteria are generally not very toxic to the host cell and do not produce tissue-damaging bacterial toxins. However, their intracellular lifestyle makes these organisms difficult to eradicate completely and chronic disease may result. Examples of diseases caused by intracellular bacteria appear in **Table 13-3**.

II. Immune Effector Mechanisms

i) Neutrophils and Macrophages

Early infections by intracellular bacteria are frequently controlled by the defensins secreted by neutrophils because these proteins can destroy the invaders before they can take refuge inside a host cell (**Fig. 13-2, #1**). Those bacteria that escape the defensins and are taken up by neutrophil phagocytosis find themselves, not in a haven for replication, but rather within a phagosome that can kill them via the powerful respiratory burst. Memory Th17 cells have an important role to play here, as the IL-17 they secrete recruits neutrophils to the site of invasion and promotes phagocytosis. Memory Th17 cells also

TABLE 13-3	Examples of Intracellular Bacteria and the Diseases They Cause
Pathogen	**Disease**
Bordetella pertussis	Diphtheria (whooping cough)
Brucella melitensis	High fevers, brucellosis
Chlamydia trachomatis	Eye and genital diseases
Legionella pneumophila	Legionnaire's disease
Listeria monocytogenes	Listeriosis
Mycobacterium leprae	Leprosy
Mycobacterium tuberculosis	Tuberculosis
Mycoplasma pneumoniae	Atypical pneumonia
Rickettsia rickettsii	Rocky Mountain spotted fever
Salmonella typhi	Typhoid fever
Salmonella typhimurium	Food poisoning
Shigella flexneri	Enteric disease

recruit activated macrophages and stimulate both their phagocytic activity and production of the IL-12 needed for Th1 cell differentiation (**#2**). For both neutrophils and macrophages, the killing of phagocytosed bacteria is frequently enhanced by certain host proteins present within the phagolysosomal membrane, and also by host enzymes in the ER or Golgi that regulate the maturation of pathogen-containing phagosomes. These enzymes are greatly upregulated in response to IFNs or LPS. In addition to phagocytosis, macrophages activated by TLR engagement produce pro-inflammatory cytokines that promote NK cell activation and Th1 differentiation (see following sections).

> NOTE: The role of TLRs in defense against intracellular bacteria has been highlighted by the fact that lipoprotein and lipoglycan components of *Mycobacteria* are readily recognized by TLR2 and TLR4. In addition, from a clinical perspective, certain SNPs in TLR5 render individuals highly susceptible to Legionnaire's disease, a form of pneumonia caused by the flagellin-expressing intracellular bacterium *L. pneumophila*.

ii) NK Cells and γδ T Cells

NK cells stimulated by macrophage-derived IL-12 detect infected host cells by their deficit in MHC class I expression (which is typically downregulated by the infection) and destroy them by natural cytotoxicity (**Fig. 13-2, #3**). In addition, activated NK cells secrete copious amounts of IFNγ, which promotes macrophage activation directly and Th1 cell differentiation indirectly. γδ T cells are also important in combatting at least some intracellular infections. Many species of intracellular bacteria (particularly the *Mycobacteria*) release small phosphorylated molecules as they attempt to colonize the host. These metabolites trigger the generation of γδ T cell effectors that either carry out cytolysis or secrete IFNγ (**#4**).

iii) CD8+ T Cells

CTLs are critical for resolving many intracellular bacterial infections. If the bacterium replicates in the cytosol of the infected cell, some of its component proteins enter the endogenous antigen processing pathway and are presented on MHC class I, marking the cell as a target for CTL-mediated destruction (**Fig. 13-2, #5**). These CTLs are generated from pathogen-specific naïve Tc cells that were activated in the draining lymph node. This Tc activation is initiated by DCs that acquired antigens derived from the degradation of a phagocytosed bacterium or a dying host cell, followed by cross-presentation of peptides from these antigens on MHC class I. Interestingly, CTLs rarely use Fas-mediated apoptosis or perforin/granzyme-mediated cytolysis to kill target cells infected with intracellular bacteria, in contrast to their destruction of virus-infected cells (see below). Rather, CTLs eliminate these targets by relying on secreted TNF and IFNγ and/or granule components with direct antimicrobial activity. Accordingly, individuals lacking the IFNγ receptor are highly susceptible to *Mycobacterial* infections.

iv) CD4+ T Cells

CD4+ T cells make a significant contribution to defense against intracellular bacteria (see **Box 13-1**), not only because of the IL-2 they secrete to support Tc differentiation but also because Th1 cells are required for macrophage hyperactivation. It is not unusual for intracellular bacteria phagocytosed by macrophages to be resistant to routine phagosomal killing, and the IFNγ produced by activated Th1 effectors hyperactivates the macrophages such that they gain enhanced microbicidal powers. The sequence of events starts when bacterial antigens either secreted by the bacteria themselves or released by necrotic infected cells are taken up by DCs. In the local lymph node, peptides from these antigens are bound to MHC class II and presented to CD4+ T cells (**Fig. 13-2, #6**). IL-12 produced by macrophages favors the differentiation of Th1 effectors, which supply the intercellular contacts (particularly CD40L) and IFNγ that drive macrophage hyperactivation. A hyperactivated macrophage produces large quantities of ROIs and RNIs that efficiently kill almost all intracellular pathogens. If the bacterium is still resistant, however, a hyperactivated macrophage may go on to participate in formation of a **granuloma** (see later) in order to contain the threat.

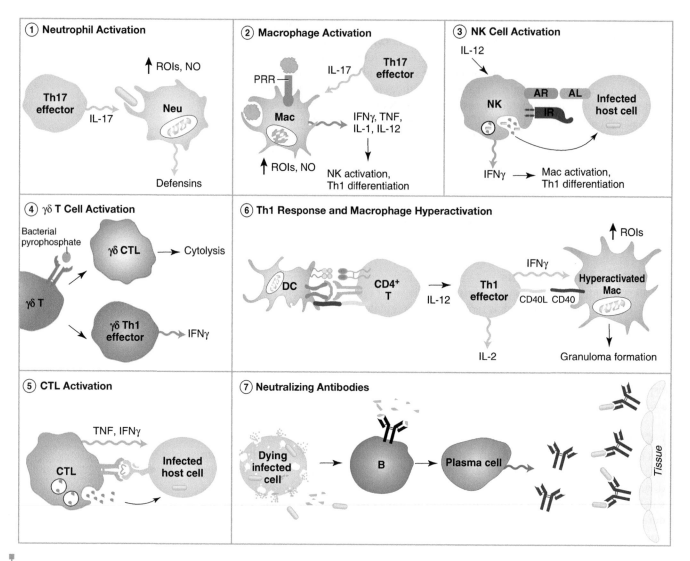

Fig. 13-2
Major Mechanisms of Immune Defense against Intracellular Bacteria

(**1**) Th17 cell-derived IL-17 recruits neutrophils to a site of infection where they capture intracellular bacteria by phagocytosis, kill them via the respiratory burst, and produce antimicrobial peptides. (**2**) IL-17 also recruits macrophages that are then activated by TLR engagement or phagocytosis of intracellular bacteria. These cells initiate phagosomal killing and secrete pro-inflammatory cytokines. (**3**) NK cells activated by IL-12 kill infected host cells by natural cytotoxicity and secrete IFNγ. (**4**) Bacterial phosphorylated metabolites activate $\gamma\delta$T cells. (**5**) CTLs recognizing bacterial peptides presented by an infected host cell kill it by releasing toxic granule contents and/or cytokines. (**6**) Infected DCs present bacterial peptides on MHC class II to CD4+ T cells, which generate Th1 effectors in the presence of IL-12. These Th1 cells supply cytokines for the CTL response and macrophage hyperactivation. (**7**) Bacterial components released from a dying infected cell activate B cells to produce neutralizing antibodies that intercept any bacterium that is temporarily extracellular.

v) Humoral Defense

Antibodies can make an important contribution to host defense against at least some intracellular bacteria. Bacterial components released from a dying infected cell may activate B cells to produce neutralizing antibodies (**Fig. 13-2, #7**). These antibodies may bind to newly arrived bacteria or to bacterial progeny that have been released into the extracellular milieu but have not yet infected a fresh host cell. The antibody-bound bacteria are unable to enter host cells and are eliminated by opsonized phagocytosis or classical complement-mediated lysis, curbing pathogen spread.

vi) Granuloma Formation

When an intracellular pathogen like *Mycobacterium tuberculosis* is able to resist killing by CTLs and hyperactivated macrophages, the body attempts to wall off the pathogen

Box 13-1   Lessons from Leprosy

The importance of the Th1 response to defense against intracellular pathogens is clearly illustrated in human immunity to *Mycobacterium leprae* infection. Individuals who are predisposed to mounting Th2 responses (i.e., their Th cells preferentially secrete IL-4 and IL-10) and are infected with *M. leprae* suffer from a devastating form of leprosy known as *lepromatous leprosy*. The DCs in the epidermis and dermis of these patients exhibit reduced expression of the costimulatory molecule B7, which further compromises the effectiveness of the T cell response. In contrast, individuals who usually mount Th1 responses (i.e., their Th cells preferentially secrete IFNγ) and are infected with *M. leprae* present with *tuberculoid leprosy*, which is generally a less severe form of the disease. The cell-mediated immunity favored by a Th1 response is clearly more effective against this intracellular pathogen than the Th2 response that promotes humoral immunity.

Recent studies have shown that TLRs are intimately involved in the balance between lepromatous and tuberculoid leprosy. DCs and monocytes in lesions of patients with tuberculoid leprosy show strong TLR2 and TLR1 expression, but DCs and monocytes in lesions of patients with lepromatous leprosy do not. A heterodimer of TLR1/TLR2 forms a PRR that recognizes a lipoprotein of *M. leprae*. Engagement of this TLR1/TLR2 complex normally results in leukocyte production of IL-2, IL-12, TNF and IFNγ, which influence nearby DCs to induce Th1 differentiation. In the absence of normal TLR1/TLR2 signaling, however, the leukocytes tend to produce IL-10, which induces DCs to promote Th2 differentiation. Thus, a lower level of the TLR1/TLR2 complex, or a failure in the TLR1/TLR2 signaling pathway, most often favors the development of the more severe form of leprosy.

in a cellular structure called a **granuloma** that forms around the infected macrophages (**Plate 13-2**). The inner layer of a granuloma contains macrophages and CD4+ T cells, whereas the exterior layer is composed of CD8+ T cells. Eventually, the granuloma exterior becomes calcified and fibrotic, and cells in the center undergo necrosis. In some cases, all the pathogens trapped in the dying cells are killed, and the infection is resolved. In other cases, a few pathogens remain viable but dormant within the granuloma, causing it to persist. Granuloma persistence is an overt sign that the disease is becoming chronic. If the granuloma breaks down, the trapped pathogens are released back into the body to resume replication. Should the host be immunosuppressed and unable to marshal the T cells and macrophages necessary to fight this fresh assault, the pathogen may reach the blood. As the bacteria travel in the circulation, they can infect organs throughout the body and even precipitate death.

Cytokines play a critical role in granuloma formation. IL-17 production by Th17 cells is required for Th1 effector recruitment and the stimulation of IL-12 production by macrophages. Sustained IFNγ production by Th1 cells and CTLs is needed to maintain macrophage hyperactivation. TNF production by hyperactivated macrophages is crucial not only for early chemokine synthesis (to recruit leukocytes to the incipient granuloma) but also for aggregating these cells and establishing the "wall" around the invaders. IL-4 and IL-10 secreted by Th2 cells late in an adaptive response control granuloma formation, damping it down as the bacterial threat is contained. Recent work has shown that TLR signaling also influences granuloma formation and thus the

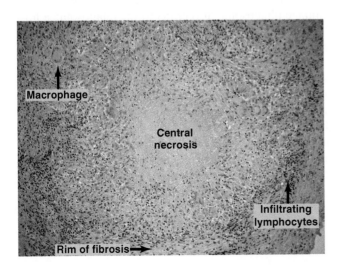

Macrophage

Central necrosis

Infiltrating lymphocytes

Rim of fibrosis

▪ **Plate 13-2**
Granuloma Cross-section

A central zone of necrosis is surrounded by activated macrophages. The rim of fibrosis, infiltrating lymphocytes, and a macrophage are indicated. *[Reproduced by permission of David Hwang, Department of Pathology, University Health Network, Toronto General Hospital.]*

TABLE 13-4	Evasion of the Immune System by Intracellular Bacteria
Immune System Element Thwarted	**Bacterial Mechanism**
Host PRRs	Produce modified PAMPs that inhibit normal signaling
	Produce modified PAMPs that trigger abnormal signaling which inhibits APCs
Phagosomal destruction	Infect a non-phagocyte
	Synthesize molecules blocking lysosomal fusion, phagosomal acidification, ROI/RNI killing
	Recruit host proteins blocking lysosome function
Hyperactivated macrophages	Block expression of host genes needed for macrophage hyperactivation
Antibodies	Spread to new host cell via pseudopod invasion
T cells	Reduce antigen presentation by APCs
	Induce DCs to produce immunosuppressive cytokines

outcome of infections by pathogens normally contained by them. For example, certain SNPs in TLR9 and NOD2 appear to increase susceptibility to *M. tuberculosis* and the TB it causes, while some TLR8 SNPs are linked to TB resistance.

III. Evasion Strategies

Evasion strategies used by intracellular bacteria are summarized in **Table 13-4**.

i) Interfere with Host PRRs

Like certain extracellular bacteria, some intracellular bacteria, including *L. pneumophila*, produce modified forms of TLR ligands such as LPS. These ligands inhibit PRR signaling and block the activation of innate leukocytes. Operating in the opposite way, *M. tuberculosis* produces a small lipoprotein that binds fiercely to host TLR2 and prolongs its signaling. This abnormal signaling inhibits IFNγ production and antigen processing by APCs, downregulating T cell responses and allowing the bacteria to persist.

ii) Avoid Phagosomal Destruction

Some intracellular bacteria avoid phagosomal killing by replicating in non-phagocytic cells. For example, *M. leprae* infects the Schwann cells of the human peripheral nervous system. Other intracellular bacteria deliberately enter phagocytes but then inactivate them or take steps to escape phagosomal killing. For example, *L. monocytogenes* accesses mouse phagocytes via host FcRs and CRs but then synthesizes a protein called *listeriolysin O* (LLO) that induces pore formation in the phagolysosomal membrane. The bacterium escapes through the pore into the relative safety of the cytoplasm. *B. pertussis* expresses a surface receptor that binds to a glycoprotein found primarily on phagocytes, promoting deliberate engulfment of the bacterium. Once inside the phagocyte, *B. pertussis* neutralizes the respiratory burst and inhibits other bactericidal activities, allowing the pathogen to persist within the host cell. When *M. tuberculosis* finds itself being engulfed in a macrophage phagosome, it recruits to the phagosome a host protein called TACO that inhibits the fusion of the phagosome to lysosomes. *M. tuberculosis* also produces NH_4^+, which reverses the acidification of phagolysosomes and promotes fusion with harmless endosomes. In addition, *M. tuberculosis* infection interferes with the expression of host genes needed for microbicidal action and macrophage hyperactivation. As a result of all these measures, *Mycobacteria* can survive within host phagosomes for long periods. Certain *Salmonella* species produce molecules that decrease the recruitment of NADPH oxidase to the phagolysosome, inhibiting ROI and RNI generation. Other intracellular bacteria block phagosomal ROIs and RNIs either by chemically neutralizing them or by synthesizing the enzymes superoxide dismutase and catalase that break down ROIs, RNIs and hydrogen peroxide.

iii) Avoid Antibodies

Some intracellular pathogens avoid the humoral response by moving directly from one host cell to another, giving antibodies no chance to bind. For example, in mice, *L. monocytogenes*

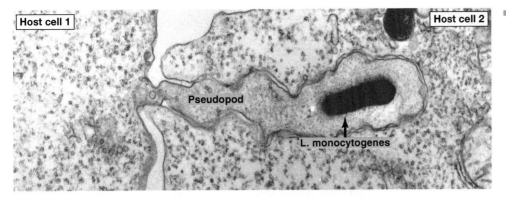

Host cell 1

Host cell 2

Pseudopod

L. monocytogenes

▶ **Plate 13-3**
Pseudopod Invasion

A pseudopod containing a *Listeria monocytogenes* bacterium is extended by an infected cell (Host cell 1) and engulfed by an uninfected neighboring cell (Host cell 2), allowing the bacterium to spread without exposure to host antibody. *[Reproduced by permission.]*

can induce the actin-based formation of a pseudopod that invaginates into a neighboring non-phagocytic cell (**Plate 13-3**). The neighboring cell engulfs the bacterium-containing pseudopod and confines it in a vacuole. The bacterium then uses LLO and phospholipases to break out of the vacuole and enter the cytoplasm of the new cell. Because the bacterium is never exposed in the extracellular milieu, it never becomes an antibody target.

iv) Avoid T Cells

Some intracellular bacteria avoid stimulating T cell responses by interfering with APC function. For example, infection of DCs by *M. tuberculosis* promotes downregulation of the expression of MHC class I, MHC class II and CD1. Antigen presentation to T cells and NKT cells is thus inhibited. *B. pertussis* alters the functions of DCs by inducing them to switch to IL-10 production, thereby suppressing the antipathogen response.

D. Immunity to Viruses

I. Disease Mechanisms

Viruses are stripped-down intracellular pathogens that consist of a nucleic acid genome packaged in a protein coat called a *capsid*. The viral genome may be DNA or RNA, and the capsid may or may not be covered in a membranous structure called an *envelope*. Most viruses enter a host cell by binding to a host surface receptor. Replication of the viral genome and synthesis of viral mRNAs follow, which may be carried out by host or viral enzymes, depending on the virus. However, all viruses lack protein synthesis machinery and rely on the host cell for viral protein translation and progeny virion assembly. Progeny virions released from an infected cell attack neighboring host cells and initiate new replicative cycles that lead to widespread dissemination of the virus. Progeny virions that reach the blood are free to spread systemically. Examples of diseases caused by viruses are given in **Table 13-5**.

Viruses cause disease both directly and indirectly. Viruses frequently kill or at least inactivate host cells, depriving the host of these cells' normal functions such that clinical symptoms appear. As well, the immune response to the viral infection frequently damages host tissues and induces inflammation, causing immunopathic disease. Clinicians classify diseases caused by viruses as either *acute* or *chronic*. When a host is initially infected with a virus, the host experiences **acute disease** in that the illness may be mild or severe (depending on the degree of pathogenicity or **virulence** of the virus) but is only short term in duration. An effective immune response removes the virus completely from the body. However, sometimes viruses are not completely eliminated during the acute infection and remain in the body to establish **persistent infections**. The ongoing low levels of viral replication associated with these persistent infections cause long-term or recurrent illnesses that are considered **chronic diseases**. In some cases, a host will experience no chronic disease symptoms at all if his/her cell-mediated immune response is effective enough to block the assembly of new virus particles. The spread of the virus to fresh host cells is halted, and the virus then persists in the body in an inactive state and does not

TABLE 13-5	Examples of Viruses and the Diseases They Cause
Pathogen	**Disease**
Adenovirus	Acute respiratory infections
Cytomegalovirus (CMV)	Pneumonitis, hepatitis
Ebola virus	Hemorrhagic fever
Epstein–Barr virus (EBV)	Infectious mononucleosis, Burkitt's lymphoma
Hepatitis viruses (HVA, HVB, HVC)	Hepatitis, cirrhosis, liver cancer
Herpes simplex (HSV)	Cold sores
Human immunodeficiency virus (HIV)	Acquired immunodeficiency syndrome (AIDS)
Human papilloma virus (HPV)	Skin warts, genital warts, cervical cancer
Human T cell leukemia virus 1 (HTLV-1)	T cell leukemias and lymphomas
Influenza virus	The "flu"
Kaposi's sarcoma herpes virus (KSHV)	Kaposi's sarcoma
Measles virus (MV)	Measles
Poliovirus	Poliomyelitis, post-polio fatigue
Polyoma virus	Infections of respiratory system, kidney, brain
Rabies virus	Rabies
Rhinovirus	Common cold
SARS (severe acute respiratory syndrome) virus	Severe acute respiratory syndrome
Vaccinia virus	Asymptomatic in most healthy humans, or mild rash and fever
Varicella zoster virus (VZV)	Chicken pox, shingles
Variola virus	Smallpox
West Nile virus (WNV)	Flu-like illness, fatigue, encephalitis

replicate. However, if the host's cell-mediated response weakens due to aging or immunosuppression, the latent virus reactivates, replicates and again causes acute disease. For example, the reactivation of latent varicella zoster virus (VZV), which causes chicken pox in young children, precipitates the painful adult skin condition known as shingles.

II. Immune Effector Mechanisms

i) Interferons and the Antiviral State

Production of the multifunctional cytokines IFNα, IFNβ and IFNγ is one of the earliest innate responses induced by viral infections. IFNα and IFNβ are secreted primarily by host cells infected with a virus, whereas IFNγ is initially produced by activated macrophages and NK cells and later on by activated Th1 cells. Any one of these IFNs can initiate a series of metabolic and enzymatic events in an uninfected host cell that results in it adopting an **antiviral state** (**Fig. 13-3, #1**). A host cell in the antiviral state can take enzymatic action to prevent an attacking virus from invading or starting to replicate. Both the transcription and translation of viral mRNAs and proteins are inhibited.

Another key source of IFNα and IFNβ is the plasmacytoid DC population described in Chapter 7. These cells are specialized in the use of endosomal PRRs, particularly TLR9, to sense viral RNA and DNA. Activated pDCs then respond rapidly with vigorous production of IFNα and IFNβ. For example, the herpes simplex viruses HSV1 and HSV2 have DNA genomes that are unusually rich in the CpG motif, which is a ligand of TLR9. Accordingly, pDCs are vital for immune defense against these viruses.

ii) NK Cells

Although CTLs are the prime mediators of the cell-mediated immunity needed to eliminate viruses (see later), there is often a 4–6-day delay before these cells can expand to sufficient numbers to complete the task. Where a virus causes downregulation of

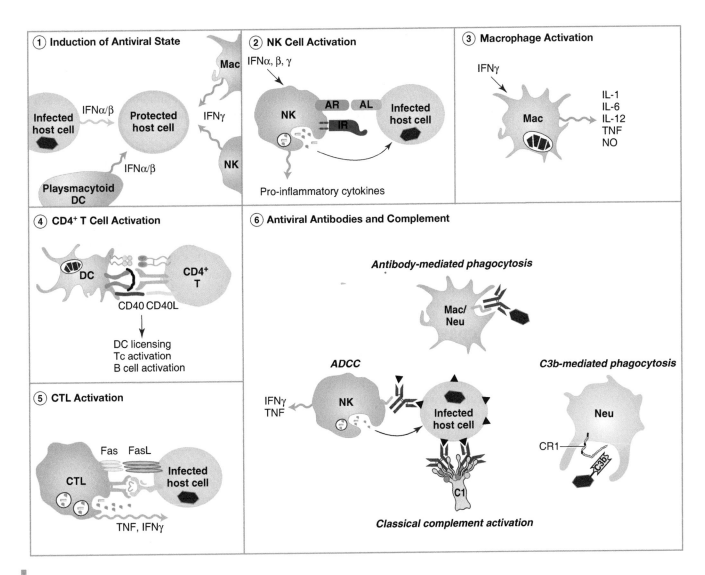

Fig. 13-3
Major Mechanisms of Immune Defense against Viruses

(1) IFNα/β secreted by infected host cells and pDCs, and IFNγ secreted by activated macrophages and NK cells, cause uninfected host cells to adopt an antiviral state. **(2)** Activated NK cells secrete cytokines and kill infected host cells that fail to express sufficient peptide-MHC class I. **(3)** Activated macrophages efficiently capture and kill viruses, and produce NO and cytotoxic cytokines. **(4)** Infected DCs, or those that have captured virions or viral products, activate CD4+ T cells, which reciprocally license DCs for Tc cell activation. **(5)** Antiviral CTLs kill virus-infected host cells by Fas killing, cytotoxic cytokines, or perforin/granzyme-mediated cytotoxicity. **(6)** Antibodies bound to a viral antigen on the surface of an infected host cell may engage FcRs on an NK cell, macrophage, or neutrophil and trigger ADCC or opsonized phagocytosis. If the bound antibody binds to C1, classical complement activation can lead to MAC-mediated destruction of the infected cell. Virus particles bound to free C3b may bind to CR1 and be taken up by opsonized phagocytosis.

MHC class I on the host cell surface, direct cytolysis of infected cells by NK cells (via natural cytotoxicity) and NK production of inflammatory cytokines can supply early defense (**Fig. 13-2, #2**). Indeed, individuals whose NK cells are not fully functional show increased susceptibility to virus infection, especially by herpesviruses. Natural cytotoxicity and inflammatory cytokine production by NK cells are stimulated by all three IFNs. NK cells are also important mediators of antiviral ADCC. The upregulation of FcR expression on both NK cells and macrophages is stimulated by IFNγ.

iii) Macrophages
Whole virions or their components may be taken into a macrophage by clathrin-mediated endocytosis or phagocytosis. Numerous viral components, including viral DNA,

dsRNA, ssRNA, envelope proteins and surface glycoproteins, serve as PAMPs that can bind to macrophage PRRs such as TLR2, TLR3, TLR4, TLR7, TLR8 and TLR9. Indeed, a TLR2 SNP that abrogates TLR signaling increases susceptibility to cytomegalovirus (CMV) infection. Macrophages activated by PRR engagement during the course of a virus infection produce copious amounts of pro-inflammatory cytokines such as IL-12 and TNF (**Fig. 13-3, #3**). The presence of IFNγ in the milieu greatly enhances this function and also allows the macrophage to express the iNOS enzyme that generates NO. This NO facilitates macrophage production of ROIs and RNIs that will aid in killing phagocytosed viruses. Macrophages can also eliminate viruses via ADCC.

> Most DCs express the same array of TLRs as macrophages and are also activated by viral PAMPs.

iv) CD4+ T Cells

TLR-stimulated DCs readily process viral proteins via the exogenous pathway and display viral peptides on MHC class II to activate naïve CD4+ T cells (**Fig. 13-3, #4**). Th cells are important for defense against most viruses because these cells both license DCs and supply IL-2 for naïve CD8+ Tc cell activation. Interaction of DCs with Th effector cells reciprocally spurs the production by the DC of pro-inflammatory mediators that recruit additional innate and adaptive leukocytes. Th cells also provide the CD40L-mediated costimulation and cytokines required for B cells to mount antibody responses to viral Td antigens.

v) CD8+ T Cells

CTLs are crucial for immune defense against most viruses. Because these pathogens replicate intracellularly, viral antigens are displayed on MHC class I on infected host cell surfaces and mark these cells as CTL targets. The effector CTLs generated from Tc cells activated in the draining lymph node return to the site of infection and kill the virus-infected cells via perforin/granzyme-mediated cytotoxicity, Fas-mediated apoptosis, or TNF and/or IFNγ secretion (**Fig. 13-3, #5**).

NOTE: Recent studies have identified small populations of CD4+ (as opposed to CD8+) cytotoxic T cells as being important for fighting persistent virus infections. In these cases, the naïve CD4+ cells within the infected host gradually lose the ability to generate Th effectors and instead generate progeny that acquire cytotoxic properties, allowing them to kill infected host cells via perforin/granzyme-mediated cytotoxicity or Fas killing. These cells are the subject of much ongoing research.

vi) Humoral Defense

Because a virus is an intracellular pathogen, it is often out of the reach of antibodies during the primary adaptive response. Nevertheless, naïve B cells may recognize viral components displayed on the surface of an infected host cell or may encounter progeny virions as they are released from an infected cell. With the appropriate T cell help, these B cells are activated and generate plasma cells and memory B cells that are usually vital for complete resolution of the infection. Late in the primary response, neutralizing antibodies are released into the circulation and block further spread of the virus. As well, in a subsequent attack, the virus will have a harder time infecting the host because the circulating neutralizing antibodies rapidly bind to the virus and bar its access to host cell receptors. Antiviral antibodies may also initiate classical complement activation. The formation of the MAC on the surface of an enveloped virus or an infected host cell kills it, and the complement components that are produced during the cascade may opsonize extracellular virions and promote their uptake by phagocytosis (**Fig. 13-3, #6**). The antiviral antibodies themselves may also serve as opsonins. Finally, antibodies that have recognized viral antigens on infected host cell surfaces may engage FcRs on phagocytes and other leukocytes (particularly NK cells) and provoke ADCC.

It should be noted that some viruses are combatted (at least in part) by B cell responses that do not require T cell help. Viruses such as vesicular stomatitis virus (VSV) have highly repetitive structures on their surfaces that induce a Ti response. Ti responses are typically faster than Td responses because a Ti response involves only a B cell and does not require

B–T cell cooperation. An antiviral Ti response can function early in an infection to minimize the spread of the virus until antibodies against viral Td antigens can be synthesized.

vii) Complement

As well as the classical complement activation that is part of the humoral response, surface components of virions can directly activate the lectin and alternative complement pathways. Opsonization of viruses by C3b (or C3d) promotes phagocytosis by neutrophils and macrophages (refer to **Fig. 13-1, #6**).

III. Evasion Strategies

Viruses with small genomes count on rapid replication and dissemination to new host cells to establish an infection before the immune system can respond. Viruses with larger genomes need more time to replicate and are transmitted more slowly. Accordingly, these latter pathogens have developed ways of interfering with various components of the host immune response that allow them sufficient time to establish an infection. Once infection has occurred, many viruses hide from the immune system. Others confront the immune response head on by interfering with host cell signaling pathways. Evasion strategies used by viruses are summarized in **Table 13-6**.

TABLE 13-6	Evasion of the Immune System by Viruses
Immune System Element Thwarted	**Viral Mechanism**
Detection	Become latent
Antibodies	Alter viral epitopes via antigenic drift or shift Express viral FcR that blocks ADCC, neutralization and/or complement activation Block B cell intracellular signaling or activation
CD8+ T cells	Infect cells with very low MHC class I expression Block MHC class I-mediated antigen presentation, including via miRNA Force pMHC internalization
CD4+ T cells	Avoid infection of DCs Interfere with MHC class II-mediated antigen presentation Force pMHC internalization
NK cells	Express viral homologs of MHC class I Increase host synthesis of HLA-E or classical MHC class I Block MICB expression via miRNA
DCs	Block DC development or maturation Block DC upregulation of costimulatory molecules Block DC expression of CCR7 Upregulate DC expression of FasL
Complement	Block convertase formation Express viral homologs of host RCA proteins Increase expression of host RCA proteins Bud through host membrane and acquire host RCA proteins
Host PRRs	Produce proteins interfering with normal PRR signaling Have a genome low in CpG motifs
Antiviral state	Block secretion of IFNs Interfere with metabolic/enzymatic events that establish the antiviral state
Apoptosis	Block various steps of extrinsic or intrinsic pathways Express homologs of death receptors and regulatory molecules Express molecules sustaining host cell survival
Cytokines/chemokines	Express competitive inhibitors of cytokines and chemokines Block cytokine/chemokine transcription Block cytokine/chemokine translation via miRNA Downregulate host cytokine/chemokine receptor expression

i) Latency

Some persistent viruses avoid removal by the immune system through **latency**. When a virus adopts a latent state, it persists in the host cell in a defective form that renders it non-infectious for a period of time. In most cases, latency involves the inactivation of viral gene transcription needed for productive infection and the subsequent expression of new viral transcripts required for latency. Reversal from latency back to productive infection requires some type of reactivation of the productive infection genes that can occur only when the host's immune system has weakened.

Different viruses achieve latency in different ways. HIV integrates a cDNA copy of its RNA genome into the DNA of its host cell in such a way that there is limited transcription of viral genes. The DNA genomes of VZV and HSV do not integrate into the host DNA but instead form a complex with host nucleosomal proteins that block transcription of productive infection genes. A similar latency mechanism operates in cases of Epstein–Barr virus (EBV) and Kaposi's sarcoma herpesvirus (KSHV) infection. However, the latency of these viruses is associated with the development of tumors in the host: B cell lymphomas and nasopharyngeal carcinomas in the case of EBV, and the AIDS-related Kaposi's sarcoma in the case of KSHV.

The structure and life cycle of HIV is described further in Chapter 15.

ii) Antigenic Variation

A common way for a virus to hide from the host immune system is to change its antigenic "stripe" over successive generations, expressing antigenically new forms of viral proteins that may not be recognized by an individual's existing memory lymphocytes or antibodies. This mechanism is most effective in long-lived hosts (like humans) that can sustain multiple re-infections, and is particularly important if the virus lacks the ability to become latent. The rapid modification of viral antigens through random mutations is known as **antigenic drift**. For example, like all RNA viruses, influenza virus cannot proofread its RNA genome during replication and thus sustains a high rate of mutation. The hemagglutinin (H) and neuraminidase (N) proteins, which are the only two viral proteins present on the surface of the influenza virion, are thus subtly different from viral generation to generation. These minor virus variants often replicate preferentially in the host, because they are not neutralized by antibodies raised against earlier strains. New influenza strains created by antigenic drift are responsible for localized influenza outbreaks. HIV is another virus that undergoes very rapid antigenic drift, even within a single infected individual. In this case, the mutations arise due to the highly error-prone reverse transcriptase involved in the replication of the HIV genome.

Almost unique to the influenza virus is its ability to undergo **antigenic shift**. The influenza virus genome exists as eight separate single-stranded RNA segments, each of which encodes a single viral protein. With such a genetic structure, two different influenza strains that simultaneously infect a single host cell can undergo a re-assortment (sometimes inaccurately called "recombination") of their genomic segments (**Fig. 13-4**). Virus particles containing new combinations of parental RNA suddenly arise, dramatically changing the spectrum of protein epitopes presented to the immune system. This antigenically novel flu virus is safe from antibodies and CTLs raised during previous exposure or vaccination, and so rapidly becomes entrenched in vulnerable hosts constituting a pandemic (see **Box 13-2**).

An *epidemic* involves an increased frequency of disease in one location. A *pandemic* is unrestricted geographically and leads to a global epidemic of disease.

NOTE: Clinical immunologists define a particular antigenic shift of an influenza virus by the identity of its hemagglutinin (H) and neuraminidase (N) molecules, since it is the presence or absence of B cell memory to these surface glycoproteins that influences the production of neutralizing antibodies. The new strains that result from antigenic shift are often referred to as "influenza virus subtypes."

iii) Interfere with Antigen Presentation

Antigen processing pathways offer many opportunities for a virus to sabotage immune responses, and a given virus can interfere at more than one step.

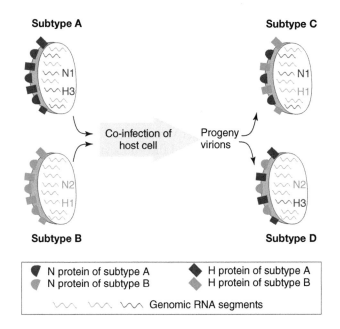

Subtype A

N1
H3

Subtype C

N1
H1

Co-infection of
host cell

Progeny
virions

N2
H1

Subtype B

N2
H3

Subtype D

- N protein of subtype A
- N protein of subtype B
- H protein of subtype A
- H protein of subtype B

Genomic RNA segments

⊪ Fig. 13-4
Principle of Antigenic Shift

The eight RNA segments of the influenza virus genome can re-assort if two different viral subtypes infect the same cell. Progeny virions acquiring various combinations of parental segments may express new constellations of proteins. In this *hypothetical* example, influenza virus subtypes A (H3N1) and B (H1N2) have contributed assorted RNA fragments that result in new progeny virus subtypes C (H1N1) and D (H3N2). Gray RNA segments encode internal viral proteins.

Box 13-2   The H1N1 Pandemic of 2009

The antigenic drift that routinely generates a new influenza virus every year is usually responsible for 3–5 million cases of severe illness worldwide and 250,000–500,000 deaths. In contrast, antigenic shifts are responsible for much more serious pandemics, as exemplified by three widespread influenza outbreaks in 1918 (H1N1), 1957 (H2N2), and 1968 (H3N2) that caused tens of millions of deaths around the globe. In this century, a global pandemic of H1N1 influenza began in April 2009. Although the exact series of genetic re-assortment events that led to the emergence of the 2009 H1N1 subtype remains undefined, it is clear that a progeny virus emerged with a hemagglutinin protein to which much of the human population was immunologically naïve. This antigenic shift combined with modern air travel allowed the novel influenza virus subtype to spread very rapidly and efficiently around the world. The potential severity of the situation caused the World Health Organization to issue a global alert, stating that the event was a "Public Health Emergency of International Concern." Fortunately, the mortality associated with this outbreak was far less than in previous pandemics, with global fatalities estimated in the tens of thousands rather than millions. Significantly, older individuals who had survived one or more of the 20th century pandemics appeared to be immune or at least somewhat resistant to the 2009 H1N1 virus. The mechanism mediating this type of cross-subtype protection after an antigenic shift is not well understood. Some immunologists believe that previous infection with the 1918 virus (for example) may have invoked some degree of long-lived protection based on immune responses to more highly conserved proteins in the virus core, such as the viral nucleoprotein or matrix protein. This hypothesis remains under investigation.

a) MHC Class I-Mediated Antigen Presentation

Different viruses avoid activating CD8+ T cells in different ways. Adenovirus blocks MHC class I synthesis in infected cells. CMV and VSV infect cells that normally have very low MHC class I expression. CMV also expresses a protein that induces deglycosylation and degradation of newly synthesized MHC class I chains. A different CMV protein associates with mature peptide–MHC class I structures that do make it to the cell surface and blocks recognition by CD8+ T cells. EBV produces viral proteins that resist proteolysis, meaning that peptides capable of fitting into the MHC binding groove are not easily generated. EBV also downregulates the expression of the TAP antigen transporter and so reduces peptide loading. Herpesviruses express small proteins that interfere with peptide binding to TAP on the cytosolic side of the ER. Other viruses express proteins that allow the peptide to bind to TAP but then trap the complex on the luminal side of the ER. HIV produces a multifunctional protein called Nef that is able to bind simultaneously to host clathrin proteins and the cytoplasmic tails of MHC class I molecules on the surface of the host cell. This Nef-mediated physical connection between MHC class I and clathrin forces the internalization and lysosomal degradation of the MHC class I molecule.

b) MHC Class II-Mediated Antigen Presentation

Viruses have also evolved myriad ways to avoid activating CD4+ T cells. Rabies virus preferentially infects neurons but is very slow to lyse these cells, meaning that viral antigens are not easily collected by APCs until well after the virus has entered the body. The adaptive response to rabies is thus delayed. CMV and adenovirus both synthesize proteins that inhibit the intracellular signaling pathways required for MHC class II expression. Other viruses express proteins that bind to MHC class II molecules and target them for proteasomal degradation. Still other viruses interfere with MHC class II presentation after the MHC molecule has entered the endocytic system. For example, a CMV protein competes with invariant chain for binding to MHC class II, and certain HPV proteins and the HIV Nef protein disrupt the acidification of the endosomal compartments necessary for peptide generation. As it can for MHC class I, HIV Nef can induce the internalization and lysosomal degradation of cell surface MHC class II molecules.

iv) Fool NK Cells

A virus that causes its host cell to downregulate MHC class I draws the attention of NK cells. CMV therefore expresses a viral homolog of MHC class I that engages NK inhibitory receptors and fools the NK cell into thinking it has detected normal MHC class I. The NK cell is not activated, and the infected host cell is not lysed. CMV also upregulates expression of the non-classical MHC class I molecule HLA-E, which can bind to NK inhibitory receptors. In contrast, the fast-replicating West Nile virus (WNV) upregulates the expression of classical host MHC class I molecules, striving to neutralize NK cells and complete its reproduction before CTLs are generated.

v) Interfere with DCs

Several viruses interfere with DC functions and thus derail T cell responses. Human T cell leukemia virus-1 (HTLV-1) infects DC precursors and prevents their differentiation into immature DCs, blocking the initiation of T cell activation. T cells that interact with the infected DCs are then infected themselves. HSV-1 and vaccinia virus infect immature DCs and block DC maturation, whereas other poxviruses induce the apoptotic death of DCs. Measles virus upregulates the expression of FasL on an infected DC, forcing it to kill any Fas-bearing T cells it encounters. Measles virus can also cause DCs to form large aggregates called *syncytia* in which the virus replicates freely and DC maturation is stymied. Ebola virus infects DCs by binding to the DC lectin DC-SIGN, and this association sequesters the virus within these cells. When CMV infects a DC, the DC becomes tolerogenic so that it anergizes, rather than activates, any naïve T cell it encounters. CMV and herpesviruses also block expression of the chemokine receptor CCR7 by DCs, preventing them from following chemokine gradients into the secondary lymphoid tissues.

vi) Interfere with Antibody Functions

Some viruses are able to interfere directly with production or effector functions of antiviral antibodies. Measles virus expresses a protein that has a negative regulatory effect on B cell activation. An attack by HSV-1 causes the infected host cell to express a viral version of FcγR that binds to IgG molecules complexed to viral antigen. The Fc portion of the antibody is rendered inaccessible by this binding so that neither ADCC nor classical complement activation can be triggered.

vii) Avoid Complement

Viruses use many of the same mechanisms as other pathogens to avoid complement-mediated destruction. Some poxviruses and herpesviruses secrete proteins that block formation of the alternative C3 convertase. Many viruses increase a host cell's expression of the RCA proteins that regulate complement activation, preventing the infected cell from undergoing MAC-mediated lysis. Other viruses express viral homologs of RCA proteins that block MAC-mediated destruction of the virion itself. Still other viruses bud through the host cell membrane and acquire its RCA proteins. The RCA proteins DAF and MIRL are acquired by HIV and vaccinia virus in this way.

viii) Interfere with Host PRRs

Like bacteria, some viruses can thwart innate leukocyte activation by interfering with PRR functions. Vaccinia virus produces proteins that either antagonize kinases or bind to adaptor proteins in TLR signaling pathways, suppressing the activation of host cell transcription needed for antiviral responses. Hepatitis C virus (HCV) synthesizes a protease that cleaves the TLR signaling mediator TRIF, thereby blocking production of IFNs. Paramyxoviruses (like measles virus) produce a protein that associates with the RLR RIG-1, inhibiting the induction of IFNβ production by viral dsRNA. Adenovirus avoids activating TLR9 by having a genome low in CpG motifs. KSHV produces an E3 ubiquitin ligase that promotes the proteasomal degradation of a factor that is needed for TLR-triggered production of IFNs.

ix) Counteract the Antiviral State

Several viruses have developed intricate mechanisms that disrupt the antiviral state. EBV expresses a soluble receptor for a growth factor essential for macrophage secretion of IFNs. In the absence of this growth factor, insufficient IFNs are produced to trigger and maintain the antiviral state. When HSV infects a cell that has already established the antiviral state, the virus expresses proteins that reverse the associated translational block, allowing viral protein synthesis to resume. Vaccinia virus and HCV also synthesize proteins that disrupt the metabolic and enzymatic events needed to maintain the antiviral state. Adenovirus expresses proteins that interfere with the activity of the host's transcription factors. KSHV produces proteins that are homologous to host transcription factors but do not permit transcription of the genes required to establish the antiviral state.

x) Manipulate Host Cell Apoptosis

Host cell apoptosis prior to completion of replication spells viral doom. Host cell apoptosis is most commonly induced by CTL degranulation, Fas/FasL interaction, or the binding of TNF to TNFR. In addition, an infected cell will sometimes be triggered to undergo "altruistic" apoptosis (death for the good of the host) by a mechanism such as **ER stress**. ER stress results when the ER machinery of a host cell is overheated by having to pump out large quantities of viral proteins. Complex viruses with large genomes have developed ways of blocking various steps of these death-inducing pathways. Adenovirus synthesizes a multiprotein complex that induces the internalization of Fas and TNFR, removing these death receptors from the cell surface and forestalling apoptosis induced by an encounter with FasL or TNF. Several poxviruses express homologs of TNFR that act as decoy receptors for TNF and related cytokines. Adenoviruses, herpesviruses and poxviruses express multiple proteins that inhibit the enzymatic cascade necessary for apoptosis. Many viruses can increase intracellular levels of host cell survival proteins that normally prevent premature apoptosis. Alternatively, a virus may express a homolog of these survival proteins that counters apoptosis.

xi) Interfere with Host Cytokines

Early in viral infections, host cells are induced to produce copious quantities of cytokines and chemokines that support antiviral responses. Viruses therefore seek to inhibit the production or action of these molecules, or their receptors. Some poxviruses alter the local cytokine milieu and make it less favorable to the cellular cooperation that underpins an immune response. Both KSHV and adenovirus express proteins that inhibit IFN-inducible gene transcription, whereas certain poxviruses express a protein that blocks IL-1 production. Herpesviruses downregulate cytokine receptor expression, and CMV disrupts the transcription of chemokine genes. Vaccinia virus secretes IFN receptor homologs that intercept IFNα and IFNγ molecules. Poxviruses synthesize a chemokine homolog that binds to chemokine receptors on host cells but blocks the chemotaxis of lymphocytes, macrophages and neutrophils.

Inhibition of IL-12 production is a major goal of many viruses since this cytokine is crucial for Th1 differentiation and thus the antiviral cell-mediated immune response.

EBV synthesizes a homolog of IL-12 that may competitively inhibit the activity of host IL-12. EBV also produces a homolog of IL-10 that suppresses IL-12 production by macrophages and IFNγ production by lymphocytes. The binding of measles virus to certain host cell receptors can also block IL-12 synthesis.

xii) Express Inhibitory miRNAs

Recent studies have revealed that many herpesviruses and some polyoma viruses express microRNA (miRNA) molecules which are not immunogenic themselves but have profound effects on antiviral immunity. Some miRNAs inhibit the expression by infected host cells of viral proteins that furnish immunodominant epitopes. Without presentation of these viral epitopes, CTLs cannot recognize infected host cells and do not kill them, allowing the virus to persist. Similarly, other miRNAs block the expression of the NK activatory ligand MICB by an infected host cell. In the absence of the MICB molecules needed to bind to the NK activatory receptor NKG2D, the NK cell may not receive sufficient activatory signaling to overcome NK inhibitory signaling; the infected host cell is spared. Still other viral miRNAs regulate the activities of innate and adaptive leukocytes by inducing or suppressing host cell expression of various cytokines and chemokines. For example, viral miRNAs may block host cell production of the chemokines IL-8, MCP-1 and/or CXCL11, preventing the recruitment of neutrophils, macrophages and activated T cells that could eliminate the threat. Other viral miRNAs block the expression of molecules that repress IL-10 synthesis, promoting the production of this immunosuppressive cytokine. Similarly, viral miRNAs that inhibit the expression of molecules blocking IL-6 expression have the effect of promoting regulatory T cell differentiation and thus the downregulation of antiviral T cell responses. Over 200 immunomodulatory viral miRNAs have been identified to date and more are emerging daily.

miRNAs are endogenous non-coding RNA molecules (~22 nucleotides in length) that are complementary to sequences in mRNAs and inhibit their translation by binding to them.

E. Immunity to Parasites

I. Disease Mechanisms

Parasites are among the biggest killers in the pathogen pantheon. Parasites include unicellular protozoans and multicellular helminth worms, both of which live within a host (*endoparasites*), as well as arthropods like ticks, fleas, lice and mites, which attach themselves to the skin or hair follicles on the exterior of a host (*ectoparasites*). These scourges claim millions of lives every year, particularly in developing countries. Some protozoans replicate extracellularly, whereas others replicate intracellularly. Helminth worms reproduce inside a host's body but outside its cells, or outside the host entirely in a location (like a water source) where access to a host is easy. Growth and maturation of the worm then occur within the host, often causing severe and long-term damage to tissues and organs. Some ectoparasites complete their entire life cycle on the host surface, whereas others attach only to feed and then detach. Other ectoparasites initially deposit their eggs on the host's skin, but the eggs later detach and mature in soil or water.

Many parasites have multistage life cycles, and each stage of a parasite may be able to infect a different host species. Parasites also frequently use vectors to infect their ultimate hosts, or serve as vectors for other types of pathogens. For example, humans contract malaria through the bite of an *Anopheles* mosquito infected with the protozoan parasite *P. falciparum*. Fleas are a vector for the Gram-negative bacterium *Yersinia pestis*, which causes bubonic plague. Parasites that do succeed in establishing an infection inside an individual may go through various life cycle stages, some of which may be intracellular and others extracellular. All these factors can create a considerable problem from a public health point of view, because a parasite that continually changes form and/or makes use of an invertebrate or animal vector is much harder to control than a pathogen that infects humans only. Both cell-mediated and humoral immunity must often be mobilized to conquer parasites. Examples of diseases caused by protozoans, helminth worms and ectoparasites are given in **Tables 13-7, 13-8** and **13-9**, respectively.

TABLE 13-7	**Examples of Parasitic Protozoans and the Diseases They Cause**
Pathogen	**Disease**
Entamoeba histolytica	Enteric disease
Leishmania donovanii	Leishmaniasis in viscera
Leishmania major	Leishmaniasis in face, ears, skin
Plasmodium falciparum	Malaria
Toxoplasma gondii	Toxoplasmosis
Trypanosoma brucei	African sleeping sickness
Trypanosoma cruzi	Chagas disease

TABLE 13-8	**Examples of Parasitic Helminth Worms and the Diseases They Cause**
Pathogen	**Disease**
Ascaris	Ascariasis
Cestoda	Tapeworms
Echinococcus	Alveolar echinococcosis
Onchocerca	African river blindness
Schistosoma	Schistosomiasis
Trichinella	Trichinosis
Wuchereria	Elephantiasis

TABLE 13-9	**Examples of Ectoparasites and the Diseases They Cause/Transmit**	
Organism	**Role in Disease**	**Disease Caused**
Acari (ticks, some mites)	Pathogen Vector	Dermatitis Lyme disease
Cimicidae (bedbugs)	Pathogen	Skin rashes
Demodex (eyelash mites)	Pathogen	Blepharitis, dermatitis
Hippoboscoidea (tsetse fly)	Vector	Elephantiasis and sleeping sickness
Oestridae (bot flies)	Pathogen	Myiasis
Phthiraptera (lice)	Pathogen	Pediculosis
Sarcoptes scabiei (mites)	Pathogen	Scabies
Siphonaptera (fleas)	Pathogen Vector	Itching, rash

II. Immune Effector Mechanisms

Different parasites evoke different types of immune responses, depending on the size and cellularity of the invader and its life cycle. In general, protozoan parasites tend to induce Th1 responses, while helminth worm infections and attacks by ectoparasites are usually handled by Th2 responses.

i) Defense against Protozoans

a) Induced Innate Defense
Many protozoan components act as PAMPs for TLRs. For example, elements of *Trypanosoma*-derived mucins, phospholipids and genomic DNA bind to TLR2, TLR4

or TLR9, respectively. Certain stages of *Plasmodium* species produce PAMPs that can activate pDCs to produce IFNs in a TLR9-dependent fashion. While some human TLR4 and TLR9 SNPs are associated with an increased risk of developing severe malaria following *Plasmodium* infection, individuals expressing particular TLR1 or TLR6 SNPs develop only mild disease.

Complement activation via the MBL-induced lectin pathway is also important for fighting *P. falciparum* and other *Plasmodium* species that cause malaria. Recent work has identified various MBL SNPs associated with different malarial states in infected individuals, including asymptomatic infection or resistance to infection entirely. Similarly, different SNPs of the complement receptor CR1 or the acute phase protein CRP affect both the frequency of malarial episodes in an infected individual as well as total parasite counts.

NOTE: Although mice genetically deficient for a single TLR do not show increased susceptibility to protozoan parasite infections, animals that lack the TLR signaling adaptor MyD88, which transduces signaling downstream of all TLRs except TLR3, are highly susceptible to infection by *Toxoplasma gondii*, *Trypanosoma cruzi* and *Leishmania major*. In contrast, humans expressing a certain polymorphism of the MyD88-like adaptor protein MAL, which transduces TLR2/TLR4 signaling, are protected against both malaria and Chagas disease (caused by *T. cruzi* infection). This observation suggests a correlation between this polymorphism and increased adaptor function.

---

b) Humoral Defense

All the effector mechanisms ascribed to antibodies for defense against extracellular bacteria (refer to **Fig. 13-1**) apply to defense against small extracellular protozoans. Antiparasite antibodies mediate neutralization, opsonized phagocytosis, and/or classical complement activation. Larger extracellular protozoans can be dispatched by ADCC mediated by neutrophils and macrophages.

c) Th1 Responses, IFNγ and Macrophage Hyperactivation

The Th1 response is critical for antiprotozoan defense because Th1 effectors are key sources of the IFNγ needed to drive macrophage hyperactivation. Like many intracellular bacteria, many protozoan parasites (e.g., *L. major*) infect or are taken up by macrophages but are not destroyed within ordinary phagosomes. Only in hyperactivated macrophages are sufficient levels of ROIs and RNIs produced to efficiently kill such parasites. In addition, TNF secreted by hyperactivated macrophages plays an important role in the control of protozoans that are still in the extracellular milieu. If all else fails and the hyperactivated macrophages cannot clear the parasite, a granuloma is formed that encompasses the infected host cells and walls off the invader.

IFNγ has several other antiprotozoan effects. This cytokine (1) is directly toxic to various forms of many protozoans; (2) stimulates IL-12 production by DCs and macrophages, which in turn triggers additional IFNγ production by NK and NKT cells; (3) induces iNOS expression in infected macrophages, resulting in the production of intracellular NO that eliminates either the parasite itself or the entire infected cell; (4) upregulates the expression of enzymes important for phagosome maturation; and (5) upregulates Fas expression on the infected macrophage surface, rendering the macrophage susceptible to Fas-mediated apoptosis when it contacts a FasL-expressing T cell. Because Th2 cytokines such as TGFβ, IL-4, IL-10 and IL-13 inhibit IFNγ production and suppress iNOS, individuals that preferentially mount Th2 responses instead of Th1 responses are highly susceptible to diseases caused by protozoan parasites.

d) CTLs and γδ T Cells

If a protozoan parasite escapes from a macrophage phagosome into the cytosol of a host cell, parasite antigens may enter the endogenous antigen processing system such that antigenic peptides are presented on MHC class I. The infected host cells then become targets for CTLs. However, perforin/granzyme-mediated cytolysis is not very effective against acute protozoan infections, and it is CTL secretion of IFNγ that is this cell type's greatest contribution to the antiprotozoan response. Similarly, IFNγ secretion by activated γδ T cells can bolster the body's defenses during the early stages of protozoan infections. Perforin/granzyme-mediated cytotoxicity becomes important for controlling chronic stages of protozoan infections.

ii) Defense against Helminth Worms

a) Induced Innate Defense

The investigation of mechanisms of innate defense against large, multicellular helminth worms is in its early stages. At least in mice, TLR4 is important for fighting the blood-dwelling trematode *S. mansonii*. Wild type mouse macrophages incubated with preparations of *Schistosoma* larvae are stimulated to produce IL-6, IL-12 and IL-10, but the production of the latter two cytokines is lost in macrophages from TLR-4-deficient mice. Studies are under way to determine the importance of TLR signaling for antihelminth worm responses in humans.

b) Th2 Responses and Humoral Defense

While Th1 responses are needed to combat protozoan parasites, Th2 responses are vital for eliminating helminth worms. For example, humans naturally resistant to *S. mansonii* express high levels of Th2 cytokines, whereas individuals susceptible to this worm exhibit increased concentrations of Th1 cytokines. The antihelminth Th2 response involves IgE, mast cells, basophils and eosinophils, a combination that does not contribute significantly to defense against any other type of pathogen. Activated CD4+ T cells are also critical for antihelminth defense because these cells differentiate into effectors supplying the Th2 cytokines and CD40L contacts required for isotype switching to IgE by B cells (**Fig. 13-5, #1**). The antiparasite IgE antibodies synthesized by the B cells enter the circulation and "arm" mast cells and basophils by binding to cell surface FcεRI. When a worm antigen engages the cell-bound IgE, the degranulation of the mast cells and basophils is triggered in close proximity to the parasite (**#2**). Histamine released by the mast cells and basophils causes the contraction of host intestinal and bronchial smooth muscles such that the parasite is shaken loose from its grip on the mucosal surface and expelled from the body. Histamine and other proteins synthesized by mast cells and basophils are also directly toxic to some helminth parasites. In addition, the vasodilation and increased vascular permeability induced by histamine allow an influx of leukocytes and circulating antibodies into the area. Circulating IgE directed against worm surface molecules may bind directly to the pathogen, attracting the attention of eosinophils expressing FcεRI molecules. The binding of the worm-bound IgE to eosinophil FcεRs triggers eosinophil degranulation and the release of substances that work directly and indirectly to kill the worm (**#3**). Some molecules degrade the skin of the worm, allowing neutrophils and other leukocytes to penetrate into its underlying tissues. These cells may also degranulate and release additional toxic proteins and peptides that kill the worm. Other molecules contained in eosinophil granules stimulate mast cells to degranulate.

Th2 cytokines are critical for eliminating helminth worms. IL-4 produced by basophils and Th2 effectors is the main cytokine driving isotype switching in B cells to IgE. IL-5 produced by Th2 cells strongly promotes the proliferation, differentiation and activation of eosinophils, and supports the differentiation of plasma cells that have undergone isotype switching to IgA production (**Fig. 13-5, #4**). Secretory IgA coats the mucosae and fends off further parasite attachment. IL-4 and IL-13 produced by basophils and Th2 cells suppress macrophage production of IL-12, inhibiting IFNγ production and hence the development of a Th1 response (which would be largely ineffective). IL-13 also stimulates bronchial and gastrointestinal expulsion responses.

iii) Defense against Ectoparasites

Ectoparasites are often arthropods that attack the exterior surface of a host. For example, the common tick is the carrier of the extracellular bacterium *Borrelia burgdorferi* responsible for Lyme disease. The bacteria are introduced into the host when the tick bites him/her to obtain a blood meal. Large numbers of basophils, eosinophils and mast cells

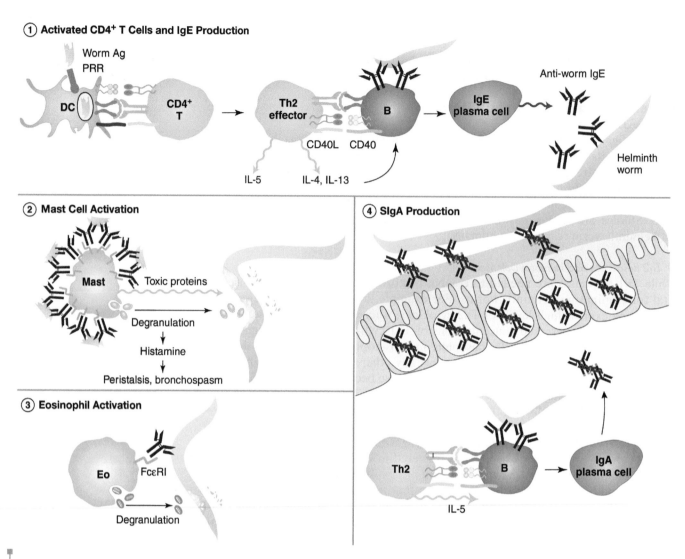

Fig. 13-5
Major Mechanisms of Immune Defense against Helminth Worm Parasites

(**1**) DCs presenting worm peptides induce CD4+ Th2 cell differentiation. Th2 effectors produce cytokines that induce activated B cells to undergo isotype switching to IgE. (**2**) Mast cells (and basophils; not shown) pre-armed with antiparasite IgE are activated by worm antigens and release histamine and toxic proteins. (**3**) Activated eosinophils bind to worm-bound antibodies via their FcεRIs and degranulate, releasing molecules that directly damage the worm surface. (**4**) IL-5 produced by Th2 cells induces isotype switching to IgA in mucosal B cells specific for worm antigens. Secretory IgA (SIgA) blocks the worm from gaining a foothold on the mucosal surface.

accumulate at the bite site to repel both the attacking bacteria and the tick. It is thought that when mast cell degranulation releases substances that increase vascular permeability, ticks have greater difficulty in locating host blood vessels. Some ectoparasites are countered by the same strategies effective against helminth worms. Antipathogen IgE bound to the surface of basophils and mast cells is critical for host defense against such invaders. For example, humans who lack adequate numbers of basophils and eosinophils develop scabies, a severe, itchy rash caused by the mite *Sarcoptes scabiei*. Much remains to be determined about the molecular details of immune responses to ectoparasites.

NOTE: The involvement of Th2 responses in defense against ectoparasites came from the unexpected finding of increased *Demodex* skin infections in mice lacking both CD28 and STAT6. CD28 is a key costimulator of Th cell activation, and STAT6 is the transcription factor required for IL-4 production by these cells.

III. Evasion Strategies

A parasite that has a multistage life cycle enjoys a wealth of opportunities to thwart the immune response. Several evasion strategies used by protozoa and/or helminth worms are described in the following sections and summarized in **Table 13-10**.

i) Avoid Antibodies

Different parasites employ different strategies to avoid antibodies. Protozoans with multiple life cycle stages can take advantage of the escape offered by antigenic variation. Just as the host mounts a humoral response to epitopes associated with one stage of the parasite, the organism may take on a totally different form and present a whole new panel of epitopes to the host's immune system. A lag in defense ensues while antibodies are produced to counter the new set of antigens. Other protozoans take a more direct approach. *L. major* hides from antibodies by sequestering itself within host macrophages. Some *Schistosome* helminths disguise themselves by acquiring a coating of host glycolipids and glycoproteins. The dense "forest" created by these host molecules blocks antibodies from binding to parasite surface antigens. Other helminths repel antibody attack by shedding parts of their external membranes, ejecting the immune complex of the parasite antigen and host antibody. Still other helminths produce substances that digest antibodies.

Trypanosoma brucei, the causative agent of African sleeping sickness, confounds antibodies by rapid antigenic variation. This pathogen can spontaneously modify its expression of its *variable surface glycoprotein (VSG)*, the molecule that is normally the main target of humoral responses to this parasite. There are hundreds of VSG genes but each trypanosome expresses only one VSG gene at a time. However, the trypanosome regularly shuts down expression of the first VSG gene and activates another, resulting in an altered glycoprotein coat that may not be recognized by antibodies raised against the first VSG protein. The trypanosomes are therefore able to outpace the immune system's ability to adapt to the change in VSG antigens, buying the time the organisms need to penetrate the blood–brain barrier and enter the central nervous system (CNS). This ability of *T. brucei* to artfully evade the immune system accounts for the near 100% mortality of untreated African sleeping sickness.

TABLE 13-10	Evasion of the Immune System by Parasites
Immune System Element Thwarted	**Parasite Mechanism**
Antibodies	Have a multistage life cycle that furnishes antigenic variation Hide in macrophages Modify parasite surface proteins to cause antigenic variation Acquire host surface proteins that block antibody binding Shed parasite membranes bearing immune complexes Secrete substances that digest antibodies
Phagocytosis	Block fusion of phagosome to lysosome Escape from phagosome into cytoplasm Block respiratory burst Lyse resting phagocytes
Complement	Degrade attached complement components or cleave Fc portions of membrane-bound antibodies Force complement component exhaustion Express homologs of RCA proteins
T cells	Inhibit Th1 response by promoting IL-10 production and decreasing IL-12 and IFNγ production Secrete proteins inducing hyporesponsiveness or tolerance of T cells Interfere with DC maturation and macrophage activation Induce downregulation of surface MHC class I and II

ii) Avoid Phagolysosomal Destruction

Helminth worms are in no danger of being captured by phagocytosis, but many protozoans have developed means of avoiding such destruction. For example, some intestinal protozoans lyse resting granulocytes and macrophages and thus minimize their chances of being engulfed in the first place. *T. gondii* blocks the fusion of macrophage phagosomes to lysosomes. *T. cruzi* enzymatically lyses the phagosomal membrane prior to lysosomal fusion and escapes to the cytoplasm of the host cell. *L. major* often remains in the phagosome but interferes with the respiratory burst.

iii) Avoid Complement

Both protozoans and helminths can take steps to avoid complement. Certain members of both groups can proteolytically remove complement-activating molecules that have attached to their surfaces, or cleave the Fc portions of parasite-bound antibodies. For example, *L. major* can induce the release of the entire complement terminal complex from its surface. Other parasites secrete molecules that force continuous fluid phase complement activation, thereby exhausting complement components. Still other parasites express a molecule that functionally mimics the mammalian RCA protein DAF.

iv) Interfere with T Cells

Members of both the protozoan and helminth groups have evolved ways of manipulating the host T cell response to favor parasite survival. For example, *P. falciparum* can promote Th cell secretion of IL-10 rather than IFNγ, resulting in downregulation of MHC class II expression and inhibition of NO production. This pathogen also expresses molecules that cause the red blood cells it infects to indirectly interfere with macrophage activation and DC maturation. *L. major* expresses molecules that can bind to CR3 and FcγR molecules on macrophages and reduce IL-12 production by these cells. The Th1 response that would kill the protozoan is thus inhibited. Nematode hookworms secrete several proteins that induce hyporesponsiveness or even tolerance in host T cells. This state of immunosuppression allows great masses of worms to accumulate in the infected host. Other filarial worms induce the APCs with which they come into contact to decrease their surface expression of MHC class I and II, and to also downregulate other genes involved in antigen presentation. The APCs cannot then participate effectively in T cell activation.

"Immunodiagnosis of *Taenia solium* taeniosis/cysticercosis" by Deckers, N. & Dorny, P. (2010) *Trends in Parasitology* 26, 137–144.

Focus on Relevant Research

Although unpleasant to deal with for both patient and physician alike, infection by the pork cestode intestinal tapeworm *Taenia solium* is relatively easy to diagnose, and adult worms that have evaded immune destruction are usually eliminated with antihelminthic medications. A much more serious condition arises when tapeworm cysts lodge in non-intestinal tissues such as the muscle, eye or brain. This condition, known as *cysticercosis*, can be extremely difficult to diagnose and may cause significant tissue damage. Arguably the most severe form of cysticercosis is *neurocysticercosis* (cysts in the brain), which can result in convulsions, permanent brain damage, or death. Neurocysticercosis caused by *T. solium* is the most common parasitic disease of the CNS worldwide, and early and accurate diagnosis of this infection could significantly reduce its global morbidity and mortality. This article by Deckers and Dorny reviews the numerous laboratory tests currently available and under development for the diagnosis of various forms of cysticercosis. The authors also describe several candidate *T. solium* molecules that may prove useful for the development of future diagnostic tools. Most of these tests are immunologically based and include techniques such as radioimmunoassay, hemagglutination, complement fixation, and ELISA. The principles of these serological techniques are illustrated in Appendix F.

F. Immunity to Fungi

I. Disease Mechanisms

Fungi are either unicellular and grow as discrete eukaryotic cells (like yeast), or are multicellular and grow in a mass (*mycelium*) of filamentous processes (*hyphae*). *Dimorphic* fungi adopt a unicellular form at one stage in their life cycle and a multicellular form at another stage. *Conidia* are haploid, non-motile fungal spores that are formed under unfavorable nutrient conditions. All fungal cells have a cell wall like bacteria but also a cell membrane like mammalian cells. Although many fungi live most of their lives in the soil, some live commensally on the topologically external surfaces of the human body. Some fungi are *dermatophytes*, filamentous fungi that infect only the skin, hair and nails. Most fungal species are not harmful to healthy humans, but when a fungus succeeds in invading the body, it usually heads for the vascular system of the target tissue. Invasion of blood vessels by a growing fungus can choke off the blood supply to the host's organs.

Fungi have recently become a more significant clinical threat due to the advent of modern protocols for organ transplantation, treatment of autoimmunity, implantation of medical devices, and chemotherapy. All these procedures call for or result in suppression of the patient's immune system, so that fungi which would normally not succeed in establishing an infection are able to do so; that is, the patients contract opportunistic infections. In particular, species of *Aspergillus*, *Candida* and *Cryptococcus* fungi have become prominent threats to immunocompromised individuals. Patients may also present with infections by one of the *Mucormycotina* group of filamentous fungi, which launch life-threatening attacks on the brain and sinuses. Diseases caused by fungal infections are generally called *mycoses*, and examples of several are given in **Table 13-11**.

> Cancer chemotherapy and organ transplantation are discussed in Chapters 16 and 17, respectively.

> NOTE: A rising concern among clinical immunologists studying fungal infections is the projected impact of global warming. With climate change will come two potential threats to the currently balanced relationship between most fungi and their mammalian hosts. First, a warmer environment will allow new fungal species to survive in previously hostile geographic areas, increasing the number of threats faced by vulnerable individuals. Second, some researchers believe that the planet is warming faster than mammals can evolve to maintain the usual difference between their own body temperature and the generally lower temperature of their surroundings. Normally, this temperature differential contributes to the resistance of healthy mammals to most fungi. As a result, global warming may open up new colonization opportunities for fungal pathogens.

II. Immune Effector Mechanisms

i) Induced Innate Immunity

Mechanisms of induced innate immunity are very important for controlling fungal infections. TLR2, TLR4, DC-SIGN, Dectin-1, Dectin-2, CR3, MBL and MR all recognize PAMPs supplied by molecules in fungal cell walls or on the fungal cell surface. These

TABLE 13-11	Examples of Fungi and the Diseases They Cause
Pathogen	**Disease**
Aspergillus species	Respiratory infections, acute and chronic pneumonias
Blastomyces dermatitidis	Blastomycosis; skin lesions, acute and chronic pneumonias
Candida species	Yeast infections, vaginitis, cystitis
Cryptococcus neoformans	Meningitis, pneumonia
Histoplasma capsulatum	Histoplasmosis; lesions in the lung
Mucormycotina	Mucormycosis; lesions in the brain, sinuses and lung; eye swelling
Paracoccidioides brasiliensis	Ulcerations of mucosae of nose and mouth
Pneumocystis (carinii) jiroveci	PCP pneumonia and lung damage
Dermatophytes	Skin, nail and hair infections

A glucan is a polymer of glucose molecules, a chitin is a polymer of N-acetylglucosamine molecules, and a mannan is a chain of mannose molecules added to a protein.

PAMPs include fungal β-glucans, chitins, mannans and oligomannosides. In particular, Dectin-1 and Dectin-2 are CLRs specialized for the recognition of fungal PAMPs. Dectin-1 binds to cell wall β-glucans, whereas Dectin-2 recognizes mannose structures common to the hyphal forms (only) of many fungi. Dectin-1 contains its own ITAM in its cytoplasmic tail, whereas Dectin-2 pairs with the FcRγ signaling chain to transduce intracellular signaling. This signaling results in host cell production of pro-inflammatory cytokines and leukotrienes instrumental in removing the invader. Dectin-1 is widely expressed on DCs, monocytes, macrophages, neutrophils and some T cells. Dectin-2 expression is more restricted, being limited to monocytes and macrophages. The importance of Dectin-1 has been highlighted by recent studies of a family in Holland, some members of which lack Dectin-1 expression and suffer from recurrent fungal infections. Individuals with deficits in the expression of other C-lectin receptors are also unusually susceptible.

Attack by a fungus often induces a host to assemble the NLRP3 inflammasome that drives IL-1 and IL-18 production. Host cell-derived DAMPs associated with fungal infections include the S100 proteins that bind to RAGE, a PRR introduced in Chapter 3. As was true for protozoan parasites, animals lacking expression of the TLR signaling adaptor protein MyD88 are very vulnerable to fungal infections. Neutrophils and macrophages activated by PRR engagement carry out vigorous phagocytosis and produce powerful antifungal defensins that induce osmotic imbalance in the fungal cells (**Fig. 13-6, #1**). Fungal cells may also trigger their own engulfment by some cell types that are not normally phagocytic, including epithelial and endothelial cells. As well as defensins, activated neutrophils and macrophages secrete copious quantities of chemokines along with IL-1, IL-12 and TNF, which are directly toxic to fungal cells. γδ T cells appear to play a significant role in antifungal defense at the mucosae (**#2**), a hypothesis based on the fact that mice engineered to lack γδ T cells show increased susceptibility to yeast infections. Activated NK cells stimulated by the presence of IL-12 contribute to fungal cell killing via TNF secretion (rather than by natural cytotoxicity) (**#3**). IFNγ produced by NK cells contributes to macrophage hyperactivation that can eventually lead to granuloma formation.

NOTE: Expression of SNPs of many of the molecules associated with antifungal responses, including TLRs, CLRs, cytokines and chemokines, leads to increased susceptibility to fungal infections. For example, a TLR2 SNP that results in increased TNF production but decreased IL-18 and IFNγ secretion by leukocytes is associated with more severe *C. albicans* infections. Similarly, TLR4 and TLR9 SNPs have been linked to *Aspergillus* infections of the lungs. SNPs in Dectin-1 and NLRP3 that impair cytokine production leave the host vulnerable to fungal infections of the mucosae and/or skin. Likewise, SNPs in MBL or MASP that alter the lectin-mediated complement activation pathway allow *Aspergillus* to set up shop, as do SNPs in TNF or TNFR1/2 that decrease their expression. Interestingly, SNPs in IFNγ and IL-4 that *increase* production of these cytokines can dysregulate the adaptive response so much that fungal invasion succeeds. Clinicians may soon be able to use all these SNPs as genetic markers to identify patients who are at high risk for developing fungal diseases if their immune systems are deliberately suppressed. Preventive treatment can then be offered to these individuals.

ii) CD4+ T Cells

T cell responses against fungi are shaped by the DCs activating them, and DCs have demonstrated a remarkable ability to distinguish among various types of fungi based on patterns of PRR engagement. Most DCs that acquire fungal antigens and experience TLR signaling undergo maturation and activate naïve T cells to generate Th1 effectors. These T cells secrete the copious quantities of IFNγ needed to complete macrophage hyperactivation (**Fig. 13-6, #4**). DCs whose MRs and Dectin-1 molecules are engaged by fungal PAMPs are stimulated to produce IL-23 and IL-10, which initiate the generation of Th17 effectors. These cells contribute mainly to protection against fungi attempting to colonize the mucosae, including *C. albicans* and some *Aspergillus* species. Antifungal Th17 cells secrete IL-17 and IL-22 that contribute to the neutrophil recruitment and proliferation crucial for fungal clearance, and support Th1 responses. Both αβ Th17 and γδ Th17 cells have been found in mice experimentally infected with *C. albicans*.

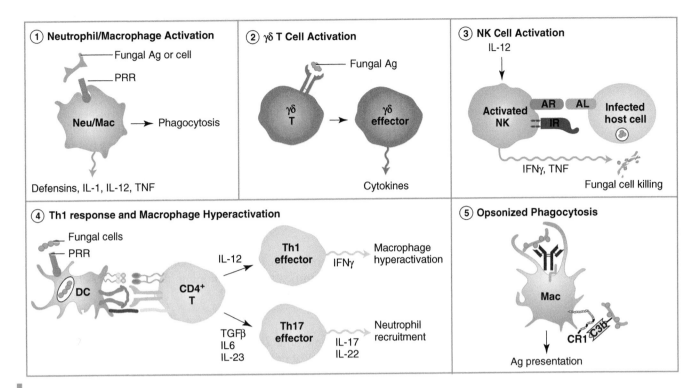

Fig. 13-6
Major Mechanisms of Immune Defense against Fungi

(1) Activated neutrophils and macrophages carry out phagocytosis of fungal cells and secrete antifungal peptides and cytokines.
(2) Activated mucosal γδ T cells generate effectors that secrete cytokines. **(3)** Activated NK cells kill fungi by secreting cytotoxic cytokines rather than by natural cytotoxicity. **(4)** Fungal PRR ligands activate DCs that drive Th1 or Th17 effector differentiation, leading to either macrophage hyperactivation and granuloma formation or neutrophil recruitment. **(5)** Fungi coated in either antifungal antibody or C3b undergo opsonized phagocytosis by macrophages (and neutrophils; not shown). Note that the structure of the fungal cell wall allows fungal cells to resist MAC-mediated lysis.

Th2 responses are comparatively rare during fungal infections and not very effective. Those patients who respond to fungi with Th2 responses instead of Th1 responses show poor resistance to these pathogens. For example, patients who mount Th1 responses against *Paracoccidioides brasiliensis* experience only mild and transient paracoccidioidomycosis (ulceration of the mucosae in the mouth and nose), whereas those who mount Th2 responses have severe disease that relapses frequently. Interestingly, the female hormone estradiol promotes Th1 responses in this context, likely accounting for the fact that paracoccidioidomycosis occurs much less often in women than in men.

iii) Humoral Defense

Conventional antibodies are thought to contribute in only a limited way to defense against fungi that manage to invade the body. Antibody-mediated opsonization may promote phagocytosis (**Fig. 13-6, #5**) and thus contribute to the clearance and presentation of fungal antigens. However, antibodies produced by a unique B cell subset called *B-1 cells* are critical for antifungal defense. In Chapters 4 and 5, we described conventional B cell production of antibodies of the IgM, IgG, IgA and IgE isotypes in response to antigen. However, the serum of normal healthy individuals contains IgM antibodies that are pre-existing and generated without the apparent need for exogenous antigen. These proteins, which are called **natural antibodies**, are produced by B-1 cells scattered in the body's periphery rather than concentrated in the bone marrow or secondary lymphoid tissues. Many natural antibodies are specific for the β-glucans and chitins in fungal cell walls.

NOTE: There is a fine balance that must be maintained during inflammatory responses to fungi to ensure host health. Acute inflammation is helpful in getting rid of the invader, but chronic inflammation that fails to clear the pathogen appears to encourage fungal persistence. Severe fungal infections can occur in patients who have started to recover function of their immune systems after immunosuppression has been used to allow transplantation, or after AIDS has been brought under control by anti-HIV drug therapy. These individuals sometimes show clinical signs of a disease labeled *immune reconstitution inflammatory syndrome (IRIS)*. As the immune system starts to reconstitute itself, it mounts an excessive inflammatory response to an opportunistic pathogen such as a fungus that has taken hold during the immunosuppression. Rather than resolving the infection, this inflammation disrupts immune system regulation and compromises fungal clearance, paradoxically making the condition worse.

III. Evasion Strategies

Many fungi adopt different morphological forms at different stages in their life cycle, affording them multiple opportunities to evade immune defense as described in the following sections and summarized in **Table 13-12**.

i) Avoid PRRs

To avoid detection by PRRs, some fungi shift their morphological form from yeast-like to hyphae to reduce the presence of detectable PAMPs. Other fungi take alternative steps to obscure their PAMPs. For example, *C. albicans* expresses a protein that covers the β-glucan molecules in its cell wall, preventing recognition by phagocytes bearing Dectin-1. Similarly, *Histoplasma capsulatum* hides its β-glucan under a layer of α-glucan, which does not bind to Dectin-1. *Aspergillus fumigatus* covers its spores in proteins such as melanin that block recognition of the conidial surface by innate leukocytes. In contrast, *Pneumocystis jiroveci* simply changes the expression of its major surface glycoproteins, confounding both PRR and antibody recognition. Lastly, the cell walls and membranes of some fungi simply lack PAMPs and other structures that might trigger PRR-mediated recognition, forestalling both phagocytosis and the binding of complement components.

Pneumocystis jiroveci is the new name of *Pneumocystis carinii*, the fungus that causes pneumonia in AIDS patients (see Ch. 15).

TABLE 13-12	Evasion of the Immune System by Fungi
Immune System Element Thwarted	**Fungal Mechanism**
Host PRRs	Shift between different morphological forms Hide lectin-binding cell wall components under another molecular layer Change expression of major surface molecules Have no LPS or peptidoglycan in cell wall
Phagocytosis	Form a very large mass Have a capsule or produce a protein that blocks phagocytosis Detoxify NO or inhibit its production Secrete factors that neutralize the phagosomal environment Alter phagosome maturation Escape a phagocyte by vomocytosis
Complement	Have a cell wall that blocks access to the cell membrane Recruit host RCA proteins to the fungal surface Produce proteases that digest complement components
Th response	Secrete molecules suppressing macrophage cytokine production and B7 expression Secrete molecules inducing ineffective Th2 response rather than Th1 response Secrete molecules blocking APC and/or T cell differentiation or proliferation Activate regulatory T cells
Antibodies	Have a multistage life cycle Have a capsule not easily recognized by antibodies Secrete molecules blocking B cell differentiation, proliferation

ii) Avoid Phagocytes

When species such as *C. albicans* and *A. fumigatus* are present in their multi-nucleate hyphal morphology, they form a mass too large to be captured by phagocytosis. In a similar vein, *Cryptococcus neoformans* forms very large polyploid "titan cells" that are encased in a thick polysaccharide capsule. The sheer size of these cells and the composition of their capsule provide a formidable barrier to phagocytosis. *C. neoformans* also secretes a small protein called App1 that is induced under conditions of low glucose, such as are routinely found in the lung (the primary site of attack of this pathogen). App1 interacts with CR2 and CR3 in such a way as to block macrophage phagocytosis of any fungal cells opsonized by C3d or iC3b, respectively.

Even when fungal cells are engulfed by phagocytes, many of these pathogens have ways of enzymatically detoxifying the NO produced by phagocyte iNOS activity. *C. albicans* secretes an unknown inhibitory factor to accomplish this task, whereas *C. neoformans* and *Blastomyces dermatitidis* establish inhibitory intercellular contacts with macrophages. *H. capsulatum* neutralizes phagosomal killing by secreting factors that alter the phagolysosomal environment and render it non-acidic. Other fungi appear to alter intracellular endosomal trafficking such that phagosome maturation is impaired. *C. neoformans* has perfected a novel route that allows it to escape from macrophage phagosomes, and indeed the whole leukocyte, without damage to either cell. The fungal cell secretes molecules that weaken the membrane of a macrophage phagosome, which then fuses with the macrophage plasma membrane. The contents of the phagosome, including the fungal cell, are then expelled from the macrophage into the extracellular milieu by exocytosis. Some immunologists have given this process the enchanting name "vomocytosis."

> "Polyploid" means that more than one set of chromosomes is contained within the nucleus of a single cell.

iii) Avoid Complement

Fungal cells are frequently opsonized by complement products. However, although fungal cells activate the complement cascade, their cell walls and recruitment of RCA proteins render them generally resistant to complement-mediated lysis. For example, *C. albicans* expresses several proteins that can recruit host RCA proteins, including Factor H and C4bp, to the fungal surface. The spores of *A. fumigatus* also produce a Factor H-binding protein. In addition, both *C. albicans* and *A. fumigatus* secrete proteases that can digest numerous host proteins, including the complement components C3b, C4b and C5.

iv) Promote a Less Effective Th Response

Many fungi produce toxins and other molecules that have immunosuppressive effects and/or promote immune deviation to an ineffective Th2 response at the expense of a Th1 response. Fungal ligands that engage certain PRRs on epithelial cells induce them to produce TSLP and IL-25. These cytokines tend to amplify Th2 responses, which are weak against fungi, and also promote iTreg cell generation. The fungus thus gains ground in winning host tolerance to itself. Infection by *C. albicans* induces host cell production of IL-10, as does a switch in fungal cell morphology by *A. fumigatus*. Other fungal molecules inhibit the transcription of genes needed for the differentiation of activated T cells. Still other fungal mediators suppress lymphocyte proliferation or macrophage cytokine production. A polysaccharide in the capsule of *C. neoformans* blocks IL-12 production by monocytes/macrophages, downregulates macrophage B7 expression, and activates regulatory T cells. Both *C. neoformans* and *H. capsulatum* (among other species) also release membrane-bound *exosomes* that contain numerous virulence factors, such as anti-oxidant proteins and capsule biosynthetic enzymes, that help to sustain the fungal infection and blunt the host's response to it. The contents of these so-called "virulence bags" remain under investigation.

v) Avoid Antibodies

The various morphologies a fungus can adopt during its life cycle may result in a constantly changing array of surface epitopes that can confound antibody recognition. In addition, even if an antibody does bind to the fungal capsule, its thick polysaccharide structure may inhibit subsequent FcR-mediated phagocytosis by innate leukocytes. Some fungi also secrete molecules that inhibit the transcription of genes needed for B cell differentiation or proliferation, or block these processes directly.

G. Prions

Prions are the pathogens that cause *spongiform encephalopathies* (SEs), which are rare, lethal neurodegenerative diseases characterized by lesions that render the brain "sponge-like." The affected host develops dementia and loses motor function control shortly before death. The major human SE diseases are called *variant Creutzfeldt–Jakob disease* (vCJD) and Kuru (the "shaking disease" of Papua New Guinea). Animal SEs include *scrapie* in sheep and *bovine spongiform encephalopathy* (BSE, or "mad cow disease") in cattle. These disorders are associated with the ingestion of infected tissues from an animal suffering from an SE. For example, a cow that consumes cattle feed made from the remains of a contaminated sheep may contract BSE, whereas a human who enjoys a hamburger made from the meat of the infected cow may eventually succumb to vCJD.

I. Disease Mechanisms

Prions are essentially transmissible proteins devoid of nucleic acid. Structurally, a prion is a conformational isomer of a normal mammalian surface glycoprotein. In the original studies of scrapie in sheep, this normal glycoprotein was denoted PrP^c (prion protein, cellular) and the altered protein was denoted PrP^{sc} (prion protein, scrapie). PrP^{res} (prion protein, resistant to proteases) is now used to denote the altered protein in any species. When PrP^{res} is introduced into a healthy animal, it acts as a template for the refolding of existing host PrP^c molecules into additional copies of PrP^{res}. The disease-causing prion thus effectively "replicates" itself in a mass conversion of the host's PrP^c molecules to the PrP^{res} conformation. The misfolded PrP^{res} protein has profoundly altered properties compared to PrP^c. As a result, the PrP^{res} protein can enter neurons in the brain and induce the degeneration of this organ that is manifested as the clinical signs of SE. Intriguingly, no other part of the body appears to be affected by the presence of PrP^{res}.

The description just given is of the infectious form of prion disease, in which there is no mutation of the PrP^c gene of the host and no change in the amino acid sequence of the affected PrP^c proteins: the disorder is purely one of protein misfolding. However, rare cases of prion disease do arise spontaneously due to a mutation of an individual's PrP^c gene that results in production of a PrP^{res} protein. As long as the tissues bearing the PrP^{res} protein are not later ingested by another animal, there is no transmission of the disease. In rare cases, however, the mutation may occur in a germ cell such that the disease is inherited by the affected animal's offspring.

NOTE: Until quite recently, the function of the normal PrP^c protein, other than to serve as a template for production of PrP^{res} proteins, was unknown. Studies in the past few years have revealed that PrP^c plays roles in several neuronal processes, including cell adhesion, neurite outgrowth, ion channel activity, and excitability. Intriguingly, the normal PrP^c protein also appears to mediate the assembly of the toxic oligomers of amyloid-β peptide that accumulate in the brains of Alzheimer's disease patients. These observations raise the possibility that therapeutic mAbs directed against the PrP^c domains that promote oligomer assembly might provide an effective treatment for Alzheimer's disease.

II. Immune Effector Mechanisms

Prion infection destroys the brain without inducing either a humoral or cell-mediated adaptive response. The host's T cells are usually tolerant to the infectious PrP^{res}

protein, as it is merely a naturally occurring self protein with a modified secondary structure. By extension, in the absence of the activation of prion-specific T cells, no Td humoral response can be mounted. Furthermore, although the "foreign" conformation of PrP^{res} might be recognized by the BCR of a B cell, the antigen itself cannot act as a Ti immunogen because it has neither the large size nor multivalency needed to activate B cells. Thus, no adaptive responses are mounted against prions. Fortunately, however, new evidence is coming to light indicating that innate immune defense against prions does exist and may help to at least slow the course of SE disease. Mice engineered to lack the transcription factor IRF3, which is important for some TLR signaling pathways, showed faster onset of prion disease than control animals. In cell cultures, cells that were infected with prions and treated to specifically inactivate IRF3 accumulated higher amounts of PrP^{res} protein, whereas cells that were engineered to express abnormally large amounts of IRF3 sustained lower levels of prion infection. Similar results have been found for mice with an inactivating mutation of TLR4. The role of innate immunity in combatting prions remains under investigation.

This brings us to the end of our description of the mechanisms of natural immune defense against pathogens. In the next chapter of this book, we discuss the "manufactured" immunity to pathogens created by vaccination.

Chapter 13 Take-Home Message

- There are six major types of pathogens: extracellular bacteria, intracellular bacteria, viruses, parasites, fungi and prions.

- Innate immunity mediated by neutrophils, NK cells, NKT cells, γδ T cells, complement and microbicidal molecules either foils the establishment of infection or slows it down until adaptive immune mechanisms can target the pathogen more effectively.

- The adaptive elements that will be most effective depend on the nature of the pathogen: extracellular versus intracellular, small versus large, fast- versus slow-replicating.

- Most extracellular entities can be coated in antibody and cleared by antibody- and complement-mediated mechanisms. Parasitic worms are targeted by IgA and IgE antibodies that prevent the worm from anchoring in the host. IgE can trigger the degranulation of mast cells, basophils and eosinophils and the release of mediators that work to expel the worm and degrade its tissues.

- Intracellular bacteria and parasites and replicating viruses must be eliminated by cell-mediated immunity. CTLs, NK cells, NKT cells and γδ T cells secrete cytotoxic cytokines and/or carry out target cell cytolysis. Macrophage hyperactivation and granuloma formation may be needed to confine persistent invaders.

- In general, Th1 and Th17 responses support cell-mediated immunity against internal threats, whereas Th2 responses are needed for humoral responses against external threats.

- Many pathogens have evolved complex strategies to evade the immune response: avoiding recognition; antigenic variation; avoiding or inactivating phagocytosis; shedding or inactivating complement components; acquiring host RCA proteins; cleaving host FcRs; inducing host cell apoptosis; and manipulating the host's immune response or cell cycle.

Did You Get It? A Self-Test Quiz

Introduction and Section A

1) Characterize the six major classes of pathogens.

2) How is disease distinct from infection?

3) What is immunopathic disease?

4) Distinguish between opportunistic and invasive pathogens.

5) Outline two ways each by which the skin and mucosae defend the body's external surfaces.

6) Name four types of leukocytes that mediate subepithelial innate defense and give examples of their effector mechanisms.

7) What is an SNP, and how can it be useful to clinical immunologists?

8) Name three pathogens that induce NLRP inflammasome assembly.

Section B

1) Distinguish between exotoxins and endotoxins.

2) Name two diseases caused by exotoxins.

3) What is endotoxic shock, and why is it considered immunopathic?

4) Describe three ways in which antibodies help protect the body against extracellular bacteria.

5) What is an antitoxin, and what does it do?

6) Describe an extracellular bacterial infection combatted by Th17 cells.

7) Outline two mechanisms each by which extracellular bacteria can evade host PRRs; antibodies; phagocytosis; complement.

Section C

1) Can you define these terms? vector, granuloma

2) How are neutrophils, NK cells, and γδ T cells helpful in combatting intracellular bacteria?

3) Name three TLRs that contribute to defense against intracellular bacteria.

4) Why is macrophage hyperactivation effective against intracellular bacteria?

5) Briefly outline the sequence of cellular and molecular events that lead from an intracellular bacterium successfully gaining a foothold in the body to CTL-mediated elimination of cells infected with that bacterium.

6) How do various subsets of CD4+ Th effectors contribute to defense against intracellular bacteria?

7) Describe how the different forms of leprosy illustrate the importance of Th responses to intracellular bacteria.

8) How can antibodies be useful for the control of intracellular bacteria?

9) Name three types of leukocytes crucial for granuloma formation and outline the role each plays.

10) Outline five mechanisms by which intracellular bacteria can evade phagosomal death.

11) Outline one mechanism each by which intracellular bacteria can avoid host PRRs; antibodies; T cells.

Section D

1) Can you define these terms? acute disease, chronic disease, persistent infection, latency, oncogenic, iNOS, syncytia

2) What is the antiviral state, and how is it induced?

3) How do pDCs and NK cells combat viruses, and why are these cells crucial for early defense?

4) Name three PRRs important for antiviral defense and the PAMP/DAMP each of these PRRs recognizes.

5) Describe three ways in which CD4+ T cells contribute to immune defense against viruses.

6) Describe three ways in which CD8+ T cells contribute to immune defense against viruses.

7) Describe four ways in which antibodies contribute to immune defense against viruses.

8) Describe two mechanisms of latency and outline how viral latency can be reversed.

9) Distinguish between antigenic drift and antigenic shift.

10) Distinguish between epidemic and pandemic.

11) Describe three ways each by which viruses can resist attack by CTLs, CD4+ T cells, and complement.

12) Outline three ways each in which viruses can counteract the antiviral state; interfere with host PRRs; inhibit the induction of host cell apoptosis; interfere with host cell cytokines.

13) Give two ways each in which viruses compromise NK cell, DC and antibody responses.

14) What is ER stress, and how can it be beneficial to a host?

15) Give three examples of how miRNA production can help a virus to persist.

Section E

1) Distinguish between ectoparasites and endoparasites.

2) What do protozoan and helminth pathogens have in common?

3) How does a multistage life cycle present a challenge for the immune system?

4) Give two examples of TLRs important for antiprotozoan defense and the PAMP/DAMP each recognizes.

5) How do antibodies combat protozoans?

6) Give four reasons why the Th1 response is crucial for antiprotozoan defense.

7) What CTL-mediated mechanism is most effective against protozoans and when?

8) What four types of leukocytes are involved in the Th2 response against helminth worms, and why are these cells important?

9) Outline three ways in which the contents of eosinophil granules combat helminths.

10) What three cytokines are important for antihelminth defense?

11) Give one example each of a disease caused by an ecto-parasite itself and an ectoparasite acting as a vector.

12) Outline two ways each by which protozoans can avoid antibodies; phagosomes; complement.

13) Describe how protozoans and helminth worms interfere with T cell responses.

Section F

1) Can you define these terms? mycelium, hyphae, dimorphic, dermatophyte, mycoses, glucan, vomocytosis

2) Why is global warming a positive development for fungi but not humans?

3) Name two CLRs vital for antifungal defense and contrast their properties.

4) Describe four mechanisms of innate defense that combat fungi.

5) Describe the contribution of Th1, Th2 and Th17 cells to antifungal defense.

6) What are natural antibodies, and what do they recognize?

7) Fungal cells are lysed by complement. True or false?

8) Outline three ways in which the structure or products of fungi promote evasion of immune responses.

9) Give three ways by which a fungus can promote a less effective Th response; avoid humoral defense.

Section G

1) Can you define these terms? SE, vCJD, BSE, PrP^{res}, PrP^c

2) What is a prion, and how does it cause disease?

3) What is the connection between PrP^c and Alzheimer's disease?

4) Give two reasons why prions are poorly immunogenic.

5) Outline two pieces of scientific evidence suggesting that prions may induce innate immune responses.

1) We noted in this chapter that a subset of B lymphocytes known as B-1 cells produces "natural antibodies" that are of the IgM isotype only and circulate in the periphery prior to antigen exposure. Keeping the idea of pattern recognition in mind, how might you argue that, despite being produced by B lymphocytes, these antibodies are actually part of the innate immune response to pathogens?

2) In many influenza epidemics and pandemics throughout history, those people who became most seriously ill or died were either the very young, the very old, or those who were immunocompromised in some way. In contrast, during the 2009 H1N1 influenza pandemic, the World Health Organization noted the following:

   a) Those who became seriously ill were often young adults in otherwise good health.

   b) Respiratory failure and shock were common causes of death.

How might differences in immune responses to different influenza strains explain these observations?

3) If you were a nefarious individual attempting to create an intracellular pathogen that could prevent the most effective immune responses directed at it by the host, what mature cell type would you design the pathogen to infect and disable?

Would You Like To Read More?

Akira, S., Uematsu, S., & Takeuchi, O. (2006). Pathogen recognition and innate immunity. *Cell, 124*(4), 783–801.

Belkaid, Y., & Tarbell, K. (2009). Regulatory T cells in the control of host-microorganism interactions. *Annual Review of Immunology, 27,* 551–589.

Biasini, E., Turnbaugh, J. A., Unterberger, U., & Harris, D. A. (2012). Prion protein at the crossroads of physiology and disease. *Trends in Neurosciences, 35*(2), 92–103.

Boss, I. W., & Renne, R. (2011). Viral miRNAs and immune evasion. *Biochimica Et Biophysica Acta, 1809*(11–12), 708–714.

Brereton, C. F., & Blander, J. M. (2011). The unexpected link between infection-induced apoptosis and a TH17 immune response. *Journal of Leukocyte Biology, 89*(4), 565–576.

Elinav, E., Strowig, T., Henao-Mejia, J., & Flavell, R. A. (2011). Regulation of the antimicrobial response by NLR proteins. *Immunity, 34*(5), 665–679.

Goldszmid, R. S., & Sher, A. (2010). Processing and presentation of antigens derived from intracellular protozoan parasites. *Current Opinion in Immunology, 22*(1), 118–123.

Hansen, T. H., & Bouvier, M. (2009). MHC class I antigen presentation: Learning from viral evasion strategies. *Nature Reviews Immunology, 9*(7), 503–513.

Horst, D., Verweij, M. C., Davison, A. J., Ressing, M. E., & Wiertz, E. J. (2011). Viral evasion of T cell immunity: Ancient mechanisms offering new applications. *Current Opinion in Immunology, 23*(1), 96–103.

Jonjic, S., Babic, M., Polic, B., & Krmpotic, A. (2008). Immune evasion of natural killer cells by viruses. *Current Opinion in Immunology, 20*(1), 30–38.

Karasuyama, H., Wada, T., Yoshikawa, S., & Obata, K. (2011). Emerging roles of basophils in protective immunity against parasites. *Trends in Immunology, 32*(3), 125–130.

Mantovani, A., Cassatella, M. A., Costantini, C., & Jaillon, S. (2011). Neutrophils in the activation and regulation of innate and adaptive immunity. *Nature Reviews Immunology, 11*(8), 519–531.

Marakalala, M. J., Kerrigan, A. M., & Brown, G. D. (2011). Dectin-1: A role in antifungal defense and consequences of genetic polymorphisms in humans. *Mammalian Genome, 22*(1–2), 55–65.

McAleer, J. P., & Kolls, J. K. (2011). Mechanisms controlling Th17 cytokine expression and host defense. *Journal of Leukocyte Biology, 90*(2), 263–270.

Modlin, R. L. (2010). The innate immune response in leprosy. *Current Opinion in Immunology, 22*(1), 48–54.

Peleg, A. Y., & Hooper, D. C. (2010). Hospital-acquired infections due to Gram-negative bacteria. *New England Journal of Medicine, 362*(19), 1804–1813.

Romani, L. (2011). Immunity to fungal infections. *Nature Reviews Immunology, 11*(4), 275–288.

Villasenor-Cardoso, M. I., & Ortega, E. (2011). Polymorphisms of innate immunity receptors in infection by parasites. *Parasite Immunology, 33*(12), 643–653.

Vaccination

The aim of military training is not just to prepare men for battle, but to make them long for it.

Louis Simpson

A. Introduction

Vaccination is a clinical application of immunization designed to artificially help the body to defend itself. A **vaccine** against infection is a modified form of a natural immunogen, which may be either the whole pathogen, one of its components, or a toxin. A vaccine does not cause disease when administered but induces the healthy host (the *vaccinee*) to mount a primary response against epitopes of the modified immunogen and to generate large numbers of memory B and T cells. In an unvaccinated individual (**Fig. 14-1**, left panel), naïve B and T cells capable of combatting an infecting pathogen are present in relatively low numbers when the pathogen is first encountered. A primary immune response is all that can be mounted so that, in many cases, the individual becomes sick until antibodies and/or effector T cells can act to clear the attacker. In a vaccinated individual (**Fig. 14-1**, right panel), a collection of circulating antibodies and an expanded army of pathogen-specific memory B and T cells have already been generated prior to a first exposure to the natural pathogen. When the natural pathogen attacks, the circulating antibodies provide a degree of immediate protection from the invader. In addition, the memory B and T cells are quickly activated, and a secondary response is mounted that rapidly clears the infection before it can cause serious illness. This type of vaccination is called *prophylactic vaccination* because it is intended to prevent disease.

NOTE: Modern air travel has made prophylactic vaccination more important than ever, because this mode of transportation has greatly facilitated the spread of contagious pathogens. For example, in 2010, an unvaccinated child who contracted measles in Europe transmitted the virus to a fellow passenger during a flight to the U.S. This passenger then attended a conference and unwittingly exposed 270 other individuals to the disease.

WHAT'S IN THIS CHAPTER?

MAK: Primer to the Immune Response. http://dx.doi.org/10.1016/B978-0-12-385245-8.00014-5

Fig. 14-1
The Principle of Vaccination

The experience of an unvaccinated individual (left panel) is compared with that of a vaccinated individual (right panel).

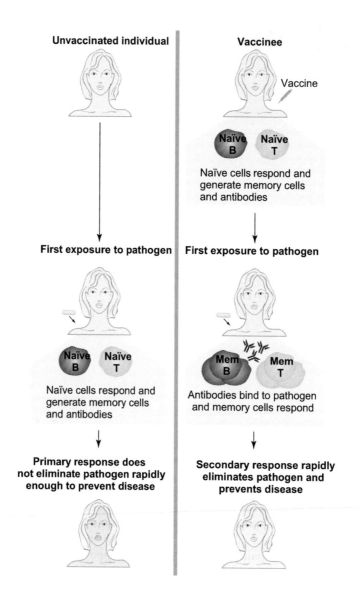

Unvaccinated individual

Vaccinee

Vaccine

Naïve B   Naïve T

Naïve cells respond and generate memory cells and antibodies

First exposure to pathogen   First exposure to pathogen

Naïve B   Naïve T

Naïve cells respond and generate memory cells and antibodies

Mem B   Mem T

Antibodies bind to pathogen and memory cells respond

Primary response does not eliminate pathogen rapidly enough to prevent disease

Secondary response rapidly eliminates pathogen and prevents disease

The effectiveness of vaccination is based on the principles of lymphocyte clonal expansion (refer to Fig. 1-6) and the differences between the primary and secondary immune responses (refer to Fig. 5-12).

Vaccination can be thought of as a form of **active immunization**, because the individual is administered a pathogen antigen and his/her body is responsible for activating the lymphocytes and making the antibodies necessary to provide defense against future assaults. In contrast, **passive immunization** is the term used to describe the transfer of protective antibodies from an immune individual to an unimmunized individual. Passive immunization is explored further in **Box 14-1**.

Today's best known vaccination success story is the global campaign of the World Health Organization (WHO) to eradicate smallpox. In 1967, the WHO began its coordination of 200,000 health workers who took 10 years to vaccinate the world's population in its remotest corners. Between 1976 and 1979, only one case of smallpox was recorded, leading to the declaration in 1980 that smallpox had been officially eradicated (**Plate 14-1**). A similar global immunization program against rinderpest is currently pushing this pathogen toward extinction (**Box 14-2**). Control of several other serious diseases, including diphtheria, tetanus, pertussis, measles, mumps and rubella, has also been achieved through vaccination, at least in the developed world (**Table 14-1**). Although mass vaccination programs do have a monetary cost, the expenses associated with a population becoming infected are much higher. According to the WHO, for every U.S. $1 million spent on global childhood vaccines, U.S. $4 million in long-term health care savings are realized. Several reports have documented the number of illnesses and deaths prevented and cost savings achieved by introducing mass immunization with effective vaccines against *Neisseria meningitidis*, *Pneumococcus*, and

Box 14-1   Passive Immunization

Passive immunization occurs when an individual receives pre-formed specific antibodies from an exogenous source. A good example of natural passive immunization is the transfer of maternal antipathogen antibodies to the developing fetus through the umbilical circulation, and later to the newborn in colostrum and breast milk. The fetus/newborn is thus protected from attack during the period when his/her own immune system is too immature to be activated. Passive immunization in a medical context can be used both pre- and post-exposure to treat an unvaccinated individual who has just been, or expects to be, exposed to a certain pathogen or toxin. Passive immunization is also helpful when no vaccine exists for a pathogen, or where the vaccine is either not 100% efficacious or not widely used. For example, someone who has suffered a dog bite may have been exposed to rabies virus. The injured person receives a preparation of antirabies virus antibodies that heads off viral replication and thus the disease. Unlike vaccination, passive immunization provides immediate protection because the 7–10 day lag necessary to mount an adaptive response has been eliminated. However, because the individual's own immune system has not been stimulated, no immunological memory is generated, and protection lasts only for days to months (rather than years). Passive immunization can also provide protection for individuals who cannot be vaccinated because they are immunodeficient or immunocompromised.

The antibodies used for passive immunization are usually derived from the pooled serum of human donors selected for high titers of the desired antibodies. Prior to serum collection, these volunteers are repeatedly immunized with a pathogen or antigen of interest to generate high levels of specific IgG antibodies; these antibodies are then recovered as a polyclonal antibody preparation. Alternatively, synthetic mAbs genetically engineered for production *in vitro* and tailored for use in humans can be employed. Antibodies suitable for passive immunization against many infectious agents, including cytomegalovirus, hepatitis A virus, hepatitis B virus, rabies virus, respiratory syncytial virus, and varicella zoster virus, are now commercially available. Antibody preparations also exist to counter several important bacterial toxins, including botulinum toxin, diphtheria toxin, and tetanus toxin, as well as the venom of certain species of insects, arachnids and snakes. These latter preparations are sometimes called "antivenoms" or "antivenins."

H. influenzae. Despite these successes, however, there are still major challenges in vaccination, especially in the developing world. Communicable diseases represent 4 of the top 10 causes of mortality worldwide and account for over 9 million deaths annually (**Table 14-2**). In some cases (e.g., measles, polio), logistical, social or economic factors prevent delivery of vaccines that are currently available and would be effective if administered. In other cases (e.g., malaria, AIDS and tuberculosis), the elusive nature of the pathogen involved has hindered the development of an effective vaccine.

Plate 14-1
WHO Declaration of Smallpox Eradication in 1980

[Reproduced by permission of the Magazine of the World Health Organization.]

TABLE 14-1 | **Reduction in Several Severe Diseases in the U.S. due to Vaccination**

Disease	Maximum Annual Cases Reported	Year Maximum Number of Cases Observed	Cases Reported in 2001	Percentage Reduction
Diphtheria	206,939	1921	2	99.99
Hib meningitis	20,000	1992	51	99.75
Measles	894,134	1941	96	99.99
Mumps	152,209	1968	216	99.86
Pertussis	265,269	1934	4788	98.20
Polio	21,269	1952	0	100.00
Rubella	57,686	1969	19	99.97
Smallpox	48,164	1901	0	100.00
Tetanus	1560	1901	26	98.34

[Adapted from Rappuoli, R. *et al.* (2002) *Science* 297, 937–939.]

TABLE 14-2 | **The WHO Top 10 Leading Causes of Death Globally and by Broad Income Group***

	Global		High-Income Countries		Low-Income Countries	
	Disease	Deaths	Disease	Deaths	Disease	Deaths
1	Ischemic heart disease	7,250,000	Ischemic heart disease	1,420,000	**Lower respiratory infections**	**1,050,000**
2	Stroke and other cerebrovascular disease	6,150,000	Stroke and other cerebrovascular disease	790,000	**Diarrheal diseases**	**760,000**
3	**Lower respiratory infections**	**3,460,000**	Cancers of trachea, bronchus, lung	540,000	**HIV/AIDS**	**720,000**
4	Chronic obstructive pulmonary disease	3,280,000	Alzheimer and other dementias	370,000	Ischemic heart disease	570,000
5	**Diarrheal diseases**	**2,460,000**	**Lower respiratory infections**	**350,000**	**Malaria**	**480,000**
6	**HIV/AIDS**	**1,780,000**	Chronic obstructive pulmonary disease	320,000	Stroke and other cerebrovascular disease	450,000
7	Cancers of trachea, bronchus, lung	1,390,000	Colorectal cancers	300,000	**Tuberculosis**	**400,000**
8	**Tuberculosis**	**1,340,000**	Diabetes mellitus	240,000	Prematurity, low birth weight	300,000
9	Diabetes mellitus	1,260,000	Hypertensive heart disease	210,000	Birth asphyxia, birth trauma	270,000
10	Road traffic accidents	1,210,000	Breast cancer	170,000	**Neonatal infections**	**240,000**

Note: Global numbers for a given disease include values from low, medium (not shown), and high-income countries.
*World Health Organization, June 2011. Entries in bold are infectious diseases. Figures are per year.

B. Vaccine Design

In designing a vaccine, one must take into consideration several factors that can affect its success. These factors include the nature of the pathogen to be combatted, the efficacy of the vaccine, and the safety of the vaccine (**Table 14-3**). These parameters are introduced here, and vaccine safety is further discussed in Section F of this chapter after specific types of prophylactic vaccines are described.

I. Nature of the Pathogen

Several characteristics of the pathogen itself can have a major influence on the success of a vaccine. Good pathogen candidates for vaccination programs are those that cause acute rather than chronic disease because these pathogens tend to invoke a vigorous

Box 14-2 Beyond Human Medicine: The Eradication of Rinderpest

Intractable pathogens can be just as horrendous a problem for animals (and thus for the people who depend on them) as for humans. Veterinary campaigns not unlike the successful initiative to wipe out smallpox can also be undertaken to eliminate such animal scourges. A good example is the Global Rinderpest Eradication Programme, which was started in Africa to stop the infection and deaths of hundreds of millions of cattle, buffalo and wild ungulates that occurred in regular epidemics of this pathogen. Rinderpest (German for "cattle plague") is a virus that causes the infected animal to develop fever, oral lesions and diarrhea that soon kill it. The disease was concentrated in Asia in the mid-1800s but outbreaks of "Cattle Plague" became common in Europe, the Middle East and Africa after the development of steam train transport and the importation of Asian and Russian cattle. In Africa, the effects were particularly devastating, as the population depended on cattle not only for food but also for moving their products to market. The first Great African Pandemic in the 1890s killed almost 90% of the cattle in Africa, and more localized rinderpest outbreaks continued to take a ghastly toll for another hundred years. The photo below shows the devastated condition of a cow infected with rinderpest in Ethiopia in 1988.

In the 1920s, veterinary immunologists in India developed a vaccine based on an attenuated rinderpest virus. This vaccine could be freeze-dried and used widely but had the disadvantage of having to be produced in live goats. In the 1960s, veterinary immunologists in Kenya found a way to produce the vaccine in cell cultures. Later refinements resulted in a vaccine that could be transported at ambient temperature for 30 days, eliminating the need for refrigeration. It was also fortuitous that rinderpest was an ideal candidate for eradication by mass vaccination because there was only one strain of rinderpest virus, no carrier state existed, and lifelong immunity to the virus developed after exposure. A series of international campaigns to vaccinate huge numbers of animals in many countries ensued in the 1970s and progress was made, but small groups of infected wild ungulates persisted in parts of Africa. As a result, a second Great African Rinderpest Pandemic raged in sub-Saharan Africa in the early 1980s. The Food and Agriculture Organization (FAO) of the United Nations, which had been involved in rinderpest control efforts since the FAO's formation in 1945, then spearheaded the Global Rinderpest Eradication Campaign, which steered away from indiscriminate mass vaccination and concentrated on those areas where the virus had been identified or was suspected to be hiding. All cattle crossing a national border were vaccinated, and community-based animal health workers immediately delivered vaccines to locations where small outbreaks occurred. Finally, in the late 1990s, after several years in which no rinderpest outbreak was reported, the virus showed its face again in the cattle of a single Sudanese tribe. Once all the animals belonging to this tribe were vaccinated, the chain of transmission was broken in this part of Africa. The very last cases of rinderpest were detected in buffaloes in Kenya's Meru National park in late 2001. After vaccination of these animals, no further reports of rinderpest were received despite intense vigilance, and despite the disruptions of natural disasters and war. On August 8, 2011, the United Nations declared that rinderpest was officially eradicated, making it the second disease in history to be completely eliminated.

Plate Box 14-2
Devastated condition of a cow infected with rinderpest in Ethiopia in 1988.

[Reproduced from Roeder, P.L. (2011) Rinderpest: The end of cattle plague. Preventive Veterinary Medicine 102: 98–106.]

immune response that leaves survivors with very long-lasting or even permanent immunity. A vaccine derived from such a pathogen is likely to induce a similar level of protective immunity. In contrast, the natural immune response to a pathogen that causes chronic disease is typically inadequate to clear the pathogen. It is therefore less likely that a vaccine derived from such a pathogen will be successful in stimulating a protective immune response in a vaccinee.

TABLE 14-3	Characteristics of a Successful Vaccine
Characteristic	**Description**
Vaccine efficacy	Stimulates vigorous and appropriate Td response leading to pathogen elimination by antibody or CTLs besides development of memory T and B cells
	Coverage of vaccinated population is typically 80–95%
Vaccine safety	No risk of causing disease
	Side effects are not worse than natural disease symptoms
Pathogen	Causes acute rather than chronic disease
	Induces immunity upon natural exposure
	Undergoes little antigenic variation
	Does not attack cells of the immune system
	Does not have an environmental or animal reservoir

Pathogens that do not exhibit a high degree of antigenic variation (e.g., measles virus) are also favored as vaccine candidates because memory B and T cells and their antibodies and effector cells will continue to recognize the pathogen in successive exposures. Conversely, pathogens that evade the immune response by undergoing extensive antigenic variation (e.g., *Trypanosoma brucei*) are poor vaccine targets. Similarly, a pathogen with many different life cycle stages and forms (e.g., the malarial parasite *Plasmodium falciparum*) can confound vaccine design because candidate antigens that remain invariant from life stage to life stage are rare. HIV presents an additional and unique challenge in vaccine design because this virus decimates an individual's CD4+ T cell population, the very cells needed to mount the protective secondary response.

The ultimate goal of a vaccination program is to completely eradicate a pathogen from a population so that, as in the case of smallpox, vaccination against the pathogen will no longer be required. The best candidates in this regard are pathogens that infect only humans, meaning that they cannot escape into an animal or an environmental **reservoir**. A reservoir in this sense is a species or environmental niche outside the human population in which a pathogen can survive. If a reservoir does exist, then once an entire human population is vaccinated and the pathogen has run out of human hosts to infect, the pathogen can retire to its reservoir. If the vaccination program is terminated at this point, there will soon be a new generation of individuals born who are susceptible to the pathogen. When the pathogen eventually emerges from its reservoir in search of fresh human hosts, it attacks the unvaccinated individuals and regains its foothold in the human population. Eradication efforts are thus thwarted.

II. Efficacy

A successful vaccine has a high level of **efficacy**; that is, it is effective in protecting the vaccinee from disease. Accordingly, the immune response that the vaccine induces must be appropriate for the elimination of the pathogen of interest. For example, an extracellular pathogen is best countered by antibodies so that the vaccine must be capable of activating B cells. Furthermore, the antibodies produced must be of an isotype known to be effective against that particular pathogen. For example, SIgA best protects the mucosae, but IgG is needed to immobilize pathogens in the blood. Similarly, if a pathogen is intracellular, the vaccine should activate CD8+ Tc cells.

The efficacy of a vaccine is often expressed as its **coverage**: the percentage of individuals vaccinated who do not experience disease after exposure to the pathogen. No vaccine has yet proved 100% effective due to the inherent genetic variation among humans, but most standard vaccines given in childhood are effective in 80–95% of a given population; that is, the coverage of these vaccines is 80–95%. It is here that the concept of **herd immunity** comes into play. If a pathogen cannot gain a foothold in a population because most members respond to the vaccine, even individuals who do not respond to the vaccine will be protected because the chance of any individual "in the herd" being exposed to the pathogen is much reduced. **Cocooning** is herd immunity on a small scale. The immediate household contacts of a vulnerable individual are immunized so that the individual is safeguarded within the cocoon of that household.

> NOTE: There has been a growing call worldwide for health care personnel to undergo mandatory seasonal vaccination against influenza virus. Voluntary immunization programs have not led to the level of coverage necessary to protect vulnerable patients such as the legions of elderly residents in nursing homes. As a result, transmission of this debilitating and frequently fatal virus by unvaccinated health care workers has been documented in numerous health care settings. A mandatory vaccination program would likely achieve the herd immunity necessary to protect patients.

Epitopes that induce an effective antipathogen response are known as *protective epitopes* and are derived from *protective antigens*. It is these protective antigens that are usually chosen to formulate a vaccine. Candidate vaccine epitopes are often first identified by the analysis of the gene and protein sequences of the antigen. An antiserum is then raised against the candidate epitope in an animal and is shown to neutralize the pathogen *in vitro*. Only if this result is positive is *in vivo* testing carried out to definitively prove that the epitope is protective and potentially useful in a vaccine. As well as bearing a protective epitope, the best protective antigens (1) are particulate and of size 20–200 nm, so that they can easily enter the lymphatic system, provoke uptake by APCs, and interact directly with B cells; (2) are repetitive in nature, so as to activate complement and cross-link the BCRs of B cells; and (3) contain PAMPs, so as to engage the TLRs of innate leukocytes and promote DC activation and licensing.

III. Safety

In addition to being effective, a vaccine needs to be safe; that is, it should have very few detrimental side effects (known as **adverse events**). Such side effects range from redness and tenderness at the injection site to high fever, seizures, pneumonia, encephalitis and even death. Balancing the severity of vaccine side effects against the incidence and severity of the disease gives different results in different parts of the world. Where disease incidence is low, as in the developed world, a vaccine that results in relatively severe adverse events is not used. In contrast, where disease incidence is much higher, as in the developing world, the risk of harm from the disease may outweigh the risk of harm from vaccine side effects.

Before any candidate vaccine can be administered to humans, both its safety and efficacy are confirmed in a series of **pre-clinical** and **clinical trials**. Pre-clinical trials test vaccines in cell cultures or animal models, whereas clinical trials test vaccines in human volunteers. In Phase I clinical trials, small numbers of human volunteers are given the vaccine and its immunogenicity, dose-response range, optimal route of administration, and adverse events are examined in the absence of any infection. Phase II and III trials further assess the vaccine's efficacy in ever-larger groups of human volunteers who receive the vaccine and then are challenged with the pathogen. Once the vaccine is approved for use in the general population, Phase IV trials continue to monitor its efficacy as well as any detrimental side effects. In addition, to collect sound scientific data on the real adverse events associated with vaccines, the U.S. Centers for Disease Control (CDC) has established the Vaccine Adverse Events Reporting System (VAERS) (see Section F of this chapter). VAERS serves as a reliable source of accurate information on vaccine safety for individuals who have concerns about immunizing themselves or their children.

Despite the preceding carefully devised structure, vaccine testing is not always straightforward. For certain pathogens, a good animal model (in which immune responses and side effects are analogous to those in humans) may not be available,

Information about current and recently completed clinical trials in the U.S., including vaccine clinical trials, can be found at http://clinicaltrials.gov/.

impeding the early stages of vaccine development. Another problem arises in situations in which it would be unethical to deliberately administer a highly pathogenic infectious agent (e.g., HIV) to challenge human volunteers who have been administered a candidate vaccine. In these cases, a clinical trial is organized in which the candidate vaccine is given to individuals already living in an area where the pathogen is endemic. The research team then tracks any changes in disease incidence relative to an otherwise identical unimmunized control group living in the same area.

NOTE: It is becoming clear that there are fundamental differences between the immune systems of humans and the mice used to test many vaccines. These differences have to be accounted for before discoveries made in mice can be safely translated to the clinic. Researchers are now joining forces to undertake the systematic analysis of the immune responses of very large numbers of humans to infection and vaccination, with the goal of designing future vaccines that are optimized for the human immune system.

C. Types of Vaccines

Table 14-4 summarizes the many different ways to construct a vaccine as well as the major advantages and disadvantages of each type. Note that the types of vaccines that work best for one class of pathogen may not work at all for another.

TABLE 14-4	Types of Vaccines and Their Pros and Cons		
Type	**Description**	**Major Pros**	**Major Cons**
Live, attenuated	Whole pathogen treated to decrease pathogenicity but maintain immunogenicity; may still replicate	Low number of doses usually very effective Minimal need for adjuvant* Supplies B and T epitopes	Cold chain required Chance of reversion of attenuating mutation
Killed or inactivated	Whole pathogen killed or inactivated to block replication but maintain immunogenicity	No possibility of reversion No cold chain required Supplies T and B epitopes	Cannot replicate so requires boosters and adjuvant Does not usually induce robust Tc responses
Toxoid	Chemically inactivated toxin of pathogen	No need to use whole organism	Effective only if disease caused solely by toxin
Subunit	Pathogen protein or polysaccharide purified from natural sources or synthesized using recombinant DNA methods	Avoids use of whole organism Can be manipulated to increase immunogenicity	Can be costly to produce May not be as immunogenic as natural pathogen component Does not usually induce robust Tc responses
Peptide	Pathogen peptide purified from natural sources or synthesized using recombinant DNA methods	Avoids use of whole organism Composition is known Very stable	Epitope size and number restricted May require coupling to a carrier protein
Recombinant DNA vector	Virus-based vector containing recombinant DNA of pathogen antigen. Vaccinee is infected with the viral vector and the pathogen DNA is transcribed and translated within the vaccinee's cells like a viral protein	Avoids use of natural pathogen Replicates like a pathogen to produce large amounts of immunogen Supplies T and B epitopes Minimal need for boosters and adjuvant	Possible side effects due to vector components Antivector antibodies raised during priming may necessitate boosting with a different vector
Naked DNA	Small plasmid-containing recombinant pathogen DNA. Plasmid is injected into a vaccinee, and the pathogen DNA is taken up by the vaccinee's cells and transcribed and translated	Easy and inexpensive to manipulate Induces B, Th and Tc responses Plasmid sequences may act as adjuvant	Not as immunogenic as protein vaccines in humans Integration of plasmid into host cell genome may induce tumorigenesis

*An adjuvant is a substance that enhances local inflammation and thus immune responses to vaccine antigens.

I. Live, Attenuated Vaccines

A **live, attenuated vaccine** consists of live, whole bacterial cells or viruses that are treated in such a way that their pathogenicity is reduced but their immunogenicity is retained. Because a viable, replicating pathogen is used, large quantities of the immunogen are produced in the vaccinee, and both the innate and adaptive arms of the immune system are readily triggered. Moreover, most attenuated pathogens supply both B and T epitopes such that both humoral and cell-mediated adaptive responses are mounted. However, although live, attenuated vaccines are generally very effective, a single dose is not usually enough to induce the long-lasting immunity enjoyed by individuals who survive natural infection with the causative pathogen. A **booster** (repeat dose of vaccine) must usually be administered to induce faster, stronger secondary immune responses and bolster protection. A disadvantage of live, attenuated vaccines compared to other vaccine types is that, in order to maintain efficacy, a **cold chain** of equipment and procedures is required to ensure that the live vaccine is kept chilled from the time of production to the moment of vaccination.

Attenuating mutations of a pathogen were originally induced by culturing the pathogen for extended periods under less than optimal conditions (i.e., conditions very different from the physiological conditions in its human host). This approach selects for organisms that have undergone mutations such that they now grow well in the suboptimal setting but can no longer cause disease in the host (although they may still be able to replicate). A virus can also be attenuated by introducing it into a species in which it does not replicate well (i.e., infection of an animal with a human virus), or by forcing the virus to replicate repeatedly in cells maintained in laboratory culture vessels, a protocol called **passaging**. The most modern methods of attenuation use recombinant DNA technology to directly mutate or delete genes encoding proteins known to contribute to pathogen virulence. Complete deletion, rather than just point mutation, of a virulence gene is preferred because it decreases the danger that an attenuated pathogen may recover its virulence through reversion of the attenuating mutation. Such reversions can lead to the vaccinee coming down with the very disease the vaccine was designed to prevent.

NOTE: In Chapter 1, we learned that groundbreaking experiments by Edward Jenner in 1796 led to the use of cowpox inoculation to protect against smallpox, initiating the modern era of vaccine science. Of course, no one at that time—including Jenner—had the faintest idea of the immunological mechanisms involved in this success. We now know that in this essentially unique case, nature has supplied a form of "live, attenuated" vaccine, because the intact, mildly pathogenic cowpox virus is able to induce a protective immune response against the related but much more deadly smallpox virus. Thankfully, despite not understanding how it worked, Jenner's keen mind and natural curiosity allowed him to harness his empirical observations to prevent untold human suffering.

II. Killed Vaccines

In a **killed vaccine,** a whole bacterium, parasite or virus is killed or inactivated by treatment with gamma irradiation or a chemical agent such as formaldehyde. Used correctly, these procedures preserve the structure of the protective epitopes but remove the pathogen's ability to replicate or recover virulence. Killed vaccines are generally more stable than live vaccines and less sensitive to cold chain disruptions. However, because the treated pathogens cannot replicate, larger amounts of these vaccines must be administered in the primary dose, raising costs. In addition, because killed vaccines are dead and therefore trigger less intense inflammation than live vaccines, the response to a given dose is weaker and frequent boosters are required. A greater problem is that killed vaccines offer only limited protection against intracellular pathogens. Because the vaccine organism is dead, it cannot actively penetrate host cells. The processing of the organism's antigens by the endogenous antigen processing pathway and presentation on MHC class I are therefore limited. As a result, Tc cell activation and

CTL generation by this route do not occur. A DC that has phagocytosed the killed vaccine may cross-present peptides derived from it, but the levels of peptide-MHC class I generated activate only limited numbers of Tc cells. Overall, killed vaccines tend to induce predominantly systemic humoral responses featuring neutralizing antibodies, a type of immune response that is not very effective against intracellular pathogens.

III. Toxoids

As described in Chapter 13, the disease caused by a pathogen may be entirely due to its production of exotoxins. Thus, vaccinating a host against the exotoxin protects the host from disease (if not infection). **Toxoids** are exotoxin molecules that have been chemically altered (usually by formalin treatment) such that they lose their toxicity but not their immunogenicity. Neutralizing antibodies generated in response to toxoid administration bind to the exotoxin and render it harmless. For example, vaccination or boosting with the tetanus toxoid protects against this disease for at least 5 years. Diphtheria is also prevented by toxoid administration.

IV. Subunit Vaccines

A **subunit vaccine** contains a protein or polysaccharide that has been purified from a pathogen and contains at least one protective epitope. A major advantage of such vaccines is that a whole organism is never used, avoiding any risk of reversion and the possibility of side effects due to irrelevant pathogen components.

i) Protein Subunit Vaccines

Some protein subunit vaccines are prepared using conventional protein purification techniques. However, these techniques are labor intensive and not very efficient, making these vaccines very costly. Recombinant DNA technology has greatly simplified the synthesis of vaccines for pathogens that are difficult or impossible to grow *in vitro*, and/or whose components are a challenge to purify in sufficient amounts from *in vivo* infections. To construct a protein subunit vaccine using recombinant DNA technology, one introduces the DNA encoding the vaccine antigen into the genome of a microorganism such as yeast or *Escherichia coli*. These microbes can easily be cultured in huge volumes in the laboratory, and the pathogen protein is synthesized in correspondingly high amounts along with other microbial proteins. A polypeptide of interest can then be isolated from among the newly synthesized proteins. The manipulation of the pathogen's DNA prior to its introduction into the microbe also allows for deliberate mutation that confers greater ease of purification or increased immunogenicity. The use of the recombinant DNA method of protein preparation also reduces the chance that potentially harmful pathogen components will be copurified with the protective antigen.

Protein subunit vaccines do have some disadvantages. Sometimes the production of a vaccine protein in a recombinant organism alters the conformation of the protein such that its stability is affected or protective conformational epitopes are no longer present. As a consequence, certain neutralizing antibodies may not be induced in the vaccinee. As well, like killed vaccines, subunit vaccines are not alive and so cannot penetrate host cells. The vaccine epitopes are thus inefficiently presented on MHC class I and can activate Tc cells only by cross-presentation, resulting in a relatively weak CTL response.

ii) Polysaccharide Subunit and Conjugate Vaccines

Subunit vaccines can also be made from pathogen polysaccharides. These vaccines are advantageous when the polysaccharide is known to be important for the virulence of the pathogen, as is true for many encapsulated bacteria. In addition, capsule polysaccharides are abundant on bacterial surfaces and thus are easy to purify, and usually do not induce harmful side effects when injected into animals or humans. However, polysaccharides are not proteins and thus do not provide T cell epitopes. Instead, polysaccharides tend to provoke Ti responses that do not generate memory lymphocytes. Indeed, many polysaccharides do not induce immune responses of any type in children under

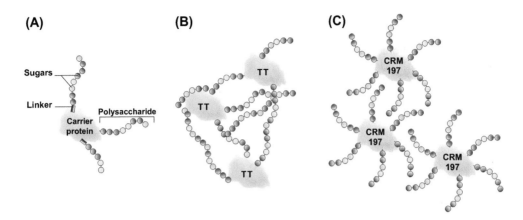

(A) **(B)** **(C)**

Sugars

Linker

Polysaccharide

Carrier
protein

TT

CRM
197

Fig. 14-2
Examples of Conjugate Vaccine Structures

(A) Pathogen-derived polysaccharide antigens of various lengths and sugar compositions can be isolated from bacterial capsules and joined either directly or through linker molecules to a carrier protein to form a conjugate vaccine. **(B)** When tetanus toxoid (TT) is used as the carrier protein, random polysaccharide-protein linking and cross-linking between polysaccharide chains give rise to a non-homogeneous, mesh-like conjugate structure that is not well defined. **(C)** To form a more homogeneous and well-defined conjugate structure, polysaccharide chains can be end-linked to the carrier protein CRM197.

2 years of age. Vaccine efficacy in these cases has been greatly increased by chemically joining a bacterial capsule polysaccharide to a carrier protein capable of supplying a T cell epitope and thus provoking a Td response. The combination of carrier protein plus polysaccharide creates a **conjugate vaccine** (**Fig. 14-2**). The diphtheria and tetanus toxoids and modified versions of these proteins are often used as the carriers in conjugate vaccines. A particularly useful new carrier is CRM197 ("cross-reacting material 197"), a modified form of diphtheria toxoid. CRM197 has proved safe and effective in humans, and is now used globally in conjugate vaccines against encapsulated bacteria such as *H. influenzae*, or species of *Pneumococcus* or *Meningococcus*. Even in infants, these vaccines induce robust Td antibody responses that show affinity maturation.

> The linked recognition of B and T cell epitopes during the mounting of a Td antibody response was described in Chapter 5 and depicted in Figure 5-6.

V. Peptide Vaccines

Some isolated peptides function as protective T and B epitopes and so can serve as vaccines. With a **peptide vaccine**, the precise molecular composition of the vaccine is known and there is no possibility of reversion to a pathogenic phenotype. In addition, due to the relatively small size of the vaccine agent, it is less likely that larger entities such as infectious agents or genomic material will copurify with the vaccine and contaminate it. Both natural and synthetic peptides have been explored as vaccine candidates, and peptides have been produced both by conventional purification methods and by recombinant DNA technology. Where a natural peptide is not immunogenic, it can be modified in the laboratory to become so. Once a protective peptide has been identified and purified, it is mixed with or linked to an adjuvant (see Section D later) that encourages its uptake by a vaccinee's APCs, including DCs. The peptide is then presented and cross-presented on MHC class II and class I to naïve Th and Tc cells, respectively.

Peptide vaccines have their own set of disadvantages. Because peptides are by definition short, the epitopes contained in such vaccines tend to be small, linear and non-conformational. However, most B cell epitopes on a whole pathogen are conformational, so that the antibodies produced in response to a peptide vaccine may offer only limited protection during a natural infection. To verify the immunogenicity of a peptide representing a B cell epitope, scientists often examine the kinetics of the binding of the candidate peptide to antibodies from the serum of a patient with the disease of interest. The verification of peptides representing T cell epitopes is more complicated because the peptide and the MHC molecule that presents it are seen as a unit by

"Th17-Based Vaccine Design for Prevention of *Strep-tococcus pneumoniae* Colonization." by Moffitt, K.L., Gierahn, T.M., Lu, Y., Gouveia, P., Alderson, M., Flechtner, J.B., Higgins, D.E., & Malley, R. (2011) *Cell Host and Microbe 9*, 158–165.

Strains of *Streptococcus pneumoniae* frequently colonize the human upper respiratory tract but do not usually cause disease unless the host immune response is weak or taxed by other infections, or the particular pneumococcal strain involved is highly aggressive. When *S. pneumoniae* does start to replicate out of control, it can cause fatal pneumonia or meningitis. Thus, this bacterium is a major cause of mortality in children worldwide, causing millions of deaths each year. Because antibiotic resistance is widespread among strains of *S. pneumoniae*, prevention of infection via immunization is a favored strategy to limit invasive pneumococcal infections. Vaccines in current use consist of pneumococcal capsular polysaccharides conjugated to a diphtheria toxoid-based carrier protein such as CRM197. However, due to (1) the generally poor antigenicity of polysaccharides; (2) the existence of more than 90 different types of pneumococcal capsules; (3) the high cost of conjugate vaccine production; and (4) the potential for emergence of bacterial strains bearing capsules not covered by a particular vaccine, clinical researchers are keen to develop a more universal protein subunit vaccine to protect against *S. pneumoniae* infection. This article by Moffitt *et al.* describes experiments aimed at identifying pneumococcal protein candidates capable of eliciting responses by CD4⁺ Th17 cells, which are key defenders against mucosal pathogens (refer to Ch. 12). Using an innovative algorithm that couples bioinformatics screening and CD4⁺ Th17 cell reactivity, the authors narrowed a list of over 2000 proteins known to be expressed from genes in the 22 known *S. pneumoniae* genomes to five candidate vaccine antigens. Each candidate antigen was highly purified and then tested in mouse immunization experiments. An example of one such experiment is shown here, in which mice were immunized intranasally with unencapsulated pneumococcal whole cell antigen (WCA; provokes a strong antibacterial immune response in mice but is not licensed for human use), or ICP47 (a herpes virus protein

not expected to induce an antibacterial response), or one of the five candidate antigens identified in the study. The mice were then challenged by intranasal administration of a clinical pneumococcal isolate. After 7 days, the numbers of bacteria colonizing the nasal cavities of the mice were determined and reported as "colony-forming units (CFU)/nasal wash." The data shown here are from Figure 2, Panel A of the article, and they indicate that immunization with candidate antigen SP 2108, SP 0148, or SP 0882 significantly reduced the number of bacteria able to colonize the nasal mucosa compared to immunization with ICP47. (*$p < 0.05$; **$p < 0.01$; ***$p < 0.001$; NS = not significant.) Other experiments evaluated the dependence of the antibacterial response on IL-17 secretion and CD4⁺ Th17 cells, as well as on mouse strain genetic background. The authors hope that these studies may provide a means of developing a more effective pneumococcal vaccine, and identifying candidate antigens for eliciting CD4⁺ Th17-mediated protection against other important mucosal pathogens.

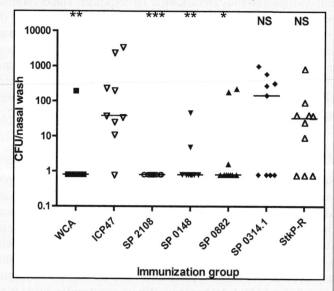

the TCR. The peptide must therefore contain not only the epitope recognized by the TCR but also amino acids allowing it to bind to MHC. Different MHC alleles may bind to and present a given peptide with varying degrees of success, causing a variation in the level of immunity induced in vaccinees. Linking the peptide to a carrier protein to form a peptide conjugate vaccine can increase its immunogenicity.

VI. DNA Vaccines

DNA vaccination involves the introduction into a vaccinee of pathogen-derived DNA sequences that direct the synthesis of immunogenic pathogen proteins *in vivo*. DNA vaccines include **recombinant vector vaccines** and **naked DNA vaccines**. Although some DNA vaccines are now licensed for veterinary immunizations, none has yet turned out to be suitable for human vaccination due to poor immunogenicity compared to the corresponding protein vaccines.

i) Recombinant Vector Vaccines

A recombinant vector vaccine uses an unrelated attenuated virus or bacterium as a vector to introduce DNA from the pathogen of interest into the vaccinee. These vectors can penetrate human cells and often replicate within them but do not cause disease in the host. The vaccine is "recombinant" because genes encoding the pathogen antigen of interest and a selectable marker (used for purification) are incorporated into the vector using recombinant DNA technology. After injection into a vaccinee, the recombinant vector infects host cells just like an unmodified virus or bacterium, and the vaccine gene(s) is transcribed and translated like a viral or bacterial component. If the host cell infected is a DC, the vaccine protein enters the endogenous antigen presentation pathway. Vaccine peptides are thus displayed on MHC class I, and Tc cells are activated. Alternatively, the vaccine protein may be released from the synthesizing host cell and activate B cells directly, or may be taken up and processed by DCs via the exogenous antigen presentation pathway and activate Th cells. Cross-presentation by DCs may also activate Tc cells.

An organism often used as a recombinant vector is the vaccinia virus. Vaccinia is a large, complex poxvirus, with many genes that are not essential for host cell invasion and replication. Consequently, these genes can be replaced in the vaccinia genome with foreign DNA encoding protective antigens from a pathogen of interest. Recombinant vectors based on poliovirus are also being explored because the natural route of infection of this virus is through the digestive tract. Thus, vectors based on poliovirus should be able to be administered orally and so induce both systemic and mucosal responses to an immunogen of interest. However, the genome of poliovirus can accept only a small amount of foreign DNA, and at least one poliovirus strain is known to be genetically unstable. Human adenovirus is now coming into more frequent use as a vector because this virus is easily cultured *in vitro,* and large amounts of foreign DNA can be inserted into its genome without affecting viral replication. Adenovirus-based vaccines also tend to be more immunogenic than vaccinia-based agents, perhaps because this virus easily enters mammalian cells. Vigorous CD8+ T cell-mediated responses are provoked by adenovirus-based vaccines.

A major problem with recombinant vector vaccines is that the vaccinee may mount an immune response against irrelevant vector components. If antivector antibodies are produced in the vaccinee following the first administration (priming), and the recombinant vector vaccine is administered a second time (boosting), the antivector antibodies may bind to the vaccine and prevent it from accessing host cells. As a result, the booster dose of pathogen antigen is never synthesized by the vaccinee's cells, and a true secondary response does not occur. This difficulty can be avoided if the priming is done with a recombinant vector vaccine but the boosting is done with a protein subunit vaccine (which does not contain the vector). Alternatively, the boosting can be done with the same pathogen antigen incorporated into a different vector.

ii) Naked DNA Vaccines

To prepare a naked DNA vaccine, one clones the gene encoding the pathogen antigen of interest into a plasmid that can replicate to a high copy number in *E. coli* but not at all in human cells. The bacteria are grown to large quantities and lysed, and the vaccine plasmid is purified and injected into the vaccinee. The DNA is said to be "naked" because it is not contained within a virion or bacterium. Once the plasmid is taken up by the vaccinee's cells, they commence production of the pathogen antigen. Protective Tc responses as well as Th and B cell responses can then be mounted. Although these vaccines have the advantage of stability, they are difficult to deliver efficiently. Additional concerns with these vaccines center around avoiding integration of the plasmid into the host cell genome, which might mutate a host gene. Such integration has been linked to tumorigenesis in experimental animals. Research is ongoing into various ways of neutralizing the dangers and enhancing the delivery of naked DNA vaccines.

D. Adjuvants and Delivery Vehicles

Researchers today try to design vaccines so that they emulate key features of pathogens, including their size, shape, molecular organization, and surface PAMPs, and so have an increased chance of being immunogenic. Live, attenuated vaccines and many recombinant vectors already retain the capacity of a pathogen to penetrate tissues, replicate in the host and induce inflammation. For these reasons, these types of vaccines are naturally highly immunogenic. However, killed vaccines, purified subunit and peptide vaccines, and naked DNA vaccines lack these properties, meaning that they need help to induce effective immune responses. In addition, protein and DNA molecules introduced into a vaccinee may encounter extracellular proteases and nucleases that degrade the vaccine before it can be taken up by APCs. To avoid these impediments to efficacy, one often administers these types of vaccines with substances called **adjuvants** and **delivery vehicles**. Adjuvants enhance local inflammation by inducing innate immune responses and thus increase the ability of the vaccine antigen to provoke an adaptive response. At the molecular level, these adjuvants often induce the upregulation of costimulatory molecules on leukocytes, or activate DCs and other APCs through PRR engagement. Delivery vehicles protect vaccine antigens from degradation, and provide a shape or surface that encourages uptake by APCs. In some cases, a single entity can act as both an adjuvant and a delivery vehicle.

> NOTE: Because prophylactic vaccines are given to healthy people to prevent disease, they must provide high benefits at low risk. Because adjuvants are components of vaccines, the adjuvant in any vaccine formulation (like its immunogen component) must be thoroughly tested in clinical trials for safety. Unfortunately, many candidate adjuvants have turned out to be too toxic for use in humans, a barrier to their inclusion in modern vaccines. As is outlined in the following section, research is ongoing into adjuvants that are effective in promoting an immune response but safe enough for human use.

I. Adjuvants

i) Alum

Although several adjuvants are available for use in experimental animals, until recently, the only adjuvant licensed for routine use in humans was alum. Alum is a gel containing salts of aluminum hydroxide or aluminum phosphate. Alum injection induces inflammation that tends to promote Th2 responses and humoral immunity rather than Th1 responses and cell-mediated immunity. However, even after 80 years of intense research, it is still not understood exactly how this agent works. It does not appear that alum administration triggers TLR signaling, unlike some other historically used substances (see later). The use of alum is not wholly without drawbacks, because it is normally injected deep into muscle. This approach ensures that vaccinees (children, in particular) are properly vaccinated but increases vaccinee discomfort. Local reactions include short-lived redness, pain, and hardening of the tissue at the injection site. Systemic reactions such as malaise, fever and aches may also occur in some cases.

> NOTE: Alum is an example of an adjuvant that not only induces inflammation but also mediates a "depot effect" in which antigen is prevented from dispersing immediately after administration. Instead, the antigen slowly and continuously leaks into the body, thereby sustaining the response. Adjuvants based on other mineral salts (such as phosphates), emulsions, or protein precipitates can also establish antigen depots. It is thought that the primary benefit of this slow antigen leakage is the maintenance of the Th responses needed to support B cell activation.

ii) Lipid- or Oil-Based Adjuvants

In the past few years, some non-alum adjuvants have been approved for use in seasonal influenza vaccines in Europe. One of these is informatively called "Adjuvant system 04" (AS04) and consists of monophosphoryl lipid A (derived from *Salmonella* LPS) adsorbed onto aluminum hydroxide. This agent, which is a TLR4 ligand, induces Th1 and antibody responses and is currently used in vaccines against human papilloma virus (HPV) and hepatitis B virus (HBV). Two other new adjuvants are MF59 and AS03, which are squalene-based oil-in-water emulsions. The squalene component triggers internalization of the emulsion by DCs, resulting in vaccine antigen uptake and vigorous Th2 and antibody responses.

iii) PRR Ligands

Several new adjuvants deliberately designed to contain PRR ligands have recently won regulatory approval in some jurisdictions. These adjuvants have been formulated to contain various combinations of bacterial proteins, carbohydrates, and/or nucleic acids known to activate innate leukocytes. In fact, PRR ligands have been playing a hidden promotional role in vaccination for as long as vaccine development has been pursued (see **Box 14-3**). More formal studies have demonstrated that the inclusion of PRR ligands in a vaccine can have powerful positive effects. For example, studies in mice have shown that antigens of *L. major* or *M. tuberculosis* suspended in oil-and-water emulsions induce only non-protective Th2 responses, but that the addition of certain PRR ligands to this emulsion can trigger protective Th1 responses. Furthermore, in preliminary studies in humans, TLR ligand-containing vaccines have been found to induce strong Th1, CTL, antibody and NK cell responses. The TLR5 ligand flagellin is a particularly attractive adjuvant candidate because this protein can be fused to a vaccine protein or peptide antigen to form a conjugate vaccine. In clinical trials of such a vaccine in which influenza HA protein was fused to flagellin, high titers of anti-influenza antibodies were induced even in elderly individuals. Similar strategies to exploit ligands of CLRs are being devised. For example, experimental vaccines have been created that are based on particles composed of the fungal cell wall glucans recognized by the CLR Dectin-1. When the vaccine antigen is

---

Box 14-3 PRR Ligands: The "Secret Ingredient" in Early Vaccine Success

Since the devising of the first whole virus vaccines, PRR ligands have played an important role in immunization successes without researchers being aware of their contribution. For example, the rabies vaccine developed in 1886 was based on the attenuated whole virus, which naturally contained the ssRNA that comprises the viral genome and is also the cognate ligand of TLR7. Similarly, the original whole cell pertussis vaccine supplied LPS (TLR4 ligand) and bacterial DNA (TLR9 ligand). It was not until the 1980s that the significance of the microbial components in such vaccines was recognized, and not until the 1990s that the PRRs binding to these components and the immunological consequences of that binding were identified. Researchers have now returned to many early vaccine formulations to understand why they worked so well. For example, the live attenuated vaccine against the virus causing yellow fever is highly efficacious, and it has now been shown that the vaccine contains ligands for RIG-1 and TLR2, -3, -7, -8 and -9. Engagement of these receptors activates multiple DC subsets, sparking intense inflammation that precedes vigorous Th1, Th2 and CTL responses against the virus. Similar investigations of vaccinia virus, which has long been used as a vaccine delivery vehicle, have revealed the importance of a viral component that binds to TLR2. DC activation is triggered that results in the generation of the desired neutralizing antibodies and memory CD8+ T cells. In addition, vaccinia virus appears to stimulate TLR-independent signaling that leads to IL-1 secretion, promoting inflammation and thus a more vigorous immune response in the vaccinee. A parallel scenario may explain why the existing **bacillus Calmette-Guérin** (BCG) vaccine against tuberculosis (TB; see **Box 14-4**) can reduce mortality caused by pathogens that co-infect a TB patient, even though the vaccine is not very effective against *M. tuberculosis* itself. The BCG vaccine contains ligands that activate TLR-2, -4, -9 and DC-SIGN, resulting in support for Th1 and Th2 responses against a range of invaders.

Another mystery solved by the recognition of the adjuvant role of PRR ligands is the effectiveness of Complete Freund's Adjuvant (CFA). CFA consists of inactivated *Mycobacteria* in an oil-in-water emulsion. This agent is used very successfully to induce antibody, Th1 and Th17 responses to specific antigens in animals but provokes severe inflammation in people and thus cannot be used for human vaccination. It turns out that CFA not only forms a site of irritation that draws innate leukocytes to the antigen but is also replete with ligands for TLRs and NOD proteins that activate these cells.

trapped inside these glucan particles, immunized animals mount robust antibody, Th1 and Th17 responses. Ligands that bind to the NLR and RLR families of PRRs are also under investigation as adjuvants, but a means of delivering them efficiently into the cytoplasm of an innate leukocyte will have to be devised.

Immunologists continue to define the features of PRR ligand adjuvants that will optimize their use in various situations. Importantly, different PRRs are expressed by different subsets of innate leukocytes, making it possible to tailor the PRR ligand included in a vaccine such that the adaptive response eventually triggered will be suited to the clearance of a particular pathogen. Scientists have also learned that incorporating more than one PRR ligand in a vaccine has synergistic effects, in that longer and more vigorous adaptive responses are mounted to lower doses of the vaccine antigen.

A major issue with all PRR ligand-containing vaccines is that the innate responses induced by PRR ligands can also promote the activation of otherwise somnolent autoreactive lymphocytes if not controlled properly. Autoimmunity is therefore a (clearly undesirable) side effect of some PRR-based vaccines. Efforts are ongoing to understand how these and other adjuvants work at the molecular level, with the goal of devising agents that have very few side effects and affect only a limited number of cell types.

II. Delivery Vehicles

Significant enhancement of vaccine-induced responses can be achieved by administering the vaccine antigen in a non-toxic delivery vehicle. The vehicle protects the vaccine molecules from protease or nuclease degradation (increasing its persistence in the tissues) and may also act as an adjuvant (inducing inflammation). Some vehicles facilitate the display of multiple molecules of the vaccine antigen on the vehicle surface, creating a multivalent form of the antigen that increases its immunogenicity. These properties have made delivery vehicles invaluable adjuncts for experimental vaccination with subunit and DNA vaccines. Several different types of delivery vehicles have been devised, some of which are starting to be used for human vaccination.

i) Liposomes

Liposomes are prepared by mixing the vaccine antigen of interest with a suspension of phospholipids under conditions that favor the formation of a spherical membranous structure. The vaccine antigen is trapped in the aqueous center of the hydrophobic liposome, which usually ranges in size from 100–10,000 nm. Liposomes are readily phagocytosed by DCs and macrophages, meaning that the antigen is rapidly processed and used to initiate T cell activation. In addition, PRR ligands can be incorporated into liposomes to serve as adjuvants. Liposomal vaccines for hepatitis A virus (HAV) and influenza are now licensed for use in some countries. A liposome-based antimalaria vaccine is currently being tested in clinical trials (see **Box 14-5** in Section G).

ii) ISCOMs

"Immunostimulating complexes" are hollow balls made up of cholesterol, phospholipid and detergent. Relatively bulky protein immunogens can easily be inserted into the interior of the ball. To the immune system, the ISCOM resembles a multivalent antigen with a shape that invites phagocytosis by APCs. In addition, the detergent in the ISCOM is a powerful adjuvant. Depending on the size of the ISCOM, it may be referred to as a nanoparticle (50 nm) or a microparticle (1000–50,000 nm). Both particle types are well within the range of natural pathogen sizes.

iii) Virosomes

Virosomes are like non-replicating "artificial viruses" that can be used to deliver vaccine antigens directly into a host cell. A virosome is basically a liposome that is covered in the envelope glycoproteins of a virus. Pathogen antigens of interest are either captured within the lumen of the virosome or are chemically cross-linked to its surface. Because of its viral envelope proteins, a virosome can bind to and "infect" host cells and deliver the antigen directly into the MHC class I antigen processing pathway. Alternatively, the virosome may be phagocytosed by an APC. A virosome vaccine that incorporates the influenza virus HA protein into liposomes has been devised for seasonal influenza immunization.

iv) Virus-like Particles

Virus-like particles (VLPs) are structures of 30–90 nm that result from the self-assembly of virus proteins without a nucleic acid genome or a lipid envelope. Sometimes it is the surface proteins of the virus of interest that form the rod-like or icosahedral VLP structure, but sometimes the viral core proteins are responsible. For example, to produce an influenza VLP, the surface H and N proteins are mixed with the viral core M1 and M2 proteins, which then self-assemble (with generic structural elements) into the VLP. Similar processes have been used to generate VLPs for HBV and HPV. In all cases, the surfaces of these VLPs have a repetitive structure that is ideal for inducing antibody production. In animal studies, VLPs have been just as effective as conventional vaccines in protecting against lethal influenza infections. In humans, VLP-based vaccines against HBV and HPV have recently been approved in some jurisdictions. Human VLP vaccines to combat viruses such as influenza virus and Norwalk virus are in clinical trials. The fact that the surface proteins of a VLP can be custom-synthesized may be a critical advantage where a vaccine has to be produced to combat a new threat. For example, a pandemic influenza virus that expresses a novel and lethal combination of H and N proteins due to antigenic shift (refer to Ch. 13) may best be fought with a VLP constructed using these exact H and N proteins.

Antigenic shift and drift in influenza virus strains were described in Box 13-2 and illustrated in Figure 13-4.

VLPs can also be constructed to contain PRR ligands as additional adjuvants. Interestingly, although VLPs cannot replicate because they lack replicases and nucleic acids, sometimes a VLP will spontaneously assemble around a fragment of RNA or DNA. This capacity offers vaccine designers the opportunity to manipulate a VLP and cause it to incorporate TLR7/9 ligands such as ssRNA or CpG sequences. In addition, proteins like the TLR5 ligand flagellin can be added to VLPs. In one study, when mice were immunized with a flagellin-containing influenza-based VLP and then infected with live influenza virus, the presence of the flagellin component promoted Th1 responses that increased the animals' IgG2a and IgG2b antiviral antibody levels over those in animals treated with empty influenza-based VLPs. Research into more and better ways to enhance VLP use with PRR ligands is under way.

E. Prophylactic Vaccines

At time of writing, there are more than 20 human infectious diseases for which safe and effective vaccines are available. In most developed countries, many of these vaccines are administered as part of a schedule of standard childhood immunizations against endemic pathogens. Other vaccines might be given only in special cases, such as if an individual is likely to encounter occupational exposure to a particular pathogen, or is traveling to an area where a pathogen uncommon in the traveler's home country is endemic. Diseases targeted by common prophylactic vaccines available in the U.S. are listed in **Table 14-5.**

The current U.S. CDC Recommended Childhood and Adolescent Immunization Schedules can be viewed online at http://www.cdc.gov/vaccines/schedules/hcp/index. Recommended vaccine schedules vary considerably by country.

Most prophylactic vaccines are designed to be administered during infancy, soon enough to prevent the onset of childhood disease but late enough to avoid the establishment of immune tolerance. Booster doses of these vaccines are usually given two to three times, with a gap of several weeks or months in between. If vaccinations are missed in infancy, an individual can receive "catch-up" vaccinations later in life. There remains some debate among immunologists (at least in the Western world) when childhood vaccination should start. Some feel that vaccination should be delayed until 6 months of age, when the infant's immune system has matured somewhat and the protection afforded by the maternal antibodies in the infant's circulation starts to fade. Others believe that it is never too soon to vaccinate against scourges such as *Bordetella pertussis*, which can easily kill a 2-month-old infant.

To circumvent the need for multiple injections and repeated clinic visits, single vaccines have been created that contain antigens from several different pathogens: these are called **combination vaccines** (**Table 14-6**), and a *valency* terminology is often

TABLE 14-5	Diseases/Pathogens Targeted by Prophylactic Vaccination in the U.S.

Targeted by Childhood Vaccination

Diphtheria

Haemophilus influenza Type b (Hib)

Hepatitis A

Hepatitis B

Human papilloma virus

Influenza

Measles

Meningococcal disease

Mumps

Pneumococcal disease

Polio

Rotavirus

Rubella

Tetanus

Varicella (Chicken pox)

Targeted by "Special Case" Vaccination

Anthrax

Cholera

Plague

Rabies

Tuberculosis

Typhoid fever

Variola (Smallpox)

Yellow fever

TABLE 14-6	Examples of Combination Vaccines

Vaccine Abbreviation	Components of Vaccine
DTaP/Hib	Diphtheria toxoid, tetanus toxoid, acellular pertussis adsorbed, *H. influenza* type b conjugate vaccine
DTaP-HepB-IPV	Diphtheria toxoid, tetanus toxoid, acellular pertussis adsorbed, hepatitis B virus vaccine, inactivated poliovirus vaccine
DTaP-IPV/Hib	Diphtheria toxoid, tetanus toxoid, acellular pertussis adsorbed, inactivated poliovirus vaccine, *H. influenza* type b conjugate vaccine
Hib-HepB	*H. influenza* type b conjugate vaccine, hepatitis B virus vaccine
HepA-HepB	Inactivated hepatitis A virus, inactivated hepatitis B virus vaccine
MMRV	Measles, mumps, rubella, varicella vaccine

used to describe them. For example, a trivalent vaccine could contain antigens from three pathogens, or antigens from three strains of the same pathogen. However, while the sparing of multiple injections is designed to bring relief to young children and their parents, combining vaccines is not always ideal. Administration of the combined measles, mumps, rubella (MMR) + varicella vaccine (MMRV vaccine; see Section E) has been found to cause more side effects in vaccinated children than does separate administration of the MMR and varicella vaccines.

The need for booster shots and their scheduling varies with the vaccine. The priming dose alone of some vaccines, particularly those composed of live, attenuated viruses, can induce very strong memory responses. In these cases, boosting may be necessary only at intervals of several years (if at all) in order to sustain life-long immunity. Alternatively, the vaccinee may naturally encounter the pathogen itself often enough to keep triggering memory responses without the need for more than a single dose of vaccine. When the nature of a vaccine is such that a booster is definitely required, it does not have to involve the same vaccine formulation as the priming dose. For example, a live, attenuated vaccine might be followed by a killed or protein subunit vaccine.

We now describe the most common prophylactic vaccines and the diseases they aim to prevent. If left unchecked, many of these disorders would hospitalize, permanently disable, disfigure or kill a significant number of people. Thousands more would be transiently incapacitated. Vaccination is therefore one of the greatest triumphs of organized public health care.

I. Anthrax

Anthrax is caused by the extracellular bacterium *Bacillus anthracis*, which occurs naturally in the soils of farms and woodlands. It commonly infects livestock and range animals but rarely humans. When *B. anthracis* does infect a human, the resulting anthrax disease takes one of three forms: inhalation (the most lethal), cutaneous, or gastrointestinal. Early symptoms range from respiratory distress to skin lesions to fever and severe diarrhea. If not rapidly treated with antibiotics, all three types of anthrax disease can lead to systemic bacterial infection and death.

> NOTE: How did *B. anthracis* come to be considered a biowarfare weapon? The bacterium has two characteristics that make it a military darling: (1) *B. anthracis* spores can survive in harsh environments for long periods, and (2) once aerosolized, the spores are readily inhaled and cause lethal respiratory infections. In 1979, an unintentional release of *B. anthracis* spores from a military lab in the former Soviet Union killed 64 people. In 1993, a religious cult in Japan unsuccessfully tried to use *B. anthracis* spores as a weapon. In 2001, *B. anthracis* spores delivered through the U.S. postal system killed 5 people and infected 17 others.

The disease caused by *B. anthracis* is due in part to the toxins it produces. Accordingly, the current licensed vaccine for anthrax, called "anthrax vaccine adsorbed" (AVA), consists of a cell-free preparation of one of the *B. anthracis* toxins. This vaccine is delivered either subcutaneously or intramuscularly. Although the protection provided by the AVA vaccine is quite strong, it is not long-lasting and the multi-dose schedule of administration is quite arduous, leading to problems with compliance among vaccinees. Thus, even where concerns about bioterrorism exist, vaccination against anthrax is recommended only for members of the military who might encounter this organism in a battlefield context; researchers culturing this organism; those working with potentially infected animals in a region of high anthrax incidence; and those handling animal hides or wools imported from countries with lax standards for spore transfer prevention.

II. Cholera

Cholera is a disease of debilitating diarrhea caused by the extracellular bacterium *Vibrio cholerae*. Transmission is by consumption of water contaminated with fecal waste from infected humans, so that in regions where sanitation is good, disease incidence is minimal. When *V. cholerae* is ingested, the bacteria multiply rapidly in the gut and produce an exotoxin that induces relentless loss of water and ions from gut epithelial cells. Life-threatening dehydration can occur within hours of infection, particularly in young children. Although cholera is easily treated with rehydration and antibiotics, these measures are not always available in developing countries, war zones, or areas afflicted by natural disasters. Where prompt treatment is lacking, the very short incubation period of the bacterium means that local outbreaks can rapidly expand into epidemics. The WHO estimates that, every year worldwide, 3–5 million cases of cholera occur that result in 100,000–120,000 deaths.

Since the 19th century, most cholera outbreaks have been due to the O1 **serotype** (strain) of *V. cholerae*. In the past, other *V. cholerae* serotypes caused mild diarrhea but did not trigger epidemics. In 1992, however, a new serotype called *V. cholerae* O139 was identified in cholera patients during an outbreak in Bangladesh. This serotype is now responsible for much of the choleric disease reported in Southeast Asia. Unfortunately, additional new *V. cholerae* serotypes have recently emerged in areas of Africa and Asia and have caused very severe cholera linked to a high death rate.

Although cholera infection induces the production of circulating antibodies recognizing the cholera exotoxin, this humoral response only mitigates and does not eliminate the disease. We now know that the most effective defense in exposed individuals is provided by mucosal SIgA, which blocks the attachment of *V. cholerae* to gut epithelial cells. Cholera is therefore an obvious candidate for a mucosally administered vaccine. However, until recently, the only cholera vaccine available was based on a killed whole bacterial cell preparation delivered by injection. Unfortunately, this vaccine offers no mucosal protection, is effective in less than 55% of a given population, protects for only 6 months, and is associated with serious adverse events. There are now two orally administered cholera vaccines that are also based on killed whole cells but result in greater coverage, confer longer-lasting protection, and have fewer side effects. However, because proper sanitation prevents cholera outbreaks in developed countries, the WHO recommends the use of these vaccines only for those living in endemic regions or traveling to them for extended periods.

III. Diphtheria, Tetanus and Pertussis

i) Diphtheria

Diphtheria is a devastating childhood disease caused by the exotoxin of the bacterium *Corynebacterium diphtheriae*. The exotoxin inhibits protein synthesis in cells of the heart and nervous system and also induces an inflammatory response in the throat that can obstruct breathing. The disease can be prevented by maintaining adequate levels of circulating antibodies to the exotoxin. The current vaccine is a formalin-inactivated exotoxin toxoid.

ii) Tetanus

Tetanus is caused by the exotoxin produced by the bacterium *Clostridium tetani*. The exotoxin attacks neurons and causes painful muscle spasms. Because the disease is caused solely by the exotoxin, neutralizing antibodies raised in response to vaccination with an exotoxin toxoid are protective.

iii) Pertussis

Whooping cough (pertussis) is a highly contagious disease caused by the bacterium *Bordetella pertussis*. *B. pertussis* gravitates to the mucosae of the bronchi and produces two exotoxins that inhibit the clearance of mucus and promote the attachment of the bacteria to the respiratory tract. As a result, severe coughing is triggered that quickly debilitates young children. This coughing can persist for weeks, leading to its monicker "the 100-day cough." Almost half of *B. pertussis* infections occur in infants, and a significant proportion of these cases require hospitalization. Potentially fatal pneumonia can occur, as well as seizures and encephalopathy leading to permanent brain damage or death. The current vaccine is made from a mixture of *B. pertussis* proteins and induces the production of neutralizing antibacterial and antitoxin antibodies. Because 75% of infants infected with *B. pertussis* catch the bacteria from a family member or close contact, cocooning is also effective in protecting children from transmission until they can be vaccinated.

Public health studies have shown that upgrading sanitary conditions is a more cost-effective way to prevent cholera outbreaks than instituting mass immunization against *V. cholerae*.

To witness the hallmark whooping cough of pertussis in a child or adult, visit http://www.cdc.gov/pertussis/ and scroll down to "Would You Know Pertussis?" to find the relevant links.

NOTE: Pneumonia kills about 1.4 million children under the age of 5 each year, mostly in South Asia and sub-Saharan Africa. While pneumonia can be treated with drugs, ensuring that children are immunized against *B. pertussis*, *Haemophilus influenzae* (Hib), *Pneumococcus*, and measles virus is the best way to prevent this serious disease.

Pneumonia is an acute respiratory infection by a bacterium or a virus that causes the alveoli of the lungs to be filled with pus and fluid instead of air. Breathing becomes painful and difficult.

iv) The Combination DTaP Vaccine

In the U.S., young children are simultaneously vaccinated against diphtheria, tetanus and pertussis through use of the combination DTaP vaccine, consisting of *B. pertussis* proteins (P) combined with the diphtheria (D) and tetanus (T) toxoids. Completion of a full series of DTaP doses at a young age is important: 83% of vaccinees receiving three doses are completely protected from these diseases, but only 36% are resistant to pertussis after one dose. In addition, the risk of adverse effects of the DTaP vaccine is greater when an individual receives his/her first dose as an adolescent or adult. A subunit vaccine for pertussis has been created but, sadly, is currently too costly for use in the developing world (where the need is greatest).

NOTE: The "a" in DTaP stands for an "acellular" preparation of *B. pertussis* proteins. The previous DTP vaccine contained whole *B. pertussis* cells, which proved to be harmful in rare cases. While the DTaP vaccine is associated with fewer side effects than the DTP vaccine, it appears to provide slightly inferior protection.

The protection against diphtheria and tetanus conferred by the childhood DTaP vaccine decreases with time, so that a booster containing only the diphtheria and tetanus toxoids (Td) is routinely recommended for adults every 10 years. More frequent Td vaccination is advisable for sewage workers and metal scrap yard workers. Parallel boosting against *B. pertussis* infection was not considered necessary in the past because infected adults and adolescents experience far fewer ill effects than do children. However, the need to provide herd immunity to protect infants means that adults, particularly those who are pregnant or spending time around newborns, are now advised to seek out a one-time booster that will combat this pathogen.

To facilitate these updated recommendations, a new combination vaccine called Tdap has recently been licensed for use as the first adult vaccine offering triple protection against tetanus, diphtheria and pertussis. Released in 2010, the Tdap vaccine contains tetanus toxoid, reduced quantities of diphtheria toxoid compared to the DTaP vaccine, and acellular pertussis vaccine. A single dose of the Tdap vaccine can be given to an adolescent or adult regardless of his/her history of vaccination with the DTaP vaccine. The individual will then need to follow up with a Td booster only every 10 years.

IV. Haemophilus Influenzae Type b

Haemophilus influenzae is an encapsulated extracellular bacterium that has nothing to do with the "flu" or the influenza virus. Different strains of *H. influenzae* bear different types of capsules. *H. influenzae* with the type "b" capsule (Hib) is particularly pathogenic to humans and causes thousands of infant deaths each year worldwide. Hib bacteria preferentially colonize the mouth and throat but can cause fatal meningitis if an individual's immune system is sufficiently depressed. Hib infection is also the second most common cause of bacterial pneumonia in children, after infection with *S. pneumoniae*.

Because neutralizing antibodies provide very good defense against *H. influenzae*, several different vaccines have been developed in which a Hib capsule polysaccharide is conjugated to a carrier protein to form a Td antigen with high immunogenicity in infants. This type of vaccine has proved very effective and has dramatically reduced the prevalence of invasive Hib disease. To date, no booster dose has been required by adolescents or adults.

Meningitis can be caused by bacteria, viruses, parasites or fungi, or may be of non-infectious origin.

V. Hepatitis A Virus

HAV is a non-enveloped single-stranded RNA virus that is contracted from the consumption of contaminated food or water. HAV infection can cause the significant liver damage characteristic of hepatitis. Fortunately, vaccination is highly effective in protecting against HAV infection. Inactivated, whole virus vaccines are available in many countries, with live, attenuated vaccines being used to a more limited extent. In the U.S., two doses of inactivated whole virus vaccine given during childhood or adolescence have proven sufficient to provide 14–20 years of protection against HAV infection. Vaccination is particularly important for those who either live in or plan to travel to places where HAV is endemic (such as the tropics). Adults with chronic liver disease and hemophiliacs should also ensure that they have been vaccinated against HAV.

VI. Hepatitis B Virus

HBV is an enveloped DNA virus that can cause severe liver damage and may also cause hepatocarcinoma. The virus is usually contracted via sexual activity or contaminated needles but can be passed from an infected mother to her fetus or breastfed newborn. HBV infection becomes chronic in about 90% of infected infants, and before the advent of the HBV vaccine, about 30–40% of chronic HBV infections in adults were believed to have resulted from transmission in early childhood. About 25% of chronically infected adults die of HBV-associated liver disease later in life. Moreover, the virus is highly infectious, meaning that chronically infected individuals can easily transmit the virus through their blood or body fluids. Adults who are companions to HBV-infected individuals, hemodialysis patients, health workers, sex workers, and scientists handling human or primate tissues or blood should take special care to ensure that they are vaccinated against HBV.

The current HBV vaccine is of the protein subunit type and consists of an HBV surface antigen produced using recombinant DNA techniques. Protective levels of neutralizing antibodies are produced in 95% of vaccinated infants, and adequate titers persist for at least 10–18 years. The HBV vaccine has been spectacularly successful in countries where HBV is endemic. For example, in 2006, a universal infant HBV vaccination program that was initiated in Shandong, China, achieved a 90% reduction in HBV infections in children aged 1–14 years.

NOTE: Despite heroic efforts, there is currently no vaccine to prevent hepatitis C virus (HCV) infection. Like HBV, this virus attacks the liver and is transmitted by contact with the blood of an infected individual. Unsafe injection using a contaminated needle is the prime culprit. HCV is not spread through shared food or water, casual kissing, or via breast milk. Symptoms can be mild or serious, and persist for a few weeks or a lifetime. The WHO estimates that 130,000–170,000 people are infected with HCV every year, and that more than 350,000 chronically infected individuals die of hepatitis C-associated liver disease annually. Oral antiviral drugs are currently the best bet for mitigating the effects of HCV infection, but the limited availability and high cost of these drugs in most developing countries mean that many infections go untreated.

VII. Human Papillomavirus

HPV is a non-enveloped DNA virus that occurs in 40 different serotypes. Infection by some of these strains is asymptomatic, but infections by others can cause genital warts or cancer of the cervix, vagina, penis, or oropharynx (tongue, tonsils, throat). Cervical cancer kills 4000 women annually in the U.S., and oropharyngeal cancer occurs in 7000 American men each year. Since it is now clear that many of these cancers are caused by HPV infection, their incidence stands to be substantially reduced by vaccination against this pathogen.

Two prophylactic vaccines are currently available, one bivalent (HPV2) and one tetravalent (HPV4). Both protect against infection with the HPV serotypes that cause the majority of HPV-associated cancers, while the HPV4 vaccine also prevents infection by the two strains that cause most genital warts. Both vaccines contain purified HPV capsid proteins and are produced using recombinant DNA techniques. Either vaccine can be administered to adolescent and adult females, whereas males should receive the HPV4 vaccine. These vaccines are most effective when given in three doses during early adolescence, so that protective antibody levels can develop before the vaccinee becomes sexually active.

Cultural biases are limiting the coverage achieved by the HPV vaccines. Despite studies proving the contrary, some parents and religious leaders still believe that vaccinating against this virus encourages early promiscuity. Some physicians will readily immunize 13–15-year-old adolescents but balk at giving the vaccine to 11-year-olds. In addition, there are health care management organizations that will not cover the expense of the three vaccine doses, making their cost a prohibitive factor for many who could benefit. Patient compliance with the three-dose schedule is also sometimes incomplete. Sadly, the groups with the highest rates of cervical cancer (low-income and/or racial minority women) have the lowest HPV vaccination rates.

VIII. Influenza Virus

The influenza virus is an enveloped RNA virus that attacks human respiratory epithelial cells and causes severe acute respiratory symptoms. Influenza virus infection is particularly problematic for very young and very old individuals whose immune systems are functioning at less than full strength. Persons suffering from type 1 diabetes, alcoholism, cardiovascular disease, or from chronic respiratory disorders such as cystic fibrosis are also at increased risk. The WHO estimates that 1 billion influenza infections occur each year worldwide, resulting in 3–5 million cases of severe disease and 300,000–500,000 deaths.

> In the U.S., influenza kills 3300–48,000 people each year, depending on the virulence of that season's virus.

There are three types of influenza viruses, A, B and C, that are distinguished by the antigenic characteristics of their structural proteins. Influenza type A is the most pathogenic to humans. In a natural influenza virus infection, neutralizing antibodies directed against important viral proteins play key roles in immune defense. However, because the influenza virus constantly changes its surface proteins due to antigenic drift and shift, it has been impossible thus far to produce a vaccine that confers lifelong protection against all influenza serotypes. Consequently, annual vaccination programs are undertaken in developed countries just prior to "flu season." How do the health authorities know which variant of the virus to target for vaccine production? In February of any given year, the WHO and the U.S. CDC study which flu viruses are emerging around the world and choose a combination of candidate strains for the production of the vaccine to be used in November. In general, these seasonal flu vaccines are formulated to contain antigens from the three virus strains considered to be most likely to circulate in the upcoming flu season. The strains chosen are often members of the influenza A H3N2 group, the influenza A H1N1 group, and the influenza B group. For example, for the 2012–2013 flu season, it was recommended that flu vaccines contain the California strain of influenza A (H1N1) virus, the Victoria strain of influenza A (H3N2), and the Wisconsin strain of influenza B virus. In the past, if the strains chosen turned out not to be responsible for the majority of flu cases in that year, the vaccine was not effective. Fortunately, the accuracy of "strain-picking" is now high enough that annual flu vaccination programs are generally quite helpful in reducing disease incidence, severity and mortality.

Two formulations of trivalent influenza vaccines are currently available: *live-attenuated influenza vaccine (LAIV),* which is administered via a nasal spray, and *trivalent inactivated influenza vaccine (TIV),* which is delivered by injection. LAIV provides superior protection when the influenza strains that are circulating in a given year do not match those used to generate that's year batch of vaccine. However,

Infants under 6 months of age cannot be vaccinated with current formulations of influenza vaccines, so that the vaccination of pregnant women and cocooning after the baby is born are vital protection strategies.

whether the vaccine is attenuated or inactivated, the fact that a whole virus is used (which by definition includes PAMPs) means that robust activation of the innate immune response occurs, often resulting in fever and aches. Patients therefore often complain that "the vaccine gave me the flu," although these effects are due to the immune response itself. That being said, according to VAERS, the TIV vaccine administered for the 2010–2011 flu season was associated with very rare but genuine febrile seizures in young children. Happily, all these youngsters eventually made a full recovery.

NOTE: One of the great concerns of immunologists with respect to influenza's ability to undergo antigenic shift is the possibility of cross-species transmission of newly emerging strains. Some known strains of influenza A that circulate among avian species or in other mammals such as pigs can also infect humans. In most cases, an influenza virus that is transmitted from an animal to a human causes clinical symptoms in the human (e.g., "bird flu" or "swine flu"), but this disease is either relatively mild or not readily transmitted from that person to other humans. However, due to the potential re-assortment of genomic segments within an animal or human that has become co-infected with two different influenza strains, a progeny virus could emerge that is both highly lethal to humans and readily spread from person to person. Because this strain would represent a radical shift from influenza A viruses previously experienced by humans as a species, the immunity of most of the world's population would be present only at the primary response level and a devastating death toll would thus likely result. Scientists are working hard to devise a "universal" flu vaccine that would induce antibodies able to recognize conserved epitopes common to any influenza virus strain. Such a vaccine would not only circumvent antigenic drift and shift, removing the need for seasonal vaccination, but might also prevent a pandemic sparked by transmission between animal reservoirs and humans.

IX. Measles, Mumps and Rubella

i) Measles

The measles virus is a highly contagious, enveloped RNA virus that initially attacks the upper respiratory tract but then spreads via the lymphatics and blood to most other tissues. Disease symptoms include fever, a characteristic rash of red spots, and temporary immune system suppression. Potentially fatal opportunistic infections may thus gain a foothold. Because the virus can remain active and contagious in the air or on infected surfaces for up to 2 hours, it is easily transmitted from person to person. Those individuals who survive a natural exposure to the measles virus usually acquire lifelong protection.

There are currently several different vaccines for measles, all of which are based on live, attenuated virus. The MCV (measles-containing vaccine) contains live, attenuated measles virus alone, whereas the combination MMR vaccine contains live, attenuated versions of the measles, mumps, and rubella viruses (see following sections).

NOTE: It is difficult to overstate the beneficial impact of immunization programs on the fight to control measles. Before the advent of widespread measles vaccination in 1980, measles killed an estimated 2.6 million people per year worldwide! Although control of this disease was quickly established in developed countries through sustained childhood immunization programs, comparable success has been elusive in low-income countries with poor health care infrastructure. Thus, despite the ready availability of these safe and cost-effective vaccines, measles is still a major leading cause of death of young children in developing countries. Indeed, measles claimed 15 young victims *per hour* according to WHO estimates in 2010. Fortunately, an intensive vaccination program implemented in many high-risk countries is starting to take effect: from 2000–2011, more than one billion young children in these countries were vaccinated against measles, and measles-related deaths have dropped by more than 70% worldwide.

The cost to immunize one child against measles is less than U.S. $1.00.

ii) Mumps

The mumps virus is an enveloped RNA virus that initially causes a respiratory infection but then travels in the blood to infect and cause swelling of the salivary glands. Severe complications include sudden and permanent deafness in one or both ears, orchitis and meningoencephalitis. Childhood vaccination provides good, but not absolute, protection against mumps.

iii) Rubella

Rubella (also known as German measles) is caused by the rubella virus, an enveloped RNA virus. In young children, rubella is a relatively mild disease characterized by a rash, low fever, malaise and mild conjunctivitis. Lifelong immunity results from childhood infection. The danger in rubella lies in its teratogenic effects on the developing fetus of a woman who never had the disease as a child. The placenta may become infected, allowing the virus to enter the fetal circulation and infect fetal organs. If infection occurs within the first trimester, the fetus may develop *congenital rubella syndrome* (CRS), which is characterized by cataracts, heart disease, deafness, and sometimes hepatitis and/or mental retardation. Since the lifetime costs of supporting individuals with CRS far exceed those for vaccination, governments have a clear incentive for ensuring their populations are protected.

iv) MMR vaccine

As introduced previously, the trivalent MMR vaccine is composed of live, attenuated forms of the measles, mumps, and rubella viruses. Two doses of this vaccine are generally sufficient to confer long-lasting protection against all three diseases. MMR vaccination induces both antibody and cell-mediated immune responses, and long-lasting memory is established. The efficacy and safety of the MMR vaccine has meant that measles is now a relative rarity in the developed world and that cases of rubella have decreased globally by 82% since 2000.

> NOTE: The Measles and Rubella (MR) Initiative is a collaboration between the WHO, CDC, UNICEF, American Red Cross and United Nations Foundation to control (if not eradicate) measles and rubella via vaccination. In April 2012, this group launched a program that aims to work with international and local health authorities to reduce global measles deaths by 95% by 2015 (compared to levels in 2000), and to eliminate measles and rubella in at least five of the six WHO regions by 2020.

X. Meningococcus

H. influenzae, *S. pneumoniae*, group B Streptococcus, and *Listeria monocytogenes* can all cause meningitis, but only the encapsulated bacterium *Neisseria meningitidis* (commonly called meningococcus) spreads readily enough to cause meningitis epidemics. Meningococcal disease most often strikes children and adolescents, and individuals with primary immunodeficiencies of complement components are particularly susceptible. Meningococcus is readily transmitted through respiratory secretions and is fatal in 50% of untreated cases. Survivors may suffer from brain damage and neurological defects leading to mental impairment, deafness or seizures. Epidemics of meningococcal meningitis occur every 7–14 years in sub-Saharan Africa, in the so-called meningitis belt. In the 2009 epidemic in this region, 14 African countries reported more than 88,000 suspected cases and over 5000 deaths. In the developed world, youth between the ages of 16 and 21 years are at the highest risk of contracting the disease. Many young adults leave home at this age to attend college, university or trade school and thus live in close proximity in residences and dormitories, facilitating bacterial transmission.

Twelve types of meningococcus exist, several of which can cause large-scale epidemics. The major miscreants are defined by their capsular polysaccharides: groups A, B, C, W135 and Y. Group A infections dominate in developing countries, particularly in Africa, whereas group B, C and Y infections are most prevalent in the Americas, Australasia and some parts of Europe. Group W135 meningococcus turns up in relatively small, localized outbreaks in various locations around the world. Regardless of group, both the capsular polysaccharide and endotoxin produced by these bacteria contribute to their virulence, and humoral immunity is the key to successful host defense.

Current vaccines are based on a preparation of capsular polysaccharides from four bacterial strains (A, C, W135, Y). These preparations may contain the capsules alone (MPSV4 vaccine) or capsules that are conjugated to a protein carrier (MCV4 vaccine). Either vaccine may be given to children older than 24 months of age, and MCV4 can be administered to infants. Boosters are required every 3–5 years for the MPSV4 vaccine, whereas boosters of the conjugate vaccine, which generates longer-lasting protective memory, are recommended at 11–12 years and 16 years of age. Mass pre-emptive vaccination against group C meningococcus in young children and adolescents has been instituted in several Western countries. In 2010, a new and highly efficacious conjugate vaccine against group A meningococcus was tested in pilot programs in three African countries within the meningitis belt. An immediate reduction in cases of bacterial meningitis was recorded, prompting the recent expansion of this vaccination program. The WHO's goal is to have this vaccine administered to large numbers of people aged 1–29 years in all 25 countries in the meningitis belt by 2016.

> NOTE: Vaccines based on the capsule polysaccharide of group B meningococcus cannot be used in humans because this bacterial polysaccharide closely resembles a polysaccharide present in human neurologic tissues. Antibodies recognizing the bacterial polysaccharide would therefore also attack the normal human tissue, a phenomenon called **molecular mimicry** (see Ch. 19). Vaccines against group B meningococcus are therefore generated on an epidemic-specific basis and contain the outer membrane protein (OMP) of the specific bacterium involved. These OMPs have no counterparts in human tissues, reducing the chance of immunopathic side effects.

XI. Plague

Plague is caused by the Gram-negative bacterium *Yersinia pestis*. *Y. pestis* naturally infects rodents (especially rats) and their fleas and is transferred to humans by flea bites. At the site of the bite, the skin becomes blistered and blackened (hence, "Black Death"). Within a week of the bite, the bacteria access the draining lymph node via the lymphatics and cause high fever and a large, painful swelling of the node known as a "bubo" (hence, *bubonic plague*). From the lymph node, the bacteria spread to the blood and organs with disastrous speed. Once the bacteria access the blood, the plague is said to be *septicemic* in form. If the bacteria reach the lungs, pneumonia occurs and *pneumonic plague* is said to be present. Unfortunately, pneumonic plague is also readily spread by the inhalation of respiratory droplets expelled by the coughing or sneezing of an infected person. In the absence of antibiotic treatment, all three forms of plague have a high fatality rate. Plague is endemic in many countries in Africa, Asia, Latin America and South America, but the risk of infection is low as long as rat-infested areas are avoided. Patients with pneumonic plague should be isolated throughout their illness to prevent air-borne transmission of the infection.

Vaccination against plague has not proved to be very helpful. Indeed, control of local rodents and elimination of their habitats is usually the most cost-effective way of preventing plague outbreaks. As a result, there is little incentive for pharmaceutical companies to pursue modern plague vaccines. The existing plague vaccine is based on formalin-killed whole bacterial cells, but it is not in common use because it has side effects of fever, headache and pain that increase in severity with repeated doses. Thus, vaccination with this vaccine is recommended only for health care workers in endemic areas, disaster relief workers, and laboratory personnel working directly with *Y. pestis*.

XII. Pneumococcus

As noted previously, *Streptococcus pneumoniae* (commonly called pneumococcus) is an encapsulated bacterium that initially colonizes the upper respiratory tract and is the leading cause of bacterial pneumonia in children. The bacterial capsule forestalls engulfment by pulmonary phagocytes so that the invaders multiply in great numbers in the lung and then spread throughout the body. Many cases of pneumonia, ear infection and meningitis are attributed to this bacterium every year.

Two types of vaccines to combat pneumococcus have been developed. The polysaccharide vaccine (PPV) contains the polysaccharides of the 23 strains of *S. pneumoniae* that represent 80–90% of disease-causing strains. However, the Ti response induced by this vaccine is ineffective in very young children. Instead, infants are given a conjugate vaccine (PCV) that contains modified diphtheria toxoid linked to the capsule polysaccharides of the 7 (PCV7) or 13 (PCV13) most common strains of *S. pneumoniae* attacking infants. When administered intramuscularly, these vaccines provoke a Td response that leads to effective protection in about 90% of vaccinated infants. Adults that have undergone splenectomy or are suffering from cardiovascular disease, diabetes or alcoholism should also ensure they are vaccinated against *S. pneumoniae*.

> NOTE: The history of the PCV7 vaccine provides a tale of caution against complacency in the face of wily pathogens. After the PCV7 vaccine came into common use in 2000, the overall incidence of invasive pneumococcal disease in children under 5 years of age dropped by 79%. In particular, the incidence of disease caused by the bacterial serotypes combatted by the vaccine decreased by 99%. However, the incidence of disease caused by bacterial serotypes *not* combatted by the vaccine rose substantially. More frighteningly, the non-vaccine bacterial strain that predominated in these infections was antibiotic-resistant. The PCV13 vaccine, which was introduced in 2010 and contains the capsular polysaccharides of six additional pneumococcus serotypes, was designed to provoke immune responses against these rising strains and thus provide wider coverage. Time will tell if still another PCV vaccine, or a vaccine based on a different approach, will be needed to fight invasive pneumococcal disease.

XIII. Polio

Poliovirus is a very infectious, non-enveloped RNA virus that occurs in three serotypes, called 1, 2 and 3. These serotypes are distinguished by their viral coat proteins, but all three cause the same disease. Poliovirus is acquired orally and replicates first in the intestinal tract but then travels in the blood to the spinal cord and CNS. In 1 in 200 cases, the virus succeeds in destroying the motor nerves and causes permanent muscle paralysis called *poliomyelitis*. Of these cases, 5–10% die when the muscles necessary to breathe are paralyzed. This devastating and incurable disease mainly affects children under 5 years of age.

Multiple doses of polio vaccine can protect a child for life. Two types of polio vaccines are currently available: inactivated poliovirus (IPV) vaccine, which contains inactivated versions of poliovirus serotypes 1, 2 and 3; and oral poliovirus (OPV) vaccine, which contains live, attenuated versions of all three serotypes. IPV is administered by intramuscular injection, requires repeated boosters, and does not induce high levels of mucosal immunity in the gastrointestinal tract. OPV is administered orally, does not require boosters, and induces significant gastrointestinal mucosal immunity. However, when the first doses of the original OPV vaccine were administered, it was discovered that there was a very small chance (one case per 2–3 million doses) that one of the attenuated viruses in this vaccine could revert and cause poliomyelitis in a vaccinee, a syndrome called **vaccine-associated paralytic polio (VAPP)**. Subsequent analysis of VAPP incidence showed that this syndrome was most likely to occur when an individual received OPV as his/her first vaccine dose, and that the probability of VAPP diminished if the individual received one or more IPV doses before his/her first OPV dose. By the mid-1990s, because the wild poliovirus had been eliminated in North America, the risk of VAPP was greater than the risk of getting the disease; accordingly, the original OPV vaccine was not recommended for use in North America, and four childhood doses of IPV were given instead.

The first IPV vaccine was known as the Salk vaccine, whereas the first OPV vaccine was known as the Sabin vaccine, the former having been developed by Jonas Salk, and the latter by Albert Sabin.

Today, to reduce the risk of VAPP, current versions of OPV vaccines require that the attenuation be based on much larger and more stable deletions of the viral genome, and that the concentration of virus used is much decreased compared to the original OPV formulation. Nevertheless, North American children continue to receive four doses of IPV vaccine. OPV stocks are held in reserve and are used only in the case of a polio outbreak. In such an emergency situation, a single OPV dose provides faster, broader protection for community members than a single IPV dose. Furthermore, the small risk of VAPP associated with the administration of a modern OPV vaccine is far outweighed by the high risk of paralysis associated with actual poliovirus infection.

In Europe, children first receive a dose of the IPV vaccine to induce a degree of systemic protection and reduce the chance of VAPP. OPV boosters are then administered to establish effective mucosal immunity. Administration of four doses of OPV vaccine alone is still recommended for vaccination programs in the few regions of the world where the wild poliovirus remains endemic. Although these programs have achieved much success, wild poliovirus can still be found in parts of Southern Asia and Africa. Sadly, some countries in which poliovirus was successfully eradicated have had it reintroduced by infected immigrants or travelers.

> NOTE: In 1988, the Global Polio Eradication Initiative was launched by the WHO, Rotary International, the CDC and UNICEF. At first, there was much optimism that polio could be eradicated globally by 2000 just as smallpox had been eradicated in 1980. Indeed, mass immunization campaigns have led to a 99% reduction in polio cases worldwide compared to 1988 levels. In 2012, poliovirus was considered endemic only in Afghanistan, Nigeria and Pakistan. However, eradication is not likely to be achieved any time soon in these areas due to uncertainties in security and political situations that restrict the activities of immunization volunteers. Because a single infected child puts all of the world's children at risk, experts acknowledge that attaining the WHO's stated goal of making the world polio-free will require coordinated and Herculean efforts by many private, governmental and public organizations to vaccinate the world's children.

Economic models estimate that the eradication of polio by 2017 could save U.S. $40–50 billion in health care costs in low-income countries.

XIV. Rabies

Rabies is caused by a slow-replicating, enveloped RNA virus that is transmitted to humans via another animal. Upon entering a human's body, the rabies virus replicates first in skeletal muscle and connective tissue. It then spreads along the peripheral nerves to the spinal cord and CNS, where it causes progressive encephalitis. The disease is inevitably fatal in humans if left untreated, and each year 55,000 people die of rabies infection around the world (most in developing countries). Indeed, a recent increase in human rabies deaths in low-income countries suggests that rabies is once again becoming a serious public health problem. While dog bites or scratches are the cause of 99% of human rabies deaths globally, bats are the source of most human rabies deaths in North America.

Modern rabies vaccines are based on killed viruses grown in human cell cultures or in chick embryos, and are safe and highly efficacious. Nevertheless, the best way to prevent rabies in humans is to vaccinate dogs, a strategy that has significantly reduced the number of human rabies deaths in several countries in Latin America. Where possible, the vaccination of wildlife, other domestic pets and stock animals can also be undertaken. Only individuals who are traveling extensively in an area at high risk for infection, or who might expect to encounter rabies in their line of work (such as wildlife workers or veterinarians), receive the human rabies vaccine prophylactically. Because the virus replicates only slowly, unvaccinated individuals bitten by a rabid animal can obtain a post-exposure shot of rabies vaccine that will still trigger an effective response against the virus. The bitten individual may also receive passive immunization with a preparation of antirabies antibodies. The WHO estimates that, every year, more than 15 million people around the world have to suffer through a painful series of rabies shots to prevent disease caused by exposure to the saliva of a rabies-infected animal. Most of those treated are young males aged

5–14 years. On the "up" side, the WHO estimates that these post-exposure regimens prevent the deaths of over 300,000 people annually. However, the cost of this post-exposure rabies treatment is out of reach for many living in the poverty-stricken areas where rabies exposure is most likely. As a result, 20,000 rabies deaths occur each year in India alone.

XV. Rotavirus

Rotavirus is a non-enveloped RNA virus that has a characteristic wheel-like shape when viewed by electron microscopy. This virus is the most common cause of severe diarrhea in children and kills over 450,000 youngsters annually around the world. Rotavirus is transmitted primarily by contact with a contaminated surface or by ingestion of contaminated food or water. About 10–15% of severe rotavirus gastroenteritis occurs in children younger than 6 months old. Although an older generation of rotavirus vaccines was withdrawn due to rare but severe detrimental effects on the intestine, newly developed vaccines have proven safe and effective, even in very young infants. The RV1 and RV5 vaccines, which are administered orally, contain one or five live attenuated bovine rotavirus strains, respectively. Even before attenuation, these organisms have a low rate of replication in the human intestine, unlike human or primate rotavirus strains.

> NOTE: Patients with PIDs, including severe combined immunodeficiency (SCID; see Ch. 15), do not normally receive live attenuated vaccines for fear that these weakened organisms will still be able to cause disease in the absence of a competent immune response. However, the first dose of rotavirus vaccine is usually given at 2 months of age, before SCID is typically diagnosed. There have been reports of infants who developed symptoms of rotavirus infection, such as dehydration and diarrhea, within a month of their rotavirus vaccination and were subsequently diagnosed SCID. As a result, scientists are racing to develop tests that can screen newborns for SCID and other PIDs so that these infants can avoid vaccination with live virus vaccines.

XVI. Tuberculosis

Tuberculosis (TB), the lung disease caused by *Mycobacterium tuberculosis*, is one of the world's leading killers. In 2010, 8.8 million people contracted TB, and 1.4 million died of this disease worldwide. In 2009, the WHO reported that 10 million children were orphaned by TB. Despite this fearsome toll, these numbers are actually a positive development, representing a 40% drop in the overall death rate due to TB since 1990.

TB is now viewed as a "co-morbidity disease," meaning that it tends to develop in individuals that are already weakened due to infection with another pathogen (such as HIV), or whose immune systems have been otherwise compromised. While healthy people who become infected with *M. tuberculosis* have only a 10% chance of coming down with "active" TB disease, those already ill due to other pathogens or factors, or are tobacco users, are far more likely to develop clinical TB. Patients with active TB are highly contagious and suffer from weight loss, loss of vigor, and coughing (often with blood). If not treated aggressively, many will eventually die of their disease. Furthermore, bacteria aerosolized by the coughing of infected individuals with active TB can readily spread to people in close proximity.

Although the BCG vaccine against *M. tuberculosis* has been available for many years, its efficacy is very low, and alternatives are continually being sought (see **Box 14-4**). The lack of an effective TB vaccine and the rise of multi-drug-resistant strains of this bacterium prompted the WHO to launch its "Global Plan to Stop TB (2006–2015)," which aims to dramatically reduce the global incidence and mortality rates of this disease. Although a major focus of this campaign is a coordinated effort to develop a new and effective TB vaccine, the WHO has also urged that the environmental conditions that favor TB prevalence and transmission (overcrowding and unsanitary housing) be addressed. In addition, rather than recommending mass vaccination programs, the WHO advises that countries focus on intensive, supervised antibiotic treatment and appropriate care of TB patients to prevent spread of the disease.

More information about the WHO's TB strategy can be found online at http://www.who.int/tb/strategy/stop_tb_strategy/en/index.html.

Box 14-4   The Challenges of Tuberculosis

M. tuberculosis is a slow-replicating intracellular bacterium that does not produce bacterial toxins. Upon initial infection, *M. tuberculosis* causes only mild inflammation in individuals with a robust immune response. However, this pathogen is extraordinarily hard to remove, so that the immune response that is provoked is prolonged. Activated T cells start to hyperactivate macrophages, which in turn secrete cytokines that damage the lungs. Granulomas that form to wall off the bacteria appear as "tubercles" (lumps) on the lungs. Eventually, the granulomas break down and release a large proportion of the bacteria. A healthy immune system can eliminate most of the invaders at this point such that the tubercular lesions calcify and become visible in X-rays as scarring on the lung (see the following Plate). Surviving bacteria in the lesions may become dormant for as long as 20 years. Indeed, 90% of individuals infected with *M. tuberculosis* remain clinically healthy. However, if the immune system is later compromised such that the bacteria can resume replication, a lesion may rupture and release millions of bacteria first into the lungs and then throughout the body. At this point, the patient is considered to have "active TB" and at risk to spread the bacteria to nearby individuals.

Researchers and clinicians have struggled for many years to develop an effective vaccine against TB. Because *M. tuberculosis* can lurk intracellularly for decades, a vaccine inducing a comprehensive immune response with long-lived memory is required. These characteristics are usually best evoked by a live, attenuated vaccine, but natural immunity to *M. tuberculosis* is still not completely understood. The only vaccine currently available to fight TB is the *bacillus Calmette–Guérin* (BCG) vaccine. BCG is a live, attenuated form of *Mycobacterium bovis*, a bacterial species that shares many antigens with *M. tuberculosis*. The original BCG strain underwent a spontaneous deletion that removed genes conferring virulence without significantly compromising immunogenicity. However, numerous mutations have since occurred in the BCG genome that have decreased its efficacy in some human populations. Indeed, for unknown reasons, protection levels for individuals vaccinated with BCG vary wildly, from zero to 80%. Scientists have been examining the genome of *M. tuberculosis* to identify genes that can be mutated to decrease the virulence or persistence of the organism without affecting its immunogenicity. However, the less virulent a mutated *M. tuberculosis* strain is, the less protective it appears to be. Researchers have also tried searching for *M. tuberculosis* antigens for use in subunit or DNA vaccines, but the lack of a reliable *in vitro* system to test for the protective efficacy of such antigens has hindered progress on this front. Moreover, no one *M. tuberculosis* antigen seems to protect against all stages of *M. tuberculosis* infection, and subunit vaccines based on bacterial products have thus far offered no more protection than the BCG vaccine. TB vaccine development has also been hampered by the lack of a sufficiently accurate animal model to use for testing. Mice are not highly susceptible to *M. tuberculosis*, and although *M. tuberculosis* infection in primates resembles that in humans, primate trials are prohibitively expensive. Another challenge is the current dearth of optimal adjuvant options. Experiments with subunit vaccines in delivery vehicles have been attempted, but none so far has offered protection superior to that induced by the venerable BCG vaccine. Lastly, because of the long latency of *M. tuberculosis* in humans, clinical trials of immense length have to be conducted to judge the efficacy of any vaccine devised. The struggle therefore continues.

Plate Box 14-4
Lung Scarring in Tuberculosis

[Reproduced by permission of Ian Kitai, The Hospital for Sick Children, Toronto.]

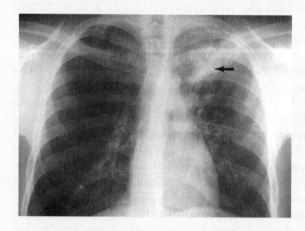

In the U.S., the need for vaccination is determined by TB experts and is restricted to individuals deemed to be at particularly high risk of TB exposure. Those living in substandard housing conditions, injection drug users, and immunocompromised patients are greatly susceptible and should be considered for vaccination. In addition, vaccination may be recommended for workers in high-risk settings such as hospitals, shelters, and correctional facilities, as well as for family members (particularly children) of TB-infected individuals.

XVII. Typhoid Fever

Typhoid fever is caused by *Salmonella typhi*, a highly invasive intracellular bacterium whose only natural host is humans. An acute infection is characterized by high fever, abdominal discomfort, malaise, and headache that can last for several weeks. Many (but not all) patients get a rash of salmon-colored spots (hence, *Salmonella typhi*). Life-threatening complications include intestinal perforation and hemorrhage. Ingested bacteria initially enter either enterocytes or M cells in the small intestine but eventually spread to macrophages throughout the body. In a small number of patients, bacteria reach the gallbladder and establish a chronic infection. These people become asymptomatic carriers who shed the bacteria in infectious form. Typhoid fever is now a rarity where good water treatment prevails, but the disease still kills hundreds of thousands in developing countries. Travelers to endemic countries should be vaccinated against typhoid fever.

The original typhoid vaccine contains inactivated whole *S. typhi* cells. This vaccine is reasonably efficacious and confers protection for 5 years but has unpleasant side effects. Two types of more modern vaccines have been developed. The first type is based on a live, attenuated *S. typhi* strain called Ty21a and is administered orally. Children in several Asian countries receive this vaccine when they attend school. However, for reasons that are unclear, the coverage afforded by this vaccine differs in different geographic areas. Other live, attenuated strains of *S. typhi* that are more immunogenic than Ty21a are under investigation for their potential as vaccines. The second type of new typhoid vaccine is a subunit vaccine containing a purified form of the bacterial capsule polysaccharide (PS) that functions as a virulence (Vi) factor for these bacteria. This ViPS vaccine is delivered intramuscularly. Although the Ty21a and ViPS vaccines are less toxic than the original, they are less efficacious and offer protection for only 2–3 years. In addition, like all polysaccharides, ViPS is not very immunogenic on its own, and a vaccine based solely on this molecule cannot be given to infants under 2 years of age. Several laboratories are therefore engaged in perfecting a conjugate vaccine in which ViPS is linked to a protein carrier. In early trials, such a vaccine has proved to be highly efficacious and to offer protection for about 4 years. The development of additional vaccines against typhoid is essential because several strains of *S. typhi* have recently become resistant to almost all antibiotics.

XVIII. Varicella (Chicken Pox)

Varicella (chicken pox) in children is caused by the varicella zoster virus (VZV). The disease is characterized by fever and an outbreak of itchy red spots all over the body. In its acute phase, the virus is highly contagious such that almost all young children in endemic areas contract chicken pox. Although most children experience relatively mild symptoms, some cases can be complicated by high fever, pneumonia or encephalitis. In any case, once the characteristic chicken pox rash has subsided, the affected individual will not usually experience it again. However, the natural immune response to VZV only suppresses viral replication and does not completely eliminate the virus from the body. Thus, latent virus from the original infection can later become reactivated and cause the painful skin disease *shingles*. This reactivation is usually associated with extreme stress or immunosuppression.

Two vaccines against varicella are now in common use, a monovalent varicella vaccine and the combination MMRV vaccine, which includes the MMR vaccine plus the varicella vaccine. The monovalent varicella vaccine that contains live, attenuated virus is highly immunogenic such that one dose is sufficient to induce antibody production in 97% of school-aged vaccinees. These antibodies persist in the circulation over several years, but it is unclear whether this persistence is due to the vaccine or to regular re-exposure to VZV in the community. A booster dose of monovalent varicella vaccine at age 4–6 years is therefore currently recommended. It is also not yet known how long vaccine-induced immunity to varicella lasts, meaning that there may be recommendations for additional booster doses in the future. Widespread vaccination of children reduces varicella in a community as a whole, which decreases the natural exposure of adults to boosting doses of the virus. It is important to determine whether several boosters of varicella vaccine received as a child can protect an adult for life, because chicken pox symptoms tend to be more severe in adults, and infection can be very dangerous to pregnant women in their third trimester. Current vaccinees are thus being carefully monitored.

The combination MMRV vaccine was formulated to reduce the number of injections endured by young children. However, an increased incidence of febrile seizures in MMRV-vaccinated infants has been reported. Although such seizures are rare and generally harmless to the vaccinee, parents made aware of this side effect may choose separate MMR and varicella vaccinations for their children.

XIX. Variola (Smallpox)

Variola virus, which causes smallpox, is no longer found in nature in any part of the world but is still feared as a bioterrorism agent. There is no effective drug for smallpox, and the length of protection afforded by the existing smallpox vaccine (based on live, attenuated vaccinia virus) is uncertain. Scientists worry that much of the world's population has not been immunized for more than 20 years, and that younger generations have never been vaccinated. Moreover, the existing vaccine is unsafe for immunocompromised individuals who, due to HIV, are now present in great numbers around the world. Even among the general population immunized in the 1960s, hundreds of complications and several deaths were recorded as a result of vaccination, representing a safety record that would be unacceptable by today's standards. Many immunologists therefore believe that a new, safer smallpox vaccine should be developed. In the meantime, most countries have opted to replenish their supplies of the existing smallpox vaccine with the goal of reserving inoculation for situations of known variola exposure. Because the virus replicates slowly, exposed individuals can be vaccinated up to 4 days later and still develop an effective adaptive response.

XX. Yellow Fever

Yellow fever is caused by a small enveloped RNA virus that is mosquito-borne and can infect both monkeys and humans. This incurable disease is prevalent in tropical climates where mosquitoes are endemic year-round. Although some patients are asymptomatic, others experience headache, fever, vomiting and nosebleeds. In severe cases, patients have fever accompanied by hepatic, circulatory and renal failure as well as severe jaundice (hence, "yellow" fever). Many severe infections are fatal. The WHO estimates that 200,000 cases of yellow fever occur each year, causing 30,000 deaths. Indeed, the number of yellow fever cases recorded globally has risen recently due to increased human activity in forests where monkeys and mosquitoes abound. In addition, climate change has increased the temperature of environments previously hostile to yellow fever virus, expanding its range.

The yellow fever vaccine is a live, attenuated virus usually given in one subcutaneous dose. Vaccination programs have been highly successful and protection can last for 30–35 years or more. An outbreak of yellow fever in a susceptible area can be prevented if 60% of the population is vaccinated. Sadly, few countries in the regions where the yellow fever virus is endemic have achieved this level of vaccination coverage. The WHO and UNICEF have been working with the governments of 12 African countries on the Yellow Fever Initiative, which seeks to vaccinate all persons older than age 9 months who live in high-risk areas. Between 2007 and 2010, 10 of these countries completed their immunization objectives. Individuals living in temperate climates receive the yellow fever vaccine only if traveling to an endemic area.

Pesticides that eliminate mosquitoes can help to control yellow fever in urban areas and allow time for mass emergency vaccination programs, but are not practical for whole forests.

F. Vaccines in the Real World

As with any medicine, a vaccine that has been shown in research labs and clinical trials to safeguard health must then have its worth proven by actual use in the real world. Within this real world, some individuals remain unimmunized against various vaccine-preventable diseases despite the urging of public health officials. In this section, we summarize some of the reasons why people choose not to vaccinate and the consequences of that decision. We also consider some of the myths and facts regarding vaccine safety, and how these may influence individuals with respect to accepting or rejecting the opportunity to vaccinate themselves or their children.

I. Reasons for Non-Vaccination

In the developing world, many people are not vaccinated because of ignorance of the benefits, the high cost of vaccines, an unreliable cold chain, a lack of sterile syringes and needles, a shortage of immunization clinics and qualified personnel, geographic distance from clinics, and suspicion of government or health care agency motives. Wars and civil conflicts block access to immunization clinics, or push vaccination programs to the bottom of a government's priority list. Sometimes a vaccine is not available because the disease it prevents afflicts only the world's poorest people, giving pharmaceutical companies little financial incentive to produce the appropriate vaccine. For all these reasons, hundreds of thousands of adults and children die of vaccine-preventable deaths each year in low-income regions.

In developed countries, failure to vaccinate is usually due to a conscious decision by seemingly well-educated adults. Some of these individuals believe that vaccination is unnecessary or prohibited by their religion, or that the risks of getting the disease are not that great, or that vaccination conflicts with a "natural" lifestyle, or that the vaccine may not have been adequately tested for safety, or that the vaccine itself causes the disease it is supposed to prevent. In addition, because of the success of vaccination in preventing various childhood scourges, many people have not witnessed the degree of devastation they can cause and so no longer appreciate the value of preventing them (refer to **Table 14-1**). Organized antivaccine groups and alternative medicine advocates argue against routine vaccination for childhood diseases, and the very small chance of a vaccine causing a severe adverse event is given a high profile by the media. More disturbingly, some health care management organizations actively support parents' decisions to forgo vaccinating their children. These actions have generated a pool of children who are not vaccinated at all, or who have not completed the required series of boosters. These children are not only themselves not protected against some very serious diseases but may also transmit them to vulnerable and unsuspecting members of their communities.

II. Consequences of Non-Vaccination

The consequences of parents refusing to vaccinate their children can be unintended and often harsh. Such choices led to the 1989–1991 measles outbreaks in high schools in the U.S. that killed 120 students. Ironically, encephalopathy following measles vaccination occurs at a rate of 1 in 10^6 vaccinees, whereas the risk of encephalomyelitis after natural measles virus infection is 1 in 10^3. In Holland in 1992, a small group of parents refused to have their children vaccinated at all for religious reasons and suffered an outbreak of 71 cases of polio, with its accompanying burden of permanent paralysis and death. Parental decisions not to vaccinate also led to severe epidemics of pertussis in Japan, the U.K. and Sweden in the early 1990s. In 2008, an American child who was intentionally not vaccinated against measles exposed over 800 people in San Diego to the virus after catching the disease during a European trip. Twelve unvaccinated children among these contacts developed measles, and $10,000 in health care costs per case were spent to contain the outbreak. In addition, parents had to disrupt their busy lives to keep their non-immune children in quarantine for 21 days. In 2009, an outbreak of 1500 cases of mumps occurred in New York and New Jersey because an unvaccinated child who traveled to Britain (where over 7000 cases of mumps were recorded that year) brought the virus back to summer camp. Similarly, several outbreaks of pertussis have recently occurred in North America due to parents' failure to either vaccinate their children in the first place, or complete the booster series. The containment of an outbreak of pertussis in Nebraska in 2008 that involved 26 initial cases eventually consumed almost 1% of the state's annual budget. In 2010, over 27,000 cases of pertussis were recorded in the U.S., the highest number since 1959.

> NOTE: In the 1970s, Japanese health authorities recommended that, rather than being given in infancy, pertussis vaccination be delayed until children were 2 years of age. It took only a few years for the annual disease statistics to go from a few hundred cases of illness due to pertussis and no deaths, to tens of thousands of cases of illness and dozens of deaths. When the recommendation to vaccinate at 2 months of age was reinstated, the numbers of cases and deaths due to pertussis rapidly returned to previous levels.

III. Vaccine Side Effects: The Facts

Vaccines prevent much misery and millions of premature deaths worldwide. Nevertheless, like all powerful medicines, vaccines carry a risk of side effects. In general, because of extensive animal and cell culture testing, such risks are low, and the associated adverse events are mild and limited to redness or pain at an injection site, sneezing or nasal congestion after intranasal administration, fatigue, or headache. In a very few instances, the adverse effects of a vaccine are more serious.

Most developed countries carry out post-licensing surveillance of adverse events that can trigger the withdrawal of a vaccine should it prove harmful in even rare circumstances. In the U.S., the VAERS was established in the 1990s by the U.S. CDC as a passive data collection system that accepts reports from the public on adverse events associated with vaccines licensed for use in the U.S. The CDC and/or the FDA monitor the VAERS data to detect new or rare adverse events, as well as increases in known side effects. These data are also scrutinized to identify patient risk factors for particular types of adverse events, or to link a specific vaccine lot with a rise in harmful side effects. [Note, however, that the VAERS data do not establish a proven cause-and-effect relationship between a given adverse event and a vaccine.]

An example of VAERS in action occurred in December 2007 when 1.2 million doses of Hib vaccine were recalled by the manufacturer due to bacterial contamination that had occurred during the manufacturing process. Routine Hib booster shots for children of 12–15 months of age were postponed, but infants at high risk for invasive Hib disease continued to be vaccinated. Adverse events were recorded by VAERS, but a review of these entries fortunately showed that no children who had received the contaminated vaccine developed a bacterial infection.

Another unusual adverse event that is still under investigation is the association between an influenza vaccine and *narcolepsy* (suddenly falling asleep at an inappropriate time). This particular vaccine was widely used during the 2009–2010 H1N1 pandemic. A Finnish study found that there was a 9-fold increased risk of narcolepsy developing in vaccinated children and adolescents between the ages of 4 and 19 years of age compared to non-vaccinated individuals of the same age. Larger populations and more countries are now under scrutiny.

> NOTE: Sometimes adverse events are manufactured by the vaccinees themselves. In 2007, 26 girls at a high school in Melbourne, Australia, reported feeling dizzy, faint and weak after receiving school-administered injection of an HPV vaccine. Four of these girls were taken to hospital. However, examination by medical professionals determined that this cluster of adverse events was likely "a mass psychogenic response to group vaccination in a school setting." In other words, a group of adolescent girls convinced themselves and each other that they were experiencing side effects that required some of them to be hospitalized, although none of them was actually ill.

Many of the current worries about vaccine side effects should be mitigated as vaccine production and purification protocols improve, and as DNA vaccines become more cost-effective to produce. *Reverse vaccinology* is the name given to the examination of the genome of an organism in order to identify novel antigens and epitopes that might constitute vaccine candidates. *Reverse genetics* allows the construction of a viral vector that is engineered to harbor a predetermined defect which blocks its ability to cause disease. These approaches are now becoming practical to use in modern vaccine design. In addition, reporting systems like VAERS drive the FDA and manufacturers to take concrete action to adjust vaccine formulations and remove any possibility of adverse events. Another recent development that may help to defuse public fears about vaccines is the new science of "adversomics," in which the specific genetic make-up of an individual is scrutinized to shed light on what vaccines or vaccine formulations might be most appropriate to minimize side effects in that person. Researchers are examining the genetic variation in the human population that affects many immune response genes with an eye to determining if there is true genetic susceptibility to specific vaccine-related adverse events. For example, in one U.S. study, aboriginal children showed more severe reactions to a measles vaccine than Caucasian children. Genetic variation has also been implicated in the patterns and levels of cytokines produced by individuals after administration of live, attenuated vaccines. Knowledge of such associations could allow physicians to better screen young patients and provide specific and appropriate advice to parents, an example of the growing movement toward "personalized medicine."

In developing countries, the biggest real "adverse event" associated with vaccination is the reuse of injection needles, a practice that spreads HIV and HCV. Needle-free vaccine delivery systems are under development.

NOTE: The overall risk of a severe reaction to any vaccine is less than 1 in a million doses. This risk must be weighed against the risk of the severe or fatal complications of contracting a disease that vaccination would prevent. For example, based on data from the CDC:

- 1 in 1000 unvaccinated individuals who contract measles will die.

- 1 in 100 unvaccinated individuals who contract polio will be paralyzed, while 1–2 in 2000 will die.

- More than half of infants under one year of age who contract pertussis will be hospitalized, and 1–2 in 200 will die.

- 1–5 in 1000 of the estimated 800,000–1.4 million individuals with chronic hepatitis B infection in the U.S. will die of liver-related disease.

IV. Links between Vaccination and Disease: Some Myths

From time to time, an association is proposed between a particular vaccine and a specific disease, generating controversy over the use of that vaccine. In the following subsections, we describe some examples of such associations and the scientific facts that have refuted them.

i) DTaP Vaccine

In the 1980s, a spurious link was drawn between the DTaP vaccine and sudden infant death syndrome (SIDS). It is true that infants of the age to receive the DTaP vaccine are also the same age as infants that die of SIDS, but proper scientific examination showed that administration of the vaccine does not increase the chance that a child will die of SIDS. In fact, statistically, it appears that children who are vaccinated are *less* likely to die of SIDS.

ii) HBV Vaccine

In the early 1990s, concerns were raised about a possible association between the HBV vaccine and multiple sclerosis (MS; see Ch. 19). However, natural infection with HBV is not a risk factor for MS, and a well-controlled clinical study carried out in 2001 showed no increased risk of MS in individuals who had received the HBV vaccine.

iii) IPV Vaccine

Early versions of the IPV vaccine used in the late 1950s and early 1960s were subsequently found to have been contaminated with a monkey virus called simian virus 40, which was derived from the cell cultures used to produce large quantities of the vaccine. Because SV40 is a cancer-causing virus in many species, fears were raised that vaccinees would develop tumors. However, extensive follow-up has shown that there has been no increase in cancer development in these IPV recipients. Refined methods of cell culture and vaccine purification have now removed the possibility of this type of contamination, so any current doubts about this form of cancer induction are unfounded.

iv) Polio Vaccine

In the 1990s, concerns were raised that previous batches of polio vaccines used for mass vaccinations in Africa in the 1950s were contaminated with HIV and responsible for introducing this insidious virus into humans. In 2001, DNA testing of preserved samples of these vaccines confirmed that no HIV sequences were present.

v) Lyme Disease Vaccine

In the U.S., well over 20,000 cases of Lyme disease occur every year, leading to significant illness and some deaths. In 1998, a protein subunit vaccine based on the causative organism *Borrelia burgdorferi* was used for an initial round of vaccinations. The vaccine succeeded in preventing the contraction of Lyme disease, and studies were under way to determine if and how often booster shots would be needed. However, some Lyme disease patient groups started to wonder whether the bacterial protein used in the vaccine was in fact inducing autoimmune arthritis and other symptoms of chronic Lyme disease. Although subsequent clinical trials proved that this was not the case, the vaccine was withdrawn in 2002. There has been little appetite on the part of drug companies to make a new vaccine, so Lyme disease is now becoming an increasingly important public health problem in some parts of the U.S.

vi) MMR Vaccine

In 1998, the medical journal *Lancet* published a paper by Andrew Wakefield and several co-authors that proposed a connection between the MMR vaccine and the increasing rate of autism in developed countries. Extensive examinations and retrospective assessments confirming that MMR vaccination does *not* increase the risk of autism did not dissuade numerous parents in the U.S. and U.K. from refusing the MMR vaccine for their children. In 2004, financial conflicts of interest on Wakefield's part were uncovered, prompting Wakefield's co-authors to remove their names from the original publication. Sadly, however, parents of autistic children, autism advocacy groups, and the antivaccination lobby continued to cite the flawed report and to circulate it widely in the uncritical popular press. The result was an avoidable tragedy: a sharp drop in MMR vaccination rates that led to outbreaks of measles, mumps and rubella in children in the U.S. and U.K. Much suffering and some deaths resulted. In addition, millions of research dollars that could have gone into finding the true cause of autism were wasted in generating the data needed to refute Wakefield's claim. In 2010, the *Lancet* officially retracted the paper, citing dishonesty and data fraud on the part of Wakefield. In 2011, it was revealed that Wakefield planned to take advantage of his manufactured MMR vaccine scare to launch his own alternative measles vaccine and other products for private profit.

vii) Multiple Vaccines

The advent of vaccines for more and more childhood diseases and the ability to give them in combination vaccines has been suggested by some commentators to "overload the immune system with antigens." This overload was proposed to be responsible for diabetes, asthma, eczema, and other forms of autoimmunity. Despite extensive scientific investigation into this claim, no supporting evidence has yet been found.

viii) Other Vaccine Components

Vaccine preparations contain various preservatives and stabilizers that have caused some parents concerns about their safety. However, intensive research has shown that these substances are eliminated rapidly from an infant's circulation and do not accumulate to toxic levels. Another issue raised has been that of *anaphylaxis* in response to vaccine components. However, in a report that examined the administration of over 7 million doses of various vaccines, only 5 potential cases of vaccine-associated anaphylaxis were identified, and none of these was fatal.

In the H1N1 influenza pandemic of 2009, when there was a shortage of conventional vaccine available in North America, some people in the U.S. refused to accept vaccines from Europe because they contained adjuvants. This resistance was based only on vague fears but persisted despite the well-documented and excellent safety record of these adjuvants and their successful use overseas. Thankfully, it turned out that adjuvants were not needed in order for the North American H1N1 2009 vaccine to induce a protective immune response. However, effective antibody responses to influenza viruses containing H5 proteins *do* require adjuvants. Should a pandemic of an H5-containing influenza virus occur in the future, an unreasoning refusal to accept administration of a fully tested, FDA-approved, adjuvant-containing vaccine could have fatal consequences.

Anaphylaxis is a severe allergic reaction that can be fatal due a catastrophic drop in blood pressure induced by excess inflammatory mediators (see Ch. 18).

G. Future Directions

I. Prophylactic Vaccines

While much ongoing basic and clinical research is devoted to improving existing prophylactic vaccines, other efforts are focused on developing vaccines for high-profile diseases such as HIV/AIDS and malaria for which there are currently no effective vaccines. Some scientists believe that one reason it has been difficult to produce good vaccines for these diseases is that the causative pathogens do not naturally establish robust, acute infections in their hosts, and so only a low-key primary response occurs. As a result, it is difficult to formulate a pathogen-derived vaccine that will induce the production of a large enough army of memory cells to mount an effective secondary response. In addition, in the case of AIDS, HIV kills billions of CD4$^+$ T cells such that the natural anti-HIV immune response is too weak to prevent the virus from establishing a permanent foothold in the body (see Ch. 15). The design of an AIDS vaccine that can induce an effective response is thus particularly challenging. With respect to malaria vaccine development, significant additional problems arise due to the complex life cycle of the causative parasite. These challenges are outlined in **Box 14-5**.

Other pathogens that are priority targets for vaccine development include: group A *Streptococci* causing flesh-eating disease and other serious ailments; bacteria and herpesviruses causing sexually transmitted diseases; enteroviruses causing childhood diarrhea; and HCV causing liver disease and hepatocarcinoma. Vaccines may also be part of the arsenal used in the near future to combat the multi-drug-resistant bacteria that have arisen with the overuse of antibiotics in developed countries, and to control the comparatively rare but potentially devastating diseases caused by the West Nile, Ebola, Hanta, SARS and dengue fever viruses. Lastly, due to security concerns and the possibility of bioterrorism, appropriate vaccines are under development to protect targeted populations and military personnel from easily transmitted organisms like anthrax.

Additional future concerns associated with vaccines have more to do with the population being vaccinated than with the vaccine itself. In general, the world's population is getting older, and how the immune systems of the elderly respond to vaccines is poorly understood. Similarly, while masses of people around the globe are malnourished and/or vitamin-deficient, many souls in highly developed countries are obese. These metabolically disruptive conditions are likely to affect how the body responds to vaccines, but little formal research has been carried out to date. To be cost-effective, vaccination programs in the future would be wise to take these factors into account and adjust vaccination protocols or vaccine components accordingly.

Box 14-5   The Life Cycle of *Plasmodium falciparum* and its Effect on Malaria Vaccine Development

Malaria kills one child in Africa every minute and costs the African economy about U.S. $2 billion each year. More than 100 countries are plagued by malaria transmission among their populace. Although increased malaria prevention and control measures are having an impact in many places, an estimated 650,000 people worldwide were killed in 2010 by this preventable and curable disease. Prevention takes the form of sleeping under an insecticide-impregnated net coupled with indoor spraying of strong insecticides. If prevention fails and malaria is contracted, the best available treatment involves artemisinin-based drugs. (Current strains of malarial parasites are resistant to the original chloroquine-based drugs.) However, the recent emergence of artemisinin-resistant strains of *Plasmodia* has spurred the WHO to strongly urge that artemisin be used only as part of a combination drug therapy. Lessons have been learned from prior experience with other nasty afflictions such as yellow fever and hookworm. In these cases, campaigns to eliminate the disease based only on vector control or drug treatment (i.e., no vaccination) all ultimately failed. An efficacious malaria vaccine will therefore be needed to buttress prevention and drug treatment efforts and achieve the worthy goal of global malaria eradication.

Malaria vaccine development has been stymied thus far by the complex, multi-host, multi-stage life cycle of the major malarial parasite *Plasmodium falciparum* (see following figure). Researchers have designed prototype vaccines targeting various *Plasmodium* stages, but none has yet proved truly efficacious. One type of vaccine is designed to prevent sporozoites from reaching the liver or producing merozoites. In theory, no RBCs are lysed, precluding the onset of clinical symptoms. A second type of vaccine is intended to target the merozoites in the RBCs. The objective is to kill the infected cells before the parasite can multiply or generate gametes. Symptoms may still arise, but they would be mild. A third vaccine type is meant to target the formation of new parasite gametes in an infected individual and to block transmission to the next host. This vaccine does not do the vaccinee much good but could reduce the spread of the disease in the community. A fourth type of vaccine uses whole, inactivated sporozoites delivered intravenously, a strategy that has been successful to a degree in animal models. However, the unintended reactivation of the vaccine sporozoites and consequent infection of the animal hosts in some trials have dulled enthusiasm for this approach. In any case, various trials of vaccine prototypes have made it clear that only a complex vaccine incorporating multiple elements will succeed in generating effective immunity against malaria. Strategies in which priming with a plasmid bearing DNA for various *Plasmodium* antigens is followed by boosting with a viral vector bearing DNA for the same *Plasmodium* antigens are currently being tested. The idea is to induce both humoral and cell-mediated responses against infected hepatocytes, infected RBCs and free sporozoites in the blood, as well as humoral responses that block merozoite entry into RBCs. Vaccines that incorporate mosquito stage antigens are also under development. The most advanced of these experimental vaccines, which targets sporozoites, is in Phase III clinical trials in seven countries in Africa as of April 2012. Early results are promising, and a formal evaluation of the efficacy of this vaccine is expected in 2014.

Artemisinin is a lactone compound derived from the sweet wormwood plant often used in traditional Chinese medicine.

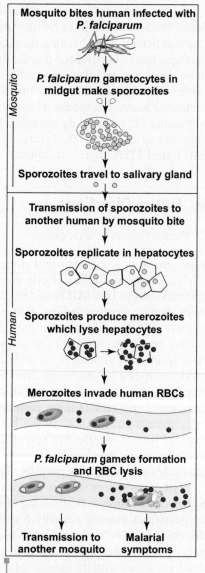

Fig. Box 14-5
Life cycle of **Plasmodium falciparum.**

II. Therapeutic Vaccines

Prophylactic vaccines are designed to prevent disease. In contrast, **therapeutic vaccines** are agents designed to cure or mitigate established disease, rather than prevent it. A therapeutic vaccine that is truly efficacious does not exist as yet, but the principles of vaccination are being explored with the goal of stimulating the immune response to eliminate tumors, cure chronic infections, suppress autoimmune disease, or relieve allergy. Some non-immunological conditions may also be mitigated by therapeutic vaccine administration in the future. Examples of these potential therapeutic vaccines are given in **Table 14-7.**

TABLE 14-7	Examples of Potential Therapeutic Vaccines		
Situation	**The Problem**	**Vaccine Agent**	**Expected Response**
Tumor therapy	Inadequate natural antitumor responses by a patient's NK cells and CTLs allow tumors to grow.	A patient's tumor cells (that express unique antigens) are stimulated *in vitro* with cytokines to promote peptide-MHC class I display. Alternatively, the patient's DCs can be loaded *in vitro* with a vaccine antigen. Inactivated tumor cells or loaded DCs are injected back into the patient.	The unique tumor antigens upregulated on the treated tumor cells, or the vaccine antigen presented by the loaded DCs, should induce effective antitumor Th, Tc and B cell responses that eliminate the tumor.
Chronic infection	Inadequate natural humoral and cell-mediated responses to a pathogen during an acute infection do not eliminate the pathogen, allowing chronic infection.	A viral antigen containing a hidden epitope expressed only in the chronic phase of an infection.	Activation of naïve lymphocytes recognizing the hidden epitope. These cells were not activated in the acute phase of the infection but may now provide the final push that eliminates the pathogen.
Indirect effects of infections	A disease is caused by an inadequate natural immune response to an underlying pathogen infection; e.g., ulcer formation and stomach cancers are caused by *H. pylori* infection.	Components of the underlying pathogen (e.g., *H. pylori*)	The immune response eliminating *H. pylori* has the effect of curing the ulcer and preventing the development of stomach cancer.
Autoimmune disease	Mechanisms of peripheral tolerance fail such that autoreactive lymphocytes are activated and destroy host tissues (see Ch. 19).	Self antigen delivered orally or delivered with a cytokine favoring immune deviation and/or activation of regulatory T cells.	Oral tolerance to the self antigen decreases the autoimmune response. Cytokine-induced immune deviation to a Th2 response or suppressive effects of regulatory T cells reduce tissue damage.
Allergy	Allergic symptoms are largely due to the production of anti-allergen* IgE antibodies that activate mast cells (see Ch. 18).	Allergen is delivered with a cytokine that favors IgG and suppresses IgE antibody production.	Anti-allergen antibodies of the IgE isotype are not produced and mast cells are not activated.
Chronic inflammation	Immunopathic inflammatory tissue damage is caused by excessive cytokine production. May contribute to cases of allergy, autoimmunity and transplant rejection.	Detoxified full-length cytokine proteins (particularly TNF or IL-1) coupled to a VLP. Alternatively, a cytokine peptide coupled to a VLP or carrier protein with non-self T cell epitope.	Anticytokine antibodies are produced that limit the effects of pro-inflammatory cytokines.
Hypertension	Dysregulation of the angiotensin proteins that govern blood pressure causes hypertension.	Peptides of angiotensins or angiotensin receptors coupled to protein carriers or VLPs.	Antibodies against angiotensins are produced that prevent these proteins from increasing blood pressure.
Neurodegenerative disease (e.g., Alzheimer's dementia)	Deposition of misfolded and aggregated amyloid beta (Aβ) peptide in the brain disrupts neuron function.	Short forms of Aβ peptides linked to non-self carrier proteins. Peptides are devoid of T cell epitopes to avoid inducing autoreactivity.	Antibodies against Aβ aggregates are produced that remove the aggregates and/or decrease their toxic properties.

*An allergen is an antigen that causes allergic symptoms.

One of the most promising therapeutic vaccination approaches today is the use of DCs as specialized delivery vehicles for experimental anticancer vaccines. This protocol calls for isolation of DCs from the cancer patient, the loading of these DCs with vaccine antigen in a culture dish, and the re-injection of the antigen-loaded DCs back into the patient. The hope is that the vaccine antigen will induce the patient's immune system to mount a more effective response against his/her tumor. However, this *ex vivo* procedure is time-consuming and expensive, and thus not suitable for routine immunization programs. Researchers have therefore experimented with ways to conjugate vaccine antigens to mAbs directed against DC surface PRRs, with the goal of provoking DCs to directly internalize the antigen-mAb-PRR complex. In mice, this approach has induced strong antigen-specific CD4$^+$ and CD8$^+$ T cell responses. A clinical trial of an HIV antigen fused to an mAb recognizing the DC PRR DEC-205 is under way.

> NOTE: An intriguing approach to maximizing the effectiveness of DCs as vaccine delivery vehicles involves *RNA interference*. All living cells naturally produce microRNA (miRNA) molecules that are complementary to mRNA sequences produced by transcription of a particular gene. As reported in Chapter 13, when an miRNA binds to the mRNA, it prevents the translation of the mRNA into protein and thereby helps to modulate levels of that protein within the cell. Researchers have devised ways to artificially introduce such RNA molecules into living cells where they act as "small interfering RNAs" (siRNAs) and can shut down the expression of a protein of interest. In the vaccination context, scientists hope to target a collection of proteins expressed by DCs whose normal function is to shut down inflammation after it is no longer needed. To this end, scientists have introduced into vaccine antigen-bearing DCs various siRNAs that disrupt the production of these negative regulators. DC-mediated inflammation therefore persists, promoting ongoing innate and adaptive responses to the vaccine antigen.

We have come to the end of our discussion of vaccines, entities that are designed to induce immune protection and reduce the incidence and severity of many devastating infectious diseases. In Chapter 15, we describe circumstances in which an individual's immune responses are profoundly compromised. We begin with a survey of primary immunodeficiency diseases (PIDs), which are determined by an individual's genetic make-up. We then turn to a discussion of AIDS, which is induced by infection with HIV.

Chapter 14 Take-Home Message

- A prophylactic vaccine is a modified form of a natural pathogen or its components that is given to an individual prior to pathogen exposure to establish expanded clones of long-lived, pathogen-specific memory lymphocytes. When the individual subsequently encounters the natural pathogen for the first time, a secondary rather than primary response is mounted such that he/she is far less likely to become seriously ill or die.

- Passive immunization with pre-formed pathogen-specific antibodies can be given to prevent disease prior to or after exposure to an infectious agent.

- Types of vaccines include live, attenuated; killed or inactivated; toxoid; subunit; peptide; recombinant DNA vector; and naked DNA vaccines.

- An efficacious vaccine induces a protective response against the pathogen in most members of the vaccinated population. A safe vaccine has side effects that pose little or no risk compared to the risk of harm from the disease

itself. Vaccine efficacy and safety are formally tested in pre-clinical and clinical trials.

- Adjuvants are used to enhance a vaccine's immunogenicity, whereas delivery vehicles protect a vaccine from degradation.

- Prophylactic vaccines may be given to children as part of a standard immunization program or to individuals in special circumstances requiring protection from a particular pathogen.

- Failure to vaccinate due to external circumstances or parental resistance or belief in a vaccine-related myth can have harsh and unintended consequences, such as the contraction of a severe but vaccine-preventable disease.

- It is particularly difficult to design prophylactic vaccines for pathogens that become entrenched in the body, have complex life cycles, or destroy elements of the immune system.

- Therapeutic vaccines are designed to cure or mitigate established disease, but none has proven highly effective to date.

Did You Get It? A Self-Test Quiz

Section A

1) Can you define these terms? vaccination, vaccine, vaccinee, prophylactic

2) Explain why vaccination prevents disease.

3) Distinguish between active and passive immunization.

4) Describe one way in which the smallpox and rinderpest eradication stories are similar and one way in which they are different.

Section B

1) Can you define these terms? coverage, herd immunity, cocooning, protective epitope

2) Give three characteristics of a pathogen that would make it a good candidate for a vaccine.

3) What is a reservoir, and how can it affect vaccination programs?

4) How is vaccine efficacy affected by whether the vaccine induces a cell-mediated or humoral response?

5) Why is *in vivo* testing an essential part of vaccine evaluation?

6) Give four characteristics of a protective antigen that would make it suitable for use in a vaccine.

7) Give four examples of vaccine side effects.

8) Why are safety concerns about a vaccine sometimes weighed differently in different parts of the world?

9) Describe the series of pre-clinical and clinical trials usually used to test a vaccine.

10) What does VAERS stand for?

11) Give two reasons why vaccine testing is not always straightforward.

Section C

1) Can you define these terms? booster, cold chain, passaging, toxoid, conjugate vaccine, recombinant vector

2) Give two advantages and two disadvantages of using a live, attenuated vaccine.

Did You Get It? A Self-Test Quiz—Continued

3) Describe three ways by which a pathogen can be attenuated.

4) Give two advantages and two disadvantages of using a killed vaccine.

5) Under what circumstances is a toxoid vaccine effective?

6) Give one advantage and one disadvantage each of a protein subunit vaccine and a polysaccharide subunit vaccine.

7) Distinguish between protein subunit vaccines and peptide vaccines.

8) Name three carrier proteins frequently used to make conjugate vaccines.

9) Give two advantages and two disadvantages of peptide vaccines.

10) Distinguish between the two current types of DNA vaccines.

11) Distinguish between subunit vaccines and recombinant vector vaccines.

12) Give one advantage and one disadvantage of naked DNA vaccines.

13) Outline two ways in which vaccine design and production have benefited from recombinant DNA technology.

Section D

1) Distinguish between adjuvants and delivery vehicles. Are they always mutually exclusive?

2) Why are adjuvants often necessary for immunizations?

3) Briefly describe three types of adjuvants and why they work.

4) What is the depot effect?

5) What is a potential disadvantage of PRR-based vaccines?

6) Describe three types of delivery vehicles and how they increase vaccination success.

7) Distinguish between virosomes and VLPs.

Section E

1) Can you define these terms? combination vaccine, IPV, OPV, VAPP, Hib, DTaP, shingles

2) Under what conditions is a booster shot of a vaccine recommended?

3) What is the advantage of using an oral vaccine to combat a pathogen like *V. cholerae*?

4) Name two diseases combatted by toxoid vaccines.

5) What organism causes the 100-day cough?

6) Distinguish between the DTaP and Tdap vaccines.

7) What is the best way to prevent pneumonia in children?

8) Name the major components of the current vaccine against invasive Hib disease, and the vaccine against HBV infection.

9) How have cultural biases affected the efficacy of the HPV vaccine?

10) Why must the influenza virus vaccine be administered yearly?

11) What are researchers seeking to prevent a future bird flu pandemic?

12) Describe the components of the MMR vaccine.

13) What is the MR Initiative?

14) Why are the MPSV4 meningococcal vaccine and the PPV pneumococcal vaccine not given to children under 2 years of age?

15) Why is no modern plague vaccine being developed?

16) What is VAPP? Describe two ways in which it can be avoided.

17) Why can the rabies vaccine be helpful *after* exposure to the virus?

18) Give three reasons why it has been so hard to develop an effective vaccine for TB.

19) Why might vaccinated adults need a booster dose of varicella vaccine?

20) Outline the current dilemma concerning smallpox vaccination.

Section F

1) Why do some parents resist vaccinating their children?

2) Give three examples of true severe side effects of vaccines.

3) Describe three proposed links between vaccines and diseases that have been disproven.

Section G

1) Why has it been so difficult to produce vaccines against HIV and malaria?

2) Give four examples of potential therapeutic vaccines.

3) What is the theory behind using DCs as delivery vehicles?

4) What is RNA interference?

Can You Extrapolate? Some Conceptual Questions

1) After mentioning her plan to become pregnant, a woman is advised by her physician to ensure that she is up-to-date on all her vaccinations. Can you think of three ways in which her adherence to this advice could later contribute to her newborn's health?

2) In this chapter, we learned that when adenovirus is used as the vector in a recombinant DNA vaccine against a particular pathogen, the secondary boost may be compromised by immune responses against vector epitopes that have the undesired effect of clearing the vector from the body. In fact, in some populations, researchers have found that adenoviral vectors are ineffective as vaccines, even for priming purposes. Why might this be?

3) In Chapter 15, we learn about various forms of primary immunodeficiency disease caused by genetic defects in either innate or adaptive immune responses. One clinical sign that may indicate the possibility of a deficient immune response is a lack of response to a vaccination. If a patient produces an antibody response to a pneumococcal polysaccharide subunit vaccine but makes no antibody response to a pneumococcal conjugate vaccine, what might this indicate about the relative ability of this patient's B and T cells to respond to antigen?

Would You Like To Read More?

Bachmann, M. F., & Jennings, G. T. (2010). Vaccine delivery: A matter of size, geometry, kinetics and molecular patterns. *Nature Reviews Immunology, 10*(11), 787–796.

Buonaguro, L., & Pulendran, B. (2011). Immunogenomics and systems biology of vaccines. *Immunological Reviews, 239*(1), 197–208.

Chen, K., & Cerutti, A. (2010). Vaccination strategies to promote mucosal antibody responses. *Immunity, 33*(4), 479–491.

Coffman, R. L., Sher, A., & Seder, R. A. (2010). Vaccine adjuvants: Putting innate immunity to work. *Immunity, 33*(4), 492–503.

Farez, M. F., & Correale, J. (2011). Immunizations and risk of multiple sclerosis: Systematic review and meta-analysis. *Journal of Neurology, 258*(7), 1197–1206.

Fay, K. E., Lai, J., & Bocchini, J. A., Jr. (2011). Update on childhood and adolescent immunizations: Selected review of US recommendations and literature: Part 1. *Current Opinion in Pediatrics, 23*(4), 460–469.

Germain, R. N. (2010). Vaccines and the future of human immunology. *Immunity, 33*(4), 441–450.

Godlee, F., Smith, J., & Marcovitch, H. (2011). Wakefield's article linking MMR vaccine and autism was fraudulent. *British Medical Journal, 342*, 7452.

Kaufmann, S. H. (2010). Future vaccination strategies against tuberculosis: Thinking outside the box. *Immunity, 33*(4), 567–577.

Klebanoff, C. A., Acquavella, N., Yu, Z., & Restifo, N. P. (2011). Therapeutic cancer vaccines: Are we there yet? *Immunological Reviews, 239*(1), 27–44.

Lai, J., Fay, K. E., & Bocchini, J. A. (2011). Update on childhood and adolescent immunizations: Selected review of US recommendations and literature: Part 2. *Current Opinion in Pediatrics, 23*(4), 470–481.

Lambert, L. C., & Fauci, A. S. (2010). Influenza vaccines for the future. *New England Journal of Medicine, 363*(21), 2036–2044.

Lambrecht, B. N., Kool, M., Willart, M. A., & Hammad, H. (2009). Mechanism of action of clinically approved adjuvants. *Current Opinion in Immunology, 21*(1), 23–29.

Lawson, L. B., Norton, E. B., & Clements, J. D. (2011). Defending the mucosa: Adjuvant and carrier formulations for mucosal immunity. *Current Opinion in Immunology, 23*(3), 414–420.

Levitz, S. M., & Golenbock, D. T. (2012). Beyond empiricism: Informing vaccine development through innate immunity research. *Cell, 148*(6), 1284–1292.

Liu, M. A. (2011). DNA vaccines: An historical perspective and view to the future. *Immunological Reviews, 239*(1), 62–84.

O'Hagan, D. T., & Rappuoli, R. (2004). The safety of vaccines. *Drug Discovery Today, 9*(19), 846–854.

Pulendran, B., & Ahmed, R. (2011). Immunological mechanisms of vaccination. *Nature Immunology, 2*(6), 509–517 1.

Rohn, T. A., & Bachmann, M. F. (2010). Vaccines against noncommunicable diseases. *Current Opinion in Immunology, 22*(3), 391–396.

Rollier, C. S., Reyes-Sandoval, A., Cottingham, M. G., Ewer, K., & Hill, A. V. (2011). Viral vectors as vaccine platforms: Deployment in sight. *Current Opinion in Immunology, 23*(3), 377–382.

Thera, M. A., & Plowe, C. V. (2012). Vaccines for malaria: How close are we? *Annual Review of Medicine, 63*, 345–357.

Immunodeficiency

"You are unwise to lower your defenses."
Darth Vader, *Star Wars*, Episode VI

The preceding chapters of this book have explained how the vigilance of our immune system prevents our bodies from falling prey to pathogen infections. In this chapter, we explore the consequences of a failure in some aspect of the immune system, a situation known as **immunodeficiency**. Immunodeficiencies can be either *primary* or *secondary*. In the case of **primary immunodeficiencies (PIDs)**, the individual is born with a genetic mutation that results in a defect in either the innate or adaptive immune response. Often these mutations affect one gene and are inherited in simple Mendelian fashion, but other PIDs have a more complex inheritance pattern involving multiple genes. The *penetrance* of these mutations, as well as environmental factors, can influence the phenotype of a given PID patient. PIDs are the subject of the first half of this chapter.

In *secondary* immunodeficiencies, the individual is born with normal immune responses but later experiences an event that damages the immune system in some way. Secondary immunodeficiencies are often the consequence of a severe infection or its treatment, cancers or cancer therapies, or trauma or malnutrition. However, the most notorious example of a secondary immunodeficiency is **acquired immunodeficiency syndrome (AIDS)**, which is caused by human immunodeficiency virus (HIV) infection. HIV/AIDS and its decimation of the immune system are the subject of the second half of this chapter.

In a population of individuals bearing a particular mutation, the "penetrance" refers to the proportion of that population that exhibits clinical symptoms caused by that mutation.

A. Primary Immunodeficiency

I. Nature of Primary Immunodeficiencies

There are now more than 180 distinct PIDs, with more identified every year. The vast majority are rare, such that the total global incidence of PIDs is about 1 in 5000 live births. In countries where consanguineous marriage is prohibited or uncommon, PID incidence is closer to 1 in 10,000 live births. PID-related mutations have been identified that affect various aspects of the generation, differentiation, proliferation, and/or effector functions of innate or adaptive leukocytes, or the function of

MAK: Primer to the Immune Response. http://dx.doi.org/10.1016/B978-0-12-385245-8.00015-7

TABLE 15-1	Major Categories of Primary Immunodeficiencies*
PID Category	**Defining Characteristics**
Innate Immunity	
Leukocyte adhesion deficiencies	Lack of extravasation of innate leukocytes (particularly neutrophils) from the circulation into inflammatory sites
Chronic granulomatous disease	Failure in phagosomal killing
Chédiak–Higashi syndrome	Failure of phagosomes or cytotoxic granules to fuse to their targets, coupled with neurologic deterioration
Defects of the IL-12/IFNγ axis	Failure in macrophage hyperactivation and granuloma formation
Defects of PRRs	Lack of activation of innate immune responses against specific pathogens
Congenital neutropenias	Reduced levels of neutrophils in the circulation, may be cyclical
Complement deficiencies	Lack of complement activation; T and B cells and phagocytes are normal
Autoinflammatory syndromes[†]	Severe local inflammation and prolonged periodic fevers with no obvious cause; autoimmune symptoms; lymphocytes are normal
Adaptive Immunity	
SCIDs and CIDs	Low numbers or absence of T cells and B cells and sometimes NK cells; lack of B cell function may be due to lack of T cell function
APC-PIDs	Lack of MHC class I or II expression, or low numbers or absence of DCs (and possibly other hematopoietic cells)
T-PIDs	Normal numbers of T, B, and NK cells, but T cells are non-functional; B cell function may be compromised
B-PIDs	B cells are absent or non-functional; T and NK cells are normal
DNA repair defects	Abnormalities of most hematopoietic cells, skewed Ig isotype balance, increased cancer incidence
Lymphoproliferative immuno-deficiency disorders[†]	Uncontrolled inflammation and proliferation of T and/or B cells

*It should be noted that different medical and immunological authorities have created different categorization schemes for PIDs.

[†]See Chapter 19 for descriptions of these disorders.

The most current information on the ever-growing list of PIDs can be obtained at www.frontiersin.org/primary_immunodeficiencies.

the complement system (**Table 15-1**). Many such mutations alter the activity of a particular enzyme or receptor and thereby cause defects in cellular interaction or regulation that have effects on immunity. The principal clinical phenotype of most PIDs is increased susceptibility to infections with organisms that are of low pathogenicity in a healthy person (i.e., opportunistic infections). Many PIDs are diagnosed in infancy, but some PIDs do not manifest symptoms until well into a patient's adult years.

Different types of PIDs are associated with different types of infections. For example, enterovirus infections are commonly seen in patients with antibody deficiencies but not in those with phagocyte defects or complement deficiencies. Conversely, infection with *Salmonella* bacteria or various fungi is most often a sign of a defect in cell-mediated immunity. In addition to infection, PID patients often present with weight loss, failure to thrive, dermatitis, and enlargement of lymph nodes, tonsils, and/or other organs. Autoimmunity, malignancies, and allergic diseases can also appear as PID symptoms depending on the immune system defect. PID severity ranges from relatively benign to life-threatening. In the worst cases, the failure in immunity allows chronic infections to take hold that can cause irreversible and sometimes fatal damage to vital organs. Because this deterioration can occur rapidly, a diagnosis of a severe

PID in a child should be addressed as quickly as possible to treat existing infections and prevent new ones.

Precise diagnosis of a PID can be a challenge because similar clinical pictures can result from very different defects, and very similar defects can produce dramatically different clinical signs. Indeed, affected members of the same family, who all carry the same mutation, can show striking variability in disease severity and range of clinical symptoms. Treatment of PIDs is briefly described in Section IV of Part A after the symptoms and causes of various PIDs are discussed.

NOTE: The PID field is an alphabet soup of acronyms. We supply here an alphabetized list of disease acronyms that the reader can refer back to as necessary.

ADA = adenosine deaminase deficiency

APAD = antipolysaccharide antibody

APECED = autoimmune polyendocrinopathy-candidiasis-ectodermal dystrophy syndrome

ALPS= autoimmune lymphoproliferative syndrome

AT = ataxia–telangiectasia

BS = Bloom syndrome

B-PID = B cell-specific PID

CGD = chronic granulomatous disease

C-HS = Chédiak–Higashi syndrome

CID = combined immunodeficiency

CVID = common variable immunodeficiency

CyN = cyclic neutropenia

DCML = DCs, monocyte, lymphocytes

IgAD = selective IgA deficiency

IgGD = Selective IgG deficiency

HIGM = hyper-IgM syndrome

HLH = hemophagocytic lymphohistiocytosis

IPEX = immune dysregulation–polyendocrinopathy–enteropathy-X-linked

LAD = leukocyte adhesion deficiency

MSMD = Mendelian susceptibility to mycobacterial disease

NBS = Nijmegen breakage syndrome

NBA = Non-Bruton's agammaglobulinemia

PNP = purine nucleoside phosphorylase

RD = reticular dysgenesis

SCID = severe combined immunodeficiency

SCN = severe congenital neutropenia

SLE = systemic lupus erythematosus

T-PID = T cell-specific PID

WAS = Wiscott–Aldrich syndrome

XLA = X-linked agammaglobulinemia

XLP = X-linked lymphoproliferation

XP = xeroderma pigmentosum

II. Primary Immunodeficiencies due to Defects in Innate Immunity

As recently as 2006, research on PIDs was concentrated almost exclusively on muta-tions in genes affecting the adaptive response. Since then, scientists have branched out to examine mutations of genes more closely associated with innate immunity and have found both new PIDs and new explanations for old PIDs. While most innate response PIDs are linked to defects in phagocytes, some stem from complement deficiencies. The reader should note that the division between innate and adaptive PIDs is not abso-lute, as a mutation that affects phagocytes may also impair lymphocyte functions.

i) Immunodeficiencies Affecting Phagocyte Responses

About 9% of PIDs are due to defects in phagocyte extravasation, activation, or func-tion, particularly that of neutrophils. These PIDs manifest at an early age with recur-rent fevers and infections by normally non-pathogenic organisms such as *Aspergillus* and *Candida* species of fungi. Most often the infection attacks the respiratory tract or the skin, but abscesses of the tissues and mouth are also common. These lesions can be fatal if antibiotics are not administered promptly.

a) Leukocyte Adhesion Deficiencies

Patients with a leukocyte adhesion deficiency (LAD) suffer from severe infections and compromised wound healing due to a lack of integrin signaling. Without functional integrins, leukocytes cannot leave the circulation and extravasate into sites of infection or injury. A clinical sign of this localized leukocyte deficiency is an absence of the pus usually found at sites of infection.

Leukocyte adhesion deficiency type I (LAD-I) is characterized by recurrent non-pyogenic bacterial and fungal infections of the mucosae. LAD-I results from autosomal recessive mutations in an integrin subunit called CD18. The CD18 protein pairs with CD11a to form the integrin LFA-1, which is primarily involved in leukocyte extravasation. However, the CD18 chain also associates with either the CD11b or CD11c protein to form the heterodimeric complement receptors CR3 and CR4, respectively. CR3 and CD4 bind to iC3b and certain bacterial PRRs, thereby facilitating phagocytosis, and also promote leu-kocyte migration. Therefore, in the absence of CD18, neutrophil phagocytosis is compro-mised, and these cells cannot extravasate into inflamed tissues. As a result, the number of neutrophils in the circulation of an LAD-I patient is almost twice that in a normal individ-ual. Few neutrophils are found in the inflammatory sites where they are needed, and those that are present function poorly. Children with severe CD18 deficiency die by age 10 years.

Leukocyte adhesion deficiency type II (LAD-II) is an extremely rare autosomal recessive disorder caused by mutations in a GDP-fucose transporter protein (CD15) essential for the selectin-mediated rolling phase of neutrophil extravasation (refer to Ch. 2). Again, the neutrophils of these patients remain in the circulation and cannot enter tissues under attack, leading to recurrent bacterial infections. Because CD15 also normally functions in metabolic reactions outside the immune system, LAD-II patients have dysmorphic facial features and suffer from growth deficits and mental retardation.

Leukocyte adhesion deficiency type III (LAD-III) is caused by autosomal recessive mutations of kindlin-3, a protein required for integrin signaling (among other roles). In addition to defective neutrophil extravasation, LAD-III patients show increased bleeding due to impaired platelet function.

b) Chronic Granulomatous Disease

Chronic granulomatous disease (CGD) is a clinically heterogeneous disorder char-acterized by a failure in phagosomal killing. Patients generally suffer from recurrent and sometimes life-threatening bacterial and fungal infections. Persistent infection results in excessive formation of granulomas in the lungs, lymph nodes, skin, ure-ters and/or bowel. CGD is caused by mutation of any one of the four subunits of the NADPH oxidase enzyme positioned in the phagosome membrane. Without a functional NADPH oxidase, there is a sharp decrease in H_2O_2 production during the respiratory burst, and killing within the phagosome is diminished. X-linked CGD is

Leukocyte extravasation was described in Chapter 2 and illustrated in Figure 2-16.

Pyogenic infections are those inducing pus production. *Pyrogenic* infections are those inducing fever.

the most common and most severe form, affecting about 70% of patients. Autosomal recessive forms of CGD also exist.

c) Chédiak–Higashi Syndrome

Chédiak–Higashi syndrome (C-HS) is an autosomal recessive disease with both hematopoietic (first described by Chédiak) and neurological (first described by Higashi) manifestations. A mutation in the *LYST* gene encoding a lysosomal trafficking regulator causes a defect in processes requiring the fusion of intracellular vesicles. Such processes include the fusion of endosomal vesicles to phagosomes, so that C-HS patients suffer from recurrent pyogenic infections with organisms that are normally eliminated by phagocytosis. In addition, the extracellular fusion of the cytotoxic granules of a CTL or granulocyte to a target cell is impaired, so that cell-mediated adaptive immunity is compromised. Interestingly, the intracellular transport of melanin is also inhibited, resulting in skin pigmentation defects and a silver-gray hair color. Chédiak–Higashi syndrome is usually fatal at a young age, although some patients can survive until age 20 or 30 years. These survivors are usually confined to a wheelchair by their neurological symptoms.

d) Defects of the IFNγ/IL-12 Axis

As discussed in Chapter 13, the control of many intracellular pathogens depends on granuloma formation by hyperactivated macrophages. Macrophage hyperactivation depends on IFNγ produced by activated Th1 cells, and Th1 differentiation depends on IL-12 produced by activated macrophages and DCs. Therefore, patients with mutations affecting IFNγ or IL-12 production or signaling suffer as infants from infections with certain environmental strains of bacteria that are normally not very harmful. For example, patients with "Mendelian susceptibility to mycobacterial disease" (MSMD) are very susceptible to infection almost exclusively by environmental mycobacteria. Diverse mutations in the IL-12p40 subunit, the IL-12Rβ1 chain, or either chain of the IFNγR cause autosomal dominant or recessive forms of MSMD. A defect in the STAT1 transcription factor that is normally activated by IL-12 or IFNγ signaling results in a more debilitating disease, with MSMD symptoms aggravated by increased susceptibility to severe virus infections.

> Environmental strains of a species are those that are commonly present in the normal healthy environment and not usually pathogenic.

e) Defects of PRRs

A defect in PRR function can lead to defective innate immunity against a particular pathogen. For example, very rare autosomal dominant mutations of the TLR signaling pathway molecules MyD88 or IRAK4 result in severe infections predominantly by pyogenic *S. pneumoniae* or *Staphylococcus* species. Mutations leading to loss of TLR3 itself or a regulator of the TLR3 signaling pathway render patients especially vulnerable to encephalitis caused by HSV infections. Patients with a relatively common autosomal recessive mutation inactivating TLR5 show increased susceptibility to infection with the flagellin-expressing pathogen *Legionella*. Other TLR deficits result in poor inflammatory responses to attack by almost any pathogen, with impaired production of acute phase proteins and little or no fever. Some TLR-deficient patients also lack robust production of IFNα/β, leaving them especially vulnerable to virus infections.

Deficiencies of PRRs outside the TLR family have also recently been identified. For example, patients lacking the CLR Dectin-1, which recognizes fungal β-glucans, suffer from an increased frequency of yeast infections. Similarly, patients without the CLR MBL cannot make good use of the lectin pathway of complement activation, leaving them susceptible to infection by a range of bacteria and fungi.

> Types of PRRs, including CLRs, NLRs and RLRs, were described in Chapter 3.

f) Congenital Neutropenias

"Neutropenia" is the term used to describe the presence of an extremely low number of neutrophils in the circulation. Congenital neutropenia may be either constant, leading to severe disease, or intermittent, resulting in a milder cyclic form of the disorder. Curiously, the exact same mutation may give rise to either constant or cyclic neutropenia, depending on other modifying factors involved in a particular patient's disease.

Severe congenital neutropenia (SCN) is characterized by recurrent bacterial and fungal infections and a dramatically reduced number of circulating neutrophils that does not fluctuate. Sporadic, autosomal dominant and autosomal recessive inheritance patterns of SCN have all been reported. Mutations in the elastase enzyme essential for neutrophil migration through the extracellular matrix underlie many SCN cases. SCN can also arise from mutation of the CXCR4 chemokine receptor, which results in an autosomal dominant syndrome long-windedly called "Warts-hypogammaglobu-linemia-infections-myelokathex." In this case, the interaction between CXCR4 and its ligand CXCL12 is compromised so that mature neutrophils in these patients are unable to leave the bone marrow and enter the circulation. B cell trafficking is also compromised, leading to a variable degree of hypogammaglobulinemia. Interestingly, these patients are also specifically susceptible to HPV infections that cause warts.

A rare autosomal recessive form of SCN is caused by mutations of the *HAX1* gene involved in intracellular signaling. A lack of HAX1 function increases the apoptosis of myeloid cells, including neutrophils. Another very rare form of autosomal recessive SCN is caused by deficiency of the endosomal adaptor protein P14. However, because the P14 protein functions in a broad range of biological processes, these SCN patients also suffer from partial *oculocutaneous albinism* and are short in stature.

Cyclic neutropenia (CyN) is characterized by recurrent infections that correlate with cyclic fluctuations in numbers of circulating neutrophils. Over a 21-day cycle, neutrophil counts in CyN patients can range from near-normal to virtually zero. During the 3–6 days when neutrophil counts are at their lowest, the patient slips from apparently normal health to a state of severe infection that leads to death in 10% of cases. Mutations in neutrophil elastase also underlie most cases of CyN, but the inheritance of CyN is always autosomal dominant.

ii) Complement Deficiencies

Deficiencies for elements of the various complement activation pathways described in Chapter 3 can cause PIDs. **Mannose-binding lectin (MBL)** is important both for the opsonization of pathogens that facilitate classical complement activation and for the direct triggering of the lectin pathway. Mutations in the MBL gene are usually autosomal dominant and relatively common, affecting 16–29% of a population depending on ethnic background. Patients with very low levels of serum MBL suffer from recurrent infections with a wide range of organisms and (for unknown reasons) are at increased risk for the development of the autoimmune disease *rheumatoid arthritis* (see Ch. 19). MBL deficiency has its greatest impact in early infancy, after maternal antibodies have dissipated and before the patient's adaptive immune system is mature enough to respond. Mutations of the MBL-associated protease MASP2 are also linked to increased susceptibility to infection.

Children lacking the earliest components (C1 and C4) of the classical complement activation pathway suffer from recurrent pyogenic infections with encapsulated bacteria. In contrast, patients deficient for C2 suffer from recurrent non-pyogenic sinopulmonary infections. For unknown reasons, patients lacking C1, C2, or C4 also have an increased incidence of the autoimmune disease *systemic lupus erythematosus* (*SLE;* see Ch.19). Patients lacking C3, or the Factor H, Factor I, or properdin components of the alternative pathway, also suffer from severe recurrent infections with pyogenic encapsulated bacteria but do not show signs of SLE. Increased infection with *Neisseria* bacteria (only) occurs in adult patients that lack any of the terminal complement components C5, C6, C7 or C8 and so cannot complete MAC formation. These latter complement component deficiencies are all rare autosomal recessive disorders and account for about 1% of all PIDs. Interestingly, patients deficient for C9 are generally asymptomatic.

NOTE: Genetic deficiency for a complement system component is not always a bad thing. In Papua New Guinea, there is a population in which many individuals have lost expression of CR1. This area of the world is also a malaria-endemic region, and the malarial parasite frequently uses CR1 to enter erythrocytes during the merozoite phase of its life cycle (refer to Box 14-5). When these CR1-deficient individuals are infected with *Plasmodium falciparum*, they do not develop severe malaria. A similar situation exists in Africa, where certain CR1 polymorphisms are associated with decreased malarial disease.

Hypogammaglobulinemia is the term used to describe abnormally low levels of Igs present in the blood.

Patients with oculocutaneous albinism cannot produce body pigments and so have pink eyes and pale white skin and hair.

iii) Autoinflammatory Syndromes

There are several inherited diseases that were originally thought to be classical PIDs because the patients suffered from recurrent bouts of severe local inflammation and prolonged periodic fevers (and so appeared to be immunodeficient). However, these symptoms occur in the absence of any obvious pathogenic cause. Recent research has revealed that the genetic defects underlying these syndromes result in the inappropriate activation of innate leukocytes mediating inflammatory responses. As such, these disorders are now thought of as a form of autoimmunity in which the innate response (rather than the more traditional adaptive response) is responsible for the immunopathology. Examples of such disorders are "familial Mediterranean fever" (FMF), "hyper-IgD with periodic fever syndrome" (HIDS), and "tumor necrosis factor receptor-associated periodic syndrome" (TRAPS). These "autoinflammatory syndromes" are discussed in Chapter 19.

III. Primary Immunodeficiencies due to Defects in Adaptive Immunity

Defects in lymphoid lineage cells mediating adaptive immunity lead to a wide range of PIDs. A mutation that results in compromised responses by both T and B cells is referred to as a *combined immunodeficiency* (CID). Because of their shared developmental path with T cells, NK and NKT cell functions are also affected by many of these mutations.

i) Combined Immunodeficiencies

a) Nature of SCIDs

Some of the best known PIDs are types of **SCID**, standing for severe combined immunodeficiency disease (**Table 15-2**). SCID cases represent about 20% of PIDs, with a prevalence of about 1 in 50,000 live births. In the 1970s, the "Boy in the Bubble" story awoke the general public to the plight of children with SCID and introduced many to the science of immunology for the first time (see **Box 15-1**). In all forms of SCID, T cell development and/or function is compromised. In some SCID cases, the B cells have an intrinsic defect. In other cases, the B cell defects are secondary to a lack of T cell help caused by the non-functional or absent T cells, so that the deficiency is still "combined." NK development and function are also impaired in many forms of SCID. In any case, no matter what their genetic defect, SCID patients generally present with chronic diarrhea, failure to thrive, and severe opportunistic infections. If left untreated, SCID is inevitably fatal, usually in early childhood but sometimes in adolescence. Even when treated so that they regain some immune system functions, SCID patients may still eventually succumb to cancers like lymphomas and leukemias.

Bone marrow transplantation is discussed in Chapter 17. The EBV-associated cancer Burkitt's lymphoma is described in Chapter 20.

TABLE 15-2	Types of SCIDs	
Name	**Defective Gene**	**Defective Process**
X-linked SCID	γc chain	Cytokine signaling
Jak3 SCID	Jak3	Cytokine signaling
IL-7Rα SCID	IL-7Rα	Cytokine signaling
ADA SCID	ADA	Nucleotide metabolism
PNP SCID	PNP	Nucleotide metabolism
Reticular dysgenesis	AK2	Nucleotide metabolism
RAG SCID	RAG-1 or RAG-2 (<1% function)	VDJ recombination
Omenn Syndrome	RAG-1 or RAG-2 (1–25% function)	VDJ recombination
Artemis SCID	Artemis	VDJ recombination and DNA repair
CD45 SCID	CD45	T cell activation due to loss of protein tyrosine phosphatase activity
Nude SCID	FoxN1	Thymic epithelial cell development

Box 15-1   "The Boy in the Bubble"

On September 21, 1971, the birth of David Vetter marked the beginning of an unprecedented experiment that gave doctors, scientists, and laypeople dramatic insight into the consequences of primary immunodeficiency. David's parents had lost a previous son to SCID, and so David was delivered by Caesarean section under "germ-free" conditions and was immediately placed in a sterile isolator. After doctors confirmed that David too suffered from SCID, they suggested that David might live in a specially designed sterile environment (his "bubble") until either a cure for SCID was found or his immune system started to function. In the meantime, researchers would learn more about SCID by studying David. David's family agreed to this plan, and David became known worldwide as "the Boy in the Bubble." Although David's immune system remained compromised, his intellectual and emotional development were normal despite his life of isolation. By age 5 years, he showed above-average speech and language abilities and attended school very successfully by telephone. In 1977, NASA presented him with a special space suit that allowed him to experience what it was like to move around outside his bubble. However, as David neared his adolescent years, he expressed a wish to live a normal life in the outside world and experience such simple pleasures as feeling the grass under his bare feet. The only hope for a cure and escape from the bubble was a bone marrow transplant from a genetically matched donor. If the donor's healthy hematopoietic cells successfully established themselves in David, they might generate the lymphocytes necessary for productive immune responses. At that time in medical history, there had to be a very good genetic match between the bone marrow donor and the recipient to prevent transplant rejection. In David's case, an exact genetic match could not be found, but he was prepared to gamble if it meant he might have his freedom. At the age of 12, David received a bone marrow transplant from his 15-year-old sister. He remained in his sterile environment while doctors waited to see if his body would accept the transplanted cells. Sadly, unbeknownst to his doctors, his sister's bone marrow cells were infected with latent Epstein–Barr virus (EBV). Once introduced into David, this oncogenic virus induced what was only later understood to be Burkitt's lymphoma. In response to David's deteriorating condition, his doctors made the decision to treat him in a hospital room. On February 7, 1984, David emerged from his bubble and was touched by his mother for the first time. He fought hard to recover from his illness but died two weeks later of cardiac arrest. In announcing David's death, his emotional doctor declared that there would be "no more bubbles." Since David's passing, multiple new strategies (including gene therapy; see Section A.IV) have been devised in an attempt to give SCID children a chance at a normal life.

b) Types of SCID

David Vetter (the "Boy in the Bubble") suffered from *X-linked SCID,* a form of the disease that accounts for about 40% of all SCID cases. X-linked SCID is caused by mutation of the X-linked gene encoding the common γ chain (γc). The γc protein is a subunit in the IL-2, IL-4, IL-7, IL-9, IL-15, and IL-21 receptors and transduces intracellular signaling initiated by the binding of the appropriate cytokine to its receptor (**Fig. 15-1A**). Because both IL-7 and IL-15 signaling are disrupted in the absence of γc, the development of T and NK cells (but not B cells) is disrupted (**Fig. 15-1B**). Boys affected with this disorder have few if any T and NK cells but normal or even increased numbers of circulating B cells. However, X-linked SCID B cells exhibit defects in isotype switching that are only partly due to an absence of T cell help and cytokine signaling.

Jak3 SCID is a disease that is clinically indistinguishable from X-linked SCID but much more rare. Jak3 SCID results from mutations in Jak3, a kinase that transduces signals downstream of γc. This disease has an autosomal recessive inheritance pattern and is characterized by a lack of T and NK cells but normal or increased B cell numbers. Again, the B cells are non-functional due to deficits in T cell help and cytokine signaling.

IL-7Rα SCID is an autosomal form of the disease that results from a mutation in the IL-7Rα chain. Only the development of T cells is affected by this mutation, but the patient's normal B cells are unable to produce Igs due to the lack of T cell help. Unlike X-linked SCID and Jak3 SCID patients, IL-7Rα SCID patients have normal numbers of functional NK cells.

ADA SCID and *PNP SCID* are two related forms of the disease that arise from autosomal recessive mutations in the genes encoding *adenosine deaminase (ADA)* and *purine nucleoside phosphorylase (PNP).* The *ADA* and *PNP* genes encode housekeeping enzymes required for the salvage pathway of purine metabolism (see **Box 15-2**). When these enzymes are non-functional, toxic metabolites accumulate that kill actively proliferating cells like lymphocytes, rendering the patient immunodeficient. Non-immune system deficits can also occur in these patients, including deafness, liver toxicity, and behavioral abnormalities.

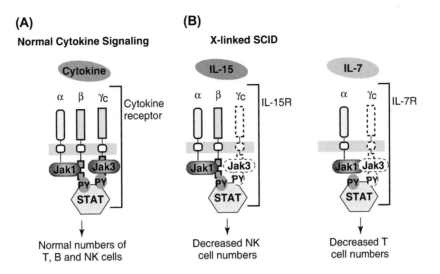

Fig. 15-1
How a Defect in Cytokine Receptor Signaling Can Lead to SCID

(A) Many cytokine receptors are multi-subunit complexes whose cytoplasmic regions associate with kinases such as Jak1 and Jak3. This association results in phosphorylation of particular tyrosine residues (PY) that in turn activate STAT transcription factors. Normal function of all these components results in normal numbers and functions of T, B, and NK cells. **(B)** In X-linked SCID, mutations of the γc chain, which is a component of the IL-15 and IL-7 receptors, alter the signaling delivered by the receptor and prevent the development of mature NK cells or T cells, respectively. B cells are usually present in normal numbers in X-linked SCID patients but are non-functional due to the lack of T cell help.

Discovered in 1972, ADA deficiency was the first known cause of SCID and accounts for about 15% of SCID cases. More than 50 distinct mutations in the *ADA* gene have now been associated with SCID, and the severity of the phenotype is determined by the level of residual activity of the mutated enzyme. Patients whose mutation depresses ADA activity to less than 0.05% of normal are severely affected, while those in whom about 1–5% of normal activity is retained have a much milder disease. Numbers of T, B and NK cells are all abnormal in ADA patients.

PNP SCID occurs much less often than ADA SCID and constitutes only 2% of all SCID cases. As little as 1.5% of normal PNP activity is sufficient for patient survival. Due to intrinsic differences in T/NK and B lymphocyte progenitors, dGTP accumulates to high levels in T/NK lineage cells in the absence of PNP but not in B lineage cells. As a result, B cell numbers are normal in PNP SCID patients. Non-immune deficits in PNP SCID patients include progressive neurological deterioration.

Reticular dysgenesis (RD) is perhaps the most severe type of SCID but extremely rare, comprising only 1–3% of SCID cases. Variable deficits in T, B, NK and other hematopoietic lineages are observed, leading to the nickname "inherited bone marrow failure." RD is caused by autosomal recessive mutations of adenylate kinase-2 (AK2), a mitochondrial enzyme essential for regulating levels of adenosine diphosphate. In the absence of AK2 function, hematopoietic cells are induced to undergo apoptosis, and the thymus and secondary lymphoid organs are therefore underdeveloped. Some patients also exhibit deafness.

RAG SCID (sometimes called "alymphocytosis") results from mutations of either RAG-1 or RAG-2. These alterations impair V(D)J recombination of the TCR and Ig genes such that mature T and B lymphocytes are absent but NK cells and other hematopoietic lineages are present.

Omenn syndrome is caused by amino acid substitution mutations in the *RAG* genes that reduce their activity to 1–25% of normal. These patients have extremely low levels of circulating B cells but a variable number of T cells.

Artemis SCID is caused by mutations of the *Artemis* gene, which is important for both V(D)J recombination and DNA repair. In the absence of Artemis, T and B cells cannot repair the double-stranded DNA cuts made by RAG-1 and RAG-2 during V(D)J recombination. Similar SCID disorders have been identified as resulting

Box 15-2   Enzymatic Functions of ADA and PNP

Normal nucleic acid degradation leads to an accumulation of purine nucleotides that are broken down into adenosine (Ado) and deoxyadenosine (dAdo), and guanosine (Guo) and deoxyguanosine (dGuo). ADA is present in all cells and converts Ado and 2′-dAdo molecules into inosine (Ino) and 2′-deoxyinosine (dIno), respectively. PNP converts Ino and 2′-dIno to hypoxanthine, and Guo and dGuo to guanine. These molecules can enter the *purine salvage* pathway (shown in green in the figure below) and are converted back to ATP and GTP that can be recycled into new purines in preparation for cell division. Thus, ADA and PNP are crucial for the recycling of elements of old purines into new purines, particularly in tissues in which cell division occurs at a rapid pace (such as the bone marrow, thymus, and lymph nodes). Without ADA/PNP function (red pathway in the figure), high levels of Ado, dAdo, Guo and dGuo accumulate and are metabolized to dATP and dGTP. These molecules are toxic and induce breaks in DNA, block normal DNA methylation, and interfere with *de novo* DNA synthesis such that cell death is triggered. Rapidly proliferating cells such as T and B lymphocytes and NK cells are most affected by dATP and dGTP accumulation, and ADA and PNP mutations therefore result in a variable loss of these cell types and clinical symptoms of SCID.

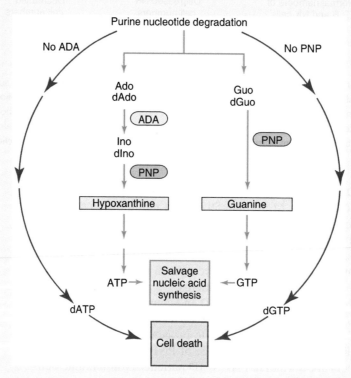

Figure Box 15-2
SCID Associated with Defects in Purine Salvage

from mutations of other proteins involved in various DNA repair pathways. These patients also often exhibit non-immune system deficits such as facial or dental abnormalities.

CD45 SCID is a very rare form of the disease that is associated with an autosomal recessive mutation in the powerful cell surface protein tyrosine phosphatase CD45. CD45 is expressed at high levels on all leukocytes and is required in particular for efficient transmembrane signaling by lymphocytes. Because CD45 is vital for T cell development and activation, CD45 SCID patients show reduced T cell numbers and defective T cell responses that compromise the responses of their normal B cell populations. NK cells are reduced in number.

Nude SCID is an autosomal recessive disorder characterized not only by immunodeficiency but also by loss of hair (*alopecia*) and malformations of the fingernails and toenails. This constellation of abnormalities closely resembles that in the immunodeficient *nude* mouse strain, accounting for the name of the parallel human disorder. Nude

SCID is caused by mutations of the *FoxN1* gene, which encodes a forkhead box transcription factor essential for the development of thymic epithelial cells. As discussed in Chapter 9, without functional cTECs and mTECs, T cell development is compromised. Without functional T cells, B cell responses are abolished although B cell numbers are normal. NK cell numbers are also normal.

ii) APC-Specific Immunodeficiencies

a) MHC Deficiencies

Patients with MHC deficiencies present with multiple severe infections that are usually fatal at a young age. The first disorder of this type to be identified was called "bare lymphocyte syndrome," since the patients' lymphocytes were devoid of the myriad MHC molecules present on the surfaces of normal cells. Modern research has revealed that mutations in subunits of the TAP transporter or tapasin impair the expression of MHC class I not only on lymphocytes but on all tissue cells. This failure to express the MHC molecules necessary for antigen presentation to Tc cells thus inhibits cell-mediated adaptive responses. Similarly, autosomal recessive mutations in RFX and CIITA, which are regulatory proteins required for MHC class II transcription, can lead to a lack of MHC class II expression on APCs, blocking Th cell activation and compromising humoral responses.

b) DC Deficiencies

Three human DC deficiency syndromes have been described to date. Individuals with a sporadic or autosomal dominant mutation in GATA2, a transcription factor important for HSC homeostasis, suffer from "DCML syndrome," in which cDCs, pDCs, monocytes, regulatory T cells, B cells and NK cells are all severely reduced in number. Curiously, conventional T cells are normal in these patients. DCML syndrome patients also show deficits in inflammatory cytokine production, leaving them highly susceptible to mycobacterial, fungal and viral infections.

The other two DC deficiency syndromes result from two different mutations of the transcription factor IRF8. These patients are highly susceptible to mycobacterial infections (only). Strikingly, although the same transcription factor is compromised, the effects of these mutations on various hematopoietic cell populations are quite different. While patients with the autosomal recessive form of this disease lack all detectable DC subsets as well as monocytes, those with the autosomal dominant form have normal hematopoietic cell populations except for reduced CD1c+ DCs. B cells, NK cells and T cells are normal. Considering the central role in adaptive immunity held by DCs, it is currently a mystery why these DC-deficient patients don't suffer from a broader range of infections.

The study of human DC deficiencies is still in its infancy, partly due to the uncertain ontogeny of these cells in humans, technical difficulties in distinguishing and isolating these fragile cells, and the fact that other APCs can substitute for some DC functions. Another obscuring factor is that a defect causing the loss of DCs often also results in the loss of other hematopoietic cells. As a result, the consequences of a deficit in, say, neutrophils (loss of innate responses to bacteria) overwhelms the consequences of a deficit in DC function (loss of T cell activation), at least initially. Research is ongoing to identify truly DC-specific PIDs.

iii) T Cell Specific Immunodeficiencies

T cell-specific PIDs (T-PIDs) are diseases in which the gene affected has its predominant impact directly on T cells as opposed to B cells or NK cells. Patients with these disorders generally have normal numbers of T cells, but these lymphocytes, particularly CTLs, show defects in activation or function. Clinically, T-PID patients exhibit increased susceptibility to a broad range of infections, stunted growth, and low Ig levels. In addition, these individuals may show signs of autoimmune disease or develop cancer. Because B cell functions are often compromised in T-PID patients due to a lack of T cell help, some clinical immunologists classify T-PIDs as SCIDs.

NOTE: "Immunodeficiency" by definition means "a lack of immune responsiveness," so it may seem odd that immunodeficiency can sometimes lead to autoimmune disease, which, by definition means "a presence of immune responsiveness to self." However, recent research has revealed the fine balance of regulatory elements controlling responses to self versus non-self, and a disruption of this balance can have unexpected consequences. In some cases where PID patients show signs of autoimmune disease, the defect is one of T cell tolerance. In other cases, the autoimmunity is due not as much to a loss of tolerance as it is to an inability to eliminate a persistent antigen, which provokes excessive and eventually chronic inflammation. In still other PIDs, the cause of the autoimmune symptoms remains a mystery.

a) Defects in TCR Signaling

Several T-PIDs result from defects in particular steps of the TCR signaling pathway. Disease-causing mutations have been identified in the CD3δ, CD3ϵ, and CD3ζ chains of the TCR complex; the Lck and ZAP70 kinases that transduce signaling downstream of the TCR; proteins regulating the Ca^{2+} flux that follows TCR triggering; and various transcription factors activated by TCR signaling. Either CD4$^+$ or CD8$^+$ or both subsets of mature T cells may be missing in these patients, but numbers of B cells and NK cells are generally normal. In the case of ZAP70 deficiency, some patients lack B cell functions, leading some clinicians to consider these children as having ZAP70 SCID.

The TCR complex and the intracellular signaling it triggers were illustrated in Figure 8-3 and Figure 8-5.

A very rare T-PID is caused by mutations in the STAT5B transcription factor activated in response to IL-2. Because IL-2 signaling is crucial for T cell homeostasis, patients lacking STAT5B are immunodeficient but also show variable signs of autoimmune disease. In addition, because IL-2 signaling normally transduces the effects of growth hormone, these patients are short in stature.

b) Coronin-1A Deficiency

A relatively new T-PID involves defects of actin polymerization. When actin polymerization is abnormal or absent, a cell cannot be activated or migrate and often undergoes apoptosis. Coronin-1A is a regulatory protein inhibiting actin polymerization in T cells. In the absence of coronin-1A, newly produced mature T cells accumulate large quantities of intracellular F-actin and cannot move easily from the thymus to the periphery. B cells and NK cells appear to be normal, likely due to compensation by other coronin family members. Clinical features not common to other PIDs have been reported in various coronin-1A-deficient patients, including psychiatric disorders and the development of chicken pox upon varicella vaccination.

G-actin is the monomeric globular form of this structural protein, whereas F-actin is the polymeric filamentous form.

c) Hyper-IgM 1 Syndrome

The name, "hyper-IgM" sounds like neither an immunodeficiency nor a T cell-related disease. However, patients with any one of the six known types of hyper-IgM syndrome (HIGM) have normal or very high levels of circulating IgM but very low or absent levels of all other isotypes. Numbers of circulating neutrophils are also very low. As a result, these patients suffer from frequent infections with opportunistic organisms. With respect to the T cell connection, only the first type of hyper-IgM syndrome (HIGM1) is caused by a genuine T cell-related defect and so is described in the following text. The other five HIGM syndromes are due to intrinsic B cell defects and are discussed later in the section on B cell-specific PIDs. **Table 15-3** summarizes the main features of all six HIGM syndromes.

HIGM1 is the most common form of hyper-IgM syndrome and is an X-linked disorder. The mutations of the *CD40L* gene that cause this form of the disease reduce the costimulatory capacity of T cells. Because the interaction between CD40L on a normal Th cell and CD40 on a B cell is vital for B cell costimulation, isotype switching, somatic hypermutation, and GC formation, the B cells of HIGM1 patients, although intrinsically normal, cannot proliferate and GCs do not form in the secondary lymphoid organs. Memory B cell generation and DC maturation are also impaired. Thus, in addition to the opportunistic infections by extracellular bacteria, fungi and parasites that

TABLE 15-3   Types of HIGM Syndrome

Type	Defective Cell Type	Defective Gene	Chromosome	Defect in Isotype Switching	Defect in Somatic Hypermutation	Defect in Cell-Mediated Immunity
HIGM1	T cell	CD40L	X	Yes	Partial	Yes
HIGM2	B cell	AID	12	Yes	Yes	No
HIGM3	B cell	CD40	20	Yes	Yes	Yes
HIGM4	B cell	AID (3′ end only)	12	Yes	No	Yes
HIGM5	B cell	UNG	12	Yes	Partial	No
HIGM-NEMO	B cell	NEMO	X	Yes	No	Yes

plague other types of HIGM patients, HIGM1 patients suffer from recurrent infections with intracellular bacteria and protozoa.

d) Hyper-IgE Syndrome

The term "hyper-IgE syndrome" again looks like it should describe a disease related primarily to B cells, but the usual primary defect is now recognized as defective Th17 cell development. The hallmark of patients with hyper-IgE syndrome is, not surprisingly, very high levels of serum IgE. IgD may also be moderately increased. As with the HIGM syndromes, it is the resulting imbalance in Ig isotype production that leads to the immunodeficiency. Children with hyper-IgE syndrome suffer from dermatitis, *Candida* infections, and *Staphylococcal* abscesses, particularly in the lungs, joints and skin. Interestingly, despite their elevated IgE, the dermatitis in these patients is quite distinct from that of patients suffering from the atopic dermatitis associated with severe allergy (see Ch. 18), and respiratory allergies are rare. In addition, although some symptoms of hyper-IgE patients resemble those of patients with CGD, the infections are by different organisms and occur in anatomically distinct sites. Lastly, many hyper-IgE syndrome patients have characteristic coarse facial features.

Most T cell subsets and B cells are normal in children with hyper-IgE syndrome, but eosinophils are greatly elevated in both blood and sputum. Most strikingly, the Th17 cells vital for mucosal defense against fungi and bacteria are missing. Numbers of neutrophils are normal in these patients, but their recruitment to sites of inflammation is impaired due to a lack of cytokines and chemokines produced by Th17 cells. Th17 cells also normally secrete IL-21, a regulator that suppresses isotype switching to IgE, so that an absence of these cells leads to excessive IgE production.

The autosomal dominant form of hyper-IgE syndrome is caused by mutations of the STAT3 transcription factor. STAT3 is essential for responses to IL-6, and IL-6 is essential for Th17 cell development. An intriguing feature of this form of the disease is that these youngsters often have double rows of teeth because their baby teeth do not fall out due to a failure in tooth root resorption. Bone density in general is reduced in these patients so that relatively minor traumas frequently result in fractures.

Patients with autosomal recessive hyper-IgE syndrome have normal bone density and tooth structures but often show signs of autoimmune disease. Mutations of TYK2, a kinase acting upstream of STAT3, have been identified in a small number of autosomal recessive hyper-IgE patients. Other cases of autosomal recessive hyper-IgE syndrome are due to mutations of DOCK8, an intracellular signal transduction adaptor protein. Like coronin-1A, DOCK8 has been implicated in actin polymerization and thus is needed for the growth, migration and adhesion of many hematopoietic cells. In addition to the usual clinical signs associated with hyper-IgE syndrome, some DOCK8-deficient patients show mental retardation and/or developmental disabilities.

Finally, in at least some hyper-IgE patients, the accumulation of IgE in the blood appears to be due to a defect in IgE catabolism that blocks normal turnover of the

Th17 cell development was described in Chapter 9.

IgE protein. Other patients may be suffering from an imbalance in Th1/Th2 cytokine secretion that promotes abnormal isotype switching to IgE.

iv) B Cell Specific Immunodeficiencies

Some PIDs are due to mutations that affect B lymphocytes only (**Fig. 15-2**), as opposed to those affecting both B and T cells, or those in which the Ig deficit is caused by a lack of T cell help. B-PIDs account for 70% of all PIDs and are relatively easy to diagnose and treat. Two broad classes of these antibody deficiencies exist: patients who do not have mature B cells and those who do. In both classes, patients are usually free of infection until age 7–9 months due to the presence of maternal antibodies. As these antibodies decrease in concentration, the patient manifests low levels of one or more serum Igs and an increased susceptibility to infections with enteroviruses, parasites, and encapsulated bacteria. The growth retardation often evident in patients with T-PIDs is not usually seen in patients with B-PIDs.

a) Congenital Agammaglobulinemias

There are several types of congenital *agammaglobulinemia*, which is an inherited lack of Ig proteins, exist. Two of the best characterized forms of this disease are described next.

Bruton's X-linked agammaglobulinemia (XLA) is a rare disorder (about 1 in 200,000) in which affected boys lack mature B cells and all antibody isotypes, and thus

Fig. 15-2

Examples of Defects Leading to B Cell-Specific Primary Immunodeficiencies

Defects at the indicated points during B cell development give rise to immunodeficiencies characterized by loss of B cell functions only.

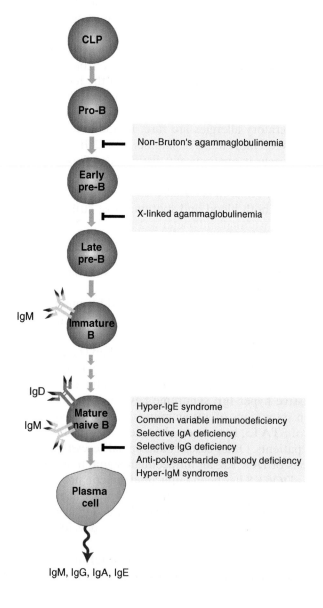

are unusually susceptible to enteroviral and bacterial infections. XLA is caused by any one of over 300 mutations in the gene encoding Bruton's tyrosine kinase (Btk). Btk, which is expressed in B cells and myeloid cells but not in T cells, is critical for the continued maturation of pre-B cells. XLA patients have greatly reduced numbers of immature B cells in the lymphoid organs and periphery but, for unknown reasons, myeloid cells are apparently normal.

Non-Bruton's agammaglobulinemia (NBA) occurs in about 10% of patients who have the clinical features of XLA but have a normal *Btk* gene. Instead, the mutations in these patients most often affect the μ heavy chain gene but may also disrupt the λ5 component of the surrogate light chain, the Igα or Igβ chain, or BLNK, a signaling adaptor acting downstream of BCR engagement. All these defects lead to a block in the pro-B to pre-B cell transition, a step earlier than the defect in XLA patients. Because the genes mutated in NBA are autosomal, this disorder can affect both boys and girls. Over 90% of patients with defects in early B cell development have a mutation in one of the NBA-associated proteins.

b) Common Variable Immunodeficiency (CVID)

Common variable immunodeficiency (CVID) is the name given to a family of diverse diseases characterized by a general impairment of humoral responses. Although both autosomal dominant and autosomal recessive inheritance patterns have been observed, most cases are sporadic. All CVID patients show profoundly decreased levels of IgA and IgG, and about 50% also lack IgM. Circulating mature B cells are present, but plasma cell differentiation and antibody production are impaired. Memory B cells may also be absent. The frequency of CVID is about 1 in 30,000, making it a fairly prevalent PID. Onset may occur in childhood, adolescence, or adulthood. The types of recurrent bacterial infections observed in CVID are similar to those in XLA. Respiratory tract infections by *H. influenzae* and *S. pneumoniae* are particularly common. Some patients show autoimmune symptoms, whereas others develop lymphomas.

Most CVID cases are related to intrinsic B cell defects, such as impaired isotype switching or somatic hypermutation, defective activation of BCR signaling, impaired DNA repair, abnormal protein tyrosine phosphorylation, or upregulated Fas expression. However, other CVID cases appear to be secondary to T cell defects, or to abnormalities in interactions between B and Th cells or between Th cells and APCs. Homozygous deletions of the *ICOS* gene, which is essential for isotype switching, can result in CVID, as can mutations of the *CD19* gene, which encodes a protein regulating the signal transduction required for B cell activation. Although the relationship is not yet clear, susceptibility to CVID has also been associated with heterozygous mutations of the TACI protein, which is a BAFF receptor.

> BAFF is a cytokine essential for B cell development (refer to Ch. 5).

c) Selective Ig Deficiencies

In patients with selective Ig deficiencies, antibodies of a particular isotype or antibodies directed against a particular pathogen structure are missing. The total level of circulating B cells is usually normal. A few of these cases are caused by straightforward deletions or mutations in the *Igh* locus that affect the constant region exons for the IgA or IgG subclasses. Other mutations that affect the synthesis of soluble antibody, or alter the regulatory or switch regions of the Ig heavy chain exons, can also result in selective Ig deficiency.

Selective IgA deficiency (IgAD) is the most common PID at a frequency of 1 in 700 people. Antibodies of both the IgA1 and IgA2 subclasses are missing. All other antibody isotypes are usually synthesized in normal amounts, although some patients may gradually go on to show signs of CVID. T cell function is normal. Gastrointestinal and respiratory infections are the most common clinical signs of IgAD, although the disease is asymptomatic in many people. Autoimmunity is not uncommon and allergy is frequent. Sporadic, autosomal dominant, and autosomal recessive forms of IgAD have been described, but the underlying genetic defects are unknown.

Selective IgG deficiency (IgGD) is characterized by the loss of production of one of the IgG subclasses and leads to recurrent upper and lower respiratory tract infections.

To date, the inheritance of IgGD has been exclusively autosomal recessive. While the underlying molecular defects are unknown in most cases, some are caused by mutations in the *Igh* locus that affect isotype switching to IgG. Other cases are associated with defects in the ICOS or CD19 genes. Transient hypogammaglobulinemia of infancy (THI) is a condition in which infants under 2 years of age exhibit abnormally low levels of IgG. This deficit is accompanied in some cases by recurrent viral infections of the upper respiratory tract. Normal IgG levels are often spontaneously restored in THI patients by 2 years of age, but an increased risk of infection persists. The underlying molecular defect is unknown.

Antipolysaccharide antibody deficiency (APAD) is a disease in which serum Igs are normal, but the affected children suffer recurrent infections because they cannot make antibodies to the polysaccharides of encapsulated bacteria. This condition often comes to light after vaccination with a polysaccharide vaccine. Mutations in the signaling pathway molecules IRAK4 and NEMO have been identified in these patients.

d) B Cell-Intrinsic HIGM Syndromes

We noted previously that the most common form of hyper-IgM syndrome (HIGM1) is due to defects in the CD40L signaling pathway in T cells. As described next, the five remaining HIGM syndromes are caused by various defects in B cell signaling pathways involved in isotype switching (refer to **Table 15-3**).

HIGM2 is caused by autosomally inherited mutations of the gene encoding activation-induced deaminase (AID), the DNA-editing enzyme that is selectively expressed in GC B cells and critical for both isotype switching and somatic hypermutation. AID expression and activation in B cells is induced by CD40 signaling. Thus, like HIGM1 patients, HIGM2 patients exhibit normal or elevated serum IgM but sharply reduced serum IgG, IgA, and IgE. However, unlike HIGM1 patients, antibody diversity due to somatic hypermutation is completely absent in HIGM2 patients. GC B cells are mature, but unlike normal GC B cells, many express IgD as well as IgM. CD40L expression and T cell functions are normal in HIGM2 patients, meaning that they do not suffer the same range of opportunistic infections as HIGM1 patients.

HIGM3 is caused by autosomal recessive mutations of the *CD40* gene. These defects block CD40-CD40L signaling and result in a disorder very similar to HIGM1 in that isotype switching and memory B cell generation are impaired. However, the defect in somatic hypermutation is more profound in HIGM3 patients than in HIGM1 patients. Like HIGM1 patients, HIGM3 patients are vulnerable to a broad range of recurrent bacterial and fungal infections.

HIGM4 patients show a phenotype similar to, but slightly milder than, that of HIGM2 patients. The AID gene is also mutated in these patients, but the defect alters only the 3′ part of the gene, so that while isotype switching is impaired in HIGM4 B cells, somatic hypermutation is not. Clinically, IgM is not as highly elevated as in HIGM2 patients, and more IgG can be detected in the blood. However, antibody responses are impaired (especially those directed against polysaccharide antigens), and HIGM4 patients therefore suffer greatly from fungal, bacterial, and mycobacterial infections.

HIGM5 patients show symptoms that most closely resemble those of HIGM2 patients: highly elevated IgM, a near-absence of IgG, enlarged lymphoid tissues, and greatly increased susceptibility to bacterial (but not fungal) infections. However, the AID gene is normal in HIGM5 patients, and the autosomal recessive defect lies in the gene encoding uracil DNA glycosylase (UNG). UNG acts downstream of AID, removing uracil residues deaminated by AID and facilitating the creation of the double-stranded DNA breaks necessary for isotype switching. HIGM5 B cells therefore show a severe defect in isotype switching and a partial defect in somatic hypermutation.

HIGM-NEMO is a mild form of HIGM that is caused by a mutation in NEMO (see Note below) and occurs in association with an X-linked disorder of ectodermal tissue development. These patients have sparse hair and abnormal or missing teeth, and lack sweat glands. Humoral responses to antigens are impaired, particularly to polysaccharide antigens. Cell-mediated immunity is also abnormal. As a result, these patients are subject to frequent fungal, bacterial, and mycobacterial infections.

> NOTE: NEMO (NFκB essential modulator), which is encoded by an X-linked gene, is part of the enzymatic IKK (IκB kinase) complex that phosphorylates IκB and induces its degradation. IκB is an inhibitor that binds to the transcription factor NFκB and holds it inactive in the cytoplasm until it is needed. NF-κB is vital for gene transcription initiated by TCR or BCR engagement by antigen.

v) Adaptive Immunodeficiencies due to Defects in DNA Repair

Immunodeficiency is also a prominent clinical feature of several diseases that are caused by defects in proteins responsible for the repair of damage to cellular DNA. Mutations in three pathways of DNA repair have been associated with immunodeficiency, but because these genes are expressed in other cell types in addition to hematopoietic cells, patients exhibit symptoms outside the immune system. Cells of patients with DNA repair defects are hypersensitive to DNA-damaging agents and show chromosomal anomalies and breakages that disrupt cell cycle progression. Lymphocytes appear to be particularly sensitive to these mutations, and both humoral and cell-mediated immune responses are impaired. For reasons that are unclear, autoimmunity is common in one of these diseases (ataxia-telangiectasia) but not in other disorders of this group. Malignancy is frequent in all these ailments due to the accumulation of unrepaired mutations in the DNA that can lead to cancerous transformation. The connection between cancer and various PIDs is explored further in **Box 15-3**.

a) Ataxia–Telangiectasia

Ataxia–telangiectasia (AT) is an autosomal recessive disease that results from mutations of the tumor suppressor gene *ATM* ("ataxia telangiectasia mutated"). A major function of ATM is to phosphorylate the tumor suppressor p53 and stabilize it in response to DNA damage, allowing p53 to arrest the cell cycle of the damaged cell and promote DNA repair. However, ATM is also involved in the pathway that re-ligates DNA during Ig isotype switching, establishing ATM's link to the immune system and accounting for the immunodeficiency observed in AT patients.

Clinically, AT is characterized by progressive cerebellar ataxia, oculocutaneous telangiectasia, and a variable CID. Recurrent severe lung and sinus infections are common and almost half of AT patients die of pulmonary failure due to pneumonia. Hypogammaglobulinemia in the form of low levels of IgA, IgE, IgG2 and IgG4 is present due to defective B cell maturation and impaired isotype switching. Thymus development is impaired, T cell responses are decreased, and circulating memory T cells are reduced in number. Patients are hypersensitive to ionizing radiation (such as that associated with sun exposure or cancer treatment) and run a 100-fold increased risk of developing lymphomas and acute lymphocytic leukemias by the age of 15 years. Autoimmunity is also common in these patients. Due to their neurological deficits, AT patients are usually confined to a wheelchair by young adulthood and die prematurely of cancer or lung infections. Although no cure for AT currently exists, clinical management of these patients has improved greatly over the past decade due to the use of better antibiotics and cancer therapies. In addition, the administration of antioxidant compounds, which theoretically should reduce the DNA damage suffered by cells of AT patients, seems to be helpful in many cases.

A disease that is very similar to AT (called "AT-like syndrome") is caused by mutations in the *MRE11* ("meiotic recombination 11") gene. Like *ATM*, *MRE11* is involved in the DNA repair pathway that also mediates Ig isotype switching.

b) Nijmegen Breakage Syndrome

Nijmegen breakage syndrome (NBS) is another rare autosomal recessive PID caused by a defect in DNA repair. NBS arises from mutations of nibrin, a protein homologous to cell cycle control proteins. Nibrin participates in a complex that acts on p53 in the same DNA damage repair pathway as ATM, and also participates in the Ig isotype

Progressive cerebellar ataxia is cerebellar degeneration leading to generalized neuromotor dysfunction. Oculocutaneous telangiectasia is abnormal dilation of blood vessels in the conjunctiva of the eyes.

Box 15-3 PIDs and Cancer Development

As described in detail in Chapter 16, the immune system is responsible for tumor **immunosurveillance**, which is the identification and killing of abnormal cells that are or could become cancerous. Because patients with PIDs by definition have compromised immune systems, they are susceptible to developing tumors. Indeed, cancer is the second greatest threat to PID patients after infections. The type of tumor developing in a PID patient depends on his/her specific genetic defect, age, and clinical history. Some genetic defects, such as those affecting the repair of DNA damage and mutations, are directly related to tumorigenesis. Other genetic defects affect the regulation of cell growth or apoptosis, or sustain chronic inflammation that promotes malignant transformation. The most common cancer developing in PID patients is non-Hodgkin lymphoma (60%), with Hodgkin lymphomas accounting for 23% and leukemias for 6% of all such malignancies. Types of cancers associated with particular PIDs are listed in the following table. (Immune responses to solid cancers are described in Chapter 16, and lymphomas and leukemias in Chapter 20.)

TABLE BOX 15-3	PIDs and Cancer
PID	**Examples of Associated Cancers**
Ataxia–telangectasia	Lymphomas, epithelial tumors, lymphoid leukemias
Nijmegen breakage syndrome	B cell lymphomas
Bloom syndrome	Lymphomas, epithelial tumors, lymphoid leukemias
Xeroderma pigmentosum	Skin cancers
CVID	Non-Hodgkin lymphoma Epithelial cell carcinomas of bladder, breast, cervix, stomach, vulva and tonsils
Hyper-IgM syndromes	Carcinomas of biliary tract, liver, pancreas
Selective IgA deficiency	Hodgkin lymphoma
SCID	Non-Hodgkin lymphoma Hodgkin lymphoma Burkitt lymphoma Leukemia Kidney and lung tumors
Wiscott–Aldrich syndrome	Diffuse large B cell lymphomas Non-Hodgkin lymphoma Leukemia Tumors of brain, smooth muscles, skin
XLP1	Burkitt's lymphoma

switching pathway involving ATM and MRE11. NBS patients can be distinguished from AT patients at birth by their bird-like facial features and microcephaly. Later on, NBS patients develop mild to moderate mental retardation that is not present in AT patients. In addition, NBS patients do not show the ataxia or telangiectasia characteristic of AT. However, like AT patients, NBS patients are noticeably short in stature; have a decreased life span; and experience recurrent, severe sinopulmonary infections due to variable CID. Hypogammaglobulinemia is common, and antibody responses to antigen are defective. NBS cells subjected to ionizing radiation display the same spectrum of chromosomal rearrangements seen in irradiated cells from AT patients. NBS patients also have an increased risk of developing cancers, particularly B cell lymphomas.

c) Bloom Syndrome

Bloom syndrome (BS) is a rare autosomal recessive disorder characterized by genomic instability and CID. BS results from a mutation in the *BLM* gene, which encodes a helicase enzyme that resolves DNA replication forks that have stalled due to DNA damage. Spontaneous chromosomal anomalies appear in BS lymphoblasts and other cell types (**Plate 15-1**). BS patients often have low levels of IgM and so experience recurrent respiratory infections that can lead to chronic lung disease. BS patients are also short in stature and highly sensitive to the sun. Due to their genomic instability, BS patients have an increased risk of developing lymphomas or lymphoid leukemias

(A)
Normal Individual

(B)
Bloom Syndrome Patient

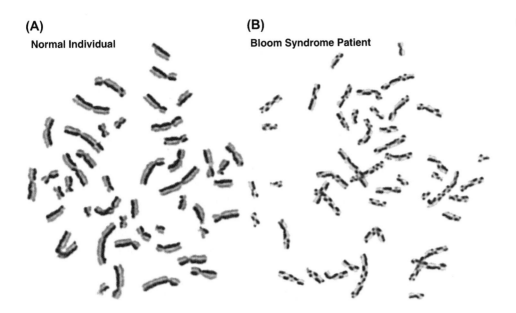

▪ **Plate 15-1**
*Bloom Syndrome Chromo-
somal Abnormalities*

Chromosomes are stained such that
genetic material from sister chro-
matids can be distinguished. Sister
chromatid exchange is rare in a
lymphocyte from a normal individual
(A) but occurs at high frequency in
a lymphocyte from a patient with
Bloom syndrome **(B)**. *[Courtesy of
the Cytogenetics Laboratory, The
Hospital for Sick Children, Toronto.]*

as young adults, and other solid tumors in later decades. Unfortunately, because of the nature of their mutations, BS patients are particularly vulnerable to the damaging side effects of cancer treatments based on radiation or chemotherapy.

d) Xeroderma Pigmentosum

Xeroderma pigmentosum (XP) is a rare disorder (1 in 250,000 live births) characterized by extreme sensitivity to the sun and a marked predisposition to skin cancer development. XP is caused by a mutation in any one of seven genes, *XPA–XPG*, involved in the same crucial DNA repair pathway. A variable defect in cell-mediated (but not humoral) responses occurs in some, but not all, XP patients. Abnormal NK-mediated natural cytotoxicity has also been reported, as has decreased IFNγ production. Cells from XP patients are highly sensitive to UV exposure *in vitro*.

vi) Lymphoproliferative Immunodeficiency Syndromes

Although it sounds counterintuitive, PIDs exist that combine unbridled lymphocyte proliferation with immunodeficiency. This phenomenon serves to demonstrate that when the various components of the immune system are not in normal balance, immune responses can become dysregulated such that normal control of cell proliferation is lost, leading to deleterious consequences for the patient. Two of these syndromes are outlined next, with more detail provided in Chapter 19.

a) Hemophagocytic Lymphohistiocytosis

The name "hemophagocytic lymphohistiocytosis" (HLH) suggests that this autosomal recessive disease is caused by a defect in innate immunity, but HLH is associated primarily with a defect in granule exocytosis by CTLs and NK cells. Patients with HLH exhibit sustained Th1 inflammatory responses that cause both T cells and macrophages to secrete copious amounts of cytokines, particularly IFNγ. Uncontrolled infiltration of activated CTLs into various organs (including the brain) causes immense tissue damage that leads to organ failure and death in the absence of immunosuppressive therapy. In addition, activated macrophages undertake excessive phagocytosis of erythrocytes, accounting for the first part of this disorder's name. However, B cell numbers and activation appear to be normal in HLH patients and there are no clinical signs of autoimmunity. Mutations in the *perforin*, *Munc13-4* and *syntaxin* genes, which all encode proteins involved in granule exocytosis, have been linked to HLH.

b) X-Linked Lymphoproliferation

X-linked lymphoproliferation-1 (XLP1) is a very curious disease of young boys in which the primary clinical manifestation is an inappropriate response to EBV infection. When XLP1 males are infected with EBV, they develop infectious mononucleosis that can be fatal due to severe liver necrosis. Excessive numbers of anti-EBV lymphocytes responding to the infection apparently contribute to the pathophysiology. Uncontrolled proliferation of B cells and CD8$^+$ T cell infiltration into the liver are observed. For unknown reasons, virtually no pathogens other than EBV appear to trigger XLP1 disease in susceptible patients. At the genetic level, the defect is often a deletion or point mutation of the *SH2D1A* gene, which encodes an adaptor protein called SAP. SAP is involved in intracellular signaling in T cells, NK cells and NKT cells, and NKT cells appear to be particularly hard hit by its loss. Because the dominant function of SAP is to support negative regulation of IFNγ secretion by T cells, XLP1 patients have high serum levels of IFNγ, and their immune responses to acute EBV infection are dramatically skewed to the Th1 type. The resulting high levels of Th1 cytokines are thought to contribute to the ensuing widespread tissue necrosis. If an XLP1 patient does not succumb to the EBV infection, he usually develops either a selective Ig deficiency or Burkitt's lymphoma. Death often occurs at a young age.

XLP2 is a disease that closely resembles XLP1 but is caused by mutations in the *BIRC4* gene encoding the "X-linked inhibitor of apoptosis" (XIAP) protein. Unlike SAP, XIAP is expressed in a wide range of cell types, so it is not clear why mutations of XIAP manifest primarily as abnormal immunity. Again, NKT cells are missing. XLP2 patients show the same rampant lymphocyte infiltration and high Th1 cytokine levels as XLP1 patients, but these symptoms do not depend on exposure to EBV.

Lastly, a PID that is similar in phenotype to XLP1 and XLP2 has been identified that is autosomal recessive rather than X-linked in inheritance. This disorder is caused by mutations of the *ITK* gene encoding a T cell tyrosine kinase.

> NOTE: There are three other lymphoproliferative syndromes called APECED (autoimmune polyendocrinopathy-candidiasis-ectodermal dystrophy syndrome), ALPS (autoimmune lymphoproliferative syndrome), and IPEX (immune dysregulation-polyendocrinopathy-enteropathy-X-linked). These congenital disorders are all due to single gene mutations (and so are PIDs) but mainly have autoimmune manifestations, and so are discussed in Chapter 19.

vii) Other Adaptive Immunodeficiencies

Two other adaptive PIDs that do not fit into the previous categories are detailed in the following sections.

a) DiGeorge Syndrome

DiGeorge syndrome is a complex disorder in which the thymus often does not develop fully or at all. Patients have characteristically abnormal facial features, including a long, narrow face; small mouth; and low-set, cupped ears. Cleft palate and cardiovascular and renal abnormalities are frequently present, and the patients suffer from recurrent infections. If the parathyroid gland is also affected, neonates can suffer fatal convulsions. In other cases, psychiatric disorders may emerge in later life. The disease is clinically very heterogeneous, and not all patients show all symptoms. Autosomal dominant, autosomal recessive, and X-linked modes of inheritance have been reported, but the precise molecular defect is unknown. The abnormality appears to affect a subset of embryonic cells that eventually develop into the thymus and parts of the face, among other structures. In most (but not all) DiGeorge syndrome patients, translocations or large deletions involving chromosome 22q11 are observed. A small number of patients have deletions of chromosome 10p13-14. With respect to immunodeficiency, the degree of the deficit is governed by how much thymic tissue is present; surprisingly little thymic tissue is sufficient for normal T cell immunity. Numbers of T cells in DiGeorge patients can range from almost nil to almost normal, while NK numbers and distribution are normal, and B cells are normal and may even be elevated in number. However, in patients who lack T cells, the reduced amount of T cell help available to B cells often compromises the humoral response. As a result, the first sign anything is amiss is usually a persistent oral fungus infection. Recurring severe infections typical of other CIDs soon follow.

b) Wiscott–Aldrich Syndrome

Wiscott–Aldrich syndrome (WAS) is an X-linked recessive, clinically heterogeneous disorder caused by mutations in the multidomain, multifunctional WAS protein (WASP). WASP is expressed constitutively by all hematopoietic cells but is particularly important for platelet and lymphocyte survival. In classical WAS patients, the mutated WASP is aberrantly expressed in lymphocytes and megakaryocytes and drives a characteristic constellation of clinical features: CID, eczema and thrombocytopenia. Both cell-mediated and humoral responses are compromised, so that patients are highly susceptible to viral, pyogenic, and opportunistic infections. Lymphomas, autoimmunity and allergy are also common in this group. T cell numbers steadily decline during early childhood, and those few cells that are left function abnormally. B cell numbers are essentially normal, but antibody responses to both protein and polysaccharide antigens are minimal. While serum IgG is normal, levels of IgM are reduced and those of IgA and IgE are increased. Numbers of monocytes and neutrophils in the blood of classical WAS patients are normal, but the chemotaxis of these cells *in vitro* is impaired. Non-classical WAS patients have only some of the preceding defects, with some showing platelet impairment without immunodeficiency, and others showing immunodeficiency without platelet impairment.

IV. Diagnosis and Treatment of Primary Immunodeficiencies

i) Diagnosis of PIDs

The initial diagnosis of PIDs is often determined by the age of onset, family history, symptoms, and nature of the recurrent infections experienced by the patient. For example, while PID patients with phagocytic defects or complement deficiencies usually suffer from infections with specific types of bacteria but not viruses, those with CVID or a selective Ig deficiency often show symptoms of infection with an enterovirus and/or a different range of bacterial species. Fungal infections are prevalent in PID patients with CVID or phagocyte defects, but not in those lacking Igs or complement. In some cases, specific physical attributes of the patient may provide clues. For example, those lacking tonsils and other secondary lymphoid tissues may have agammaglobulinemia, while the presence of ataxia may point to AT.

NOTE: The past 20 years have seen a huge increase in the number of PIDs recognized and a revolution in their diagnosis. Not only has the identification of the genetic defects leading to various PIDs exploded, but the recognition that sporadic severe infections, autoimmunity, chronic inflammation and malignancy may also point to an immunodeficiency disease has increased the complexity of PID diagnosis. Until recently, general practitioners had been advised to consider a diagnosis of PID only if a patient displayed one or more "warning signs" on a list that was dominated by a focus on recurrent infections with multiple and/or unusual organisms. Such infections might include fungal infections of the mouth or skin, bacterial infections of the ear or sinuses, pneumonia, abscesses, or septicemia. However, subsequent analyses have indicated that this approach is sufficient to diagnose only some PIDs. Indeed, just one severe infection with a common pathogen may be evidence of a PID, and ignoring the possibility simply because the infection is not recurrent could put an infant's life at risk. This scenario is particularly problematic with respect to PRR defects, which can be associated with life-threatening *S. aureus* or *S. pneumoniae* infections. In addition, the traditional list of warning signs does not include autoimmunity or malignancy, which may be the problem that brings the child to the clinic in the first place. For example, children with lymphoproliferative immunodeficiency disorders usually do not present with infections at all, at least initially. Similarly, patients with Omenn syndrome generally first come to medical attention because of severe inflammation of the gut and skin rather than infections, and it is lymphomas that drive ADA SCID and NBS patients to the hospital. Thus, recognition of the diverse presentation of PIDs is critical to implementing the appropriate treatment as quickly as possible.

An initial diagnosis of a particular PID is followed up with laboratory tests that confirm its immune hallmarks, such as neutropenia, or a combined lack of T and B cells, or a deficit in one or more serum antibody isotypes. When one is assessing these factors, it is important to compare patient samples with results obtained from normal age-matched controls. In addition, because most PID patients are infants or young children, it must be borne in mind when measuring serum antibody levels that the immunoglobulin proteins in an infant's circulation during the first 3 months of life are of maternal origin, acquired either *in utero* or via breastfeeding. These maternal antibodies may effectively mask any *bona fide* antibody deficiency suffered by the child. Moreover, it can take several years for even some healthy individuals to produce normal levels of some antibodies, particularly IgA.

Additional functional assays are used to test whether a patient's leukocytes can become activated after *in vitro* stimulation, or to analyze which cytokines they produce. The ability to mount B cell responses to Ti antigens can be tested by administering a polysaccharide vaccine and looking for specific antibody production. Similarly, responses to Td antigens can be tested by measuring titers of antibodies specific for the tetanus toxoid or diphtheria toxoid used in the DTaP vaccine (refer to Ch. 14). Patients with TLR defects can often be identified by treating their isolated blood leukocytes with a TLR ligand and measuring IL-6 and TNF production. Diagnosis of CGD can be confirmed by an *in vitro* fluorescent assay of NADPH oxidase activity in isolated phagocytes. A defect in the Fas-mediated apoptosis of a patient's cells *in vitro* suggests a diagnosis of ALPS, whereas a deficit in CTL- and/or NK-mediated cytotoxicity may be evidence of C-HS or HLH. Specific assays for the activities of ADA or PNP may reveal impaired function of these enzymes. LAD-I is confirmed when flow cytometric analysis of leukocytes reveals a lack of expression of CD18, LAD-II by a dearth of CD15, and IL-7Rα SCID by a deficit in surface IL-7R. Flow cytometry can also be helpful in detecting the absence of a specific cell subset, whereas karyotyping may reveal the presence of a 22q11 or 10p13-14 deletion associated with DiGeorge syndrome. A relatively new PCR-based technique allows the diagnosis of SCID by screening for by-products of V(D)J recombination in the TCR loci. These by-products are called *TCR excision circles* (TRECs), and a low number or absence of TRECs indicates a defect in the generation of mature T cells.

NOTE: Most pediatricians view SCID as a medical emergency. Even though it might take several weeks to firm up the diagnosis, a newborn at risk or suspected of having a form of SCID must be closely watched in the first few days and months of life to decrease his/her exposure to infection, institute prophylactic antibiotic treatment, and make sure no live, attenuated vaccines are administered. To decrease SCID-related tragedies, many states in the U.S. have implemented mass screening of newborns for SCID by looking for TRECs in dried blood spots. Such a program was started in Wisconsin in January 2008, and plans are in place to have screening programs established in all states over the next several years.

Much research continues to be devoted to identifying the mutated genes responsible for various PIDs. Knowing the genetic defect can allow a precise diagnosis, and thus a prediction of what the prognosis may be for the patient and the chances of disease for other family members. Genetic counseling of identified carriers can then be implemented, as appropriate. Prenatal diagnosis of a PID also becomes possible, allowing parents to make informed reproductive choices. However, one should bear in mind that not all changes to a particular DNA sequence are detrimental; some may in fact be benign polymorphisms, complicating this approach to diagnosis. In the future, when gene therapy (see later) becomes truly practical, knowing which alleles actually cause disease, and whether it will be helpful to replace them, will be of paramount importance.

ii) Treatment of PIDs

PIDs were first identified in the early 1950s, and the first treatment of PID patients by Ig replacement therapy (see later) was implemented in 1952. Attempts to reconstitute the immune systems of PID patients by bone marrow transplantation were begun in

1957, leading to more refined hematopoietic cell transplantation techniques that are now the standard of care for several PIDs. For some PIDs, the treatment can address only the symptoms of the disease, and is focused on appropriate prophylactic or therapeutic use of antiviral, -bacterial, or -fungal agents. Specific immunization against the organisms prominent in the disease may also be warranted, as may the boosting of the patient's nutrition. More innovative therapies are now attempting to target the root cause of a PID: the genetic defect or its immediate consequences. Sometimes, however, even when the molecular defect is known, the treatment of symptoms is still the only or most effective approach. Therapy for deficiencies of the terminal complement components fall into this category. In addition, understanding and treating the molecular defect do not always translate into clinical success. For example, in the case of hyper-IgE syndrome associated with impaired neutrophil recruitment, IFNγ treatment improved the *in vitro* chemotaxis of the neutrophils of several patients but did not bring clinical relief when applied *in vivo*. Antibiotics to combat *Staphylococcal* infections therefore remain the most effective treatment for hyper-IgE syndrome, with surgery to remove entrenched abscesses.

Next, we briefly discuss some conventional and innovative approaches to treating various PIDs.

a) Immunoglobulin Replacement

The deficit in Ig production associated with many PIDs can be treated on an ongoing basis with **intravenous immunoglobulin (IV-IG)** or **subcutaneous immunoglobulin (SC-IG) replacement therapy**. The IV-IG or SC-IG preparation most often used is derived from pooled plasma obtained from thousands of healthy blood donors and contains 10% IgG (of all specificities) in a form that reduces the chances of complement activation. Moreover, because this plasma contains significant concentrations of natural antibodies, antibody-deficient PID patients who receive IV-IG or SC-IG are well protected against common pathogens. SC-IG is sometimes more convenient for a patient (as when traveling) and generally has a lower rate of adverse reactions than IV-IG. It should be reiterated, however, that Ig replacement therapy generally has very little effect on those pathogens usually combatted by cell-mediated immunity, meaning that a PID patient may show a reduction in bacterial and fungal infections but still suffer from frequent virus infections.

> Recall that antibodies are IgM molecules which recognize frequently recurring PAMPs and are present in the blood even in the absence of overt infection or immunization.

Recent research has indicated that the benefits of IV-IG/SC-IG therapy may extend beyond mere antibody replacement because these antibody preparations also seem to have an effect on DCs. For example, DC maturation is impaired in XLA patients, but when these individuals are treated with IV-IG, their DCs become capable of completing maturation and can initiate T cell responses. It is thought that the natural antibodies present in IV-IG may stimulate DC maturation. Other studies have indicated that IV-IG also induces proliferation and Ig synthesis by the B cells of some CVID patients in a T cell-independent manner and reduces leukocyte overproduction of inflammatory cytokines.

> IV-IG therapy is a more general form of the passive immunization technique described in Chapter 14.

b) Enzyme Replacement

Some PIDs are due to an enzyme deficiency and can be treated by enzyme replacement therapy in which the patient is injected with a stabilized form of the missing protein. This strategy works only if enzyme function in the blood is sufficient to mitigate the disease, and if the protein does not actually have to enter the hematopoietic cells affected. Many ADA-SCID patients have been successfully treated using enzyme replacement therapy. Purified bovine ADA protein is covalently conjugated to polyethylene glycol (PEG) and injected intramuscularly 1–2 times per week on a continuing basis. The ADA–PEG conjugate is very effective in reducing levels of adenosine and deoxyadenosine in the serum (it does not enter cells) and is not toxic to the patient. Within 3–4 months of treatment, thymus development often resumes, and B and T cell numbers slowly increase. However, because ADA-PEG must be continuously supplied to avoid resumption of clinical signs, this therapy is extremely expensive. In addition, responsiveness to the drug has been observed to steadily decrease in some patients over the long term.

> PEG conjugation prolongs the half-life of a protein injected into the body and ensures that it cannot enter cells.

Enzyme replacement has also been used to treat human PNP deficiency. Injection of a PEG-conjugated form of purified PNP results in the processing of any inosine, deoxyinosine, guanosine, and deoxyguanosine released from cells into the blood, reducing the chance of uptake by lymphocytes. As is true for ADA–PEG, this method of therapy is safe and efficacious but very expensive.

c) Hematopoietic Stem Cell Transplantation

An HSCT is a refined type of bone marrow transplant discussed further in Ch. 17.

Many types of PID patients, including those with MHC deficiencies, or ALPS, WAS, XLP, IPEX, HLH or ADA–SCID, are now being cured with a hematopoietic stem cell transplant (HSCT) from an HLA-identical donor. (The HLA molecules expressed by the transplanted hematopoietic cells must be identical to those in the patient to prevent an attack on the foreign cells by the patient's immune system; see Ch. 17.) When successful, this procedure allows a patient to completely replace his/her immune system and subsequently lead a normal life. Indeed, prenatal diagnosis of many PIDs has made it possible to carry out an HSCT *in utero*, and the earlier the transplant is undertaken, the greater the chance of success. Up to 60–80% of PID patients treated with HSCT survive and go on to develop new T cells after 3–4 months. Indeed, for infants with SCID, HSCT using cells from an HLA-identical donor has a success rate of 95%, and the survival of older SCID patients after HSCT is still 70%. For other PID patients, if T cell function is only partially restored and B cell production and/or function is consequently still lacking, the deficit in Ig production can be treated with IV-IG or SC-IG therapy. It should be noted, however, that HSCT will not address non-immune symptoms associated with a PID. For example, the neurological symptoms of C-HS patients persist long after their immune systems are restored by HSCT.

d) Gene Therapy

Somatic gene therapy is applied only to somatic cells—that is, cells whose DNA is not passed on to the next generation.

If a PID is known to be caused by the mutation of a single gene, it may be possible to attempt *somatic gene therapy*, in which a normal copy of the faulty gene is introduced into the somatic cells of the patient. Single gene PID mutations generally affect a relatively discrete cell population, meaning that the therapy becomes practical to implement. Importantly, introduction of the normal gene has the potential to permanently cure the defect, especially if the cells receiving the new gene are long-lived and/or self-renewing (like hematopoietic progenitors).

Somatic gene therapy was first attempted with two ADA–SCID patients in 1990 (see **Box 15-4**). Since then, there have been several trials of gene therapy in which the patient's own HSCs are isolated and cultured in the lab, and a normal copy of the gene

---

Box 15-4   Gene Therapy for ADA Deficiency: The Ashanti DeSilva Story

Ashanti DeSilva lived to become a healthy adult due to the success of the first somatic gene therapy trial for a PID. Ashanti was diagnosed with ADA deficiency in 1988 at the age of 2 years and started enzyme replacement therapy with PEG-ADA right away. This treatment was successful at first, in that it increased Ashanti's T cell numbers and allowed her to resist some infections. However, by the time she was 4 years old, her immune system was not responding as well to the drug, her resistance to infection was down, and she was becoming very weak. Moreover, the option of bone marrow transplantation was ruled out due to the lack of a suitable donor. In an act of desperation, Ashanti's parents allowed her to be treated experimentally with a retroviral vector expressing the normal ADA transgene. The ADA-retroviral vector was introduced into her isolated peripheral blood lymphocytes, and the modified cells were injected back into her circulatory system. The hope was that the vector would integrate the normal ADA transgene into her leukocytes, that the transgene would be expressed to generate the normal ADA enzyme, and that the activity of this enzyme would reduce her excessive blood adenosine and deoxyadenosine concentrations. Unlike a bone marrow transplant from a donor, the cells would be Ashanti's own and should not be rejected. As a result of this ground-breaking therapy, Ashanti's T cell counts became normal after 6 months, and her cell-mediated and humoral immune responses improved steadily for 2 years despite the fact that only 20–25% of her leukocytes contained the corrective transgene. As a precaution, Ashanti continued on a low dose of PEG-ADA, but the dose did not have to be increased as she aged. She was able to attend school, undertake many of the activities of a normal childhood, and grew into an otherwise healthy, active, independent adult. The dramatic improvement in Ashanti's condition following her gene therapy was the first real confirmation of the feasibility of this approach to cure human PIDs.

is introduced into these cells via a retroviral vector capable of integrating itself into a host cell genome. Similar attempts to cure other forms of SCID (PNP, Jak3, X-linked SCID) as well as CGD and WAS have been made, and success in the form of significant clinical benefit has been achieved in a sizeable number of cases. However, although this strategy of *transgene* expression works quite well, early efforts revealed the dangers of using unmodified retroviral vectors, in that several PID patients developed leukemia due to the insertion of the vector close to a *proto-oncogene*. Current research into somatic gene therapy is focused on finding vectors that still mediate expression of the normal transgene in the patient's hematopoietic cells and cure the PID but do not promote *insertional mutagenesis*. Vectors that are self-inactivating or non-integrating are under investigation, as are vectors that contain sequences designed to protect from mutation genes neighboring the genomic integration site. At an even more basic level, researchers are seeking ways to correct the mutated gene itself without having to introduce a vector-based transgene. Custom-designed nucleases can be created that target very specific sequences within a host genome, nipping out the mutation and allowing DNA recombination mechanisms within the cell to copy a normal donor DNA template and fix the gene permanently. This approach has minimal risks of oncogenesis but at the moment is very inefficient and has some unexplained side effects.

> A transgene is a gene from one organism that has been transferred either naturally or artificially into another organism. A proto-oncogene is a normal gene that can become a cancer-causing oncogene due to mutation or abnormally increased expression.

> NOTE: After the success of Ashanti DeSilva's somatic gene therapy for a PID, a flurry of trials was launched for a variety of genetic disorders. Sadly, harsh lessons were learned about the need for improved safety of vectors and treatment protocols. In 1999, Jesse Gelsinger was an 18-year-old patient with a relatively mild metabolic disorder caused by the mutation of a particular enzyme needed to metabolize nitrogen. When his liver was infused with an adenovirus-based vector expressing the enzyme transgene, his immune system mounted a severe inflammatory response that quickly led to lung failure, then multiple organ failure, and then death. Investigation of this tragedy revealed that Jesse's liver status on the day he received the adenoviral vector should have precluded its administration, a clear violation of the trial protocol. As a result, multiple scientists were suspended from clinical research for several years. The inquiry into this case also revealed that many somatic gene therapy trials were under-reporting adverse events. Stricter protocols for risk assessment and disclosure in somatic gene therapy trials are now in place.

B. HIV Infection and Acquired Immunodeficiency Syndrome (AIDS)

As described in Part A of this chapter, primary immunodeficiencies result from genetic defects of the immune system that are congenital. In contrast, secondary immunodeficiencies are caused by non-genetic, external events and so can be considered to be "acquired." Part B of this chapter focuses on the most infamous form of secondary immunodeficiency, namely acquired immunodeficiency syndrome (AIDS) caused by infection with HIV. The sociologic and economic aspects of the AIDS epidemic are beyond the scope of this book and have been well examined elsewhere. Instead, we concentrate on the biology of HIV and how it destroys the adaptive response to cause AIDS. The reader should note that, in addition to HIV infection, there are several other important causes of secondary immunodeficiency, as briefly summarized in **Box 15-5**.

I. Overview of HIV/AIDS

i) Impact of AIDS

AIDS first became known to the medical community in the early 1980s, when rare tumors and cases of an unusual pneumonia appeared in ostensibly healthy young men. After several years of intensive research during which the rapid spread of AIDS made it a lethal worldwide epidemic, HIV was identified as the causative agent of this scourge.

Box 15-5   Non-HIV Causes of Secondary Immunodeficiency

Severe metabolic disturbances, trauma, infection with certain viruses other than HIV, diseases like diabetes can all cause secondary immunodeficiency. Immunosuppressive drugs taken to treat conditions such as transplant rejection, hypersensitivity, autoimmune disease, or cancer can also depress the immune system. Weakened immune responses are naturally a function of being very young in age or very elderly. Perhaps surprisingly, the most common cause of secondary immunodeficiency is malnutrition, which generally impairs the body's ability to mount innate and adaptive responses (see the table below). Interestingly, the lack of a particular nutrient can have a very specific effect on immunity. For example, recent work has revealed that vitamin C is needed to strengthen mucosal barriers, while vitamin D bolsters the ability of macrophages to combat intracellular pathogens. Patients with secondary immunodeficiency no matter how caused are left vulnerable to opportunistic bacterial and viral infections that can be lethal if not promptly treated with proper nutrition and antibiotics or antiviral drugs.

TABLE BOX 15-5   Some Non-HIV Causes of Secondary Immunodeficiency

Being very young:
- Decreased numbers of neutrophils, NK cells, splenic B cells and limited functionality of neutrophils
- Reduced TLR signaling, complement components, cytokines
- Immature secondary lymphoid organs and MALT
- Reduced B cell costimulatory molecule expression and low levels of immunological memory

Being very old:
- Decreased generation of macrophages and neutrophils
- Reduced cell-mediated immunity
- Impaired proliferation of stimulated lymphocytes and reduced capacity to respond to new antigens

Malnutrition:
- Impaired function of mucosal barriers and macrophages
- Reduced cell-mediated immunity

Metabolic disruption:
- Reduced chemotaxis and impaired phagocytosis
- Decreased lymphocyte proliferation and impaired memory B cell responses
- Increased T cell anergy

Immunosuppressive agents:
- Impaired mucosal barrier function
- Decreased pro-inflammatory cytokine production, reduced chemotaxis and impaired phagocytosis
- Reduced numbers of lymphocytes and neutrophils

Trauma or surgery:
- Loss of local mucosal barrier function
- Impaired T cell responses

UV or cosmic radiation, low oxygen, other detrimental environmental conditions:
- Decreased numbers of lymphocytes
- Impaired antibody responses
- Reduced cell-mediated immunity

Although much has been learned about the biology of HIV over the past three decades, this complex and wily virus still frustrates the best efforts of researchers to produce an effective vaccine. Progress has been made in recent years in reducing the number of new HIV infections by educating the public on how to avoid acquiring and transmitting the virus. In addition, the advent of effective antiretroviral drugs has rendered AIDS a chronic disease in countries where these drugs can easily be obtained. However, while new infections fell from 4.3 million in 2006 to 2.7 million in 2010, 1.8 million deaths due to AIDS were still reported in that year and 34 million people were living with HIV infection (**Table 15-4**). Approximately equal numbers of men and women are infected, and, due to mother-to-child transmission during pregnancy, childbirth or breastfeeding, about 10% of all victims are children under 15 years of age. Sadly, half of all AIDS victims are now women and children. The grip of AIDS is much worse in lower-income countries

TABLE 15-4	The Global HIV/AIDS Epidemic		
Region	**Total Number of Persons Living with HIV/AIDS in 2010**	**New HIV Infections during 2010**	**Deaths due to AIDS in 2010**
Sub-Saharan Africa	22,900,000	1,900,000	1,200,000
South & Southeast Asia	4,000,000	270,000	250,000
Latin America	1,500,000	100,000	67,000
Eastern Europe & Central Asia	1,500,000	160,000	90,000
North America	1,300,000	58,000	20,000
Western & Central Europe	840,000	30,000	9,900
East Asia	790,000	88,000	56,000
Middle East & North Africa	470,000	59,000	35,000
Caribbean	200,000	12,000	9,000
Oceania	54,000	3,300	1,600
Total	**34.0 million**	**2.7 million**	**1.8 million**

With data from: UNAIDS World AIDS Day Report, 2011 (http://www.unaids.org/en/media/unaids/contentassets/documents/unaidspublication/2011/JC2216_WorldAIDSday_report_2011_en.pdf)

where public education campaigns to control transmission can be hampered politically and where a lack of regular, affordable access to antiretroviral drugs makes the disease more likely to be acute rather than chronic. To make matters worse, strains of HIV have recently been identified that are resistant to multiple antiretroviral drugs. Thus, although battles against HIV are being won, the war against AIDS is far from over.

> Of all persons currently infected with HIV, 60% live in sub-Saharan Africa, whereas <5% live in North America.

ii) Types of HIV

There are two known human immunodeficiency viruses: HIV-1 and HIV-2. HIV-1 can be found anywhere in the world but is most prevalent in the Western hemisphere. HIV-2 occurs almost exclusively in Western Africa. Both HIV-1 and HIV-2 are retroviruses belonging to the lentivirus class and are *cytopathic*, meaning that they usually kill the cells they infect. A typical lentivirus persists in its host despite the humoral and cell-mediated immune responses directed against it, and clinical disease is detected only after a long latency period. Although both HIV-1 and HIV-2 also infect macrophages and DCs, it is the destruction of CD4+ T cells by these viruses that causes AIDS. Because it has been more extensively studied, the remainder of this chapter focuses on the biology of HIV-1.

iii) HIV Transmission

HIV is transmitted primarily via the transfer of body fluids. An HIV virion in such fluids most often accesses another individual's body by breaching the mucosae during sexual contact or via intravenous drug injection involving contaminated needles. As mentioned previously, HIV can also be transmitted from mother to child before or during birth or by breastfeeding. Merely touching, kissing or being sneezed on by an HIV-infected person is not enough to acquire the virus. HIV infection is easily prevented if elementary precautions are taken, such as the use of condoms, gloves, screened or cloned blood products, and clean needles and syringes. With respect to mother-to-child HIV transmission, treatment of the mother with antiretroviral drugs during pregnancy, delivery and after the birth can substantially reduce viremia in the mother and thus the risk of transmission to the child. The infant can also be given antiretroviral drugs for the first 6 weeks of life. Unfortunately, many women and infants who could benefit from these drugs do not have access to them.

iv) HIV Life Cycle

An illustration and description of the life cycle of an HIV virion following its infection of a CD4+ T cell is shown in **Figure 15-3**. The end of the cycle is marked by the budding

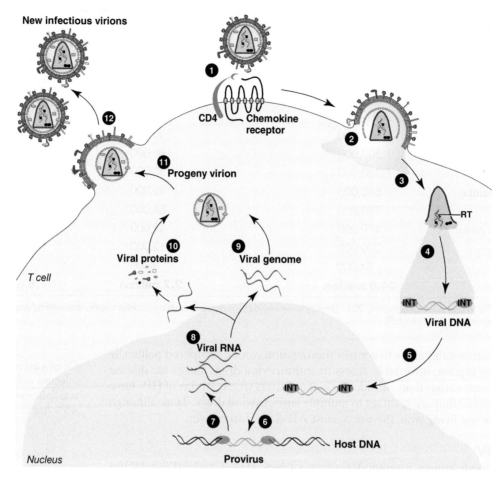

Fig. 15-3
HIV Life Cycle

(**1**) The HIV virion uses its gp120 protein (see **Fig. 15-4**) to bind to CD4 and a chemokine receptor (either CCR5 or CXCR4) on the surface of a host T cell. (**2**) The viral envelope fuses to the host cell membrane, and the capsid is everted into the cytoplasm, where it is uncoated by host cell proteases (**3**). The viral enzyme *reverse transcriptase (RT)* commences reverse transcription of the viral RNA genome into a DNA copy that becomes associated with a viral *integrase (INT)* enzyme (**4**). The viral DNA-INT complex is imported into the cell nucleus (**5**), where INT mediates insertion of the viral DNA into the host cell genome (**6**). This integrated viral DNA is called proviral DNA or the *provirus*. The provirus may remain untranscribed in the host cell genome for a considerable time without disturbing the infected cell. However, when the cell is activated, transcription of the provirus is triggered (**7**), and new copies of the viral RNA are produced and exported out of the nucleus (**8**). Some viral RNA molecules become the genomic RNA of new progeny viruses (**9**), while others are translated to generate the viral proteins necessary to associate with the new genomes and assemble the new progeny virions (**10**). The progeny virions bud through the host cell membrane (**11**), acquiring an envelope containing not only newly synthesized viral envelope proteins but also host cell surface proteins (**12**). The progeny virions can then proceed to infect fresh host T cells.

of newly formed progeny virions through the T cell membrane (**Plate 15-2**). The infectious progeny then spread through the body in search of fresh host cells to infect, and the CD4+ T cell that has supported the generation of these virions dies. It is the resulting systemic loss of CD4+ T cells that leaves the victim vulnerable to fatal opportunistic infections and tumorigenesis.

v) HIV-1 Structure and Genome

The HIV-1 virion has a multilayered structure consisting of the *envelope*, the *matrix* and the *capsid*. The viral genome is contained within the capsid at the center of the virion. Like all retroviruses, the HIV genome contains the three principal genes *gag*, *pol* and *env*. These genes encode three HIV polyprotein precursors, called Gag, Pol and Env, from which the major structural and functional HIV proteins are derived. The genome also contains *tat* and *rev* genes that encode replication regulatory factors common to all

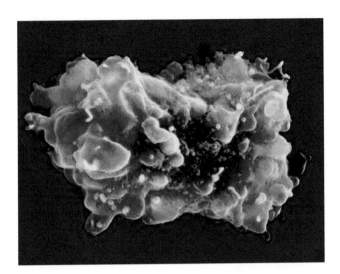

Budding HIV virions appear light blue. *[Courtesy of Boehringer Ingelheim Pharma KG, photo Lennart Nilsson, Albert Bonniers Förlag AB.]*

lentiviruses. In addition, HIV has four small genes that encode the accessory proteins Vif, Vpr, Vpu and Nef. The basic structure of the HIV virion is illustrated and further described in **Figure 15-4**. The HIV proteins and the genes from which they are derived are summarized in **Table 15-5**.

II. Molecular and Clinical Events of HIV Infection

i) Viral Tropism

The *tropism* of a virus refers to the range of host cells the virus can infect. Although the predominant target of HIV is the CD4$^+$ T cell population, the virus can attack a wide range of other cell types that express CD4, including macrophages, LCs and dermal DCs. However, unlike HIV-infected T cells, HIV-infected macrophages and DCs are quite resistant to the cytopathic effects of HIV and are not killed in great numbers. Nevertheless, the ability of these cells to carry out phagocytosis, chemotaxis and cytokine secretion is compromised by HIV infection. Infected and ineffective macrophages and DCs thus represent a significant reservoir within the host where HIV can safely reproduce.

The entry of HIV into cells requires not only that the virus bind to CD4 but also that it bind to either the CCR5 or CXCR4 chemokine receptor. CCR5 is expressed by CD4$^+$ T cells, DCs and macrophages, whereas CXCR4 is expressed by CD4$^+$ T cells and DCs but not by macrophages. Strains of HIV that bind to CCR5 only are known as **R5 viruses** and predominantly infect memory T cells and macrophages (and low numbers of DCs). Strains of HIV that bind to CXCR4 only are known as **X4 viruses** and predominantly infect resting T cells (and low numbers of DCs). HIV strains that can bind to both CCR5 and CXCR4 are called R5X4 strains. In 90% of HIV patients, the virions generated during the early stages of infection are exclusively R5 in character, a property that facilitates efficient entry of the virus at body portals because cells in these anatomical regions typically express CCR5. In particular, uptake of the virus by CCR5$^+$ memory T cells in these locations appears to be critical for establishing an infection. In some individuals, as the infection progresses and mutations accumulate in the viral genome, the original R5 strain gives rise to an R5X4 or X4 strain. Because X4 viruses usually replicate more rapidly than R5 viruses, the development of AIDS is accelerated during the later stages of the infection.

ii) HIV Entry and Replication

When the HIV envelope protein gp120 binds to CD4, the conformation of gp120 is altered such that a region capable of binding to chemokine receptors is exposed. The

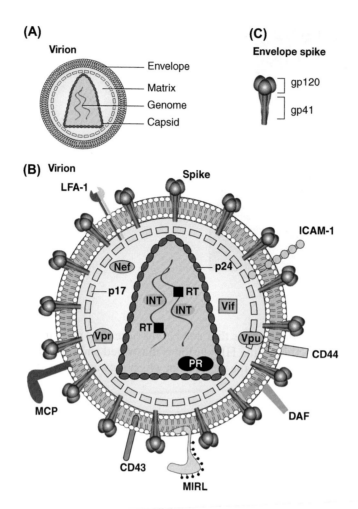

Fig. 15-4
Structure of the HIV Virion

(**A**) The HIV virion has three basic layers: the envelope, matrix and capsid. The spherical matrix surrounds the cone-shaped capsid and also supports the phospholipid bilayer constituting the viral envelope. The capsid protects the two single strands of RNA that make up the HIV genome. (**B**) The matrix is composed primarily of the viral structural protein p17, which is anchored to the viral envelope and creates the scaffolding around which the envelope is wrapped. Embedded in the envelope are various viral proteins as well as host cell proteins such as the adhesion molecules ICAM-1, CD44, CD43 and LFA-1, and the RCA proteins DAF, MIRL and MCP. The most prominent viral envelope protein is the "spike" made up of the gp41 and gp120 glycoproteins (see panel C). Within the matrix below the envelope are the viral accessory proteins Vif, Vpr, Vpu and Nef. The matrix surrounds the capsid, which is composed of ~1200 molecules of the viral structural protein p24. The genomic viral RNA molecules within the capsid are associated with viral reverse transcriptase (RT) and integrase (INT), which are required for provirus generation and integration into the host cell genome. A viral protease (PR) within the capsid cleaves the Gag polyprotein into p17 and p24, and the Pol polyprotein into RT and INT. (**C**) The gp41 and gp120 glycoproteins are produced by the action of a host protease on the viral Env polyprotein. This cleavage leaves the membrane-bound gp41 protein non-covalently associated with the non-membrane-bound, globular gp120 protein. Three gp41–gp120 units come together to form each spike. It is the carbohydrate-rich gp120 component that binds to host cell CD4 and CCR5.

interaction of the CD4–gp120 complex with CCR5 or CXCR4 further adjusts the conformation of the entire envelope spike such that gp41 is brought into contact with the host cell membrane. Gp41 inserts into the host cell membrane and promotes its fusion to the viral envelope, allowing the virus to efficiently transfer into the cell's interior. HIV can also gain access to T cells by making use of a DC that has captured the virus and sequestered it in a specialized intracellular vacuole. If the HIV-loaded DC makes contact with a CD4+ T cell, an intercellular interface called an *infectious synapse* is formed that facilitates the rapid transfer of the virus from the DC into the T cell. Once an HIV virion has entered a T cell, the genomic viral RNA undergoes reverse transcription to create the DNA **provirus**. After the provirus integrates into the host cell DNA, the virus is said to be in its **latent form**. Infected T cells in this *preactivation stage* do not transcribe the provirus to any meaningful extent and very few progeny

TABLE 15-5	HIV Genes and Major Proteins		
Gene	**Polyprotein**	**Protein**	**Function**
Structural/Enzymatic			
gag	Gag	p17 matrix protein	Acts as scaffolding for HIV envelope
		p24 capsid protein	Protects RNA genome
pol	Pol	Reverse transcriptase (RT)	Transcribes viral RNA into viral DNA
		Integrase (INT)	Inserts viral DNA into host DNA to form provirus
		Protease (PR)	Cleaves Gag and Pol polyproteins into smaller functional proteins
env	Env	gp41 transmembrane envelope protein	Part of envelope spike that promotes fusion with host cell membrane
		gp120 non-transmembrane envelope protein	Part of envelope spike that binds to host cell receptors to initiate viral entry
Regulatory			
tat	–	Tat	Sustains host cell transcription of proviral DNA
rev	–	Rev	Promotes viral mRNA transport to host cytoplasm and translation
Accessory			
vif	–	Vif	Promotes viral cDNA synthesis by inhibiting a host antiviral protein
vpr	–	Vpr	Facilitates transport of viral DNA into host cell nucleus for integration as provirus; induces cell cycle arrest; required for expression of non-integrated viral DNA
vpu	–	Vpu	May promote progeny virion budding
nef	–	Nef	Acts in several ways to promote host cell survival and activation that support viral DNA synthesis Forces internalization and degradation of MHC class I and MHC class II

virions are formed. However, if the T cell is stimulated by either TCR engagement or cytokine binding, intracellular signaling is triggered that initiates new transcription within the host cell. Along with various host genes, the HIV regulatory genes start to be expressed. Expression of the HIV genes encoding structural and accessory proteins as well as enzymes then commences, followed by progeny virion production.

iii) HIV Antigenic Variation

The reverse transcriptase (RT) enzyme that makes the proviral DNA is highly error-prone, meaning that it introduces mutations at a relatively high rate as it copies the RNA genome. This high mutation rate, combined with the rapid replication of the virus, provides plenty of DNA synthesis cycles and thus numerous chances for mutations to be introduced. HIV infections are therefore characterized by extreme antigenic variation. Those virions whose antigenic variation allows them to evade neutralizing antibodies and antigen-specific CTLs are called **genetic escape mutants**. Although a single HIV clone initiates a new infection, distinct HIV isolates with a range of genomic RNA sequences can soon be found within an individual patient.

Several broad groups or *clades* of antigenically distinct types of HIV-1 have been defined based on nucleotide differences in the *env* and *gag* genes. At the protein level, the most variable HIV molecule is gp120. At the clinical level, some minor differences in transmissibility and disease progression have been noted among HIV-1 clades, but all cause premature death in the absence of antiretroviral drug treatment. The ability of viruses of different clades to recombine if they co-infect the same individual can suddenly generate a new viral genome with sequence characteristics of both parental viruses, further expanding antigenic variation.

A clade is a group of related organisms descended from a common ancestor.

iv) Progression of HIV Infection

HIV is most often introduced into the body by sexual contact, and the rectal and vaginal mucosae are particularly vulnerable to HIV attack. In the earliest, asymptomatic stages of a primary HIV infection, the virus attaches to DCs and infects macrophages and CD4+ T cells resident in the rectal or vaginal lamina propria. In addition to spreading the virus locally, these virus-bearing DCs and infected T cells carry the virus to resting CD4+ T cells in the local lymph nodes. Within these lymph nodes, progeny virions that have attracted the attention of complement components bind to CRs expressed on the surfaces of FDCs in the GCs and remain "trapped" on the exterior surfaces of these cells. Macrophages and T cells that come into contact with these FDCs then may also become infected. The infected cells release vast numbers of fresh virions that spread to uninfected T cells in lymphoid and non-lymphoid tissues, so that viral replication proceeds at an exponential rate.

From a clinical point of view, within 2–6 weeks of exposure, an HIV-infected individual may experience acute fever and an illness similar to infectious mononucleosis (**Fig. 15-5**). The number of virions present in the circulation (the *viral load*) is high, and significant levels of p24 can be detected in the blood. It is during this time that the individual, who may not suspect HIV infection at all, is most contagious.

At the cellular level, macrophages and resting T cells are infected on a massive scale during the early weeks of infection, and proviruses integrate into the genomes

Fig. 15-5
Clinical and Immunological Course of HIV Infection

(**A**) Overview of clinical course of HIV infection. (**B–D**) These three graphs show the progression of several clinical and immunological parameters over a typical time course of 1 month to 11 years. (**B**) Relative levels of p24 viral protein and virus particles in the blood (viremia). (**C**) Relative levels of antiviral IgM and IgG antibodies directed against p24 and gp120. (**D**) Relative levels of CD8+ CTLs and CD4+ T cells.

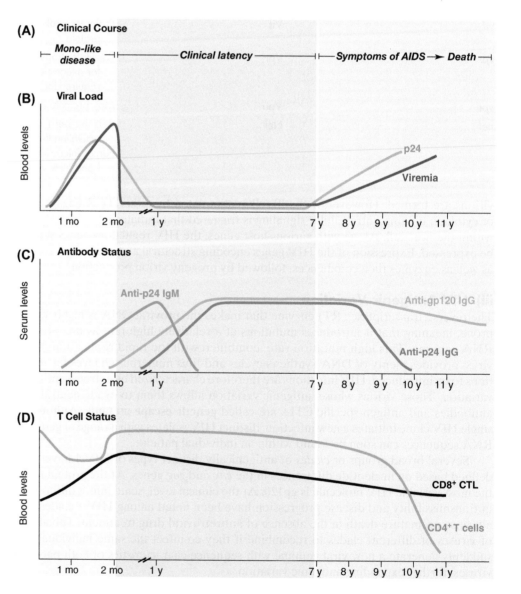

of these cells. The rate of viral replication slows as the patient enters the next phase of infection. During this phase, whenever an infected CD4$^+$ T cell or macrophage is stimulated, the latent provirus is transcribed and translated, and progeny virions are produced and released systemically. This ongoing viral replication occurs at a low enough rate that the innate and adaptive responses are able to control the virus, but they still cannot clear it from the body. Studies at the molecular level have revealed that, instead, the anti-HIV CTLs and neutralizing anti-HIV antibodies generated at this stage exert selection pressure on the virus to mutate. The result is the generation of genetic escape mutants that display extensive antigenic variation and stay one step ahead of the immune system. Clinically, the patient is not overtly ill, and high viral loads are not seen in his/her blood. CD4$^+$ T cell numbers hold relatively steady during this time due to the tremendous power of the bone marrow to produce new hematopoietic cells. Thus, although the virus is active during this phase, the patient experiences a period of "clinical latency" that may extend for many years (refer to **Fig. 15-5**).

As long as an HIV patient's **T4 count** remains above 200 CD4$^+$ T cells/mm^3 blood (about 25% of normal), the infected individual continues to show no clinical signs of illness. Eventually, however, there is a collapse in the CD4$^+$ T cell population followed by a drop in anti-HIV antibody titers and a decline in the number of anti-HIV CTLs (refer to **Fig. 15-5**). Viral p24 reappears in the blood, and viremia increases sharply due to the production of ~10^{10} virions/day! As the disease approaches its late stages, lymph node architecture degenerates, and the virus that was trapped on FDC surfaces is freed. The viral load continues to mount as the T4 count falls to below 200 CD4$^+$ T cells/mm^3 blood. The destruction of CD4$^+$ T cells, FDCs and the lymph node microenvironment drastically compromises the patient's ability to mount adaptive immune responses.

Clinical latency ends and symptoms of AIDS appear, which can range from mild to severe opportunistic infections and/or the development of malignancies like *Kaposi's sarcoma* (**Table 15-6**). Kaposi's sarcoma is an unusual tumor of connective tissues (**Plate 15-3**) and is caused not by HIV but by *Kaposi sarcoma herpesvirus*, a virus that gains a foothold only after HIV destroys the bulk of the CD4$^+$ T cell population. Two other frequent features of very ill AIDS patients are **cachexia** (wasting syndrome) and pneumonia caused by infection with the fungal pathogen *Pneumocystis (carinii) jiroveci*. When a patient's T4 count drops to <50 CD4$^+$ T cells/mm^3, death is imminent in the absence of antiretroviral drug therapy.

A normal T4 count is 800 CD4$^+$ T cells/mm^3 of blood.

Recall that *Pneumocystis jiroveci* was originally known as *Pneumocystis carinii*. Both terms appear in the research literature.

NOTE: HIV's ability to evade complete immunological clearance is partly due to its capacity to spread widely and establish latency soon after the onset of infection. Recent research has indicated that this process is facilitated by the existence of the two pools of CD4$^+$ T memory cells described in Chapter 9: the abundant Tem cells that patrol the peripheral tissues such as the mucosae and can flock to sites of inflammation, and the smaller pools of Tcm cells that are confined to the secondary lymphoid tissues such as lymph nodes. The HIV virions that first enter the body via the mucosae encounter a large population of CD4$^+$ Tem cells expressing CCR5. Since the virus initially has an R5 character with respect to tropism, these cells are readily infected and killed off. However, Tem cell numbers are continually replenished by the differentiation of Tcm cells in the secondary lymphoid organs. In contrast to Tem cells, Tcm cells express mainly CXCR4 and only very low levels of CCR5, and so are initially protected from widespread HIV infection. Newly differentiated Tem cells derived from these proliferating Tcm cells are recruited to sites of HIV infection, providing a fresh supply of uninfected host cell targets for progeny virions to attack. As a result, the virus is able to sustain a very high level of replication and spreads quickly to large numbers of uninfected Tem cells. Moreover, because Tem cells are relatively long-lived, once they are infected, they have the potential to harbor latent HIV for extended periods and thus act as a long-term viral reservoir. Within this reservoir, the virus slowly but steadily continues to acquire new mutations that change its character from R5 to R5X4 or X4. Tcm cells in the secondary lymphoid tissues then become vulnerable to infection and are also destroyed, contributing to the complete breakdown in infection control that marks the onset of AIDS.

TABLE 15-6	Common Opportunistic Infections and Neoplasms in AIDS Patients	
Type of Infection/Disease	**Causative Agent**	**Tissue Affected**
Candida albicans	Yeast	Oral mucosae, esophagus, lungs, bloodstream
Cryptococcus neoformans	Yeast	Central nervous system
Cryptosporidium species	Protozoan	Intestines
Cytomegalovirus	Herpesvirus	Retina, intestines, and others
Herpes simplex virus I	Herpesvirus	Brain, esophagus, lungs
Mycobacterium avium complex	Bacterium	Bloodstream and others
Mycobacterium tuberculosis	Bacterium	Lungs and others
Pneumocystis jiroveci (originally Pneumocystis carinii)	Fungus	Lungs
Progressive multifocal leukoencephalopathy	JC polyomavirus	Brain
Salmonella species	Bacterium	Bloodstream
Toxoplasma gondii	Protozoan	Brain
Type of Neoplasm	**Causative Agent**	**Tissue Affected**
Kaposi sarcoma	Kaposi sarcoma herpes virus (KSHV)	Skin, mucosae
Cervical cancer	Human papillomavirus (HPV)	Cervix and surrounding tissues
B cell lymphoma	Epstein–Barr virus (EBV)	Brain

Plate 15-3
Kaposi's Sarcoma
[Reproduced by permission of the CDC/Steve Kraus.]

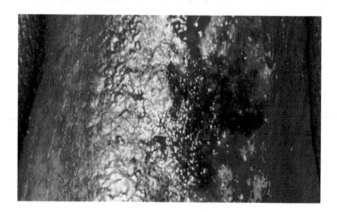

v) Resistance to HIV Infection and Delayed Disease Progression

Although rare, some groups of individuals appear to be able to either resist HIV infection when directly exposed to the virus, or retain their health for a lengthy or indefinite period after infection has occurred.

The best known group that is truly resistant to HIV infection (the virus cannot enter host cells at all) is composed of persons carrying a mutation in the host CCR5 gene required for viral entry. A mutated allele of CCR5 called CCR5Δ32 has a 32-base pair deletion in the receptor nucleotide sequence that renders it non-functional. The vast majority of individuals homozygous for CCR5Δ32 do not appear to be susceptible to HIV-1 infection, do not make detectable levels of anti-HIV antibodies, and indeed suffer no health deficits at all. Furthermore, many CCR5/CCR5Δ32 heterozygotes who become infected with HIV experience at least a 2-year delay in the onset of AIDS.

Other rare individuals suffer HIV infection (the virus can enter host cells) but are able to control it such that they are resistant to disease progression, viremia and AIDS. These individuals are variously known as "long-term non-progressors" (LTNPs) or "chronic suppressors," among other similar terms. LNTPs are identified by their long-term maintenance of a stable T4 count, a variable but generally low level of viremia, and a significant delay in the onset of full-blown AIDS in the absence of any antiretroviral drug treatment. It is currently estimated that 2–15% of HIV-positive individuals are LTNPs. In terms of immune responses, some LTNPs appear to retain large populations of HIV-specific CTLs and/or HIV-specific Th cells capable of unusually vigorous Th1 responses, while others produce high levels of neutralizing antibodies. Some LTNPs who have been infected for 2–3 years also generate "broadly neutralizing" antibodies that can target a larger range of HIV variants than the neutralizing antibodies found in other HIV-infected individuals. These additional antibodies in LNTPs help to foil HIV's attempts at genetic escape.

About 5–10% of LTNPs form a small subset of patients that are variously called "elite controllers," "elite suppressors," "HIV controllers" or "aviremic controllers." These individuals, who comprise less than 1% of all HIV-positive individuals, show no viremia after infection and are able to maintain this state independently of any drug treatment, at least for a while. T4 counts are typically high and relatively stable, and HIV infection generally has little impact on the health of an elite controller for an extended period. However, for reasons that are not yet clear, some of these patients eventually lose control over the virus and experience some progression of the disease, although the complete onset of AIDS is still rare. It is believed that elite controllers express certain protective MHC class I alleles that present peptides derived from regions of HIV proteins (especially Gag) which are conserved by the virus because they are required to maintain a normal level of replication. This antigen presentation facilitates the early activation of CD8$^+$ T cells able to kill HIV-infected host cells, containing viral spread and limiting viral variants to those that replicate only slowly. This reduced viral burden allows the immune system to "get ahead" of the virus and establish control of the infection.

III. Immune Responses During HIV Infection

Although HIV is not a highly infectious pathogen, it has been phenomenally successful. Part of this success stems from the fact that HIV spreads quickly, undergoes extensive antigenic variation, and establishes latency in large numbers of host cells. We have already outlined some aspects of the antibody and CTL responses to HIV that are commonly tracked as indicators of clinical progression, and mentioned how these responses exert ongoing selection of genetic escape mutants which enjoy a survival advantage. As a result of this continual push/pull between the virus and the adaptive response, whenever the virus comes out of latency, the immune system tends to be one step behind in controlling it. We will now examine in more detail the effects of HIV on the innate and adaptive leukocytes tasked with mounting the body's defense against HIV, as well as several sophisticated mechanisms employed by this virus that permit it to both evade host immune responses and destroy hematopoietic cells.

i) γδ T cells

Our increasing understanding of the major role of innate immunity in protecting us against pathogen assaults has sparked intense research into the activities of innate leukocytes during HIV infection. In particular, γδ T cells have come under scrutiny, since these cells are ideally positioned at the mucosae, precisely where HIV first attacks. As described in Chapter 11, γδ T cells in healthy individuals can respond immediately to non-peptide antigens, without the need for antigen presentation or the clonal expansion needed for an adaptive response. The γδ Th effectors produced secrete antimicrobial factors as well as chemokines and cytokines. Recent work has suggested that a chemokine released by γδ T cells competes with HIV for binding to CCR5, perhaps forestalling infection for a time. Other cytokines produced by γδ T cells induce inflammation and activate additional T cell subsets, particularly mucosal memory αβ Th17 cells and HIV-specific CD8$^+$ αβ CTLs. Concomitantly, γδ CTLs are produced that can also carry out cytolysis of infected cells. The importance of γδ T cells is highlighted by the fact that many AIDS patients have a deficit in this cell type in the blood, and the degree of deficit correlates with disease progression. Moreover, any γδ T cells present in HIV-infected individuals appear to be anergic. However, it is not yet clear if HIV actually infects γδ T cells, or whether the observed changes to the γδ T cell population precede the assault on the CD4+ αβ T cell population. Interestingly, a group of HIV-infected patients that was able to suppress replication of the virus without antiretroviral drugs showed an increase in γδ T cell numbers.

ii) Th1 and Th17 Responses

At least in the initial stages of an HIV infection, HIV-specific Th cells play a key role in keeping viral loads under control. At the mucosae, IL-17 and IL-22 produced by activated Th17 cells induce antimicrobial peptide production and recruit neutrophils to the site of attack. In parallel, the cytokine help supplied by Th1 cells supports the differentiation of HIV-specific CTLs that kill infected macrophages, DCs and T cells and thus reduce viremia. However, in most HIV-infected individuals, the virus soon mutates to evade these CTLs and is free to infect large numbers of resting Th cells. HIV interferes with the expression of survival genes in these cells and promotes their apoptosis; those few Th cells that survive appear to be anergized. In addition, as noted in Chapter 13, HIV's Nef protein binds simultaneously to host clathrin proteins and the cytoplasmic tails of MHC class II molecules, forcing the internalization of the latter and their lysosomal degradation. As a result of all these deficits, only weak (if any) anti-HIV Th responses are mounted later during the infection, which also compromises CTL responses.

> Mechanisms by which T cell anergy is induced were discussed in Chapter 10.

iii) CTL Responses

As mentioned previously, HIV-1 replication and viremia are initially held at bay in part by vigorous CTL responses. These responses are usually directed against epitopes in gp120 or in highly conserved regions of the Pol, Nef and Gag proteins. HIV-specific CTLs destroy HIV-infected cells mainly by perforin/granzyme-mediated cytolysis rather than by cytotoxic cytokine secretion. However, in most cases, the anti-HIV CTL response eventually fails because of the catastrophic loss of Th cells. The genetic escape mutants generated by the continuously mutating virus present new epitopes, requiring the activation of more and more naïve CD8++ Tc cells. In the absence of sufficient Th help, these cells are not adequately stimulated. In addition, the presence of HIV frequently induces a naïve CD8++ Tc cell to express CD4 on its surface, marking it for HIV-mediated destruction. HIV also forces the upregulation of Fas expression on a CTL, making it vulnerable to apoptosis induced by FasL-expressing CD4+ T cells. Finally, the HIV Nef protein helps to dampen the anti-HIV CTL response by forcing internalization and degradation of surface MHC class I. The resulting reduction in pMHC on the surface of an infected cell inhibits its recognition by CD8++ CTLs.

iv) Antibody Responses

As described earlier, due to the extreme antigenic variability of HIV, neutralizing antibodies generated early in the course of HIV infection often do not recognize epitopes of the progeny virions generated later in the infection. Indeed, as most HIV infections progress, the viral strains that emerge appear to have been selected for complete resistance to antibody neutralization. Intriguingly, recent work has shown that HIV-infected persons also produce *non*-neutralizing anti-HIV antibodies that attempt to control the virus. The Fc regions (rather than the Fab regions) of these antibodies facilitate phagocytosis, NK-mediated ADCC, and complement-mediated lysis. By activating these mechanisms of innate immunity, these non-neutralizing antibodies help to contain the virus in the early stages of infection. Indeed, some patients deemed to be elite controllers or LTNPs have higher than normal levels of non-neutralizing antibodies, implying that these Igs are important for the suppression of HIV infection.

> Neutralizing antibodies block the entry of virus into host cells, whereas non-neutralizing antibodies facilitate FcR-mediated effector functions of innate leukocytes and complement activation.

Although HIV does not replicate in B cells, the virus has a severe impact on B cell function. In addition to binding to CD4 and chemokine receptors on host cells, gp120 can bind to a site in the V region of the Ig heavy chain (outside the antigen-binding cleft) in a BCR and non-specifically activate multiple clones of B cells. As a result of this extraneous B cell expansion, only about 20% of the antibodies produced in response to HIV infection are actually directed against HIV. The non-specific antibodies spur the formation of circulating immune complexes and promote autoimmunity in many HIV-infected persons. More importantly, included among the B cell clones activated non-specifically by gp120 are those directed against many bacterial and fungal pathogens. The eventual clonal exhaustion of these B cells robs the patient's immune system of a key weapon needed to defend against these pathogens later in the infection. In addition to the loss of these B cell clones, the decimation of the CD4+ Th cell population impairs the activation of any remaining clones of naïve anti-HIV B cells.

v) NK Cells

Although NK cells are resistant to HIV infection *in vitro*, HIV has significant detrimental effects on NK cells *in vivo*. There is some evidence that extracellular Tat protein released by an HIV-infected cell may interfere with the natural cytotoxicity exerted by an NK cell that has bound to this cell as a target. The NK cell shows an altered profile of activatory and inhibitory receptors, and loses the ability to synthesize perforin-containing granules. An NK cell's capacity for ADCC is also lost as HIV infection progresses. Thus, even though HIV downregulates MHC class I expression on infected T cells (creating a deficit that should act as a call to arms for NK cells), the warriors are disabled and the infected T cells escape death.

vi) Cytokines

HIV infection results in an abnormal profile of cytokine secretion that contributes to many AIDS symptoms. In the early stages of the disease, HIV-infected persons have elevated blood levels of TNF, IL-1, IL-2, IL-6 and IFNα. TNF in particular is a powerful activator of the HIV provirus and induces the wasting syndrome associated with AIDS. IL-1 has been linked to AIDS-associated fever and dementia.

vii) Complement

HIV blocks complement-mediated defense in three ways. First, as new progeny virions bud through the host cell membrane, they incorporate host RCA proteins into their envelopes and can thus block the deposition of the MAC. Second, the HIV envelope has multiple binding sites that recruit the soluble RCA protein Factor H to the virion surface. Factor H then inhibits alternative complement activation. Third, HIV infection downregulates expression of CRs on host cells, so that when a complement component binds to an infected cell, there is a decreased likelihood that the cell will be removed by opsonized phagocytosis.

IV. Efforts to Develop HIV Vaccines

i) Overview

The impact and incurability of AIDS made the development of an HIV vaccine a global priority for much of the past three decades. In the 1990s, because the virus had been identified so rapidly, there was much optimism that an effective vaccine could be readily produced using modern biotechnology. Unfortunately, in a perfect reflection of HIV's ability to avoid provoking a protective response during natural infection, this pathogen's antigenic variation and multiple strategies for evading and destroying the human immune system have greatly hampered vaccine development. It remains unclear which epitopes of the virus should be included in a vaccine, what type of response (humoral or cell-mediated or both) will be most effective, where (mucosally or systemically) the response will have to be induced, and how strong the response will have to be. Many traditional vaccine trials have now been undertaken, and all have failed. Nevertheless, the small number of people who have been continually exposed to HIV but never infected, and the LTNPs and elite controllers who have the virus but experience a long delay in developing AIDS, represent hope for the eventual production of an HIV vaccine that might induce a similar level of protective immunity in the general population. Study of the genetics and immune responses of these individuals and how they cope with a virus that has evolved to very quickly replicate, mutate, spread, and adopt latency may provide clues as to the nature of a vaccine that could help an individual's immune system defeat, or at least stay ahead, of this pathogen.

HIV's ability to elude the natural immune response also allows it to escape vaccine-induced immunity.

Interestingly, the most recent research in monkeys has demonstrated that there is a short period following mucosal transmission of *simian immunodeficiency virus (SIV)*, the monkey equivalent of HIV, when the virus is unusually susceptible to control by MALT T cells. The second wave of HIV vaccine development is thus focused on identifying chinks in the armor of HIV that can be exploited in a non-traditional way. If this avenue also fails, it may still be possible to devise a vaccine that induces an immune response that cannot prevent HIV infection but can target epitopes that the virus needs for efficient replication. Such responses may perhaps control the infection and thereby limit its progression to AIDS.

NOTE: In 2003, the Global HIV/AIDS Vaccine Enterprise (http://www.hivvaccineenterprise.org/) was established among AIDS researchers to coordinate efforts and thereby accelerate the development of a prophylactic vaccine for HIV/AIDS. The Enterprise involves private sector and government scientists, public health officials, advocacy groups and funding agencies. Its mandate is to promote the sharing of strategies, resources and results. The Enterprise is also actively seeking ways to overcome regulatory obstacles and increase numbers of trained personnel in developing countries.

ii) Barriers to HIV Vaccine Development

Obviously, the destruction of the very cells responsible for responding to a vaccine antigen constitutes a huge hurdle to HIV vaccine development, but there are other barriers. The extreme variability of HIV presents enormous challenges to the development of an effective vaccine. An HIV immunogen has yet to be identified that induces antibodies or CTLs capable of recognizing a broad range of primary viral isolates from a multitude of patients. Neutralizing antibodies recognizing one clade usually do not recognize another. In addition, focusing on one epitope tends to accelerate the emergence of genetic escape mutants. Moreover, it seems that the most conserved epitopes in HIV tend to be the least immunogenic, and those that are immunogenic are subject to extensive antigenic drift. These factors do not bode well for producing a vaccine that will induce protective immunity in most individuals within a population, let alone between populations.

Another technical challenge in HIV vaccine development is related to the animal systems used to grow the virus and test candidate vaccines. Laboratory-maintained strains of HIV do not behave exactly like primary HIV isolates obtained from patients, making it difficult to extrapolate results. Similarly, although monkey and mouse models have been very helpful in investigating certain aspects of HIV biology, infections in these models differ significantly from natural HIV infections of humans. Again, any promise of a vaccine tested in these models might not extend to the human situation.

In view of the failure of vaccines based on the induction of systemic immunity to HIV, some researchers have recently begun to examine the possibility of immunizing HIV-infected persons to preferentially activate γδ T cells and thus bolster mucosal responses. As noted previously, there is some recent evidence from animal studies that stimulation of innate leukocyte activation at the mucosae correlates with a slowing of AIDS progression. Other scientists are investigating ways to induce greater production of non-neutralizing antibodies able to trap HIV in mucus, mediate ADCC, or inhibit passage of the HIV virion across epithelial cell barriers. Indeed, in 2009, a trial of a potential anti-HIV vaccine called RV144 generated some surprising findings supporting these approaches. First, the RV144 vaccine, which contained two immunogens that were both unsuccessful when tried alone, conferred a modest degree of protection against HIV infection in humans when combined. Second, this protection was achieved without inducing either a strong neutralizing antibody or CTL response against the virus. Instead, these patients had significant levels of non-neutralizing anti-HIV antibodies whose protective effect correlated with the ability of their Fc domains to bind to FcRs or CRs expressed by innate leukocytes. Consistent with these results, HIV-infected individuals who express certain FcR polymorphisms have been found to exhibit a delay in AIDS progression. It is thought that these particular forms of the receptor bind preferentially to IgG2, which mediates phagocytosis more efficiently than other IgG subclasses.

V. HIV/AIDS Treatment

Until an effective HIV vaccine is developed, the best weapon the world has against AIDS is antiretroviral drug therapy. These drugs are responsible for the dramatic reductions in AIDS progression and AIDS-related deaths seen in the developed world since the late 1990s. There are at present six classes of licensed antiretroviral drugs that act on proteins essential for viral spread. Suppression of viral replication by any one of these inhibitors should reduce the number of HIV replicative cycles and thus the patient's viral load. However, due to HIV's propensity for genetic escape, the use of only one antiretroviral drug at a time soon selects for a strain of HIV resistant to that drug, re-escalating the attack on the immune system. Clinicians therefore often treat HIV patients with **highly active antiretroviral therapy (HAART)**, a regimen that features combinations ("cocktails") of three or more antiretroviral drugs from at least two different drug classes. As well as firmly shutting down viral replication for an extended period, this multidrug approach greatly decreases the chance of the virus generating a substrain that is simultaneously resistant to all drugs used in the cocktail. The patient's immune system then gains some breathing room so that at least some restoration of HIV-specific T cell responses, as well as responses against other pathogens, can occur. To determine the exact course of treatment for a given HIV-infected person, both his/her immunological status and clinical presentation are taken into account. These factors vary widely among patients, with some individuals being HIV-infected but asymptomatic or showing only mild disease, and others suffering from moderate or severe opportunistic infections, and sometimes cancers.

> Administration of IL-7 or IL-2 can further increase the T4 count of HAART-treated patients.

The HAART regimen has been highly successful in extending the lives of HIV patients, but only if the patient has been assiduously compliant with the prescribed regimen. Because treatment must be life-long, the quality of that life is also becoming an increasingly important issue. A large number of antiretroviral drugs now exists, meaning the tolerability of a drug's side effects and the convenience of its administration are being considered when therapy programs are formulated. Combination pills containing multiple antiretroviral drugs have been devised not only to reduce the chance that a drug-resistant strain will emerge but also to make compliance simpler and improve quality of life.

i) Classes of Antiretroviral Drugs

a) Protease Inhibitors

Protease inhibitors (PRIs) are small molecules that work by competitively binding to the active site of the HIV PR. Without PR function, the production of progeny viruses is stymied because the Gag and Pol polyproteins of the virus are not cleaved. Although PRIs can be highly effective, therapy with older drugs in this class entails the taking of numerous pills that have unpleasant side effects, decreasing patient compliance. Newer generations of PRIs require fewer pills, have fewer side effects, and show greater efficacy against the virus.

> Information on current drugs used for HAART can be found at http://aidsinfo.nih.gov/contentfiles/ApprovedMedstoTreatHIV_FS_en.pdf.

b) Nucleoside RT Inhibitors

Nucleoside RT inhibitors (NRTIs) work on the principles of competitive inhibition and premature DNA chain termination during viral replication. The inhibitors are analogs of the usual deoxynucleosides that the RT joins together to synthesize DNA. Occupation of the active site of RT by the inhibitor blocks the elongation of the growing viral DNA chain. In addition, the presence of large numbers of molecules of the inhibitor makes it hard for the RT to find its natural substrate. Unfortunately, patients taking NRTIs also experience unpleasant side effects.

c) Non-Nucleoside RT Inhibitors

Unlike nucleoside inhibitors, the non-nucleoside inhibitors (NNRTIs) act on the RT enzyme at locations distant from the active site. Although these inhibitors are diverse in structure, they all induce major conformational changes to the RT molecule that disrupt its enzymatic activity. These drugs are better tolerated by patients than either PRIs or NRTIs. The most modern NNRTIs induce two conformational changes, so that if the virus acquires a mutation allowing it to circumvent the first conformational change in its RT enzyme, it will still be blocked by the second conformational change.

A commonly prescribed HAART regimen includes two NRTIs plus either an NNRTI, a PRI, or an ISTI.

d) Integrase Strand Transfer Inhibitors

Integrase strand transfer inhibitors (ISTIs) block the function of HIV INT, thereby interfering with the integration of the HIV cDNA into the host genome. These drugs tend to act faster than NNRTIs in reducing viral load and increasing T4 count. Side effects on the CNS and metabolism are also milder.

e) Chemokine Receptor Inhibitors

The apparent good health of individuals who naturally lack expression of CCR5, and the resistance of these persons to HIV infection, has made CCR5 a highly promising therapeutic target. Several agents have been produced to block the access of HIV to CCR5 and thus prevent cell entry. However, major problems with toxicity and efficacy have arisen with all but one of these agents. In 2007, maraviroc became the first chemokine receptor inhibitor to be approved for the treatment of HIV patients. When maraviroc binds to CCR5, gp120 cannot gain access to this receptor, and the virus cannot enter the cell. When used in combination with other antiretroviral drugs, maraviroc slows down AIDS progression but cannot cure it. In addition, because HIV can sometimes use CXCR4 to invade T cells (especially in the later stages of many infections), testing for which chemokine receptor is targeted by the virus in a given patient is required prior to maraviroc treatment. This powerful drug can cause serious liver damage, so that careful and monitored dosing only to patients who can benefit is essential.

f) Fusion Inhibitors

Enfuvirtide is the only fusion inhibitor licensed to date. This drug is a synthetic peptide that blocks the conformational change of the HIV envelope spike needed for viral fusion with the host cell membrane and cell entry. Two practical disadvantages of enfuvirtide are that it requires cold storage and must be delivered by subcutaneous injection twice daily. Enfuvirtide is therefore usually recommended only for patients whose viral isolates show resistance to all other classes of inhibitors.

ii) Limitations of HAART

Despite the successes achieved with HAART, clinicians have come to the realization that it will be extremely difficult to completely eliminate HIV from a patient's body. The antiretroviral drugs currently available are effective only on actively replicating HIV and do not eliminate latent virus lurking in the genomes of resting cells. HIV can maintain latency for years inside resting T cells, resting macrophages and in other inaccessible body reservoirs such as the CNS. Moreover, HAART apparently does not stop *all* viral replication. The highly mutable nature of HIV means that the strain lingering in the patient is constantly evolving and will likely develop resistance to existing HAART drugs. New approaches are needed that can completely suppress viral replication and attack viruses in reservoirs without harming normal cells. For example, scientists are searching for novel agents that can specifically degrade viral RNA in the hope that this strategy will eliminate or incapacitate the virus wherever it hides.

Timing of HAART treatment has also been an issue—when to start, and whether to give drugs steadily after initiation or to interrupt the dose schedule. Traditionally, implementation of HAART does not occur at HIV diagnosis but is delayed until an asymptomatic patient's T4 count is 350–500. Some studies in the early 2000s hinted that intermittent administration of HAART drugs to these patients was at least as effective and perhaps even more efficacious than steady treatment. This approach was attractive because significant clinical problems have arisen from the extended use of HAART, including asthma; drug allergies; excessive inflammatory responses to opportunistic infections; and cardiac, renal and/or hepatic complications. In addition, HAART drugs are very expensive. Thus, a therapy plan that would allow a patient to take fewer pills while maintaining effective treatment would be beneficial on both fronts. Unfortunately, larger-scale trials showed that patients who underwent interrupted HAART administration experienced a greater number of non-AIDS side effects compared to patients who maintained steady HAART treatment. Nevertheless, while significant physical ailments are associated with extended HAART, this treatment is usually considered preferable to the premature death caused by an untreated HIV infection.

> NOTE: At the 19ᵗʰ International AIDS Conference held in Washington, D.C., in July 2012, a treatment approach was showcased that holds considerable promise for many HIV-infected individuals. Rather than waiting for their T cell counts to decline significantly, a group of HIV-positive persons were started on antiretroviral drugs very soon after diagnosis, when their T4 counts were normal and their blood HIV levels were still very low. These people were maintained on this treatment for 3 years until the virus was completely undetectable. At time of writing, these individuals remain healthy in the absence of further drug treatment. Scientists speculate that these people may not be truly cured but may have become LTNPs by virtue of their early therapy. Needless to say, these individuals are being carefully monitored to determine the extent of their triumph over HIV.

iii) Future Treatment Alternatives

The limitations of antiretroviral drug therapy have spurred clinicians and scientists to search for alternative approaches. This chapter's Focus on Relevant Research highlights attempts to use gene therapy and stem cell transplantation to outwit HIV.

We have reached the end of our discussion of primary and secondary immunodeficiencies. In Chapter 16, we describe how a healthy immune system strives to fight cancer development and how tumors evade both immune surveillance and elimination by immune system effector cells.

"Hematopoietic-Stem-Cell-Based Gene Therapy for HIV Disease" by Kiem, H.P, Jerome, K.R., Deeks, S.G., & McCune, J.M. (2012) *Cell Stem Cell* 10, 137–147.

Focus on Relevant Research

For nearly two decades, the medical community has been aware of the existence of patients who have been infected with HIV and yet neither exhibit high levels of circulating virus nor develop HIV-related disease. As noted in the main text, some of these individuals carry a mutation in the CCR5 gene, a key coreceptor for the cellular entry of HIV. Nature, it seems, has provided clues to blocking HIV infection, and indeed the CCR5 inhibitor maraviroc was developed using this line of reasoning. Unfortunately, to date, no drugs or vaccines have proven effective in preventing or curing HIV infection. However, in a much publicized clinical report, Hütter and colleagues described an HIV-positive patient with acute myeloid leukemia who was apparently "cured" of HIV infection following the transplantation of stem cells from a donor homozygous for the CCR5 mutation (Hütter et al., 2009, *New England Journal of Medicine*, 360, 692–698). This patient, who has been subsequently referred to as "The Berlin Patient" and is the only person at this time to have been cured of AIDS, has served as a proof-of-concept for HIV treatment strategies that couple gene therapy and stem cell transplantation. It is these approaches that are reviewed here in the article by Kiem *et al*. As shown here, Figure 3 from this article depicts a generalized approach to modifying stem cells for HIV resistance. In addition to techniques that replace the normal *CCR5* gene with its HIV-unfriendly allele, the authors review the use of targeted nucleases such as zinc-finger nucleases (ZFNs), TAL effector nucleases (TALENs), and hom-

ing endonucleases (HEs) that can directly destroy genes essential for HIV replication. Major challenges lie ahead, as indicated in the diagram by the thin lines leading to undesirable outcomes. However, investigators hope that this area of research will foster the development of more permanent means of treating HIV disease than are currently available.

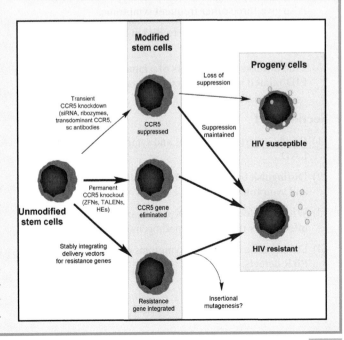

Chapter 15 Take-Home Message

- Immunodeficiencies can be primary or secondary.

- Primary immunodeficiencies (PIDs) are due to genetic defects that compromise innate and/or adaptive leukocytes or the complement system.

- Secondary immunodeficiencies are the consequences of an external cause, such as infection with HIV or other viruses, trauma, cancer therapy, immunosuppression or malnutrition.

- Historically, PIDs were diagnosed mainly on the basis of recurrent infections, but it is now clear that a single severe infection with a common pathogen may be a diagnostic sign.

- PIDs are usually treated by restoration of the missing immune component via intravenous or subcutaneous immunoglobulin replacement therapy, enzyme replacement therapy, or hematopoietic stem cell transplantation. Future treatments may involve somatic gene therapy.

- HIV is a cytopathic RNA retrovirus causing AIDS in human hosts.

- The interaction of the HIV envelope protein gp120 with CD4 and the host chemokine receptors CCR5 or CXCR4 allows the virus to enter macrophages, DCs and CD4+ T cells.

- The HIV provirus is integrated into the host cell genome. When the infected host cell is stimulated, viral transcription and translation produce virions that bud from the host cell prior to its death.

- HIV preferentially attacks DCs, macrophages and CD4+ T cells. B cells and CD8+ T cells respond to HIV infection, but the virus-specific antibodies and CTLs produced are ultimately insufficient to contain the virus.

- Clinically, an HIV-infected individual experiences a mononucleosis-like illness followed by an asymptomatic period of clinical latency that may last for several years. Death is usually caused by opportunistic infections or malignancy.

- Many barriers continue to thwart the development of a successful HIV vaccine, including HIV antigenic variation, T cell depletion, and a lack of animal models. New approaches based on the resistance of some individuals to HIV infection are under investigation.

- Treatment with antiretroviral drugs extends the life of an infected individual but does not eliminate the virus. Approaches based on gene therapy and stem cell transplantation have shown promise but are still experimental.

Did You Get It? A Self-Test Quiz

Introduction–Section A.I

1) Distinguish between primary and secondary immunodeficiencies.

2) How does consanguineous marriage affect PID incidence and why?

3) What is the principal clinical phenotype of most PIDs? Also give three other frequent symptoms.

4) Why is a diagnosis of severe PID in a child considered a medical emergency?

5) True or false: All members of a family carrying the same PID-related mutation show the same clinical phenotype.

Section A.II

1) Briefly describe the causes and symptoms of two types of LAD.

2) Distinguish between CGD and C-HS in terms of causes and symptoms.

3) Why do defects in the IFNγ/IL-12 axis often result in infection by intracellular pathogens?

4) Describe the causes and symptoms of three PIDs involving PRR defects.

5) Give two differences and two similarities between SCN and CyN.

6) Which complement deficiencies are linked to autoimmune disease?

Section A.III

1) How are T, B, NK and NKT cells affected in SCID patients?

2) What was the theory behind giving a bone marrow transplant to the "Boy in the Bubble"?

3) Why does loss of γc chain signaling lead to SCID?

4) Why do patients with ADA SCID and PNP SCID have non-immune deficits? Give two examples.

5) Distinguish between Artemis SCID, RAG SCID and Omenn syndrome.

6) What cell type is compromised in Nude SCID, and how does this lead to disease?

7) Give two examples each of APC-specific PIDs and T cell-specific PIDs.

8) What factors are slowing research into DC deficiency diseases?

9) Which HIGM syndrome is *not* a B cell-specific PID and why?

10) Explain why hyper-IgE syndrome is most often a T cell-specific PID rather than a B cell-specific PID.

11) Can you replicate **Figure 15-2** from memory?

12) Name four causes of CVID.

13) What is the basis for the slight phenotypic differences between HIGM2 and HIGM4 patients?

14) Why do mutations in DNA repair pathways result in PIDs?

15) Name two PIDs in which the patient's cells are abnormally sensitive to sun exposure.

16) Briefly describe three lymphoproliferative syndromes considered to be PIDs.

17) In which lymphoproliferative PIDs are NKT cells especially hard hit?

18) What is the principal anatomical defect in patients with DiGeorge syndrome?

Section A.IV

1) How has the diagnosis of PIDs evolved over the past several years?

2) Name three laboratory tests that might be used to confirm a PID diagnosis.

3) What are TRECs, and how are they useful for PID diagnosis?

4) Name three specific benefits of IV-IG/SC-IG therapy.

5) Describe how ADA–SCID patients are treated with enzyme replacement therapy. What are two drawbacks to this approach?

6) Name three PIDs that can be treated with HSCT.

7) Give two "pros" and two "cons" of somatic gene therapy.

Part B Introduction–Section B.I

1) Give four non-HIV causes of secondary immunodeficiency.

2) Is the number of people currently living with AIDS worldwide closer to 300,000, 3,000,000 or 30,000,000?

3) True or false: More men than women are infected with HIV.

4) Approximately what percentage of all HIV-infected persons are children?

5) Can you define these terms? lentivirus, cytopathic

6) Which cell type is the primary target of HIV? What other cell types does it infect?

7) Give three methods by which HIV is transmitted between individuals.

8) What is a provirus, and how does it function in the generation of progeny virions?

9) Sketch the structural layers of the HIV virion and label each. Include the structure of the HIV envelope spike.

10) Describe the three polyprotein precursors of HIV and the structural or functional proteins derived from each.

11) Name four host-derived proteins that appear in the HIV envelope.

Section B.II

1) Can you define these terms? tropism, infectious synapse, genetic escape mutant, clade, viral load

2) Why are macrophages and DCs a reservoir for HIV?

3) What molecules are required for HIV entry into a host cell?

4) Give two differences between X4 and R5 HIV strains.

5) HIV is said to be predominantly "R5 or X4 in character." Which character of the virus is prevalent in 90% of HIV patients at diagnosis? At the later stages of HIV infection?

6) How does gp41 facilitate entry of HIV into a cell?

7) What is the preactivation stage of HIV infection, and how is it terminated?

8) Why are HIV infections characterized by extreme antigenic variation?

9) Give a clinical and biological description of the three main phases of HIV infection.

10) What is the T4 count in a healthy individual? In an HIV-infected person in clinical latency? In a symptomatic AIDS patient?

11) Give three symptoms of advanced AIDS.

12) What group of people is truly resistant to HIV infection and why?

13) Distinguish between neutralizing, broadly neutralizing, and non-neutralizing antibodies.

14) Explain at the molecular level why LTNPs enjoy a delay in AIDS onset. Do the same for elite controllers.

Section B.III

1) Why has HIV been such a successful pathogen?

2) Why might γδ T cells help to slow the spread of HIV soon after infection?

3) Give two effects of HIV on Th cells that contribute to the eventual failure of the Th response.

4) How does HIV blunt the effectiveness of the CTL response?

5) Is the humoral response effective against HIV? Explain.

6) What effects does HIV have on B cell functions?

7) How does HIV interfere with the NK cell response?

Did You Get It? A Self-Test Quiz—Continued

8) Which cytokine is a powerful activator of the HIV provirus?

9) Give two ways by which HIV blocks complement-mediated defense.

Section B.IV

1) What factors remain to be determined before an effective HIV vaccine can be designed?

2) What is the purpose of the Global HIV/AIDS Vaccine Enterprise?

3) Give three barriers to HIV vaccine development.

4) Outline two novel approaches to inducing non-traditional immunity to HIV.

Section B. V

1) What is HAART and why is it used?

2) Name six classes of antiretroviral drugs and describe how they work.

3) Give two limitations of the antiretroviral drugs currently in use.

4) Why was the Berlin Patient's situation unique?

5) How might the procedure used to treat the Berlin Patient be generalized to treat other HIV-infected persons?

Can You Extrapolate? Some Conceptual Questions

1) In Chapter 3, we emphasized the fact that PRRs function as receptors recognizing molecular patterns present on a broad range of microbes. However, mutations of genes involved in PRR signaling often lead to infections with specific pathogens. Why might this be the case?

2) If an individual had one of the following genetic deficiencies, what major immune response mechanisms would he/she lack, and what might be the general consequences?

   a) Defective secretory component (SC) structure

   b) Defective J chain structure

   c) Mutation in the peptide-binding site of an HLA-A class I molecule (not affecting disulphide bond structure)

   d) Mutation in the $\beta2$ microglobulin molecule that prevents intrachain disulphide bonding

3) Without human intervention, HIV might well have become a victim of its own success by now. Where people do not modify their behavior to avoid becoming infected or do not receive the benefits of antiretroviral drug therapy, HIV spreads widely in hosts of all ages and is highly lethal. Thus, over time, a virus with these properties should end up wiping out the host population it needs to infect to survive and reproduce. Fortunately, human intervention has prevented this scenario from unfolding with HIV/AIDS, but if it had not, what adaptations might HIV have been forced to undergo to avoid completely killing off its host?

4) Rare individuals lack susceptibility to HIV infection because they do not express the CCR5 chemokine receptor that is the normal cellular port of entry for the virus. This protection from HIV infection is good news for these individuals, but if CCR5 is important for normal health, how would you explain the observation that CCR5-deficient individuals do not apparently suffer from any health deficits at all?

Would You Like To Read More?

A. Primary Immunodeficiency

Aloj, G., Giardino, G., Valentino, L., Maio, F., Gallo, V., Esposito, T., et al. (2012). Severe combined immunodeficiencies: New and old scenarios. *International Reviews of Immunology, 31*(1), 43–65.

Arkwright, P. D., & Gennery, A. R. (2011). Ten warning signs of primary immunodeficiency: A new paradigm is needed for the 21st century. *Annals of the New York Academy of Sciences, 1238*, 7–14.

Booth, C., Gaspar, H. B., & Thrasher, A. J. (2011). Gene therapy for primary immunodeficiency. *Current Opinion in Pediatrics, 23*(6), 659–666.

Chapel, H. (2012). Classification of primary immunodeficiency diseases by the International Union of Immunological Societies (IUIS) expert committee on primary immunodeficiency 2011. *Clinical & Experimental Immunology, 168*(1), 58–59.

Conley, M. E., Dobbs, A. K., Farmer, D. M., Kilic, S., Paris, K., Grigoriadou, S., *et al.* (2009). Primary B cell immunodeficiencies: Comparisons and contrasts. *Annual Review of Immunology, 27*, 199–227.

Cunningham-Rundles, C. (2011). Key aspects for successful immunoglobulin therapy of primary immunodeficiencies. *Clinical & Experimental Immunology, 164*(Suppl. 2), 16–19.

de Vries, E. European Society for Immunodeficiencies (ESID) members. (2012). Patient-centred screening for primary immunodeficiency, a multi-stage diagnostic protocol designed for non-immunologists: 2011 update. *Clinical & Experimental Immunology, 167*(1), 108–119.

Ferrua, F., Brigida, I., & Aiuti, A. (2010). Update on gene therapy for adenosine deaminase-deficient severe combined immunodeficiency. *Current Opinion in Allergy & Clinical Immunology, 10*(6), 551–556.

Kaveri, S. V., Maddur, M. S., Hegde, P., Lacroix-Desmazes, S., & Bayry, J. (2011). Intravenous immunoglobulins in immunodeficiencies: More than mere replacement therapy. *Clinical & Experimental Immunology, 164*(Suppl. 2), 2–5.

Netea, M. G., & van der Meer, J. W. (2011). Immunodeficiency and genetic defects of pattern-recognition receptors. *New England Journal of Medicine, 364*(1), 60–70.

Ochs, H. D., Oukka, M., & Torgerson, T. R. (2009). TH17 cells and regulatory T cells in primary immunodeficiency diseases. *Journal of Allergy & Clinical Immunology, 123*(5), 977–983.

Pessach, I. M., & Notarangelo, L. D. (2011). Gene therapy for primary immunodeficiencies: Looking ahead, toward gene correction. *Journal of Allergy & Clinical Immunology, 127*(6), 1344–1350.

Samarghitean, C., & Vihinen, M. (2009). Bioinformatics services related to diagnosis of primary immunodeficiencies. *Current Opinion in Allergy & Clinical Immunology, 9*(6), 531–536.

B. HIV Infection and Acquired Immunodeficiency Syndrome (AIDS)

Ackerman, M. E., Dugast, A. S., & Alter, G. (2012). Emerging concepts on the role of innate immunity in the prevention and control of HIV infection. *Annual Review of Medicine, 63*, 113–130.

Allers, K., Hutter, G., Hofmann, J., Loddenkemper, C., Rieger, K., Thiel, E., *et al.* (2011). Evidence for the cure of HIV infection by CCR5Δ32/Δ32 stem cell transplantation. *Blood, 117*(10), 2791–2799.

Carrington, M., & Walker, B. D. (2012). Immunogenetics of spontaneous control of HIV. *Annual Review of Medicine, 63*, 131–145.

Chakrabarti, L. A., & Simon, V. (2010). Immune mechanisms of HIV control. *Current Opinion in Immunology, 22*(4), 488–496.

Elhed, A., & Unutmaz, D. (2010). Th17 cells and HIV infection. *Current Opinion in HIV & AIDS, 5*(2), 146–150.

Goulder, P. J., & Watkins, D. I. (2008). Impact of MHC class I diversity on immune control of immunodeficiency virus replication. *Nature Reviews Immunology, 8*(8), 619–630.

Kwong, P. D., & Wilson, I. A. (2009). HIV-1 and influenza antibodies: Seeing antigens in new ways. *Nature Immunology, 10*(6), 573–578.

Mascola, J. R., & Montefiori, D. C. (2010). The role of antibodies in HIV vaccines. *Annual Review of Immunology, 28*, 413–444.

Picker, L. J., Hansen, S. G., & Lifson, J. D. (2012). New paradigms for HIV/AIDS vaccine development. *Annual Review of Medicine, 63*, 95–111.

Verkoczy, L., Kelsoe, G., Moody, M. A., & Haynes, B. F. (2011). Role of immune mechanisms in induction of HIV-1 broadly neutralizing antibodies. *Current Opinion in Immunology, 23*(3), 383–390.

Wilkin, T. J., & Gulick, R. M. (2012). CCR5 antagonism in HIV infection: Current concepts and future opportunities. *Annual Review of Medicine, 63*, 81–93.

Tumor Immunology

Chapter 16

Do what you can, with what you have, where you are.

Theodore Roosevelt

Over 8 million new cases of cancer occur annually throughout the world. At the start of 2012, the American Cancer Society predicted that 1,638,910 Americans would be newly diagnosed with some kind of malignancy, and that 577,190 Americans would die of their cancers during the year. As well as mortality, cancer imposes a huge economic burden. The direct costs associated with the treatment of cancer patients in the U.S. (including drugs, hospital care and home care) are estimated to exceed U.S. $50 billion annually, even before the lost productivity of stricken workers is taken into account. Although research has revealed much about the origin and nature of cancers, the role of immune responses in dealing with them remains unclear. Many immunologists believe that the immune system tries to protect the host against cancer by acting as a "tumor surveillance" mechanism. However, although cancers are clearly a threat to the host, the immune system does not eliminate all tumors promptly and some malignancies appear to induce tolerance rather than an immune response.

A. Tumor Biology

I. Carcinogenesis

i) Tumors versus Cancers

Throughout the life of a healthy organism, body cells divide, differentiate and die in a carefully controlled manner. When the cells of a tissue undergo unusual division that serves no useful function for the host, an abnormal tissue mass called a **neoplasm** ("new growth") or **tumor** ("swelling") may be created. All tumors can be classified as either **benign** or **malignant**, and a malignant tumor is a *cancer*. A benign tumor is relatively slow growing and contains cells that are well differentiated and well organized (**Plate 16-1A**). A benign tumor is also securely encapsulated such that its cells cannot break away from the cell mass and enter the blood. Benign tumors do not normally cause death, but if they do, they do so indirectly by compressing or damaging an adjacent organ. In contrast, cells comprising malignant

MAK: Primer to the Immune Response. http://dx.doi.org/10.1016/B978-0-12-385245-8.00016-9

423

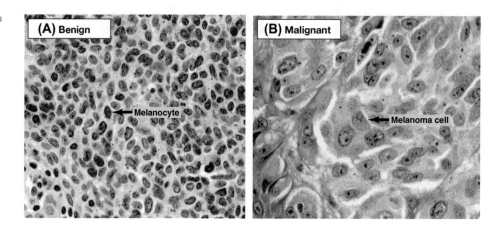

tumors are often poorly differentiated, the mass is disorganized, and encapsulation is rare (**Plate 16-1B**). These growths are lethal to the host unless they are completely removed or killed.

The transformation of a cell from normal to malignant occurs in a multistep process called *carcinogenesis*. During carcinogenesis, mutations accumulate in the genes controlling cell proliferation and programmed cell death, resulting in deregulated growth. In normal cells, multiple mechanisms exist that rapidly and accurately repair mutated DNA, or kill the cell if repair is impossible. Cancer results only when these mechanisms fail, or environmental factors increase the rate of mutation such that the repair mechanisms cannot keep up. Deleterious mutations then accumulate, leading to inappropriate gene expression that drives malignant transformation. *Carcinogens* are agents or substances, such as radiation or certain chemicals, that induce the genetic mutations and/or gene expression deregulation leading to cancer (**Table 16-1**). In addition, about 15% of human cancers appear to be linked to infection with a particular pathogen (**Table 16-2**). Chronic inflammation, with all its cytokine mediators and the ROIs and RNIs produced by activated leukocytes, is also thought to increase the chance that a cell will sustain a carcinogenic mutation (see later).

TABLE 16-1	Examples of Chemical and Radioactive Carcinogens
Carcinogen	**Tumor Association**
Chemical Carcinogens	
Alcohol	Cancers of the liver, esophagus, larynx
Aluminum	Lung cancer
Asbestos	Cancers of the gastrointestinal tract, peritoneum, lung
Benzene	Leukemia
Cadmium compounds	Lung cancer
Diethylstilbestrol	Cancers of the cervix, vagina, breast, testis
Nickel compounds	Cancers of the lung, nasal sinus
Silica crystals	Lung cancer
Soot	Lung cancer
Tobacco	Cancers of the lung, esophagus, larynx, pancreas, liver
Radioactive Carcinogens	
Plutonium-239 and decay products	Bone, liver, lung cancers
Radium-224 and decay products	Bone cancer
Radon-222 and decay products	Lung cancer
Thorium-232 and decay products	Lung cancer, leukemia
Iodine-131 and decay products	Breast and thyroid cancers, leukemia

TABLE 16-2	Examples of Pathogens Associated with Carcinogenesis
Pathogen	**Tumor Association**
Epstein–Barr virus	Burkitt lymphoma, Hodgkin lymphoma
Helicobacter pylori	Stomach cancer
Hepatitis B virus	Liver cancer
Hepatitis C virus	Liver cancer
Human papillomavirus	Cancers of the cervix, penis
Kaposi's sarcoma herpesvirus	Kaposi's sarcoma
Human T cell leukemia virus	Leukemia

NOTE: At the clinical level, human cancers are described as **sporadic** or **familial**. Most cancers are sporadic, in that the tumorigenic mutations occur in a somatic cell of a tissue and do not arise from alterations to a person's germ cells. In rare cases, the germ cells of an individual exhibit an accumulation of tumorigenic mutations that causes the host to be genetically predisposed toward a particular malignancy. If cancers then arise in the affected individual's descendants due to this inherited genotype, the malignancies are said to be familial.

ii) Steps in Tumor Development

Traditionally, the establishment of a tumor has been described as a process of *clonal evolution* that proceeds in four steps: *initiation, promotion, progression* and *malignant conversion*. In the initiation step, the DNA in the **target cell** or *cell of origin* of the tumor experiences a mutation that confers a growth or survival advantage. During promotion, the target cell is exposed to a stimulus that induces its selective proliferation such that a *preneoplastic* clone of genetically altered cells develops. However, unlike initiation, promotion is completely reversible, so that if the promoting stimulus is removed, the clone will undergo **regression**. Progression involves the introduction of additional genetic instabilities that drive the preneoplastic clone to become a *neoplastic* clone. Neoplastic clones contain cells that have a significant growth advantage over normal cells and are on the threshold of malignancy. Malignant conversion is achieved when a neoplastic cell acquires a transforming mutation that allows invasive growth and confers resistance to normal death signals. The highly mutated and malignant cell then generates progeny cells that grow in a totally deregulated manner and form the tumor. As a result of this succession of mutations, a tumor is usually genetically heterogeneous, and its component cells can vary widely in their growth rates, a fact that can complicate treatment.

iii) Metastasis

Malignant tumors are also frequently *metastatic*, meaning that cells from the original tumor mass (the **primary tumor**) gain additional mutations that allow them to break away and spread via the blood to nearby or distant secondary sites. Once established and growing in their new locations, these secondary tumors are called **metastases**. Metastasis occurs in four broad steps, the first of which is called the *epithelial-mesenchymal transition (EMT)*. During the EMT, a tumor cell changes its shape and acquires the ability to detach from the primary tumor mass and move on its own through the extracellular matrix (ECM) of the tissue. Loss of expression of the adhesion molecule E-cadherin is a key marker of the EMT. In the second step of metastasis, the tumor cell pushes through the endothelial cell layer lining a local efferent lymphatic or blood vessel and enters the circulation. However, it is estimated that only 0.01% of such cells survive in this new environment, meaning that most of them die at this point. For those rare tumor cells that survive in the circulation, the third step of metastasis is the use of integrin-mediated adhesion to adhere to a blood vessel in a new tissue site. The tumor cells extravasate through the blood vessel wall and into the ECM of the underlying tissue. In the last step of metastasis, if the tumor cell can survive in the new tissue

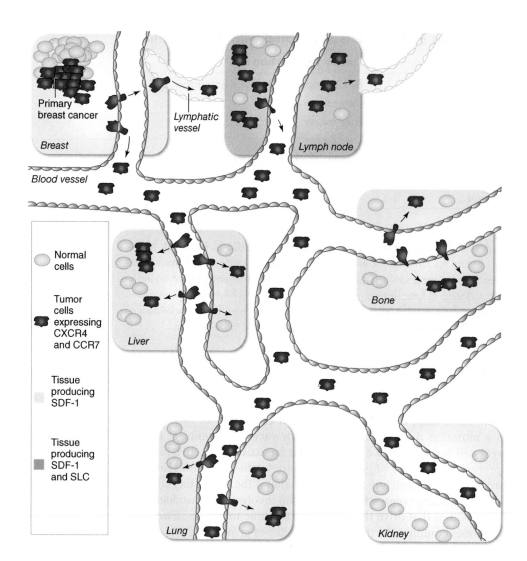

Fig. 16-1
The Role of Chemokines in Metastasis

Cells in some primary breast cancers express abnormally high levels of the chemokine receptors CXCR4 and CCR7. If these cells are circulating in the blood or lymphatics, they are drawn to leave the circulation wherever the cells of a tissue express high levels of the chemokines SDF-1 and/or SLC, which are ligands for CXCR4 and CCR7, respectively. SDF-1 is produced by resting cells in the lungs, liver, bone marrow and lymph nodes, whereas SLC is synthesized solely by lymph node cells. Accordingly, it is in these tissues that breast cancer metastases are most commonly found.

environment, it begins to divide in this location and establishes a secondary tumor. Studies of metastasis have shown that the migration of tumor cells is not random. Different types of cancer cells establish secondary tumors in different tissues, depending on the chemokines produced by a given tissue and the chemokine receptors expressed by the malignant cells. An example is illustrated in **Figure 16-1**.

iv) Cancer Stem Cells

Not all cells in a tumor have an equal capacity to initiate new tumors. In fact, in animal models, only a small fraction of the total cancerous cell population is capable of establishing a new tumor if transplanted into a suitable recipient. These findings have led to the *cancer stem cell hypothesis*, in which a small number of malignant cells (that arose from a single mutated cell) are responsible for sustaining the bulk of the tumor. In this scenario, the vast majority of cells in the tumor are derived from **cancer stem cells (CSCs)** and proliferate rapidly but cannot establish a new tumor on their own. As well as this ability to establish a new tumor, CSCs also have the HSC-like properties of self-renewal, a slow rate of cell division, and the capacity to give rise to more differentiated "progenitor" cells.

How are CSCs thought to be generated? In some cases, a CSC can result from the malignant transformation of a normal tissue-specific stem cell, the cell type responsible for maintaining a particular tissue or organ throughout the life of an organism. The long life span of a tissue-specific stem cell allows it ample time to accumulate the many mutations required for malignant conversion. These growth-deregulating changes are passed on to daughter progenitor cells that then undergo uncontrolled cell division, resulting in cancer expansion. In other cases, a CSC arises because a cancer-causing gene has been activated in a progenitor cell that has already differentiated. This progenitor then appears to "de-differentiate" and re-acquire stem cell-like properties, particularly the ability to self-renew and establish new tumors.

Because of their slow cell division, CSCs may escape destruction by current anticancer drugs, which kill only rapidly proliferating cells. Thus, CSCs may survive to re-establish the primary tumor or distant secondary growths after the treatment has stopped and could account for the recurrence of many cancers. If a surface marker unique to CSCs can be found, it may be possible to isolate these cells and study them more fully, and perhaps devise new therapeutic strategies to kill them.

NOTE: The cancer stem cell hypothesis is just that at the moment—a hypothesis that is not yet completely proven. While the evidence is very strong that hematopoietic malignancies like leukemias are driven by true (and very rare) CSCs, some immunologists believe that the situation may not be exactly parallel for solid tumors. CSCs have been reportedly identified in human tumors of the brain, skin, colon and breast, but other scientists argue that there still are no *direct* data showing that an unmanipulated solid tumor contains any CSCs at all. Since it is ethically and technically challenging to perform the type of assay in humans that could identify CSCs *in situ*, this question may remain unresolved for some time.

II. Tumorigenic Genetic Alterations

Mutations to three major classes of genes involved in controlling cell proliferation and death are at the root of most cancers. These classes are *DNA repair genes*, *oncogenes* and *tumor suppressor genes* (**Table 16-3**).

Databases listing the many hundreds of genes whose disruption is associated with cancer can be found at http://www.binfo.ncku.edu.tw/TAG/GeneDoc.php and http://www.tumor-gene.org/TGDB/tgdb.html.

i) DNA Repair Genes
The protein products of DNA repair genes detect and fix mutations to other genes. If a **DNA repair gene** does not function properly, the chance is increased that genes involved in the control of proliferation, differentiation or apoptosis will acquire deregulating mutations that lead to malignant transformation. As noted in Chapter 15, the skin cancers in PID patients with xeroderma pigmentosum are due to defects in genes involved in the pathway that repairs DNA damaged by exposure to UV radiation. Similarly, the lymphomas and leukemias developing in PID patients with ataxia telangiectasia are due to mutations in the ATM gene. ATM encodes a kinase important for the repair of double-strand DNA breaks caused by ionizing radiation or toxic drugs. Examples of other genes important in various DNA repair pathways are *MSH2*, *BRCA1* and *BRCA2*. MSH2 is frequently mutated in colorectal cancers, whereas mutations of BRCA1 and BRCA2 are associated with familial breast and ovarian cancers.

ii) Oncogenes
The term **oncogene** ("cancer gene") refers to a normal cellular gene that is altered or deregulated in a way that *directly* contributes to malignant transformation. Once an oncogene is identified, its normal cellular counterpart is referred to as a *proto-oncogene*. Most proto-oncogenes are positive regulators of cell proliferation, such as growth factors and their receptors, intracellular signaling molecules, and transcription factors. A proto-oncogene gains oncogene status when a mutation causes it to become constitutively activated such that it drives perpetual cell division. Important oncogenes include *Ras*, *Myc*, *epidermal growth factor receptor (EGFR)*, and *HER2*. Ras encodes an intracellular signal transducer and is mutated in many colorectal, pancreatic, lung

TABLE 16-3	Examples of Genes Altered in Various Tumors
Gene Name	**Tumor Association**
DNA Repair Genes	
ATM	Childhood leukemia, lymphoma
BRCA1	Breast, ovarian cancer
BRCA2	Breast, ovarian cancer
MSH2	Colorectal cancer
NBS1	Leukemia, breast cancer, multiple myeloma
XPC	Melanoma in xeroderma pigmentosum
XRCC1	Breast cancer
XRCC3	Melanoma
Oncogenes	
Bcr-Abl	Leukemia
HER2	Breast cancer
Jun	Bone, skin cancer
Met	Renal cell cancer
Myc	Breast, colon, lung cancer
PKC	Skin cancer
Raf	Liver, lung cancer
Ras	Colorectal, pancreatic, breast, skin cancer
Tumor Suppressor Genes	
APC	Colorectal cancer
p53	Multiple advanced cancers
PTEN	Multiple advanced cancers
Rb	Retinoblastoma
VHL	Renal cell cancer
WT1	Renal cell cancer

When a gene is amplified, it is present in a cell's genome in an increased number of copies relative to the norm.

and skin cancers. Myc encodes a transcription factor and is frequently amplified in cells of breast, colon and lung tumors. EGFR is the name given to a family of receptors that bind to epidermal growth factor and have tyrosine kinase activity. Alterations to EGFR molecules are associated with various types of tumors, particularly lung and brain cancers. HER2 is a member of the EGFR family, but its amplification is most often associated with sporadic breast cancers.

iii) Tumor Suppressor Genes

Unlike oncogenes, **tumor suppressor genes (TSGs)** usually encode negative regulators of cell proliferation or survival. Because the normal function of a TSG protein product is to discourage tumor development, a TSG that undergoes a "loss-of-function" mutation allows malignant transformation. Because diploid cells have two copies of any TSG, mutation of one of them is usually not sufficient to cause malignant conversion. However, if the remaining normal copy of the TSG is later inactivated, the individual becomes predisposed to tumor formation. Examples of important TSGs are *p53*, *PTEN* and *Rb*. The p53 gene encodes a transcription factor that functions as a master regulator of cell cycle progression and cell death. Normal p53 function causes cells that have suffered irreparable DNA damage to undergo growth arrest, senescence, or death. When p53 is mutated, DNA-damaged cells can proliferate uncontrollably. Mutations of p53 have been found in close to 50% of all human cancers. PTEN is a TSG whose normal function is to negatively regulate several key pathways driving cell survival, proliferation and metabolism. In the absence of PTEN, cells may escape a scheduled death and go on to proliferate instead. PTEN is the second most commonly mutated gene in human cancers. The normal function of the Rb protein is to block the division of cells with genetic abnormalities. Rb was discovered through studies of the development of retinoblastoma (retinal cancer) in young children.

B. Tumor Antigens

As mentioned previously, some scientists believe that one of the functions of the immune system is to carry out **immunosurveillance** and routinely seek out and destroy tumor cells. Such vigilance is thought to account for cases of "spontaneous regression" in which a tumor disappears apparently on its own. Even several decades ago, spontaneous regression was attributed to successful immune responses against unknown antigens expressed by tumor cells. Modern studies of mice deficient for various components of the immune system have also supported the concept of immunosurveillance, as these mutant animals show increased frequencies of spontaneous and carcinogen-induced tumors. A parallel exists in humans, since AIDS patients suffer from an increased frequency of cancers, particularly those associated with oncogenic virus infection. However, other investigators question the validity of immunosurveillance and point to the fact that the more common malignancies, such as colon, breast and lung cancers, do not occur at increased frequency in immunodeficient or immunosuppressed humans. These researchers also argue that the virus-associated cancers in AIDS patients may arise due to a failure in antiviral defense rather than in tumor immunosurveillance. Moreover, the fact that most cancers develop in immunocompetent individuals clearly shows that if immunosurveillance exists, it fails on a regular basis.

Despite these differing views, it has generally been assumed that genetic mutations associated with carcinogenesis can lead to the expression by a tumor cell of macromolecules that might be immunogenic. When these *tumor antigens* have been identified, they have turned out to be macromolecules that are abnormal in structure, concentration or location, or are normal macromolecules expressed at an anomalous time during the life of an animal. Although most tumor antigens are proteins, unusual carbohydrates that may serve as tumor antigens can arise when cells experience carcinogenic mutations in genes that control various aspects of carbohydrate synthesis and modification. Regardless of their protein or carbohydrate nature, tumor antigens generally fall into two classes: **tumor-associated antigens (TAAs)** and **tumor-specific antigens (TSAs)**. The reader is cautioned here that, despite much research effort, only a handful of *bona fide* TAA and TSAs have been isolated to date.

I. Tumor-Associated Antigens (TAAs)

A TAA of a tumor cell is a normal protein or carbohydrate expressed in a way that is abnormal relative to its status in the healthy, fully differentiated cells in the surrounding tissue of origin. In other words, a TAA is almost always a case of the "right molecule expressed at the wrong concentration, place and/or time."

i) "Wrong Concentration"

Genetic modifications that deregulate the expression of a normal gene can result in a tumor cell that expresses a normal protein at much higher levels (up to 100-fold) than those present on a healthy cell (**Fig. 16-2A**). A specific example is the protein product of the oncogene HER2. The HER2 protein forms heterodimers with other EGFR chains to generate a receptor that mediates the stimulation of the cell by epidermal growth factor (EGF) and related growth factors. In a healthy individual, HER2 is expressed at low levels by various types of epithelial cells. However, in the cells of some breast cancers, the HER2 protein is greatly overexpressed due to amplification of the HER2 gene. The same phenomenon occurs in a small proportion of prostate, colon, lung and pancreatic cancers. In all these tumors, circulating growth factors lead to excessive activation of the HER2 molecules present on the tumor cells, resulting in uncontrolled proliferation. Another good example of a "wrong concentration"

Fig. 16-2
Types of Tumor-Associated Antigens

TAAs are encoded by normal genes but are expressed at an abnormal concentration (**A**), place (**B, C**), or time (**D**). In contrast, TSAs are derived from mutated genes in the tumor cell genome (not shown).

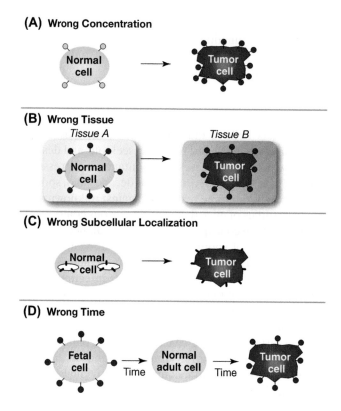

(A) Wrong Concentration

Normal cell → Tumor cell

(B) Wrong Tissue

Tissue A   Tissue B

Normal cell → Tumor cell

(C) Wrong Subcellular Localization

Normal cell → Tumor cell

(D) Wrong Time

Fetal cell → Time → Normal adult cell → Time → Tumor cell

TAA is CD20, a surface marker that is expressed only at low levels on all normal B cells but is highly overexpressed by the malignant B cells comprising several types of lymphomas (see Ch. 20). In other cases of "wrong concentration," a protein that is usually expressed only on a rare cell type (and thus whose total concentration in the body is only very low) is present at an abnormally high concentration because a cell of this type has become malignant and proliferated extensively.

ii) "Wrong Place"

Some TAAs arise when a protein whose expression is normally restricted to cells of tissue A is expressed inappropriately by tumor cells of tissue B; the protein appears in the "wrong place" (**Fig. 16-2B**). The **cancer-testis antigen** family of proteins furnishes some good examples of "wrong place" TAAs. The cancer-testis proteins are expressed solely in spermatogonia and spermatocytes in healthy individuals but are found on other cell types when they become cancerous. The best-studied examples of this class of TAAs are the MAGE proteins, which were first discovered in melanomas. "Wrong place" TAAs also include several that are expressed in the appropriate tissue type but show abnormal subcellular localization (**Fig. 16-2C**). For example, a melanocyte-specific protein called TRP-1 constitutes a TAA when it appears in the plasma membrane of melanoma cells instead of in the membranes of the intracellular vesicles with which it is normally associated.

iii) "Wrong Time"

Some TAAs are **embryonic antigens**, proteins whose expression is normally restricted to fetal cells but which are abnormally expressed in adult cells that have undergone malignant transformation (**Fig. 16-2D**). The gene encoding the embryonic antigen is silent in normal adult tissues but reactivated in tumor cells. An example of an embryonic TAA is *carcinoembryonic antigen (CEA)*, which is normally expressed only in the liver, intestines and pancreas of the human fetus but is highly associated with colon, breast and ovarian cancers in the adult. Similarly, *alpha-fetoprotein (AFP)* is normally expressed only in fetal liver and yolk sac cells but is strongly linked to liver and testicular cancers in the adult.

II. Tumor-Specific Antigens (TSAs)

TSAs are new macromolecules that are unique to the tumor and are not produced by any type of normal cell. Because of their non-self nature, TSAs should constitute true immunogens capable of provoking an immune response. However, for reasons that are elaborated in this section, very few of them do. The cancer cells expressing these TSAs therefore continue to grow unchallenged.

Macromolecules constituting TSAs may be localized on the tumor cell surface or in its interior. The new molecular structures may be derived from a simple amino acid substitution that alters the linear sequence and/or conformation of a protein, or from a more drastic internal rearrangement, or from deletion of parts of a gene. Unique proteins may also be produced when a chromosomal translocation fuses DNA sequences from two different genes, creating a new gene encoding a novel fusion protein composed of parts of two different proteins. TSAs may also be derived from DNA sequences that are not mutated but that are not transcribed in the normal way. For example, new proteins can arise from the use of a cryptic transcription or translation start site, an alternative reading frame, transcription of a gene in the opposite direction, or transcription/translation of a pseudogene sequence. Other TSAs are abnormal carbohydrates, produced when the gene encoding an enzyme involved in carbohydrate biosynthesis has undergone a transforming mutation. Certain mucins and gangliosides, which are altered in some types of melanomas and brain, breast, ovary and lung cancers, are carbohydrate TSAs. Finally, the viral proteins expressed by tumor cells infected with an oncogenic virus are considered to be TSAs.

Many TSAs are the mutated protein products of genes controlling cell proliferation or apoptosis. For example, mutations of the TSG p53 result in a non-functional protein that cannot induce cell cycle arrest or apoptosis in response to DNA damage. Another example occurs in certain head-and-neck cancer patients whose tumor cells show a mutation in the pro-apoptotic gene caspase-8. This mutation leads to the addition of numerous amino acids to the caspase-8 protein, an alteration that interferes with its ability to induce cell death. In some melanoma patients, the tumor cells have a mutation in the cell cycle control gene Cdk4 that results in uncontrolled cell proliferation. Finally, mutated forms of EGFR are expressed by several different types of cancer cells. Abnormal signaling driven by the binding of EGF to the altered receptor drives tumor expansion.

NOTE: Prostate cancer is among the leading causes of cancer-related death in men. For early detection of this malignancy, physicians have moved away from older, less sensitive physical exams and currently utilize the *PSA blood test*. PSA stands for "prostate-specific antigen," which is actually an enzyme called "kallikrein-related peptidase-3." The PSA protein is a 33 kD protease that is secreted into semen by epithelial cells of the prostate gland. The normal role of PSA is to liquefy both semen and the mucus protecting the cervix, and thereby increase sperm mobility. However, an elevated level of PSA in a man's blood is believed to be indicative of prostate cancer, making it a TAA in this context. Such blood elevations of PSA, even when subtle, can be detected by highly sensitive immunoassays. This sensitivity has generated controversy, however, because elevated blood PSA can also be seen in non-cancerous conditions such as benign prostate hyperplasia and prostatitis. A doctor who fails to rule out these more innocuous diagnoses may mistakenly subject his/her patient to unnecessary and potentially dangerous surgery, chemotherapy, and/or radiation to treat a cancer that doesn't exist.

Various immunoassays used for antigen detection are illustrated in Appendix F.

C. Immune System Function in the Tumor Microenvironment

I. Interplay Between the Immune Response and Tumors

i) Antitumor Immunity

The immune system can work to prevent carcinogenesis in the first place or strive to eliminate the tumor cells in an established cancer. With respect to prevention, a tumor

Plate 16-2
*Tumor-Infiltrating
Lymphocytes in Melanoma*
[Reproduced by permission of
Danny Ghazarian, Princess
Margaret Hospital, University
Health Network, Toronto.]

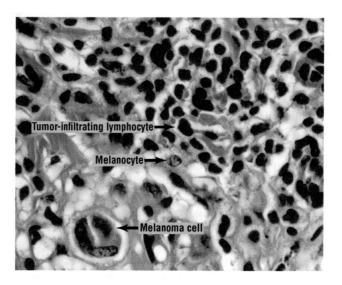

may never form at all if the immune system responds successfully to an infection by an oncogenic virus or resolves acute inflammation before it can become chronic. With respect to tumor cell elimination, most immunologists feel that various components of innate and adaptive immunity do indeed respond to cancer cells and kill them, a belief held despite the difficulties researchers continue to experience in identifying TAAs and TSAs. Indeed, extensive studies over the past two decades have confirmed that mice genetically engineered to lack a component of the immune system are more susceptible to both spontaneous tumors and cancers induced by viral or chemical carcinogens. When the regression of a tumor is induced by an obvious immune response, the process is called **tumor rejection**.

Evidence that leukocytes respond to the presence of a tumor has also been gained from histological studies. In samples acquired from cancer patients, collections of lymphoid cells can often be found within the tumor tissue itself. When such cells are mature CD4+ or CD8+ T cells or B cells, they are often referred to as **tumor-infiltrating lymphocytes (TILs)** (**Plate 16-2**). Additional lymphoid lineage cells such as NK and NKT cells can also penetrate tumors and so are sometimes referred to as TILs. Other scientists label tumor-infiltrating NK and NKT cells as well as tumor-infiltrating myeloid cell types as "tumor-associated" cells. Importantly, extensive leukocyte infiltration in a tumor has been associated with a better prognosis in patients with melanoma or ovarian cancer.

ii) Evasion of Antitumor Immunity

Although it is clear that immune responses are mounted against tumor cells, the prevalence of cancer in the world's human population indicates that the immune system does not always succeed in preventing or controlling tumorigenesis. Just as pathogens have evolved various means of evading or thwarting immune responses (refer to Ch. 13), tumor cells often possess or develop mechanisms that allow the cancer some degree of escape from the immune system. In addition, sometimes the activities of the leukocytes that do manage to respond to the cancer end up promoting tumor cell growth rather than suppressing it. Complicating the situation further is the fact that molecules produced by non-cancerous cells in the surrounding tumor microenvironment may influence the behavior of both leukocytes and tumor cells. All these considerations mean that the outcome of an encounter between leukocytes and cancer cells within a tumor site is by no means certain.

Two forms of escape from local immune responses are thought to be associated with all tumor microenvironments, regardless of which leukocytes respond to the malignancy. First, the abnormal properties of tumor vasculature can make effective immune responses difficult. Unlike a normal tissue, the capillaries that wind in and out of a tumor mass and support malignant cell growth are disorganized and frequently lack expression of integrins, hindering leukocyte extravasation into the tumor site. Second, many human cancer patients exhibit elevated levels of plasma TGFβ, a cytokine that was first named for its ability to promote the malignant transformation of fibroblasts. TGFβ stimulates **angiogenesis** within the tumor, which increases the delivery of oxygen and nutrients to the growing tumor. TGFβ has also been shown *in vitro* to block the functions of CTLs and NK cells and to suppress antibody synthesis.

In addition to having an effect on local immune responses, the presence of a cancer in the body may also downregulate immunity in a systemic way. The existence of such general tumor-associated immunosuppression is supported by two lines of evidence. First, most cancer patients are highly prone to opportunistic infections. However, within weeks of removal of a tumor, the patient's immune system is often restored to full function. Second, animals that bear tumors respond poorly to antigens that induce vigorous immune responses in healthy animals.

The next few sections cover in more detail how particular immune system mechanisms attempt to eliminate tumor cells, while other mechanisms seemingly protect malignant cells or promote their growth. How the tumor cells themselves can neutralize immune responses against them is also described.

II. Inflammation

The involvement of inflammation in cancer is multilayered and its effects complex. While acute inflammation has mostly antitumorigenic effects, chronic inflammation is thought to be a key promoter of carcinogenesis. Some immunologists now believe that inflammation is such an important part of tumorigenesis that most cancers would be preventable if the chronic inflammation supporting them was eliminated. Obviously, the prevention of inflammation would be a much more economical approach for protecting the general public than the current practice of treating the disease only after it has become entrenched enough to diagnose. These concepts are the subject of much ongoing research.

i) Acute Inflammation

Acute inflammation in the local tumor environment can be provoked in various ways. In some cases, the stress of a rapidly growing mass in a tissue causes normal cells in the area to upregulate their surface expression of DAMPs or to release them into the immediate milieu. These DAMPs can bind to the PRRs of innate leukocytes, inducing their activation. In other cases, the premalignant or malignant cells themselves express or release DAMPs that trigger an inflammatory response (**Fig. 16-3**). Other DAMPs,

Angiogenesis is a term used to describe the generation of new blood vessels.

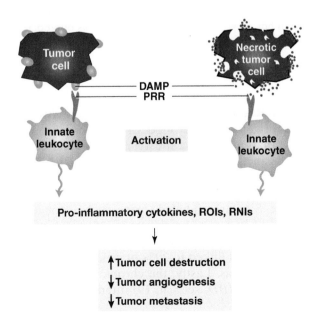

► Fig. 16-3
The Role of Tumor-derived DAMPs in Triggering Acute Inflammation

Intact tumor cells may express surface DAMPs that engage the PRRs of an innate leukocyte (left). Dying tumor cells may release DAMPs that can also bind to leukocyte PRRs (right). In both cases, acute inflammation may be induced, with the production of cytokines, ROIs and RNIs that work to destroy the tumor and its supporting infrastructure.

such as HSPs, HMGB1 and ATP, may be present in the debris of necrotic tumor or normal cells. Neutrophils and macrophages activated by HSPs play key roles in tumor rejection because these cells release the major pro-inflammatory cytokines IFNγ, IL-12 and TNF. TNF was originally identified by its ability to directly induce the hemorrhagic necrosis of a tumor. IFNγ and IL-12 drive Th1 and NK cell responses and support CTL generation, and also have powerful anti-angiogenic effects that prevent the tumor from building up the blood vessels needed for its growth. In addition, IFNγ induces the formation of a collagen capsule around a tumor mass, a natural barrier against metastasis. HMGB1 and ATP combine to activate TLR4-bearing leukocytes, resulting in formation of the NLPR3 inflammasome and copious production of IL-1. The assembly of this inflammasome in DCs and the binding of the resulting IL-1 to IL-1R molecules on the surface of CTLs is critical for supporting their subsequent attack on tumor cells. IL-1 can also promote the differentiation of tumor-fighting γδ Th17 cells.

Cytokines are not the only anticancer substances released by leukocytes participating in an acute inflammatory response to a tumor. Activated eosinophils and other granulocytes as well as macrophages release ROIs and RNIs that can kill cancer cells and damage tumor vasculature.

NOTE: Not all acute inflammation reduces tumorigenesis. *Therapy-induced inflammation* is acute inflammation associated with the necrotic death of both normal and tumor cells in a patient treated with certain chemotherapy drugs or radiation. The tissue trauma caused by these "heavy hammer" treatments induces the body to mount a wound healing-like response to DAMPs that can promote the cross-presentation of internal TAA/TSAs and thus enhance antitumor responses. On the other hand, this type of inflammation also triggers the expression of pro-survival genes whose protein products may allow a few cancer cells to survive the treatment. The expression of pro-angiogenic genes that stimulate tumor angiogenesis may also be activated. In the worst case scenario, the continued expression of pro-survival genes in residual cancer cells can foster the emergence of drug- or radiation-resistant tumor cells that establish an even more dangerous cancer. These considerations have caused clinicians to explore the judicious use of anti-inflammatory drugs during cancer treatments. Such approaches include drugs that can (1) inhibit the transcription factors and signal transducers that are activated by inflammation and support tumor cell growth; (2) reduce tumor-promoting cytokines/chemokines while increasing tumor-inhibiting cytokines/chemokines; and (3) block the activities of certain leukocytes while sparing the functions of others. There are even examples in which an anti-inflammatory drug has been used prophylactically in an effort to prevent a cancer-prone person from developing the disease. For instance, aspirin has been given to individuals at high risk of developing colon, breast or prostate cancer, although a benefit has been seen only in those having a genetic polymorphism linked to high lymphotoxin production. Since the long-term use of aspirin has severe detrimental side effects, this approach should be recommended only when the patients have been pre-screened for the presence of this lymphotoxin polymorphism.

Lymphotoxin is a TNF-related cytokine with TNF-like properties (see Appendix E).

ii) Chronic Inflammation

As far back as the 1860s, pathologist Rudolph Virchow suggested that chronic inflammation might contribute to tumorigenesis. Many modern day clinicians now support this view, having recognized that most of the lifestyle, dietary and environmental risk factors associated with cancer predisposition can be linked in some way to chronic inflammation. For example, chronic infections are closely linked to 20% of malignancies, and pathogen persistence is a prime driving force behind an unstoppable inflammatory response. Similarly, inhaled toxins, such as tobacco smoke and asbestos particles, are well known to trigger chronic inflammation in the respiratory tract and have been linked to 30% of cancers. In the same fashion, improper diet leading to obesity, which is linked to 20% of cancers, results in chronic inflammation in the liver and pancreas. Lastly, the chronic inflammation associated with *inflammatory bowel disease (IBD;* see Ch. 19) often precedes the development of colon cancer.

NOTE: Not all chronic inflammation promotes tumorigenesis: for unknown reasons, patients who suffer from the inflammatory symptoms associated with rheumatoid arthritis or psoriasis actually have a neutral or decreased cancer risk (see Ch. 19).

What components of chronic inflammation are actually tumor-promoting? The ROIs and RNIs generated by activated leukocytes help to destroy infected normal cells in the short term, but if these reactive molecules persist in the local microenvironment, they start to exert genomic stress on surrounding normal cells, promoting tumor initiation. Similarly, the cytokines and growth factors produced by activated leukocytes concentrate the immune attack where it is needed and drive lymphocyte expansion, but can also promote the proliferation of preneoplastic cells that have gained a growth advantage. Other leukocyte-derived chemical mediators produced to promote wound healing increase the angiogenesis that supports tumor growth, while molecules generated to increase the migration of innate leukocytes through the ECM have the downside of facilitating tumor cell invasiveness. Consequently, molecules found at high levels in inflammatory sites, such as TNF, IL-1 and IL-12, may both enhance antitumor responses and promote tumor progression. In addition, as described later, other activities of the leukocyte populations recruited during prolonged inflammatory responses may promote tumor progression, rather than suppress it. Thus, whether or not a chronic inflammatory response supports tumor growth may depend on its overall duration; the balance of activated leukocytes; tumor cells and stromal cells in the immediate microenvironment; and the cytokines, chemokines and other modulatory molecules produced by all these cells.

NOTE: Clinical researchers have long known that the inhalation of tiny particles of silica or asbestos can cause lung cancer. Both of these substances are vigorous activators of the NLRP3 inflammasome. As a result, the lungs of an individual breathing air polluted with silica or asbestos eventually become sites of chronic inflammation, establishing a microenvironment that favors tumorigenesis. Scientists have shown that NLRP3-deficient mice forced to breathe asbestos-contaminated air exhibit much less lung inflammation than control animals. The specific relevance of the NLRP3 inflammasome to human lung cancers is still under investigation.

III. γδ T Cells

Mice that are engineered to lack γδ T cells and are exposed to carcinogens develop skin tumors with increased frequency and rapidity compared to normal mice exposed to the same substances. This observation suggests that γδ T cells resident in the epithelial layers of the body may guard against malignancies. Whether γδ T cells perform such a function in humans and to what extent are under debate, since most common human adult cancers (such as tumors of the lung, breast, colon and prostate) arise from the transformation of an epithelial cell that occurs despite the presence of γδ T cells. Nevertheless, many researchers believe that at least some γδ T cells mount antitumor responses following the binding of tumor antigens to their γδ TCRs. These responses may be enhanced by a type of costimulatory signaling triggered by the binding of the stress molecules MICA and/or MICB to the NK activatory receptor NKG2D. MICA and MICB are frequently expressed on human cancer cell surfaces, and many γδ T cells express NKG2D. Once fully activated by tumor-derived ligands, antitumor γδ Th and γδ CTLs are generated that kill tumor cells via cytotoxicity, and produce IFNγ and other cytokines which influence αβ T cells. The IL-17 produced by γδ Th17 cells stimulates the production of IFNγ by antitumor αβ Th1 cells and CTLs. In addition, as described in Chapter 11, certain γδ T cell subsets respond to HSPs, which may be present in the debris of necrotic cells and can serve as DAMPs activating the innate response. However, the involvement of HSPs in antitumor γδ T cell responses *in vivo* awaits definitive confirmation.

The antitumor activities described previously have been attributed to conventional γδ T cells that function to protect the host. However, there is some evidence that a γδ T

The expansion of γδ T cells can be induced by amino-bisphosphonates, which are anticancer drugs used to treat myeloma and lymphoma.

cell subset exists that has regulatory functions similar to those of iTreg cells (although these γδ T cells do not express CD25 or Foxp3). Such regulatory γδ T cells have been found infiltrating human breast, renal and prostate cancers, and appear to have a pro-tumorigenic effect in that they are capable of shutting down the activities of conventional antitumor αβ T cells. In animal models, this regulatory γδ T cell population inhibits antitumor immune responses via secretion of IL-10 and TGFβ, and can also block DC maturation. The importance of these cells in human cancers is under investigation.

IV. Macrophages

In Chapter 2, we introduced macrophages as a single population of leukocytes. In Chapter 3, we concentrated on the functions of these cells as inflammatory phagocytes, and in Chapter 7 on their functions as APCs in immune responses. These classical macrophages, which are activated by IFNγ and PAMPs/DAMPs, express high levels of pro-inflammatory cytokines (especially IL-12), and engage in ADCC, have recently been designated as *M1 macrophages.* These macrophages are believed to perform all the activities we have previously described to promote antitumor immune responses. However, researchers studying the inflammation present in tumor and transplant sites have identified a second type of macrophage, dubbed the *M2 macrophage. In vitro* analyses of M2 macrophages have revealed that they downregulate their expression of MHC class II, stop producing IFNγ, and start producing IL-4, IL-10 and IL-13. These characteristics have spurred scientists to theorize that M2 macrophages are generated to balance the pro-inflammatory actions of M1 macrophages and thus help to maintain homeostasis. Within tumor sites, the total macrophage population present often appears to be dominated by cells of the M2 phenotype, such that their overall effect is the promotion, rather than suppression, of cancer growth. Specifically, the activities of M2 macrophages have been linked to the support of tumor angiogenesis as well as cancer cell invasiveness and metastasis. The derivation of M1 and M2 macrophages and their differing roles within the tumor microenvironment are addressed in this chapter's "Focus on Relevant Research."

V. Dendritic Cells

Obviously, as the lynchpins of adaptive responses, DCs are critical for antitumor immunity. In addition to their activation of Th and Tc cells (and the activation of B cells that Th cell activation permits), DCs produce large amounts of IL-12 and IFNs that can activate macrophages, NK cells and other leukocytes to attack tumor cells. As noted previously, DCs whose PRRs are engaged by the appropriate DAMPs undergo inflammasome assembly that leads to copious production of IL-1. The pro-inflammatory properties of this cytokine help to support acute inflammation capable of eliminating tumor cells and also stimulate IL-17 production and the generation of antitumor γδ T cells.

However, even DCs have been found to sometimes have a pro-tumorigenic effect. The IL-1 produced by DCs can accumulate to a concentration that promotes both chronic inflammation and angiogenesis, favoring cancer growth. IL-1 also stimulates the recruitment to a tumor site of *myeloid-derived suppressor cells (MDSCs)* that can evolve into tumor-suppressive M2 macrophages. A little less obvious is the potential role of DCs themselves in tumor angiogenesis. In response to inflammatory peptide mediators released in the tumor microenvironment, immature DCs are attracted to the tumor site, where growth factors can induce some DCs to take on the characteristics of endothelial cells that help to build the tumor's blood supply.

VI. Neutrophils

Unlike macrophages, neutrophils are not prominent infiltrators of tumors, despite the fact that tumors and *tumor-associated macrophages (TAMs)* frequently secrete

"Paired Immunoglobulin-Like Receptor-B Regulates the Suppressive Function and Fate of Myeloid-Derived Suppressor Cells" by Ma, G., Pan, P., Eisenstein, S., Divino, C., Lowell, C., Takai, T., & Chen, S. (2011) *Immunity*. 34, 385–395.

Focus on Relevant Research

The cover of this textbook bears an artist's rendering of a tumor microenvironment in which tumor-associated macrophages (TAMs) have infiltrated among the cancer cells. These TAMs are derived from "myeloid-derived suppressor cells" (MDSCs), which are bone marrow-derived myeloid progenitors that appear in sites of inflammation. After infiltrating a tumor, MDSCs seem to acquire the potential to differentiate into several different types of innate leukocytes, including M1 macrophages (which have antitumor properties) and M2 macrophages (which promote tumor progression). Prior to the research reported in this article, the mechanisms regulating the differentiation of M1 versus M2 TAMs from MDSCs were completely unknown. In this article, Ma *et al.* identify a mouse macrophage cell surface receptor called PIR-B (encoded by the *Lilrb3* gene) as a key regulator of MDSC differentiation. They demonstrate that, within the tumor microenvironment, an MDSC that expresses PIR-B becomes a tumor-promoting M2 macrophage, whereas an MDSC that lacks PIR-B expression becomes a tumor-destroying M1 macrophage. Ma *et al.* established this finding in part by using "depletion-reconstitution" experiments, an example of which is shown here in selected panels taken from Figure 4 of the article. First, tumor-bearing mice were either left untreated (control; "CT") or were depleted of their endogenous MDSCs ("Depletion"). The depleted mice were then reconstituted with either wild-type (WT) MDSCs or MDSCs harvested from mice lacking expression of PIR-B (*Lilrb3⁻/⁻* MDSCs). Tumors from these mice were isolated, and the percentage of immunosuppressive Treg cells present (top panel) as well as the tumor weights (middle panel) were determined. Both Treg cell numbers and tumor weights were significantly reduced in mice reconstituted with PIR-B-deficient MDSCs, suggesting that the generation of immune response-promoting M1 macrophages dominated when PIR-B was not expressed. In contrast, when the MDSCs did express PIR-B, the expansion and activities of Treg cells were enhanced and tumor weights were increased, suggesting that the generation of immunosuppressive M2 macrophages was favored. Tumor vascularization (as determined by anti-CD31 immunohistochemistry; bottom panel) was also reduced in mice reconstituted with PIR-B-deficient MDSCs, further indicating that such MDSCs can give rise to TAMs with the antitumor M1 phenotype. The article goes on to describe several other experiments that collectively suggest mechanisms by which MSDC differentiation into M1/M2 macrophages is controlled in the tumor microenvironment. The authors also present data supporting the design of novel therapies that could theoretically be used to enhance antitumor immune responses.

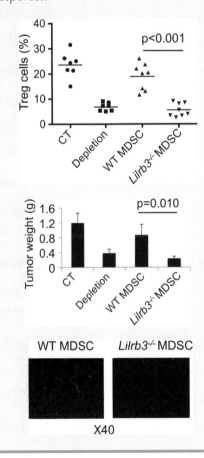

chemokines that attract these innate leukocytes. However, recent research has revealed that, like the differentiation of M1 versus M2 TAMs, low numbers of *tumor-associated neutrophils (TANs)* may be present that fall into two opposing subsets called *N1* and *N2 neutrophils*. N1 neutrophils tend to have an antitumor phenotype, producing copious amounts of ROIs and RNIs that kill tumor cells directly. In the absence of the immunosuppressive cytokine TGFβ, N1 neutrophils also increase their production of hydrogen peroxide and TNF, undertake Fas-mediated killing, and, most importantly, stimulate antitumor CTL activity. Researchers are now searching for agents that will drive the differentiation of all TANs within a cancer patient toward the antitumor N1 phenotype.

In contrast to N1 neutrophils, N2 neutrophils tend to be found in microenvironments high in TGFβ. These cells have lower direct cytotoxic capacity, release fewer TNF and ROIs, and are less able to stimulate CTL activity. Furthermore, N2 neutrophils routinely synthesize molecules that can encourage tumor angiogenesis and dissolve the ECM to ease tumor cell migration. The tumor-promoting properties of N2 cells may explain why cancers often secrete neutrophil-attracting chemokines. Indeed, it has been shown in mouse models that a general depletion of neutrophils can actually inhibit cancer growth.

> NOTE: The ROIs and RNIs produced by neutrophils are a double-edged sword. Although these chemical radicals can kill large numbers of tumor cells, they may also increase the genetic instability of any survivors, allowing these cells to acquire new mutations that accelerate cancer progression.

VII. NK Cells

NK cells are thought to be key effectors of anticancer immunity. As described in Chapter 11, NK cells are activated when the signals transduced through their activatory receptors overpower the signals generated by their inhibitory receptors. As previously noted, human tumor cells often lose expression of the MHC class I molecules that would normally act as inhibitory ligands but do express high levels of the activatory ligands MICA and/or MICB. Thus, activatory receptors such as NKG2D are engaged, but NK inhibitory receptors are not, so that the NK cell delivers a cytotoxic "hit" to the tumor cell either through the perforin/granzyme mechanism or Fas-FasL ligation. NK cells also carry out sustained IFNγ production that is critical for the inhibition of metastasis in mouse models.

However, precisely because mature NK cells are such potent tumor cell killers, cancers often develop various mechanisms to thwart them. The downregulation of MHC class I by many malignancies, which should invite elimination by NK cells, can be offset in some cases by tumor cell expression of non-classical HLA-E, HLA-F or HLA-G molecules. These MHC class Ib proteins can bind to NK inhibitory receptors and prevent an NK-mediated assault on the tumor. Significantly, HLA-G expression on cancer cells is upregulated by IL-10, an immunosuppressive cytokine often found in the tumor microenvironment. In addition, at least in mice, cancers can also manipulate immature NK cells to hobble the immune system. In mouse models, certain leukemic cells have been found to produce soluble factors that cause immature NK cells to suppress DC functions via cell-to-cell contact. These DCs are then not able to activate antitumor T cells, crippling the adaptive response. Whether the same phenomenon occurs in humans is under study.

VIII. NKT Cells

In some tumor microenvironments, immature DCs that capture TAA/TSAs may mature in the local site and present non-peptide antigens derived from these TAA/TSA molecules on CD1d, enabling these DCs to activate NKT cells. Although it is not yet certain that NKT cells actually recognize such antigens, mutant mice engineered to lack NKT cells show increased susceptibility to chemical carcinogens. Conversely, the transfer of functional NKT cells back into these mutants protects them from carcinogenesis. Another line of evidence supporting an antitumor role for NKT cells is the observation that administration of the prototypic NKT antigen αGalCer to tumor-bearing animals inhibits cancer progression and metastasis. These findings indicate that activated NKT cells do provide early and important defense against tumor establishment. As described in Chapter 11, activated NKT cells rapidly express a broad collection of cytokines, including copious amounts of IFNγ. This early burst of IFNγ stimulates the antitumor activities of NK cells, causing them to produce IFNγ in a sustained manner. As noted earlier, IFNγ has an important antimetastatic effect, at least in mice. In addition to stimulating NK cells, activated NKT cells can carry out direct perforin-dependent cytolysis of tumor cells *in vitro*, although it has been difficult to demonstrate this capacity *in vivo*.

In contrast to the antitumor functions of type 1 NKT cells, a pro-tumorigenic role has been proposed for the rare type 2 NKT cell subset based on studies of tumorigenesis in animal models. Type 2 NKT cells appear to be able to suppress immune responses to developing cancers. This activity is independent of the antitumor suppression mediated by CD4$^+$CD25$^+$ Treg cells. The relevance of type 2 NKT cells to tumor promotion in humans is under investigation.

The NKT cells discussed here are the more common type 1 NKT cells. The differences between type 1 and type 2 NKT cells were outlined in Chapter 11.

IX. αβ T Cells

Responses to cancers by αβ T cells depend on whether their TCRs can be engaged by TAAs and/or TSAs, and whether that engagement triggers T cell activation. For tumors expressing TAAs, central and peripheral T cell tolerance will have already been established, and the malignant cells will be seen as self. This tolerance will have to be "broken" in order for these TAAs to become visible to the immune system, with the added complication that self tissues expressing that TAA may come under attack. In the case of tumor cells expressing TSAs, the new macromolecules may indeed constitute antigenic structures that become targets for adaptive immunity. Notably, while the initial assault on a tumor by αβ T cells is largely in response to epitopes of TSA/TAAs derived from the surface or interior of the tumor cells themselves, the nature of the targets may change with time. The T cell attack on the malignant mass frequently exposes new tumor antigens as the cells and surrounding tissues are degraded. Some of these antigens, which may have previously been hidden from the immune system, are taken up by DCs and presented as new pMHC epitopes to draw additional clones of naïve T cells to the fight. In general, the broadening of an immune response due to the increased accessibility of new antigens is called **epitope spreading**. Epitope spreading is thought to be important for sustaining the immune response against a tumor long enough for rejection to be achieved.

NOTE: Once drawn into a tumor microenvironment by local inflammation, infiltrating T cells may encounter a hotbed of TSA production due to the inherent genetic instability of cancer cells. *Microsatellites* are short repeated sequences of DNA present in all normal cells. In cells with defects in DNA repair, microsatellites may be duplicated or deleted such that the cell exhibits microsatellite instability. Tumors with high microsatellite instability have been found to generate unique TSAs that can be recognized by TILs, which then remain in the tumor site and initiate a strong adaptive antitumor response. Indeed, histologically, cancers whose component cells show high levels of microsatellite instability are often heavily infiltrated with TILs. Tertiary lymphoid tissue may also develop near or within these tumors, more evidence that a local immune response is occurring.

In terms of cellular dynamics, it is believed that antitumor responses by αβ T cells commence when immature DCs that have acquired TAA/TSAs from a tumor microenvironment mature as they migrate to the local lymph node. Within the node, these mature DCs may activate naïve antitumor αβ Th cells, becoming licensed DCs that may activate naïve antitumor αβ Tc cells via cross-presentation. Studies in humans and mice have provided convincing evidence that both CD4$^+$ Th and CD8$^+$ Tc cells contribute to effective tumor rejection. Researchers have established many cloned T cell lines with antitumor activity. Under the appropriate conditions, the transfer of such a T cell line to a tumor-bearing mouse can reduce the size of its malignancies. Not surprisingly, the primary function of activated antitumor CD4$^+$ Th cells is believed to be to supply help to naïve antitumor CD8$^+$ Tc cells and thus promote the generation of antitumor CTLs. Such CTLs make a major contribution to the control of human cancers, with IFNγ production being a crucial activity.

In addition to supporting the antitumor CTL response, Th1 and Th2 effectors release cytokines that recruit and further stimulate innate leukocytes such as neutrophils, eosinophils, macrophages and NK cells. Th17 cells produce IL-17, IFNγ, CXCL8 and granulocyte-macrophage colony stimulating factor (GM-CSF) which promote neutrophil recruitment, activation and prolonged survival at inflammatory sites (including in the tumor microenvironment). Furthermore, IL-17 supports the generation and functions of CTLs. The importance of IL-17's antitumor functions has been illustrated in studies of a mouse strain that is particularly susceptible to colon cancer development. These mutants experience more rapid tumor growth and greater metastasis if they are genetically engineered to also lack IL-17.

Naturally, because αβ T cells are so important for the adaptive antitumor response, cancers often develop ways to evade assaults by these cells so that they may continue to grow in the host. New mutations acquired by tumor cells, or products that they secrete, can allow them to escape immunosurveillance, suppress immune responses, and/or thwart leukocyte effector functions. In such cases, the tumor appears to induce immune tolerance to itself. This apparent tolerance is frequently mediated by cytokines such as IL-10 and TGFβ. In the presence of these immunosuppressive molecules, any TAA/TSAs acquired by an immature DC are taken up in a microenvironment that does not support DC maturation. T cells that encounter these DCs are thus more likely to be anergized than activated. Accordingly, patients with colorectal cancer or lymphoma often have a worse prognosis if they also have elevated plasma IL-10 levels. *In vitro*, IL-10 treatment decreases MHC class I expression on melanoma cell lines, inhibits TAP function, suppresses DC activity, and blocks CTL-mediated lysis of tumor cell lines. In addition, some cancers can force the downregulation of TCR-associated signaling molecules in TILs that have invaded them, blocking their attack. It is unknown how the cancer accomplishes this feat, but the establishment of an inhibitory intercellular contact is suspected. Other tumor cells upregulate their expression of ligands for negative T cell regulators such as CTLA-4 and PD-1. Engagement of these molecules shuts down any T cells attacking the tumor regardless of antigenic specificity.

The negative regulation of T cell activity exerted by CTLA-4 and PD-1 was described in Chapter 9.

Although it is likely that an evolutionary imperative of T cells is to fight cancers, some of their effector functions can have pro-tumorigenic effects. For example, Th1 cells produce IL-2, which is essential for the generation of Treg cells. If these Treg cells suppress the responses of conventional antitumor T cells, cancer expansion may be promoted. Studies have shown that tumor rejection is enhanced when Treg cells are depleted from mice prior to their transplantation with cancer cells. *In vitro,* Treg cells can non-specifically inhibit the antitumor responses of not only Th cells and CTLs but also NK cells. In human cancer patients, the presence of Treg cells has been linked to reduced survival in some, but not all, cases. (On the other hand, sometimes Treg cells appear to decrease tumor-promoting inflammation, making their overall contribution to tumorigenesis hard to predict.)

The cytokines and chemokines produced by Th2 and Th17 cells can also sometimes support cancer cell survival. As noted in Chapter 9, the cytokines produced by Th2 cells tend to suppress Th1 and Th17 cell differentiation, reducing antitumor populations of these cells. The IL-17 produced by any Th17 cells that are present can induce the synthesis of molecules such as vascular endothelial growth factor (VEGF) that promote angiogenesis, as well as IL-6, which stimulates tumor cell growth. In the case of ovarian cancers, Th17 cells have been associated with the progression of these malignancies because the neutrophil-recruiting chemokines produced by Th17 cells attract the pro-tumorigenic N2 subset.

X. B Cells

It remains unclear whether B lymphocytes actively participate in immunosurveillance and tumor rejection. Antitumor antibodies are not often detected in cancer patients, and those that are present do not seem to be effective. However, in certain mouse cancer models, animals depleted of B cells show an increase in tumorigenesis, implying that B cells do indeed contribute to antitumor immunity, at least in some contexts. As described later in this chapter and in Chapter 20, antitumor antibodies produced experimentally can be used successfully for antitumor immunotherapy.

Evidence that B cells can have pro-tumorigenic effects has emerged from animal studies. Mice that are depleted of B cells and treated with carcinogens show a delay in the conversion of pre-malignant cells to fully malignant cells. B cells have also been implicated in the production of cytokines that inhibit Th1 responses and support the establishment of a microenvironment favoring tumor cell growth. Cancer patients do produce low levels of antibodies to TAAs/TSAs, and these antibodies can form immune complexes that may bind to the FcRs of innate leukocytes. If this interaction results in persistent leukocyte activation, the sustained inflammation that results may be pro-tumorigenic. Finally, the B-1 subset of B cells that resides primarily in the peritoneum has been found to promote metastasis in mouse models of melanoma. This effect is mediated by intercellular contacts that result in tumor cells upregulating adhesion molecules facilitating migration.

NOTE: Some immunologists view the immune system's involvement in antitumorigenesis as a three-phase process of **cancer immunoediting**. In the first phase, dubbed "elimination," innate and adaptive leukocytes detect and destroy incipient tumor cells when their numbers are low and before a clinically detectable malignancy forms. The second phase is "equilibrium," in which the tumor cells are not all killed but are kept under control by the immune system such that the cancer may or may not become clinically significant. If the cancer does cause illness but the patient is treated appropriately, he/she may appear to have beaten the threat and may be healthy for an extended period. However, this patient may later suffer a relapse when his/her immune system is weakened or suppressed. During equilibrium, the tumor cells present are deemed to be highly immunogenic in that they readily attract leukocyte attention. However, this attention by the immune system also exerts selective pressure on the tumor cells, much as the application of an antibiotic promotes the selection of drug-resistant bacteria. Such selection can then lead to the third "escape" phase, in which newly acquired mutations allow some tumor cells to become completely resistant to control by the host's leukocytes and inexorably generate a clinically detectable malignant growth. Tumor cells in the escape phase are deemed to be either poorly immunogenic because they are no longer recognized by the host's immune system, or are immunosuppressive and produce substances that actively shut down the host's immune response. The involvement of the host immune system in shaping the nature of a cancer during each of these three phases is viewed by some immunologists as a form of "editing"; hence, the adoption of the term "cancer immunoediting."

D. Conventional Cancer Therapy

The first step in cancer therapy is almost always the surgical removal of the complete primary tumor, if possible. This approach is clearly not an option for very diffuse cancers (such as leukemias), in which the tumor cells circulate throughout the body, nor for many metastatic cancers, in which tumors may be present in multiple small and hidden sites. Neither is surgery advisable when the tumor is located such that the efforts to remove it would irreparably damage a vital body structure, as in many types of brain cancers. In these cases, clinicians turn to chemotherapy and/or radiation therapy. Both these techniques predominantly affect rapidly dividing cells while sparing the resting or slowly dividing cells that comprise the majority of normal cells in the body. Chemotherapy and/or radiation therapy are also often used in conjunction with surgery in an effort to kill any tumor cells that were missed and prevent re-establishment of the cancer.

I. Chemotherapy

Chemotherapy is the use of pharmaceutical drugs to kill tumor cells in a cancer patient. The underlying principle is that a chemotherapeutic agent primarily affects only those cells that are growing faster than most normal cells, or cells that have a metabolic imbalance. Chemotherapeutic drugs work by directly or indirectly damaging the replicating DNA of the dividing tumor cell, inhibiting DNA synthesis, preventing cell division, or blocking the access of the tumor cells to a necessary growth factor. Most chemotherapy drugs are alkylating agents, antimetabolites, glucocorticoids or plant alkaloids. Because combinations of two or more chemotherapeutic drugs often work synergistically, patients are usually treated with "cocktails" containing anywhere from two to six agents, depending on the type and stage of the tumor. The use of more than one agent also greatly reduces the chance of the cancer becoming resistant to drug treatment, since the malignant cells would have to acquire resistance-conferring mutations to multiple drugs at once.

Unfortunately, by its very nature, conventional chemotherapy is not very specific, and fast-growing normal cells such as those in bone marrow are also damaged. Immunity to pathogens therefore plunges. Cells of the gastrointestinal tract and hair follicles are also highly proliferative, so that chemotherapy almost inevitably causes nausea, vomiting and hair loss. In addition, because the liver and kidneys tend to accumulate these powerful drugs and their metabolites, hepatic and nephrotic toxicity are common side effects. To address these issues, scientists have developed molecular agents designed to specifically target tumor cells. Several of these agents have demonstrated higher efficacy and lower toxicity for particular cancers than conventional chemotherapeutics.

II. Radiation Therapy

The principle of radiation therapy is that photons and sub-nuclear particles emitted by radioactive substances generate ROIs that cause severe damage to the DNA of tumor cells with which they interact. The tumor cells then die when they attempt to initiate DNA replication. Radiation oncologists can use sophisticated radiation machines to externally deliver "hits" of damaging energy to tumors that are hidden deep within the body. Alternatively, internal delivery (**brachytherapy**) can be achieved by implanting a metal "seed" containing the radioactive material either directly in the tumor or very close to it. In this latter method, high doses of radiation are delivered for short periods of time, decreasing detrimental side effects on patients. Most radiation side effects are due to the impact of radiation on fast-growing normal tissues, such as the gastrointestinal mucosae, the skin and the bone marrow. Nausea, vomiting, hair loss and bone marrow suppression (failure to generate new hematopoietic cells) are all commonly observed in radiation-treated cancer patients.

> NOTE: Researchers are examining the way tumor cells are killed by chemotherapy and radiation and the implications of the form of their deaths for antitumor immunity. Chemotherapy and radiation tend to induce the *apoptotic* death of rapidly dividing cells, meaning that few DAMPs are released into the surrounding microenvironment to alert cells of the immune system. If these therapies are successful in killing 100% of the cancer cells, then the lack of an immune response to the dead cells does not matter: the cancer is killed, mission accomplished. However, fully one out of every two human cancers bears a mutation in p53, the master tumor suppressor that normally induces the growth arrest or apoptotic death of cells with DNA damage. Thus, in the absence of p53 to mediate apoptotic death, chemotherapy and radiation therapy are often less than completely effective treatments. In these cases, any residual cancer cells remaining after the treatment will persist in an environment relatively lacking in stimuli that would induce innate or adaptive responses, allowing the tumor cells to grow unchecked. Thus, scientists are looking for ways to treat cancers such that necrotic, rather than apoptotic, death is induced. As described in Chapter 2, necrotic cell death is messy and floods the immediate area with DAMPs, which then recruit and activate innate leukocytes. Moreover, if immature DCs acquire TAAs/TSAs from tumor cells undergoing necrotic death, the chance of an antitumor adaptive response being mounted greatly increases, which could bolster treatment efficacy.

A list of chemotherapeutic drugs currently in use for the treatment of particular malignancies can be found at http://www.cancer.org/treatment/treatmentsandsideeffects/guidetocancerdrugs/index.

The U.S. National Institutes of Health (NIH) list over 2500 clinical trials for chemotherapeutic drugs at http://clinicaltrials.gov/search/open/intervention=chemotherapy.

E. Immunotherapy

Although chemotherapy and radiation therapy have improved the survival of many cancer patients, there are several problems with these approaches. The high levels of drugs and radiation applied in an effort to achieve sufficient killing power at a tumor site may encourage the rise of drug-resistant and radiation-resistant populations of tumor cells. In addition, systemic or local toxicity can be significant because there is often no way to confine the drug or radiation treatment precisely to the tumor site. This problem is compounded if the tumor has metastasized. Another concern is that the biology of the CSCs thought to drive tumorigenesis appears to render them resistant to current modes of chemotherapy and radiation therapy. Clinical researchers have thus turned to the immune system and devised various forms of **immunotherapy** that can theoretically kill cancer cells with high specificity. The idea is to harness components of the immune system such as antibodies, T cells and/or cytokines to selectively destroy tumor cells. Immunotherapies are already being used to complement conventional cancer treatments and are starting to replace them in select circumstances.

The obvious antigens to target to induce antitumor immunity are the TSAs expressed by cancer cells, and researchers were originally optimistic that they could induce clinically effective antibody and T cell responses specific for such antigens. However, it has proved extremely challenging to identify TSAs, and although TAAs might serve as suitable targets, TAA-directed antibodies and lymphocytes have the potential to attack healthy tissues. As alternatives to TAA/TSA-based therapies, researchers are testing strategies based on cytokine secretion. Also under examination are agents that disrupt tumor angiogenesis or block the supply or action of a required tumor growth factor. Another hope for the future is that unique antigens may be identified on CSCs that can be used to generate antibodies or T cells capable of destroying them.

I. Antibody-Based Immunotherapies

Several novel cancer therapy techniques have taken advantage of the specificity and ease of manipulation of monoclonal antibodies (mAbs). Such *antibody-based immunotherapies* usually involve the laboratory production of a tumor-specific mAb that is then administered to a cancer patient. These clinical mAbs are often produced by immunizing mice with the human antigen of interest and then generating either *chimeric* or *humanized* immunoglobulins using genetic engineering techniques. A higher percentage of human protein sequences within the mAb structure generally correlates with increased mAb function and resistance to premature clearance from the patient's body. Some antitumor mAbs are used without any further modification to the Ig protein, in which case their antitumor effects are due to standard antibody effector functions. Other antitumor mAbs are covalently linked to a cytotoxic molecule (such as a toxin, drug or radioisotope) that can kill a cell on contact. These therapeutic mAbs are called **immunoconjugates** and act as a drug delivery mechanism that specifically targets tumor cells. Hypothetically, the use of a TSA-directed immunoconjugate ensures that normal cells are spared while the toxic agent is targeted to cancer cells in both the primary tumor and metastases.

A monoclonal antibody is derived from a single B cell clone. The properties and production of mAbs were described in Chapter 5.

i) Unconjugated mAbs That Target Tumor Cells
a) mAbs against Tumor Antigens
A therapeutic unconjugated mAb that binds to a TAA/TSA on the surface of a tumor cell may in turn be bound by the FcRs of a macrophage, neutrophil, eosinophil or NK cell. The tumor cell is then destroyed by phagocytosis, ADCC or complement-mediated lysis. Alternatively, an unconjugated mAb may be used to sequester growth factors needed by a tumor to continue its expansion, or to interfere with angiogenic factors and thus reduce the blood supply to the malignant mass. In other cases, mAb binding may be employed to coat a tumor cell and make it more attractive to DCs and other APCs, increasing the chances of inducing an adaptive response to the tumor. Interestingly, cross-presentation achieved by such mAb administration can induce CTL responses against more than one

Fig. 16-4

Examples of Immuno-therapy with Monoclonal Antibodies

(**A**) Unconjugated mAbs specific for a TAA/TSA can be used to induce the death of tumor cells by ADCC. (**B**) MAbs specific for a TAA/TSA and conjugated to a toxin like Ricin A (or to a cytotoxic drug; not shown) can be used to directly kill tumor cells. (**C**) MAbs specific for a TAA/TSA and conjugated to a radioisotope like ^{90}Y can be used to directly kill tumor cells by radioactivity. (**D**) MAbs specific for a TAA/TSA and conjugated to an enzyme can be used to kill tumor cells by turning an inactive pro-drug into an active cytotoxic drug. (**E**) MAbs specific for a TAA/TSA and conjugated to a cytokine can be used to link a tumor cell to an effector leukocyte bearing the receptor for that cytokine. The effector cell is then in a prime position to kill the tumor cell.

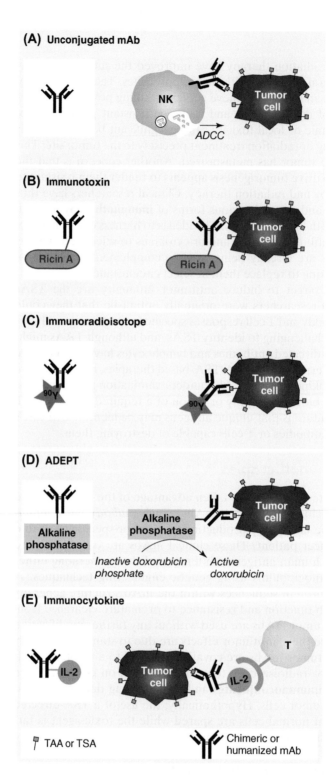

tumor antigen; that is, a response is mounted against a different tumor antigen that is not targeted by the mAb but is present on the same tumor cell.

Unconjugated mAbs can be used on their own (or in combination with chemotherapy or radiation therapy) to combat human cancer *in vivo* in those rare cases in which a TAA/TSA can be identified on the tumor cell, and a mAb can be raised against that TAA/TSA in mice (**Fig. 16-4A** and **Table 16-4**). For example, a successful antitumor mAb treatment has been devised for the 25% of breast cancer patients whose tumors overexpress HER2. The drug Herceptin (trastuzumab) is a humanized mAb that binds to the HER2 protein and both promotes NK cell–mediated ADCC

TABLE 16-4	Examples of Unconjugated Monoclonal Antibodies Used for Immunotherapy		
Antibody Name	**Specificity**	**Mode of Action**	**Cancers Treated**
Alemtuzumab (CAMPATH-H)	CD52	Facilitates killing of CD52-bearing B cells	Chronic B cell lymphocytic leukemia
Bevacizumab (Avastin)	VEGF	Prevents tumor angiogenesis	Lung cancer Metastatic colorectal cancer HER-negative breast cancer
Cetuximab	EGFR	Facilitates killing of tumor cells overexpressing EGFR	Metastatic colorectal cancer Advanced head-and-neck cancers
‡Farletuzumab	Folate receptor	Facilitates killing of tumor cells overexpressing folate receptor	Ovarian cancers
Ipilimumab	CTLA-4	Blocks CTLA-4–mediated inhibition of T cell–mediated antitumor responses	Metastatic melanoma
Ofatumumab	CD20	Facilitates killing of CD20-bearing B cells	Refractory chronic B cell lymphocytic leukemia
Panitumumab	EGFR	Facilitates killing of tumor cells bearing EGFR	Metastatic colorectal cancer
‡Ramucirumab	VEGF receptor	Prevents tumor angiogenesis	Breast, gastric and liver cancers
Rituximab	CD20	Facilitates killing of CD20-bearing B cells	B cell lymphoma Chronic B cell lymphocytic leukemia
Trastuzumab (Herceptin)	HER2	Blocks proliferative signaling in tumor cells and promotes NK killing	HER2+ breast cancer HER2+ gastric carcinoma
‡Zalutumumab	EGFR	Facilitates killing of tumor cells overexpressing EGFR	Head-and-neck cancers

‡Promising but not yet approved for general clinical use.

and interferes with HER2-mediated signals driving breast cancer cell proliferation. Similarly, cetuximab is a chimeric mouse/human mAb that promotes the killing of colorectal cancer cells or head-and-neck carcinoma cells expressing EGFR. Two other useful mAbs called rituximab and ofatumumab are directed against the B cell surface marker CD20. These agents are now widely used for treatment of various types of B cell lymphomas in which the cancerous lymphocytes greatly overexpress CD20. Other B cell lymphomas that overexpress CD52 rather than CD20 can be treated with alemtuzumab.

Cancers of hematopoietic cells and their treatment are discussed in more detail in Chapter 20.

NOTE: The name of a mAb conveys a great deal of information about what species it was derived from, whether the protein is chimeric (e.g., a mouse Ig's constant region is replaced with the human form) or humanized (e.g., a mouse Ig's variable regions are replaced with the human form), and its target (a specific tumor type, pathogen or disease). For example, trastuzumab, which is used to treat human breast cancer, is composed of the elements "tras" = trade name prefix; "tu" = tumor target; "zu" = humanized Ig; and "mab" = monoclonal antibody. Updated naming conventions for mAbs can be found at http://www.ama-assn.org/ama/pub/physician-resources/medical-science/united-states-adopted-names-council/naming-guidelines/naming-biologics/monoclonal-antibodies.page.

As well as mAbs targeting tumor cells, mAbs have been developed to attack the tumor vasculature. These antibodies target molecules which are expressed at higher levels on tumor blood vessels than on normal blood vessels. Other mAbs bind to factors like VEGF, preventing this angiogenic factor from binding to its corresponding receptor on endothelial cells, and thereby impairing the ability of the tumor to build new blood vessels. Another mAb-based approach that has shown promise at the basic research level involves the induction of tumor cell apoptosis. TRAIL is a TNF-related molecule that triggers apoptosis by engaging its receptor TRAIL-R, and mapatumumab is an agonistic humanized mAb that

An *antagonistic* mAb shuts down the activity of the surface protein it binds to, while an *agonistic* mAb stimulates its activity.

mimics TRAIL activity by binding to TRAIL-R and inducing cell death. Tumor cells expressing TRAIL-R that bind to mapatumumab are forced to commit suicide. Pre-clinical testing of the specificity and usefulness of this mAb and related agents is under way.

b) Bi-Specific mAbs

Researchers have genetically engineered immunoglobulin genes to encode antibody proteins with one Fc region but two different Fab sites. Sometimes, each Fab site recognizes a different TAA/TSA, increasing the chance that the antibody will bind tightly to a tumor cell expressing both antigens. The antibody then triggers the usual Fc-mediated effector functions that eliminate the tumor cell. In other cases, one of the Fab sites recognizes a TAA/TSA while the other binds to the CD3 molecules of the TCR complex. The engagement of CD3 by such a mAb activates a T cell regardless of its antigenic specificity. A tumor cell encountering an anti-TSA/anti-CD3 bi-specific mAb is thus attacked by effector functions activated through the Fc region of the mAb as well as by effector functions of any T cell brought into contact with the tumor cell by the mAb. An example is an agent called catumaxomab, which is a bi-specific antibody that uses its two different Fab sites to simultaneously bind to CD3 and epithelial cell adhesion molecule (EpCAM), a TAA upregulated on breast cancer and colon cancer cells. The Fc region of this mAb then engages the FcRs of innate leukocytes. The effectiveness of catumaxomab in killing tumor cells *in vitro* and *in vivo* has led to its approval for the treatment of certain malignancies in Europe.

A spin-off of bi-specific antibodies is a class of molecule called a "bi-specific T cell engager" (BiTE), in which an Fab site of an anti-CD3 mAb is directly linked to the Fab site of an anti-TAA mAb. No Fc region is present, and thus no complete mAb is involved. BiTE proteins are translated from a synthetic gene in which the required Fab coding sequences are linked together at the nucleotide level. Any T cell bound by a BiTE is activated immediately when the other end of the BiTE engages a tumor cell. A CD3-CD19 BiTE is currently in clinical trials for the treatment of advanced non-Hodgkin lymphoma, a lymphoid cell cancer that frequently expresses the B cell marker CD19.

ii) Conjugated mAbs That Target Tumor Cells

As introduced previously, immunoconjugates are chimeric proteins in which mAbs are linked either chemically or via genetic engineering to lethal effector molecules such as toxins, drugs, radioisotopes, enzymes and cytokines. The idea is that the specificity of the mAb brings the effector molecule directly to the tumor site, sparing normal tissues. Sometimes the immunoconjugate is internalized by the tumor cell, and the toxic portion kills the cell by disrupting intracellular survival pathways. In other cases, the effector molecule is delivered within the cancerous mass or its vasculature but remains external to the individual tumor cell. Some examples of immunoconjugates are given in **Table 16-5**.

TABLE 16-5	Examples of Conjugated Monoclonal Antibodies Used for Immunotherapy			
Antibody Name	**Specificity**	**Conjugate**	**Mode of Action**	**Cancers Treated**
‡Brentuximab vedotin	CD30	Auristatin E	Facilitates killing of CD30-bearing tumor cells via conjugated antimitotic agent	Hodgkin lymphoma Anaplastic large cell lymphoma
Gemtuzumab ozogamicin (Mylotarg)	CD33*	Calicheamicin	Facilitates killing of CD33-bearing hematopoietic cells via conjugated cytotoxic antibiotic	CD33+ acute myeloid leukemia
Ibritumomab tiuxetan (Zevalin)	CD20	90Y	Facilitates killing of CD20-bearing B cells via radioisotope yttrium-90	B cell lymphoma
Tositumomab (Bexxar)	CD20	131I	Facilitates killing of CD20-bearing B cells via radioisotope iodine-131	B cell lymphoma
‡Trastuzumab-emtansine	HER2	Emtansine	Facilitates killing of HER2-bearing tumor cells via conjugated antimitotic agent	HER2+ breast cancer

*CD33 is an adhesion protein found on leukemic cells and immature myeloid cells.
‡Promising but not yet approved for general clinical use.

As attractive as the immunoconjugate concept is, it faces two major challenges. First, there are very few known TSAs to raise a mAb against, making it difficult to construct truly tumor-specific agents. Second, the preparatory research in animal models frequently overestimates the efficacy and underestimates the toxicity of a drug or toxin in humans. Despite these difficulties, researchers are pursuing several therapeutic approaches based on immunoconjugates.

a) Immunotoxins and ADCs

When the molecule linked to a mAb is a toxin or drug, the immunoconjugate is known as an *immunotoxin* or an *antibody-drug conjugate (ADC)* (**Fig. 16-4B**). For example, mAbs have been joined to bacterial toxins such as pseudomonas exotoxin (PE) and diphtheria toxin (DT) as well as to the "A" subunit of the plant toxin ricin. Once internalized by the tumor cell, the toxin inhibits protein or nucleic acid synthesis or damages the genomic DNA. Many immunotoxins are extremely potent, requiring only one molecule of toxin per tumor cell to kill. This efficacy means that very few immunotoxin molecules have to reach the tumor site to have a therapeutic effect. Indeed, significant clinical promise has been demonstrated by an immunotoxin in which modified PE toxin is linked to part of a mAb recognizing CD22. However, it has been difficult to control the immunogenicity and half-life of immunotoxins, and to achieve adequate penetration of solid tumors. More recently developed ADC preparations include classes of drugs that are antibiotics (e.g., calicheamicin) or interfere with the mitosis of tumor cells (e.g., auristatin E). The ADC brentuximab vedotin, which targets CD30 on tumor cells, has been approved for the treatment of some lymphomas.

b) Immunoradioisotopes

When a radioisotope such as iodine-131 or yttrium-90 is linked to a mAb, the resulting immunoconjugates are called *immunoradioisotope*s (**Fig. 16-4C**), and the use of these agents to treat cancer patients is called *radioimmunotherapy (RIT)*. For example, a yttrium-90-labeled mAb has been constructed that binds to the CEA protein often expressed by colon, pancreatic, breast, ovarian and thyroid cancers. The targeting of this immunoconjugate to these tumor cells has been encouragingly precise, and effective clinical responses have been observed in cancer patients. Two immunoradioisotopes based on murine mAbs recognizing CD20 have been approved for the treatment of certain lymphomas, and others are in clinical trials. Interestingly, these patients do not seem to mount immune responses against the mouse mAb proteins.

One difficulty with RIT is that the entire body (and particularly the bone marrow) is exposed to the radioisotope while the mAb is circulating to find its target. Researchers are developing administration strategies that more quickly concentrate the immunoradioisotope at the tumor site so that it does minimal damage to normal tissues. Another drawback to RIT is that radioisotope released from the mAb during catabolism can accumulate in the liver or kidney, damaging these organs.

c) ADEPT

Another type of immunoconjugate is created when a mAb is linked to an enzyme and is used for *antibody-directed enzyme/pro-drug therapy (ADEPT)* (**Fig. 16-4D**). In this situation, an antitumor mAb is conjugated to an enzyme capable of converting an inert pro-drug into an active cytotoxic drug. For example, alkaline phosphatase conjugated to a mAb is capable of converting inactive doxorubicin phosphate to the active anticancer drug doxorubicin. The immunoconjugate is given to the patient in advance of the pro-drug, giving the mAb–enzyme complex time to specifically localize in the tumor site and to be cleared from other areas of the body. The pro-drug is then administered so that when the toxic metabolite is generated, it is (theoretically) confined to the tumor mass where the mAb–enzyme is bound. In practice, it is sometimes difficult to clear the mAb–enzyme complex from all non-cancerous tissues. Another challenge is that, since the immunoconjugate in this case contains an enzyme (which is a relatively large protein), the immunoconjugate may itself be immunogenic. An immune response may be induced that inactivates the enzyme before it can catalyze pro-drug conversion.

d) Immunocytokines

Some immunoconjugates are *immunocytokines (ICKs),* chimeric proteins in which an anti-TAA/TSA mAb is linked to a cytokine that either has antitumor properties or can stimulate leukocytes in the tumor site (**Fig. 16-4E**). (ICKs are not internalized like immunotoxins.) Here, the plan is to prolong the half-life of the cytokine and bring it directly to the malignant mass. For example, one well-studied ICK contains IL-2 fused to a mAb directed against EpCAM. Use of this immunoconjugate in both humans and mice has shown promise for the control of colon cancer cell metastasis. ICKs containing IL-2 have also worked well in some T cell leukemia patients.

> NOTE: Antibodies are traditionally viewed as binding to extracellular epitopes only, and therefore not useful for targeting intracellular proteins. However, recent research in mouse cancer models has demonstrated that at least some mAbs are internalized by tumor cells, whereupon the mAb can bind to an intracellular antigen and inhibit tumor cell growth from the inside. If this finding withstands further scrutiny, it may become possible to target both the internal and external mediators of a cancer cell's extraordinary survival and proliferation.

iii) Immune Blockade with mAbs

Unconjugated mAbs can be used to establish an **immune blockade**, in which regulatory mechanisms that could restrain an antitumor T cell response are inhibited. Researchers speculate that one reason certain cancers are allowed to progress is that the T cell responses against them are eventually brought to a natural halt by negative regulators such as CTLA-4. As described in Chapter 9, CTLA-4 is expressed on the surfaces of T cells about 2 days after activation. For a conventional T cell, the binding of CTLA-4 to the B7 ligands it shares with CD28 shuts down TCR signaling, induces inhibitory signaling, and damps down the cell's ability to respond. For a regulatory T cell, engagement of CTLA-4 enhances its suppressive effects on conventional T cells, also contributing to response shutdown. Thus, a mAb that binds to CTLA-4 and prevents it from interacting with its B7 ligands may help to sustain an existing antitumor T cell response. A huge advantage of this type of approach is that one does not have to know the identity of a TAA/TSA to affect the T cell response to it. Indeed, in the absence of CTLA-4-mediated negative regulation, antitumor T cell responses have been found to proceed much more vigorously. As a result, the anti-CTLA-4 mAb ipilimumab has now been approved as an adjunct treatment for advanced prostate cancers and melanomas, and indeed is the only therapy so far to prolong the survival of patients with metastatic melanoma. However, immune blockade comes with an important caveat: interference with the negative regulatory mechanisms controlling conventional T cell responses may also unleash autoreactive T cells in the periphery. Accordingly, some melanoma patients treated with ipilimumab to achieve a CTLA-4 blockade have shown signs of autoimmune disease. Careful dosing and monitoring of these patients is therefore required to achieve the right balance between inducing tumor regression and avoiding autoimmunity.

> NOTE: Two interesting clinical issues have arisen as the CTLA-4-mediated immune blockade approach has made the long trek to the clinic. Ipilimumab treatment is effective for only a small subset of melanoma patients, carries the risk of significant side effects, and is currently very expensive. The search is thus on for some way that clinicians could use to identify those patients who would benefit most from ipilimumab treatment. A *biomarker* is a molecule expressed by a patient that can accurately predict the response of that patient to treatment with a particular therapeutic agent and is most often related to the pathway affected by that agent. Indeed, biomarkers are sought for all oncology therapies that help only subsets of patients. To date, a reliable biomarker for responsiveness to ipilimumab has yet to be isolated.
>
> The other interesting clinical issue regards the criteria used to determine whether a patient has responded to an anticancer agent. Immune blockade appears to take effect much more slowly than conventional chemotherapeutic drugs or radiation therapy. A tumor in a patient treated with immune blockade may even increase in size before it finally starts to

regress. As a result, a patient evaluated at a conventional time after treatment initiation might be rated as "not responding" to treatment, when a longer observation period would reveal a slow-developing response. Many clinicians are now pushing to have the formal clinical response criteria revised in order to account for the tardy success of immune blockade agents.

Other modulators of T cell responses are currently under investigation as potential therapeutic targets for immune blockade. As described in Chapter 9, PD-1 is primarily a negative regulator of TCR signaling in effector T cells in an inflammatory site. However, unlike CTLA-4, PD-1 is also expressed by NK cells and B cells, and so also controls the power of these leukocytes. Many tumor cell types upregulate their expression of PD-1 ligands in an effort to engage PD-1 on antitumor T cells, B cells and NK cells and thus forestall an immune assault. An antagonistic mAb that binds to the extracellular domain of PD-1 and establishes a PD-1 immune blockade has therefore been effective in boosting antitumor T cell responses in small numbers of cancer patients. This mAb may also enhance antitumor NK cells and B cell responses, although this has yet to be formally established. Moreover, this agent is well tolerated and does not induce the autoimmune symptoms experienced by patients treated with CTLA-4 blockade. More extensive clinical trials are under way to examine this mAb's effects on melanomas as well as on renal, colon and lung cancers.

Another way to indirectly establish an immune blockade may be to use agonistic mAbs recognizing T cell costimulatory molecules such as OX40 and CD40. Early clinical trials examining such mAbs have found that these agents not only stimulate the responses of antitumor T cells but also have the beneficial effect of shutting down regulatory T cell differentiation. The suppressive actions of these latter T cell subsets that might promote tumor cell growth are thus curtailed.

NOTE: In an approach parallel to the immune blockade methods described above, mAbs have been exploited to enhance innate antitumor responses. Some cancer cells overexpress NK activatory ligands not seen on normal cells, making these tumor cells potential NK cell targets. However, if the tumor cell also expresses inhibitory ligands that can bind to an NK cell's inhibitory receptors, the activatory receptor signaling triggered by the tumor cell's activatory ligands may be counterbalanced and the NK cell attack thwarted. Researchers are currently investigating the possibility of using antagonistic mAbs that can bind to tumor cell inhibitory receptors and block their ability to shut down NK cell activation. Any tumor cell overexpressing activatory ligands would then be vulnerable to death by NK-mediated cytotoxicity. One such antagonistic mAb is currently in early clinical trials.

II. Cancer Vaccines

i) Pathogen-Based Cancer Vaccines

When a cancer is associated with a particular pathogen infection, it may be possible to devise a prophylactic vaccine that can block the infection and thus prevent subsequent tumorigenesis. At the moment, the most common pathogen-based cancer vaccines are those directed against the oncogenic viruses. It is still debated whether these viruses are themselves carcinogenic, since there is very little solid evidence that these pathogens are solely responsible for the genetic deregulation seen in these malignancies; that is, not every individual infected with one of these pathogens develops a cancer. However, it is generally agreed that infected cells expressing oncoviral antigens often later become malignant. Many scientists now believe that it may be the chronic inflammation associated with persistent infection by these pathogens that tips the balance toward carcinogenesis. In any case, oncoviral antigens are logical targets for cancer vaccine development, and therapies designed to target them can be quite successful. For example, vaccines containing proteins of the HBV virus reduce not only the incidence of hepatitis but also liver cancer. Similarly, a vaccine against HPV has been proven to decrease the incidence of cervical cancer.

Many cancer vaccines are designed using the same approaches underlying antipathogen vaccines (refer to Ch 14).

ii) TAA/TSA-Based Cancer Vaccines

Because the immune system does mount responses against cancers, scientists have been trying for many years to use the principles outlined in Chapter 14 to devise vaccines that could enhance the natural antitumor response to tumor antigens in a way that would cure cancers. Such vaccines are therapeutic rather than prophylactic, as they are administered when a patient already has a malignancy in an attempt to generate or boost an antitumor T cell response *in vivo*. The vaccine antigen, which may be a TAA or a TSA, can be introduced into the patient using various vaccination approaches. For example, the vaccine could contain the isolated antigenic protein or peptides derived from it. Alternatively, DNA encoding the antigen could be delivered into the patient directly or by using an expression vector such as a recombinant virus.

a) TAA Vaccines

In theory, a vaccine directed against a TAA might provoke an attack against normal cells. In reality, many TAAs are expressed in non-vital organs, and the limited immune destruction of healthy cells in these locations does not threaten the survival of the host. Indeed, studies of animal cancer models have demonstrated the potential of this approach. For example, TRAMP ("transgenic adenocarcinoma mouse prostate") mice spontaneously develop prostate cancer in a manner that closely approximates the development and progression of prostate cancer in humans. Using a recombinant DNA immunization protocol based on the use of the prostate TAA "prostate stem cell antigen" (PSCA), researchers were able to induce a dramatic reduction in tumor size in TRAMP mice (**Plate 16-3**). A similar approach in humans with various types of cancers has not yet been widely successful in clinical trials, although there have been some encouraging results. Small numbers of melanoma patients immunized with melanoma TAAs have shown some regression of their tumors without significant damage to healthy tissues. In addition, some colon cancer patients have shown clinical improvement after vaccination with CEA, an embryonic TAA that is not normally expressed on adult tissues. However, none of these vaccines has proved to be efficacious in the few large Phase III clinical trials that have been carried out to date.

b) TSA Vaccines

Two TSAs that have been considered as vaccine candidates are the proteins encoded by the oncogene Ras and the TSG p53. In theory, distinctive peptides from the oncogenic versions of these proteins should be presented on the surface of tumor cells and

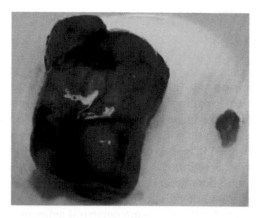

Plate 16-3
Successful Therapeutic Vaccination against a Prostate TAA in Mice

TRAMP mice are a transgenic model of prostate cancer where the animals spontaneously develop large prostate tumors. Age-matched groups of TRAMP mice were vaccinated using a prime-boost vaccination strategy in which the mice were immunized with DNA encoding the prostate TAA mPSCA and boosted using Venezuelan Equine Encephalitis virus particles encoding the same antigen. Prostate tumors were isolated after sacrifice at 340 days. A representative tumor in an unvaccinated TRAMP mouse is shown on the left as a control, while a much smaller cancer from an mPSCA-vaccinated mouse is shown on the right. [*Reproduced from Graya A. et al. (2009) Vaccine, Vol. 27, Suppl. 6:G52–G59.*]

should act as targets for CTLs. *In vitro* assays have demonstrated the presence in mice of CTLs capable of responding to purified mutated Ras peptides (when appropriately presented). Unfortunately, however, these anti-Ras CTLs did not respond to intact tumor cells. In addition, although immune responses against mutated p53 peptides occur in experimental animals, they have not been detected in human cancer patients.

Another type of TSA that may represent a cancer vaccine antigen of the future is the tumor-associated carbohydrate antigen. These abnormal carbohydrates are generally very specific to cancer cells, meaning that a response targeting them would spare normal cells. In addition, such TSAs tend to be expressed on a wide range of cancers, making it practical to create a vaccine based on one or more of them. However, most carbohydrate TSAs are Ti antigens, meaning that they cannot induce an immune response strong enough to eliminate a cancer on their own. In an effort to increase the immunogenicity of carbohydrate TSAs, scientists are experimenting with altering their structures and/or developing carbohydrate subunit vaccines in which the antigenic molecule is coupled to a protein carrier.

iii) Overcoming Barriers to Cancer Vaccine Success

Many clinicians believe that the current failure of Phase III cancer vaccine trials is due to the age and stage of the patients involved. Most of the patients recruited to these cancer vaccine trials are already in the advanced stages of their disease and therefore often severely immunocompromised. Without the ability to mount adaptive responses against TAA/TSAs, there is little chance that a positive trial outcome will be seen. From an ethical standpoint, these patients are the best candidates to take on the risk of a clinical trial because they have no other therapeutic options. On the other hand, the negative results of these trials may not be an accurate reflection of the therapeutic potential of this approach for treatment of the disease at an earlier stage.

Another difficulty with TAA/TSA cancer vaccines is that many TAAs and TSAs are simply not very immunogenic. Moreover, as discussed previously, the DCs present in a tumor site may not be able to activate naïve antitumor T cells due to a lack of DAMPs and/or the presence of tumor-secreted immunosuppressive cytokines. Anergization (rather than activation) of responding TAA/TSA-specific lymphocytes may thus be induced. Indeed, in a few cases, administration of a TAA/TSA vaccine has actually enhanced cancer cell growth.

One way to improve the effectiveness of cancer vaccines may be to deliver them using DCs that have been forced to mature *in vitro*. For example, immature DCs isolated from a mouse can be treated *in vitro* with the growth factor GM-CSF to trigger maturation. The mature DCs are then mixed with tumor antigens that may be purified TAAs or TSAs, or contained in a whole tumor cell lysate. In a process called *electroporation*, the culture is subjected to a mild electrical current that opens tiny pores in cell membranes. The tumor antigens enter the mature DCs such that these cells are said to be "pulsed" or "loaded." The loaded mature DCs are then administered as a vaccine to a tumor-bearing, syngeneic mouse in the hopes of sparking an anticancer response. Because electroporation introduces the tumor antigens directly into the DC cytoplasm, peptides derived from these antigens are presented on MHC class I and can provoke a CD8+ T cell response. If some of the tumor antigens enter the DC via clathrin-mediated endocytosis, the relevant peptides may be presented on MHC class II such that CD4+ T cells are also activated. A variation on the preceding approach is in early clinical trials in humans. The patient's own tumor cells and immature DCs are isolated, the tumor cells are lysed, and the lysate is used to load the patient's immature DCs. These DCs are stimulated to mature *in vitro* and are then returned to the patient with the goal of provoking an enhanced T cell response. Although the earliest results using this approach were disappointing, researchers are working steadily to improve them. In particular, some trials are examining personalized protocols, in which the vaccine antigen is based on peptides derived from the patient's own tumor.

Another way to ensure that the tumor antigens making up a cancer vaccine are ultimately presented by mature rather than immature DCs may be to administer the vaccine using HSPs as an adjuvant. Because HSPs are DAMPs that are naturally released

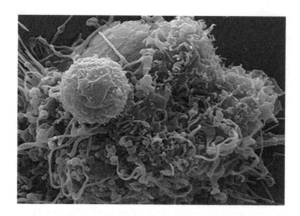

by stressed cells, their administration may promote DC maturation *in vivo* and induce
leukocytes to secrete pro-inflammatory cytokines. In this microenvironment, TAA/
TSA-specific lymphocytes that encounter the mature DCs presenting TAA/TSA peptides should be activated rather than anergized. An adaptive response should then be
mounted against the tumor. This approach has met with considerable success in mouse
models and has now moved into the early clinical trial stage in humans.

In addition to electroporation as a method of introducing TAA/TSAs into DCs,
researchers have been exploring the possibility of loading DCs via phagocytosis of
nanoparticles. Lysates of tumor cells (potentially containing myriad TAA/TSAs) can
be encapsulated in polymer nanoparticles that are begging to be phagocytized by DCs
(**Plate 16-4**). *In vitro* at least, DCs that engulf such nanoparticles gain an enhanced
capacity to activate tumor-specific CD8+ T cells. Translation of these findings to the
in vivo situation is ongoing.

III. Adoptive T Cell Cancer Therapies

In contrast to cancer vaccines, which are a form of active immunization, **adoptive T cell
cancer therapy** is a form of passive immunization in which samples of the cancer patient's
T cells (including rare antitumor T cells) are isolated, activated *in vitro*, and transfused
back into the patient. The hope is that these cells will attack and destroy the tumor and
prevent its recurrence. The transfused T cell population received by **adoptive transfer**
may be polyclonal and contain T cells specific for a variety of (unknown) tumor antigens,
or may be an enriched population selected for a particular T cell surface marker phenotype, or may be a single clone selected by co-culture with APCs loaded with a known
tumor antigen. The T cells harvested from the cancer patient's peripheral blood or draining lymph node often include Th cells and Treg cells as well as CD8+ T cells. Alternatively,
TILs present in the tumor tissue itself may be isolated. The harvested T cells are greatly
expanded in number in culture and activated by incubation with cytokines and APCs.
The activation of these T cells may be enhanced by exposure to agonistic anti-CD3 and/
or anti-CD28 mAbs. The expansion/activation process can take several weeks, and, in the
meantime, the cancer patient is treated with chemotherapeutic drugs that deplete his/her
body of lymphocytes to make space for the incoming population of T cells. The activated
T cells are then transfused into the patient, whereupon they home to the tumor site. The
transfused patient is then monitored for signs of tumor regression.

There has been some clinical success with adoptive T cell therapy in a small number of cases. For example, in one study of melanoma patients, T cells that had infiltrated each tumor were isolated, expanded in the laboratory, and transferred back
to the original patient. Most of these patients subsequently showed accumulations of
melanoma-reactive T cells in their peripheral blood and enjoyed some degree of tumor
regression. Early clinical trials are also under way to explore adoptive T cell cancer
therapy as a treatment for HER2 breast cancer. However, this approach is currently
extremely expensive, labor-intensive, lengthy, and difficult to scale up to the Phase III
clinical trial level. Researchers are exploring ways to overcome these barriers.

IV. Cytokine-Based Therapies

Although it might be expected that cytokines should have antitumor effects, it has been surprisingly difficult to exploit them. Many cytokines also have tumor-promoting effects, most are too rapidly cleared from the body to be effective, and almost all have turned out to be too toxic for use in humans, at least in their pure forms. Scientists have speculated that cytokines might be effective if the dose could be delivered directly to the tumor in a sustained manner. The ICK approach described previously is one attempt to remedy this problem because the antibody that is conjugated to the cytokine will guide it to the tumor. Another method may be to use a transfection approach in which tumor cells isolated from an individual are genetically engineered to express a cytokine gene. When these modified tumor cells are returned to the same individual, the cells home back to the primary tumor and commence secreting the cytokine. This approach has been shown to work in animal models. For example, mice injected with tumor cells engineered to express GM-CSF show increased DC activity and develop strong antitumor Th1 and Th2 responses. The production of NO and anti-angiogenic factors by activated macrophages is also enhanced. In human melanoma patients, trials have been carried out in which the patient's tumor cells were isolated, engineered *in vitro* to express GM-CSF, and returned to the patient. The modified cells homed back to the primary tumor and created a local milieu of GM-CSF that induced the infiltration of the tumor by activated lymphocytes, neutrophils and eosinophils. These leukocytes then worked together to destroy the tumor vasculature. In addition, vigorous antibody responses against melanoma surface antigens and CTL responses against intracellular epitopes were induced. Clinical trials are ongoing to refine and expand the utility of this currently labor-intensive therapy.

We have reached the end of our discussion of whether and how the immune system attempts to fight cancer. We move now from a field in which researchers try to induce immune responses to sustain life to one in which they try to suppress them to sustain life: human tissue transplantation.

Chapter 16 Take-Home Message

- A benign tumor contains well-differentiated cells, grows slowly, and is not usually life-threatening. A malignant tumor or "cancer" is composed of poorly differentiated cells, grows in an uncontrolled and invasive way, and is lethal if not treated.

- TAAs are normal cellular proteins expressed at abnormal concentration, tissue location or developmental stage in a tumor, whereas TSAs are macromolecules unique to the tumor and not expressed by any normal cell.

- Immunosurveillance by the immune system may detect TAAs and TSAs and implement defense against tumors via the actions of NK, NKT and T cells; inflammatory cells; and/or cytokines such as TNF, IFNγ and IL-12.

- Tumor cells may actively evade or suppress immune responses by losing expression of a TSA to which an immune response has been initiated; secreting immunosuppressive cytokines; promoting regulatory T cell-mediated suppression of antitumor responses; inhibiting T cell signaling; expressing non-classical MHC class I molecules that shut down NK cells; and/or having a vasculature that discourages T cell extravasation.

- Immunotherapies are being explored that can specifically kill tumor cells or induce an augmented antitumor immune response. These strategies include the use of mAbs in unconjugated and immunoconjugate forms, cancer vaccines, adoptive T cell therapy and engineered cytokines.

Did You Get It? A Self-Test Quiz

Section A

1) Can you define these terms? neoplasm, metastasis, metastases, proto-oncogene

2) Distinguish between the terms "tumor" and "cancer."

3) Give three differences between benign and malignant tumors.

4) What alterations to rates of cell survival, division and apoptosis might lead to the growth of a tumor?

5) Give three examples each of chemicals, radioisotopes and pathogens linked to cancers. Specify the tumor types involved.

6) Distinguish between sporadic and familial cancers.

7) Describe the four steps of carcinogenesis.

8) Why is metastasis not random?

9) Briefly outline the cancer stem cell hypothesis.

10) Name the three classes of genes commonly mutated in cancers and describe how each leads to tumorigenesis. Give an example of each class of gene.

Section B

1) Can you define these terms? spontaneous regression, tumor antigen

2) Give two reasons why some researchers believe in the effectiveness of immunosurveillance.

3) Give two reasons why other researchers doubt the effectiveness of immunosurveillance.

4) What are the two major classes of tumor antigens?

5) Complete the following: "A TAA is almost always a case of the right molecule expressed at the wrong ____, ____ or ____."

6) Give one example each of a cancer-testis antigen and an embryonic antigen.

7) How is a TSA different from a TAA?

8) Describe four ways a TSA can arise.

9) Give two examples of how expression of a TSA can lead to cancer.

Section C

1) Can you define these terms? angiogenesis, TILs, TAMs, TANs, microsatellite

2) What is tumor rejection?

3) Give three reasons why immune responses against tumors may not be effective.

4) Outline two lines of *in vivo* evidence that suggest tumors actively evade or block immune responses.

5) Describe three ways in which tumor cells might suppress or evade the immune system.

6) How might acute inflammation help to combat tumorigenesis?

7) Describe the pros and cons of therapy-induced inflammation.

8) Outline three ways chronic inflammation promotes tumorigenesis

9) Describe the potential contributions of γδ T, NKT, NK and DCs to anticancer responses.

10) Distinguish between M1 and M2 macrophages in terms of phenotype and effects on tumors.

11) Distinguish between N1 and N2 neutrophils in terms of phenotype and effects on tumors.

12) Describe two antitumorigenic effects of αβ T cell responses.

13) What is epitope spreading, and how might it help during an anticancer response?

14) What contribution does the humoral response make to anticancer responses?

15) Briefly outline the three phases of the concept of cancer immunoediting.

Section D

1) Describe the three conventional modes of cancer treatment.

2) Why do chemotherapy and radiation therapy have such devastating side effects?

3) Chemotherapy and radiation therapy do not appear to be effective against cancer stem cells. Why not?

Section E

1) Why might immunotherapy be an improvement over current conventional approaches for treating cancers?

2) Give two reasons why it has been difficult to induce clinically effective anticancer responses with immunotherapy.

3) Describe the two major ways in which monoclonal antibodies are used for immunotherapy.

4) Why are clinical mAbs humanized?

5) Distinguish between bi-specific mAbs and BiTES.

6) Give four examples of immunoconjugates and describe how each functions.

7) Give an example of an immune blockade and how it might be used to treat cancer.

8) Describe the two basic types of cancer vaccines. Are they effective? If not, why not?

9) Outline three barriers that might be responsible for the ineffectiveness of current cancer vaccines.

10) Describe three methods with the potential to improve cancer vaccine efficacy.

11) What is adoptive T cell cancer therapy, and how does it work?

12) Describe two types of cytokine-based immunotherapies.

Can You Extrapolate? Some Conceptual Questions

1) A primary tumor is identified in a patient and is found upon biopsy to express a particular chemokine receptor. What might be the effect of treating this patient with a mAb against this chemokine receptor?

2) Imagine a scenario in which a large cohort of cancer patients was monitored over time for susceptibility to chronic infections. Also imagine that half of these cancer patients were treated with a reagent that created a PD-1 immune blockade. If patients in this latter group were observed to have fewer chronic infections than the untreated group, how might you explain this?

3) A patient has a cancer that is linked to a mutation in a DNA repair gene. How might this mutation affect the long-term success of immunotherapy with a mAb specific for a particular TAA?

Would You Like To Read More?

Baskar, R., Lee, K. A., Yeo, R., & Yeoh, K. W. (2012). Cancer and radiation therapy: Current advances and future directions. *International Journal of Medical Sciences, 9*(3), 193–199.

Chow, M. T., Moller, A., & Smyth, M. J. (2012). Inflammation and immune surveillance in cancer. *Seminars in Cancer Biology, 22*(1), 23–32.

Clevers, H. (2011). The cancer stem cell: Premises, promises and challenges. *Nature Medicine, 17*(3), 313–319.

Grivennikov, S. I., Greten, F. R., & Karin, M. (2010). Immunity, inflammation, and cancer. *Cell, 140*(6), 883–899.

Liu, X. Y., Pop, L. M., & Vitetta, E. S. (2008). Engineering therapeutic monoclonal antibodies. *Immunological Reviews, 222,* 9–27.

Maniati, E., Soper, R., & Hagemann, T. (2010). Up for mischief? IL-17/Th17 in the tumour microenvironment. *Oncogene, 29*(42), 5653–5662.

Mantovani, A., Cassatella, M. A., Costantini, C., & Jaillon, S. (2011). Neutrophils in the activation and regulation of innate and adaptive immunity. *Nature Reviews. Immunology, 11*(8), 519–531.

Ogbomo, H., Cinatl, J., Jr, Mody, C. H., & Forsyth, P. A. (2011). Immunotherapy in gliomas: Limitations and potential of natural killer (NK) cell therapy. *Trends in Molecular Medicine, 17*(8), 433–441.

Pantic, I. (2011). Cancer stem cell hypotheses: Impact on modern molecular physiology and pharmacology research. *Journal of Biosciences, 36*(5), 957–961.

Pardoll, D. M. (2012). The blockade of immune checkpoints in cancer immunotherapy. *Nature Reviews.Cancer, 12*(4), 252–264.

Pillay, V., Gan, H. K., & Scott, A. M. (2011). Antibodies in oncology. *New Biotechnology, 28*(5), 518–529.

Poschke, I., & Kiessling, R. (2012). On the armament and appearances of human myeloid-derived suppressor cells. *Clinical Immunology, 144*(3), 250–268.

Prasad, S., Cody, V., Saucier-Sawyer, J. K., Saltzman, W. M., Sasaki, C. T., Edelson, R. L., Birchall, M. A., & Hanlon, D. J. (2011). Polymer nanoparticles containing tumor lysates as antigen delivery vehicles for dendritic cell-based antitumor immunotherapy. *Nanomedicine, 7*(1), 1–10.

Vermeulen, L., de Sousa e Melo, F., Richel, D. J., & Medema, J. P (2012). The developing cancer stem-cell model: Clinical challenges and opportunities. *Lancet Oncology, 13*(2), e83–89.

Vesely, M. D., Kershaw, M. H., Schreiber, R. D., & Smyth, M. J. (2011). Natural innate and adaptive immunity to cancer. *Annual Review of Immunology, 29,* 235–271.

Weiner, L. M., Murray, J. C., & Shuptrine, C. W. (2012). Antibody-based immunotherapy of cancer. *Cell, 148*(6), 1081–1084.

Transplantation

War is the only game in which it doesn't pay to have the home-court advantage.

Dick Motta

A. The Nature of Transplantation

The term "transplantation" typically brings life-saving surgeries to mind: kidney transplants, heart and lung transplants, and skin grafts for burn victims. These procedures are called *solid organ transplants*. However, the **bone marrow transplants (BMTs)** and **hematopoietic stem cell transplants (HSCTs)** used to treat malignancies like leukemias are also tissue transplants, as are blood transfusions. In all these cases, the term "transplantation" refers to the replacement of a patient's non-functional tissue or cells with healthy tissue or cells from a donor. Indeed, tissue transplantation is now the preferred treatment for many serious disorders of the heart, lung, kidney, liver, intestine, pancreas, and hematopoietic system. Statistics on the frequencies of the common solid organ transplants in the U.S. are given in **Table 17-1**, while two more unusual types of transplants are described in **Box 17-1**. The use of BMT and HSCT to treat various hematopoietic cancers is discussed in Chapter 20.

Unless the donor and recipient of a transplant are genetically identical, the donated tissue is viewed by the immune system of the recipient as non-self. Without drug-induced suppression of the recipient's immune system, the donated tissue is destroyed by recipient leukocytes in an immune response called **graft rejection (Plate 17-1)**. Researchers have defined specific terms to describe the genetic relationship between an organ donor and a prospective recipient. Individuals who are **syngeneic** are identical at all genetic loci, such that transplantation between them may be undertaken without fear of graft rejection. Individuals who are **allogeneic** have different alleles at various genetic loci so that transplants between such persons will be rejected in the absence of immunosuppression. In the simplest of transplant scenarios, the donor and recipient are the same patient, as in the case of a burn victim whose own healthy skin is used to replace a damaged section of skin elsewhere on his/her body. This

The terms "transplant" and "graft" are often used interchangeably.

Current statistics on transplantation in the U.S. can be found on the website of the Organ Procurement and Transplantation Network (OPTN): http://optn.transplant.hrsa.gov.

MAK: Primer to the Immune Response. http://dx.doi.org/10.1016/B978-0-12-385245-8.00017-0

TABLE 17-1	Snapshot of U.S. Solid Organ Transplantation Statistics*	
Transplant Type	**Number of Patients on the Waiting List as of December 7, 2012**	**Number of Transplants Performed in 2011**
Kidney	94,652	16,813
Liver	15,962	6,342
Lung	1,617	1,822
Heart	3,372	2,322
Pancreas	1,212	287
Intestine	258	129
Heart/Lung	49	27
Kidney/Pancreas	2,145	795
Total	**116,524**	**28,537**

*Data from the U.S. Organ Procurement and Transplantation Network (OPTN), accessed December 12, 2012.

Box 17-1   Two Unusual Types of Transplantation

Advances in transplantation medicine have allowed surgeons to transfer what might be considered by some to be rather unusual tissues. Due to its critical physiological function, some clinicians consider the gut microbiota to be an "organ." This "organ" is hugely disrupted in individuals suffering from *Clostridium difficile* infection (CDI), and these patients suffer from severe diarrhea, abdominal pain, nausea and fever. A last-ditch method for treating these stubborn infections is the transplantation of fecal microbiota (FMT). In an FMT, large numbers of gut microbes are isolated from a fecal sample of a healthy donor. The sample is filtered before introduction into the gut of a CDI patient by enema, transcolonic infusion, or nasogastric infusion. The administration of these fresh commensal organisms swamps the pathogenic invader, restores a normal spectrum of gut microbes, and relieves the patient's symptoms. The FMT approach is highly effective and has been in use for over 50 years for the treatment of colitis linked to CDI, and researchers are now starting to apply FMT to the treatment of other, non-pathogen-related gut disorders. For example, several hypersensitivities and autoimmune diseases (see Ch. 18 and 19, respectively) are linked to abnormalities of the microflora, suggesting that FMT may constitute a therapeutic avenue for these disorders as well. Although a healthy individual might see FMT as a revolting procedure (what some clinicians call "the yuck factor"), patients with severe gut disorders have unreservedly embraced FMT for the clinical relief it offers.

Another relatively unusual area of transplantation is that of the human face. The first human facial transplant was performed in 2005, and over a dozen such transplants have been carried out in adults to ameliorate conditions such as severe burns, trauma, or reconstruction after tumor removal. These patients must remain on a life-long regimen of immunosuppressive drugs, which leaves them at risk for complications such as drug toxicity, infection and tumorigenesis. The seriousness of these adverse events has raised an ethical dilemma for physicians, since a damaged face, while psychologically difficult to live with, is not usually life-threatening. In an effort to remove the need for life-long immunosuppression, clinical researchers are avidly seeking ways to induce tolerance to the grafted facial tissue. More on inducing graft tolerance appears later in this chapter.

type of syngeneic, self–self tissue transfer is called an **autologous graft**. Syngeneic grafts that involve the transfer of tissue between two genetically identical individuals (such as between human identical twins or two mice of the same inbred strain) are called **isografts**. Conversely, tissue transfers between genetically different members of the same species (such as two humans who are not identical twins) are called **allografts**. **Xenografts** are tissue transfers between members of two different species, as in the case of a pig organ transplanted into a human.

NOTE: In immunological circles, the relationship between the MHC genotypes of the individuals of interest is often of overriding importance, so that the terms "syngeneic" and "allogeneic" may sometimes be used to denote genetic identity or non-identity at the MHC only.

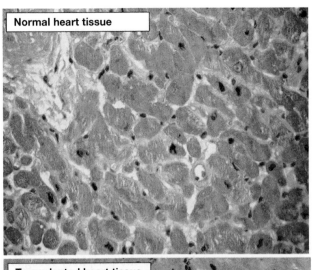

Normal heart tissue

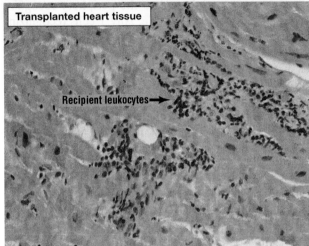

Transplanted heart tissue

Recipient leukocytes→

Plate 17-1
Graft Rejection

Histological sections of a normal heart (top) and a transplanted heart in which the muscle tissue has been infiltrated by mononuclear cells (bottom). *[Reproduced by permission of Jagdish Butany, University Health Network/Toronto Medical Laboratories, Toronto General Hospital.]*

B. The Molecular Basis of Graft Rejection

The strong, rapid responses that are principally responsible for allograft rejection are due to the recognition of allelic differences in antigens encoded by the *major histocompatibility (MHC) loci.* Thus, the term **allorecognition**, as it is most commonly used, refers to recipient immune responses mounted against donor MHC. Certain genes outside the MHC also show a low degree of allelic variation and may also be involved in graft rejection responses. These latter genes came to be called **minor histocompatibility loci** based on their influence in solid organ transplant situations, in which allelic differences at these loci result in slower, milder antigraft responses relative to those triggered by MHC differences. In the case of BMT or HSCT, differences at minor histocompatibility loci can cause rapid rejection of the transplanted hematopoietic cells.

I. Immune Recognition of Allogeneic MHC Molecules

How does an allogeneic MHC molecule (allo-MHC) in a graft provoke lymphocyte responses? For B cells, the mechanism of allo-MHC recognition is straightforward. The allo-MHC molecules of the donor often have a slightly different conformation than recipient MHC molecules, so that the recipient has B cells with BCRs that recognize allo-MHC as they would any incoming non-self protein. The recipient's B cells are activated and produce **alloantibodies** directed against the allo-MHC molecules displayed on the graft surface. For T cells, the situation is more complex because TCRs recognize epitopes composed of peptide bound to MHC. There are two mechanisms that account for T cell responses to allo-MHC: **direct allorecognition** and **indirect allorecognition**.

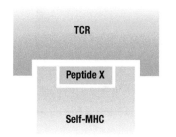
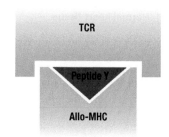

i) Direct Allorecognition

Direct allorecognition occurs when a recipient's T cells recognize pMHCs involving not *self*-MHC, as would occur during a pathogen attack, but *allo*-MHC molecules expressed by cells of the graft. What is the basis for this apparently contradictory recognition of non-self MHC? T cells are selected in the thymus through the mechanisms of central tolerance such that the lymphocytes released to the periphery recognize complexes of [self-MHC + X], where X is a non-self peptide. Any T cells recognizing [self-MHC + Y], where Y is a self peptide, are either eliminated in the thymus or silenced in the periphery by the mechanisms of peripheral tolerance (refer to Ch. 10). In a transplant situation, the cells of an allograft are covered with allo-MHC molecules loaded with a wide variety of peptides. Most of these peptides are considered "self" with respect to both the donor and recipient because they are derived from proteins that are invariant (monomorphic) in a given species. Thus, from the recipient's point of view, the pMHCs in the transplanted tissue can be described as [allo-MHC + Y]. However, T cells do not distinguish the individual components of such complexes: they recognize the overall shape. If the conformation of an [allo-MHC + Y] epitope looks like some combination of [self-MHC + X], a recipient T cell specific for [self-MHC + X] will be activated by a donor cell bearing [allo-MHC + Y] (**Fig. 17-1**). The response of the recipient's T cells to the allograft is then really a form of cross-reactivity that occurs when peptide presented on allo-MHC looks like non-self peptide presented on self-MHC.

How does allo-MHC in a grafted tissue such as a kidney manage to activate cross-reactive T cells, since naïve T cells are usually activated only in lymph nodes? During transplantation, donor DCs expressing allo-MHC travel along with the donated organ and are introduced into the recipient. These donor DCs can migrate out of the graft into the recipient lymph node draining the transplant site. Naïve recipient Th and Tc cells expressing cross-reactive TCRs can then be activated by *donor* DCs within the node and generate T cell effectors as well as memory T cells. The Th effectors support the activation of antigraft B cells and Tc cells within the lymph node, and the differentiated progeny of these cells home back to the graft and attack it.

> NOTE: T cell responses to allogeneic cells are much more intense than T cell responses to pathogens, and the mechanism of direct allorecognition explains why. A vast array of peptides are presented simultaneously and continuously by the allo-MHC molecules of a graft, allowing for the simultaneous and continuous activation of a wide range of naïve T cell clones. In contrast, a pathogen attack results in the presentation of a relatively small collection of pathogen-derived peptides that stimulates only a limited number of naïve T cell clones. It has been estimated that, while only 1 in 10,000 T cell clones responds to any given pathogen, as many as 1 in 50 T cell clones can be activated by allo-MHC.

ii) Indirect Allorecognition

In indirect allorecognition, naïve recipient T cells are activated by *recipient* DCs that have acquired peptides derived from MHC proteins of the donor (**Fig. 17-2**). These proteins may have been shed into the surrounding recipient tissues when cells of the allograft died, or the recipient APCs may have entered the grafted tissue and acquired shed donor proteins. Most of these donor proteins are encoded by genes that are

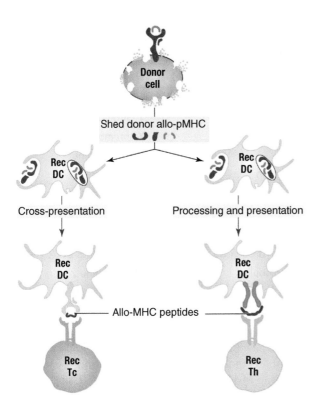

➤ **Fig. 17-2**
Indirect Allorecognition

In a transplantation site, spent or damaged donor cells shed allo-pMHC complexes into the graft and surrounding tissue. These proteins are taken up by recipient (Rec) DCs and either cross-presented on MHC class I, or presented conventionally on MHC class II. Because the peptide is derived from a molecule that is not expressed in the recipient, the pMHC combinations on the surfaces of these DCs are seen as "non-self" by recipient Th and Tc cells. These T cells may become activated and attack the graft.

monomorphic within a species, so that peptides derived from these donor proteins are seen as "self." If acquired by recipient DCs, these self peptides are presented on self-MHC and do not provoke an immune response because the naïve recipient T cells recognizing this combination were deleted during the establishment of central tolerance. However, some of the shed donor proteins will be allo-MHC molecules, and peptides derived from allo-MHC may be seen as non-self. If these non-self peptides are acquired by recipient DCs, they may be presented and cross-presented on self-MHC to naïve Th and Tc cells, respectively, in the local lymph node. Effector Th cells are generated that home back to the allograft and recognize complexes of allo-MHC peptide bound to self-MHC that are presented by additional recipient APCs. If the recipient and donor are syngeneic at some MHC class II loci, then donor APCs from within the grafted tissue may also present allo-MHC peptides that can be recognized by Th cells. Similarly, if the recipient and donor are syngeneic at some MHC class I loci, Tc cells stimulated by indirect allorecognition in the lymph node can generate CTLs that home back to the transplant site and attack the allograft.

II. Immune Recognition of Minor Histocompatibility Antigens

As mentioned previously, within a given species, subtle allelic variation may exist in genes outside the MHC. In the case of a transplant between individuals who are allogeneic at such a locus, the proteins encoded by the donor allele and the recipient allele will differ slightly in amino acid sequence. If this difference leads to graft rejection, this protein is considered to be a **minor histocompatibility antigen (MiHA)**. The corresponding gene is then classified as a *minor histocompatibility gene*. Where an organ donor and recipient are allogeneic at an MiHA locus, antigen processing of MiHA proteins in donated tissue (by either donor or recipient APCs) gives rise to **minor H peptides** that may combine with MHC molecules on recipient APCs or graft cells to form pMHC structures that look like [self-MHC + X] to recipient T cells. Thus, even when an organ donor and recipient are MHC-matched, graft rejection can occur due to differences in MiHAs. Some MiHAs result from allelic variation in "housekeeping" genes within a population. Housekeeping genes are expressed in virtually all cell types

TABLE 17-2	Examples of Human and Murine Minor Histocompatibility Antigens	
MiHA Name	**Protein/Gene**	**Gene Localization**
Human		
HY-B7	SMCY	Y chromosome
HY-A1	DFFRY	Y chromosome
HY-DRB3	DBY	Y chromosome
HA-2	Class I myosin	Chromosome 6
HB-1	HB-1	Chromosome 5
Mouse		
COI	Cytochrome oxidase	Mitochondrial DNA
ND-1	NADH dehydrogenase	Mitochondrial DNA
β2m	β2-Microglobulin	Chromosome 2
H-YK^K	SMCY	Y chromosome

Humoral responses to MiHAs are uncommon because the conformational differences involved are usually too small to trigger recognition by the BCRs of recipient B cells.

of a species and generally encode proteins that maintain basal metabolic functions. Other MiHAs result from differences in the expression of tissue-specific genes, such as certain mitochondrial proteins and proteins that differ between males and females of a species. Examples of human and murine MiHAs are given in **Table 17-2**, and T cell responses to MiHAs are illustrated in **Figure 17-3**.

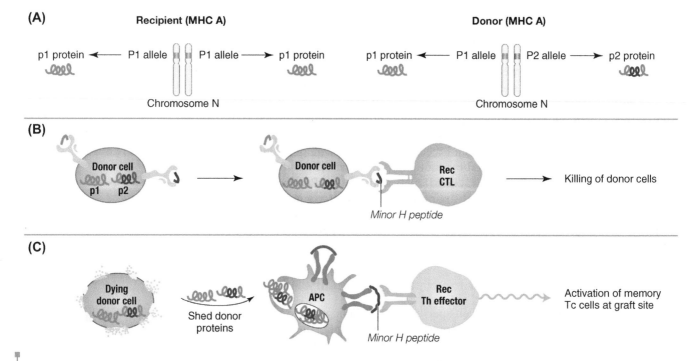

Fig. 17-3
Recognition of Minor Histocompatibility Antigens

(**A**) In this hypothetical example, a minor histocompatibility antigen is encoded by the housekeeping gene P, which is found on chromosome N and is co-dominantly expressed. A transplant has taken place between a donor and a recipient, both of whom express MHC class I molecule A (MHC A). The transplant recipient is homozygous for allele P1 of the housekeeping gene, which gives rise to p1 proteins. The transplant donor is heterozygous for the housekeeping gene, and so produces p1 and p2 proteins, which differ in the peptides they give rise to upon processing (gray vs. red). (**B**) In the transplant site, the donor cells present peptides from both p1 and p2 proteins on MHC A. However, because the p2 peptide is not present in the recipient, the p2-MHC A combination is perceived as non-self by the recipient's CTLs. The donor cells are killed. (**C**) Dying donor cells shed both p1 and p2 proteins, which serve as sources of p1 and p2 peptides, respectively. Provision of cytokine help for Tc cell-mediated graft rejection response arises when recipient Th effector cells recognize a p2-derived minor H peptide presented by MHC class II on either a donor or recipient APC.

NOTE: Why is the response to an MiHA weak compared to that invoked by allo-MHC? An MiHA expressed by a donor cell will give rise to a very limited number of minor H peptides with the potential to form only a few pMHC structures that recipient T cells will recognize as foreign. Indeed, the proportion of recipient T cells responding to an MiHA difference is usually no greater than that responding to a pathogen-derived antigen. In contrast, as described earlier, allo-MHC molecules in a graft form pMHC complexes with a large number of different peptides, leading to the activation of many different cross-reactive T cell clones.

C. Solid Organ Transplantation

I. Immunology of Solid Organ Transplant Rejection

Transplantation of solid organs is a traumatic event from the body's point of view. Tissues are surgically resected, and blood vessels in both the graft and the transplant site are damaged. All these events initiate responses from innate and adaptive leukocytes.

i) Innate Responses

Upon transplantation of a tissue, adhesion molecules are upregulated in the injured endothelium of the graft. Even in a syngeneic transplant situation, these molecules promote the infiltration of the graft by activated recipient macrophages, NK cells, and other inflammatory cells that produce large amounts of IL-1 and IL-6. Traditionally, because these events are non-specific and do not usually result in complete loss of the grafted tissue, they have not been considered graft rejection *per se*; rather, they have been deemed to play an important role in setting the stage for graft rejection. However, recent work has elevated the importance of the innate response during transplantation, spurred in part by the observation that deletion of attacking recipient T cells alone does not spare a renal or intestinal allograft from rejection mediated by infiltrating macrophages and/or eosinophils. Similarly, in the absence of immunosuppression, heart transplants in experimental mice lacking all lymphoid cells are soon infiltrated by innate leukocytes secreting high levels of pro-inflammatory cytokines. These observations indicate that the innate response has effects on graft rejection which need to be elucidated if therapeutic manipulation of the adaptive response to achieve graft tolerance is to be successful.

TLR signaling triggered during transplantation can play an important role in graft rejection. Recent work at the molecular level has shown that engagement of TLRs on innate leukocytes by PAMPs/DAMPs reduces graft acceptance in experimental animals, and that administration of TLR ligands can trigger rejection even in animals treated to induce T cell tolerance to a graft. Conversely, the survival of heart or skin allografts is increased in mice deficient for MyD88, a signal transduction adaptor in the TLR signaling pathway. Interestingly, the tissues most often exposed to pathogens (and thus most likely to mount innate responses featuring TLR engagement) are those in which acute rejection occurs most often and in which it is the most difficult to establish graft tolerance. These tissues include the intestine, lung and skin. Scientists speculate that TLR signaling triggered by the presence of DAMPs associated with the transplantation surgery may promote the differentiation of Th1 effector cells that secrete pro-inflammatory mediators supporting graft rejection. In addition, TLR signaling may inhibit the generation of regulatory T cells that can block the attack of conventional antigraft T cells.

NOTE: In Chapter 16, we described the phenotypic distinctions of M1 and M2 macrophages in the context of cancer. To recap, M1 macrophages are generally associated with initial antitumor responses, while M2 macrophages predominate at later stages with activity that may promote tumor growth. A similar dichotomy exists in transplantation situations. M1 macrophages are prominent in a transplant site soon after the new organ is introduced, but M2 macrophages, with their properties that encourage wound healing and angiogenesis, gradually come to dominate the macrophage population in this locale as the organ settles into its new life.

ii) Adaptive Responses

Graft rejection proper is caused by adaptive immune responses against alloantigens in the incoming organ. Once the organ is transplanted, both donor APCs (including immature DCs) from within the graft and recipient APCs that subsequently infiltrate the donated organ will eventually travel to the local lymph node. Because of DAMPs in the inflammatory milieu surrounding the graft, any DCs that have captured alloantigens are induced to mature. Donor DCs can then activate naïve T cells via direct allorecognition, while recipient DCs mediate indirect allorecognition. Both donor and recipient DCs can be involved in the presentation of peptides derived from MiHAs. In all these cases, the DCs present pMHC structures that naïve recipient Th cells can recognize.

> NOTE: Sometimes the alloreactive T lymphocytes activated during a transplant are memory T cells. How could this be if the patient has never received a graft before? Memory T cells can be generated in cases in which a naïve T cell was primed *prior to allograft transplantation* by exposure to a pathogen providing [self-MHC + X] structures that happen to resemble [allo-MHC + Y] structures present in the graft. These cross-reactive antipathogen memory T cells then attack the incoming transplant.

Once a recipient's Th cells are activated, they generate Th effectors that provide T cell help for the activation of naïve antigraft B cells and Tc cells in the lymph node. The activated B cells generate plasma cells that produce antibodies directed against the allo-MHC proteins of the graft. These alloantibodies enter the circulation and travel to the graft, where they kill its cells via classical complement activation and/or ADCC. Although the alloantibodies produced may be directed against either MHC class I or MHC class II of the donor, the majority of the clinical damage will be due to the alloantibodies that recognize MHC class I because, unlike MHC class II, these proteins are widely expressed on most cell types (including the cells making up the graft). Tc cells that become activated generate CTLs that home to the graft, recognize allo-MHC present on graft cells, and kill these cells using their multiple cytotoxic mechanisms. If the recipient and donor share identity at some MHC class I loci, CTLs specific for minor H peptides will also be able to attack graft cells directly. In addition, CTLs directed against minor H peptides complexed to self-MHC class II may attack recipient APCs that have infiltrated the graft, and, in so doing, collaterally damage graft cells. Lastly, Th effectors can home to the allograft and release damaging cytokines, particularly IFNγ.

II. Clinical Graft Rejection

Clinically, the rejection of a solid organ transplant is classified as one of four types: *hyperacute*, *acute cellular*, *acute humoral*, or *chronic*. The clinical and histological features of these forms of graft rejection and their proposed underlying mechanisms are summarized in **Table 17-3**.

i) Hyperacute Graft Rejection

Hyperacute graft rejection (HAR) is the destruction of a graft almost immediately after transplantation. Solid organs are vascularized structures, and HAR is usually caused by the presence in the recipient of *pre-formed* antibodies that recognize antigens on the endothelial cells of blood vessels within the transplanted organ. When these pre-formed antibodies bind to these antigens, they rapidly initiate complement activation via the classical pathway such that the vascular network of the graft is destroyed *within minutes* by MAC deposition. In general, no cellular infiltration is seen.

The chance of HAR is greatest when a recipient possesses pre-formed anti-ABO blood group antibodies. Although **ABO antigens** are defined as antigenic structures present on an individual's erythrocytes (see Section F on blood transfusions, later in this chapter), they are also expressed on vascular endothelial cells, making them a major target within donated tissue. HAR can also be triggered by pre-formed antibodies recognizing non-self MHC. Such antibodies are common in the circulation of prospective

TABLE 17-3	Types of Rejection of Solid Organ Transplants			
	Hyperacute (HAR)	**Acute Cellular (ACR)**	**Acute Humoral (AHR)**	**Chronic (CGR)**
Time for graft rejection	Within minutes	Within days or weeks	Within days or weeks	Within months
Targets on donor cell	ABO blood group antigens; allogeneic MHC	Allogeneic MHC	Allogeneic MHC	Allogeneic MHC
Mechanism	Pre-formed antibodies that induce death by MAC formation	Direct allorecognition by recipient CTLs that exert cytotoxicity and secrete cytokines	Indirect allorecognition by recipient CD4+ T cells that promote alloantibodies production	Indirect allorecognition; cytokine-induced ischemia
Cellular infiltration	None	Lymphocytes	Neutrophils	Lymphocytes and neutrophils
C4d staining	Yes	No	Yes	Yes
Fibrosis	No	No	Yes	Yes
Treatment	Pre-transplant plasmapheresis; no effective treatment once rejection is under way	Immunosuppressive drugs, steroids	Immunosuppressive drugs, post-transplant plasmapheresis	None effective

recipients who have been previously exposed to allo-MHC molecules, as in the case of individuals who have had a previous graft, blood transfusion or pregnancy. In the transplantation world, patients who harbor pre-formed antibodies likely to cause HAR are said to be "highly sensitized" or "hyperimmunized." The chance of HAR of a graft can be greatly decreased by careful screening of the recipient and potential donor for blood group and MHC mismatches (see later). In addition, the recipient may undergo a pre-transplant procedure called **plasmapheresis** in which the recipient's blood is passed through a machine that removes all antibodies before returning the blood to the recipient. If the levels of the offending antibody are successfully diminished to below those required for HAR, the donated organ is less likely to be rejected, at least initially.

ii) Acute Graft Rejection

Two types of acute graft rejection, **acute cellular rejection (ACR)** and **acute humoral rejection (AHR)**, generally occur within a few days or weeks of transplantation. These types of rejection are distinguished by their underlying mechanisms.

a) Acute Cellular Rejection

ACR is almost always the result of direct allorecognition of mismatched donor MHC by alloreactive recipient CTLs. The greater the number of MHC differences between the donor and recipient, the more Tc clones are mobilized and the faster the rejection mediated by effector CTLs. The CTLs destroy the graft both by secreting destructive cytokines such as TNF and by initiating perforin/granzyme-mediated cytotoxicity. Histologically, ACR is typically characterized by lymphocyte infiltration into the transplanted organ and inflammation of its blood vessels.

The reader may wonder whether NK-mediated cytotoxicity also contributes to ACR, as the incoming organ may not express the self-MHC molecules needed to bind to NK inhibitory receptors on the recipient's NK cells. Indeed, any NK cell that does not express any inhibitory receptor specific for any of the MHC class I molecules expressed by an allogeneic graft can be activated by graft cells. However, within the recipient NK cell population as a whole, different NK cell subsets will express different collections of inhibitory receptors, and the proportion of the recipient NK cell population activated by the graft will vary depending on the degree of MHC matching between the donor and recipient. Moreover, the strength of any such response will depend on the levels of NK activatory ligands expressed by graft cells. Studies have indicated that direct NK killing of graft cells does not contribute significantly to ACR, but that these leukocytes have an important indirect role. NK cells appear to infiltrate solid organ grafts *before* T cells get there and may set up conditions that promote T cell-mediated ACR. For example, even if only a few of these infiltrating NK cells become activated by graft cells, they may produce sufficient TNF and IFNγ to enhance DC maturation and promote Th1 cell differentiation.

> NOTE: Paradoxically, because NK cell activation depends on a balance of inhibitory and activatory signaling, NK cells can sometimes promote graft acceptance rather than ACR-mediated graft rejection. If activated antigraft recipient T cells in a site of transplantation express stress ligands, NK cells that are not adequately restrained by inhibitory ligands may kill them. NK cells may also destroy donor-derived DCs that have come along for the ride in the graft, reducing the chance that these cells might present donor antigens to recipient T cells and spark rejection. In addition, some NK cells have been found to block DC functions and/or produce IL-10, damping down the activation and proliferation of any nearby antigraft T cells.

ACR can also be promoted by a loss of regulatory T cell functions. In the absence of Treg cell-mediated suppression, a recipient's conventional T cells may be activated by pMHCs displayed by the graft and attack it. What precipitates a loss of Treg cell functions? Some of these cells may be killed by NK-mediated cytotoxicity, as NK cells have been observed to kill Treg cells that express activatory ligands. In addition, when alloreactive Th17 and Th1 cells infiltrate a graft, they produce IL-17 and IFNγ, respectively. These cytokines can stimulate various cells to produce IL-6, which in turn inhibits the differentiation of Treg cells.

b) Acute Humoral Rejection

AHR is usually due to humoral responses initiated via indirect allorecognition by recipient Th cells. The cytokines produced by the resulting Th effector cells support the activation of antigraft B cells and the synthesis of antigraft alloantibodies. These antibodies destroy graft cells via ADCC or classical complement activation. The histological features of AHR include neutrophil (but not lymphocyte) infiltration into the graft, vascular inflammation, fibrosis, and necrosis of blood vessel walls. AHR can also be detected by staining for C4d, a degradation product of the complement component C4b. C4d tends to accumulate on capillaries in a graft undergoing AHR, whereas ACR is generally negative for C4d staining. The effects of AHR can be mitigated if the recipient undergoes post-transplantation plasmapheresis to remove the alloantibodies.

iii) Chronic Graft Rejection

Chronic graft rejection (CGR) of solid organs is defined as the loss of allograft *function* several months after transplantation. The transplanted organ may still be in place, but persistent immune system attacks on the allo-MHC expressed by its component cells have gradually caused the organ to cease functioning. Precisely what causes CGR is unknown, but the most important clinical predictor of CGR is a prior, transitory episode of acute rejection. Work at the molecular level has implicated IL-17 and other Th17-related cytokines in both prior acute rejection episodes and the ensuing CGR.

Both alloantibodies and cell-mediated responses are involved in CGR, and the indirect pathway of allorecognition appears to be particularly important. In response to allogeneic donor peptide presented on recipient APCs, recipient memory T cells are activated and differentiate into Th effectors. These Th effectors produce cytokines that either directly damage the grafted tissue or constitute T cell help for recipient B cells producing alloantibodies. Over half of chronically rejected organs exhibit C4d deposition in the capillaries of the transplant. In a critical part of CGR that is characteristic but poorly understood, activated recipient T cells also induce monocytes and macrophages that have infiltrated the graft, as well as the endothelial cells comprising the graft's blood vessels, to produce numerous growth factors and cytokines. These molecules drive both smooth muscle cell proliferation and the synthesis of extracellular matrix (ECM) proteins. The proliferating smooth muscle cells effectively shrink the lumens of the graft's blood vessels, causing *ischemia*. The increase in ECM proteins also results in fibrosis, further compromising graft function. Thus, CGR is histologically characterized by fibrosis, collagen deposition, and a loss of blood flow within the graft.

Ischemia is a deficiency of blood in a tissue that is caused by constriction or obstruction of local blood vessels and results in a reduced supply of oxygen to the tissue.

III. Graft-Versus-Host Disease (GvHD) in Solid Organ Transplantation

Transplantation, although a lifesaving procedure, can be a double-edged sword. Sometimes donor leukocytes traveling along with a transplanted organ attack recipient tissues, such that **graft-versus-host disease (GvHD)** is said to have occurred. The epithelial cells of the skin, liver and intestine of the transplant recipient are the prime targets of GvHD. As a result, the recipient may start to lose weight as well as liver and/or lung functions. Onset can be acute (and sometimes fatal) or chronic. Although GvHD rarely occurs in solid organ transplants, it is a huge problem for patients undergoing HSCT since, by definition, the transplant is of donated immunocompetent cells that disperse throughout the body. These aspects of GvHD are discussed in more detail in Section E of this chapter.

D. Minimizing Graft Rejection

Graft rejection can be minimized by a combination of approaches that rely on MHC matching, alloantibody analysis and immunosuppression. Researchers are also currently experimenting with treatments designed to increase a recipient's immunological tolerance of an incoming graft. As illustrated in **Table 17-4**, the proportion of transplant patients surviving for an extended period varies with organ type, as does the speed with which the transplanted organ is rejected. These variations may reflect differences in the type and strength of immune responses occurring in different tissue locations in the body. The derivation of the transplanted organ can also influence graft acceptance/ rejection. Various sources of donated human organs are outlined in **Box 17-2**, and the possibility of using animal organs for human transplantation is explored at the end of this section in **Box 17-3**.

I. HLA Typing

i) Rationale

In order to minimize the intensity of a transplant recipient's adaptive response to a donated organ, clinicians attempt to ensure the closest possible match between the MHC alleles of the recipient and those of the donor. Thus, the first step in a transplantation protocol is to determine the MHC haplotype of the prospective recipient. Since MHC molecules in humans are known as HLAs (human leukocyte antigens), this procedure is known as **HLA typing**. Collectively, close to 10,000 alleles of HLA-A, -B, -C, -DR, -DQ and -DP molecules are now known.

HLA typing is carried out by examining the collection of HLA alleles expressed on a recipient's leukocytes. Once a donor organ becomes available, the HLA haplotype

TABLE 17-4	Transplant Patient Survival Rates in the U.S.*		
Transplant Type	**1-Year Patient Survival (%)**	**3-Year Patient Survival (%)**	**5-Year Patient Survival (%)**
Kidney	96	91	85
Liver	87	78	72
Lung	83	62	47
Heart	88	79	72
Pancreas	94	90	82
Intestine	79	59	47
Heart/Lung	66	50	39
Kidney/Pancreas	95	90	85

*Data from the Organ Procurement and Transplantation Network (OPTN), based on transplants carried out between 1997 and 2004. Accessed December 14, 2012.

Box 17-2   Types of Human Organ Donation

Originally, organs for transplantation were harvested only from persons whose hearts had stopped beating (*cadavers*). Unfortunately, the organs of a cadaver remain healthy for only a few minutes or hours before the lack of oxygen kills their cells. Persons who suffer "brain death" (but retain heart and lung function) may also be organ donors. With life support machinery, a brain-dead donor's organs remain healthy while family consent to organ donation is obtained, MHC haplotypes are matched, and the donated organ is transferred to the recipient's hospital. However, only a small percentage of deaths involve brain death, and the number of patients on transplant waiting lists greatly exceeds the organ supply. Clinicians have thus increasingly reached out to "living related" donors, who are healthy relatives of the patient who donate an organ or tissue under circumstances in which they are highly likely to survive after the surgery. For example, most adults can live quite cheerfully with only one kidney, and the amazing regenerative powers of the liver mean that individuals can often donate sections of this organ with minimal risk to themselves. There are three advantages to living related donors: (1) family members will likely share at least some MHC alleles with the prospective recipient, increasing the chance that the transplant will be successful; (2) the transplant can be planned in advance and proceed without delay; and (3) organs from living donors have a better chance of surviving for longer in the recipient than the organs of cadavers. Because of fears of undue pressure or calls for financial compensation, offers of organs from living, *non*-related donors are usually declined. At least for kidney transplants, the use of living related donors has meant that reliance on kidneys from cadavers has plunged: from 1988 to 2012, the percentage of kidney transplants in the U.S. that involved cadaveric donors dropped from 80% to 65%. Furthermore, current U.S. data show that, whereas 90% of patients receiving a kidney from a living donor survive for at least 5 years after their transplant, only 82% of patients receiving a cadaveric kidney do the same. Similar improvements in rates of long-term survival for patients receiving a lung or liver transplant have recently been achieved using organs from living donors instead of cadavers.

Box 17-3   Xenotransplantation

Transplantation is now the cure of choice for many human disorders. However, the supply of donated organs generally remains stuck at about one-third of the number of patients on the waiting list. Artificial hearts are in development and are undergoing clinical testing but are not yet a standard alternative to cardiac transplantation, and no artificial lungs exist. Xenotransplantation has been proposed to prolong the survival of waiting list patients by temporarily installing an animal organ until a suitable human organ becomes available. Although chimpanzees are close genetic cousins to humans, these animals are an endangered species and expensive to maintain. The use of primate species also raises the concern that a primate pathogen might be transferred to humans. Researchers have thus turned to a breed of "mini-pigs," which are readily available, can be raised in a pathogen-free environment, and have organs comparable in size to human organs. However, the vigorous rejection of pig organs by the human immune system is a formidable problem. This rejection occurs because pig (but not human) cells express glycoproteins with a terminal sugar that cross-reacts with certain human IgM antibodies. These antibodies are normally present in human blood due to B cell responses induced by a commensal bacterium that colonizes the human gut. The glycoproteins of this bacterium end in a sugar very similar to that on the ends of the pig glycoproteins. In the case of a pig–human transplant, the human antibodies directed against the bacterial sugars immediately recognize the sugar on the pig organ endothelial cells and activate the human complement system, destroying the graft via HAR. Various strategies have been tried to remove or inhibit the antisugar antibodies and to engineer the expression of human RCA proteins on pig cells. Researchers are also developing a genetically engineered mini-pig that cannot synthesize the offending sugar. Unfortunately, even if the HAR problem is solved, the pig graft may still be rejected due to acute or chronic rejection. Another problem is that these approaches do not address the possibility that a virus infecting a mini-pig organ might make the jump to infecting humans upon transplantation. All these matters remain under investigation.

of the donor is similarly determined so that HLA mismatches with the recipient can be evaluated. Since only 1 in 400,000 unrelated persons are fully matched at the major HLA loci, almost all transplants (except those involving twins or siblings) are done under conditions of at least some HLA mismatching. Depending on the type of transplant being performed and the alternatives open to the patient (i.e., dialysis in the case of kidney failure), the doctors involved decide what degree of mismatch, if any, is acceptable before proceeding with the transplant. In addition, clinicians are now frequently attempting to match donor and recipient MiHAs (to avoid recipient T cell activation by these minor antigens), as well as MICA, MICB and KIR molecules (to avoid activation of NK cells and γδ T cells). However, despite these measures, substantial

TABLE 17-5 Comparison of Methods of HLA Typing

Type of Assay	Requirements of Sample	Requirements of Assay	Time to Results	Advantages	Disadvantages
CDC	Viable peripheral lymphocytes	Typing sera containing defined anti-HLA antibodies Complement Vital dye	3–4 hrs	Low cost Rapid Identifies HLA alleles that are actually expressed as proteins	Labor-intensive Limited specificity Variable reproducibility Affected by patient's health status
PCR-SSP	Any nucleated cells	Allele-specific HLA primers	3–4 hrs	Low cost	Not suitable for automation Allele level typing requires additional testing
PCR-SSOP	Any nucleated cells	Generic HLA primers	3–4 hrs	Can be automated	Allele level typing requires additional testing
SBT	Any nucleated cells	Locus-specific HLA primers	1–3 days	Allele level typing determined directly	High cost due to expensive instrumentation

numbers of transplants must be conducted under conditions of significant mismatching of HLA and other molecules. Fortunately, mismatched transplants of heart, lung, liver, intestine, pancreas and kidney can now be safely carried out due to the control of graft rejection by modern immunosuppressive drugs.

ii) Techniques of HLA Typing

Two main approaches can be used to ascertain HLA haplotypes: serological and DNA-based. Serological typing methods such as the *complement-dependent cytotoxicity (CDC)* assay were established in the early days of clinical transplantation, while common DNA-based typing methods have evolved more recently in step with advances in DNA manipulation technology. Standard examples of these methods are described in the next several sections and summarized in **Table 17-5**.

a) Serological Approach

Traditional HLA typing is carried out using serological methods based on the detection of HLA molecules present on the surfaces of the donor's and recipient's lymphocytes. In the CDC assay, leukocytes from the individual to be typed are isolated from peripheral blood and mixed in an array of microwells with defined collections of anti-HLA antibodies. The anti-HLA antibodies used are obtained from individuals exposed to allogeneic cells via a previous transplant, transfusion or pregnancy. Also added to the wells are complement and a visible dye (such as trypan blue) that is excluded from viable cells. In any given well, if the cells express HLA alleles recognized by the added anti-HLA antibodies, the cells are perforated by classical complement activation such that the dye can enter the cells and turn them blue. HLA mismatching of a prospective organ donor and transplant recipient is thus revealed by different patterns of reactivity across the same array of antibodies.

An advantage of serological HLA typing over newer DNA-based methods (see the following section) is that it automatically confirms that a particular HLA allele is not only present in an individual's genome but is actually expressed as a protein on the surface of the individual's cells. Comparisons of HLA genotypes (as determined by DNA-based methods) with HLA phenotypes (as defined by serological methods like CDC) have revealed that some HLA alleles are not actually expressed, leading to their designation as "null" or "blank" alleles. Null HLA alleles will not provoke an immune response, and so can be discounted in assessing the degree of HLA matching between a prospective organ donor and recipient.

The CDC assay is further detailed and illustrated in Appendix F.

b) DNA-Based Approach

Most HLA typing today relies on highly sensitive and accurate analyses of DNA, many of which are based on the *polymerase chain reaction (PCR)*. PCR is a widely used

The mechanics of the PCR reaction are further described and illustrated in Appendix F.

technique in molecular biology in which specific oligonucleotide primers are added to a miniscule sample of DNA to drive the rapid amplification of particular DNA sequences. When PCR is used for HLA typing, the sample DNA is usually acquired from peripheral blood lymphocytes. PCR has two advantages over serological methods for HLA typing: (1) unlike the live cells required for serology-based techniques, DNA can easily be stored and recovered for repeat analyses; and (2) large numbers of DNA samples can be screened simultaneously using automated equipment. The complete nucleotide sequences of most known human MHC class I and II alleles have been defined, so that the HLA-specific primers required for PCR-based methods are commercially available. As a result, clinicians can now very quickly and specifically identify the HLA genotype of a given individual.

Three different DNA-based methods, each involving an initial PCR step, are now commonly used to define the HLA haplotype of prospective transplant recipients and donors. These techniques are called "PCR followed by sequence-specific oligonucleotide probing" (PCR-SSOP); "PCR using sequence-specific primers" (PCR-SSP); and "sequence-based typing" (SBT). Each of these methods is briefly outlined here, and PCR-SSOP and PCR-SSP are further described and illustrated in Appendix F.

For PCR-SSOP, the use of generic HLA primers allows PCR amplification of any and all HLA alleles present in a DNA sample obtained from the cells of a prospective donor or recipient. The amplified HLA sequences are then split into multiple aliquots and tested for complementary binding to HLA allele-specific probes. Hybridization of a particular probe to the recipient's DNA confirms that the recipient expresses that HLA allele. This method is best suited for typing large numbers of samples in batches, although recent technical refinements allow its use for smaller numbers of samples.

For PCR-SSP, aliquots of an individual's DNA are separately mixed with a series of HLA allele-specific primers under PCR reaction conditions. The amplification products are then detected by gel electrophoresis. If PCR amplification occurs in a particular sample, it indicates that the primer was able to bind to the individual's DNA and that he/she possesses that particular HLA allele. This method is well suited for deceased donor typing, where time is of the essence to recover transplantable organs before they deteriorate. However, because a given allele-specific primer may in fact bind to the DNA encoding several closely related HLA molecules, PCR-SSP is not recommended for situations in which a precise distinction among HLA alleles is required. In addition, this method is not suitable for large batch screening.

For SBT, aliquots of an individual's DNA are mixed with "locus-specific" primers that allow amplification of exons 2 and 3 of all alleles encoded by the MHC class I loci and exon 2 of all alleles encoded by the MHC class II loci. As noted in Chapter 6, these exons encode the regions of highest polymorphism in the corresponding HLA proteins. The exact DNA sequence of these exons in an individual's HLA molecules is determined by standard DNA sequencing, and each allele's precise identity is established by comparison to a database of known HLA sequences. Such fine specificity is usually necessary for HLA typing before an HSCT and for confirming newly identified HLA alleles.

II. Alloantibody Analysis

i) Alloantibody Screening

Transplant survival is greatly decreased if the recipient has circulating pre-formed alloantibodies that could potentially mediate HAR. Prior alloantibody screening can determine whether anti-HLA antibodies are present in the recipient's serum and allow the physician to select the donor accordingly. Traditionally, a recipient's serum was tested for the presence of anti-HLA antibodies by assessing the serum's reactivity against panels of cells, the so-called *panel reactive antibody (PRA)* test. The CDC assay described previously or methods based on the enzyme-linked immunosorbent assay, or ELISA (see Appendix F) can also be employed. However, although these methods are relatively cheap, they require days or weeks to yield results. More importantly, because there are literally thousands of MHC molecules and no test cell expresses just one HLA allele at a time, these types of testing do not reveal the exact specificity of the anti-HLA antibodies in the recipient's serum. Today, molecular methods allow rapid screening of recipient serum for the presence of antibodies against individual HLA alleles. In such tests, various HLA molecules are separately synthesized *in vitro* using recombinant DNA technology. These HLA molecules are then individually coated onto synthetic beads and mixed with samples of the recipient's serum to allow the binding of antibodies that recognize a particular HLA allele. The antibodies are often linked to fluorescent markers so that results can be obtained using flow cytometry. The pattern of serum reactivity with the collection of HLA-coated beads indicates the range of anti-HLA antibodies possessed by the recipient and thus the HLA haplotypes that should be avoided when considering potential donors for this recipient.

ii) Cross-Matching

Once a potential donor is identified for a given prospective transplant recipient, a **cross-matching** test is carried out to confirm that the recipient does not possess any pre-formed antibodies that could attack a graft from that particular donor. A CDC assay can be helpful in this context, since it is a functional test that can predict the fate of the cells in a graft. T cells and B cells are separately isolated from the blood of a prospective donor to enable testing for recipient antibodies against donor MHC class I and class II molecules, respectively. The isolated cells are then mixed with exogenous complement plus the prospective recipient's serum. If the recipient possesses antidonor alloantibodies, these serum Igs bind to the donor lymphocytes and activate the complement, lysing the donor cells. A *positive cross-match* is said to have been identified. However, if antidonor alloantibodies are indeed absent, no donor cells are lysed, and a *negative cross-match* is said to have been established. The transplant can then proceed immediately. If circumstances are such that a transplant has to proceed between a recipient and donor exhibiting a positive cross-match, the recipient undergoes pre-transplant plasmapheresis to remove the pre-formed alloantibodies. Additional cross-matching assays may be carried out post-transplantation to monitor a recipient for the reappearance of antibodies that could jeopardize the long-term acceptance of the donated organ.

Flow cytometry is also used to perform cross-matching. Lymphocytes from the prospective donor are incubated with the recipient's serum, giving any alloantibodies present a chance to bind to the cells. The cells are washed, and antihuman Ig antibody tagged with a fluorescent marker is added. The tagged cells are passed through a flow cytometer, and the intensity of the fluorescence, which correlates with the concentration of alloantibodies, is measured. Flow cytometric cross-matching is more sensitive but more costly than CDC cross-matching, and in some transplant centers is reserved for recipients that have high levels of anti-HLA antibodies or have previously suffered a graft rejection. In some cases, the flow cytometric analysis of a particular cross-match may be positive but the CDC result is negative, indicating that the alloantibodies in question bind only weakly to the donor's cells. In these cases, the alloantibodies are likely not an impediment to successful transplantation, and the recipient does not have to wait for another donor organ to become available.

Solid phase methods in which donor HLA itself is fixed onto beads and is used to screen recipient serum are now starting to be used for cross-matching.

III. Immunosuppressive Drugs

Graft rejection can be combatted by immunosuppressive drugs designed to derail the alloreactive response to graft antigens. These drugs typically disrupt the normal functions of innate and/or adaptive leukocytes and generally fall into the following categories: corticosteroids, which dampen inflammatory responses; antiproliferatives, which inhibit cell division, particularly that of lymphocytes; calcineurin inhibitors, which block calcineurin, a phosphatase required for IL-2 transcription; mTOR inhibitors, which block mTOR (molecular target of rapamycin), a phosphatase involved in cell proliferation and protein synthesis; and T cell regulators, which specifically downregulate T cell functions. The most commonly used drugs of these classes, how they work in controlling graft rejection, and their associated side effects are outlined in **Table 17-6**.

TABLE 17-6   Immunosuppressive Drugs Commonly Used to Prevent Transplant Rejection

Drug	Nature	Mechanism	Outcome	Common Side Effects
Alemtuzumab	Humanized mAb recognizing CD52	Binds to T cells and promotes their death by ADCC, complement or direct induction of apoptosis	T cells are depleted	Chills, nausea, muscle spasms, fever
Azathioprine	Nucleoside drug precursor	Converted to a purine analog that inhibits DNA replication and transcription in lymphocytes	T and B cells cannot proliferate; may induce apoptosis	Nausea, vomiting, fatigue, skin rashes, anemia, infection
Belatacept	Extracellular domain of CTLA-4 fused to Fc region of human IgG1	Binds to activated T cells and shuts off costimulation	Responses of T to allogeneic cells are dampened	Anemia, neutropenia, headache, peripheral edema, PTLD, brain infection
Cyclosporine A (CsA)*	Fungal compound	Inhibits a phosphatase essential for interleukin transcription, especially IL-2	Responses of T and B cells to allogeneic cells are dampened	Cardiovascular disease, diabetes, toxicity to kidneys and CNS, periodontal swelling, dysregulated hair growth
Fludarabine	Nucleoside drug precursor	Converted to a purine analog that inhibits DNA replication and transcription in lymphocytes	T and B cells cannot proliferate; may induce apoptosis	Fatigue, fever
Malononitrilamide (MNA)	Low molecular weight molecule	Blocks pyrimidine synthesis and thus DNA replication in lymphocytes‡; inhibits intracellular signaling required for Ab and selectin expression	T and B cells cannot proliferate; Ab production and lymphocyte extravasation are inhibited	Nausea, chills, dizziness, headache
Mycophenolate mofetil (MMF)	Derivative of mycophenolic acid	Blocks purine synthesis and thus DNA replication in lymphocytes‡	T and B cells cannot proliferate	Nausea, vomiting, diarrhea, low leukocyte and platelet counts
Prednisone	Corticosteroid	Inhibits cytokine production; reduces influx of leukocytes into a graft; blocks T cell activation and proliferation	Leukocyte numbers and activation in graft site are reduced; responses of T cells to allogeneic cells are dampened	Hypertension, hyperglycemia, hyperlipidemia, osteopenia, opportunistic infections, weight gain, psychosis
Sirolimus (rapamycin)	Bacterial compound	Blocks enzyme required for signaling leading to IL-2-induced proliferation; blocks pro-inflammatory cytokine and growth factor expression	T cells cannot proliferate; B cell activation and antigen uptake by DCs are also inhibited	Anemia, low leukocyte and platelet counts, dysregulated lipid metabolism
Tacrolimus (TAC)	Bacterial compound	Same as CsA	Same as CsA but more effective	Similar but milder than those of CsA

*Use has been largely replaced by tacrolimus.
‡Other cell types use an alternative pathway to synthesize purines and pyrimidines and so are not affected by these drugs.

NOTE: In addition to the immunosuppressive drugs described in **Table 17-6**, researchers are continually developing new kinds of agents (mAbs, fusion proteins and small-molecule inhibitors) in the hopes of finding some that are effective in controlling rejection but are less toxic than traditional drugs. Four such agents are belimumab, atacicept, bortezomib and eculizumab, which all target molecules in signaling pathways important for B cell survival and/or plasma cell activation. These drugs are currently in clinical trials to assess their ability to reduce AHR. Also under investigation for their capacity to maintain immunosuppression are new agents designed to target molecules important in T cell signaling pathways, such as CD40, CD28, CD2, IL-2R, and Jak kinase. Several of these agents have shown promise in extending the long-term survival of transplant recipients.

Modern immunosuppressive drugs have made transplants between unrelated individuals possible, saving the lives of countless patients. However, the same immune system machinery that is blocked to avoid graft rejection is also normally responsible for tumor surveillance and dealing with pathogens. Thus, an immunosuppressed patient is liable to develop malignancies or come down with an opportunistic infection. In addition to these risks, the various drugs used for immunosuppression are often associated with serious medical complications. Osteoporosis is a frequent concern, as is the development of **post-transplant lymphoproliferative disease (PTLD)** (see **Box 17-4**). Physicians thus must attempt to maintain a balance between transplant survival and the patient's tolerance of detrimental side effects.

Box 17-4 Post-Transplant Lymphoproliferative Disease

Post-transplant lymphoproliferative disease (PTLD) was first recognized as an important complication of solid organ transplants and HSCTs in the mid-1980s. This disorder is clinically very heterogeneous but generally involves malignancies of B lineage cells, with the most common manifestation being diffuse large cell lymphoma (DLCL; see Ch. 20). PTLD often results when B cells of a recipient are infected with Epstein–Barr virus (EBV) and proliferate uncontrollably due to the immunosuppression necessary to prevent graft rejection. The EBV infection may have been latent in the recipient prior to transplantation, or it may have been acquired with the transplanted organ if the donor was EBV+. In other cases, the recipient may become infected by EBV well after the transplant has taken place. PTLD is especially likely to develop if the recipient's therapeutic immunosuppression is focused on T cells, since the anti-EBV T cell response necessary to control the virus is weakened. Interestingly, a transplant recipient who is initially EBV− is at greater risk of developing PTLD than an EBV+ recipient because, if the virus does attack, the immune system of the EBV− individual is capable of mounting only a primary response. In contrast, the EBV+ individual will already have anti-EBV memory lymphocytes available to mount a secondary response capable of shutting down viral replication.

Sadly, PTLD is not uncommon, occurring in about 10% of all transplant recipients. This number has steadily increased following the introduction of immunosuppressive drugs that block the proliferation and function of T cells. In retrospect, immunologists have realized that it was PTLD that killed David Vetter, "The Boy in the Bubble" (refer to Box 15-1). The BMT given to David in an attempt to set him free from his isolator was infected with EBV, a fact unknown to the donor and David's doctors. The only effective treatment currently available for PTLD is to reduce the degree of immunosuppression the patient is experiencing, while keeping a careful eye out for signs of graft rejection. Anti-B cell therapy, such as the use of rituximab, may also be warranted in some cases.

NOTE: Both transplant recipients and HIV/AIDS patients show an increased incidence of cancer, such that this disease is now a major cause of death in both populations. Intriguingly, despite the fact that these two groups have very different lifestyle-related risk factors for tumorigenesis, the types of malignancies they develop are the same: lymphomas, non-melanoma skin cancers, Kaposi's sarcoma, and ano-genital cancers. Researchers have concluded that the common influence allowing the development of a specific collection of tumors in these two groups is immunosuppression: therapeutic in the case of transplant recipients and pathogenic in the case of AIDS patients.

IV. Induction of Graft Tolerance

The ultimate objective of transplantation researchers is to induce permanent tolerance to an allogeneic graft in the absence of immunosuppressive drugs. Some scientists are seeking ways to interfere with the metabolism, proliferation and/or migration of attacking lymphocytes, whereas others are investigating approaches designed to anergize or delete only those recipient T cells that might attack the graft. These objectives overlap with new approaches to treating allergies (see Ch. 18) and autoimmune diseases (see Ch. 19), two other situations in which unwanted immune responses cause disease. This use of strategies that are designed to specifically downregulate disease-causing lymphocytes rather than shutting down the entire immune system is in line with the goal of the Immune Tolerance Network to make immunotherapy a practical reality for a broad range of diseases.

i) Bone Marrow Manipulation

If thymocytes capable of recognizing particular allo-MHC alleles could be negatively selected in a recipient's thymus, central tolerance to a particular donor's tissue would be achieved. The first step in implementing such a strategy is to establish a state of **mixed chimerism** in a recipient's bone marrow such that both recipient and donor HSCs carry out hematopoiesis. To achieve mixed chimerism, the recipient receives an HSCT from the donor prior to the solid organ transplantation procedure. As is described in more detail in Section E, to prevent the destruction of the donor HSCs by recipient T cells, the patient is first treated with a regimen of chemotherapy and/or irradiation such that peripheral lymphocytes are destroyed but HSCs are preserved. The infused donor HSCs can then safely travel to the recipient's bone marrow and engraft in this location alongside the recipient's HSCs. Hematopoiesis in the recipient then generates progenitors of both recipient and donor origin. These progenitors seed the depleted recipient thymus, giving rise to thymic DCs that express donor MHC alongside thymic DCs expressing recipient MHC. Developing thymocytes then undergo negative selection following encounters with both donor and recipient pMHCs. Some immunologists say that the thymocytes become "educated" to recognize both donor and recipient MHC molecules as self. Thus, when lymphocytes in the reconstituted recipient encounter a transplanted solid organ from the same donor, they are tolerant to the MHC differences expressed by the donor cells, and graft rejection is less likely to occur. Some success with this approach has been achieved both in animal models and in early clinical trials involving patients undergoing corneal or β-islet cell transplantation.

ii) Thymic Manipulation

Some researchers have tried to induce central T cell tolerance to alloantigens by introducing donor cells directly into the thymus of an animal whose mature peripheral T cells have been depleted by irradiation or antilymphocyte antibody treatment ("global depletion"). The presence of donor pMHC structures on cells in the thymus should result in the deletion of thymocytes with the potential to mediate allorecognition of a graft from this same donor. In addition, the presence of alloantigen in the thymus may induce the generation of nTreg cells recognizing alloantigenic peptides presented by recipient APCs. Evidence is accumulating that nTreg cells specific for alloantigen can migrate to the periphery and suppress rejection of grafts expressing the alloantigen. Scientists have begun experimenting with techniques that facilitate the fresh isolation of nTreg cells, their expansion *in vitro*, and their adoptive transfer into a transplant recipient.

iii) Immune Blockade

As described in Chapter 16, artificial interference with the delivery of T cell regulatory signals is known as establishing an immune blockade. In theory, the advantage of an immune blockade in a transplant situation is that, rather than targeting all T cells, the effect should be largely focused on T cells activated by an encounter with specific antigens of the graft. In practice, it seems that no one blockade agent on its own can induce permanent tolerance to a transplant. In animal models, a combination of blockade

Anergy and clonal deletion were described in Chapter 10, and the Immune Tolerance Network was discussed in Box 10-1.

Pre-transplant treatment of a recipient with chemotherapy and/or irradiation is known as **myeloablative, reduced intensity,** or **non-myeloablative** conditioning, depending on the degree of destruction of peripheral lymphocytes (see Section E).

As introduced in Chapter 16, adoptive transfer is the transfer of mature lymphocytes from one individual into another, most often by intravenous or peritoneal infusion.

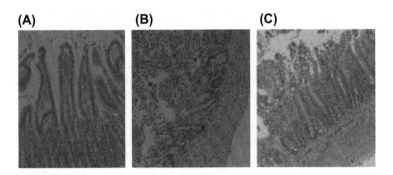

(A)   **(B)**   **(C)**

Plate 17-2
Immune Blockade of MAdCAM-1 to Reduce Graft Rejection

Rats that received intestinal grafts were left untreated, or were treated with antibody directed against MAdCAM-1 (mucosal addressin cell adhesion molecule-1) to establish an immune blockade. Histological sections of the transplanted intestines in each case were examined after 6 days. (**A**) A syngeneic control graft shows almost normal villus architecture. (**B**) In the absence of antibody treatment, acute rejection of an allogeneic graft occurred, as indicated by the presence of marked lymphocyte infiltration and destruction of villus architecture. (**C**) In rats that received an allogeneic graft and were treated with immune blockade, lymphocyte infiltration was minimal and villus architecture was better maintained. [*From Yoshiyuki Ihara, Shuji Miyagawa, Toshimichi Hasegawa, Takuya Kimura, Hengjie Xu, Masahiro Fukuzawa (2007)* Transplant Immunology 17 (4), 271–277.]

agents has demonstrated some temporary improvement in graft survival, but much work remains to optimize this approach.

Immune blockades have been established experimentally in mice by injecting anti-B7, anti-CD40L or anti-ICOS mAbs that inhibit the delivery of T cell costimulatory signaling. Alternatively, a CTLA-4 blockade similar to that used to prolong T cell responses against tumor cells (refer to Ch. 16) may be employed. In the transplantation context, a soluble fusion protein made up of CTLA-4 plus the constant region of an Ig molecule has been generated to promote CTLA-4 function and downregulate the responses of recipient antigraft T cells. An alternative blockade approach seeks to reduce the number of T cells infiltrating a graft. For example, the administration of an antibody recognizing the mucosal addressin MAdCAM-1 to rats that had received an intestinal graft decreased the number of recipient T cells infiltrating the transplanted tissue (**Plate 17-2**). These methods remain under investigation.

iv) DC Manipulation

Some immunologists are attempting to invoke graft tolerance by generating tolerogenic DCs that can induce T cell anergy (refer to Ch. 10). The idea is to deliver donor antigens to immature DCs *in vivo* under conditions that prevent DC maturation. The DCs should then be capable of anergizing any T cells they encounter, preventing an alloreactive response. In one study, researchers biochemically linked a donor antigen of interest to a mouse mAb that recognized a DC-specific marker. Administration of this mAb to a recipient mouse allowed the mAb to convey the donor antigen to immature DCs in lymphoid tissues and facilitate antigen internalization without triggering maturation. Alternatively, since immature DCs routinely process apoptotic cells to acquire self antigens without triggering maturation, isolated apoptotic donor cells have been injected intravenously into mice to promote the acquisition of donor antigens.

Another approach that has been explored in animals is to isolate a transplant recipient's DCs, culture them *in vitro* under conditions that would tolerize them, and return them to the transplant recipient. These DCs would then be expected to anergize, rather than activate, any naïve allo-MHC-specific T cells they encounter. These DCs might also trigger the differentiation of regulatory T cells that could suppress alloreactive T cells. There is some evidence from animal studies that this approach can work, and its translation to human organ transplantation is ongoing. Thus far, it seems that culture *in vitro* with IL-10 is the best way to produce DCs that can generate subsets of regulatory T cells promoting graft tolerance.

v) Manipulation of Induced Regulatory T Cells

Transplantation researchers are experimenting with strategies to produce large quantities of purified iTreg cells and manipulate their activation and function in the hopes of increasing graft tolerance. *In vivo*, iTreg cells normally occur in low numbers, but recent work suggests that methods which can trigger the expansion of these cells *in vivo* or *in vitro* followed by adoptive transfer may be of real therapeutic benefit.

a) In Vivo Generation of iTreg Cells

Several different approaches have been found to have very similar outcomes in terms of *in vivo* expansion of regulatory T cell populations recognizing donor pMHCs (donor-specific Treg cells). Treating a transplant-bearing individual transiently with anti-CD3 antibodies or antithymocyte antibodies both improves graft survival and induces tolerance. The same effect can be achieved with short-term use of either the anti-CD52 mAb alemtuzumab (discussed in Ch. 16 as an antileukemia/lymphoma drug), or mAbs directed against the integrin VLA4 or the adhesion molecule LFA1. It is believed that these agents initially kill most T cells in the recipient (including those attacking the graft) and then subsequently promote the preferential differentiation down the iTreg pathway of the cells that subsequently repopulate the recipient's T cell compartment. These donor-specific iTreg cells are activated by the presence of donor pMHC on graft cells and can suppress the activity of any residual conventional recipient T cells attacking the graft. It should be noted that the beneficial effects of these methods depend on their judicious use, as long-term administration of such powerful agents can impose general immunosuppression that can leave a recipient vulnerable to infections and cancer.

Another approach to *in vivo* iTreg cell generation involves rapamycin administration. Although this prototypic mTOR inhibitor acts as a general T cell killer and immunosuppressant if employed over the long term, short-term use of this drug can promote the growth and survival of iTreg cells. In contrast, calcineurin-based immunosuppressants have been found to decrease the number of circulating regulatory T cells in transplant recipients, as do mAbs directed against the T cell surface molecules IL-2R and CD2. More information is needed on which agents can best eliminate conventional T cells while supporting the generation and expansion of iTreg cell populations *in vivo*.

b) In Vitro Generation of iTreg Cells

Despite the promise of the preceding techniques, sometimes the *in vivo* production of regulatory T cells is insufficient to ensure graft survival. Ideally, transplant surgeons would like to be able to have a ready supply of regulatory T cells "in a bottle" to give to their patients. If iTreg cells could be reproducibly generated under defined conditions, they could be administered in predetermined numbers to recipients at the time of transplant and could be made available to faltering patients after transplant if necessary. New hope for iTreg cell manipulation in this way may lie in the results of one recent mouse study in which *in vitro*-generated iTreg cells were used to block graft rejection. Researchers isolated total T cells from mice that would later receive skin allografts and incubated these cells with donor cells so that a population of alloantigen-specific T cells was generated in culture. The investigators then used a retrovirus to introduce a Foxp3 transgene into the alloantigen-specific T cells such that these cells stably overexpressed this transcription factor and became iTreg cells. It was found that the manipulated iTreg cells could be expanded to huge numbers in culture, and that when they were later introduced into the animals that received a skin allograft, graft rejection was successfully prevented and no immunosuppressive drugs had to be given. Grafts from a different donor (whose cells had not been used to generate the alloantigen-specific T cell population) were rejected normally.

In vitro iTreg cell generation and manipulation also show promise in the human context. In one study, human T cells induced *in vitro* to overexpress a human Foxp3 transgene for 7–12 days acquired regulatory capacity. This approach, in which conventional T cells are converted into regulatory T cells, erases the difficulties associated with isolating rare nTreg or iTreg cells *in vivo*. Human transplant recipients may therefore soon benefit from the administration of *in vitror*-generated alloantigen-specific iTregs. However, several daunting challenges remain. In humans, a gene expression signature of Foxp3+CD25+ is not exclusive to regulatory T cells, meaning that alloantigen-specific effector T cells that can attack the graft might be expanded inadvertently. In addition, it is not yet clear whether these *in vitro*-generated iTreg cells are exclusively alloantigen-specific: it is possible that large numbers of such iTreg cells roaming the body might not only dampen graft rejection but also shut down antipathogen and antitumor responses, putting the recipient at risk. Lastly, it is unknown whether *in vitro*-induced iTreg cells retain their suppressive character permanently, raising the possibility that a cadre of antidonor effector T cells might eventually emerge. Should both specificity for alloantigen and T cell regulatory character be lost, anti-self effector T cells might be produced that could cause autoimmune disease.

> NOTE: Some transplant immunologists maintain that populations of regulatory B cells exist that contribute to graft tolerance by promoting the differentiation of alloantigen-specific regulatory T cells. The nature of these helpful "Breg" cells is under investigation.

vi) Mast Cell Manipulation

There is currently no evidence that mast cells participate in graft rejection, but they do appear to play a role in graft tolerance. In one study, researchers successfully established the acceptance of skin grafts in normal mice by treating the animals with anti-CD40L antibody to establish an immune blockade. However, when the same approach was tried in mast cell-deficient mice, the skin grafts were rejected. Analysis of these animals showed an unexpected link between mast cells and regulatory T cells, in that the IL-9 produced by regulatory T cells at the site of a graft supported the differentiation of mast cells. These mast cells in turn produced a protease called MCP6 (mast cell protease-6) that degrades IL-6. As discussed previously, Treg cell differentiation is inhibited by the presence of IL-6 in the immediate microenvironment. Increased levels of IL-6 in the skin grafts under study correlated with their death by ACR, presumably because fewer Treg cells were present to block the destructive effects of antigraft conventional T cells. In a mast cell-rich environment, more MCP6 and less IL-6 would theoretically be present, preserving Treg cells. Thus, strategies that promote mast cell expansion in the graft microenvironment might foster graft tolerance.

E. Hematopoietic Stem Cell Transplantation

I. The Nature of HSCT

Hematopoietic stem cell transplants (HSCTs) are carried out when a patient has a disease involving hematopoietic cells. In the past, such ailments could be treated only by the painful and risky transplantation of whole bone marrow (BMT). Although BMT is still sometimes used to restore the hematopoietic systems of cancer patients subjected to high dose chemotherapy, other conditions are usually treated by HSCT, which involves the much less stressful collection of peripheral blood. Prior to the HSCT procedure, the donor's blood is enriched for HSCs by giving the donor repeated injections of purified growth factors known to drive HSC proliferation. The HSCs then "spill out" from the bone marrow into the blood and are harvested by **leukapheresis**, a process in which blood passes out of a patient's body into a machine that collects leukocytes but returns erythrocytes and other blood components to the patient's circulation. The harvested leukocytes (including the increased numbers of critical HSCs) are then frozen in preparation for the transplant. Meanwhile, the prospective recipient undergoes pre-transplant conditioning (see Section E.II) to partially empty his/her hematopoietic compartment of leukocytes. The conditioned recipient then intravenously receives a donor cell infusion consisting of the thawed donor leukocyte preparation with its increased concentration of HSCs. The conditioning of the recipient prolongs the survival of the infused donor leukocytes, allowing the donated HSCs among them to repopulate the recipient's immune system.

NOTE: An alternative to bone marrow and peripheral blood as sources of HSCs is *umbilical cord blood (UCB)*. The first transplant involving UCB was carried out in 1988, and the growing success and acceptance of this procedure for allogeneic HSCTs has now led to the performance of over 20,000 such transplants worldwide. To date, more than an estimated 600,000 UCB units have been banked in almost 50 special UCB cryopreservation facilities around the globe. The ease of collecting UCB and its tendency to cause less severe graft-versus-host disease (see Section E.III) compared to adult hematopoietic cell preparations have made it a popular commodity. In addition, because the UCB of a given individual can be frozen at birth, should that individual require an HSCT later in life, an autologous transplant can be performed. A UCB HSCT is thus a rare example of a truly syngeneic transplant in humans, where the matching of all alleles, including HLA and ABO, is exact. One drawback to UCB is that, because it is collected from newborns who have yet to be exposed to many foreign antigens, most T cells present are naïve cells rather than memory cells, and therefore less easily activated. Thus, immune system reconstitution in recipients of UCB transplants may be delayed compared to a conventional BMT or HSCT, both of which contain mature lymphocytes. UCB transplant recipients are therefore more susceptible to potentially fatal post-transplant infections. Researchers are investigating factors that can affect UCB transplant success, such as the number of transplanted cells, timing of the transplant, and degree of HLA mismatching, in order to maximize the benefits gained from this approach.

II. Avoidance of Graft Rejection in HSCT

If an HSCT must be performed between a donor and recipient who exhibit HLA mismatches, the recipient undergoes pre-transplant conditioning consisting of a regimen of chemotherapy and/or irradiation designed to kill the leukocytes in his/her hematopoietic cell compartment. There are two objectives to such conditioning. First, the destruction of leukocytes eliminates alloreactive recipient lymphocytes, allowing the incoming donor HSCs in the HSCT to survive and repopulate the recipient's immune system. Second, if the HSCT is being given to treat a patient with a hematopoietic cancer such as a leukemia, the conditioning regimen is also used to reduce the patient's tumor burden. A clinician chooses the type of conditioning to be employed for each patient based on his/her age, type of cancer, and overall health status. If the preparatory regimen destroys most or all of the cells in the recipient's hematopoietic system, it is known as **myeloablative conditioning**. Myeloablative conditioning is very effective in preventing a graft-destroying response by the recipient's lymphocytes but also leaves the patient unable to mount any immune responses or generate new blood components until his/her hematopoietic system is successfully reconstituted by the HSCT. As a result, myeloablative conditioning is often poorly tolerated by patients (particularly older ones) and is considered very toxic. At the other end of the conditioning spectrum is **non-myeloablative conditioning**, which results in only partial destruction of the recipient's hematopoietic cells. These regimens are less severe (and so better tolerated by patients) but also frequently less successful in preventing graft rejection. More recently, newer chemotherapeutic drug options have allowed the evolution of an intermediate approach known as **reduced intensity conditioning**. These regimens come closest to the ultimate goal of achieving graft acceptance with minimal toxic side effects.

Pre-transplantation conditioning of recipients permits a clinician to broaden the pool of potential transplant donors to include unrelated, allogeneic individuals. Indeed, in the absence of recipient conditioning, an HSCT would require near-identity in donor and recipient HLA haplotypes to prevent graft rejection by the recipient's lymphocytes. HSCT donors would thus be effectively limited to the patient's HLA-matched siblings, and very few such transplants would be possible.

Hematopoietic cancers, including leukemias, lymphomas and myeloma, are discussed in Chapter 20.

NOTE: The ABO blood group compatibility that is important for solid organ transplants is much less relevant for HSCTs due to the lack of a graft vasculature to protect. Indeed, about 40% of allogeneic HSCTs are performed under conditions of ABO incompatibility.

Do recipient NK cells contribute to the rejection of HSCTs? In our previous discussion of solid organ transplants, we noted that NK cells appear to play little direct role in ACR. However, in the HSCT context, NK cells can attack and destroy allogeneic hematopoietic cells that are missing self MHC class I. These observations suggest that hematopoietic cells are inherently more sensitive to some forms of immune assault than are somatic cells within solid organs and tissues. This increased sensitivity of allogeneic hematopoietic cells to NK cell-mediated attack explains the phenomenon of "hybrid resistance," the name originally given to the once-mysterious observation that the F1 offspring of the mating of two homozygous parental strains of mice of two different MHC haplotypes were able to reject parental BMTs. Since the cells of these offspring expressed both parental MHC haplotypes, the parental BMTs should have been accepted as "self" based on the expected reactivity of the offspring's lymphocytes. Indeed, solid organ grafts from either parental strain were accepted as "self" by the F1 offspring. The fact that the parental BMTs were rejected flew in the face of the classical dogma that graft rejection was strictly due to the recognition by T and B cells of non-self pMHC structures on the graft. Scientists later theorized that the BMTs must have been "missing some self" that made the parental bone marrow cells vulnerable to attack by activated NK cells.

III. Graft-Versus-Host Disease (GvHD) in HSCT

i) Nature of GvHD in HSCT

As mentioned earlier, GvHD occurs when donor cells in the transplanted tissue attack recipient cells and tissues. Because the transplanted tissue in an HSCT consists of immunocompetent leukocytes, GvHD is a significant issue for HSCT recipients. In particular, the HLA mismatching in allogeneic HSCT has the potential to lead to acute and severe GvHD that can be fatal. Once activated by recognition of an MHC or MiHA mismatch, alloreactive donor T cells undergo clonal expansion. Pro-inflammatory cytokines and chemokines are released that recruit donor and recipient macrophages and other cell types that do not discriminate between donor and recipient targets. TNF secreted by these cells is thought to be responsible for the metabolic wasting associated with many cases of GvHD. However, the bulk of tissue destruction in GvHD is caused by Fas- or perforin/granzyme-mediated cytotoxicity exerted by donor CTLs. Donor Th cell responses directed against MiHA presented on MHC class II may also contribute. Interestingly, donor NK cells appear to play an important opposing role in this context. As noted previously, NK cells do not generally attack allogeneic cells in solid organs, but they will kill allogeneic hematopoietic cells. Thus, donor NK cells may destroy allogeneic recipient DCs, helping to reduce attacks by donor T cells on recipient tissues and thereby acting to prevent, rather than cause, GvHD.

GvHD in HSCT is especially hard on the thymus, where recipient cTECs and mTECs required for thymocyte differentiation are prime targets for destruction by alloreactive donor T cells. As a result of this damage, the rate of naïve T cell production in an HSCT recipient drops markedly for 6–12 months after the transplant, leaving the recipient vulnerable to infection and tumorigenesis.

NOTE: There is a very important form of GvHD that is actually helpful when the recipient of an HSCT is a leukemia patient. The disease in these individuals is caused by the uncontrolled proliferation of malignant leukocytes, so that the ultimate goal of treatment is to destroy these cells completely. In the early 1990s, clinicians observed that the donated immunocompetent cells in an allogeneic HSCT could recognize and kill any allogeneic leukemic cells that had survived prior chemotherapy and irradiation treatment and remained in the patient. This beneficial outcome became known as the **graft-versus-leukemia (GvL)** effect. Subsequent work has confirmed that the GvL effect can make a significant contribution to the successful treatment of hematopoietic cancers, particularly when patients have undergone reduced intensity conditioning that might allow the survival of residual cancer cells. The GvL effect is discussed in more detail in Chapter 20.

ii) Minimizing GvHD in HSCT

In the absence of clinical intervention, acute GvHD occurs in about 75% of HSCT recipients with a mismatch at a single MHC locus, and in 80% of recipients with three mismatches. Even among recipients completely matched at all MHC loci, about 30–50% will develop clinically significant GvHD due to MiHA mismatches. Clinicians are thus searching for ways to reduce GvHD while achieving efficient immune system reconstitution. A major issue is that the conditioning regimens used to prepare recipients for HSCT also promote GvHD. The tissue trauma caused by the relatively high doses of drugs used to fully or partially empty the recipient's hematopoietic compartment promotes an inflammatory response and the production of cytokines that both cause additional damage and encourage donor T cells to attack recipient cells. Thus, establishing the least amount of conditioning that can be used in a given situation may help to minimize GvHD. Moreover, because much of the damage of GvHD is due to cytokines, scientists are exploring ways to block the production or action of these molecules using various antibodies or antagonists. Indeed, concrete reductions in GvHD in animals have been observed when TNF is neutralized. In addition, because of the particular vulnerability of TECs to GvHD, researchers have been looking at the administration of FGF7 (fibroblast growth factor-7) as a protective measure. FGF7 promotes the proliferation of thymic stromal cells *in vitro*, and clinical trials of its utility *in vivo* are under way.

Future strategies to combat GvHD in HSCT may be based on the prevention of donor T cell recognition of recipient antigens, or the anergization of donor T cells. Alternatively, the adoptive transfer of regulatory T cells may be useful. In one human study, HSCT recipients with chronic GvHD were infused with donor-specific nTreg cells that had been isolated and expanded *in vitro*. GvHD in these individuals was modestly reduced after treatment, and the engraftment of the donor tissues was improved. In another investigation using animal models, lethal GvHD could be prevented if the HSCT was accompanied by nTreg cell transplantation. Although some researchers would like to move this approach into clinical trials and optimize this therapeutic avenue for human use, significant challenges in its safe application remain. For example, there is a very real risk that GvHD might be initiated or exacerbated if conventional T cells are not fully eliminated from the Treg cell preparation, a task that is currently very difficult and costly. In addition, it generally takes 2–3 weeks to isolate Treg cells from a donor and expand them sufficiently *in vitro* for infusion into an HSCT recipient. A recipient who is experiencing severe GvHD may not survive long enough to be treated with Treg cells. Unfortunately, it is currently impossible to predict which HSCT recipients will develop severe enough GvHD to justify the risks of Treg administration, or to determine in advance how large a dose of Treg cells will be needed to counter the disease. These considerations preclude the preventative administration of *in vitro*-generated donor Treg cells. Nevertheless, despite these hurdles, clinical researchers are optimistic that adoptive transfer of Treg cells will eventually become a useful tool in the HSCT armamentarium.

IV. Immune System Reconstitution after HSCT

Because HSCTs consist mainly of HSCs and progenitor cells, it takes a significant amount of time for these cells to completely rebuild the recipient's immune system, particularly if the patient has undergone myeloablative conditioning. Although innate immunity, at least in the form of responses by myeloid and NK cells, is recovered relatively quickly after a transplant, it takes much longer to produce the mature functional T and B cells needed to power the adaptive response. Thus, an HSCT recipient is particularly vulnerable to infection and tumorigenesis during the reconstitution period. The faster and more extensively reconstitution is completed, the better the outcome for the patient. The age of the recipient, the degree of his/her thymic involution, the degree of donor T cell depletion, the percentage of donor T cell engraftment, whether GvHD has occurred, and whether the recipient's thymus has been damaged by immunosuppressive (or other) drug administration, all influence the degree of success of reconstitution. In adult patients who receive an HSCT depleted of donor T cells, it can take up to 2 years to restore normal thymopoiesis. Moreover, if severe donor T cell depletion has been undertaken to reduce GvHD, the lag in the reconstitution of the adaptive response may be extended by an additional 6–12 months. Such recipients are at a high risk for increased morbidity and mortality due to opportunistic viral infections. Recipients of an HSCT derived from UCB are also at increased risk of infection because, as noted earlier, the naïveté of the T cells in newborns means that adaptive defense is only at a primary response level. Investigators are attempting to find ways to increase the speed and safety of HSCT reconstitution.

Focus on Relevant Research

"Assessing the Safety of Stem Cell Therapeutics" by Goldring, C.E.P., Duffy, P.A., Benvenisty, N., Andrews, P.W., Ben-David, U., Eakins, R., French, N., Hanley, N.A., Kelly, L., Kitteringham, N.R., Kurth, J., Ladenheim, D., Laverty, H., McBlane, J., Narayanan, G., Patel, S., Reinhardt, J., Rossi, A., Sharpe, M., and Park, B.K. (2011) *Cell Stem Cell* 8, 618–628.

The constant improvements to HSCTs devised by clinical researchers offer hope to many patients with diseases mediated by dysregulated immune responses, including those suffering from autoimmune disorders, hematopoietic cancers or hypersensitivities. Other scientists working with types of stem cells other than HSCs are striving to offer the same hope to patients afflicted with a wide range of non-immunological diseases. The goal is to coax tissue-specific stem cells to generate a rudimentary healthy tissue in a patient in need of a new organ, an approach that can be viewed as a new type of "transplantation."

Stem cells were originally tapped for use in regenerative medicine to replace a diverse range of damaged tissues and organs. Subsequently, the potential of stem cell-based therapies to solve a broader range of medical issues has headlined news reports. The popular media, however, have paid only limited attention to the daunting challenges surrounding the pre-clinical development, trial design, standardization, and safety of therapeutic stem cell applications. This review article familiarizes the reader with the numerous hurdles faced by stem cell biologists. Six central issues are explored: (1) Stem cells are more difficult to manufacture consistently than a conventional small-molecule drug. (2) Stem cells can be genetically unstable. (3) The dosing and pharmacokinetics of stem cells used in a therapeutic context are difficult to control. (4) Stem cells, especially those that are introduced intravenously, have the potential to migrate to unintended body sites. (5) Allogeneic stem cells may be rejected by the patient's immune system. (6) Stem cells have the potential to form tumors. As a result of these and many other considerations, complex regulatory algorithms have been proposed to guide the translation of stem cell-based therapies from the laboratory to the clinic. This review provides a striking illustration of the tortuous roadmap that stem cell-based therapies must follow to reach the clinic.

F. Blood Transfusions

One of the most common tissue transplants is blood transfusion, in which a preparation of blood cells (but not serum) donated by healthy volunteers is injected into the bloodstream of injured or diseased individuals or those undergoing surgery. However, the RBCs of different individuals frequently express different antigens, so that two individuals may be of incompatible blood types. If the blood cells of an incompatible individual are transfused into a recipient by mistake, a **transfusion reaction** results because the recipient's circulation contains pre-formed antibodies specific for donor RBC antigens. These antibodies immediately attack the transfused RBCs and induce their complement-mediated destruction, an event referred to as *hemolysis*. Clinically, the symptoms of a transfusion reaction may be as mild as a headache or wheezing, or as serious as tissue necrosis. In the most severe cases, the recipient's blood pressure falls and the blood vessels constrict, resulting in renal failure and shock. In the U.S., transfusion reactions account for about 1 patient death per 500,000 units of blood transfused.

The majority of severe transfusion reactions result when blood of an incorrect ABO blood type is transfused into a recipient. Because of differences in the terminal sugars on the erythrocyte surface antigens A and B (**Fig. 17-4**), these proteins constitute Ti antigens in a person of an incompatible blood type. During the development of central tolerance, an

A transfusion reaction is a systemic response by a recipient's body to the administration of blood from an incompatible donor.

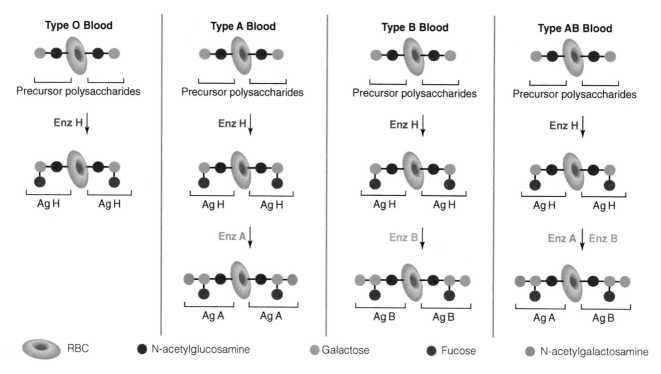

Fig. 17-4
Structures of ABO Blood Sugars

All individuals express a precursor polysaccharide containing N-acetylglucosamine and galactose on the surfaces of their RBCs. All individuals also express a transferase enzyme (Enz H) that attaches fucose to the precursor to form the H antigen. The RBCs of individuals possessing blood type O have only this type of complex sugar on their surfaces. Individuals of blood type A express an additional transferase (Enz A) that attaches N-acetylgalactosamine to the H antigen to form the A antigen. Individuals of blood type B express a different transferase (Enz B) that attaches galactose to the H antigen to form the B antigen. Individuals of blood type AB express both Enz A and Enz B and so have RBCs expressing both the A and B antigens.

individual of blood type A loses the B lymphocytes recognizing antigen A (due to negative selection) but retains the B cells recognizing antigen B. Unfortunately, certain common intestinal bacteria express glycoproteins with epitopes closely resembling those of antigens A and B. Therefore, long before any transfusion, the B cells in the type A individual that recognize antigen B are activated by cross-reaction with the bacterial epitopes, and anti-B antibodies are synthesized and released into the circulation. These cross-reactive antibodies are then maintained at significant levels by the ongoing presence of the bacteria. Thus, when a transfusion of blood of type B is attempted in the individual of blood type A, the pre-formed cross-reactive antibodies in the recipient's circulation immediately attack the transfused RBCs: a transfusion reaction results. **Table 17-7** provides a summary of blood type compatibility between recipients and donors of various ABO blood types based on the RBC blood group antigens and antibodies involved in each case.

Platelets, which are transfused into patients with blood clotting disorders and other hematological diseases, also express ABO antigens. However, hemolytic transfusion reactions triggered by platelet ABO incompatibility are relatively rare and seldom fatal. Thus, until recently, clinicians did not bother to routinely screen these blood products for ABO compatibility before administering them to a patient. Today, more and more clinics are trying to maximize patient safety by transfusing only ABO-identical blood components of any type.

In the U.S., transfusion reactions can be reported to the CDC's National Healthcare Safety Network. These "hemovigilance" data are shared nationally as part of a larger "biovigilance" program to improve patient safety.

NOTE: Transfusion-related GvHD is the term given to the situation in which donor leukocytes transferred with the transfused blood attack recipient tissues. The incidence of transfusion-related GvHD is very rare in developed countries because, prior to transfusion, leukocytes are removed from the donated blood by either irradiation or physical depletion. In countries where leukocyte removal is not practiced, 5% of cases of transfusion-related GvHD result in death.

TABLE 17-7 Blood Group Antigens and Antibodies

Blood Type of Recipient	Genotype	Sugars Added by Transferases	Antigens on RBCs	Anti-ABO Antibodies in Circulation	Blood Type of Compatible Donors
A	AA or AO	Fucose; NAGA*	A	Anti-B	A, O
B	BB or BO	Fucose; GAL†	B	Anti-A	B, O
AB	AB	Fucose; NAGA; GAL	A and B	None	A, B, AB, O
O	OO	Fucose	H	Anti-A and anti-B	O

*N-Acetylgalactosamine.
†Galactose.

The ABO blood group antigens are not the only antigens capable of provoking transfusion reactions. A prospective recipient should also be screened for the expression of the highly immunogenic *Rh antigen* in order to avoid **Rh disease** (see **Box 17-5**). A severe anti-Rh antibody response may occur in a transfusion situation if a recipient is Rh⁻ but the donated blood is Rh⁺, and if the Rh⁻ recipient possesses circulating anti-Rh antibodies due to prior exposure to Rh⁺ cells during pregnancy or tissue transplantation. There are other minor blood group antigens that may be mismatched between donors and recipients but physicians do not usually screen for these alloantigens because they are not highly immunogenic. Any immune responses provoked in recipients by these mismatches are mild and of little clinical significance.

We have come to the end of our discussion of transplantation and move next to a description of allergy and other hypersensitivities. Many of the concepts learned in Chapters 16 and 17 overlap those relevant to the control of hypersensitivities (Chapter 18) and autoimmune diseases (Chapter 19). Clinical researchers hope that the lessons learned in one discipline will have direct application to the others.

Box 17-5 Passive Immunization to Prevent Rh Disease

Rh is a protein expressed on the RBCs of many, but not all, humans. In Rh disease (also known as hemolytic disease of the newborn, or erythroblastosis fetalis), the immune system of a mother whose RBCs do not express Rh (Rh⁻) destroys the RBCs of her fetus that does express Rh (Rh⁺) if she has *previously* been pregnant with an Rh⁺ fetus. During a first pregnancy in which the Rh⁻ mother is carrying an Rh⁺ baby, very few fetal cells get into the mother's circulation prior to birth, and only naïve maternal anti-Rh lymphocytes are present. Any anti-Rh immune response mounted against the first Rh⁺ fetus is weak and not particularly harmful. However, the traumatic process of birth propels many more fetal cells into the maternal circulation. These fetal cells provoke a vigorous response by the maternal immune system that eliminates the fetal RBCs and generates anti-Rh memory cells. If a subsequent pregnancy also involves an Rh⁺ fetus, the few fetal cells that make it into the maternal circulation prior to birth will activate memory B cells. Maternal plasma cells are generated that produce large amounts of anti-Rh antibodies capable of crossing the placenta and destroying RBCs in the fetus.

Rh disease can be prevented by passively immunizing an Rh⁻ woman with an anti-Rh antibody preparation early during her first pregnancy and again shortly after the birth of her first child. The exogenous anti-Rh antibodies bind to the Rh antigen on any fetal RBCs that accessed the maternal circulation during birth, clearing them before they can interact with naïve anti-Rh B cells in the mother. Virtually no memory anti-Rh B cells are generated, so that if the fetus in the next pregnancy is Rh⁺, the risk of fetal damage is reduced. As insurance, Rh⁻ women are given anti-Rh antibodies throughout and after subsequent pregnancies.

Chapter 17 Take-Home Message

- Allogeneic MHC molecules on transplanted tissue trigger strong immune responses that may cause rapid graft rejection, whereas molecules encoded by allogeneic minor histocompatibility loci usually trigger weak antigraft responses.

- Direct allorecognition occurs when the TCRs of recipient T cells specific for complexes of [self-MHC + non-self peptide] cross-react with complexes of [non-self MHC + self peptide] expressed by cells in a graft.

- Indirect allorecognition occurs when allogeneic MHC molecules shed from a graft are processed by recipient APCs and presented to recipient Th cells.

- Hyperacute graft rejection of a solid organ transplant is very fast and mediated by pre-formed alloantibodies.

- Acute graft rejection of a solid organ transplant is fast and mediated by CTLs (acute cellular rejection) or B cells (acute humoral rejection).

- Chronic graft rejection of a solid organ transplant is slow and mediated by cytokines and other molecules induced by CTL- and Ab-induced graft destruction.

- In GvHD of solid organ transplants, donor lymphocytes in the allograft attack recipient tissues. Donor NK cells do not contribute to GvHD in these transplants.

- MHC matching between a recipient and donor improves transplant and patient survival and is optimized by tissue typing techniques.

- ABO blood group matching between recipient and donor and testing of the recipient's serum for graft-specific alloantibodies minimize the chance of HAR during solid organ transplantation.

- HSCT involves the transfer of healthy HSCs from a donor to a recipient depleted of hematopoietic cells.

- The risk of GvHD is significant in allogeneic HSCT due to collateral transfer of donor mature T cells. Donor NK cells do contribute to GvHD in HSCTs.

- Donor CTLs and NK cells are responsible for the beneficial GvL effect.

- Long-term allograft acceptance is currently established using immunosuppressive agents. The induction of specific graft tolerance in a recipient may be achieved in the future by adoptive transfer of regulatory T cells.

- Blood transfusions are a common form of transplantation and require ABO blood group compatibility for success.

Did You Get It? A Self-Test Quiz

Section A

1) One type of tissue transplantation involves solid organs. What are two other types?

2) What is graft rejection, and why does it occur?

3) Distinguish between syngeneic and allogeneic transplants.

4) Distinguish between isografts, allografts, and xenografts.

5) What does FMT stand for, and how does it work?

Section B

1) Define "allorecognition" as the term is commonly used in the transplantation context.

2) Slower, milder graft rejection is due to differences in what type of genetic locus?

3) What are the two mechanisms that account for T cell responses to allo-MHC, and what are the consequences of these responses?

4) Why are T cell responses to an allogeneic transplant much stronger than T cell responses to a pathogen?

5) What are minor H peptides, and why do they provoke a graft rejection response?

6) Give two examples of minor histocompatibility loci.

7) Why are humoral responses not usually a component of rejection due to minor histocompatibility antigen differences?

8) How does the strength of a T cell response to a minor histocompatibility antigen compare to that of a T cell response to a pathogen? Explain.

Section C

1) Can you define these terms? hyperimmunization, plasmapheresis, ischemia

2) What elements of the innate response contribute to the rejection of a solid organ transplant?

3) Name the four types of clinical graft rejection and state their rates of onset.

4) Contrast the underlying mechanisms of the four types of clinical graft rejection.

5) How might NK cells contribute to ACR, and how might they counteract ACR?

6) Name two ways regulatory T cells in a transplant site might be lost.

7) C4d staining is a feature of what type of graft rejection?

8) What is the most important clinical predictor of CGR?

9) What are the histological features of CGR, and how are they caused?

10) How is graft-versus-host disease caused, and to what degree does it affect solid organ transplantation?

Section D

1) Give two reasons why organs from living related donors are preferable to cadaveric organs for human transplantation.

2) Name three types of antigens that are now being typed to match organ donors and recipients.

3) Compare the advantages/disadvantages of serological and DNA-based HLA typing.

4) What is a null HLA allele, and how does it affect the choice of donor?

5) Briefly outline the differences between the three most common methods of DNA-based HLA typing.

6) Describe three potential difficulties associated with the use of animal organs for human transplants.

7) Distinguish between alloantibody screening and cross-matching.

8) What is a positive cross-match, and how is it determined?

9) What are the pros and cons of the use of immunosuppressive drugs in a transplant situation?

10) Briefly describe three types of commonly used immuno-suppressive drugs and how they work.

11) Why are clinicians seeking new types of immunosuppressive drugs?

12) What is PTLD, and how is it caused?

13) Why would induction of graft tolerance be beneficial for a transplant recipient?

14) What is mixed chimerism?

15) What is adoptive transfer, and how is it achieved?

16) Describe four ways in which graft tolerance might theoretically be induced.

17) Outline how regulatory T cells might be manipulated to increase graft acceptance.

Section E

1) For what types of disorders is HSCT a treatment option?

2) What is leukapheresis, and why is it helpful for HSCT?

3) What are the advantages and disadvantages of using umbilical cord blood for HSCT?

4) What is the difference between myeloablative and non-myeloablative conditioning?

5) How does the requirement for HLA matching between donor and recipient compare for HSCT versus a solid organ transplant? ABO matching?

6) Briefly explain the phenomenon of "hybrid resistance."

7) Why is GvHD more important in HSCT than in solid organ transplants?

8) What is the graft-versus-leukemia effect, and why is it useful?

9) Describe three measures that may help to reduce GvHD in HSCT?

10) What factors affect immune system reconstitution after an HSCT?

11) Outline four general issues with stem cell therapeutics that are raising safety concerns.

Section F

1) What is a transfusion reaction?

2) What is the mechanism underlying ABO blood group incompatibility?

3) Sketch out a table defining the ABO blood groups and indicating who may give blood to whom and why.

4) What is "hemovigilance," and why might it be helpful?

5) What is transfusion-related GvHD, and how might it be prevented?

6) An Rh-negative mother is having her second child. What measure should she take to prevent Rh disease in her baby and why?

Can You Extrapolate? Some Conceptual Questions

1) Consider the following xenotransplantation scenario: A solid organ transplant is carried out between a recipient and donor of different species. The basic immune response mechanisms in these species are very similar and their MHC molecules are of identical function, but these proteins are extremely different in structure. Assuming that the basic mechanisms of "xenograft recognition" are similar to those mediating allograft recognition (i.e., direct allorecognition and indirect allorecognition), what might you expect to observe with respect to

   a) innate immune responses?

   b) direct "xenorecognition" responses (i.e., the mechanism involved is like direct allorecognition)?

   c) indirect "xenorecognition" responses (i.e., the mechanism involved is like indirect allorecognition)?

2) **Table 17-4** shows transplant survival rates for various types of solid organ transplantation. The 5-year patient survival rates for kidney, liver, heart, pancreas, and kidney/pancreas transplants are all greater than 70%. In contrast, the 5-year patient survival rates for lung, heart/lung and intestine transplants are all less than 50%. With the immunological basis of allogeneic organ rejection in mind, what possible effect might the anatomical location of a graft have on its relative chances of survival?

3) In Section F on transfusion reactions, a Tip refers to a system for reporting these adverse events in patients. How would you compare this system and one of a similar nature discussed in Chapter 14 on vaccination? Do the terms "hemovigilance" and "biovigilance" used in this chapter apply to the system described in Chapter 14?

Would You Like To Read More?

Burrell, B. E., & Bishop, D. K. (2010). Th17 cells and transplant acceptance. *Transplantation*, *90*(9), 945–948.

Cavazzana-Calvo, M., Andre-Schmutz, I., Dal Cortivo, L., Neven, B., Hacein-Bey-Abina, S., & Fischer, A. (2009). Immune reconstitution after haematopoietic stem cell transplantation: Obstacles and anticipated progress. *Current Opinion in Immunology*, *21*(5), 544–548.

de Kort, H., de Koning, E. J., Rabelink, T. J., Bruijn, J. A., & Bajema, I. M. (2011). Islet transplantation in type 1 diabetes. *BMJ*, *342*, 217.

Dunn, P. P. (2011). Human leucocyte antigen typing: Techniques and technology, a critical appraisal. *International Journal of Immunogenetics*, *38*(6), 463–473.

Edinger, M., & Hoffmann, P. (2011). Regulatory T cells in stem cell transplantation: Strategies and first clinical experiences. *Current Opinion in Immunology*, *23*(5), 679–684.

Eng, H. S., & Leffell, M. S. (2011). Histocompatibility testing after fifty years of transplantation. *Journal of Immunological Methods*, *369*(1–2), 1–21.

Fishbein, T. M. (2009). Intestinal transplantation. *New England Journal of Medicine*, *361*(10), 998–1008.

Gill, R. G. (2010). NK cells: Elusive participants in transplantation immunity and tolerance. *Current Opinion in Immunology*, *22*(5), 649–654.

Gokmen, M. R., Lombardi, G., & Lechler, R. I. (2008). The importance of the indirect pathway of allorecognition in clinical transplantation. *Current Opinion in Immunology*, *20*(5), 568–574.

Howell, W. M., Carter, V., & Clark, B. (2010). The HLA system: Immunobiology, HLA typing, antibody screening and crossmatching techniques. *Journal of Clinical Pathology*, *63*(5), 387–390.

Lees, J. R., Azimzadeh, A. M., & Bromberg, J. S. (2011). Myeloid derived suppressor cells in transplantation. *Current Opinion in Immunology*, *23*(5), 692–697.

Murphy, S. P., Porrett, P. M., & Turka, L. A. (2011). Innate immunity in transplant tolerance and rejection. *Immunological Reviews*, *241*(1), 39–48.

Nankivell, B. J., & Alexander, S. I. (2010). Rejection of the kidney allograft. *New England Journal of Medicine*, *363*(15), 1451–1462.

Regateiro, F. S., Howie, D., Cobbold, S. P., & Waldmann, H. (2011). TGF-beta in transplantation tolerance. *Current Opinion in Immunology*, *23*(5), 660–669.

Roncarolo, M. G., Gregori, S., Lucarelli, B., Ciceri, F., & Bacchetta, R. (2011). Clinical tolerance in allogeneic hematopoietic stem cell transplantation. *Immunological Reviews*, *241*(1), 145–163.

Shilling, R. A., & Wilkes, D. S. (2011). Role of Th17 cells and IL-17 in lung transplant rejection. *Seminars in Immunopathology*, *33*(2), 129–134.

Thiruchelvam, P. T., Willicombe, M., Hakim, N., Taube, D., & Papalois, V. (2011). Renal transplantation. *BMJ*, *343*, 7300.

Wood, K. J., Bushell, A., & Jones, N. D. (2011). Immunologic unresponsiveness to alloantigen *in vivo*: A role for regulatory T cells. *Immunological Reviews*, *241*(1), 119–132.

Immune Hypersensitivity

Chapter 18

Govern a great nation as you would cook a small fish. Do not overdo it.

Lao Tzu

The immune response is usually viewed as helpful because it protects against pathogen attack. In most cases, the secondary response to a pathogen is so effective that the individual does not get sick at all. In this chapter, we examine what happens when a primary response is followed by a secondary response that hurts rather than helps the individual. Such disorders are called **immune hypersensitivities**, and immunologists classify them into four types based on their underlying mechanisms: **type I, IgE-mediated hypersensitivity**; **type II, direct antibody-mediated cytotoxic hypersensitivity**; **type III, immune complex-mediated hypersensitivity**; and **type IV, delayed type hypersensitivity**. Key features of these disorders are summarized in **Table 18-1** and discussed in detail in the following sections.

All hypersensitivities develop in two stages: the **sensitization** stage and the **effector** stage. The sensitization stage is the primary immune response to an antigen, whereas the effector stage is a secondary immune response. In this context, hypersensitivity (HS) is defined as any excessive or abnormal secondary immune response to an antigen. A first exposure to an antigen causes most individuals to mount a normal primary response that is followed by normal secondary response upon a subsequent exposure. It remains unknown why a first exposure to an antigen results in a primary response that sensitizes some individuals, who then experience a hypersensitivity reaction upon a subsequent exposure to the same antigen.

A. Type I Hypersensitivity: IgE-Mediated or Immediate

I. What is Type I HS?

Type I hypersensitivity is what most people think of as "allergy." Allergies occur in individuals who express IgE antibodies directed against certain innocuous antigens in the environment. These antigens, which

MAK: Primer to the Immune Response. http://dx.doi.org/10.1016/B978-0-12-385245-8.00018-2

TABLE 18-1	Types of Hypersensitivity (HS) and their Key Characteristics			
Type	**Type I HS**	**Type II HS**	**Type III HS**	**Type IV HS**
Common name(s)	IgE-mediated HS Immediate HS Allergy, atopy	Direct antibody-mediated cytotoxic HS	Immune complex-mediated HS	Delayed type HS Cell-mediated HS
Primary immune system mediator	Antibody (IgE)	Antibody (IgG or IgM)	Antibody (IgG or IgM)	Effector T cells, macrophages
Time to symptoms	<1–30 min	5–8 hr	4–6 hr	24–72 hr
Mechanism	Allergens cross-link IgE bound on mast cells and basophils and induce degranulation	IgG or IgM bind to cell-bound antigen; cell is destroyed by phagocytosis, complement activation or ADCC	Immune complexes trigger complement activation; phagocyte FcR engagement leads to release of lytic mediators	Effector T cells produce IFNγ and other cytokines promoting macrophage hyperactivation
Examples	Asthma, hay fever, eczema, hives, food allergies, anaphylaxis	Hemolytic anemias, Goodpasture's syndrome	Arthus reaction, aspects of rheumatoid arthritis (RA) and systemic lupus erythematosus (SLE)	Lesions of TB and leprosy, poison ivy, farmer's lung

In North America, IgE-mediated allergies affect more than 25% of the population.

In the U.S., over 5000 deaths and 500,000 hospitalizations each year are attributed to asthma attacks, and the annual cost of caring for asthma patients exceeds $6 billion.

are commonly referred to as **allergens**, are typically soluble proteins that are components of larger particles such as pet dander or tree pollen (**Table 18-2**). Most people encountering such antigens produce IgM, IgG or IgA antibodies that successfully clear the antigens without causing any symptoms. However, in individuals who make IgE antibodies to these antigens, reactions are triggered that lead to side effects which can range from itching and swelling to breathing difficulties and even shock or death. The response to the allergen is generally very rapid and occurs within 30 minutes of the encounter, so that type I HS is also known as "immediate" HS. Why only some people produce IgE antibodies to allergens is a mystery.

Another term for allergy is **atopy**, so that clinicians speak of "atopic patients" and "atopic reactions." There are two types of atopic reactions: systemic and local. A *systemic* atopic response is called **anaphylaxis** and affects the entire body (see later). In a local atopic reaction, the allergic symptoms depend on the anatomical location of the affected tissue and are generally confined to that site. For example, a local IgE-mediated response to an allergen in the nose is manifested as *atopic rhinitis* (hay fever), while a local response in the airway and lungs that results in the inflammation of these tissues is called *atopic asthma*. In the skin, local atopic responses take the form of *atopic dermatitis* (eczema) or *atopic urticaria* (hives). It is not understood why some antigens cause localized reactions, whereas others have systemic effects.

TABLE 18-2	Examples of Common Allergens	
Allergen Name*	**Scientific Name of Source**	**Common Name of Source**
Amb a 2	*Ambrosia artemisiifolia*	Ragweed
Api m 1	*Apis mellifera*	Bee venom
Ara h 2	*Arachis hypogea*	Peanut
Bet v 1	*Betula verrucosa*	Birch tree pollen
Can f 1	*Canis familiaris*	Dog dander
Der p 1	*Dermatophagoides pteronyssinus*	House dust mite
Pen a 1	*Penaeus aztecus*	Shrimp
Phl p 5	*Phleum pratense*	Timothy grass

*Allergens are named according to the first three letters of the genus of the organism from which the antigen is derived combined with the first letter of the species and a number indicating the order of discovery. For example, Amb a 2 is the second allergen derived from **Amb**rosia **a**rtemisiifolia.

NOTE: Allergy testing in common parlance is the process by which the allergen irritating a particular patient is identified. One or more of several different assays may be used. Enzyme-based immunoassays seek to detect the presence of IgE antibodies in a patient's serum. A positive result indicates that the patient is atopic (assuming that pathogen-related causes of IgE elevation are ruled out). This type of diagnostic method is safe for the patient because he/she never comes in direct contact with the allergen and so avoids any chance of anaphylaxis. However, this test does not identify the specific allergen. In addition, in some cases, most of the patient's IgE is bound to mast cells, as opposed to circulating in the blood, so that the patient's allergy may go undetected by this serum-based approach. To avoid the latter difficulty and to identify a specific allergen, a skin prick test is most often used. Typically, the skin surface of a patient's forearm or back is pricked in different spots with a collection of needles, each coated at the tip with a different diluted purified allergen. The development of a skin eruption at the site of a particular prick indicates that the individual is allergic to the corresponding allergen. While this method is low cost and fast (results within 20 minutes), some individuals may show positive skin prick tests for allergens to which they are not clinically allergic (have no troubling physical symptoms). Nevertheless, the relative severity of the skin reactions provoked by the various test allergens may help to identify those that are causing overt allergy in the patient.

The most modern atopy diagnosis method, which is not yet in common use, involves microarray analysis. A very small volume of a patient's serum is applied to a microscopic solid surface upon which tiny amounts of a vast array of purified allergens are fixed in known positions. The specific binding of antibodies in the patient's serum to proteins in the array is then quantitated relative to the appropriate controls to determine which among the allergens may be the culprits causing the patient's symptoms. This method offers unparalleled discrimination among allergens, but it takes some skill to master the technique and interpret its results reliably.

II. Mechanisms Underlying Type I HS

Like all hypersensitivities, type I HS has a sensitization stage and an effector stage.

i) Sensitization Stage

The sensitization stage of a type I HS reaction is illustrated and described in detail in **Figure 18-1**. Th2 effector cells generated in response to pMHC complexes containing allergen peptides produce copious amounts of IL-4, IL-5 and IL-13 that influence the isotype switching of allergen-activated B cells toward IgE (rather than IgG or IgA). Some of these IgE antibodies bind to allergen in its site of penetration, whereas others bind to high-affinity FcεRI receptors expressed on the surfaces of mast cells in the immediate area. The remaining IgE antibodies enter the circulation and eventually bind to FcεRI molecules on the surfaces of basophils and mast cells in the blood and tissues. As a result, the mast cells and basophils are soon coated with allergen-specific IgE and become bombs waiting to be triggered by a subsequent encounter with the allergen. These sensitized mast cells and basophils can remain "armed" in this way for an extended period.

At this point, the reader may well ask, "But if allergens are generally innocuous, what supplies the DAMPs that induce allergen-capturing DCs to mature and activate naïve anti-allergen T cells?" Researchers speculate that allergens in fact are associated with some kind of cellular stress or damage that provokes a primary response in everyone. However, in non-atopic individuals, this response takes the form of harmless IgG production rather than IgE. In an atopic person, the lymph nodes contain abnormally high local concentrations of IL-4 that promote Th2 effector cell differentiation. These cells then secrete the Th2 cytokines that promote isotype switching to IgE in activated B cells and profoundly influence the effector actions of other leukocytes responding to the allergen. The Th2 phenotype of the response appears to be inextricably linked to allergy, since type I HS reactions to allergens in mouse models can be mitigated by inducing immune deviation to Th1.

Because they harm, rather than help, an individual, anti-allergen antibodies are sometimes described as being "pathological."

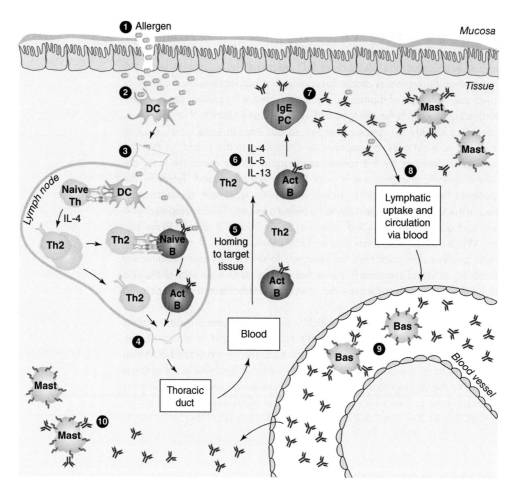

Fig. 18-1
Sensitization Stage of Type I Hypersensitivity

Allergen breaches a mucosal barrier (**1**) and is taken up by an immature DC (**2**), which conveys it to the local lymph node (**3**). Naïve T cells recognizing pMHC (derived from the allergen) presented by the now mature DC are activated and, in the presence of cytokines such as IL-4, differentiate into Th2 effectors that supply help to allergen-specific B cells. Activated B cells and Th2 cells exit the lymph node (**4**) and commence the expression of specific homing receptors that direct them to travel via the thoracic duct and blood back to the site where the allergen first entered the body (**5**). The Th2 cells supply cytokines (**6**) that influence differentiating plasma cells to switch to IgE production (**7**). In the local tissue, IgE may bind directly to the allergen or may bind to high affinity FcεRs expressed on mast cells, sensitizing them in readiness for the effector stage of type I hypersensitivity (see **Fig. 18-2**). Excess IgE is taken up into local lymphatics (**8**), enters the blood circulation, and travels to distant tissues. When the IgE encounters basophils in the blood (**9**) and mast cells in other tissues (**10**), it binds to FcεR molecules on these cells, sensitizing them as well.

ii) Effector Stage

If an allergen enters the body of an atopic person a second time during the period that mast cells and basophils are armed with anti-allergen IgE, the effector stage is triggered. The effector stage takes place in two phases: the **early phase reaction** and the **late phase reaction**. Because sensitized mast cells are already present in the target tissue, the soluble mediators released by these cells in response to allergen binding are the main drivers of the early phase reaction. Sensitized basophils must be recruited from the blood by chemokines released by the allergen-activated mast cells, and so make their contribution to the late phase reaction. Other leukocytes drawn to the site of allergen accumulation, particularly eosinophils, also play their roles during the late phase reaction. Although all type I HS reactions have the same underlying bi-phasic mechanism, the symptoms seen vary widely because the mediators released affect different cell types in different locations. A list of mediators contributing to type I HS appears in **Table 18-3**.

TABLE 18-3	**Major Mediators Contributing to Type I Hypersensitivity**	
Mediator	**Cellular Source**	**Promotes**
Histamine	Mast cell and basophil granules	Vasodilation, vessel permeability, bronchial smooth muscle contraction, mucus production, itching, sneezing
Serotonin	Mast cell and basophil granules	Vasodilation, bronchial smooth muscle contraction
Chemotactic factors	Mast cell and basophil granules	Chemotaxis of eosinophils and neutrophils
Proteases	Mast cell and basophil granules	Mucus production, basement membrane digestion, increased blood pressure
Cytokines and growth factors	Mast cell, basophil synthesis	Mobilization and activation of immune system cells; initiation and maintenance of inflammatory response
Platelet-activating factor (PAF)	Mast cell membrane breakdown; eosinophil granules; synthesis by neutrophils, macrophages	Platelet aggregation and degranulation, pulmonary smooth muscle contraction
Leukotrienes	Mast cell membrane breakdown; eosinophil granules; synthesis by neutrophils, macrophages	Vasodilation, increased vessel permeability, bronchial smooth muscle contraction, mucus production
Prostaglandins	Mast cell membrane breakdown; synthesis by neutrophils, macrophages	Platelet aggregation, pulmonary smooth muscle contraction
Major basic protein	Eosinophil granules	Mast cell degranulation, smooth muscle contraction, death of respiratory epithelial cells
Eosinophil-derived neurotoxin	Eosinophil granules	Death of myelinated axons and neurons
Eosinophil cationic protein	Eosinophil granules	Death of respiratory epithelial cells

a) Early Phase Reaction

As illustrated in **Figure 18-2**, the early phase reaction of type I HS is mediated primarily by the degranulation of sensitized mast cells in the target tissue. Upon repeat entry of the allergen, it is captured by the IgE molecules bound to FcεRs on the mast cell surface. When several of these IgE molecules are engaged by allergen, the FcεRs are effectively cross-linked and trigger the degranulation of the mast cell. The symptoms induced by the release of granule contents depend on the localization of the mast cells and the particular effects the mediators have on the local tissues. Mast cells are abundant in the skin; in the loose connective tissue surrounding blood vessels, nerves and glandular ducts; and in the mucosae. In the lungs, mast cells are often found surrounding the blood vessels as well as in the bronchial connective tissues and alveolar spaces.

Mast cell degranulation most likely evolved to combat parasites, and so induces coughing, sneezing, tearing of the eyes, scratching of the skin, and/or cramping of the gut and diarrhea, all of which are designed to expel these types of pathogens. The fact that fast-acting mediators are *pre-formed* and stored in mast cell intracellular granules accounts for the "immediate" nature of type I HS responses and the rapid onset of initial symptoms. Symptoms are then sustained for several hours by the action of newly generated mast cell mediators that require some type of synthesis for their formation, including cytokines (particularly TNF, IL-1 and IL-6) and chemokines. Membrane breakdown products of spent mast cells can also serve as inflammatory mediators.

How do the mediators released during the early phase reaction cause allergic symptoms? Histamine and platelet activating factor (PAF) bind to their specific receptors on the smooth muscle cells supporting the blood vessels and induce them to relax, expanding the diameter of the blood vessel lumen (vasodilation) and increasing blood flow to the local area. Simultaneously, histamine and leukotrienes induce the contraction of the endothelial cells lining the blood vessels (increased vessel permeability), creating opportunities for cells and plasma proteins such as complement components to leak out of the circulation into the tissues. The action of histamine on sensory nerves causes the itching of eczema and the sneezing of hay fever. Histamine also induces the increased mucus secretion in the bronchioles that is characteristic of asthma. PAF

Perhaps the best known mediator of the early phase reaction of type I HS is histamine.

491

Fig. 18-2
Effector Stage of Type I Hypersensitivity: Early Phase Reaction

An allergen enters a tissue where sensitized mast cells are present (**1**) and binds to the antigen-binding sites of IgE molecules that are bound through their Fc regions to FcεRs on the mast cell surface. Cross-linking of these receptors then triggers degranulation (**2**). Histamine and other pre-formed mediators are released (**3**), as are multiple cytokines, chemokines and growth factors (**4**). When a spent mast cell breaks down, enzymatic digestion of its membrane components generates mediators such as platelet-activating factor (PAF) and others (**5**). Together, all these molecules combine to produce the tissue-specific symptoms typical of an allergic response (**6**). Chemotactic factors released during mast cell activation induce the expression by endothelial cells of new adhesion molecules (**7**), encouraging the extravasation of additional leukocytes such as eosinophils and sensitized basophils.

Leukocyte extravasation was discussed in Chapter 2 and illustrated in Figure 2-16.

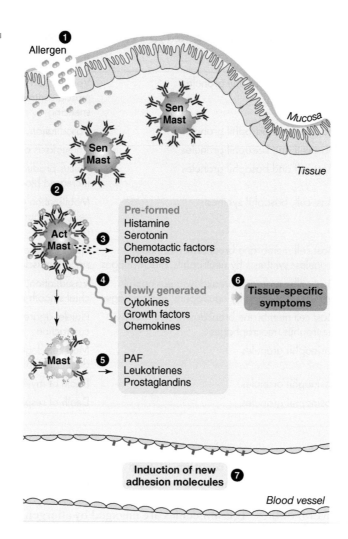

binds to endothelial cells, induces smooth muscle relaxation, and triggers the activation of platelets, which then release additional inflammatory mediators.

b) Late Phase Reaction
About 4–6 hours after the initiation of a type I HS reaction, the late phase reaction occurs. As illustrated in **Figure 18-3**, leukocytes such as eosinophils, sensitized basophils, macrophages and neutrophils exit the circulation in response to chemotactic factors released during the early phase reaction. These cells migrate into the allergen-contaminated tissue and are activated to carry out their destructive effector functions. Eosinophils are particularly prominent during the late phase reaction, and airway epithelial cells are especially sensitive to the damaging proteins they produce. Indeed, the clinical symptoms of asthma can be attributed mainly to eosinophil activation.

III. Examples of Type I HS

i) Localized Atopy
Most people experience allergies as a local type I HS reaction affecting a specific target tissue such as the skin or bronchial passages. The sensitized mast cells that are triggered in these instances usually lurk among the epithelial cells lining the target tissue. It remains unclear why some atopic people have a tendency to develop one particular clinical manifestation (such as urticaria), while others tend to develop another affliction (such as asthma).

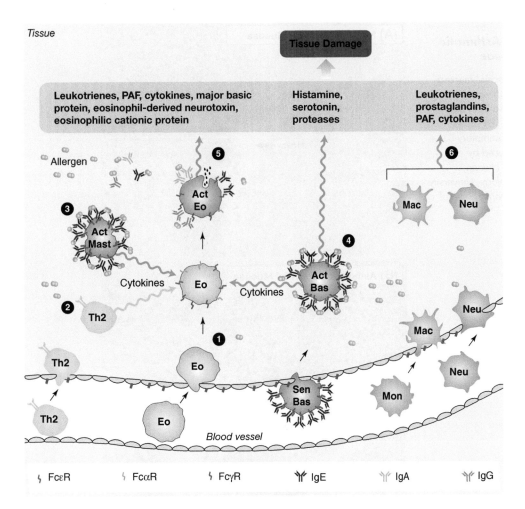

Fig. 18-3
Effector Stage of Type I Hypersensitivity: Late Phase Reaction

Chemotactic factors released during the early phase reaction induce the expression of new adhesion molecules on the endothelium of local blood vessels so that leukocytes in the circulation are drawn into an allergen-polluted tissue (**1**). Th2 effectors (**2**) as well as activated mast cells (**3**) and basophils (**4**) secrete cytokines that stimulate eosinophil chemotaxis and activation (**5**). The activated eosinophils express FcεR as ell as FcαR and FcγRs. Upon cross-linking of their Ig-bound FcRs by allergen, the eosinophils and basophils release the indicated mediators, which contribute to tissue damage and reinforce eosinophil activation. Macrophages and neutrophils (**6**) that have extravasated into the site in response to inflammatory cytokines and chemokines also secrete mediators that can cause tissue damage.

a) Allergic Rhinitis (Hay Fever)

Allergic rhinitis, the prototypical example of a type I HS, results when airborne allergens such as ragweed pollen or mold spores are inhaled. Sensitized mast cells resident in the upper respiratory tract, the conjunctiva of the eyes, and the nasal mucosa are triggered to degranulate and release pro-inflammatory mediators in these locations. The mediators cause the characteristic symptoms of hay fever: coughing, tearing and itching of the eyes, sneezing, and the blockage of nasal passages. About 20% of individuals in the developed world suffer from allergic rhinitis.

"Hay fever" is the name farmers in the early 1800s gave to the symptoms they experienced when they harvested their hay crops. We now recognize this affliction as allergic rhinitis caused by seasonal exposure to airborne pollen grains in grassy fields.

b) Atopic Asthma

Atopic asthma is a type I HS reaction that occurs in the lower respiratory tract. About 10–20% of children and adults living in developed countries suffer from atopic asthma. Inhalation of antigen triggers degranulation of sensitized mast cells resident in the nasal or bronchiolar mucosae. The resulting activation of eosinophils and release of pro-inflammatory mediators induce the production of copious amounts of mucus that constrict the bronchioles (sometimes severely). The patient soon complains of tightness in

Plate 18-1
Normal versus Asthmatic Bronchial Mucosae

Compared with bronchial mucosa from a non-atopic individual **(A)** bronchial mucosa from an asthmatic individual **(B)** shows an infiltration of inflammatory leukocytes. [Reproduced by permission of David Hwang, Department of Pathology, University Health Network, Toronto General Hospital.]

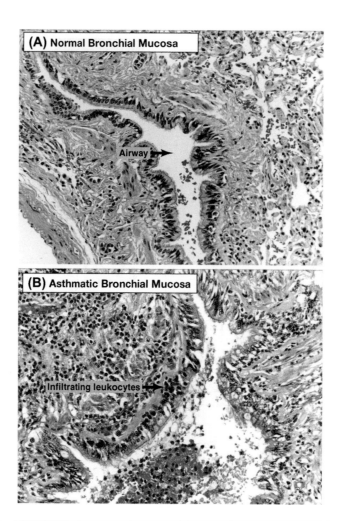

the chest and begins to wheeze or gasp for air. This increased propensity to experience symptoms that result in narrowing of the airway is termed *airway hyper-responsiveness* and is a hallmark of atopic asthma. Asthma can be fatal if an acute attack totally blocks the airway.

Histologically, the airway of an asthmatic patient appears chronically inflamed with mast cells, eosinophils, lymphocytes and neutrophils (**Plate 18-1**). In addition, the basement membrane of the airway is increased in thickness, the bronchiolar smooth muscle layer is enlarged, and increased mucus is present. Over 50 distinct inflammatory mediators are associated with asthma symptoms, and high levels of pro-inflammatory cytokines are found in the lung secretions of asthmatic patients.

c) Atopic Urticaria (Hives)

Atopic urticaria is a type I HS reaction in which sensitized skin mast cells degranulate and release mediators that cause swollen, reddened patches on the skin known as hives or "wheal and flare reactions" (**Plate 18-2**). The whitish wheal in the center of the hive is composed of leukocytes that have escaped the blood vessels due to the increased permeability of these channels. The flare is the ring of redness seen surrounding the wheal due to increased blood flow into this area. The intense itching and pain of hives are caused by the stimulation of skin nerve endings by histamine. Eventually, the hives may become confluent such that the reaction covers a large area of the body. As well, urticaria is frequently accompanied by prominent swelling beneath the mucosal and cutaneous layers in the site of allergen exposure. Allergies to latex, hair chemicals, food additives, insect bites and some drugs are associated with acute urticaria.

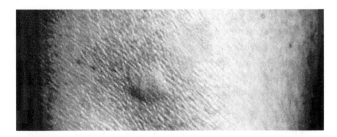

Plate 18-2
Atopic Urticaria (Wheal and Flare Reaction)

A hive typical of atopic urticaria has a raised, circular area in its center (wheal) due to the infiltration of leukocytes into the site. The outer ring of redness (flare) is caused by increased blood flow into the area that has been facilitated by local vasodilation. *[Reproduced by permission of the Mayo Foundation for Medical Education and Research.]*

d) Atopic Dermatitis (Eczema)

Atopic dermatitis is a type I HS reaction in the skin that results in excessive dryness and an itchy rash that is more scaly than in urticaria. Eczematous lesions may affect different parts of the body at different ages. Atopic dermatitis tends to be more chronic in nature than urticaria and is often associated with respiratory allergies later in life. Individuals with atopic dermatitis also tend to be more susceptible to skin infections because the barrier function of the skin is compromised by the eczematous lesions.

ii) Systemic Atopy: Anaphylaxis

Anaphylaxis is a type I HS response with striking systemic consequences. Clinically, anaphylaxis is a form of extreme shock (hence, "anaphylactic shock") that can kill within minutes of exposure to the triggering antigen. Anaphylactic shock is most frequently observed in individuals sensitized to insect stings, peanuts, seafood or penicillin.

During anaphylaxis, large quantities of inflammatory mediators and vasodilators are released into the circulation by activated mast cells and basophils, causing rapid dilation of blood vessels throughout the body. Respiration immediately becomes difficult and is followed by a dramatic drop in the victim's blood pressure and extensive edema in the tissues. Patients have been known to report a "feeling of doom" at this point. The lungs may fill with fluid, the heart may beat irregularly, and control of the smooth muscles of the gut and bladder is often lost. Constriction of the bronchioles may cause lethal suffocation of the victim unless treatment with epinephrine (adrenaline) is started immediately. Sometimes the clinical course is bi-phasic, in that severe symptoms initially appear and then seem to resolve for 1–3 hours. Symptoms then return with a vengeance and can kill the patient if treatment is not sought again immediately.

iii) Food Allergies

Food allergies result from IgE-mediated reactions to allergenic proteins in consumed foods. "Intolerances" to milk and alcohol can resemble allergies in some of their symptoms but are not IgE-mediated and are therefore not allergies. The health impact of food allergies also has a cultural component, depending on the diet of the population involved. Proteins that are important as food allergens in one culture may be of no significance in another culture if that population does not eat food containing that protein. That being said, 90% of food allergies in any culture have been linked to peanuts, soy, milk, eggs, wheat and fish.

Food allergies are manifested in a variety of ways. Some food allergens cause a type of urticaria that takes the form of burning or itching of the tongue, lips and throat. Severe mucosal edema of the mouth and pharynx may occur. Other food allergens do not affect the mouth but trigger sensitized mast cells in the gut. Mediators released by these mast cells act on the gut smooth muscles, causing them to contract. Vomiting, nausea, abdominal pain, and cramping and/or diarrhea may result. Large numbers of eosinophils may infiltrate into the gastric and intestinal walls. In other cases, the

The "Inform All" database is a useful U.K.-based collection of information on food allergens [http://foodallergens.ifr.ac.uk].

Latex allergy can be considered a food allergy, since latex shares antigenic determinants with bananas, avocados, grapes and corn.

mediators may increase the permeability of the gut mucosae such that a food allergen enters the circulation. Depending on where it ends up, the allergen may induce an asthmatic response or urticaria or eczema in a site distant from the mouth or gut. While the only proven therapy for a food allergy is elimination of that food from the diet, sensitivity to a food allergen often diminishes with age. Thus, a child may appear to "outgrow" his/her allergy with time.

NOTE: A whole generation of children has now grown up with an acute awareness of food allergies, thanks to the relatively recent surge in numbers of youngsters with peanut allergies. In the U.S., about 1% of the population is allergic to peanuts, and the reactions experienced by these individuals if they ingest even the smallest quantities of these nuts are particularly severe. In general, anti-peanut antibodies are overwhelmingly of the IgE isotype. There are 12 different peanut allergens defined by experts in the field, and studies of these proteins have indicated that their glycosylation pattern is the key to the allergenicity of many of them. The uptake of certain carbohydrates can suppress a DC's ability to secrete the Th1-promoting cytokine IL-12. Thus, a DC that has acquired a peanut antigen containing such carbohydrates may be influenced to promote a harmful (in this context) Th2 response instead of a more benign Th1 response. In addition, certain components of peanut extract activate complement, thereby generating the anaphylatoxin C3a that helps to precipitate anaphylactic shock. Children with a peanut allergy should always be within easy reach of an epinephrine injection.

IV. Determinants Associated with Type I HS

i) What Makes an Antigen an Allergen?

Despite intensive study, it has not been possible to identify a single characteristic common to all allergens. They are diverse in their structures and biochemical properties, enter the body in different ways, act at different concentrations, and interact with different molecules or cell types once within the body. It is also not known why an atopic individual might be allergic to cat dander but not dog dander, nor why allergy to birch tree pollen is common but allergy to pine tree pollen is extremely rare.

The "Allergome" database contains the gene and protein sequences of many allergens, along with their secondary structures and associated immunological and clinical consequences [http://www.allergome.org].

In theory, any protein should have the potential to be an allergen; in practice, very few of them are. Of the over 12,000 protein families recognized by biochemical experts, only 255 are associated with allergies. All types of primary and secondary protein structures are represented among allergens, implying that there must be structural characteristics beyond these simple features that act as the determinants defining a particular molecule as an allergen. For example, 20 different protein families are represented among food allergens. The one feature shared by most of these molecules is the ability to either bypass or resist the digestive system such that sufficient protein remains in the body in a form able to provoke an immune response. Interestingly, many food and airborne allergens are able to bind to lipids, a type of interaction that may help to protect them from digestion by the gut or expulsion by the respiratory system. In addition, some dietary proteins maintain a particulate rather than soluble form in the gut, so that rather than being absorbed by intestinal epithelial cells in a relatively nonimmunogenic way, they are taken up by the M cells of the GALT. Protein glycosylation is also a structural determinant of allergenicity because it plays a role in protein stability and can affect immunogenicity.

The structural features of proteins or the molecules with which they associate can influence the character and strength of innate immune responses, which in turn may determine the likelihood of an allergic reaction being provoked. For example, in-depth examination of some of the globe's most common allergens, such as Der p 2 (derived from house dust mites), has revealed that some of these molecules structurally resemble the lipids that normally bind to the LPS-binding moiety of the TLR4 signaling complex. The engagement of TLR4 on DCs by such molecules can influence these cells to promote Th2 responses, which may be manifested as allergic reactions. Examination of several other airborne allergens has shown that they can activate the TLR2 signaling complex in a similar manner, leading to allergic inflammation in the lung.

In addition to TLR engagement, NLR engagement can lead to allergic responses. For example, house dust mites contain proteases that are able to activate the NLRP3 inflammasome in human keratinocytes, causing these cells to release IL-1 and IL-18. These pro-inflammatory cytokines may create a microenvironment in the skin that promotes the sensitization of atopic individuals to house dust mite allergens.

NLRs and inflammasomes were described in Chapter 3.

NOTE: The allergenicity of a given antigen may be influenced by how it is first encountered. When an antigen is inhaled, some proportion of it will normally reach the digestive tract. Thus, both food antigens and many airborne antigens undergo biochemical processing in the maternal gut before entering the breast milk. In addition, as discussed in Chapter 10, breast milk is replete with IL-10 and TGFβ, both of which suppress pro-inflammatory cytokines and promote regulatory T cell generation. Potential allergens are thus passed on to neonates in a form and environment that discourages the development of allergy. Indeed, epidemiological studies have demonstrated that, in general, breast-fed children develop fewer allergies than bottle-fed children. Accordingly, rather than counseling breast-feeding mothers to avoid eating foods correlated with a high incidence of allergy, clinicians are now starting to consider the possible benefits of exposing very young infants to food and environmental antigens via their mother's milk. Indeed, such exposure to cereal antigens before the age of 6 months decreases the chance that the child will develop a wheat allergy later on in life. In line with this reasoning, the incidence of peanut allergy in children is lower in cultures in which peanuts are consumed throughout pregnancy and early childhood compared to cultures in which peanuts are traditionally avoided. It may be that the Western habit of eschewing allergens during pregnancy has inadvertently made children in these countries more allergy-prone.

Neonatal exposure to an allergen via the skin increases, rather than decreases, the chances of later developing atopy. This observation emphasizes the unique qualities of oral tolerance (refer to Ch. 10).

ii) What Makes an Individual Atopic?

The development of atopy clearly depends on an individual's genetic background because particular alleles of certain polymorphic genes are strongly associated with a predisposition to atopy. However, studies of atopy in identical twins have revealed that both twins are allergic in only 60% of cases. Thus, other non-genetic factors must also be involved. As noted previously, the circumstances under which an individual first encounters an allergen can affect whether that person becomes sensitized. Similarly, the environment in which an individual is raised can influence the competence of the immune system and how it responds to an allergen. Furthermore, for unknown reasons, a genetically and environmentally predisposed person may still not respond to an allergen unless that allergen is encountered in the context of a triggering event. The major determinants thought to play a role in atopy are summarized in **Figure 18-4** and discussed in the following sections.

a) Genetic Determinants

The familial nature of allergy was first noticed in the mid-1800s. In a family in which both parents are atopic, 50% of the children will have allergies, compared to only 19% of children in a family with no history of atopy. Several examples of the many chromosomal regions and genes that are thought to contribute to allergy in humans are shown in **Table 18-4**. The chromosomal region 5q31–33 is of special interest because it contains a cluster of genes that encode cytokines promoting Th2 differentiation. Patients with certain polymorphisms in this region have more Th2 cells in the relevant tissues than non-allergic individuals, and these Th2 cells produce greater than normal amounts of IL-4. For example, single nucleotide polymorphisms (SNPs) in the IL-4Rα and IL-13 genes are strongly correlated with asthma. Another cytokine linked to atopy is TSLP, which we discussed in Chapter 10 as a powerful inducer of Th2 cell differentiation. Increased TSLP levels are present in the inflamed tissues of individuals with atopic dermatitis, allergic rhinitis and asthma. Furthermore, specific polymorphisms of the TSLP gene (located in chromosomal region 5q22) are associated with elevated IgE, increased eosinophils in the blood and tissues, and airway hyper-responsiveness.

As introduced previously, studies of allergen structure suggest that innate responses initiated by allergen engagement of TLRs can play a key role in creating

Fig. 18-4
Determinants Favoring Atopy

The indicated genes and environmental factors may work alone or in concert to predispose an individual to atopy. Exposure to one or more of the indicated triggering events can result in a clinically relevant response to an allergen that enters the body.

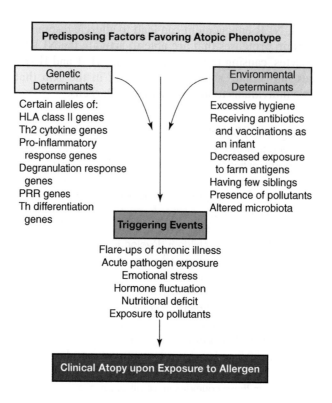

a tissue microenvironment favoring allergic sensitization. In line with this thinking, researchers have identified polymorphisms in genes encoding TLRs or the proteins regulating their expression that appear to correlate with predisposition to atopy. For example, in one study, specific SNPs in the TLR2 gene were linked to asthma. In a different study, the HSCs of infants at high risk for atopy showed lower levels of TLR2, TLR4 and TLR9 expression, and these HSCs gave rise to significantly more basophils and eosinophils when induced to differentiate *in vitro* than did HSCs from infants not at risk for atopy. Examinations of other PRRs, including the mannose receptor and Dectin-2, have shown that these receptors can mediate DC internalization of allergens in a way that preferentially leads to Th2 responses. Immunologists therefore expect to uncover polymorphisms in these genes that can also influence an individual's predisposition to atopy.

TABLE 18-4	**Examples of Human Genes Potentially Contributing to Atopy**	
Chromosomal Region*	**Candidate Gene(s)**	**Proposed Role or Function**
3q27	Bcl-6	Regulation of Th2 responses
5q31–33	IL-3, IL-4, IL-5, IL-9, IL-13, TGFβ	Immune deviation to Th2; isotype switching to IgE; activation of eosinophils, basophils and mast cells
6p21–22	HLA-DR2, HLA-DR4, HLA-DR7, TAP TNF	Antigen processing and presentation Inflammation
11q13	FcεRI	Mast cell degranulation
12q14–24	IFNγ, iNOS, SCF, mast cell growth factor	Regulation of IL-4 transcription and thus isotype switching to IgE; inflammatory response; mast cell stimulation
14q11–13	TCRA, TCRD NF-κB	T cell recognition of allergen Transcription factor driving pro-inflammatory gene expression
16p	IL-4Rα	Component of IL-4 receptor
5q22.1	TSLP	Stimulation of IgE production, eosinophil development, and airway hyperresponsiveness

*p and q indicate the short and long arms of a chromosome, respectively. Thus, "3q27" means "chromosome 3, long arm, 2nd block, 7th band" as seen in standard karyotyping.

b) Environmental Determinants

In the 1800s in Europe, the group of disorders that we now call "allergies" were known as "the rich and noblemen's disease." Even today, clinicians continue to report a steady increase in the frequency of atopy in the developed world compared to the less-developed world. It is not clear why an elevation of living standard increases the risk of atopy. One factor relevant to current times may be the emphasis in more developed cultures on energy conservation in the home. The upgrading of windows, insulation and carpeting may have decreased draftiness but has also tremendously reduced the exchange of inside and outside air. As a result, occupants of such homes may suffer increased exposure to house dust mites and other such allergens. Another factor may be the exposure in more developed countries to higher levels of pollutants that promote allergen-specific IgE production and increase sensitization rates.

Atopy may also be increasing in the developed world due to an obsession with "germs." In more advanced countries, childhood infections are prevented by vaccination or rapidly resolved with antibiotics, and personal and food hygiene are fastidious. As a result, a child may not encounter sufficient pathogens in infancy to establish the appropriate populations of Th effector cells and regulatory T cells needed to maintain a healthy balance of pro- and anti-inflammatory responses. This concept forms the basis of the **hygiene hypothesis**, which posits that increased hygiene leads to underexposure to pathogens and a consequent imbalance in the generation and functions of Th1, Th2, Th17, and regulatory T cells. The hygiene hypothesis is outlined in **Box 18-1**, and recent research on this concept is the subject of this chapter's "Focus on Relevant Research."

Box 18-1 The Hygiene Hypothesis

Where exposure to germs is limited, the immune system gets fewer chances to mount the Th1 and Th17 responses necessary to combat intracellular pathogens. Regulatory T cell development may also be compromised, and T cell differentiation may be biased toward Th2 development. Many immunologists believe that this imbalance may create a predisposition to atopic disease. Consistent with this idea, in both developed and less-developed countries, a history of substantial childhood infections (rather than the presence of specific pollutants or allergens) correlates strongly with resistance to atopic disease. For example, a child who is one of several siblings, or who attends a day care center at an early age, has a decreased likelihood of developing skin allergies, hay fever, or asthma. On the other hand, a child who receives copious quantities of antibiotics and/or is vaccinated against numerous pathogens during early childhood is at increased risk for allergies. In one study of Swedish children, the group that had the highest frequency of atopy (25%) also had the highest rate of childhood vaccination and antibiotic use. In the group that rarely used antibiotics and was vaccinated only against tetanus and polio, the frequency of atopy was 13% and asthma incidence was reduced almost 4-fold. Another example is the surge in atopy that occurred in the former East Germany after the fall of the Berlin Wall in 1989. People of this region rapidly increased their standard of living but also increased their risk for several atopic diseases.

What is the underlying basis for the hygiene hypothesis? The placenta is a unique microenvironment that secretes Th2 cytokines and other factors designed to maintain tolerance of the growing fetus. As a result, T cells in the cord blood of newborns are largely biased to the Th2 phenotype. Serum IFNγ levels are reduced, and Th1/Th17 cells are decreased in frequency. With the growing infant's normal exposure to pathogens, the Th1/Th17 populations are built up at the expense of the Th2 population, and the finely tuned balance among these cell subsets necessary for properly controlled and comprehensive immune responses is restored. In the absence of significant infection by pathogens (due to antibiotics or vaccinations), the balance remains tipped in favor of Th2 responses and thus atopy.

As attractive as the hygiene hypothesis is, it is likely too simplistic. Individuals infected with helminth worms, which strongly induce Th2 responses characterized by IgE production and eosinophilia, do not suffer from an increase in atopic disease. Also, a simple shift in the balance between Th1/Th17 responses and Th2 responses should be associated not only with an increase in Th2-mediated atopy but also with a decrease in Th1/Th17-mediated autoimmune diseases, but such a decrease has not been observed. Indeed, the incidence of Th1/Th17-associated autoimmune diseases such as type I diabetes and inflammatory bowel disease is on the rise in the Western world (see Ch. 19). Like atopy, autoimmune diseases are more prevalent in developed countries than in developing ones, and in urban as opposed to rural environments.

Scientists now theorize that it is not the development of Th1/Th17 responses *per se* that is important for suppressing atopy so much as the generation of regulatory T cells and an adequate capacity to produce anti-inflammatory cytokines. Indeed, helminth-infected individuals have elevated levels of IL-10, while allergic individuals show decreased IL-10. IL-10 and TGFβ are known to suppress both cytokine secretion and T cell proliferation, and IL-10 also inhibits mast cell degranulation. Thus, it may be that numerous childhood infections help to establish a robust regulatory T cell population and a powerful anti-inflammatory response that combine to suppress allergic and autoimmune responses. If these capacities are not fully developed during childhood, a predisposition to atopy or autoimmune disease might arise.

"Infection, Inflammation, and Chronic Diseases: Consequences of a Modern Lifestyle" by Ehlers, S., & Kaufmann, S.H.E. (2010) *Trends in Immunology* 31, 184–190.

Focus on Relevant Research

As discussed in **Box 18-1**, the "hygiene hypothesis" suggests that early exposure to certain bacterial, viral, and helminth antigens conditions the immune system in a way that downregulates chronic inflammation. Such exposure routinely occurs in a rural setting or in an area with poor sanitation, but is much less common in highly developed countries and urban environments. In this article, Ehlers and Kaufmann summarize the proceedings of the 99th Dahlem Conference that took place in Berlin in 2010. At this conference, experts from diverse fields discussed and debated the hygiene hypothesis, presenting data derived from molecular, cellular, genetic, and animal model systems. Further research is clearly needed to explain why an excessively sanitary lifestyle is not necessarily a good thing, and to identify key factors contributing to healthy immune system homeostasis.

c) Triggering Events

Many allergic responses appear to be associated with a triggering event. Such events include a flare-up of a chronic illness, an acute pathogen infection, emotional stress, fluctuating hormone levels, a nutritional deficit, or exposure to a pollutant. The common factor here may be a weakening of the immune system that leaves the body more accessible to allergen entry and sensitization. If an infection is the trigger, then a molecule from the pathogen itself may be acting as the allergen, and the infected individual mounts an IgE response to the offending protein. The response successfully clears the pathogen but leaves the individual with atopic symptoms. For example, many flare-ups of atopic dermatitis are associated with the presence of certain skin fungi, and IgE antibodies directed against proteins from these organisms are found in the sufferer's circulation.

NOTE: The contribution of the gut microbiota to the prevention of atopy is becoming a hot topic. In one study of intestinal microbiota in children at high risk for developing allergies, those who had a decreased diversity of intestinal microbes also showed higher levels of allergen-specific IgE and an increased incidence of allergic symptoms. Similar findings have been reported for mice, in that germ-free animals are more susceptible than normal mice to the chemical induction of colitis and asthma. Interestingly, a major player in these atopic reactions may be NKT cells. As described in Chapter 11, these cells are activated by lipid antigens presented on CD1 molecules and secrete copious amounts of the Th2 cytokines IL-4 and IL-13. In germ-free mice lacking a diversity of commensal organisms, excessive numbers of NKT cells accumulate in the lamina propria of the colon and lung, where they can be readily activated by allergen administration. It seems that exposure to a wide variety of microbes at a very young age is needed to establish a mucosal NKT population that will not respond to innocuous antigens. Immunologists are now coming to believe that events that interfere with the health of the gut microbiota, such as a significant gastrointestinal infection or excessive antibiotic use, can play a role in predisposing an individual to atopy.

V. Therapy for Type I HS

The appropriate therapy for type I HS depends on an individual's circumstances and exposure to the allergen. Sometimes the easiest way to prevent an allergic response is to simply minimize contact with the allergen. Antibiotics can reduce the presence of pathogens that trigger or exacerbate flare-ups of particular allergies, while emollients can help to rehydrate dry skin and thus reduce the chance of allergen entry. However, such basic approaches are not always effective, and many patients must resort to more involved therapies.

i) Antihistamines

Because a principal mediator of symptoms released during the effector phase of an allergic response is histamine, molecules that mitigate its effects are often used to treat relatively mild allergies such as hives and hay fever. Antihistamines work by binding to histamine receptors on target organs so that histamine released by degranulating mast cells and basophils cannot trigger symptoms. Antihistamines are normally taken at the onset of symptoms, but if an individual knows that he/she is about to be exposed to an allergen, the prior use of antihistamines can prevent symptom development.

ii) Lipoxygenase Antagonists

Leukotrienes generated in both the early and late phase reactions of the allergic response mediate or promote many aspects of atopy, particularly bronchoconstriction and eosinophil infiltration. Because lipoxygenase antagonists block the generation of leukotrienes from innate leukocyte membranes, the inhalation of these agents can provide relief from asthma symptoms and improve pulmonary function. Lipoxygenase antagonists can also be given orally and usually have minimal side effects.

iii) Bronchodilators

Acute asthma attacks are often treated by inhalation of rapidly acting bronchodilators (delivered using inhalers or "puffers"). These agents are frequently aerosolized drugs that block mast cell degranulation and induce smooth muscle relaxation. However, bronchodilators do not address the underlying inflammatory response.

iv) Corticosteroids

Corticosteroids have powerful anti-inflammatory effects because they indirectly inactivate a wide range of transcription factors. Both cytokine production and the expression of adhesion molecules required for the entry of inflammatory cells into target tissues are thus inhibited. Low-dose corticosteroids are often used in a cream form to treat atopic dermatitis, as nose drops to treat rhinitis, or as an inhaled agent to treat asthma. Corticosteroids have very few side effects when used at a low dose to treat atopic conditions, but high-dose, long-term or oral use of corticosteroids can have severe consequences, including general immunosuppression.

v) Cromones

Cromones are effective anti-inflammatory drugs that are thought to block mast cell degranulation, inhibit cytokine release by macrophages and eosinophils, and/or impair leukocyte extravasation. Intranasal cromones can alleviate symptoms of allergic rhinitis and asthma, whereas oral cromones are useful for treating food allergies. Cromones tend to have fewer side effects than corticosteroids.

vi) Anti-IgE Immunotherapy

Omalizumab (Xolair) is a humanized IgG mAb that binds specifically to human IgE. Administration of this therapeutic mAb reduces free IgE in the blood and interstitial fluid. It does not bind to IgE that has already been bound to FcεRI expressed by mast cells, basophils or APCs because the required binding site in the C_H3 domain of the IgE molecule is inaccessible. This mAb was introduced as an asthma medication and has been helpful in many cases of moderate and severe asthma. However, it has also been linked more recently to unexpectedly serious side effects in some asthma patients, including cardiac failure, thrombosis, and cerebrovascular disorders, and so remains under governmental evaluation for this application. In contrast, Xolair is well tolerated by patients with chronic urticaria and is quite effective in reducing symptoms in these patients. It is unclear why this mAb appears to work differently in different types of atopy.

vii) Allergen-Specific Immunotherapy

Allergen-specific immunotherapy (abbreviated by many in this field as SIT) is a newer name for a therapeutic approach that was commonly called *hyposensitization* or *desensitization*. In this procedure, an allergy sufferer receives subcutaneous injections of ever-increasing amounts of the purified allergen every week or month for a period of up to 3–5 years. The idea is to induce the individual to produce harmless anti-allergen IgG antibodies that can capture the allergen and prevent it from binding to the individual's pathological anti-allergen IgE antibodies. In many cases, the atopic individual eventually ceases to show an allergic reaction to the allergen. An advantage to the SIT approach is that it can be long-lasting and free of the side effects that plague drug or chemical treatments. In addition, SIT is currently the only form of allergy therapy that works by altering the immune response mechanism responsible for the patient's hypersensitivity, rather than just mitigating atopic symptoms. However, the technique does have limitations. SIT works well for individuals with atopic rhinitis or urticaria, or who are allergic to insect venom, but is not as effective for those suffering from asthma or eczema. SIT is not usually started until after a child's fifth birthday, and careful monitoring is necessary because induction of anaphylaxis has been observed in rare cases.

> Anti-allergen IgG antibodies induced by SIT are sometimes called "blocking antibodies" because they prevent the binding of IgE antibodies to the allergen.

NOTE: In some cases, SIT can be performed by sublingual allergen administration rather than injection, much to the relief of small children and their parents. However, the effectiveness of this route is often less than that of subcutaneous injection. Oral allergen administration, which was originally of great interest to clinical immunologists because of the known properties of oral tolerance, has not been effective to date. Injection of allergens into lymph nodes works in the lab but is inconvenient and painful for patients. Skin patches that work via painless absorption of the allergen through the skin are under investigation.

The end result of successful SIT is the apparent conversion of a harmful IgE anti-allergen response to a harmless IgG response, but the underlying mechanism is still not completely understood. Whether the procedure achieves its goal may depend on the final balance of T cell subsets present and their functionality. One hypothesis is that continuous dosing with allergen may alter the number and characteristics of DC subsets. Some of these DCs may then promote the differentiation of allergen-specific Th1/Th17 cells at the expense of Th2 cells. Whereas Th2 cells secrete cytokines (such as IL-5) favoring mast cell activation and IgE production, Th1 cells produce cytokines (such as IFNγ) that inhibit isotype switching to IgE and promote switching to IgG. Another theory is that allergen-specific Th2 cells may be anergized by repeated stimulation with allergen under conditions in which costimulation is minimal. As a corollary, SIT may also gradually increase the number and/or activation of regulatory T cells that express high levels of IL-10 and TGFβ. *In vitro*, these immunosuppressive cytokines promote isotype switching to IgG and dampen T cell, eosinophil and mast cell functions. It has also been suggested that, by reducing the ability of APCs to engage in IgE-mediated allergen presentation to T cells, SIT decreases allergen-induced T cell activation that would otherwise contribute to the late phase reaction. In reality, all these mechanisms may make a contribution to SIT.

NOTE: Some immunologists consider the various proteins used for SIT to be allergy "vaccines" because an individual is intentionally exposed to an immunogen (i.e., the allergen) in a form that is harmless and induces a more favorable immune response when the immunogen (allergen) is later encountered in its natural form. In the case of a standard vaccine of the type described in Chapter 14, the result is the mitigation of disease due to a stronger, faster response when the natural pathogen is first encountered. In the case of an allergy vaccine, the result is the prevention of atopic symptoms due to the mounting of a weaker and/or non-sensitizing response when the true allergen is first encountered.

Until recently, the only sources of allergen components suitable for use in SIT have been extracts of natural allergen sources. These extracts are often poorly characterized and inconsistent from batch to batch. Modern molecular analysis tools have now made it possible to accurately determine the secondary structures as well as the protein and nucleotide sequences of allergens, allowing the reproducible production of allergen proteins through recombinant DNA technology. For example, purified recombinant grass pollen antigens and recombinant birch pollen antigens used for SIT have successfully alleviated atopic symptoms in individuals allergic to these substances. These types of discoveries have opened up promising new avenues for SIT. Some of these novel approaches, which are in various stages of translation from the lab to the clinic, are outlined in **Box 18-2**.

Box 18-2   New Approaches to Allergen-Specific Immunotherapy

---

(1) Hypoallergenic allergen derivatives. Recombinant DNA technology allows creation of an allergen in which a B cell epitope that tends to induce anti-allergen IgE antibody production is deliberately mutated. This modification reduces the chance that an IgE response will be mounted against the altered allergen when it is delivered to a patient receiving SIT. Importantly, the modified allergen retains the T cell epitopes required for a T cell response that promotes a shift from anti-allergen IgE production to anti-allergen IgG production. Anti-allergen IgE production is thus decreased or even completely suppressed in the patient. A subsequent encounter with the natural allergen provokes a strong Th1 response and production of anti-allergen IgG rather than IgE antibodies. Because allergens modified in this way do not induce IgE-related side effects, higher doses can be administered than would be used in conventional SIT. The first successful clinical trials using this approach were reported in the mid-2000s and used recombinant hypoallergenic birch pollen to relieve symptoms in atopic individuals.

A different approach to allergen manipulation is to alter the amino acid structure of an allergen such that it does not form oligomers. Oligomers are characteristic of many native allergens and are apparently responsible for binding to IgE and subsequent basophil activation. SIT allergens that are altered to resist oligomer formation reportedly elicit both Th1 and Th2 responses, in contrast to the sole induction of Th2 responses by the native allergen.

(2) Fusion proteins. Recombinant DNA technology has also been used to generate a fusion protein (FP) in which pieces of the Fel d 1 cat allergen were joined to a part of an HBV protein. This HBV fragment has already been shown to be a safe carrier protein that is fully able to induce T cell help. When injected into experimental animals, the recombinant FP antigen provoked a strong IgG response. The anti-FP IgG antibodies produced blocked the binding of anti-FP IgE to the native Fel d1 antigen, greatly reducing mast cell activation.

(3) B cell epitopes. Isolated allergen peptides that are part of or close to the natural allergen epitopes that bind to IgE can be generated in the lab. Such peptides have been coupled to large carrier proteins and used to vaccinate atopic individuals. In some cases, allergen-specific IgG responses were induced that blocked the IgE response and its accompanying inflammation.

(4) TLR ligands. Another allergy vaccination technique exploits the innate response to alter T cell responses through DCs. For example, suppose a natural allergen is a ligand for TLR2 or TLR4, and the engagement of these receptors on an atopic individual's DCs results in a Th2 response that generates allergic symptoms. Vaccination of this individual with an altered form of the allergen conjugated to a repeated CpG motif should cause the allergen to bind to TLR9 molecules on a DC instead of TLR2 or TLR4. The DC is then influenced to direct any anti-allergen T cell it encounters down the Th1 differentiation pathway, avoiding atopic symptoms. When the vaccinated individual is later exposed to the unmodified allergen in the natural environment, his/her memory T cells should continue to mount Th1 responses (rather than Th2 responses), keeping the individual symptom-free. This technique of reducing allergenicity has been tried in mice, where an allergen coupled to a repeated CpG motif successfully induced a Th1 response and decreased symptoms. However, the outcomes of the limited number of human trials of this approach have been mixed.

B. Type II Hypersensitivity: Direct Antibody-Mediated Cytotoxic Hypersensitivity

I. What is Type II HS?

During a type II HS response, clinical damage is sustained when the pathological antibodies bind directly to antigens on the surfaces of cells and induce their lysis. In contrast to the disease-causing IgE antibodies of type I HS, those mediating type II HS are mainly of the IgM or IgG isotype. In some cases of type II HS, the pathological antibodies attack leukocytes or RBCs, which are mobile cells. In other cases, the antibodies bind to cells that are "fixed" as part of a solid tissue. The antigen recognized may be a foreign entity that has become "stuck" in some way on the surface of a mobile or fixed cell, or may be an autoantigen. In the latter case, the pathological antibodies are **autoantibodies** that, due to a failure in tolerance mechanisms, are free in the periphery to bind to self epitopes. The type II HS reaction that ensues in this situation is then manifested as part of an autoimmune disease (see Ch. 19). Other examples of type II HS include some forms of anemia, blood transfusion reactions, reactions to certain drugs, some platelet disorders, and some types of tissue transplant rejection.

II. Mechanisms Underlying Type II HS

The mechanisms of cell lysis and tissue damage involved in type II HS are the same as those induced when IgG or IgM antibodies bind to pathogens. The interaction of the pathological antibody with the antigen triggers classical complement activation that results in C3b deposition on the host cell. In the case of RBCs, which are free in the circulation, some cells will be killed by MAC deposition, while others will succumb to complement receptor-mediated phagocytosis or FcR-mediated ADCC. In the case of a cell in a solid tissue like a kidney, the cell is "fixed" and cannot be engulfed by phagocytes, causing them to experience "frustrated phagocytosis." The phagocytes eventually degranulate and release toxic contents that kill the tissue cells. MAC-mediated lysis and FcR-mediated ADCC can also damage fixed cells. The resulting release of DAMPs into the microenvironment surrounding the damaged cells further activates innate leukocytes and increases local inflammation. Examples of type II HS reactions against mobile and fixed cells are illustrated in **Figure 18-5A and 18-5B**, respectively.

III. Examples of Type II HS

i) Hemolytic Anemias

Hemolytic anemia is the lytic destruction of RBCs (*hemolysis*) caused by a pathological antibody in a type II HS reaction (refer to **Fig. 18-5A**). The destruction may occur within the blood vessels or within the spleen or liver, and may be due to autoantibodies or alloantibodies.

a) Autoimmune Hemolytic Anemias

Autoimmune hemolytic anemias occur when the individual makes antibodies directed against epitopes on his/her own RBCs. The autoantibodies involved are classified as being either "warm" or "cold," depending on the temperature at which they show optimal reactivity when tested in the laboratory. Warm autoantibodies are most potent around 37°C and have reduced effects at lower temperatures, whereas cold autoantibodies bind effectively only below 37°C. Most autoimmune hemolytic anemias are due to warm autoantibodies. In acute onset cases, the anemia strikes in a sudden, potentially life-threatening way. Treatment of warm autoimmune hemolytic anemias usually involves glucocorticoids. For patients with cold autoantibodies, the onset of winter may be enough to trigger episodes of acute hemolysis. The extremities of these patients feel cold and turn blue due to the lack of RBCs available to transport oxygen to these tissues. The primary therapy for these people is to avoid exposure to cold temperatures.

b) Alloimmune Hemolytic Anemias

Alloimmune hemolytic anemias occur when an individual has circulating antibodies directed against foreign RBC antigens and is then exposed to the corresponding allogeneic RBCs. The interaction of the pre-formed antibodies with the allogeneic RBCs may induce hemolysis, and the resulting clinical symptoms constitute a type II HS reaction. Two of the best known examples of type II HS are transfusion reactions and Rh disease (refer to Ch. 17). Transfusion reactions occur when the incoming donor RBCs encounter naturally occurring anti-ABO antibodies in the recipient. Rh disease occurs

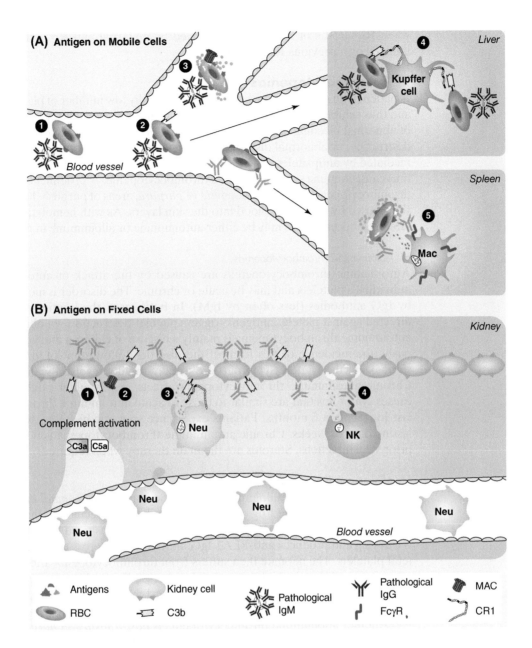

Fig. 18-5
Examples of Type II Hypersensitivity

Type II HS is mediated by IgM or IgG antibodies directed against antigens (which may be self antigens) on either mobile or fixed host cells. **Panel A** depicts a case of autoimmune hemolytic anemia, in which an individual produces IgM or IgG antibodies that recognize a surface antigen on his/her own RBCs (**1**). Within a blood vessel, the binding of a pathological IgM antibody triggers the activation of the classical complement pathway and initiates the deposition of C3b on the RBC surface (**2**). The cascade proceeds to MAC formation and RBC lysis (**3**). If an RBC complexed to IgM avoids MAC-mediated destruction in the blood vessel and makes it to the liver, its destruction can be mediated by resident Kupffer cells via C3b-opsonized phagocytosis (**4**). RBCs bound by pathological IgG antibodies are usually destroyed in the spleen by resident macrophages employing FcγR-mediated ADCC (**5**). **Panel B** depicts a case of type II HS in which pathological IgG antibodies recognize an antigen on the surface of kidney cells. These antibodies initiate classical complement activation and the production of chemokines (**1**) that promote neutrophil extravasation into the tissue. The deposition of C3b on the fixed kidney cells leads to their direct destruction by MAC-mediated lysis (**2**). Neutrophils attempt CR1-mediated phagocytosis of C3b-coated kidney cells but cannot engulf these fixed targets. Instead, they experience "frustrated phagocytosis" that causes them to release their cytotoxic contents toward the kidney cells (**3**). Antibody-bound kidney cells may also be damaged by either NK cells (**4**) or neutrophils employing FcR-mediated ADCC.

when Rh+ RBCs of a fetus are destroyed by maternal anti-Rh antibodies that were induced by a previous Rh+ pregnancy.

ii) Thrombocytopenias

A patient with *thrombocytopenia* has an abnormally low number of platelets in the blood and thus exhibits impaired blood clotting. Thrombocytopenia is the most common cause of abnormal bleeding and is caused by decreased platelet production, increased platelet destruction, or abnormal distribution of platelets within the body. Type II HS reactions mediated by antiplatelet antibodies can contribute to increased platelet destruction and thus cause immune system-mediated thrombocytopenia. A prominent clinical feature of thrombocytopenias is the development of *purpura*, areas of purplish discoloration on the skin caused by leakage of blood into the skin layers. As with hemolytic anemias, type II HS thrombocytopenias may be either autoimmune or alloimmune in nature.

a) Autoimmune Thrombocytopenias

Autoimmune thrombocytopenias are caused by the attack of autoantibodies on an individual's platelets and may be acute or chronic. The disorder is mediated most often by IgG antibodies (less often by IgM). In both cases, the binding of autoantibodies directed against platelet antigens triggers platelet destruction via phagocytosis. Acute autoimmune thrombocytopenia is mainly a disease of children and young adults. Curiously, the incidence of this form of the disorder follows that of viral infections and so peaks in the winter. Symptoms are usually rapidly resolved without intervention. Chronic autoimmune thrombocytopenia affects mainly older adults. This disorder causes symptoms that are similar to those of acute autoimmune thrombocytopenia but last longer than 6 months. Patients experience sporadic bleeding, with each episode lasting days or weeks. Chronic autoimmune thrombocytopenia is not associated with prior viral infections. Steroids are the main therapy when required.

b) Alloimmune Thrombocytopenias

A good example of an alloimmune thrombocytopenia is *neonatal thrombocytopenia*, a rare disorder mediated by maternal alloantibodies specific for the platelet surface antigen PLA1. A pregnant, PLA1− woman becomes sensitized by the PLA1+ fetus she is carrying and produces anti-PLA1 IgG antibodies that cross the placenta and destroy fetal platelets. The neonate then suffers from thrombocytopenia after birth and develops purpura. The principal danger of this disease lies in the potential for intracranial bleeding affecting the neonatal brain. Standard therapy for neonatal thrombocytopenia usually involves giving the newborn a platelet transfusion as well as corticosteroids.

Another alloimmune thrombocytopenia is *post-transfusion purpura*, which occurs when PLA1− individuals receive PLA1+ platelets during a blood transfusion. Although some PLA1− individuals naturally harbor antibodies recognizing PLA1, post-transfusion purpura occurs most often in PLA1− patients who have been sensitized to PLA1 during prior pregnancies or transfusions. The patient's anti-PLA1 antibodies initiate platelet destruction, resulting in the sudden onset of thrombocytopenia. Luckily, this disease is usually self-limited, but intracranial hemorrhage can be a concern. Therapy may involve steroids or plasmapheresis to remove the pathological antibodies.

iii) Antibody-Mediated Rejection of Solid Tissue Transplants

An example of type II HS against a fixed cellular target is hyperacute graft rejection (HAR; refer to Ch. 17). HAR occurs within minutes or hours of organ transplantation when the recipient has pre-existing alloantibodies directed against MHC molecules expressed on cells of a donated organ. These antibodies are usually present because of a previous pregnancy, organ or bone marrow transplant, or blood transfusion.

iv) Goodpasture's Syndrome

Goodpasture's syndrome is an autoimmune disease (see Ch. 20) caused by autoantibodies that recognize a collagen protein found in the basement membranes of the glomeruli in the kidney and the alveoli in the lungs. The autoantibodies trigger classical

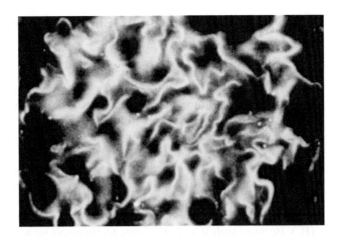

Plate 18-3
Immunofluorescent Detection of Anticollagen Antibodies in the Kidney in Goodpasture's Syndrome

Fluorescently labeled anti-Ig antibodies are used to detect autoantibodies that have bound to collagen throughout the glomerular basement membrane of the kidney, resulting in a smooth, linear binding pattern. *[Reproduced by permission of William G. Couser, University of Washington, and Michael P. Madaio, Temple University School of Medicine.]*

complement activation that damages epithelial and endothelial cells in the target organs, causing lung hemorrhage and inflammation of the renal glomeruli. Patients present with transient kidney dysfunction, bleeding in the lungs, and blood in the sputum and urine. Although permanent lung damage is rare, the damage to the kidneys can be severe and long-lasting and may result in renal failure if left untreated. The worst cases of Goodpasture's syndrome are fatal due to lung hemorrhage and respiratory failure. In **Plate 18-3**, fluorescently labeled anti-IgG antibodies have been used to highlight the typical linear deposits of anticollagen IgG along the basement membrane of the renal glomerulus in a Goodpasture's syndrome patient. Therapy for Goodpasture's syndrome usually involves a combination of plasmapheresis, corticosteroids and immunosuppressive drugs.

v) Pemphigus

Pemphigus is an autoimmune disease (see Ch. 20) characterized by potentially fatal blistering of the skin and mucosae that promotes dehydration and infection. This disorder is caused by autoantibodies (usually IgG or IgA) that attack adhesion proteins called *desmogleins*. Desmogleins "glue" keratinocytes together to form intact upper epidermal layers, and do the same for mucosal epithelial cells to form mucosae. Autoantibody binding to desmogleins not only induces separation of the epidermal or mucosal layers but also allows the release of a protease that causes very painful blisters (**Plate 18-4**). These blisters are exceedingly fragile, and just touching the affected skin can cause it to peel off, leaving the individual vulnerable to infections. Patients are usually middle-aged or elderly, and corticosteroid administration is the standard treatment.

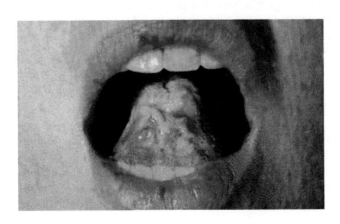

Plate 18-4
Mouth Blisters in Pemphigus

Mouth blisters result when proteases are released from mucosal layers following the binding of autoantibodies. *[Reproduced by permission of Vijay Chaddah, Grey Bruce Health Services, Owen Sound, Ontario.]*

507

C. Type III Hypersensitivity: Immune Complex-Mediated Hypersensitivity

I. What is Type III HS?

During a normal humoral response against a soluble antigen, molecules of antigen and specific antibody bind together to form complexes. These complexes remain small and soluble because complement component C1q binds to the Fc regions of the participating antibodies. This binding has two effects: (1) the classical complement cascade is triggered and leads to clearance of the antibody-bound antigen by phagocytes, and (2) the deposition of C3b interferes with any extension of the antigen–antibody lattice structure. However, inefficient removal or unchecked expansion of these complexes can allow the antigen–antibody lattice to become a large and insoluble **immune complex (IC)**. Because such ICs are too large to phagocytose and clear from the blood, they often become lodged in narrow channels in the body and provoke an immune response that collaterally damages surrounding cells. The clinical outcome of this response depends on the site of IC deposition but commonly involves inflammation of the renal glomeruli (glomerulonephritis), blood vessels (vasculitis), or joints (arthritis). The detrimental immune responses underlying these conditions are known as type III HS reactions.

Many type III HS reactions are clinical complications of infections with certain pathogens, such as those causing meningitis, malaria or hepatitis. A bacterium, parasite or virus may supply large amounts of an antigen that persists after the individual's immune system has dealt with the actual pathogen. Circulating antibodies may bind to these persistent antigens and form large ICs that are deposited in various tissues, causing symptoms that are distinct from those due to the pathogen itself. Some drug or environmental "allergies" may also be due to type III HS reactions. In these cases, the symptoms persist as long as the individual is taking the offending drug, or is exposed to the questionable environmental antigen, because the pathological antibodies induced by the presence of these agents form immune complexes with them, leading to IC deposition. Type III HS is also often found in patients expressing autoantibodies. The autoantibodies combine with soluble autoantigens, which might be proteins, glycoproteins or even DNA, that are naturally and abundantly present at all times. Finally, in rare cases, a cancer patient may make antibodies to tumor antigens shed into the blood by cancer cells. Unless the tumor resolves naturally or is forced to do so by medical intervention, exposure to such antigens is continuous and relatively long term. The ICs that form between the antibodies and tumor antigens may then cause type III HS symptoms.

NOTE: One of the first type III hypersensitivities was originally recognized in the early 1900s. Immunologists noticed that some patients who received passive immunization of serum preparations containing large doses of antidiphtheria or antitetanus antibodies (given to protect against these diseases) experienced a systemic inflammatory illness that they termed "serum sickness." It was later surmised that these individuals mounted a humoral response to the incoming non-human serum proteins, including to the antibodies themselves, that resulted in the formation and tissue deposition of large ICs. Now that progress in vaccine science has allowed active immunization to largely replace passive immunization, this original form of serum sickness is not often seen. However, the term is sometimes still used to describe the type III HS reactions experienced by some individuals in response to drug treatment or environmental antigen exposure.

II. Mechanism Underlying Type III HS

The mechanism by which ICs cause type III HS is illustrated in **Figure 18-6**. Immune complexes that get stuck in narrow body channels induce inflammation and activate complement, resulting in MAC-mediated damage to endothelial cells that allows ICs to penetrate into the underlying tissue. Complement activation then generates not only C3b that coats tissue cells but also anaphylatoxins that summon mast cells. The activated mast cells release

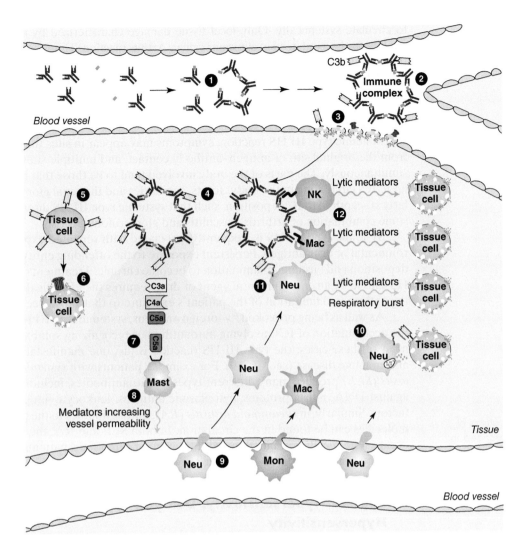

Fig. 18-6
Type III Hypersensitivity

Type III HS results when antigen-antibody pairs in the blood (**1**) cross-link to form insoluble ICs that get stuck in a narrow body channel such as a capillary (**2**). The presence of the ICs activates complement, coating nearby endothelial cells in C3b and triggering their MAC-mediated lysis (**3**). The endothelial layer becomes leaky, and ICs enter the underlying tissue (**4**). Additional complement activation is triggered that results in the coating of tissue cells with C3b (**5**) followed by their lysis (**6**). Anaphylatoxins generated during complement activation draw mast cells to the site and activate them (**7**). These cells in turn secrete mediators (**8**) that increase blood vessel permeability and promote leukocyte extravasation (**9**). Tissue cells coated in C3b are then destroyed by the degranulation of CR1-bearing neutrophils and macrophages that experience frustrated phagocytosis (**10**). Similar damage is done to bystander cells via cytotoxic molecules released by neutrophils that bind to ICs via C3b-CR1 interactions (**11**). Macrophages, NK cells and neutrophils that bind to ICs via their FcRs also release lytic mediators that collaterally kill tissue cells (**12**).

mediators which, along with the anaphylatoxins, increase vascular permeability and facilitate leukocyte extravasation. Tissue cells are then destroyed by phagocyte degranulation and release of cytotoxic molecules. Once again, DAMPS are released from these damaged cells that further activate innate leukocytes and prolong local inflammation.

III. Examples of Type III HS

Type III HS reactions are classified by whether the inflammation they induce appears locally or systemically.

i) Localized Type III HS

In a localized type III HS reaction, the ICs involved are deposited only where antibody and antigen first encounter one another during the effector phase; the ICs do not get a chance to circulate systemically. Only local tissue damage characterized by pain, redness, and swelling is observed in what clinicians call an **Arthus reaction**. This reaction typically takes the form of localized vasculitis due to deposition of ICs in dermal blood vessels. Localized type III HS reactions in patients are comparatively rare, although researchers sometimes induce Arthus reactions in animals to study the mechanisms underlying this type of HS.

ii) Systemic Type III HS

In a systemic type III HS reaction, symptoms may appear in sites that are far removed from the original site of antigen–antibody contact, and multiple sites may be affected simultaneously. The parts of the body involved tend to be those that inadvertently trap large ICs circulating in the body. Joints, capillaries and the renal glomeruli are particularly susceptible to IC deposition, such that systemic type III HS disease often involves some combination of arthritis, vasculitis and glomerulonephritis.

Systemic type III HS is frequently associated with repeated exposure to an environmental or drug antigen. Persistent exposure to the offending entity can cause the IC deposition and ensuing inflammation to become chronic. Treatment of chronic type III HS caused by an environmental agent or drug requires the identification of the triggering antigen and limitation of the patient's exposure to that substance.

As well as being provoked by foreign antigens, systemic type III HS can result from the accumulation of ICs involving autoantibodies recognizing soluble self antigens. In most of these cases, the type III HS reaction is just one manifestation of a complex autoimmune disease (see Ch. 19). For example, patients with *systemic lupus erythematosus (SLE)* produce many different types of autoantibodies, including those directed against DNA, nucleoproteins, cytoplasmic antigens, leukocyte antigens, and clotting factors. Similarly, in *rheumatoid arthritis (RA)*, autoantibodies to the patient's own IgG molecules can be found in the circulation. In both SLE and RA, the deposition of ICs occurs systemically so that, along with their other symptoms, patients experience joint inflammation, vasculitis and, in the case of SLE, kidney disease.

> The autoimmune features of SLE and RA are described in more detail in Chapter 19.

D. Type IV Hypersensitivity: Delayed Type or Cell-Mediated Hypersensitivity

I. What is Type IV HS?

Type IV HS reactions do not occur until about 24–72 hours after exposure of a sensitized individual to the antigen (hence, "delayed-type" hypersensitivity or DTH). Whereas type I, II and III HS are all antibody-mediated, type IV HS results primarily from the tissue-damaging actions of effector Th cells, CTLs and macrophages. The delay in this type of HS is due to the time required for T cell activation and differentiation, cytokine and chemokine secretion, and for the accumulation of macrophages and other leukocytes at the site of exposure. Common examples of type IV HS include *chronic DTH reactions, contact hypersensitivity,* and *hypersensitivity pneumonitis,* all of which are described in the following sections. Another type IV HS reaction is the cell-mediated response to autoantigen in certain autoimmune diseases. In addition, some clinicians consider some forms of chronic graft rejection to be type IV HS reactions because an ongoing cell-mediated response causes immunopathological damage to the transplant recipient.

II. Examples of Type IV HS and their Mechanisms

i) Chronic DTH Reactions

Chronic DTH reactions are initiated by antigens derived from agents that are unusually resistant to elimination by the immune system. Such agents include persistent intracellular pathogens (e.g., those causing tuberculosis, leprosy, leishmaniasis), certain non-infectious agents (e.g., in silicosis, berylliosis), and some unknown agents (e.g., in Crohn's disease, sarcoidosis). An example of how a chronic DTH reaction to a persistent pathogen develops is shown in **Figure 18-7**. Macrophages activated by the capture

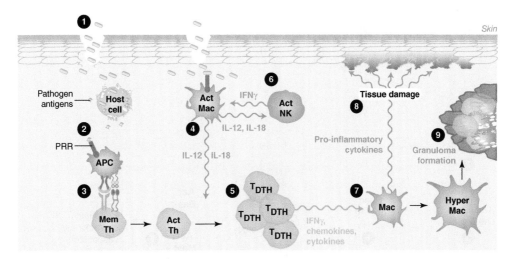

Fig. 18-7
Type IV Hypersensitivity: Chronic DTH Reaction

A pathogen has penetrated the skin (**1**) and has infected a host cell. Pathogen antigens released by the infected cell are taken up by an APC (**2**), processed, and presented to memory Th cells (**3**), which are then activated. At the same time, macrophages activated by the pathogen produce IL-12 and IL-18 (**4**), which promote the differentiation of the Th cells into T$_{DTH}$ effectors (**5**). The cytokines secreted by activated macrophages also stimulate NK cells to secrete large amounts of IFNγ (**6**), which acts on the macrophages to further upregulate their production of IL-12 and sustains T$_{DTH}$ cell generation. In the presence of persistent antigen, the T$_{DTH}$ cells produce IFNγ and other cytokines and chemokines that recruit additional macrophages to the site and activate them (**7**). These macrophages secrete cytokines that lead to the damage of host keratinocytes (**8**). As the response persists, the macrophages may become hyperactivated and initiate granuloma formation (**9**).

of a pathogen antigen produce cytokines that promote the differentiation of Th1 effectors that are called, in this context, **T$_{DTH}$ cells**. In the presence of the persistent antigen, these T$_{DTH}$ cells produce IFNγ and other cytokines and chemokines that recruit additional leukocytes to the site and activate them. Hypersensitivity arises when activated macrophages (and sometimes NK cells) relentlessly secrete pro-inflammatory cytokines, leading to the damage of host keratinocytes. Clinically, the skin over the site of antigen entrenchment becomes red and inflamed. As the response persists, the macrophages may become hyperactivated and initiate granuloma formation around entities that cannot be eliminated. If a granuloma forms in an organ such as the liver or lung, it can cause a lesion that may interfere with that organ's function, leading to serious liver disease or respiratory failure. Corticosteroid treatment is most often used to treat chronic DTH reactions.

Granuloma formation was discussed in Chapter 13 and shown in Plate 13-2.

ii) Contact Hypersensitivity

Contact hypersensitivity (CHS), sometimes called "contact dermatitis," is a secondary immune response to a small, chemically reactive molecule that has bound covalently to self proteins in the uppermost layers of the skin. Examples of CHS include the patchy rash and intense itching that follow a plunge into a patch of poison oak or poison ivy, and the local skin irritations experienced by individuals sensitive to drugs, metals, cosmetics, or industrial or natural chemicals. The alteration of self proteins by the binding of a CHS antigen present in these substances generates a "non-self" entity that can be thought of as a **neo-antigen** ("new" antigen). Some CHS neo-antigens are created when the chemically reactive molecule oxidizes self proteins. In other cases, a metal may form a stable metalloprotein complex that is particularly provocative to macrophages. Sometimes a chemical will have to be metabolized in the liver before the reactive component is available for neo-antigen formation. In any case, the neo-antigen induces skin cells to release mediators drawing leukocytes to the area, particularly memory Th and Tc cells. As illustrated in **Figure 18-8**, Th effectors and CTLs derived from these memory lymphocytes make a major contribution to the tissue damage associated with CHS. Mast cells also appear to have an important role, because natural mouse mutants lacking mast cells show impaired CHS reactions. NKT cells may participate in CHS

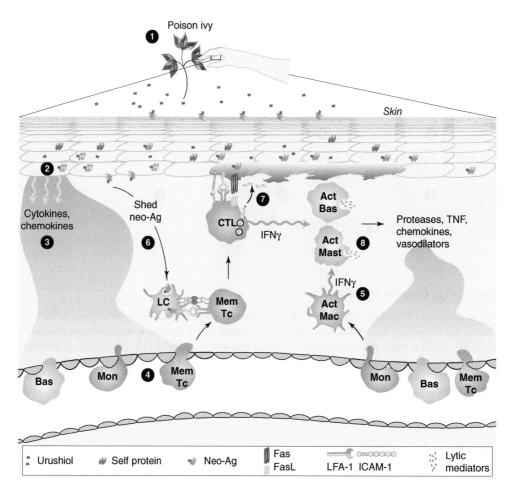

Fig. 18-8
Type IV Hypersensitivity: Contact Hypersensitivity

A misguided individual has touched a poison ivy plant (**1**), which contains the CHS antigen urushiol. Upon contact with human skin, urushiol molecules penetrate the keratinocyte layer and bind covalently to reactive groups present on local self proteins, forming a neo-antigen (**2**). The presence of the neo-antigen induces skin cells to release numerous cytokines and chemokines (**3**) that draw leukocytes from the circulation (**4**) into the affected area where they activate macrophages to secrete IFNγ (**5**). Neo-antigen shed by skin cells (**6**) is taken up by Langerhans cells (LCs), and peptides derived from it are cross-presented to memory Tc cells that were generated during the sensitization stage. CTL effectors are generated that use their mechanisms of cytotoxicity to damage skin cells displaying neo-antigen-derived pMHCs (**7**). CTLs also secrete IFNγ that stimulates the degranulation of activated basophils and mast cells (**8**), which release additional damaging mediators that perpetuate local inflammation.

through their secretion of IL-4, which may stimulate a B cell response to the neo-antigen. Finally, recent studies have shown that the damage caused by CHS is worse if an individual is deficient in regulatory T cells, implying that these cells are necessary for balancing the responses of effector T cells to neo-antigens.

The primary mode of treatment of CHS is avoidance of the inciting antigen. If exposure does occur, corticosteroid cream may be applied to the affected area.

iii) Hypersensitivity Pneumonitis

Hypersensitivity pneumonitis (HP) is a type IV HS reaction in the lung caused by prolonged exposure to an inhaled antigen. These antigens, which include certain chemicals, microbial components, plant components and even rodent urinary proteins, are acquired in the form of aerosolized organic dust particles of size <5 μm—i.e., small enough to reach the distal airway. Regardless of the nature of the antigen, the resulting type IV HS response usually occurs in the same three clinical stages: the *acute* phase, the *subacute* phase and the *chronic* phase.

Both the innate and adaptive immune responses contribute to HP pathogenesis. When a sensitized individual inhales the offending antigen, macrophages are activated in the lungs either because these cells ingest the antigenic particles directly or because of complement activation in the local environment. Within 48 hours, chemokines secreted by the macrophages promote an influx of neutrophils and other granulocytes. In some cases, the interaction of the antigen itself with TLR2 expressed by neutrophils can promote the recruitment of these leukocytes. Somewhat later, memory Th and Tc cells are drawn to the site, where they commence differentiation into CTLs and Th1/Th17 effectors secreting copious cytokines. Additional chemokines and other inflammatory mediators are contributed by activated eosinophils. At this acute stage of the disease, the affected individual likely experiences influenza-like symptoms (high fever, chills, body aches, malaise) that resolve quickly if exposure to the antigen ends. If a formal diagnosis of HP is made, corticosteroids may be given to dampen the inflammatory response. However, if acute HP goes unrecognized and exposure to the antigen continues, the subacute phase ensues during which macrophages become hyperactivated, secrete large amounts of IFNγ, and initiate the formation of a granuloma around the antigen in the lung. Inflammatory mediators promote the synthesis of collagen, which leads to airway fibrosis. The fever may be gone and overt symptoms may be rare for several days or weeks, but fatigue, cough and anorexia then reappear and progress steadily. If there is still no clinical intervention and exposure continues, the chronic phase of HP sets in, and the lung tissue is damaged by the same mechanisms as operate in chronic DTH responses. In particular, activated macrophages in the lungs secrete large amounts of TGFβ that encourage fibrosis. At this point, the damage to the lungs may be irreversible. The patient suffers from a slow progressive decline characterized by dyspnea (shortness of breath), anorexia, weight loss, weakness and fatigue. Even with corticosteroid treatment, the disease may eventually be fatal.

The first cases of HP were described in 1713 as an occupational disease of grain farmers.

As mentioned previously, the antigens causing HP are proteins that can be derived from microbes, plants or animals and are often associated with occupational exposure. In the past, workplace exposure to antigens triggering HP gave rise to afflictions labeled as "farmer's lung" and "cheese washer's lung," and now the workplace theme continues in modern times in the form of "machine operator's lung" and "yacht-maker's lung" (**Table 18-5**). However, other modern HP cases have been linked to non-occupational exposure to inhaled antigens, giving rise to disorders with monickers like "sauna-taker's disease," "hot tub lung," and even "gerbil keeper's lung." Interestingly, only 5–15% of all individuals exposed to environmental antigens capable of inducing HP actually develop the disease. Those who are exposed to these inhaled antigens but remain disease-free mount a harmless IgG response with no cellular involvement. In some instances,

TABLE 18-5	Examples of Occupational Hypersensitivity	
Occupation	**Antigen**	**Disease**
Hay/silage/grain farming	Fungal species in crop	Farmer's lung
Tobacco farming	Fungal species in crop	Tobacco worker's disease
Vintner	Mold on grapes used for wine-making	Wine grower's lung
Pheasant farmer	Pheasant antigen in feathers	Pheasant rearer's lung
Cheese maker	Bacterial growth on cheese	Cheese washer's lung
Grain mill worker	Wheat weevil in crop	Wheat weevil disease
Research lab worker	Rodent urinary protein	Rodent keeper's lung
Furrier	Animal fur antigen	Furrier's lung
Metalwork machine operator	Bacteria in metal-working fluid	Machine operator's lung
Professional musician	Bacterial contamination of trombone	Trombone player's lung
Wood worker	Tree dust or mold contaminating the trees to be processed	Wood worker's lung
Polyurethane worker	Chemical dust from polyurethane foam used to build yachts	Yacht-maker's lung

[Adapted from Zacharisen, M. C., & Fink, J. N. (2011). Hypersensitivity pneumonitis and related conditions in the work environment. *Immunology & Allergy Clinics of North America, 31*(4), 769–786.]

individuals with HP recall having had an acute respiratory infection just prior to the onset of symptoms, but the external factors that influence HP susceptibility remain obscure. There is some evidence that polymorphisms in the TNF gene, TAP genes, and in genes encoding components of the proteasome involved in MHC class I antigen presentation may confer genetic predisposition to HP.

NOTE: As awareness of the harmful consequences of inhaled antigens has grown, the use of safety precautions and protective garb has increased to prevent workers from breathing in noxious particles. Accordingly, the number of reported cases of HP has plummeted in the U.S. in the past 5 years.

This concludes our description of immune hypersensitivity. We move now to a discussion of autoimmune disease, one of the great mysteries of immunology. It is still unknown why the immune system sometimes attacks an individual's own tissues and how such attacks can best be prevented.

Chapter 18 Take-Home Message

- The development of hypersensitivity occurs in two stages: the sensitization stage and the effector stage. The sensitization stage is a primary immune response to an antigen and has no clinical consequences. The effector stage is a secondary immune response that is deleterious to the individual.

- Type I hypersensitivity (allergy or atopy) is mediated by IgE antibodies specific for antigens that are normally non-pathogenic (allergens).

- In the sensitization stage of type I HS, a first exposure to allergen triggers the production of allergen-specific IgE that binds to FcεRs of mast cells and basophils. In the early phase of the effector stage, re-exposure of the sensitized mast cells to the allergen triggers the immediate release of pre-formed mediators via degranulation as well as new synthesis of inflammatory molecules. Eosinophils, sensitized basophils and other leukocytes drawn to the site mediate the late phase of the effector stage via granule release and the production of additional mediators.

- The major mediators characteristic of type I HS induce vasodilation, smooth muscle contraction, increased blood vessel permeability, and leukocyte recruitment.

- Local atopic responses can result in dermatitis, urticaria, asthma or rhinitis.

- Anaphylaxis is a systemic atopic response that can be life-threatening and requires immediate epinephrine administration.

- Food allergies are atopic responses to consumed allergens.

- Type II hypersensitivity is mediated by direct antibody-mediated cytotoxicity. IgG or IgM antibodies bind to antigenic epitopes on fixed or mobile cells and trigger ADCC, complement activation and/or phagocytosis.

- Type III hypersensitivity occurs when soluble antigen binds to IgG or IgM to form large insoluble immune complexes. Deposition of these complexes in narrow body channels triggers inflammatory responses that damage underlying tissues.

- Type IV hypersensitivity is mediated by effector T cells, macrophages and other leukocytes that infiltrate a site of antigen exposure and induce a delayed form of inflammatory tissue damage.

Did You Get It? A Self-Test Quiz

Introduction and Section A.I–II

1) Define "immune hypersensitivity."

2) Name the four types of immune HS.

3) All immune HS develop in what two stages?

4) Give two names used to refer to type I HS.

5) Describe the sequence of events making up the sensitization stage of type I HS.

6) What difference in cytokine production by atopic and non-atopic persons is relevant to type I HS?

7) Distinguish between the two phases of the effector stage of type I HS.

8) Give three examples each of pre-formed and newly generated mediators relevant to type I HS.

9) Describe three effects of mediator release that lead to allergic symptoms.

10) What contribution do eosinophils make to type I HS?

Section A.III–V

1) Distinguish between localized and systemic atopy.

2) What are the clinical names for the following? hives, eczema, hay fever

3) What is airway hyper-responsiveness, and what type of atopy is it associated with?

4) What is a "wheal and flare reaction," and what causes it to develop?

5) Distinguish between atopic dermatitis and atopic urticaria.

6) Why is anaphylaxis a medical emergency?

7) In a person with a food allergy, which body sites may show atopic symptoms? What are these symptoms?

8) All allergens share a common structure that targets mast cells. True or false?

9) How are DCs instrumental in determining whether an allergic reaction will occur?

10) How does breast-feeding reduce the chance of later developing an allergy?

11) What three factors may combine to determine whether a person will have an atopic response to an allergen?

12) Give three examples of genes associated with atopy and discuss why they may be relevant.

13) Outline the hygiene hypothesis.

14) Give three examples of triggering events with respect to atopic responses.

15) What are the roles of the microbiota and NKT cells in atopy?

16) Give four examples of commonly used allergy therapies and how they work.

17) What is SIT, and what is the rationale behind using it for allergy therapy?

18) Give two examples of experimental approaches to SIT and explain how they might work.

Section B

1) How do pathological antibodies cause type II HS? Is IgE relevant?

2) What cytolytic mechanisms are operating in type II HS responses against mobile cells? Against fixed cells?

3) Give two examples each of autoimmune and alloimmune type II HS disorders and describe the epitopes targeted by the pathological antibodies.

4) Why is hyperacute graft rejection considered to be a case of type II HS?

5) What protein is targeted in Goodpasture's syndrome? In pemphigus?

Section C

1) What is an immune complex?

2) How does the triggering of type III HS differ from that of types I and II HS?

3) Give three sources of antigen that might trigger type III HS.

4) What is an Arthus reaction?

5) Give two examples of a systemic type III HS disorder.

Section D

1) What is the main difference between type IV HS and types I–III?

2) Why are type IV HS responses comparatively delayed in onset?

3) Outline the mechanism underlying chronic DTH reactions.

4) Outline the mechanism underlying CHS.

5) Outline the mechanism underlying hypersensitivity pneumonitis.

Can You Extrapolate? Some Conceptual Questions

1) You decide to try a new food that you have never eaten before. Soon after, you feel unwell and experience unpleasant gastrointestinal symptoms. A friend suggests that you must be allergic to that food. Would you agree or disagree? Explain.

2) For each of the following type I hypersensitivity therapies, which numbered mechanism(s) in which figure(s) in the main text would be affected?

   a) antihistamines

   b) lipoxygenase antagonists

   c) bronchodilators

   d) corticosteroids

   e) cromones

   f) allergen-specific immunotherapies

3) You will learn in Chapter 19 that many autoimmune diseases have symptoms that are either systemic, meaning that they affect many sites in the body, or local, meaning that they affect only very specific cell types or tissues. As we have seen in this chapter on hypersensitivity, the IgM and IgG antibodies that mediate type II and III hypersensitivities are often directed against self antigens, and so play a role in autoimmunity. Would you expect type II hypersensitivity to correlate with systemic or local autoimmunity, and why? How about type III hypersensitivity?

4) The neoantigens that trigger contact hypersensitivity draw memory cells (Th and Tc) into the skin where they mediate tissue damage. Would you expect these memory T cells to be Tem or Tcm? (Hint: You may want to refer to Ch. 9.)

Would You Like To Read More?

Besnard, A. G., Togbe, D., Couillin, I., Tan, Z., Zheng, S. G., Erard, F., et al. (2012). Inflammasome–IL-1–Th17 response in allergic lung inflammation. *Journal of Molecular Cell Biology*, 4(1), 3–10.

Boguniewicz, M., & Leung, D. Y. (2011). Atopic dermatitis: A disease of altered skin barrier and immune dysregulation. *Immunological Reviews*, 242(1), 233–246.

Boyce, J. A., Bochner, B., Finkelman, F. D., & Rothenberg, M. E. (2012). Advances in mechanisms of asthma, allergy, and immunology in 2011. *Journal of Allergy & Clinical Immunology*, 129(2), 335–341.

Claas, F. H., & Doxiadis, I. I. (2009). Management of the highly sensitized patient. *Current Opinion in Immunology*, 21(5), 569–572.

Cosmi, L., Liotta, F., Maggi, E., Romagnani, S., & Annunziato, F. (2011). Th17 cells: New players in asthma pathogenesis. *Allergy*, 66(8), 989–998.

Di Meglio, P., Perera, G. K., & Nestle, F. O. (2011). The multitasking organ: Recent insights into skin immune function. *Immunity*, 35(6), 857–869.

Karp, C. L. (2010). Guilt by intimate association: What makes an allergen an allergen? *Journal of Allergy & Clinical Immunology*, 125(5), 955–960.

Mari, A., Scala, E., & Alessandri, C. (2011). The IgE-microarray testing in atopic dermatitis: A suitable modern tool for the immunological and clinical phenotyping of the disease. *Current Opinion in Allergy & Clinical Immunology*, 11(5), 438–444.

Masilamani, M., Commins, S., & Shreffler, W. (2012). Determinants of food allergy. *Immunology & Allergy Clinics of North America*, 32(1), 11–33.

Metz, M., & Maurer, M. (2009). Innate immunity and allergy in the skin. *Current Opinion in Immunology*, 21(6), 687–693.

Valenta, R., Campana, R., Marth, K., & van Hage, M. (2012). Allergen-specific immunotherapy: From therapeutic vaccines to prophylactic approaches. *Journal of Internal Medicine*, 272(2), 144–157.

Valenta, R., Ferreira, F., Focke-Tejkl, M., Linhart, B., Niederberger, V., Swoboda, I., et al. (2010). From allergen genes to allergy vaccines. *Annual Review of Immunology*, 28, 211–241.

Zacharisen, M. C., & Fink, J. N. (2011). Hypersensitivity pneumonitis and related conditions in the work environment. *Immunology & Allergy Clinics of North America*, 31(4), 769–786.

Autoimmune Diseases

We have failed to grasp the fact that mankind is becoming a single unit, and that for a unit to fight against itself is suicide.

Havelock Ellis

A. What is an Autoimmune Disease?

In the early days of immunological studies, scientists believed that the immune system would simply be incapable of mounting anti-self immune responses because they would harm the body. In 1900, Paul Ehrlich labeled this concept as "horror autotoxicus," meaning that the immune system should have a "horror of" (and therefore avoid) being "autotoxic." However, even at that time, there were isolated reports suggesting that autoreactivity could definitely occur. Clinical and experimental evidence for this phenomenon slowly accumulated in the 1950s, and by the 1960s, it was generally accepted that several diseases, including *systemic lupus erythematosus (SLE)* and *multiple sclerosis (MS),* resulted from "**autoimmunity**." It has since become clear that autoimmunity itself is not unusual and is not the same as "**autoimmune disease**." In fact, every healthy person is autoimmune to a limited degree, as demonstrated by the low levels of anti-self antibodies that can be found in all individuals. The concentrations of these autoantibodies can even increase as a consequence of inflammation or infection without the individual experiencing any clinical effects. Autoimmune disease arises only when autoimmunity causes clinical damage to a self tissue or disrupts its normal function. Currently, there are about 100 different autoimmune diseases that vary widely in incidence and phenotype. The U.S. National Institutes of Health (NIH) estimate that over 20 million Americans collectively suffer from these often chronic and debilitating diseases.

Autoimmune diseases are generally considered to be disorders of the adaptive response. The clinical symptoms associated with these afflictions are due to the activation and effector functions of autoreactive lymphocytes responding to self proteins acting as **autoantigens**. Activated autoreactive Th cells, particularly Th1 and Th17 effectors, release cytokines such as IFNγ and IL-17. These cytokines act on other

WHAT'S IN THIS CHAPTER?

MAK: Primer to the Immune Response. http://dx.doi.org/10.1016/B978-0-12-385245-8.00019-4

leukocytes (granulocytes, monocytes, macrophages) as well as non-hematopoietic cells (endothelial cells, fibroblasts) and induce them to secrete additional cytokines and chemokines. These mediators, which include IL-1, IL-6 and TNF, precipitate tissue damage and chronic inflammation.

Autoreactive Th cells also supply T cell help for **autoantibody** production by autoreactive B cells. Sometimes autoantibodies cause autoimmune disease by directly deregulating or disrupting the function of the *target tissue* expressing the specific self antigen under attack. For example, in *myasthenia gravis (MG)* and *Graves' disease (GD)*, the binding of autoantibodies to particular cell surface receptors causes dysfunction of the muscles or thyroid gland, respectively. In other cases, such as *Goodpasture's syndrome (GS)*, autoantibody binding leads to the destruction of the targeted cells and damage to nearby tissues that is manifested as type II hypersensitivity (refer to Ch. 18). Type III hypersensitivity may also contribute to autoimmune disease when autoantibodies bind to a self antigen and form large insoluble immune complexes that travel throughout the body in the circulation. If these complexes lodge in the body's narrow channels, inflammatory tissue damage can be triggered at distant sites where the self antigen is not even expressed. SLE furnishes a good example of an autoimmune disease with a type III hypersensitivity component.

In addition to the destruction caused by autoantibodies, autoreactive Tc cells receiving help from autoreactive Th cells can generate CTLs that kill healthy cells expressing the self antigen. The cytokines released by these CTLs also damage surrounding cells and help to sustain the activation of the APCs required for autoreactive Th responses. This cell-mediated tissue destruction caused by autoreactive Th effectors and CTLs constitutes a form of type IV hypersensitivity. The autoimmune form of *diabetes mellitus* is considered by many immunologists to be an example of such an affliction.

NOTE: Clinicians have traditionally categorized autoimmune diseases into two broad classes: (1) *organ-specific* autoimmunity, in which a particular anatomical site is targeted for immune destruction, and (2) *systemic* autoimmunity, in which the immune response is not restricted to a particular organ or tissue. However, this categorization is based on the clinical manifestations of the disease rather than on the expression pattern of the targeted self antigen. In some cases, an antigen may be ubiquitously expressed, but the autoimmune response to it may occur in only one organ. Why the attack is limited in this way is not yet clear.

In Sections B through E of this chapter, we discuss possible mechanisms and determinants of autoimmune disease, followed by descriptions of some of the better known examples of these disorders and the therapies used to treat them. In Section F, we present some examples of **autoinflammatory disorders**, diseases that are like autoimmune diseases in that they are mediated by the immune system in the absence of a pathogen, but unlike them in that no autoantigen recognition or adaptive response is involved. Instead, autoinflammatory disorders are characterized by uncontrolled inflammation associated with dysregulation of the innate response.

B. Mechanisms Underlying Autoimmune Diseases

How does an autoimmune disease arise? Negative selection during lymphocyte development removes the vast majority of autoreactive lymphocyte clones during the establishment of central tolerance. Any autoreactive clones that do escape to the periphery are normally controlled by the mechanisms of peripheral tolerance (refer to Ch. 10). Thus, four things must happen for an autoimmune disease to develop:

(1) An autoreactive lymphocyte must escape elimination by central tolerance mechanisms and be released to the periphery;

(2) The escaped autoreactive lymphocyte must encounter its specific self antigen in the periphery;

(3) The peripheral tolerance mechanisms designed to regulate autoreactive lymphocyte responses must fail; and

(4) The response by the autoreactive lymphocyte must result in clinical damage.

In the sections that follow, we describe several mechanisms that are believed to contribute to the activation of autoreactive lymphocytes in the periphery and thus set the stage for the onset of autoimmune disease. These mechanisms, which may not act in a mutually exclusive way, can affect almost any aspect of adaptive immunity.

> The processes by which developing B and T lymphocytes are negatively selected were described in Chapters 5 and 9, respectively.

NOTE: While most autoimmune diseases are due to inadequacies in peripheral tolerance, some are caused by rare defects in central tolerance mechanisms. For example, as introduced in Chapter 15, *autoimmune polyendocrinopathy candidiasis ectodermal dystrophy (APECED)* is a disease that results from mutations to a transcriptional regulator called *autoimmune regulator (AIRE)*. In the absence of AIRE function, there is a deficit in the expression by mTECs of certain self antigens during negative selection in the thymus (refer to Ch. 9). As a result, T cells bearing TCRs specific for these self antigens cannot be eliminated and are released to the periphery, where they attack multiple tissues. As well as abnormal T cell functions, APECED patients exhibit hypoparathyroidism and adrenal insufficiency due to attacks by organ-specific autoantibodies.

I. Inflammation

Inflammation, whether induced by infection or tissue damage or stress, is thought to play a key role in the development of autoimmune disease. In addition, as is described in **Box 19-1**, putative inflammation-related links between cancer and autoimmune disease have also been found.

i) Infection-Induced Inflammation and Autoimmunity

Many episodes of autoimmune disease appear to occur soon after infection with a pathogen. One school of thought is that the inflammatory milieu that arises in the course of a pathogen infection can "break" peripheral tolerance. As described in Chapter 10, two key mechanisms contribute to the maintenance of peripheral tolerance. First, if an autoreactive T cell binds to its cognate pMHC presented by a tolerogenic DC, the autoreactive T cell is anergized due to a lack of costimulation and no longer responds to the antigen. Second, any autoreactive T cells that are not anergized and attempt to initiate activation are soon suppressed by regulatory T cells. Both of these measures may be undermined when pathogen invasion generates a local inflammatory environment. At the mechanistic level, the DNA and other components of infecting microbes, plus any host DNA released upon pathogen-induced cell death, constitute PAMPs/DAMPs that can engage the PRRs of immature DCs accumulating in the target tissue. These DCs commence maturation, upregulate costimulatory molecules, and migrate to the local lymph node where they display pMHCs derived from both pathogen antigens and self antigens to naïve T cells, including to autoreactive T cells. A naïve autoreactive T cell may thus be activated because the DC it encounters is mature. Furthermore, the large amounts of pro-inflammatory cytokines secreted by the mature DC may allow any autoreactive T cell effectors to escape the control of regulatory T cells. These autoreactive T cell effectors migrate back to the target tissue expressing the self antigen and destroy tissue cells either directly or by facilitating autoantibody production. Because the self antigen is continuously present as part of a tissue, the immune response cannot mop up the antigen in the same way as it would a pathogen. The unwanted autoimmune attack is thus perpetuated and causes an autoimmune disease.

Recent research has indicated that the non-pathogenic commensal microbes protecting the mucosae may also occasionally contribute to autoimmune disease during an infection. Like all microbes, commensal organisms bear common molecular patterns, which some immunologists call **microbiota-associated molecular patterns (MAMPs)**. It is speculated that some of these MAMPs may bind to certain PRRs at the same time as PAMPs or DAMPs engage others, enhancing inflammation. Thus, although they cannot normally promote the activation of autoreactive lymphocytes on their own,

MAMP-expressing commensal organisms may contribute to the inflammatory milieu that encourages the activation of these cells.

Support for the connection between infection-induced inflammation and auto-immune disease initiation is growing. Upregulated PRR expression has been found on mature APCs in damaged tissues of patients with organ-specific autoimmune diseases, and the persistence of PAMPs in a patient's tissues following an infection has been linked to later development of autoimmune symptoms. This association between PAMPs and autoimmune disease has been intensively investigated in animal models, where clinical signs of autoimmune disease do not emerge if DCs are prevented from accumulating in the target tissue. In mice induced to develop *experimental autoimmune encephalitis (EAE)*, which resembles MS in humans, the severity of the resulting disease is much worse if the animals are infected with bacteria 7 days after EAE induction. This worsening of autoimmune symptoms does not occur in TLR2-deficient mice, which lack the capacity to respond to the PAMPs of the bacteria. These observations are in line with the experience of human cancer patients who also have the autoimmune disease *psoriasis (PS)*. When these patients are treated with an anti-cancer mAb that triggers TLR7/8 signaling, their PS symptoms intensify.

NOTE: Infection does not always correlate with exacerbated autoimmune symptoms. In developed countries, where better sanitation and antibiotics keep a lid on pathogens, the incidence of several autoimmune diseases is much higher than in less-developed countries where infections are more common (particularly infections with helminth parasites). As noted in our discussion of the hygiene hypothesis, many immunologists believe that infections with pathogens that promote Th2 responses (which are associated with anti-inflammatory cytokine

Box 19-1   The Relationship between Cancer and Autoimmunity

Many autoimmune disease patients have a higher risk than the general population of developing cancer. Lymphomas, which are solid tumors composed of cancerous lymphocytes (see Ch. 20), were the first malignancies identified in autoimmune disease patients, but leukemias and tumors of the kidney, lung, breast and GI tract also seem to occur with increased frequency. Some of these cancers do not appear until as long as 20 years after the initial diagnosis of the autoimmune disease. Intriguingly, the converse is also true, in that many cancer patients have a propensity to develop clinical features of autoimmunity (particularly autoantibody production) many years after the diagnosis of their malignancy.

How could autoimmunity lead to cancer? It is thought that the destructive chronic inflammation associated with many autoimmune diseases may increase the level of carcinogenic molecules (such as ROIs) in the microenvironment. These reactive substances could act on a nearby cell in an early stage of carcinogenesis and push it down the path to malignant conversion. Alternatively, the relentless proliferation of an autoreactive lymphocyte activated by abundant self antigen might provide more opportunities for oncogenic mutations to occur and accumulate.

How might a cancer lead to autoimmunity? Lymphocytes directed against tumor-related antigens can sometimes be found in the blood of cancer patients. Although some of the antigens targeted by these lymphocytes may be TSAs, others may be TAAs, such as cell cycle control proteins and tumor suppressors (see Ch. 16). Self tissues expressing these proteins might then come under autoimmune attack. Such a mechanism is thought to account for the onset of *vitiligo* in melanoma patients. Vitiligo is a relatively rare autoimmune disease in which autoimmune destruction of melanocytes leaves patients with patches of abnormally white skin. Some researchers believe that the immune system's attack on the malignant melanoma cells targets a TAA also expressed by normal melanocytes. Indeed, the development of vitiligo in a melanoma patient often presages an improved prognosis. In addition to TAA recognition, the physical disruption associated with a growing tumor may expose previously hidden autoantigens that then can be recognized by the immune system. Lastly, the therapies used to rid cancer patients of their tumors may induce autoimmune disease onset, or may make a patient more vulnerable to infection with a pathogen associated with autoimmune disease.

Sometimes an autoimmune disease is linked to a *decreased* risk of cancer. For example, although MS patients have a greater chance of developing CNS tumors (the site of the autoimmune attack), they are less likely to develop other types of malignancies. Similarly, patients with RA and SLE are often at an overall lower risk of developing cancers. The reasons why some autoimmune diseases reduce cancer risk while others increase it are not clear, although some researchers have speculated that, in certain cases, the increased activity of the immune system in autoimmune disease patients might ramp up immunosurveillance and thus more efficiently eliminate cells in the process of malignant transformation.

secretion and Treg cell generation) can have suppressive effects on immune hypersensitivities and autoimmune diseases. Evidence to support this hypothesis has been gathered from mouse studies in which animals induced to develop an autoimmune disease but also infected with a helminth parasite produce cytokines and Treg cells that suppress autoreactive Th17 cells. Thus, it seems that infections which trigger inflammation associated with Th1/Th17 responses may indeed exacerbate autoimmune disease symptoms, whereas infections that are handled by Th2 responses may assuage autoimmune disease symptoms.

The hygiene hypothesis was discussed in Chapter 18.

ii) Damage/Stress-Induced Inflammation and Autoimmunity

Even in the absence of a pathogen infection, DC maturation promoting autoimmune disease may occur if tissue cells experience significant stress or physical trauma. Cells that have become necrotic due to mechanical injury, transformation or other assaults frequently release stress molecules that constitute DAMPs with effects on DCs. For example, *in vitro*, HSPs alone can induce DCs to mature, produce pro-inflammatory cytokines, and upregulate MHC class II and B7 molecules. Intracellular molecules like ATP that can activate inflammasomes may also contribute, as can the binding of host DNA or RNA to TLR7 or TLR9. Again, autoreactive T cells that interact with DCs induced to mature by this inflammation may recognize pMHCs presented by these cells and initiate an autoimmune response that leads to clinical symptoms.

Another contributor to autoimmune disease in this context may be a failure to resolve any normal, acute inflammation that has played a role in healing a stressed or damaged tissue. This return to homeostasis requires not only the dialing down of the activities of inflammatory leukocytes but also the actions of molecules that actively inhibit proteases produced by these cells. For example, one such helpful molecule appears to be TSLP, which induces the expression of a particular protease inhibitor needed to block *neutrophil elastase* (NE) activity. In one study, researchers used a chemical agent to establish a mouse model of autoimmune intestinal inflammation. Wild-type mice experienced severe tissue damage soon after administration of the agent but then were able to fully recover after its removal. However, mice that had undergone genetic deletion of TSLP and were treated with the chemical agent could not resolve the intestinal inflammation and eventually died. Further investigation showed that TSLP normally binds to a ligand expressed on intestinal epithelial cells and triggers the expression of the required NE inhibitor. In the absence of TSLP, the inhibitor was not expressed and NE activity continued, perpetuating autoimmune tissue damage. Thus, non-hematopoietic cell types and their products may be critical for supporting the healing processes that prevent damage/stress-induced inflammation from persisting and promoting autoimmune disease.

Thymic stromal lymphopoietin (TSLP) was discussed in the context of peripheral tolerance in Chapter 10. TSLP-stimulated DCs can induce naïve T cells to differentiate into nTreg cells.

II. Molecular Mimicry by Pathogen Antigens

In addition to triggering inflammation, pathogens may also contribute to the onset of an autoimmune disease via **molecular mimicry**. Molecular mimicry is deemed to occur when a component of a pathogen bears an epitope that resembles an epitope derived from a self antigen. Thus, some T and/or B cells in an individual may bear antigen receptors that recognize both the self epitope and the pathogen epitope. In the absence of infection by the relevant pathogen, such potentially autoreactive lymphocytes are held in check by peripheral tolerance. However, if the pathogen infects the individual, the presentation of a relatively large amount of the cross-reactive pathogen epitope in the inflammatory milieu created by the infection may break peripheral tolerance and trigger activation of the autoreactive lymphocyte. An attack on host tissues expressing the self epitope then ensues. Several examples of pathogen amino acid sequences that resemble regions of self proteins thought to be involved in human autoimmune diseases are shown in **Figure 19-1**.

Both T cell- and B cell-mediated responses can be triggered by molecular mimicry. The first step leading to T cell molecular mimicry is the acquisition of a pathogen or its components by an immature DC and the induction of DC maturation. As illustrated in **Figure 19-2**, a mature DC presents pathogen peptides on MHC class II to naïve Th

Autoimmune Disease	Cross-reacting sequences	Origin of peptide
Rheumatic fever	QK M RRD LE E	Human myosin
	KG L RRD LD A	*Streptococcus* cell wall protein
Multiple sclerosis	V V H F F K N I V	Human myelin basic protein
	V Y H F V K K H V	Epstein-Barr virus protein
Graves' disease	DA FG GVY S	Human thyroid stimulating hormone receptor
	DA LG NV TS	*Yersinia enterocolitica* outer membrane protein
Systemic lupus erythematosus	P P P G M R P P	Human nuclear spliceosome protein
	P P P G R R P	Epstein-Barr virus protein
Rat equivalent of multiple sclerosis	YG SL PQ KS QR TQ DE N	Rat myelin basic protein
	YGC LLP RN PR TE DQ N	*Chlamydia pneumoniae* protein
Mouse equivalent of inflammatory bowel disease	N IIS DA	Mouse heat shock protein
	NAAS IA	*Mycobacterium* protein

Fig. 19-1
Examples of Amino Acid Sequences Potentially Involved in Molecular Mimicry

Several autoimmune diseases in humans may arise from molecular mimicry between pathogen protein sequences and host proteins of similar sequence. Similar mimicry has been observed in rodent models of autoimmune diseases. Cross-reacting sequences are given in single-letter amino acid code, and amino acids that share identity are highlighted. *[With information from (1) Rohm A. P. et al. (2003) Mimicking the way to autoimmunity: an evolving theory of sequence and structural homology. Trends in Microbiology 11, 101; (2) Quinn A. et al. (1998) Immunological relationship between the class I epitopes of Streptococcal M protein and myosin. Infection and Immunity 66, 4418; (3) Zhe Wang et al. (2010) Identification of outer membrane porin F protein of Yersinia enterocolitica recognized by antithyrotropin receptor antibodies in Graves' disease and determination of its epitope using mass spectrometry and bioinformatics tools. J Clin Endocrinol Metab 95, 4012; (4) James J.A. et al. (2001) Systemic lupus erythematosus in adults is associated with previous Epstein–Barr virus exposure. Arthritis and Rheumatism 44, 1122.]*

cells bearing TCRs that recognize not only these pMHC epitopes but also a structurally similar pMHC derived from a self antigen expressed by healthy host cells. The activation of these autoreactive Th cells results in the licensing of the DC and the generation of Th effector cells that supply help for the activation of naïve autoreactive Tc cells. The activated Tc cells in turn generate CTLs whose effector actions eliminate the pathogen but also damage healthy tissues routinely presenting the self pMHC. In the case of B cell-mediated molecular mimicry, a pathogen may supply B cell epitopes that activate naïve B cells (which receive help from Th effectors). The plasma cell progeny of these B cells produce antibodies that eliminate the pathogen but may also recognize similar B cell epitopes present on non-infected host cells. Healthy tissues are damaged as the antibodies attack host cells displaying the cross-reactive epitope.

The molecular mimicry theory is supported by the identification of T cells and antibodies that respond to both pathogen and self antigens *in vitro*. In addition, extensive work in rodent models has demonstrated that autoimmune symptoms can be triggered under highly controlled conditions by exposing the animals to infectious agents identified as potential sources of cross-reactive epitopes. However, the definitive experiments to prove molecular mimicry in humans are not likely to be conducted because they would involve the highly unethical injection of a pathogen protein or peptide into a healthy individual with the expectation that he/she would develop an autoimmune disease. As a result, it remains unproven that this type of attack is responsible for clinical autoimmune disease in humans.

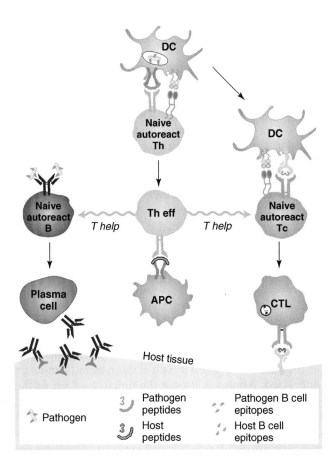

ⵀ▪ **Fig. 19-2**
Cellular Model of Molecular Mimicry

A pathogen can supply B and T cell epitopes that mimic epitopes on host tissues. If a naïve autoreactive B cell is activated by such an epitope, its plasma cell progeny produce antibodies that eliminate the pathogen but may also attack host tissues expressing the cross-reactive self epitopes. Similarly, if a mature, licensed DC presents epitopes composed of pathogen peptides plus MHC that resemble self peptide plus MHC, naïve autoreactive Th and Tc cells recognizing these structures may be activated. Th effector cells and CTLs are generated that help eliminate the pathogen but also attack host tissues.

III. Defects in Immune System Components

i) Abnormalities in Regulatory T Cells

Because regulatory T cells play a key role in maintaining peripheral tolerance, abnormalities in the numbers or functions of these cells can lead to autoimmune disease. Particularly implicated are the Treg cells that primarily use intercellular contacts to suppress the activation of effector T cells. For example, failure of Treg cells to develop in an animal, or artificial depletion of Treg cells that already exist, allows activated autoreactive T cells to escape normal shutdown and to cause autoimmune symptoms. In humans, some autoimmune diseases have been associated with normal numbers of Treg cells of decreased suppressive activity and/or anti-inflammatory cytokine secretion, or with decreased numbers of Treg cells of normal suppressive activity. A very rare disease called *immune dysregulation, polyendocrinopathy, enteropathy X-linked (IPEX)*, which has a significant autoimmune component, is caused by mutations in the Foxp3 gene required for Treg cell development. IPEX patients exhibit an array of endocrine anomalies and the abnormal infiltration of lymphocytes into the skin, pancreas and intestinal mucosa. In most cases, IPEX patients have an absence or severe deficiency of Treg cells. The same is true for *scurfy* mice, which have a natural mutation of the Foxp3 gene. Because IL-2 is crucial for upregulating Foxp3 in Tregs, mutations of the IL-2Rα gene can also cause an IPEX-like disease.

IPEX was introduced in Chapter 15 as a primary immunodeficiency disease with an autoimmune component.

NOTE: Abnormalities in conventional T cells can be associated with autoimmune disease, although the precise mechanisms are unknown. In some cases, the defects may cause the autoreactive T cells to adopt a "pre-primed" state such that they can be activated more easily than normal T cells. For example, T cells from individuals with SLE have been found to display greater density and more rapid activation of cell surface signaling molecules. In other cases, the autoreactive T cells may not respond to the negative T cell regulators CTLA-4 and PD-1. Indeed, PD-1-deficient mice show SLE-like symptoms. Alternatively, effector T cells in an individual may be resistant to Treg cell-mediated suppression or the effects of anti-inflammatory cytokines.

Negative regulation of T cell activation by CTLA-4 and PD-1 was discussed in Chapter 9.

ii) Abnormalities in APCs

At least in animals, some autoimmune diseases are associated with aberrations in the generation and/or function of APCs. In one rodent autoimmunity model, the activated macrophages present showed enhanced powers of migration and pro-inflammatory cytokine secretion that were linked to increased non-specific damage to self tissues. It was hypothesized that this tissue damage exposed otherwise sequestered self antigens and resulted in the unintended activation of autoreactive lymphocytes. In another rodent autoimmunity model, the abnormal DCs present showed reduced expression of MHC class II and B7. Although this deficit meant that naïve autoreactive T cells were less likely to be activated, it also impeded the activation of the regulatory T cells necessary to block autoimmune disease development. Accordingly, other studies have shown that the onset of autoimmune symptoms in susceptible rodents is hastened, rather than suppressed, if DCs are treated such that the interaction between B7 and CD28 is blocked.

iii) Abnormalities in B Cells

As described in Chapter 5, the random process of V(D)J recombination during B cell development in the bone marrow naturally generates some B cells with BCRs recognizing self antigens. The majority of these autoreactive B cells are eliminated by negative selection in the bone marrow during the establishment of B cell central tolerance. Autoreactive B cells that escape negative selection undergo receptor editing in a second effort to generate a non-autoreactive BCR. B cells that fail this step, escape to the periphery, and have their BCRs engaged by self antigen usually undergo Fas-mediated apoptosis or are anergized. Abnormalities in the function of any of the proteins associated with these various B cell maturation checkpoints can result in the survival of autoreactive B cells that would normally have been removed. Although evidence of such abnormalities in human patients is sparse, studies of mouse models have revealed a broad spectrum of mutations affecting B cell survival, development and activation that can give rise to an autoimmune phenotype. For example, SLE-like symptoms develop in mice bearing mutations that decrease the expression or function of Fas or its ligand FasL.

iv) Abnormalities in Cytokine Production

Autoimmune disease has also been linked in animal models to overexpression or deficiency of particular cytokines as well as to defects in cytokine signaling. Such abnormal skewing of cytokine production can alter Th cell differentiation and create an imbalance of Th cell subsets that then leads to further abnormalities in cytokine levels. For example, overproduction of IL-12 by APCs can trigger a susceptible mouse to develop autoimmune disease because the excessive IL-12 increases the number of Th1 cells present. The production of IFNγ and IL-2 by these Th1 cells drives macrophage activation and consequently TNF-mediated inflammation. Similarly, an overproduction of IL-6, TGFβ and/or IL-23 can skew Th differentiation toward Th17 cells and away from Treg cells. The resulting enhanced secretion of IL-17 and IL-6 may then persistently drive inflammation. Indeed, IL-17 and IL-6 are increased in the inflammatory lesions of human MS and RA patients, and in a mouse model of RA, inhibition of IL-17 or IL-6 blocks the development of autoimmune symptoms. The implication is that overproduction of Th1 or Th17 cytokines, or of any cytokine promoting Th1 or Th17 cell differentiation, may facilitate the development of autoimmune disease.

Although IL-1, IL-6 and IL-17 are the cytokines most often implicated in autoimmune disease onset, others can play a role. Some RA patients show increased amounts of IL-15 and chemokines in inflamed tissues. Similarly, the IL-12-related cytokine IL-27 has been associated with glomerulonephritis in human SLE patients. Exactly how these cytokines contribute to these autoimmune diseases is not yet clear. Sometimes the influence of a cytokine on an autoimmune disease can change as the disease progresses. For example, some RA patients treated with anti-TNF antibodies to clear TNF from their inflamed joints initially experience clinical relief but later show signs of exacerbated disease. It seems that the same pro-inflammatory cytokines that promote damage to self tissues also trigger regulatory mechanisms that slowly gain control and shut down autoreactive lymphocytes. This observation has important implications for the management of patients treated with therapies that target cytokines.

> NOTE: Th17 cells produce IL-17 as part of a healthy anti-pathogen adaptive response, but the normal functions of this cytokine have effects that can promote autoimmune disease. For example, IL-17 can
>
> 1) Stimulate osteoclastogenesis, which results in the enhanced generation of macrophage-like cells that destroy bone.
>
> 2) Induce the production of IL-6 and stress response molecules by stromal cells, driving inflammation.
>
> 3) Increase VEGF production, which promotes angiogenesis that facilitates the recruitment of innate leukocytes into joints.
>
> 4) Inhibit the proliferation of intestinal epithelial cells, preventing the healing of tissue damaged by inflammation.
>
> 5) Promote B cell activation, supporting autoantibody production.

v) Inappropriate Engagement of PRRs

Some autoimmune disease symptoms have been linked to abnormal responses to self nucleic acids. As described in Chapter 3, TLR7 and TLR9 are endosomal PRRs that recognize ssRNA and DNA, respectively. In a healthy individual, free self nucleic acids are degraded by serum nucleases before they can make it to the endolysosomes where TLR7 and TLR9 are localized. Thus, the triggering of inflammation by these receptors is limited to when they are engaged by the foreign nucleic acids of invading pathogens. However, if conditions in a host somehow permit self nucleic acids to access TLR7/9, receptor engagement can drive abnormal inflammation that contributes to autoimmune disease symptoms. Immunologists believe that TLR7/9 engagement by self nucleic acids may occur when self RNA or DNA becomes complexed to a host protein that protects it until the complex is captured in a TLR7/9-expressing endolysosome. In addition, if inflammation has already been provoked by pathogen invasion, the intracellular trafficking of TLR7/9 molecules between the ER (site of synthesis) to the endolysosomes (site of function) increases. This increased trafficking offers more opportunities for the nascent TLR7/9 molecules to encounter free self nucleic acids, triggering inappropriate inflammatory signaling. Researchers suspect that this type of mechanism may contribute significantly to the pathology of SLE and RA.

vi) Defects in the Complement System

Individuals with mutations of various complement components often exhibit systemic autoimmune symptoms. Without the function of the C1, C2 or C4 complement components, the level of C3b is suboptimal, and so the clearance of immune complexes is reduced. However, immune complexes involving pathogens accumulate only transiently in complement-deficient individuals because the invader is eventually cleared by other components of the immune system. In contrast, because self antigens by definition are present continuously, immune complexes involving self antigen can persist in complement-deficient individuals. In the absence of C3b to speed their removal, these immune complexes coalesce to form large networks that can become trapped in narrow body channels. These complexes can then trigger systemic autoimmunity in the form of a type III hypersensitivity reaction.

IV. Epitope Spreading

Epitope spreading is thought by many researchers to drive the progression of autoimmune diseases and to contribute to recurrent flare-ups. As introduced in Chapter 16,

epitope spreading is the term describing the phenomenon in which the immune system expands its response beyond the immunodominant epitopes first recognized by T and B cells. In the context of autoimmunity, the original self-reactive response may damage tissues such that new epitopes that were originally "cryptic" (hidden) become accessible. These epitopes would previously have been either totally sequestered from the immune system, or processed by DCs in amounts insufficient to activate T cells, so that lymphocytes recognizing these epitopes would not have been deleted during the establishment of central tolerance nor anergized in the periphery. Upon damage-induced exposure, molecules containing the cryptic epitopes become available for the first time for uptake by DCs. In an inflammatory milieu, autoreactive T and B lymphocytes directed against these epitopes may be activated, expanding the autoreactive attack to additional self tissues where these proteins are expressed. There might then be virtually no limit to the duration of the autoimmune response because new cryptic epitopes might be continually revealed as tissue destruction proceeded.

There is convincing evidence from mouse models to support the epitope spreading theory. In mice prone to autoimmune diabetes, the initial damage to the insulin-producing pancreatic islet cells arises from an autoimmune attack on the islet enzyme *glutamic acid decarboxylase (GAD)*. As the disease progresses, however, lymphocytes directed against different islet cell proteins become activated. Similarly, in mouse models of MS, autoreactive responses are mounted against a series of cryptic CNS epitopes in a reproducible sequence that can be attributed to epitope spreading.

Direct evidence for epitope spreading in human autoimmune diseases is less firm, but one example may occur in *pemphigus (PG)*. Patients with PG exhibit blistering of the mouth and skin due to an autoimmune attack on the desmoglein proteins needed to "glue" epithelial layers together. Blistering in the mouth almost always precedes blistering in the skin because the autoantibodies inflicting the initial damage recognize desmoglein-3, which is expressed in the mouth mucosa. It is not until the attacks on desmoglein-3 expose epitopes on the related skin isoform desmoglein-1 that autoantibodies against this latter protein are produced and skin blistering commences. The autoimmune response thus appears to spread from epitopes of desmoglein-3 to include epitopes of desmoglein-1. Other potential examples of epitope spreading in human autoimmune diseases are discussed with specific disorders in the next section.

C. Examples of Autoimmune Diseases

An alphabetical listing (by acronym) of 20 of the most common autoimmune diseases is presented in **Figure 19-3 (Parts 1 and 2)**. In this figure, the "Dominant sex" has been indicated because some autoimmune diseases occur predominantly in either males or females. "Disease pattern" indicates whether the clinical course of a particular autoimmune disease is considered acute or chronic, and whether it involves relapses and remissions (rather than a steady progression). In the sections that follow, we provide more detailed information on 12 of the better-studied human autoimmune diseases.

I. Systemic Lupus Erythematosus (SLE)

SLE is a systemic autoimmune disease that affects the skin, joints, kidney, lung, heart and brain. A pattern of relapse and remission is common, with an unpredictable frequency of flare-ups. SLE patients exhibit a characteristic rash that gives the disease its name. The rash is red in color (erythematous) and is concentrated on the cheeks such that some patients have a "wolfish" appearance (*lupus* is Latin for "wolf") (**Plate 19-1**). Multiple elements of the immune system may be disrupted in SLE patients such that these individuals are usually vulnerable to opportunistic infections.

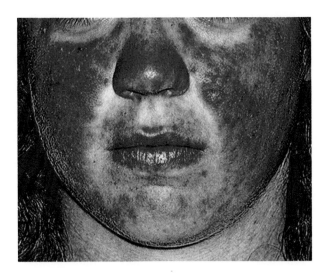

■- **Plate 19-1**
Characteristic Rash of SLE

[From Habif, T.P. Clinical Dermatology, 4th ed. New York: Mosby, 2004, pp. 592–606.]

A signature feature of SLE is the production of high levels of "anti-nuclear" autoantibodies, which are directed against dsDNA and small nuclear proteins. Other autoantibodies in SLE patients recognize non-nuclear entities such as IgG, complement components, and membrane phospholipids. All these autoantibodies form immune complexes that accumulate first in the blood and eventually in the target tissues, triggering damaging inflammation. The production of these autoantibodies has been linked to abnormalities in B cell development and activation. Increased numbers of B cells at all stages of differentiation can be found in the circulation of SLE patients, and these B cells tend to be more sensitive than normal B cells to the effects of cytokines.

With respect to T cells, SLE patients show an increase in Th17 cells in the peripheral blood, along with elevated levels of IL-17 in the circulation. SLE patients also have higher serum levels of IL-10 than do normal individuals. Both IL-17 and IL-10 are known to stimulate B cell activation, proliferation and differentiation. SLE patients have decreased numbers of Treg cells and these cells show reduced functionality. In addition, effector T cells in SLE patients are resistant to Treg cell-mediated suppression. When SLE patients are treated with corticosteroids, their Treg cell numbers slowly rise again and autoimmune disease symptoms are mitigated.

NOTE: SLE may be due in part to a defect in neutrophil NETS. As described in Chapter 3, neutrophils can release their DNA to form an extracellular net in which a pathogen can become entrapped. In a healthy individual, these NETS are eventually degraded by DNAse I once they have served their purpose. However, in at least some SLE patients, this DNAse I function is impaired so that the NETS are not degraded and self DNA persists at unusually high levels in the extracellular environment. This continued presence of self DNA may eventually induce the generation of anti-nuclear antibodies. When such autoantibodies, as well as others directed against protein components of the NETs, bind to their cognate antigens, they can form large immune complexes that can contribute to SLE symptoms.

II. Rheumatoid Arthritis (RA)

RA is caused by an autoimmune attack on antigens expressed in the synovial tissue and cartilage of the joints. In the early phase of RA, the patient typically experiences morning stiffness in the affected joints. As the RA becomes severe, cartilage destruction and bone erosion may deform the digits (**Plate 19-2**), and inflammation of the cervical spine may lead to progressive crippling. Activated macrophages and DCs extravasate from the local blood vessel into the joint and produce large quantities of pro-inflammatory cytokines, particularly TNF. Ligaments, tendons and bones may suffer degradation

Autoimmune disease		Dominant sex	Disease pattern	Tissue(s) affected	Self antigen (if known)
APS	Anti-phospholipid syndrome	F	C	Blood clots at multiple sites	Glycoproteins of prothrombin activator complex
ARF	Acute rheumatic fever	F ≅ M	C	Heart muscle and valves, kidney, CNS, joints	Myosin protein in heart muscle
AS	Ankylosing spondylitis	M	C	Tendons, bones ligaments, joints	Fibrocartilage-derived Ag
CD	Crohn's disease	M = F	C R/R	Walls of colon and small intestine	?
GBS	Guillain–Barré syndrome	M = F	A	Peripheral nerves	Neuronal glycolipids and gangliosides
GD	Graves' disease	F	C R/R	Thyroid gland	TSHR and other thyroid gland proteins
GS	Goodpasture's syndrome	M	A	Kidney, lung	Collagen in basement membrane
HT	Hashimoto's thyroiditis	F	C	Thyroid gland	TSHR and other thyroid gland proteins
ITP	Immune thrombocytopenia purpura	M = F (children) F (adults)	A (children) C (adults)	Platelets	Platelet membrane glycoproteins
MG	Myasthenia gravis	F (30–50 yrs) M (70–80 yrs)	C RR	Muscles	Acetylcholine receptors

Fig. 19-3 (Part 1)
Examples of Human Autoimmune Diseases

Several features of human autoimmune diseases are summarized. "Dominant sex" indicates whether males (M) or females (F) are affected most often. "Disease pattern" describes whether the course of the disease is acute (A), chronic (C), or follows a relapsing/remitting pattern (R/R). (Some autoimmune diseases are manifested in more than one pattern.) "Tissues affected" specifies those tissues that characteristically come under autoimmune attack. "Self antigen" is the molecular target of the autoimmune attack, if known.

Plate 19-2
Joint Deformation in Rheumatoid Arthritis

The digits of patients with RA may show joint deformation due to cartilage destruction and bone erosion. *[Reproduced by permission of Rae Yeung, The Hospital for Sick Children, Toronto.]*

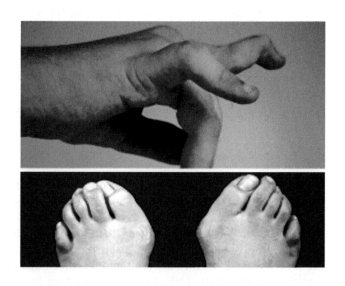

Autoimmune disease		Dominant sex	Disease pattern	Tissue(s) affected	Self antigen (if known)
MS	Multiple sclerosis	F	C R/R	Brain, spinal cord	Myelin basic protein, oligodendrocyte proteins
PG	Pemphigus	M=F	C	Mucosae, skin	Desmoglein proteins in skin and mucosae
PM	Polymyositis	F	C R/R	Muscles	Aminoacyl tRNA synthetases, dsDNA, small nuclear proteins
PS	Psoriasis	M = F	C R/R	Skin	?
RA	Rheumatoid arthritis	F	C	Joints, tendons, ligaments, bone	Synovial and cartilage proteins, IgG
SLE	Systemic lupus erythematosus	F	R/R	Skin, joints, kidney, lung, heart, brain	dsDNA, small nuclear proteins, IgG, complement
SS	Sjögren syndrome	F	C	Exocrine glands (lacrimal, salivary)	Ro52 E3 ligase, ribonuclear proteins
T1DM	Type 1 diabetes mellitus	M = F	C	β-islet cells of pancreas	GAD, insulin, other β-islet cell antigens
TTP	Thrombic thrombocytopenia purpura	F	A R/R	Platelets	von Willebrand factor (clotting)
UC	Ulcerative colitis	M = F	C R/R	Inner wall of colon	?

Fig. 19-3 (Part 2)
Examples of Human Autoimmune Diseases

due to the action of proteases secreted by the activated macrophages. The blood vessels within the inflamed joint soon take on the characteristics of HEVs capable of facilitating the extravasation of lymphocytes. Eventually, low numbers of activated CD4+ Th effectors (particularly Th17 cells) and CD8+ CTLs infiltrate RA joints and produce cytokines such as TNF and IL-17 that help to perpetuate the inflammation. IL-17 is found in the synovial fluid of RA patients and induces the production of IL-6 by synovial cells. While Treg cells are increased in number in RA patients, particularly in the joint synovial fluid, their functionality is decreased.

RA synovial tissues contain GCs that are "ectopic," meaning that these structures have developed in the wrong tissue. The plasma cells in these abnormal GCs produce autoantibodies directed against antigens in the synovial membrane and cartilage. A distinctive (but not exclusive) feature of RA is the presence in a patient's serum of **rheumatoid factor**. Rheumatoid factor is not a single entity but rather a collection of autoantibodies that are directed against the patient's own IgG molecules.

III. Psoriasis (PS)

Psoriasis (from a Greek word meaning "the itch") is an autoimmune disease characterized by reddened skin lesions that develop a covering of silver scaly skin cells. The lesions may occur either in localized patches or cover large sections of the body. Patients generally suffer from persistent itching and flaking of the skin, and often epidermal thickening. The disease can strike at any time, affects males and females equally, and can adopt a relapsing/remitting pattern that does not get worse with time. Flare-ups may have no apparent cause. While not a great threat to physical health, PS can cause the sufferer much discomfort and sometimes emotional or self-esteem problems.

The scaling of PS is due to the presence in the affected region of abnormal keratinocytes with an accelerated growth rate. The root cause of the abnormality is unknown, but most researchers suspect that activated autoreactive T cells may be releasing cytokines that stimulate keratinocyte growth. Skin biopsies from PS patients have revealed the presence of Th17 and Tc17 effector cells as well as high levels of IFNγ and IL-17. Abnormal regulation of TNF and IL-1 may also contribute to PS development. Some PS patients have autoantibodies directed against keratinocyte proteins or IL-1, but the significance of these antibodies to the disease is unclear.

IV. Acute Rheumatic Fever (ARF)

Acute rheumatic fever primarily affects cells in the heart muscle, heart valves, kidney and CNS but also compromises the joints. ARF follows a relapsing/remitting pattern with major clinical signs of fever, a distinctive rash, carditis, arthritis and neurological effects. Some patients present with very rapid heartbeat or even acute cardiac failure, giving ARF a relatively high mortality rate. In many cases, the autoimmune symptoms of ARF appear 2–6 weeks after infection with certain virulent strains of group A Streptococcal bacteria, including those causing "strep throat." These bacteria express a cell wall protein called M antigen that researchers believe resembles an epitope found in the human heart protein *myosin* (refer to **Fig. 19-1**). Thus, ARF may be established most often by molecular mimicry. In the presence of large quantities of M antigen, autoreactive lymphocytes that are specific for human heart myosin epitopes (and that are normally held in check by peripheral tolerance mechanisms) may break tolerance and become activated by epitopes of the bacterial M antigen. Once activated, these lymphocytes then attack heart tissues as well as Streptococci expressing the M antigen epitope. Indeed, antibodies recognizing the M protein are present at high levels in patients with ARF, and subsets of T cells isolated from these patients recognize pMHCs containing M protein peptides *in vitro*. Furthermore, relapses of ARF are often triggered by a subsequent infection with Streptococcus.

Rheumatic fever is another example of a disease in which epitope spreading contributes to its persistence. Once the inflammation and tissue damage associated with this autoimmune disease become chronic, the response is dominated by lymphocytes directed against previously hidden epitopes in collagen and laminin, as opposed to the original myosin epitopes.

V. Type 1 Diabetes Mellitus (T1DM)

Diabetes results when the body either does not produce enough insulin or its cells cannot respond to it normally. As a consequence, glucose cannot get into cells properly and blood sugar levels rise. Symptoms of diabetes include excessive hunger and thirst, increased urination, weight loss, fatigue and blurred vision. If blood glucose levels rise too far, the patient can fall into a life-threatening diabetic coma. When diabetes is controlled inadequately for an extended period, the elevated sugar in the blood damages the vascular endothelium in multiple tissues. This damage can result in blindness, potentially fatal heart attacks and strokes, nerve deterioration, kidney failure, or tissue necrosis necessitating limb amputation. There are two major types of diabetes mellitus: *type 1 diabetes mellitus (T1DM)*, in which the pancreas does not produce enough insulin, and *type 2 diabetes mellitus (T2DM)*, in which tissue cells become resistant to insulin and the patient exhibits glucose intolerance.

T1DM, which accounts for about 20% of diabetic patients, has long been considered an autoimmune disease caused by lymphocyte attacks on the insulin-producing β-islet cells of the pancreas. The pancreas is invaded by numerous leukocytes, including macrophages, B cells, and CD4+ and CD8+ T cells. Antibodies and CTLs directed against various β-islet cell antigens combine forces to destroy the islets, leading to a failure in insulin production. Antibodies and T cells recognizing epitopes of the β-islet enzyme GAD or the tyrosine phosphatase IA-2 are often found in T1DM patients, as are antibodies recognizing insulin. While T1DM patients have normal numbers of Treg cells, the functionality of these cells is decreased. In addition, effector T cells in these patients are resistant to Treg cell–mediated suppression.

Until recently, T2DM was not considered an autoimmune disease but rather a consequence of obesity. A hallmark of T2DM is chronic inflammation in the visceral adipose tissue that is characterized by the infiltration of macrophages and particular subsets of Th cells and CTLs. These cells produce the cytokines driving insulin resistance. However, researchers studying mouse models of T2DM have determined that B cells play a key role in the development of this disease because they produce pathogenic IgG antibodies and promote the activation of the infiltrating inflammatory macrophages and T cells. Indeed, obesity-prone mouse strains engineered to lack B cells do not develop T2DM even after they become obese. Similarly, obesity-prone animals that possess B cells but are treated with anti-CD20 mAb develop less severe autoimmune disease. These findings have a parallel in humans, in that autoantibodies have recently been detected in some T2DM patients. All these factors suggest that T2DM may be a genuine autoimmune disease in at least some patients, albeit a disorder whose incidence can be greatly reduced by adopting a healthy lifestyle.

VI. Multiple Sclerosis (MS)

Multiple sclerosis is an autoimmune disease that primarily affects the brain and spinal cord. The autoimmune attack is directed against the myelin sheath surrounding the nerve axons, as well as against cells called *oligodendrocytes* that make the myelin. Axon function may be lost due to demyelination, ultimately leading to the inhibition of nerve impulse transmission. MS patients thus suffer from widespread motor weakness and sensory impairments. The nerve fibers in the brains and spinal cords of MS patients show sites of demyelination called *plaques* where normal tissue has undergone hardening (*sclerosis*) (**Plate 19-3**). The disease course of MS varies widely: some patients show very few symptoms, whereas others are severely disabled within months. A relapsing/remitting pattern that persists for years is common, but some patients experience chronic, progressive disability.

MS is thought to stem from the activation of autoreactive T cells that can recognize pMHCs involving peptides derived from proteins of the myelin sheath. Three such proteins are *myelin basic protein (MBP)*, *proteolipid protein (PLP)* and *myelin oligodendrocyte glycoprotein (MOG)*. Many scientists believe that at least some cases of MS may result from molecular mimicry in which an epitope of an EBV protein closely resembles an epitope in MBP (refer to **Fig. 19-1**). Epitope spreading also contributes, in that it appears to shape the course of MS in many patients. Autoreactive responses

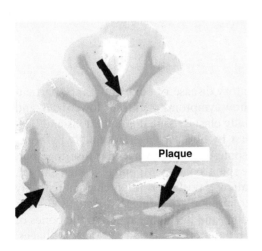

Plate 19-3
Plaque Formation in Multiple Sclerosis

The arrows indicate areas of demyelination (plaques) in the white matter of the brain. *[Reproduced from McAlpine's Multiple Sclerosis (Fourth Edition), 2006, pg. 557–599, Chapter 12, "The pathology of multiple sclerosis." Fig. 12-8A. Hans Lassmann, Hartmut Wekerle.]*

Plaque

to the dominant myelin epitope occur early in the disease but fall off later in favor of increasing responses to cryptic myelin epitopes.

No matter how they are activated, autoreactive T cells in MS patients soon generate effector T cells that apparently upregulate certain adhesion molecules that allow them to cross the blood-brain barrier, along with activated B cells and other inflammatory cells. Accordingly, elevated numbers of Th1 and Th17 cells have been found in the cerebrospinal fluid of MS patients. The IFNγ and IL-17 secreted by these cells likely make a major contribution to MS-associated inflammation. Although Treg cells are normal in number and even increased in the CNS of MS patients, they are defective in function.

Autoreactive CTLs responding to various peptides presented by oligodendrocytes produce cytokines and proteases that collectively damage the myelin sheath. These cytokines induce **microglia** (macrophage-like cells resident in the brain) and infiltrating neutrophils to secrete additional pro-inflammatory mediators and cytokines that further contribute to myelin destruction. Nitric oxide produced by these phagocytes blocks nerve conduction pathways and contributes to structural damage. Sites of demyelination have been found to contain elevated levels of complement products as well as autoantibodies directed against MBP, PLP or MOG.

Remission of MS symptoms is thought to occur when anti-inflammatory cytokines and growth factors produced by cells in the inflammatory infiltrate offset the autoimmune attack and allow the oligodendrocytes to remyelinate the damaged nerves. However, over time, the build-up of scar tissue around the nerves, coupled with the accumulating assaults on the oligodendrocytes, may prevent remyelination so that the patient is not able to fully recover from an MS episode. If the sclerosis becomes severe, the patient enters the chronic progressive stage of MS.

NOTE: In 2009, controversy in the MS field was sparked by a study suggesting that MS was associated with a condition called "chronic cerebrospinal venous insufficiency" (CCVI). The study's authors hypothesized that MS, rather than being an autoimmune disease, was the result of abnormal constrictions in blood vessels draining the brain. Within a very short time, clinics around the world began offering MS patients a procedure involving balloon angioplasty, which is commonly used to open blocked cardiac vessels in heart attack patients. This treatment to open vessels in the necks of MS patients became known as "liberation therapy," and the Internet was soon flooded with articles written by proponents and opponents of the procedure. While individual MS patients have anecdotally sung the praises of liberation therapy, proper clinical studies have shown no benefit that appears to warrant the health risks involved with an invasive procedure of this nature. Most government agencies have therefore not granted approval for liberation therapy. As a result, many desperate MS patients have traveled abroad to obtain the procedure, sometimes submitting to dubious treatment conditions and almost always incurring great personal expense. Controversies such as that surrounding liberation therapy highlight the importance of carefully controlled research trials, as well as the need for patients to exercise caution and critical thinking when presented with any new (but unproven) therapy.

VII. Ankylosing Spondylitis (AS)

Ankylosing spondylitis is a chronic inflammatory disease of bone and joints, particularly of the spine. Affected individuals first show symptoms of chronic lower back and hip pain that can persist for years. The normally elastic tissue in the tendons and ligaments of the joint is eroded and replaced first by fibrocartilage and finally by bone. When this bone replacement occurs in the spine, the lower vertebrae fuse together, causing irreversible and serious damage to the spinal column. Mobility is compromised, chest expansion may be reduced, and the patient's posture is often characteristically altered. The target tissue that undergoes autoimmune attack in AS appears to be the fibrocartilage supporting the sites where tendons and ligaments attach to the bones of the joints. Affected AS joints contain elevated numbers of plasma cells, macrophages, lymphocytes (particularly CD8$^+$ T cells) and mast cells. The serum of AS patients often shows elevated IgA, IL-10 and acute phase proteins but, curiously, not autoantibodies.

VIII. Autoimmune Thyroiditis: Graves' Disease (GD) and Hashimoto's Thyroiditis (HT)

There are two major types of autoimmune thyroiditis: *Graves' disease (GD)* and *Hashimoto's thyroiditis (HT)*. In GD, the autoimmune attack on the thyroid gland causes it to become hyperactive and overproduce the hormone *thyroxine*, which controls the body's metabolic rate. In HT, the autoreactive response against the thyroid gland results in insufficient thyroxine production. The putative pathogeneses of GD and HT are illustrated in **Figure 19-4**.

In GD patients, autoantibodies directed against *thyroid-stimulating hormone receptor (TSHR)* overstimulate its signaling, resulting in overproduction of thyroxine and clinical hyperthyroidism. GD patients exhibit hand tremors, insomnia, weight loss and rapid heart beat. As well, the thyroid glands of GD patients are infiltrated by T cells that respond to pMHCs derived from TSHR and other thyroid autoantigens. Sometimes autoreactive T cells of unknown specificity infiltrate the intraocular muscles and orbital tissues around the eyes of GD patients. This infiltration induces inflammation that causes these tissues to swell, resulting in a distinctive bulging of the eyes known as *Graves' ophthalmopathy*.

In HT patients, the autoantibodies bind to TSHR (in a region different from that targeted in GD) and block TSHR signaling. Insufficient thyroxine is produced, which results in clinical hypothyroidism. HT patients complain of depression, fatigue, weight gain, and dry rough skin. In severe cases, the thyroid gland becomes greatly enlarged in size (*goiter*) as it attempts to make large amounts of thyroxine precursors to compensate for the thyroxine deficiency. In some HT cases, NK cells and autoreactive CTLs infiltrate the thyroid gland, facilitating gland destruction via ADCC and perforin/granzyme-mediated cytotoxicity.

IX. Myasthenia Gravis (MG)

Myasthenia gravis is a rare autoimmune disease that results in severe and specific muscle weakness. Onset may be early or late, and clinical presentation is heterogeneous. The disease is usually chronic in character, and a relapsing/remitting pattern is common. Weakness in the facial muscles may cause a patient to have difficulty talking and swallowing, whereas respiratory muscle weakness can lead to life-threatening breathing difficulties. MG is caused by autoantibodies that recognize an *acetylcholine receptor* expressed on muscle cells. Acetylcholine is a chemical messenger necessary for the transmission of nerve signals. The binding of the autoantibodies to the acetylcholine receptor prevents the binding of acetylcholine and thus interferes with the transmission of electrical impulses across the neuromuscular junction between a neuron and a muscle cell. Without transmission of this signal, the muscle is unable to contract.

X. Guillain–Barré Syndrome (GBS)

Guillain–Barré syndrome is a rare disease resulting from an acute autoimmune attack on the peripheral nerves. Autoantibodies directed against gangliosides and glycolipids attack neurons in the peripheral nerves, inducing acute inflammation that demyelinates the nerve fibers and reduces electrical impulse transmission. The patient may first notice tingling in the feet or hands that rapidly (within hours) spreads up or down the body. Blurred vision, clumsiness, fainting and swallowing difficulties may occur. Within days or weeks, the muscle weakness can progress to extensive paralysis. In the worst cases, the respiratory muscles are paralyzed, and the patient has to be put on a respirator. Serum TNF is usually elevated, and lymphocytes and macrophages infiltrate the peripheral nerves. Despite the serious clinical picture of GBS, the vast majority of

Fig. 19-4 ─────────
Putative Pathogenesis of Autoimmune Thyroiditis

(A) Top panel: Overview of the structure of the normal thyroid gland, showing the thyroid follicles composed of thyroid cells and colloid. Bottom panel: Close-up view of the microenvironment surrounding several thyroid follicle cells. In a normal individual, thyroid-stimulating hormone (TSH) produced by the anterior pituitary gland binds to thyroid-stimulating hormone receptors (TSHR) expressed by thyroid follicular cells. These cells routinely make the inactive precursor thyroglobulin and store it in the colloid of the thyroid follicles, where it is iodinated. In response to TSHR signaling, the iodinated thyroglobulin is imported back into the thyroid follicular cells, followed by cleavage to yield thyroxine. Thyroxine is released into the local blood vessel and then the circulation to control the body's metabolic rate. **(B)** In GD patients, autoantibodies overstimulate TSHR signaling, resulting in overproduction of thyroxine and clinical hyperthyroidism.
(C) In HT patients, autoantibodies inhibit TSHR signaling, resulting in a lack of thyroxine and clinical hypothyroidism.

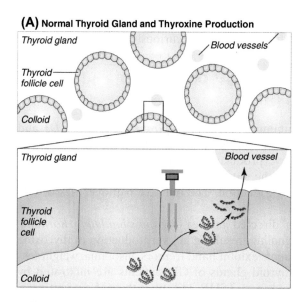

(A) Normal Thyroid Gland and Thyroxine Production

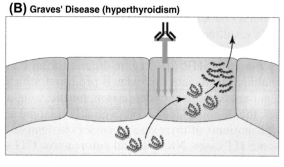

(B) Graves' Disease (hyperthyroidism)

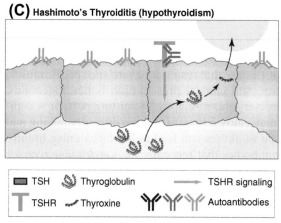

(C) Hashimoto's Thyroiditis (hypothyroidism)

patients recover without treatment. GBS may be a case of molecular mimicry because, in two-thirds of cases, onset occurs following vaccination against or recovery from a gastrointestinal or respiratory infection, in particular with *Campylobacter jejuni*. Antibodies that recognize both the LPS of *C. jejuni* and human nerve gangliosides have been isolated from GBS patients.

XI. Sjögren Syndrome (SS)

Sjögren syndrome is a systemic autoimmune disease that primarily strikes post-menopausal women. Autoantibodies produced in these patients attack exocrine glands such as the lacrimal glands in the eyes and the salivary glands in the mouth, causing these structures to stop production of their usual secretions. As a result, the signature clinical signs of SS are a constellation of complaints known as *sicca symptoms* (sicca, meaning "dry"). These include dry eyes, leading to blurry vision; dry mouth, leading to swallowing difficulties; and dry nose and skin. Patients may also exhibit RA-like arthritis or SLE-like symptoms. Although SS is a relatively benign disorder in most patients, 5% present with lymphoproliferation that can progress to lymphoma.

Infiltrating activated T cells can be found in the lacrimal and salivary glands of SS patients, and antiribonucleoprotein antibodies abound in the serum. Antithyroid antibodies and rheumatoid factor may also be present. The most recent evidence suggests that Ro52, an enzyme crucial for protein ubiquitination and the regulation of apoptosis, may be an important autoantigen in SS.

XII. Inflammatory Bowel Disease (IBD): Crohn's Disease (CD) and Ulcerative Colitis (UC)

Inflammatory bowel disease is actually a family of autoimmune diseases affecting the large and/or small intestines. The two major types of IBD are *Crohn's disease (CD)* and *ulcerative colitis (UC)*. CD affects all layers of the wall of the colon and the small intestine, whereas UC affects only the innermost layer of the wall of the colon. In CD, there may be healthy patches of bowel interspersed with diseased patches, whereas in UC, the entire colonic lining is usually affected. Both CD and UC are characterized by chronic, relapsing/remitting inflammation of the intestinal lining that causes loss of appetite, weight loss, fatigue, diarrhea and fever. The scarring and swelling associated with IBD can result in obstructions that lead to abdominal cramps and vomiting. Both CD and UC patients have an increased risk of later developing colon cancer. Susceptibility to IBD has a strong genetic component, but environmental factors also appear to be involved, particularly exposure to industrial pollution or tobacco smoke.

Elevated levels of TNF and IL-17 are frequently present in the mucosal lesions of CD patients, and the blood and intestinal lamina propria of these patients contain significant numbers of Th1 and Th17 cells. In contrast, UC is associated with elevated Th2 cytokines. Numbers of fully functional Treg cells in the blood are decreased in UC patients but normal in CD patients. It is unclear why the Treg cells in CD patients cannot prevent the autoimmune disease symptoms. Although autoantibodies are sometimes detected in IBD patients, B cells do not appear to play a large part in this family of diseases.

IBD is thought to arise initially from inappropriate inflammatory responses to unknown antigens furnished by commensal organisms normally present in the gut (and thus considered "self"). Accordingly, many immunologists are starting to consider the IBD family of diseases as entities that straddle the line between classical autoimmune diseases and autoinflammatory disorders (see Section F of this chapter). While some cases of CD clearly involve NOD2 mutations and thus have a genetic component, others appear to be triggered mainly by an imbalance in the microbiota that leads to persistent inflammation. It is this inflammation that may slowly expose self antigens which later come under attack by autoreactive T cells.

> Irritable bowel syndrome (IBS) is not a member of the IBD family. IBS is a milder, non-autoimmune disease of colonic muscle contraction and is not characterized by intestinal inflammation.

> The overall health costs associated with IBD in the U.S. have been estimated at more than $1.7 billion.

NOTE: Patients with IBD require close monitoring of their disorder. Researchers have recently discovered that certain DAMPs released by infiltrating neutrophils and other phagocytes are good markers of the intensity of intestinal inflammation. For example, measurement of levels of various members of the S100 family of stress proteins is now coming into more frequent use as a disease assessment tool.

D. Determinants of Autoimmune Diseases

Earlier in this chapter, we described the basic events that must occur before an autoimmune disease can develop. In this section, we discuss several variables or *determinants* that influence whether or not any of these events occur. First, the genetic make-up

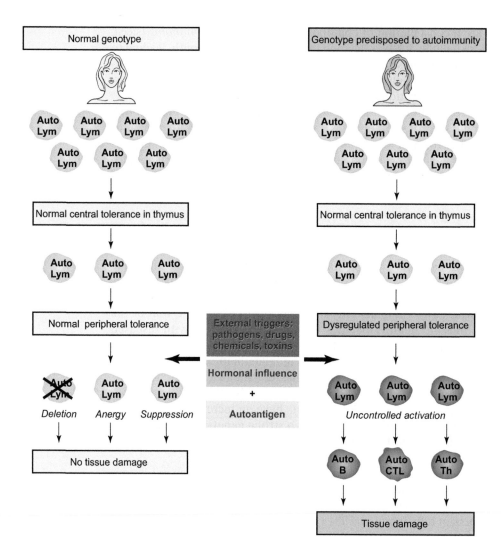

Fig. 19-5
Determinants Influencing the Fate of Autoreactive Lymphocytes

In an individual with a normal genotype, most autoreactive lymphocytes are eliminated by negative selection during the establishment of central tolerance. Those autoreactive cells that escape into the periphery are kept under control by the mechanisms of peripheral tolerance. An encounter with autoantigen, even in the presence of one of the indicated triggers, almost always results in clonal deletion or anergization. In addition, regulatory T cells may act to suppress autoreactive lymphocytes. However, in an individual with an autoimmune disease-prone genotype, normal central tolerance is usually in place, but the mechanisms of peripheral tolerance are compromised. Autoreactive lymphocytes that escape to the periphery and encounter autoantigen in the presence of a trigger are activated uncontrollably and may resist suppression by regulatory T cells. The effector functions of these autoreactive cells then can cause tissue damage and clinical symptoms that constitute "autoimmune disease." *[Adapted from Bach J.F. (2002). The effect of infections on susceptibility to autoimmune and allergic diseases. New England Journal of Medicine 347, 911–920.]*

of an individual plays a critical role in determining his/her susceptibility to autoimmune disease. Allelic variation in genes encoding proteins important for immune system function can result in the failure of peripheral tolerance mechanisms. However, mere possession of an autoimmune-prone genotype is usually not enough for overt disease to appear. For many autoimmune diseases, the actual manifestation is linked to an external trigger such as an encounter with a particular pathogen, drug, chemical or toxin. Alternatively, a change in hormonal status may serve as an internal trigger of autoimmunity in genetically predisposed individuals. An overview of how genetics, external triggers and hormonal influences can set the stage for an autoimmune disease appears in **Figure 19-5**.

The involvement of multiple determinants in the occurrence of autoimmunity makes it challenging to elucidate how these diseases develop in humans. Unlike

Box 19-2 Population Trends in Human Autoimmune Diseases

Many autoimmune diseases appear to vary in incidence by region or ethnic group. For example, Southern European and Asian countries have a lower incidence of T1DM and MS than do Northern European countries, and SLE incidence is higher in natives of African and Caribbean countries than in Europeans and North Americans. Within the U.S., African-Americans and Hispanics have a higher risk of developing SLE than do Caucasians but a lower risk of T1DM and MS. Such variation may sometimes be due to the uneven prevalence in particular ethnic groups of a gene linked to a given autoimmune disease. Similarly, differences in diet or in the occurrence of a triggering pathogen or chemical agent due to geographic factors may influence the frequency of an autoimmune disease. In other cases, however, the reasons for variation in autoimmune disease incidence by region or ethnic group are not obvious.

It is also curious that the incidence of certain prototypical autoimmune diseases, including CD, T1DM and MS, has rapidly escalated in developed countries since the 1970s. Some researchers believe that this increase may be related to the hygiene hypothesis described in Chapter 18. To recap, this theory posits that excessive zeal in maintaining sanitary living conditions may skew the balance of pro- and anti-inflammatory influences (particularly relative numbers of Th1, Th2 and Th17 cells) in a child's developing immune system such that a predisposition to allergy or autoimmunity arises. In this light, the observation that SLE incidence is much higher in Black Americans than in West Africans (who share ethnicity but not environment) makes sense. However, contradictory evidence also exists, and although some animal studies have supported the hygiene hypothesis with respect to autoimmunity, others have not. The rise in autoimmune disease in the developed world thus remains a conundrum.

laboratory mice, humans are genetically outbred and subject to a diverse array of external conditions. Sometimes a trend in a particular autoimmune disease can be observed if the population being studied is restricted by geographical location or ethnic background, as outlined in **Box 19-2**.

I. Genetic Predisposition

The most convincing evidence that predisposition to autoimmune disease depends on genetics comes from studies of identical twins. When one identical twin develops autoimmune symptoms, there is a 12–60% chance that the other twin will develop the same ailment. In contrast, the chance of the same autoimmune disease developing in two fraternal twins is only about 5%. Inherited predisposition to autoimmunity is also seen in families without twins. Strikingly, in these cases, the actual form of the autoimmunity may vary in different members of the same family. It is also very common for patients suffering from one type of autoimmune disease to later develop another autoimmune disease. These observations suggest that what is often inherited is a general susceptibility to autoimmunity rather than a single defect associated with a particular autoimmune disease.

The number and identity of the genes involved in human predisposition to autoimmunity are not precisely known, and different individuals have different collections of predisposing alleles. Moreover, not every human who has a particular predisposing allele experiences autoimmune disease even when exposed to the same environmental influences, and the course of a given autoimmune disease can vary greatly from person to person. For example, the increased frequency of T1DM in close relatives of a diabetic individual is consistent with a genetic basis for this disorder; however, the disease can skip whole generations.

Due to their direct involvement in T cell responses, the most important genes conferring predisposition to autoimmunity are the HLA genes (**Table 19-1**). These genes were also the first to be identified in early studies of the linkage between specific genes and particular autoimmune diseases. Perhaps one of the best illustrations of HLA association with human autoimmunity can be found in AS. Over 90% of Caucasians with AS express one of a small collection of unusual alleles belonging to the HLA-B27 family. It should be noted, however, that an association between an autoimmune disease and a particular HLA molecule may not hold true for all populations. For example, an allele called HLA-DRB1*04 is tightly linked to RA development in Caucasians but

TABLE 19-1	Examples of HLA Alleles Associated with Autoimmune Diseases		
Autoimmune Disease	**HLA Allele**	**Autoimmune Disease**	**HLA Allele**
Ankylosing spondylitis	HLA-B27	Myasthenia gravis	HLA-B7, B8, DR2, DR3, DR7
Crohn's disease	HLA-DRB1	Polymyositis	HLA-DR3
Goodpasture's syndrome	HLA-DR2	Rheumatoid arthritis	HLA-DR4
Graves' disease	HLA-DR3	Sjögren syndrome	HLA-DR3
Hashimoto's thyroiditis	HLA-DR3, DR5	Systemic lupus erythematosus	HLA-DR2, DR3
Multiple sclerosis	HLA-DR2	Type 1 diabetes mellitus	HLA-DR3, DR4, DQ2, DQ8

not in Hispanic or African populations. To complicate matters further, some HLA proteins confer protection against the development of autoimmune disease even when another HLA molecule linked to increased autoimmune susceptibility is present. For example, HLA-DQ8 is strongly associated with the development of T1DM. However, if an HLA-DQ8-expressing individual also expresses HLA-DQ6, T1DM is far less likely to develop. How this protection works is not known.

A non-HLA gene of relevance to several autoimmune diseases is Runx-1, which encodes a subunit of a transcription factor called CBF. CBF binds to a DNA motif (called the "Runx-1 binding site") that occurs throughout the human genome. Runx-1 binding sites that have undergone mutations have been pinpointed as autoimmune susceptibility loci. For example, mutations in a Runx-1 binding site on chromosome 2 are associated with SLE, and mutations of a Runx-1 binding site on chromosome 17 have been linked to RA. Sometimes the gene(s) associated with an autoimmune disease are unknown, but a region of a chromosome may have been identified as containing a susceptibility locus for that disorder. For example, AS is associated with specific genetic susceptibility regions that have been mapped to chromosomes 1, 2, 6, 9, 10, 16 and 19. Researchers are working to isolate the culprit genes in these regions.

New links between human autoimmunity and non-HLA genes have recently been found using *genome-wide association studies (GWAS)*. These approaches allow the identification of large numbers of *single nucleotide polymorphisms (SNPs)* associated with a given disease in which multiple genes play complex roles (see Note below). In the case of autoimmune diseases, many genes identified by GWAS and SNP analyses have been directly related to immune system functions. Hundreds of allelic variants of non-HLA genes involved in inflammation, pathogen recognition, complement activation/function, autophagy, antigen clearance, or T cell differentiation or regulation are now linked to autoimmunity (**Table 19-2**). For example, SNPs linked to SLE development occur in the CR1, HSP70 and TNF genes (among others). Similarly, in human

TABLE 19-2	Examples of Non-HLA Genes Associated with Autoimmune Diseases
Autoimmune Disease	**Gene**
Ankylosing spondylitis	Unknown genes on chromosomes 1, 2, 6, 9, 10, 16 and 19
Crohn's disease	NOD2, cation transporters
Graves' disease	CTLA-4
Guillain–Barré syndrome	FcγRII, TNF
Hashimoto's thyroiditis	CTLA-4
Rheumatoid arthritis	IL-1, IL-1R, MBL, TNFR2, ICAM-1, IFNγ, FcRs
Systemic lupus erythematosus	CR1, HSP70, FcγRII, FcγRIII, IL-6, IL-10, TNF, C1, C4, TNFR2
Type 1 diabetes mellitus	Insulin, CTLA-4, IL-12

RA patients, particular alleles of the IL-1, IL-1R and IFNγ genes occur at increased frequency. Polymorphisms of TLR4 and TLR9 are associated with CD, as is a particular mutation of NOD2. T1DM has been linked to a polymorphism of TLR2.

NOTE: Genome-wide association studies rely on improved DNA sequencing technology that allows the comparison of mass numbers of DNA sequences at once. To perform a GWAS, researchers obtain DNA samples from two large groups of individuals: a well-documented control population and a population of interest whose members share a particular phenotypic trait, such as an autoimmune disease. These DNA samples are subjected to automated gene sequencing such that several hundred thousand SNPs may be detected scattered throughout the collective genomes of the population of interest. Any allelic variants occurring at increased incidence among large numbers of persons with the same disease may be causally linked to disease development. A crucial result that has emerged from GWAS efforts is that a significant number of the genetic alterations associated with a particular disease are found in non-protein coding areas of the genome; these regions are likely involved in regulating gene expression. Another key finding has been that the presence in an individual's genome of even several of these variants is no guarantee that he/she will have the disorder in question, emphasizing the importance of external factors and triggers in disease onset.

Despite the challenge of sorting through the large amounts of data generated by GWAS, this approach is thought to have distinct advantages over previous methods. In traditional genetic association studies, researchers use what they know about the biology of a given disease to identify a small number of candidate genes which might be associated with disease development. To support this association, the researchers look for allelic variants of these genes that are over-represented in the patient population relative to the general population. However, because the biology of a complicated disease is often not completely understood, only a limited number of gene disease associations may be possible using traditional methods, and some of them may be misleading. With GWAS, disease occurrence can be linked to specific genetic alterations in the absence of any pre-existing functional correlation between a given gene and a given disease. In other words, no candidate gene has to be identified before the study can begin, and no prior knowledge of the biochemical pathway associated with the disease is required.

Despite their profusion, the biological significance of the myriad autoimmune disease-linked genetic variations identified to date is unclear. Especially puzzling is the fact that inheritance of a putative disease-associated allele does not lead to autoimmune symptoms as often as might be expected. Indeed, it has been difficult to prove that any of these polymorphisms actively contribute to the development of an autoimmune disease in humans. However, a rare exception is described in the Focus on Relevant Research box below.

II. External Triggers

External stimuli, including chemical agents and pathogens, show tantalizing links to autoimmune disease onset or flare-ups in both animal models and humans. However, convincing proof that an encounter with an environmental stimulus actually triggers the initial onset of human autoimmunity is lacking.

i) Chemical Agents

Certain chemical and pharmaceutical agents appear to precipitate particular autoimmune symptoms. The use of hair dyes, smoking, glue sniffing, and exposure to silica dust or other toxins have been linked to episodes of RA, SLE, HT and GD. Heavy exposure to paint thinners has been blamed for some cases of MS, whereas a weak association has been noted between pesticide exposure and RA. Excess iodine intake may be a triggering factor for some cases of HT and GD, and an episode of allergic rhinitis may bring on GD in certain patients. Exposure to UV radiation has been linked to flare-ups of SLE. Other autoimmune diseases may initiate in response to drug treatment. For example,

"Human ITCH E3 Ubiquitin Ligase Deficiency Causes Syndromic Multisystem Autoimmune Disease" by Lohr, N.J., Molleston, J.P., Strauss, K.A., Torres-Martinez, W., Sherman, E.A., Squires, R.H., Rider, N.L. Chikwava, K.R., Cummings, O.W., Morton, D.H., & Puffenberger, E.G. (2010) *American Journal of Human Genetics* 86, 447–453.

The discovery of genes responsible for autoimmune diseases is difficult because (1) unrelated humans are highly variable genetically; (2) the immune system is very complex; and (3) it is impractical and unethical to "knock out" genes in human subjects as is routinely done to create mutant mice. Given these issues, the best way to identify a gene associated with an autoimmune disease is to have access to large numbers of related humans suffering from the same autoimmune disease of interest, and a suitable animal model of that disease. In this article, Lohr *et al.* use just this approach to pinpoint the human gene responsible for Syndromic Multisystem Autoimmune Disease (SMAD). By comparing DNA sequences from several patients in a large family with a history of SMAD, these researchers identified a single base insertion in the ITCH gene (as shown here in Figure 4, panel C from the paper). The ITCH gene encodes a ubiquitin ligase important for the normal ubiquitination of proteins destined for proteasomal processing and antigen presentation. The identified mutation causes a frameshift that leads to the introduction of a downstream stop codon (TGA). This codon blocks the translation of a complete ITCH protein, resulting in the loss of ITCH function and consequently the development of SMAD. Bolstering this link between ITCH and the autoimmunity observed in this family, gene-targeted mice deficient for murine ITCH showed parallel physiological deficits.

The genetic linkage analysis described in this report was possible only because the researchers had access to DNA samples from a highly inbred population of Old World Amish patients for which extensive and accurate family pedigrees were available. Such strategies coupling epidemiological and molecular tools have proven powerful in the hunt for disease-related genes.

```
                    ITCH codon 132
CTT GGA GGT GAC AAA GAG CCA ACA GAG ACA ATA GGA GAC TTG TCA ATT TGT CTT GAT GGG CTA
Leu Gly Gly Asp Lys Glu Pro Thr Glu Thr Ile Gly Asp Leu Ser Ile Cys Leu Asp Gly Leu
                                         c.394_395insA
CTT GGA GGT GAC AAA GAG CCA ACA GAG ACA AAT AGG AGA CTT GTC AAT TTG TCT TGA TGG GCT
Leu Gly Gly Asp Lys Glu Pro Thr Glu Thr Asn Arg Arg Leu Val Asn Leu Ser ***
```

thiol-containing drugs and sulfonamide derivatives, as well as certain antibiotics and non-steroidal anti-inflammatory drugs, appear to trigger the onset of PG. Anti-TNF agents prescribed as anti-inflammatory medications have also been associated with the appearance of antinuclear antibodies. Several pharmaceuticals have been linked to the initiation of an SLE-like syndrome in a small proportion of patients, and the administration of other drugs prescribed to combat high blood pressure or irregular heartbeat has been followed by the onset of SLE-like symptoms. Fortunately, in these cases, the disease is often mild and of short duration, and usually resolves when the drug is withdrawn.

ii) Infections

As mentioned earlier in this chapter, infections with certain pathogens appear to provoke the initiation or intensification of some autoimmune diseases in genetically predisposed individuals. This connection can be attributed to molecular mimicry or chronic inflammation, or both. Severe bacterial or viral infections frequently seem to trigger an increase in autoreactive antibodies or CTLs that can lead to a flare-up of quiescent autoimmune disease or an exacerbation of existing symptoms.

With respect to viruses, flare-ups of SLE have been associated with infections by a wide range of these pathogens, including EBV, CMV and human parvovirus. The development of GBS often follows infection with HSV-1, CMV or EBV, and infections with respiratory viruses may trigger GD or acute *immunopathic thrombocytopenic purpura* (*ITP*; refer to **Fig. 19-3**). The onset of acute ITP may also be preceded by varicella infection. The vacuolar appearance of muscle cells in patients with *polymyositis* (*PM*; refer to **Fig. 19-3**) suggests that a virus (possibly EBV) may trigger this autoimmune disease. Some researchers believe that, especially in children, T1DM is triggered by a virus (particularly rubella virus or Coxsackie B4 virus); however, this theory is controversial. Whether MS is triggered by a viral infection is also unclear. About 30% of new MS cases emerge following infection with some kind of virus, and relapses in some MS patients seem to be associated with adenovirus or GI infections. As stated earlier, molecular mimicry may exist between human MBP and an EBV protein, but no one virus has been consistently found in MS patients. EBV infection has also been linked to RA, SS and PS, while HCV infection is associated with CD, PS and *anti-phospholipid syndrome* (*APS;* refer to **Fig. 19-3**).

Infections with various bacterial species have also been associated with human autoimmune diseases. The most striking example is the development of ARF following recovery from infection with particular strains of group A Streptococci. Similarly, episodes of AS and the onset of GBS frequently follow an infection with *C. jejuni.* GD has been linked to *Yersinia enterocolitica* infections, and mycoplasma infections are thought to sometimes precipitate the onset of RA or CD. A connection between *Helicobacter pylori* and ITP has been proposed because some ITP patients infected with this organism show improvement in their ITP symptoms following antibiotic treatment. SS patients often show alterations of the normal microbial flora of the mouth, but whether these changes cause SS or are consequences of the reduced salivation in these individuals is unclear.

Parasite infections may contribute to some human autoimmune diseases. For example, severe autoimmune damage to the heart muscle occurs in 30% of patients with Chagas disease, which is caused by the protozoan parasite *Trypanosoma cruzi* (refer to Ch. 13). Several *T. cruzi* antigens feature epitopes highly similar to those of host antigens in the heart, resulting in a cross-reactive attack. Similarly, high titers of antibodies directed against *Toxoplasma* antigens have been found in patients with two other rare autoimmune diseases.

III. Hormonal Influences

Many autoimmune diseases show a gender bias. For example, women account for 88% of SLE patients and 90% of SS patients, and RA affects three times as many women as men. GD and HT are also predominantly found in women. In contrast, AS and GS patients are usually males. These findings suggest that sex hormones can play a major role in inducing the onset of autoimmune disease in genetically predisposed individuals. Scientists hypothesize that hormone expression may somehow "reactivate" previously tolerized lymphocytes so that they can attack self tissues. For example, in a mouse model of SLE, the administration of estrogen was shown to block B cell tolerization and to result in autoimmune symptoms. In human SLE patients, estrogen metabolism is often abnormal. Furthermore, flare-ups of SLE are frequently associated with changes in hormonal status, such as during pregnancy or the initiation of hormone replacement therapy. Similarly, significant numbers of HT and GD patients first develop their disease in the postpartum period, a time of major hormonal changes. Intriguingly, pregnant RA and MS patients frequently experience an improvement in their autoimmune symptoms during their last trimester. It is thought that the changing hormonal environment associated with pregnancy may upregulate the production of regulatory T cells that are meant to help protect an allogeneic fetus but may also suppress autoreactive lymphocytes.

The glucocorticoid hormones responsible for the "fight or flight" response to acute stress may also be relevant to autoimmune disease etiology. Animals that produce lower levels of these hormones show increased susceptibility to autoimmunity. Indeed, when experimentally exposed to a stress such as hypoglycemia, SLE patients produce lower levels of serum glucocorticoids than do normal individuals. Psychological stress may also be associated with symptoms of autoimmune disease. For example, marital difficulties, job stress and economic hardship have all been tentatively linked to the onset of RA or PG. Scientists speculate that, in normal individuals subjected to stress, the production of glucocorticoids may be increased to restrain autoreactive lymphocytes. In individuals predisposed to autoimmunity, this increase in glucocorticoids may not occur or is insufficient to prevent autoimmune disease.

E. Therapy of Autoimmune Diseases

Both conventional and immunotherapeutic strategies are used to treat autoimmune diseases. Conventional therapies include anti-inflammatory and immunosuppressive drugs that are helpful but usually only alleviate symptoms and do not address the underlying cause of the autoimmunity. There are also concerns about the potentially toxic side effects of long-term use of these drugs. Other conventional treatments can either blunt the effects of autoantibodies or mechanically remove them from a patient's blood but, again, are not a cure. Immunotherapies use aspects or components of the immune system to ameliorate disease. These approaches generally attempt to eliminate or control the autoreactive lymphocytes or autoantibodies causing the damage. However, the development of these agents has been slow largely because results obtained in animal models have rarely translated well to the human situation. Lack of efficacy of a candidate agent or even toxicity and exacerbation of autoimmunity have been reported in some clinical trial participants. In addition, the results of human clinical trials must be cautiously interpreted because many human autoimmune diseases have a relapsing/remitting pattern. Thus, the observed resolution of an autoimmune symptom may or may not be the direct result of the candidate treatment. Such hurdles have made the development of new therapeutics for autoimmune diseases a tortuous process. Current conventional and immunotherapeutic approaches to autoimmune disease treatment as well as several experimental approaches are discussed in the following sections.

I. Conventional Therapies

The mainstays of conventional autoimmune disease treatment are anti-inflammatory agents, immunosuppressive drugs and technologies that non-specifically control auto-antibodies. For autoimmune diseases that have a strong association with a particular pathogen infection, antibiotics or vaccines may be beneficial. For an autoimmune disease that has a disease-specific symptom, a treatment to alleviate that symptom may improve a patient's quality of life. For example, the excessive blood clotting in patients with APS can be treated with anti-coagulant drugs. Similarly, individuals with T1DM can regulate their blood sugar with insulin injections, and MS patients can be given anti-spasmodic agents to reduce muscle spasms.

i) Anti-Inflammatory Drugs

Although there is no known cure for the arthritis associated with RA or the spinal fusion of AS, various treatments aimed at relieving the inflammation driving the bone and joint symptoms of these disorders can be applied. Traditionally, such patients are treated with anti-inflammatory agents such as aspirin and ibuprofen. The administration of corticosteroids that block the transcription of TNF, IL-1 and IL-17 can also help. Such agents improve joint function in RA and AS, decrease relapses of MS, reduce inflammation in SLE, ameliorate muscle weakness in MG and PM, control bleeding in GS, and soothe the blistering of PG. However, prolonged use of large doses of corticosteroids can have detrimental side effects, including decreased resistance to infection and development of osteoporosis (brittle bones).

ii) Immunosuppressive Drugs

Drugs that inhibit lymphocyte proliferation are used to treat many autoimmune diseases, including severe cases of RA, MG, PM, SS, GS, IBD and PG. Some of these drugs were described in Table 17-6 as agents used to suppress transplant rejection. However, these drugs are non-specific in their effects such that all activated lymphocytes, including non-autoreactive T and B cells needed to protect the autoimmune patient from infection or cancer, may be affected. Thus, prudent use of these immunosuppressive therapies is required to avoid increasing the risk of life-threatening infections or tumorigenesis. For example, many RA patients benefit from treatment with leflunomide, a malononitrilamide that reduces patient symptoms without increasing opportunistic infections. Cyclosporine A can be helpful in increasing the platelet count in ITP patients, and mycophenolate mofetil has relieved symptoms in some SLE cases.

iii) Non-Specific Control of Autoantibodies

Plasmapheresis can be used to mechanically remove all antibody proteins (including autoantibodies) from a patient's blood. Patients with *thrombic thrombocytopenia*

Plasmapheresis was described in Chapter 17 as a method of removing alloreactive antibodies from a patient prior to organ transplantation.

purpura (TTP; refer to **Fig. 19-3**) receive the greatest benefit from plasmapheresis, whereas patients with MS, GBS, MG, ITP, GS and PG show a more modest clinical improvement.

Another method used to non-specifically control autoantibodies is infusion of a very high dose of *intravenous immunoglobulin (IV-IG)*. As described in Chapter 15, IV-IG is a preparation of antibodies of multiple specificities pooled from a group of healthy donors. IV-IG administration has successfully reduced autoimmune symptoms in many cases of MS, SLE, RA, MG, GBS, ITP and PM. Recent work suggests that IV-IG may reduce the inflammation associated with autoimmune disease by damping down the responses of innate leukocytes. As discussed in Chapter 4, immunoglobulin proteins are glycosylated in a variety of ways, and may or may not bear sialic acid moieties in their Fc regions. IgG antibodies that have non-sialylated Fc regions and engage their cognate antigen bind preferentially to FcγRs constitutively expressed by innate leukocytes. This binding normally triggers pro-inflammatory signaling, as described in Chapter 5. However, IgG antibodies that have sialylated Fc regions do not bind to FcγRs but rather to DC-SIGN, the CLR described in Chapter 3, as being abundantly expressed on DC surfaces. Importantly, this binding can occur regardless of whether the IgG has engaged its cognate antigen. DC-SIGN engagement by the sialylated Fc region of the IgG triggers the DC to express an FcγR called FcγRIIB. Unlike activatory FcγRs, when FcγRIIB is engaged by either a sialylated or non-sialylated IgG molecule, inflammatory signaling is inhibited rather than promoted. As a result, when IV-IG is infused into a patient, the level of FcγRIIB expression relative to that of other FcγRs increases, causing inhibitory signaling to dominate activatory signaling; inflammation therefore decreases. The fact that sialylated antibodies are relatively scarce in any plasma preparation accounts for the previously mysterious observation that very high doses of IV-IG have to be used to achieve a therapeutic effect for autoimmune disease patients.

II. Immunotherapies

Current immunotherapeutic strategies for the treatment of autoimmune diseases are summarized in **Table 19-3**. Although deliberate tinkering with the immune system affects responses to pathogens and tumor cells and thus leaves the patient vulnerable to potentially serious infections and/or cancer development, the hope is that responses to autoantigens will also be inhibited, providing the patient with relief from autoimmune symptoms. In the future, it may be possible to treat an autoimmune disease with exquisite precision by establishing tolerance to the relevant autoantigen (see **Box 19-3** below).

i) Cytokine Blockade

New agents have been developed that block the harmful inflammation associated with autoimmune diseases while causing fewer side effects than traditional anti-inflammatory or immunosuppressive drugs. Many RA and AS patients, whose disease is largely due to excessive TNF, have benefitted from treatment with anti-TNF mAbs or agents that block access to TNFR. Patients experience decreased inflammation in their affected joints and enjoy an improved quality of life. Anti-TNF agents are also helpful in cases of IBD and PS. An interesting corollary to anti-TNF therapy is that Treg cells, which appear to be present in normal numbers but inactive in at least some RA patients, recover their ability to inhibit cytokine production by effector T cells.

RA patients that fail to respond to measures countering TNF are thought to have disease driven by IL-17 or IL-1. In a healthy person, the binding of IL-1 to its receptor IL-1R triggers inflammation that is modulated by the competitive binding to IL-1R of a natural IL-1R antagonist protein called IL-1Ra. *Anakinra* is a synthetic form of IL-1Ra that binds to IL-1R and blocks its function, thereby reducing inflammation. Joint symptoms in many RA patients have been relieved by anakinra treatment. With respect to IL-17, a humanized mAb that blocks IL-17 signaling has been examined in clinical

TABLE 19-3	Examples of Immunotherapeutics Used to Treat Autoimmune Diseases*		
Target	**Effector Molecule**	**Mechanism**	**Diseases Treated**
Extravasation			
CD2 (LFA-2)	LFA-3–IgG fusion protein	T cell costimulatory blockade; autoimmune disease CC promotion	PS
CD49d (VLA-4)	Anti-VLA-4 mAb	Disruption of T cell trafficking	IBD, MS
CD11 (LFA-1)	Anti-CD11 mAb	Disruption of T cell trafficking	PS
α4β1 and α4β7 integrins	Anti-integrin mAbs	Disruption of leukocyte extravasation	MS, UC, CD
Cytokine Blockade			
TNF	Anti-TNF mAb or TNFR blocking agent	Inhibition of TNF signaling	RA, AS, IBD, PS, CD
IL-1	IL-1R antagonist	Inhibition of IL-1 signaling	RA
IL-6	Anti-IL-6R mAb	Inhibition of IL-6 signaling	RA and other forms of arthritis
IL-12	Anti-IL-12 mAb	Inhibition of IL-12 signaling	PS, CD
IL-17	Anti-IL-17 mAb	Inhibition of IL-17 signaling	PS, RA
IL-23	Anti-IL-23 mAb	Inhibition of Th17 cell differentiation	PS, CD
Cytokine Administration			
DCs	IFNβ	Inhibition of autoantigen presentation, T cell migration, Th17 cell differentiation	MS
DCs	IL-10, TGFβ	Inhibition of DC maturation	PS
Targeting of TLRs			
TLR4	Anti-TLR4 mAb	Inhibition of TLR4 signaling	IBD
TLR7/9	TLR7/9 antagonist	Inhibition of TLR7/9 signaling	PS
Targeting of Autoreactive T Cells			
CD52	Anti-CD52 mAb	T cell depletion	MS, SLE, ITP, RA
CD3	Anti-CD3 mAb	T cell anergization; iTreg induction	T1DM, UC, MS
CD25 (IL-2Rα chain)	Anti-CD25 mAb	Inhibition of T cell activation	PG, PS, ITP, T1DM
CD4	Anti-CD4 mAb	T cell anergization	RA, PS
CD28	Soluble CTLA-4 protein	T cell costimulatory blockade	MS, SLE, RA, PS
Targeting of Autoreactive B Cells			
CD20	Anti-CD20 mAb	B cell depletion	SLE, RA, MG, ITP
CD22	Anti-CD22 mAb	B cell depletion	SLE, RA, MG, ITP
CD40L	Anti-CD40L mAb	B cell costimulatory blockade	SLE, ITP
CD52	Anti-CD52 mAb	B cell depletion	MS, SLE, ITP, RA
CD257 (BAFF)	Anti-CD257 mAb	Inhibition of B cell development	SLE

*Includes investigational treatments.

trials as a therapy for PS and RA patients, with some encouraging results. In addition, because IL-23 is a cytokine crucial for the differentiation of Th17 cells, researchers have treated PS and CD patients with an antagonistic anti-IL-23 mAb. Again, some clinical benefit has been observed in early trials.

Lastly, in some groups of RA patients, it is believed that other cytokines such as IL-15, GM-CSF or IL-6 may drive the disease. Accordingly, a mAb that targets IL-6R and blocks the pro-inflammatory effects of IL-6 signaling has been developed. In mouse models of arthritis, this anti-IL-6R mAb inhibits Th17 cell differentiation and upregulates

Treg cell production. In humans, this anti-IL-6R mAb was recently proven in clinical trials to be efficacious for the relief of RA symptoms. Early success in limited numbers of patients with SLE, PM, AS or CD has also been achieved. Studies are currently under way to examine the effectiveness of this agent for the treatment of other autoimmune diseases.

Cytokines important for the generation of various Th effector cell subsets were discussed in Chapter 9.

ii) Cytokine Administration

Although some cytokines play a role in driving autoimmunity, others can be effective in countering it. For example, large numbers of MS patients have been successfully treated with IFNβ. The original rationale for this approach was that IFNβ would control the unknown virus that was thought to be triggering MS. However, it now appears that IFNβ may reduce MS relapse rates by opposing the effects of pro-inflammatory cytokines on DCs and by downregulating the expression of MHC class II that facilitates autoantigen presentation. In addition, IFNβ is a powerful inhibitor of the matrix metalloproteinases used by activated autoreactive T cells to invade brain tissue. Another more recently proposed hypothesis is that IFNβ decreases the expression of cytokines and transcription factors important for Th17 cell differentiation, but this theory remains under debate.

Administration of the immunosuppressive cytokines IL-10 and TGFβ is beneficial for some autoimmune disease patients (particularly for those with PS), although toxicity has been an issue for others. IL-10 and TGFβ block the maturation and migration of DCs and inhibit the expression of pro-inflammatory cytokines by APCs. In patients benefitting from treatment with these cytokines, inflammation was suppressed, the interval between relapses was lengthened, and the number of relapses was reduced.

iii) Targeting of Leukocyte Extravasation

The ability of leukocytes to exit the circulation and migrate within tissues to sites where their effector functions are needed is crucial for effective immune responses against pathogens. However, this same capacity for extravasation allows autoreactive lymphocytes to reach tissues targeted by a particular autoimmune disease. Immunologists concerned with mitigating autoimmune symptoms have therefore sought ways to prevent leukocytes from accessing target tissues. Several different mAbs directed against integrin proteins have been developed to block leukocytes from binding to their required counter-receptors during extravasation. Interestingly, some of these anti-integrin antibodies also downregulate the expression of costimulatory molecules by T cells, helping to block the activation of autoreactive lymphocytes. For example, an anti-LFA-1 mAb has been helpful for the treatment of some cases of PS. However, the use of a related agent that initially showed promise for the treatment of MS, RA and IBD was suspended due to significant toxicity. Similarly, anti-VLA-4 mAbs have been beneficial in some cases of IBD and MS but have been unacceptably toxic in other cases. Newer mAbs targeting the α4 subunit of the α4β1 and α4β7 integrins have recently been approved for the treatment of MS, UC and CD.

Leukocyte extravasation was described in Chapter 2 and illustrated in Figure 2-16.

iv) Targeting of TLRs

The observation that many autoimmune diseases appear to correlate with infections has led researchers to wonder whether the binding of PAMPs/DAMPs to particular TLRs contributes to patients' symptoms. If so, anti-TLR mAbs or other inhibitors of TLR signaling might be helpful as therapies for these disorders. To test this hypothesis, scientists deleted specific TLR genes in mouse models of a particular autoimmune disease and evaluated whether autoimmune symptoms were abrogated. For example, in a mouse model of autoimmune arthritis, the genetic deletion of TLR4 or treatment with a TLR4 antagonist reduced Th17 responses and ameliorated joint symptoms. Similarly, in a rat model of IBD, treatment with either a synthetic TLR4 antagonist or an anti-TLR4 mAb decreased autoimmune symptoms. With this success in mind, a humanized anti-TLR4 mAb is undergoing clinical trials for the treatment of human IBD. Parallel development of antagonists or mAbs directed against TLR2, TLR7/9 and TLR8 is in progress.

v) Targeting of Autoreactive T Cells

Some immunotherapeutic approaches are focused directly on inhibiting autoreactive T cells (keeping in mind that activated, non-autoreactive T cells are often affected as well). Significant success has been achieved with the anti-CD52 mAb introduced in Chapter 16. Binding of this mAb to an autoreactive T cell dooms it to ADCC- or complement-mediated destruction. The symptoms of autoimmune diseases as diverse as RA, ITP, MS and SLE have been ameliorated following treatment with anti-CD52 mAb. Another approach that has shown promise is the use of a modified anti-CD3 mAb to anergize T cells in T1DM patients. Long-term positive effects on the preservation of pancreatic islet cells have been observed. Other mAbs investigated as potential therapies include anti-CD3 antibodies that promote the generation of iTreg cells; anti-CD4 mAbs that block T cell effector functions; and anti-CD25 mAbs that bind to the IL-2Rα chain and inhibit the effects of IL-2. Anti-CD25 mAbs have proved useful for the treatment of some cases of PG, PS and ITP, whereas anti-CD4 mAbs have achieved limited success in trials for the treatment of RA and PS.

Another strategy in this context is to interfere with the costimulation of an autoreactive T cell. A soluble form of the CTLA-4 protein that negatively regulates CD28

Box 19-3   Induction of Tolerance as Therapy for Autoimmune Diseases

There were originally high hopes for new autoimmune disease therapies based on "vaccinating" patients with their own autoantigens under tolerizing conditions. Whereas a conventional vaccination is intended to induce an effective immune response, autoimmune "vaccination" is designed to anergize or exhaust autoreactive lymphocytes. Hypothetically, such elimination or inactivation of lymphocytes could be achieved if the patient is exposed to the offending autoantigen either in a benign context or via an unusual route of administration. The induction of oral tolerance has thus been examined as a potential tactic, since iTreg cells induced by the administration of an oral antigen usually secrete IL-10 and/or TGFβ and thus suppress the activation of any nearby effector T cells, regardless of antigenic specificity. In theory, the identity of the autoantigen causing the autoimmune disease does not have to be known for oral tolerance to help mitigate autoimmune disease symptoms. However, although the feeding of an antigen to induce oral tolerance has been successful in rodents, it has been ineffective in humans. For example, no efficacy was demonstrated in a group of MS patients given oral MBP, or in a group of RA patients given oral collagen.

Tolerization through low-dose antigen administration has also been tried as a means of preventing autoimmune disease progression. The U.S. Diabetes Prevention Trial (2003) examined whether small, regular doses of insulin could prevent the development of T1DM in individuals who were at high risk but still healthy at the start of the trial. The thought was that the continuous administration of insulin might induce tolerance to it and thus reduce β-islet cell destruction. Unfortunately, although successful in rodents, this strategy did not prevent or delay overt diabetes in humans. On the other hand, administration of a protein vaccine containing the islet-associated antigen GAD coupled to alum was able to preserve residual insulin secretion in some T1DM patients.

DNA and peptide vaccination strategies have also been explored in animal models as a possible means of inducing tolerance to autoantigens. For example, in a mouse model of MS, paralysis was reversed if the affected animals were given DNA encoding an MBP peptide. In humans, researchers have vaccinated MS patients with low doses of purified peptides that are tolerogenic versions of the MBP peptides associated with autoimmunity. Some of these vaccinated MS patients have enjoyed a modest clinical improvement. RA patients treated with autoantigen peptides have shown a switch to production of IL-10 from TNF, while SLE patients treated with autoantigen peptide have exhibited a reduction in anti-dsDNA autoantibodies. Vaccination with altered peptides derived from insulin and desmoglein-3 are under investigation for the treatment of T1DM and PG, respectively. Of interest, immunologists have recently become aware of the relationship between the dose of autoantigen administered and the risk of severe side effects. A patient who receives repeated high doses of autoantigen may eventually develop tolerance to it but is also more likely to suffer a severe hypersensitivity reaction. A repeated low dose of autoantigen has fewer side effects but may not induce sufficient tolerance to result in clinical improvement. It seems that a program of slowly escalating doses of autoantigen achieves the best balance between induction of tolerance and avoidance of debilitating side effects.

Another currently experimental method of inducing tolerance to autoantigens involves promoting the interaction of tolerogenic, self antigen-presenting DCs with autoreactive T cells. In the same vein, murine B cells have been modified so that they present autoantigen peptides on MHC class II in a tolerogenic context. When these approaches have been tried in animal models, a diminution of disease symptoms has been seen that is thought to depend on the induction of Treg cells. A similar approach called *autologous Treg cell therapy* is under investigation in humans. A patient's own regulatory T cells are isolated and cultured *in vitro* to expand their numbers before infusion back into the individual. The rationale is that this army of Treg cells should suppress the activities of autoreactive T cells. To date, this approach has yielded promising results in clinical trials involving children with T1DM.

signaling has been used to treat patients with PS, RA, SLE or MS, with some benefit being observed in some cases. Other researchers have devised a fusion protein composed of part of an IgG molecule and the extracellular domain of LFA-3, the CD2 receptor. This agent simultaneously binds to CD2 on T cells and FcγRs on phagocytes and NK cells, promoting the death of the T cells by ADCC. Improvement in some PS patients treated with this agent has been noted.

A caveat with the non-specific targeting of T cells is that, as well as affecting autoreactive, anti-pathogen and anti-tumor T cells, the activation and/or function of regulatory T cells may be impeded. Thus, autoimmune disease symptoms may be exacerbated rather than mitigated by this type of therapy.

In 2007, the annual sales in the U.S. of antibodies for clinical use topped $27 billion.

vi) Targeting of Autoreactive B Cells

Autoreactive B cells not only differentiate into plasma cells producing damaging autoantibodies but may also serve as powerful APCs driving the activation of autoreactive memory T cells. Antibody-mediated depletion of autoreactive B cells has therefore been explored for the treatment of several autoimmune diseases. The binding of mAbs directed against B cell surface molecules induces the destruction of B cells (both normal and autoreactive) by either complement-mediated cytolysis, ADCC, or the induction of apoptosis. Accordingly, the administration of mAbs recognizing the B cell markers CD20 or CD22 has been of clinical benefit to some SLE, RA, MG and ITP patients. Only mild side effects have been observed, and increased infection due to a general impairment of the humoral response does not appear to be a problem. Anti-CD52 mAb treatment also been helpful for some autoimmune disease patients because CD52 is found on B cells as well as T cells. Treatment with anti-CD40L mAb has been investigated as a means of interfering with the CD40/CD40L engagement that is critical for antibody production. In some SLE patients treated with anti-CD40L mAb, titers of circulating autoantibodies have been decreased and glomerulonephritis mitigated. While anti-CD40L mAbs have also been used to ameliorate symptoms in some ITP patients, unacceptable toxicity has occurred in other cases. Lastly, a mAb directed against the B cell development factor BAFF has recently been approved for SLE treatment.

F. Autoinflammatory Disorders

I. Nature

Autoinflammatory disorders (or syndromes) are a group of inflammatory diseases that occur in the absence of infection or tissue damage that could supply recognizable PAMPs or DAMPs. The major clinical signs of these disorders are recurrent or periodic bouts of fever, rash, and swelling of the joints, with other symptoms varying by disease. While these syndromes are relatively rare, they are often first seen at a very young age, and their effects can be quite severe. In the past, it was assumed that these patients were experiencing repeated infections by undetermined pathogens, leading clinicians to diagnose them as suffering from an underlying primary immunodeficiency disease. However, over the past few years, genetic analyses combined with a greater understanding of innate response mechanisms has clarified their nature. While these patients may indeed have alterations in genes encoding immune system components, rather than being loss-of-function mutations impairing the adaptive response, they are instead usually gain-of-function mutations leading to uncontrolled activation of innate responses. Thus, these individuals do not suffer from immunodeficiency but instead experience "hyperactive innate immunity" that causes systemic inflammatory disease. Some immunologists think of these disorders as examples of autoimmune diseases, categorizing them this way because the patient's immune system is responsible for the symptoms. However, since there are no identifiable autoantigens involved in these disorders, nor any activation of autoreactive T or B lymphocytes, other immunologists believe these syndromes fall outside the traditional definition of "autoimmune disease."

| TABLE 19-4 | **Examples of Autoinflammatory Disorders** |

Autoinflammatory Disease	Clinical Characteristics	Component(s) of the Innate Response Affected
Familial Mediterranean Fever (FMF)	Autosomal recessive; childhood onset Most common in persons of Middle Eastern descent Characterized by periodic fevers of 1–3 days, abdominal and chest pain, skin rash, arthritis Kidney failure can be a late complication	Mutations in the pyrin protein lead to deregulation of the NLRP3 inflammasome in activated monocytes and granulocytes
TNF Receptor-Associated Periodic Syndrome (TRAPS)	Autosomal dominant; onset at any age Characterized by fevers lasting from 7–30 days, severe localized inflammation, abdominal pain, arthritis, spreading rash, eye edema	Mutations in TNFR1 result in uncontrolled stimulation of the pro-inflammatory TNFR1 signaling pathway
Hyper-IgD with Periodic Fever Syndrome (HIDS)	Autosomal recessive; onset in infancy Most common in persons of Northern European descent Characterized by fevers of 3–7 days, abdominal and joint pain, rash, elevated serum IgD	Mutations in a cholesterol biosynthesis enzyme; mechanism remains unclear
Cryopyrinopathies:	Autosomal dominant; variable onset Characterized by urticaria-like skin rash, fever, chills, joint pain, eye redness	Mutations in NLRs lead to inflammasome hyperactivation, excessive IL-1 production
i) Familial Cold Autoinflammatory Syndrome (FCAS)	FCAS is the mildest cryopyrinopathy; onset at around 6 months of age; often triggered by cold temperatures	Mutations in NLRP3, NLRP12
ii) Muckle–Wells Syndrome (MWS)	MWS is a slightly more severe cryopyrinopathy; fevers last longer and occur regardless of environmental temperature; damage to inner ear and kidneys may occur	Mutations in NLRP3
iii) Neonatal Onset Multisystem Inflammatory Disease (NOMID)	NOMID is the most severe cryopyrinopathy; onset as young as 6 weeks; may result in meningitis, loss of hearing and vision, abnormal bone growth and mental retardation; 20% of untreated cases are fatal	
Deficiency of Interleukin-1 Receptor Antagonist (DIRA)	Autosomal recessive; onset in infancy/childhood Most common in persons of Dutch or Puerto Rican descent; may be fatal Characterized by pain and swelling of bone tissue, bone deformities, skin rash	Mutations in the negative regulator IL-1Ra lead to uncontrolled IL-1R signaling
Blau Syndrome	Autosomal dominant; onset by age 4 years Characterized by skin granulomas, arthritis, eye inflammation	Mutations in NOD2 lead to its constitutive activation
Behcet's Disease	Autosomal recessive; onset in young adulthood Most common in persons of Middle Eastern, Japanese or Asian descent	Genes involved are currently unknown

Some autoinflammatory disorders, including the cryopyrinopathies introduced in Chapter 3 and the TRAPS and FMF syndromes mentioned in Chapter 15, are due to an activating mutation in a single gene associated with inflammation. Others appear to be more complex disorders caused by alterations to several genes. However, in more than half of patients with an autoinflammatory disease, no known mutation has been detected. Much research therefore remains to be done to properly dissect these afflictions. The characteristics of some better known autoinflammatory disorders are summarized in **Table 19-4**.

Other conditions exist that resemble the cryopyrinopathies, in that the clinical symptoms are caused by excessive activation of the NLRP3 inflammasome, but these diseases are not caused by genetic mutations. Instead, persistent inflammation is triggered by the local deposition of organic crystals that act as DAMPs in specific tissue sites. For example, gout is caused by the deposition of uric acid crystals in the joints, which then become swollen and painful. Another example is atherosclerosis, in which cholesterol crystals accumulate in the artery walls and trigger the inflammation that damages these vessels.

NOTE: In Chapter 15, we introduced autoimmune lymphoproliferative syndrome (ALPS) as a primary immunodeficiency with an autoimmune component. As in the case of autoinflammatory disorders, the various forms of ALPS involve genetic defects that result in uncontrolled activity of the immune system in the absence of any autoantigen or pathogen. However, ALPS causes dysregulation of lymphocyte proliferation as opposed to inflammation, and so is clearly a disease of adaptive, rather than innate, immunity.

Each of the various ALPS subtypes is due to a defect in cellular apoptosis that results in the selective hyperproliferation of B cells, CD8+ T cells, and DN thymocytes. It is not known why other cell subsets are not affected. The expanding lymphocytes accumulate in the lymphoid organs in a non-malignant way, causing lymphadenopathy and hepatosplenomegaly. Due to the B cell hyperproliferation, levels of serum IgG and IgA are increased, and autoantibody production is found in almost all patients. Many of these circulating antibodies are directed against erythrocytes and platelets, leading to anemia and thrombocytopenia, respectively. ALPS patients also have an increased risk of developing tumors of various types.

ALPS is very rare. It is sometimes diagnosed within the first year of life and almost always by age 5 years, and patients are classified by their underlying genetic mutations. ALPS Ia patients (the majority) have an autosomal dominant mutation in the gene encoding the Fas death receptor (CD95), while ALPS Ib patients have a mutation in FasL. ALPS II patients have mutations in the genes encoding caspase-10 and caspase-8, which are enzymes vital for the apoptotic cascade downstream of Fas ligation. The genes mutated in many other ALPS patients remain to be identified.

Lymphadenopathy means "swelling of the lymph nodes." Hepatosplenomegaly means "swelling of the liver and spleen." Thrombocytopenia means "reduced platelet count."

II. Therapies

The front line treatment of autoinflammatory disorders often involves the use of agents that block TNF signaling (such as anti-TNF mAb) or IL-1 signaling (such as anakinra), and so reduce inflammation. Another useful approach is anti-IL-6 mAb administration, which inhibits the differentiation of Th17 cells while promoting that of Treg cells. Anakinra has also been effective in relieving symptoms in some patients with T2DM, suggesting that, in at least some cases, this mainly metabolic disease may have an autoinflammatory component.

NOTE: A recent discovery that may have significant implications for the treatment of the cryopyrinopathies is that their associated recurrent inflammation leads to deposition in the tissues of fibrils derived from a misfolded acute phase protein called *serum amyloid A*. This symptom, termed *systemic amyloidosis*, is a serious complication of these diseases and other autoinflammatory disorders. Aggressive treatment with anti-inflammatory agents can shrink these deposits in animal models, raising hopes that a similar strategy may be effective in humans.

We have come to the end of our discussion of autoimmune diseases and autoinflammatory disorders, topics with many questions remaining. We move to another subject about which there are still several mysteries: cancers of immune system cells. Whereas Chapter 16 addressed the role of the immune system in fighting tumorigenesis, Chapter 20 discusses what happens when immune system cells themselves undergo malignant transformation.

Chapter 19 Take-Home Message

- An autoimmune disease is a pathophysiological state in which an individual's tissues are damaged as a result of an attack by the immune system. A large number of autoimmune diseases have been identified that vary widely in their symptoms.

- Four events are required for the development of an autoimmune disease: (1) an autoreactive lymphocyte must escape elimination by central tolerance mechanisms and be released to the periphery; (2) the escaped autoreactive lymphocyte must encounter its specific self antigen in the periphery; (3) the peripheral tolerance mechanisms designed to regulate autoreactive lymphocyte responses must fail; and (4) the response by the autoreactive lymphocyte must result in clinical damage.

- The breaking of peripheral tolerance may be due to inflammation; molecular mimicry by pathogen antigens; defects in immune system cells, cytokines or complement; and/or epitope spreading.

- Activated autoreactive Th, Tc and B cells produce cytokines, CTLs and autoantibodies, respectively, that cause tissue damage. This damage may be localized or systemic, depending on the nature and distribution of the self antigen.

- Many of the peripheral tolerance defects underlying autoimmune diseases involve proteins encoded by polymorphic genes. Thus, an individual's genotype, particularly the identity of his/her HLA alleles, may predispose him/her to developing an autoimmune disease.

- In someone genetically prone to autoimmunity, the balance may be tipped toward autoimmune disease by triggers such as pathogen infection, toxin exposure or altered hormone levels.

- Some autoimmune disease therapies are aimed at alleviating symptoms, whereas other approaches seek to control or eliminate autoreactive lymphocytes.

- Autoinflammatory disorders are characterized by periodic bouts of inflammation in the absence of any lymphocyte-mounted attack on a pathogen or self antigen. These disorders are thought to be mediated by dysregulated innate responses.

Did You Get It? A Self-Test Quiz

Section A

1) How do autoimmunity and autoimmune disease differ?

2) Describe the roles of Th cells, B cells and CTLs in causing the tissue damage associated with autoimmune disease.

3) Give two examples of how hypersensitivities can be involved in autoimmune diseases.

4) Distinguish between organ-specific and systemic autoimmune diseases.

Section B

1) What four events must occur for autoimmune disease to develop?

2) Why is the mechanism underlying APECED unusual for an autoimmune disease?

3) How does infection-induced inflammation contribute to autoimmune disease?

4) What are MAMPs, and how might they be related to autoimmune disease?

5) How does tissue damage/stress-induced inflammation contribute to autoimmune disease?

6) How can non-hematopoietic cells help to prevent autoimmune disease?

7) Why might a cancer arise in a patient with an autoimmune disease?

8) Why might an autoimmune disease arise in a cancer patient?

9) What is the theory of molecular mimicry? Give two examples.

10) Give five examples of immune system component defects that might favor autoimmunity.

11) How is epitope spreading thought to perpetuate an auto-immune response? Give an example.

Section C

1) Can you define these terms? relapsing/remitting, rheumatoid factor, sclerosis

2) Give two examples of proteins that are targets of autoantibodies in systemic lupus erythematosus.

3) What is the primary target tissue in rheumatoid arthritis?

4) Rheumatic fever and Guillain–Barré syndrome are often cited as examples of molecular mimicry. Why?

5) Give two examples of autoimmune diseases driven by epitope spreading.

6) Distinguish between type 1 and type 2 diabetes mellitus.

7) How is the tissue damage of multiple sclerosis caused, and what is the clinical outcome?

8) What is liberation therapy, and why is it controversial?

9) Compare the clinical features and mechanisms of the two types of autoimmune thyroiditis.

10) Compare the clinical features and mechanisms of myasthenia gravis and Guillain–Barré syndrome.

11) Compare the clinical features and mechanisms of the two major types of inflammatory bowel disease.

Did You Get It? A Self-Test Quiz—Continued

Section D

1) What three major determinants govern whether an individual develops autoimmunity?

2) Why do some autoimmune diseases appear to vary by geographic location?

3) Give five examples of genes linked to autoimmune diseases.

4) You have an autoimmune disease and so do many members of your extended family. How might you be able to advance research into the cause of your disease?

5) Give three examples of chemical agents and the autoimmune diseases to which they may be linked.

6) Give three examples of pathogens and the autoimmune diseases to which they may be linked.

7) How might hormonal changes trigger episodes of autoimmune disease? Give two examples.

8) How might stress trigger episodes of autoimmune disease?

Section E

1) Give three examples of conventional approaches used to treat autoimmune diseases.

2) Describe two ways in which the effects of autoantibodies can be non-specifically controlled.

3) Cytokines can either be administered or their effects blocked to treat autoimmunity. Give an example of each circumstance and its therapeutic rationale.

4) What is anakinra, and what is its role in autoimmune disease therapy?

5) What is the rationale behind the treatment of MS patients with IFNβ?

6) Why might disabling leukocyte extravasation be helpful in treating autoimmune diseases? Give two examples.

7) Why might targeting TLRs be useful for treating autoimmune diseases?

8) Give three examples each of how autoreactive T and B cells can be targeted to treat autoimmunity.

9) What is a major disadvantage of current immunotherapeutic strategies?

10) Give two examples of potential autoimmune disease treatments based on inducing tolerance.

Section F

1) How do autoinflammatory disorders differ from classical autoimmune diseases?

2) What is the nature of the "deficiency" in those autoinflammatory disorders that are considered primary immunodeficiencies?

3) In what way are gout and atherosclerosis unlike the majority of autoinflammatory disorders?

4) Describe a major therapeutic approach used to treat autoinflammatory disorders.

Can You Extrapolate? Some Conceptual Questions

1) In some cases of autoimmune disease such as Guillain–Barré syndrome, the onset of symptoms may be connected to recent vaccination. What mechanism might account for this association? In what way might you decrease the chances of such an association between vaccination and an autoimmune disease?

2) If a genetic polymorphism was found in human autoimmune disease patients that correlated with severely decreased levels of apoptotic death among developing T cells in the thymus, how might you explain this disease association?

3) Many autoimmune diseases involve pathology caused by hypersensitivity responses. Would you categorize the following autoimmune diseases as having a hypersensitivity component, and if so, is it of type II or III character?

   a) Acute rheumatic fever

   b) Myasthenia gravis

   c) Systemic lupus erythematosus

   d) Hashimoto's thyroiditis

   e) Graves' disease

Would You Like To Read More?

Buckner, J. H. (2010). Mechanisms of impaired regulation by CD4(+)CD25(+)FOXP3(+) regulatory T cells in human autoimmune diseases. *Nature Reviews. Immunology*, *10*(12), 849–859.

Chan, A. C., & Carter, P. J. (2010). Therapeutic antibodies for autoimmunity and inflammation. *Nature Reviews. Immunology*, *10*(5), 301–316.

Chervonsky, A. V. (2010). Influence of microbial environment on autoimmunity. *Nature Immunology*, *11*(1), 28–35.

Cho, J. H., & Gregersen, P. K. (2011). Genomics and the multifactorial nature of human autoimmune disease. *New England Journal of Medicine*, *365*(17), 1612–1623.

Doherty, T. A., Brydges, S. D., & Hoffman, H. M. (2011). Autoinflammation: Translating mechanism to therapy. *Journal of Leukocyte Biology*, *90*(1), 37–47.

Franks, A. L., & Slansky, J. E. (2012). Multiple associations between a broad spectrum of autoimmune diseases, chronic inflammatory diseases and cancer. *Anticancer Research*, *32*(4), 1119–1136.

Gelfand, E. W. (2012). Intravenous immune globulin in autoimmune and inflammatory diseases. *New England Journal of Medicine*, *367*(21), 2015–2025.

Getts, M. T., & Miller, S. D. (2010). 99th Dahlem conference on infection, inflammation and chronic inflammatory disorders: Triggering of autoimmune diseases by infections. *Clinical & Experimental Immunology*, *160*(1), 15–21.

Haskins, K., & Cooke, A. (2011). CD4 T cells and their antigens in the pathogenesis of autoimmune diabetes. *Current Opinion in Immunology*, *23*(6), 739–745.

Hu, X., & Daly, M. (2012). What have we learned from six years of GWAS in autoimmune diseases, and what is next? *Current Opinion in Immunology*, *24*(5), 571–575.

Liao, W. W., & Arthur, J. W. (2011). Predicting peptide binding to major histocompatibility complex molecules. *Autoimmunity Reviews*, *10*(8), 469–473.

Maloy, K. J., & Powrie, F. (2011). Intestinal homeostasis and its breakdown in inflammatory bowel disease. *Nature*, *474*(7351), 298–306.

Mills, K. H. (2011). TLR-dependent T cell activation in autoimmunity. *Nature Reviews. Immunology*, *11*(12), 807–822.

Munz, C., Lunemann, J. D., Getts, M. T., & Miller, S. D. (2009). Antiviral immune responses: Triggers of or triggered by autoimmunity? *Nature Reviews. Immunology*, *9*(4), 246–258.

Murdaca, G., Colombo, B. M., & Puppo, F. (2011). The role of Th17 lymphocytes in the autoimmune and chronic inflammatory diseases. *Internal & Emergency Medicine*, *6*(6), 487–495.

Nimmerjahn, F., & Ravetch, J. V. (2010). Antibody-mediated modulation of immune responses. *Immunological Reviews*, *236*, 265–275.

Ombrello, M. J., & Kastner, D. L. (2011). Autoinflammation in 2010: Expanding clinical spectrum and broadening therapeutic horizons. *Nature Reviews Rheumatology*, *7*(2), 82–84.

Sabatos-Peyton, C. A., Verhagen, J., & Wraith, D. C. (2010). Antigen-specific immunotherapy of autoimmune and allergic diseases. *Current Opinion in Immunology*, *22*(5), 609–615.

Sheffield, V. C., Stone, E. M., & Carmi, R. (1998). Use of isolated inbred human populations for identification of disease genes. *Trends in Genetics*, *14*(10), 391–396.

Wing, K., & Sakaguchi, S. (2010). Regulatory T cells exert checks and balances on self tolerance and autoimmunity. *Nature Immunology*, *11*(1), 7–13.

Zepp, J., Wu, L., & Li, X. (2011). IL-17 receptor signaling and T helper 17-mediated autoimmune demyelinating disease. *Trends in Immunology*, *32*(5), 232–239.

Hematopoietic Cancers

Where there is the need for a controller, a controller of the controller is also needed.

Tadeusz Kotarbinski

A. Overview of the Biology and Treatment of Hematopoietic Cancers

I. What Are Hematopoietic Cancers?

In Chapter 16, we discussed how the immune system attempts to deal with cancerous cells. In this chapter, we examine the cancers that arise from the malignant transformation of immune system cells. We have chosen to call these tumors "hematopoietic cancers" (HCs) to distinguish them from the non-hematopoietic cancers (NHCs) described in Chapter 16. HCs account for about 8–10% of all cancer diagnoses in the developed world and a similar percentage of cancer deaths. The tumor biology we described in Chapter 16 remains mostly relevant here despite the fact that cancers of the immune system concern hematopoietic cells that are inherently mobile and not fixed like those of body organs. However, there are unique aspects to the biology of HCs that make them a fascinating area of study in their own right. For the purposes of this book, we focus on the three main types of HCs: **leukemias, myelomas** and **lymphomas**. The development of a leukemia is called *leukemogenesis*, while that of a myeloma is *myelomagenesis*, and that of a lymphoma is *lymphomagenesis*.

Leukemias are tumors that arise from the transformation of a hematopoietic cell in the blood or a hematopoietic precursor in bone marrow (BM). In the latter case, the cancerous progeny of the transformed cell usually make their way into the blood. Thus, leukemias most often occur as "liquid tumors" that are manifested as greatly increased numbers of myeloid, lymphoid or (more rarely) erythroid lineage cells in the blood and BM. Myelomas are tumors of fully differentiated plasma cells that are present either as solid masses or as dispersed clones in the BM, blood or tissues. Unlike normal plasma

MAK: Primer to the Immune Response. http://dx.doi.org/10.1016/B978-0-12-385245-8.00020-0

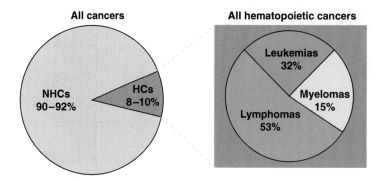

Fig. 20-1
Relative Frequencies of Hematopoietic Cancers in North America

Based on current rates of diagnosis, hematopoietic cancers (HCs) comprise 8–10% of all human malignancies. The three major classes of HCs are leukemias, myelomas and lymphomas. Lymphomas are sub-classified as either Hodgkin lymphomas (HLs) or non-Hodgkin lymphomas (NHLs). NHCs, non-hematopoietic cancers. *[With data from Howlader et al., (eds). SEER Cancer Statistics Review, 1975–2009 (Vintage 2009 Populations), National Cancer Institute. Bethesda, MD, http://seer.cancer.gov/csr/1975_2009_pops09/, based on November 2011 SEER data submission, posted to the SEER website, April 2012.]*

TABLE 20-1	**Hematopoietic Cancer Terminology**	
HC Term	**Target Cell**	**Target Cell Description**
Myeloid or myelogenous*	Myeloid cell	Any cell of the myeloid lineage
Myeloblastic	Myeloblast	Myeloid precursor
Myelocytic	Myelocyte	Mature myeloid cell
Lymphoid	Lymphoid cell	Any cell of the lymphoid lineage
Lymphoblastic	Lymphoblast	T or B precursor cell
Lymphocytic	Lymphocyte	Mature T or B cell

*Historical term still in use for some malignancies.

The U.S. National Cancer Institute web page http://seer.cancer.gov/ and the American Cancer Society web page http://www.cancer.org/research/cancerfactsfigures/index are excellent resources for cancer statistics in this country.

cells, which do not divide after they differentiate, myeloma cells continue to proliferate in an uncontrolled way and synthesize large amounts of Ig chains. Lymphomas develop from transformed lymphocytes that may travel through the body but most often form distinct stationary masses of lymphocytes. These "solid tumors" are generally found in the lymph nodes, thymus or spleen and are classified as either **Hodgkin lymphomas** or **non-Hodgkin lymphomas**. The relative frequencies of these HCs in North America are illustrated in **Figure 20-1**.

As was true for the NHCs described in Chapter 16, the specific cell that undergoes malignant transformation is known as the **target cell** or *cell of origin*. For HCs, the target cell may be either a developing hematopoietic precursor or a mature cell type. The specific terminology used by hematologists to describe these malignancies is based on the nature of the target cell involved and is outlined in **Table 20-1**.

NOTE: Much of this text has focused on the benefits of a properly functioning immune system with its myriad component leukocyte subsets. However, having too many of these beneficial cells, as occurs in HCs, can be as dangerous as having too few of them. Inappropriate overproduction of leukocytes can drain a body's nutritional resources and set up a competition between proliferating malignant cells and differentiating effector cells trying to protect the body against infection. In addition, the accumulation of leukocytes in various body sites or the deposition of their products in these locations can damage vital organs because, by their very nature, leukocytes secrete biologically active cytokines and hydrolytic enzymes. Accordingly, the hallmark symptoms of patients suffering from HCs are fever, bleeding problems, fatigue, weight loss, and increased susceptibility to opportunistic infections.

II. Hematopoietic Cancer Carcinogenesis

Like NHCs, HCs are clonal in nature and arise when a target cell sustains multiple genetic alterations to DNA repair genes, oncogenes and/or tumor suppressor genes (TSGs). Also like NHCs, many HCs appear to be driven by cancer stem cells (CSCs). In fact, researchers were studying a form of HC when they found the first evidence pointing to the existence of CSCs. These investigators discovered that a very rare cell type in the mouse leukemia they were examining retained the dual capacity of HSCs to either self-renew or differentiate. Importantly, mouse transplantation studies showed that only these leukemic stem cells were capable of transferring the leukemia to a susceptible recipient. The bulk of the leukemia cells were more differentiated than stem cells and unable to self-renew or establish new malignancies in recipients. Today, researchers are actively working on defining the characteristics of CSCs in a variety of HCs.

Cancer stem cells were discussed in Chapter 16.

Although many HCs show relatively subtle genetic aberrations such as small chromosomal deletions or point mutations, an estimated 50% of leukemias and lymphomas are associated with major chromosomal disruptions. These disruptions frequently take the form of translocations involving the abnormal exchange of genetic material between two different chromosomes. Because the same translocation appears in many patients with the same type of leukemia or lymphoma, these translocations are called **recurring chromosomal translocations** (see **Box 20-1** and **Box 20-2**). Not surprisingly, recurring translocations associated with HC development affect chromosomes in regions where genes regulating cellular growth, differentiation and apoptosis are located. Sometimes the exchanged chromosomal fragments come together in such a way that a new genetic entity (which may be a functional gene) is created. If this gene is transcribed and translated to produce a new protein, that protein (or parts of it) may constitute a TSA from the perspective of the immune system.

In addition to translocations, exposure to environmental carcinogens, radiation or chemicals can promote HC development (just as was true for NHCs). For example, cigarette smoking and exposure to industrial solvents can increase the risk of leukemogenesis. A higher incidence of leukemia has also been recorded in firefighters exposed to burning plastics, as well as in populations living close to the sites of the atomic bomb explosions in Japan in 1945 and the nuclear reactor meltdown in Chernobyl in 1986. Sadly, leukemias can develop as secondary tumors following intensive radiation or chemotherapy applied either as a treatment for a primary cancer (HC or NHC) or in preparation for a BMT or an HSCT.

III. Clinical Assessment and Treatment of HCs

i) Characterization and Diagnosis

HCs are usually categorized by their grade and whether they cause acute or chronic disease. High-grade HCs grow aggressively and cause acute disease that kills rapidly (in the absence of treatment) due to the accumulation of cancerous cells and the failure of the BM to maintain normal hematopoiesis. These types of malignancies often result when early precursor cells (*blasts*) of a particular hematopoietic lineage become transformed. Medium- and low-grade HCs tend to cause chronic disease that has a slower course and symptoms that may appear to come and go. The transformed cells in these malignancies are often quite well differentiated or even mature.

The 5-year survival rate of HC patients has improved greatly over the past several decades (**Table 20-2**). Much of this success is attributable to aggressive chemotherapy and radiation treatment regimens as well as the increased use of immunotherapies and HSCTs. However, more accurate diagnoses of malignant conditions have also helped to improve clinical outcomes. One of the methods now commonly used to evaluate HCs is *immunophenotyping*, which involves preparing cells from the patient's BM and blood and staining these cells with tagged antibodies specific for surface marker proteins. The particular array of markers expressed on a given cancer cell allows an oncologist to more precisely identify the cell type that originally underwent transformation. Another tool of the modern HC diagnostic trade is *fluorescence in situ*

Box 20-1 Recurring Chromosomal Translocations

Recurring translocations are responsible for many HCs. In the example shown in the following figure, **panel A** depicts one chromosome each of the hypothetical chromosome pairs 1 and 2 at metaphase. These chromosomes sustain breaks in their DNA (**panel B**) that allow the reciprocal transfer of genetic material between chromosome 2 and chromosome 1 (**panel C**). The resulting translocated chromosomes (**panel D**) may contain existing genes that have been disrupted or repositioned, or new genes created by the fusion of sequences from both chromosomes. In the example shown in **panel E**, the translocation positions a hypothetical transcription factor originally on chromosome 1 downstream of a hypothetical gene expression enhancer normally situated on chromosome 2. Alternatively, the translocation may result in the fusion of these sequences. Deregulated expression of the transcription factor induced by the enhancer may help to drive malignant transformation. It should be noted that not every reciprocal translocation is itself sufficient to cause an HC, and not every person bearing one of these translocations is fated to develop an HC. Often the translocation is only an initiating event, and subsequent mutations to oncogenes, TSGs and/or DNA repair genes are needed to complete malignant conversion.

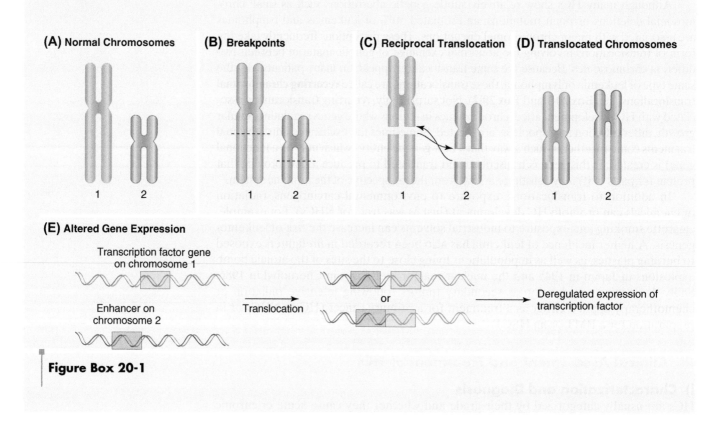

Figure Box 20-1

hybridization (FISH), used for the examination of tumorigenic chromosomal translocations. In FISH, probes specific for a chromosomal region of interest are synthesized by incorporating an easily detected fluorochrome into a DNA fragment of complementary sequence. One or more of these fluorescently tagged probes can then be applied to whole cells or to a chromosomal spread on a microscope slide to visualize one or more whole chromosomes or a specific region of a single chromosome. Major aberrations become clearly visible, pointing the observer to a probable diagnosis. Karyotyping by microarray-based *comparative genomic hybridization (CGH)*, which involves hybridizing fluorescently labelled DNA from test and normal samples to several thousand oligonucleotide probes, is becoming a popular and less costly means of detecting chromosomal anomalies. Finally, the identification of unique RNA patterns expressed by malignant cells has increasingly been used to definitively classify HC subtypes. All these techniques help the physician to arrive at a more accurate diagnosis of the HC. The treatment is then chosen that is most appropriate to an individual case and therefore offers the best chance of extending survival.

Box 20-2 Examples of Reciprocal Translocations Leading to HCs

Examples of reciprocal translocations associated with HCs are given in the following table. In a designation such as t(12;22) (p13;q11), the "t" indicates "translocation," while the "12" and "22" in the first set of parentheses refer to the two chromosomes involved listed in ascending order. In the second set of parentheses, p13 indicates that the break in chromosome 12 took place in the first block, third band (originally defined by karyotypic staining methods) of the short ("p") arm of the chromosome. (The "p" stands for the French word "petit," meaning small.) Similarly, q11 indicates that the break in chromosome 22 took place in the first block, first band of the long ("q") arm. These breakpoints are sometimes expressed as 12p13 and 22q11.

TABLE BOX 20-2	Examples of Reciprocal Translocations Leading to HCs		
Hematopoietic Cancer	**Translocation**	**Genes Involved (Proteins Encoded)**	
Acute myeloid leukemia	t(8;21)(q22;q22)	ETO (DNA binding protein)	CBF2A (transcription factor subunit)
	t(12;22)(p13;q11)	TEL (transcription repressor)	MN1 (nuclear protein)
Chronic myelogenous leukemia	t(9;22)(q34;q11)	Abl (tyrosine kinase)	Bcr (kinase/GTP exchange protein)
	t(9;12)(q34;p13)	Abl (tyrosine kinase)	TEL (transcription repressor)
B cell acute lymphoblastic leukemia	t(8;14)(q24;q32)	c-Myc (transcription factor)	Igh (Ig heavy chain)
	t(12;21)(p13;q22)	TEL (transcription repressor)	CBF2A (transcription factor subunit)
T cell acute lymphoblastic leukemia	t(1;7)(p32;q35)	TAL1 (transcription regulator)	TCRB (TCRβ chain)
	t(8;14)(q24;q11)	c-Myc (transcription factor)	TCRA (TCRα chain)
Chronic lymphocytic leukemia	t(14;19)(q32;q13)	Igh (Ig heavy chain)	Bcl-3 (transcription coactivator)
Myeloma	t(11;14)(q13;q32)	Cyclin D1 (cell cycle regulator)	Igh (Ig heavy chain)
	t(4;14)(p16;q32)	FGFR3 (growth factor receptor)	Igh (Ig heavy chain)
Non-Hodgkin lymphoma	t(14;18)(q32;q21)	Igh (Ig heavy chain)	Bcl-2 (anti-apoptosis protein)
	t(3;14)(q26;q32)	Bcl-6 (transcription repressor)	Igh (Ig heavy chain)

TABLE 20-2	Five-Year Survival Rates for Hematopoietic Cancer Patients			
	1950–1954	**1975–1977**	**1987–1989**	**2002–2008**
Leukemia	10%	34%	48%	59%
Myeloma	6%	25%	33%	43%
Hodgkin lymphoma	30%	72%	85%	88%
Non-Hodgkin lymphoma	33%	47%	59%	72%

[With data from Howlader et al., (eds). SEER Cancer Statistics Review, 1975–2009 (Vintage 2009 Populations), National Cancer Institute. Bethesda, MD, http://seer.cancer.gov/csr/1975_2009_pops09/, based on November 2011 SEER data submission, posted to the SEER website, April 2012.]

ii) Chemotherapy and Radiation

Standard chemotherapy and radiation therapy are the first modes of treatment offered to newly diagnosed HC patients. The response to these treatments is assessed by examining smears of cells taken from the patient's blood and/or BM. Malignant HC cells often morphologically resemble the blast stages of normal hematopoietic cells and so are termed "blast-like" (**Plate 20-1**). In a healthy individual, such blast-like cells account for up to 5% of cells developing in BM but are virtually absent from the blood, so that the presence of cells of this morphology in the blood is often a sign of cancer. After an HC patient has been treated, if more than 5% of cells in a BM smear are still blast-like and blasts are still present in a peripheral blood smear, the treatment has failed and there has been no clinical response. A patient whose HC does not respond to treatment at all is said to have *refractory disease*. A *partial response* to treatment means that between 5% and 20% of cells in the BM smear may be blast-like, but the number of blast-like cells in the blood smear has been reduced to almost zero. In addition, neutrophil and platelet counts are close to normal and the patient does not require a blood transfusion. A *complete response* means that <5% blast-like cells remain in the patient's post-treatment BM smear and that the blood smear is essentially clear of blast-like cells. If these levels hold over a period of at least 4 weeks following treatment, the patient's disease is said to be in *remission*. A *relapse* is the reappearance of disease in a patient who was previously in remission, and is often due to the survival and proliferation of CSCs. In general, a patient who has not suffered a relapse of the same tumor for a period of 5 years is said to be a *long-term survivor*.

Rarely, remission occurs due to a natural resolution of an HC in the absence of treatment.

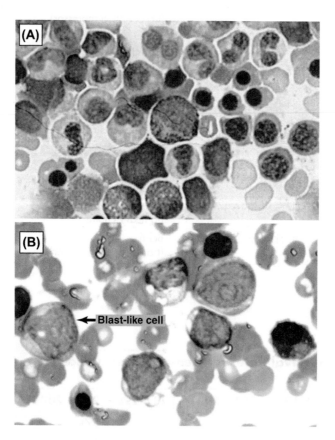

Plate 20-1
Comparison of Normal and HC Bone Marrow Aspirates

(**A**) In a normal bone marrow sample, only about 5% of hematopoietic cells have a blast-like morphology (large nucleus). (**B**) In a bone marrow sample from a patient with acute leukemia, many more than 5% of cells are blast-like. *[Reproduced by permission of Doug Tkachuk, Lifelabs, Ontario, Toronto.]*

iii) Immunotherapy

Immunotherapy has greatly contributed to recent progress in treating HCs. Monoclonal antibodies directed against molecules expressed predominantly on the surfaces of tumor cells are now used as therapeutic agents on a regular basis. For example, the tumor cells of many B cell lymphomas and leukemias show elevated expression of the surface marker proteins CD20 and/or CD22, so that mAbs directed against these molecules can be used to induce tumor cell death by either complement activation or ADCC. Conjugation of these antibodies to toxins or radioisotopes to form immunotoxins or immunoradioisotopes (refer to Ch. 16) can also be used to kill cancerous cells bearing the appropriate markers. Examples of immunotherapies specific to a particular type of HC are included in the following sections.

Examples of immunotherapeutic mAbs that can be used for treatment of either an HC or NHC were summarized in Tables 16-4 and 16-5.

NOTE: A mAb that is used for immunotherapy not only binds to its antigen with its Fab sites but also uses its Fc region to interact with FcRs expressed by innate leukocytes. This interaction typically triggers the innate leukocytes to undertake phagocytosis or ADCC, or activates complement. However, which activity is induced and to what degree depend on the FcR subtypes engaged and the specific polymorphisms of these receptors. For example, the success of treatment of non-Hodgkin lymphoma with rituximab, an anti-CD20 mAb, is heavily influenced by the types of FcRs expressed by the patient's leukocytes. Individuals expressing particular FcγR polymorphisms that are associated with increased IgG binding are much more likely than patients expressing other FcR alleles to benefit from rituximab. In addition, expression by the patient's leukocytes of an FcR polymorphism that encourages leukocyte effector functions, such as the ADCC mediated by NK cells, can enhance rituximab effectiveness. Such observations suggest that the sequence of the Fc region of a therapeutic mAb should be engineered to ensure that the mAb binds to the most useful FcR for a given circumstance, and that it does so with maximum affinity.

iv) Hematopoietic Stem Cell Transplants

Hematopoietic stem cell transplants can provide a cure for some HCs. Depending on the malignancy, the availability of potential donors, and the overall health status of the patient, either autologous or allogeneic HSCT may be used. Prior to receiving an HSCT, clinicians choose chemotherapy and radiation therapy conditioning regimens that they feel are most appropriate for the individual patient. As described in Chapter 17, these protocols range from the weakest, most easily tolerated form (non-myeloablative conditioning) to the strongest, most toxic form (myeloablative conditioning). The goal is to apply conditioning that is mild enough to be withstood by the patient but severe enough to destroy as many cancerous hematopoietic cells as possible. In addition, the conditioning regimen should ideally kill any immunocompetent effector cells in the recipient that might mount an attack on the incoming donor leukocytes and prevent their engraftment.

a) Autologous HSCT

In an autologous HSCT, a patient's own hematopoietic cells are used for reconstitution of his/her immune system. The HC patient first undergoes a round of conditioning that kills the rapidly dividing HC cells and so leads to remission. The patient's normal HSCs, which are quiescent, are spared. The patient is then treated with an agent such as *granulocyte colony-stimulating factor (G-CSF)* to mobilize the HSCs into the peripheral blood. The blood is collected, and the HSCs are purified by leukapheresis and other techniques that depend on surface marker expression. The purified HSCs are stored at −150°C while the patient undergoes a round of more intense conditioning that empties the patient's BM and periphery of virtually all hematopoietic cells. The patient is then infused with his/her stored HSCs to reconstitute a hopefully cancer-free immune system. Alternatively, if umbilical cord blood of the patient was stored at birth, it can be used as a source of cells for an autologous HSCT.

b) Allogeneic HSCT

In an allogeneic HSCT, donated hematopoietic cells are used to reconstitute the immune system of an appropriately conditioned HC patient. Most often, these transplants involve leukocytes prepared from the peripheral blood or BM of a donor who is matched with the patient at the HLA loci but differs at various minor histocompatibility loci. The donated leukocyte preparation contains not only HSCs but also immunocompetent donor lymphocytes and NK cells. As introduced in Chapter 17, these leukocytes recognize differences between molecules expressed by donor and patient cells and therefore destroy residual HC cells in the patient. This beneficial killing, which was discovered first in leukemia patients and therefore dubbed the **graft-versus-leukemia (GvL)** effect, accounts for the observation that relapse rates are lower in many HC patients who receive an allogeneic HSCT compared to those who receive an autologous HSCT. Subsequent work has shown that the same benefits can be realized in some lymphoma and myeloma patients, and even in some cases of solid tumors like metastatic renal cell carcinomas; metastatic colon carcinomas; neuroblastomas; and ovarian, pancreatic and prostate cancers. This wide spectrum of positive effects has prompted clinicians to talk of a broader **graft-versus-cancer** (GvC) effect.

The GvL (or GvC) effect first came to light when clinicians repeatedly observed that leukemia patients who received an allogeneic HSCT depleted of *all* donor T cells (in an effort to reduce potentially fatal GvHD) suffered a relapse of their disease. This relapse of leukemia was not observed if the patient received an HSCT that had been only mildly depleted such that it still contained residual donor T cells. It was deduced that, although the residual donor cells might attack allogeneic non-cancerous cells and thus contribute to GvHD, they could also turn their alloreactivity against any leukemic cells arising in the recipient after the HSCT. Originally, it was believed that these responses were due solely to the activities of alloreactive donor CTLs, but it is now apparent that alloreactive CD4+ T cells and NK cells are also important mediators of the GvC effect. CD4+ donor T cells kill cancerous recipient cells via Fas-mediated apoptosis, while CTLs use the perforin/granzyme system to carry out direct cytotoxicity. Cytokines produced by both effector T cell subsets contribute to tumor cell eradication as well as leukocyte recruitment. Although the donor lymphocytes appear to be responding mainly to major and minor histocompatibility differences expressed by hematopoietic cells of the recipient, some CTL responses against specific leukemic cell TAA/TSAs have recently been identified.

With respect to the involvement of NK cells in the GvC effect, the donor NK cells in an HSCT are activated when the KIRs they bear are not sufficiently bound by recipient MHC class I molecules to block NK cytotoxicity. Recipient cells that do not express the correct MHC class I ligands for these KIRs are killed. Such recipient cells will include not only leukemic cells but also recipient T cells responsible for graft rejection and recipient DCs that facilitate GvHD. Graft survival is improved and leukemic cells are killed, benefitting the patient on two fronts. As noted in Chapter 17, in the HSCT context, NK cells attack only other hematopoietic cells and do not kill recipient cells in non-hematopoietic tissues to any great extent. However, cancer cell killing by NK cells occurs on a more modest scale than that by T cells, and the NK cell contribution to the GvC effect is most obvious when an HSCT has been depleted of donor T cells.

Graft-versus-host disease (GvHD) was described in Chapter 17.

NK cell activatory and inhibitory receptors, including killer Ig-like receptors (KIRs), were described in Chapter 11.

NOTE: B cells may also be important for the GvC effect. Alloantibodies have been associated with GvHD, although it is not clear whether these antibodies are also involved in GvC activity. However, some HC patients who received an allogeneic HSCT and were later treated with anti-CD20 mAb to control severe GvHD experienced an almost immediate relapse of their malignancy, suggesting that B cells were mediating a GvC effect. In addition, in at least some HSCT-treated leukemia patients, antibodies have been identified that recognize antigens associated with the leukemic cells. Investigation of the humoral response's contribution to the GvC effect is ongoing.

Today, when deciding whether to treat a patient with allogeneic HSCT, a physician must weigh the value of GvHD prevention against the value of the GvC effect. Ideally, clinicians aim to preserve the donor T and NK cell responses to recipient *hemato-poietic* cells (maintaining GvC) while preventing the donor T responses to recipient *epithelial* cells (preventing GvHD). The cell preparations used for such HSCTs may be derived from BM, cord blood or peripheral blood. Judicious use of patient conditioning regimens and immunosuppressive drugs may be required to prevent excessive GvHD while preserving the killing of tumor cells by GvC and preventing a relapse of disease. An important consideration is that the earlier in disease progression the HSCT is undertaken, the more powerful the GvC effect appears to be.

Immunotherapy in which deliberate mismatching of donor NK cell KIRs and recipient MHC class I is invoked to trigger GvC has been working well in clinical trials of its use for treating certain acute leukemias. In one of the original studies establishing the validity of the GvC effect, HC patients received an HSCT from an unrelated donor along with treatment to prevent GvHD. The HSCT donors exhibited either KIR compatibility or incompatibility with their recipients. Patients who received KIR-incompatible HSCTs enjoyed both longer survival and a decreased probability of relapse compared to those who received KIR-compatible HSCTs. Most encouragingly, patients with myeloid HCs who were treated with KIR-incompatible HSCTs were alive and disease-free more than 4 years after the transplant. In addition to KIR-incompatible HSCT on its own, there is evidence from animal models that NK cell-mediated killing of leukemic cells in an HSCT recipient can be enhanced if the recipient is also treated with anti-KIR antibodies that activate donor NK cells. Translation of these approaches to humans is continuing.

v) HC Vaccines

Another innovative approach to the treatment of HCs may be vaccination with proteins or peptides representing antigens specific to the malignancy (i.e., a TSA). *In vitro* testing has confirmed that T cells recognizing such TSAs can be generated in vaccinated patients. Moreover, as is described in the following sections, clinical responses to TSA-related vaccination have been observed in some early clinical trials, particularly those involving leukemia patients.

The rest of this chapter outlines the clinical and genetic features of the major subtypes of leukemias, myelomas and lymphomas, and provides examples of various treatment options.

B. Leukemias

The majority of leukemias are broadly classified by whether the transformed cell is derived from the myeloid or lymphoid lineage and by whether disease onset and course are acute or chronic. The four major classes of leukemias are *acute myeloid leukemia (AML), chronic myelogenous leukemia (CML), acute lymphoblastic leukemia (ALL),* and *chronic lymphocytic leukemia (CLL).* The relative frequencies of these disorders in North America are given in **Figure 20-2**. Other leukemias include those involving transformed cells of erythroid, megakaryocytic or NK cell lineages. Acute leukemias can strike both children and adults, whereas chronic leukemias tend to arise in individuals 50 years of age and older. The overall incidence of leukemia is relatively low compared to NHCs and accounts for only 3% of all malignancies. Nevertheless, in persons under 20 years of age, acute leukemia is the leading fatal cancer.

The distribution of frequencies of the various types of leukemias is very different in adults and children. While ALL accounts for 12% of adult leukemias, it constitutes over 75% of leukemias in children.

I. Acute Myeloid Leukemia (AML)

AMLs are acute cancers of the myeloid lineage. The target cells are often early stage myeloblasts (**Plate 20-2**), but the disease is biologically very heterogeneous. Although AML mainly affects patients over 60 years of age but also accounts for 20% of childhood leukemias.

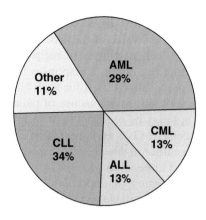

Fig. 20-2
Relative Frequencies of Leukemias in North America

The relative frequencies of the major types of human leukemias are shown. AML, acute myelogenous leukemia; CLL, chronic lymphocytic leukemia; CML, chronic myelogenous leukemia; ALL, acute lymphocytic leukemia. Numbers are based on estimated cases for 2012. *[With data from Howlader et al., (eds). SEER Cancer Statistics Review, 1975–2009 (Vintage 2009 Populations), National Cancer Institute. Bethesda, MD, http://seer.cancer.gov/csr/1975_2009_pops09/, based on November 2011 SEER data submission, posted to the SEER website, April 2012.]*

Anemia means "low red blood cells." Thrombocytopenia means "low platelets." Malaise means "weakness." Pallor means "pale skin."

i) Clinical Features

Like most leukemia sufferers, AML patients present with anemia and thrombocytopenia that are manifested as malaise, fatigue, pallor, dizziness, shortness of breath, bleeding, easy bruising, fever, and weight loss. Bone pain is present in rare cases. The high numbers of leukemic cells disrupt hematopoiesis in the BM such that the BM is said to have "failed." Respiratory infections are common due to the resulting impairment of innate immunity. *Hepatosplenomegaly*, which is caused by an accumulation of leukemic cells in the liver and spleen, may be observed. The leukemic cells may also migrate to **extramedullary tissues**, a term used by hematologists to refer to tissues and organs in the body that are neither lymphoid tissues nor BM.

ii) Genetic Aberrations

About half of AML cases show gross chromosomal abnormalities that are easily detected. In some of these patients, a whole chromosome has been gained or lost, but why these aberrations result in AML is not known. Most other cases of AML involving significant genetic anomalies are due to chromosomal translocations, and over 200 such aberrations have been identified. Not surprisingly, the genes affected are usually involved in cell proliferation, differentiation and/or apoptosis. Other abnormalities alter important signaling pathways or the BM microenvironment, and thereby disrupt hematopoiesis.

Plate 20-2
Acute Myeloid Leukemia

A bone marrow aspirate of an AML patient shows the presence of several myeloblasts. *[Reproduced by permission of Doug Tkachuk, Lifelabs, Ontario, Toronto.]*

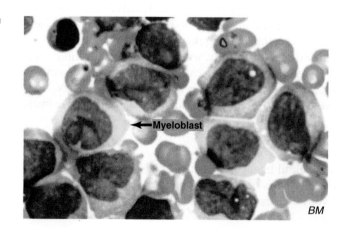

A common recurring translocation called (t8;21)(q22;q22) leads to AML because a part of the gene encoding the CBF2A subunit of the transcription factor CBF is fused to part of the ETO gene encoding a DNA binding protein. The failure to produce normal CBF2A impairs the normal function of CBF, which is to regulate many genes involved in myeloid cell development, proliferation and function. As a result of the translocation, myeloid precursors developing in the BM get "stuck" at an early stage of development and never receive the appropriate signals to stop proliferating and start differentiating. The excess cells spill into the blood and are detected as AML. Another subset of AML cases involves disruption of the TEL gene, which encodes a transcriptional repressor. The recurring translocation t(12;22)(p13;q11) fuses part of the TEL gene to part of the MN1 gene. The TEL gene is required for hematopoiesis in the BM and MN1 encodes a transcriptional coactivator.

In about 50% of AML cases, the leukemic cells do not display a gross chromosomal anomaly. An interesting subgroup of these AML cases is associated with the primary immunodeficiencies *Bloom syndrome* and *ataxia–telangiectasia*, which both involve hereditary defects in DNA repair genes. A second subgroup comprises patients with *Li–Fraumeni syndrome*, most of whom have germline mutations of the p53 TSG. Other patients with essentially normal karyotypes have point mutations or small deletions of the p15 gene essential for controlling cell cycle progression; the nucleophosmin gene involved in ribosome synthesis and intracellular transport of these organelles; or the FLT3 gene, which encodes a receptor with tyrosine kinase activity that is involved in myeloid differentiation and stem cell survival.

Bloom syndrome and ataxia–telangiectasia were described in Chapter 15.

iii) Treatment

Despite steady advances in cancer therapy since the late 1960s, the prognosis for AML patients remains poor. Although better clinical management of anemia, bleeding and infections, as well as careful monitoring of cardiac and other complications, has helped to improve the survival of AML patients, they still have a high rate of relapse despite apparently successful initial treatment.

AML patients diagnosed with very high leukemic cell counts need to be treated immediately with large doses of standard chemotherapeutics (refer to Ch. 16). Several new types of drugs are also in limited use, including agents that inhibit DNA replication, decrease DNA methylation, or block the cyclin-dependent kinases required for cell division. Chemotherapy of AML patients is often followed by HSCT. About 50% of AML patients receiving an allogeneic HSCT are alive and disease-free 5 years after the transplant, and the survival rate of pediatric patients has reached 60% in clinical trials in developed countries. Newer protocols of reduced intensity conditioning implemented prior to transplantation have contributed to this success, as has more individualized care. Clinicians are also currently exploring ways of further enhancing the GvC effect to reduce GvHD-related toxicity and decrease the rate of cancer recurrence after transplant. Some AML patients have been treated with an autologous HSCT, but the rate of relapse has been relatively high.

Most cases of AML have been refractory to the mAb-based therapies helpful in other types of HCs (see later). This difficulty most likely stems from the pronounced heterogeneity of AML: only when the myriad underlying genetic defects are defined will it be possible to design effective targeted agents. That being said, cases of AML in which the leukemic blasts express the surface marker CD33 can be killed if they internalize an immunotoxin composed of anti-CD33 mAb conjugated to a toxic drug. The success of therapy with this anti-CD33 immunotoxin has been enhanced by co-treatment with an agent designed to block the expression of Bcl-2, a critical anti-apoptotic protein that promotes cell survival. Parallel approaches can be expected to improve the prognosis of individual AML patients as additional molecules that can be targeted by mAbs are identified.

Most recently, scientists have sought to identify TAA/TSAs expressed by AML leukemic cells that could be used for cancer vaccines—that is, could induce an immune response to the malignant cells in the patient. Some of the numerous chromosomal translocations identified in AML patients generate antigenic fusion proteins that represent unique TSAs. Other AML-associated TSAs arise from duplication mutations in the FLT3 or nucleophosmin genes. A difficulty is that most of these antigenic proteins are expressed by relatively small subgroups of AML patients, meaning that strategies based on them would have only limited application. An exception is WT1 (Wilm's tumor 1), an oncogenic TAA that is upregulated in AML stem cells compared to normal HSCs. Clinical trials have established that immunization of AML patients with a cancer vaccine containing WT1 peptide or WT1 DNA can induce WT1-specific Th and CTL responses. Most encouragingly, early data indicate that some patients enjoy periods of remission or even complete eradication of their AML disease after WT1 vaccination. Similar promising results have been obtained in AML patients vaccinated with a modified peptide called PR1, which is derived from a protein aberrantly overexpressed by AML cells. Mucin-1 (MUC1), another oncogenic TAA often overexpressed by AML cells, is also under investigation as a vaccine candidate for AML patients.

Peptide and DNA vaccination approaches were discussed in Chapter 14.

> NOTE: Wilm's tumor antigen 1 (WT1) is a transcription factor frequently overexpressed by leukemic cells and in solid tumors. WT1 is expressed by normal HSCs in the BM but not by normal somatic cells. Due to this highly restricted pattern of tissue expression, this TAA is practically a TSA, and so an attractive target for immunotherapy and cancer vaccination.

II. Chronic Myelogenous Leukemia (CML)

Chronic myelogenous leukemia (CML) is characterized by increased numbers of mature and immature granulocytes in the blood. Although CML accounts for a mere 0.3% of all cancers, it represents about 10% of adult leukemias. This malignancy usually strikes adults with a median age of 50 years, and men are affected almost twice as often as women. CML accounts for less than 5% of childhood leukemias.

i) Clinical Features

The onset of CML is relatively benign, and the disease can remain "silent" in many patients for 3–5 years. This stage of the disease is known as the *chronic phase*, in which blood smears show increased numbers of precursor and mature granulocytes (**Plate 20-3**). However, the disease eventually progresses to an *accelerated phase* and finally to a *blast crisis*. In the accelerated phase, there is a sharp increase in the numbers of immature leukemic cells in the blood or BM. Splenomegaly is often present. In the blast crisis, CML blast cells make up more than 30% of cells in the blood or BM. Once the blast crisis has started, the leukemic cells start to infiltrate extramedullary tissues, and an untreated patient survives only a few months. Symptoms of blast crisis CML include weight loss, abdominal discomfort, bleeding, weakness, lethargy and night sweats. Histological examination of the blood and BM usually reveals mild anemia with increased numbers of

Plate 20-3
Chronic Myelogenous Leukemia: Chronic Phase

Peripheral blood smear of a CML patient showing increased numbers of precursor and mature (indicated) granulocytes. *[Reproduced by permission of Doug Tkachuk, Lifelabs, Ontario, Toronto.]*

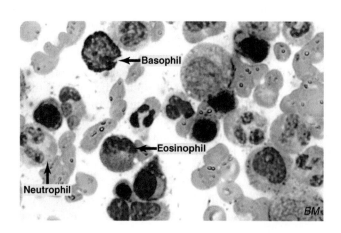

mostly granulocyte precursors but also B lymphoid (rarely T lymphoid), erythroid and megakaryocytic lineage cells, plus connective tissue-forming cells. This spectrum of leukemic cell types indicates that the original transformation event must have taken place in the BM in a very early progenitor that retained its ability to divide and differentiate into multiple hematopoietic lineages. As a result, despite the name of this disorder, cells of either the myeloid or lymphoid lineage may predominate in the blast crisis phase. Indeed, phenotypically, blast crisis CML can bear a striking resemblance to either AML or ALL.

ii) Genetic Aberrations

The tumor cells in over 90% of CML patients possess the *Philadelphia* (Ph) *chromosome*, whose derivation by a recurring translocation between chromosomes 9 and 22 is illustrated in **Figure 20-3**. The product of the translocation is an abnormal tyrosine kinase termed the *Bcr-Abl protein*, which directly and indirectly activates other genes involved in cell division, cell survival and myeloid differentiation. A less frequent recurring chromosomal translocation found in *atypical CML* is t(9;12)(q34;p13). This fusion event joins the Abl kinase gene to the TEL gene described earlier for AML.

iii) Treatment

The prognosis of CML patients varies according to the phase of the disease at diagnosis, with patients in blast crisis having the shortest survival times. In the past, CML

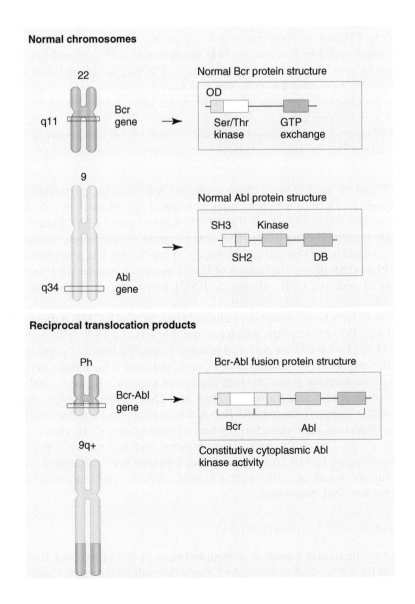

Normal chromosomes

22
q11

Bcr
gene

Normal Bcr protein structure

OD

Ser/Thr
kinase

GTP
exchange

9

Normal Abl protein structure

SH3   Kinase

SH2   DB

q34

Abl
gene

Reciprocal translocation products

Ph

Bcr-Abl
gene

Bcr-Abl fusion protein structure

Bcr   Abl

9q+

Constitutive cytoplasmic Abl
kinase activity

Fig. 20-3
The Philadelphia (Ph) Chromosome

The Ph chromosome arises from a reciprocal translocation that occurs between chromosomes 9 and 22. Normally, the Bcr gene on chromosome 22 encodes a protein with serine–threonine kinase and GTP exchange activities as well as an oligomerization domain (OD). The normal Abl gene on chromosome 9 encodes a protein that possesses tyrosine kinase activity, SH2 and SH3 protein interaction domains, and a DNA-binding domain (DB). During leukemogenesis, a t(9;22) (q34;q11) reciprocal translocation occurs in which the chromosomal breakpoints are located in the Bcr and Abl genes. One of the products of this translocation is a shortened version of chromosome 22 (the Ph chromosome) containing a new fusion gene made up of parts of the Bcr and Abl genes. The chimeric protein encoded by the Bcr-Abl fusion gene has constitutive Abl tyrosine kinase activity and acts in the cytoplasm due to regulatory effects exerted by the Bcr moiety.

patients were treated with either chemotherapy or the antiproliferative cytokine IFNα. Although IFNα administration is effective in more than 50% of chronic phase CML patients, it has significant toxic side effects, and it is still not clear exactly how it works.

In the early 2000s, the drug imatinib mesylate (Gleevec or Glivec) was introduced to treat CML patients whose tumor cells bear Bcr-Abl or TEL-Abl translocations. Imatinib mesylate is a *tyrosine kinase inhibitor (TKI)* that can be administered orally and potently inhibits the activity of the chimeric Abl kinase driving the proliferation of most CML leukemic cells. Other kinases active in normal cells, such as various Src family kinases crucial for lymphocyte signaling, may also be inhibited *in vitro* by this drug. However, at the doses of imatinib mesylate used to treat CML patients, systemic side effects on normal cells are not usually observed. Long-term imatinib mesylate administration has proved highly efficacious for both adults and those rare children with early chronic phase CML. However, TKI treatment is not a cure for CML, and control of the disease requires continuous and long-term drug administration. As a result, in 20% of imatinib mesylate-treated CML patients, the leukemic cells eventually develop resistance to this drug. Subsequent treatment with newer "second generation" TKIs can then be employed and may be effective for a while. Sadly, however, most TKI-treated CML patients eventually experience a fatal relapse of their disease due to the continued presence of CML stem cells. Unlike most CML leukemic cells, which express Bcr-Abl and undergo rampant proliferation, CML stem cells are quiescent (do not enter the cell cycle) and do not express Bcr-Abl, rendering them resistant to TKI treatment.

> NOTE: Despite its toxicity, IFNα administration has won a new place for itself in today's CML treatment regimens. Scientists speculate that, because IFNα stimulates normal HSCs to undergo cell division, it may have the same effect on CML stem cells; that is, IFNα may force these cells to enter the cell cycle, rendering them vulnerable to the effects of various chemotherapeutic drugs and TKIs that target proliferating cells. Moreover, IFNα treatment has been found to extend the lives of CML patients whose leukemic cells have developed resistance to TKIs. Still other recent evidence suggests that IFNα may increase a patient's antitumor CTL and NK cell populations. All these effects have ensured that IFNα continues to be part of the armamentarium used against CML.

Allogeneic HSCT can be used to treat chronic phase CML patients, particularly those who do not respond to TKIs, or who are under 30 years of age. More than half of chronic phase CML patients who receive such transplants achieve over 5 years disease-free survival, and some are cured completely. The slow progress of the disease at this stage may be partly responsible for this success. However, if the disease has progressed to the accelerated or blast crisis phase, the chance of HSCT success is reduced to about 15%. In the rare cases of pediatric CML, allogeneic HSCT is considered the first-line treatment because it can potentially cure the patient.

Most recently, researchers have undertaken clinical trials in which CML patients have been vaccinated with Bcr-Abl peptide, which presumably acts as a TSA. Although responses by anti-CML CTLs have been detected in some cases, the Bcr-Abl peptide used has proven to be only weakly immunogenic in most patients. This failure may be due to the fact that the junction point creating the fusion gene in any given CML patient can vary slightly in amino acid sequence. Thus, the TSA formed is unique to each tumor and may differ from any "standard" Bcr-Abl sequence chosen as the basis of a peptide vaccine. Clinicians have therefore turned to immunizing CML patients with peptide vaccines derived from the more immunogenic and less variable WT1 oncoprotein, with encouraging results. Approaches that enhance the recruitment of APCs to the vaccine site and boost the antigen presentation capacities of these cells are also under investigation for CML treatment.

III. Acute Lymphoblastic Leukemia (ALL)

ALL is characterized by increased numbers of lymphoblasts in the blood and BM. Cases of ALL account for 1.5% of all cancers. ALL can affect adults but occurs most often in children of age 3–5 years. Indeed, in industrialized countries, this disease is the most frequent form of cancer in children. The only firm cause identified for ALL is exposure to ionizing radiation, as occurred in Japan in 1945 and in Chernobyl in 1986. Other links sometimes cited as causes of ALL (exposure to electromagnetic fields or industrial pollutants) have yet to be definitively proven. Interestingly, like allergies and autoimmune diseases, ALL is more common in developed countries than in less-developed ones. Some researchers therefore speculate that skewed T cell and inflammatory responses to common infections (as proposed in the hygiene hypothesis) may be to blame for at least some cases of childhood ALL. However, to date, no infectious agent has been directly linked to ALL.

> The hygiene hypothesis in the contexts of allergy and autoimmune disease was discussed in Chapter 18 and Chapter 19, respectively.

i) Clinical Features

In general, ALL patients first present with fever, fatigue, weight loss, infections, bleeding, dizziness, easy bruising and joint pain. High numbers of lymphoblast-like leukemic cells are found in the blood, BM and extramedullary tissues. Most cases of ALL involve cells of the B lineage, although less common forms involve T lineage cells. Patients with B cell ALL (B-ALL) often have masses in the abdomen that can compromise kidney function. Patients with T cell ALL (T-ALL) may have a mass in the chest that can lead to wheezing and cardiac complications if it is sufficiently large. CNS involvement, characterized by leukemic cell invasion of the spinal cord and brain, is rare at ALL presentation but can be a source of relapsed disease if not addressed promptly. Symptoms of CNS involvement include headaches, nausea, lethargy and irritability.

A diagnosis of ALL is reached when analysis of the patient's BM smear reveals that at least 20% of cells present are leukemic lymphoblasts with large nuclei and minimal cytoplasm (**Plate 20-4**). Under the microscope, cases of B-ALL and T-ALL are morphologically alike so that immunophenotyping is used to distinguish them. B-ALL cells usually exhibit surface expression of CD19 and CD20, whereas T-ALL cells tend to express CD2, CD3, CD5 and CD7.

ii) Genetic Aberrations

Some of the genetic changes leading to ALL development are similar to those found in AML and CML patients. The tumor cells in about 25% of adult and 5% of childhood ALL cases exhibit a variant of the chimeric Bcr-Abl protein first described for CML. About 30% of childhood ALL cases show the t(12;21)(p13;q22) translocation in which the gene encoding the CBF transcription factor subunit CBF2A is translocated to the TEL transcriptional repressor locus. Other recurring chromosomal translocations leading to ALL involve the introduction of gene fragments into either the Ig loci or the TCR loci. In cases of mature cell B-ALL, the c-Myc gene is inserted into the Igh locus such that c-Myc becomes constitutively activated and deregulates the expression of genes involved in cell proliferation, apoptosis and differentiation. Similarly, the insertion of the c-Myc gene into the TCRA locus results in cases of T-ALL.

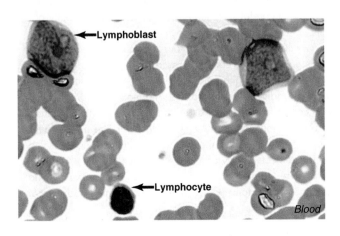

Plate 20-4
Acute Lymphoblastic Leukemia

Bone marrow aspirate of an ALL patient. A lymphoblast with its large nucleus and minimal cytoplasm is indicated, as is a normal mature lymphocyte. *[Reproduced by permission of Doug Tkachuk, Lifelabs, Ontario, Toronto.]*

Other genetic aberrations associated with ALL involve too many or too few chromosomes in the leukemic cells. In many cases of precursor cell B-ALL in children, the leukemic cells show more than 50 chromosomes per cell (normal cells have 46 chromosomes). These cases have a relatively favorable prognosis. Trisomy (three copies of a chromosome, rather than the usual two) for chromosomes 4, 10 or 17 is particularly common in childhood ALL. Other ALL leukemic cells show fewer than 45 chromosomes; these cases have a relatively poor prognosis. Intriguingly, with or without a chromosomal abnormality, more than half of T-ALL cases show inactivating mutations in the gene encoding the important cell fate-determining protein Notch1.

iii) Treatment

Chemotherapy is the first line of treatment for ALL, and modern drug regimens are quite successful in pediatric cases. However, the situation is not as rosy for adults with ALL. Although current chemotherapy strategies induce remission in 65–90% of treated adult ALL patients, the toxicity associated with these regimens is high, the majority of patients eventually suffer a relapse, and the survival rate is poor. Allogeneic HSCT is a viable option for relapsing ALL patients, but the success rate is low in adults.

Molecularly targeted therapies are now in more frequent use for adult ALL treatment. Administration of TKIs such as imatinib mesylate has been very successful for ALL patients whose leukemic cells bear the Ph chromosome. In addition, new classes of drugs have been developed that inhibit essential kinases and other molecules in the metabolic pathways that cancer cells use to circumvent unfavorable nutrient or oxygen conditions in their immediate microenvironment. Other agents block the pathways that tumor cells manipulate to resist normal death signals and thus survive. Finally, novel drugs targeting anti-apoptotic proteins like Bcl-2 are under investigation for ALL treatment. Although mAb-based immunotherapy has been explored for ALL, it has not been as effective as it has for other types of leukemias. Trials at various stages are examining the use of immunotoxins involving mAbs directed against CD20, CD52, CD19 or CD3 in adult ALL patients. One promising avenue may be TSA vaccination because the leukemic cells of 70% of all ALL patients overexpress WT1. The potential of WT1 to act as an ALL vaccine antigen is suggested by the observation that HSCT-treated ALL patients who mount CTL responses against WT1 show a decrease in their burden of leukemic cells and a lower risk of relapse.

NOTE: The differential diagnosis of ALL can be a challenge, as can be appreciated from the following hypothetical scenario. An 8-year-old boy is brought to his pediatrician because of an ongoing fever, wheezing breath sounds, generalized malaise, loss of appetite, and pallor. Upon initial review of these symptoms, the pediatrician first considers that the child might have an infection because, from a purely statistical standpoint, an infection is the most likely culprit. For example, a child with infectious mononucleosis ("mono") caused by EBV shows exactly the list of symptoms set out here. However, this doctor knows that leukemias are the leading cause of cancer death in children, and, being a careful sort, she also considers the possibility of a childhood HC. Fever, loss of appetite, and malaise are often the first signs of an HC in children. Her young patient is also bruised, which may have arisen through rough childhood play but could also be caused by thrombocytopenia, which is commonly associated with childhood HCs. Similarly, the child's pallor may be an outward manifestation of severe anemia, which is also a feature of HCs. The doctor decides that additional testing for ALL, the most common cancer in children, is warranted and orders a complete differential blood cell count and a BM aspirate. Sadly, the results confirm that the patient does indeed have ALL. However, the child's family does not despair because the pediatrician tells them that the cure rate for this malignancy is very high (about 90%) when this leukemia is detected early and treated promptly.

This scenario highlights the importance of keeping an HC in mind when confronted by what at first may seem like a mundane range of symptoms. Correct diagnosis of childhood ALL can avoid a potential tragedy.

IV. Chronic Lymphocytic Leukemia (CLL)

CLL arises from the transformation of a peripheral (mature) blood lymphocyte. As its name suggests, CLL is a chronic form of lymphocytic leukemia and is characterized by a prolonged disease course that is ultimately fatal. Clinicians consider this disease to be incurable due to the low rate of response of CLL patients to treatment and their high relapse rates. CLL mostly affects individuals over 60 years of age and occurs about twice as often in men as in women. Curiously, although CLL represents about 35% of adult leukemias in North America, it accounts for fewer than 5% of adult cases in Asia.

i) Clinical Features

Some CLL patients have no clinical symptoms of this disease at diagnosis and are only identified when a blood analysis is conducted to investigate other health concerns. These patients are considered at "low risk" and generally survive at least another 10 years even without treatment. Other CLL patients show splenomegaly or hepatomegaly and are considered to be at "intermediate risk." These patients live an average of 7 years following diagnosis. Patients considered at "high risk" have more advanced disease and display *lymphadenopathy* (swelling of the lymph nodes) and hepatosplenomegaly. Bacterial and viral infections are frequent in high-risk CLL patients. Furthermore, the chance of a CLL patient developing an NHC is twice that of the general population. To add insult to injury, about 20% of all CLL patients develop some type of autoimmune disease such as autoimmune hemolytic anemia or ITP (refer to Ch. 19). As a result of all these difficulties, CLL patients with advanced disease can expect to live only 2–4 years after diagnosis.

CLL is mainly a disorder of the differentiation and/or proliferation of B lineage cells. B cell production in the BM and B cell functions in the periphery are disrupted. Two subgroups of CLL patients have been identified at the molecular level, each accounting for about half of all cases. In one subgroup, the malignant B cells have Ig genes showing evidence of somatic hypermutation; the prognosis for these patients is relatively favorable. In patients in the other subgroup, the Ig genes of the malignant cells have not undergone somatic hypermutation; these patients have a poor prognosis. Alterations to the normal levels of many cytokines, the kinase ZAP70, and the transcription factor Bcl6 have also been noted in both subgroups. Most likely as a result of these B cell abnormalities, Th cell responses to antigens are also impaired, culminating in reduced antibody responses in CLL patients. About 5% of CLL cases involve transformed mature T lymphocytes. Patients with this HC respond only weakly to chemotherapy and have a very poor prognosis.

ii) Genetic Aberrations

Leukemic cells in about 80% of CLL cases show overt cytogenetic alterations. The most common mutation, found in over 50% of CLL patients, is a deletion at 13q14. This deletion affects the micro-RNA genes miR15 and miR16, which are involved in tissue-specific gene regulation. Of note, the Rb TSG found in this chromosomal region is normal in these patients. Another chromosomal deletion found in 20% of CLL cases occurs at 11q23 and removes the ATM TSG. The disease course is particularly aggressive in 11q23 CLL patients. About 15% of CLL patients have trisomy 12. In half of trisomy 12 CLL cells, a translocation at t(14;19)(q32;q13) joins the Igh locus to that encoding the transcriptional coactivator Bcl-3. Overexpression of the Bcl-3 gene stimulated by elements in the Igh locus may in turn drive excessive expression of survival genes, causing lymphocyte accumulation.

Atypical CLL occurs in about 15% of patients. In these cases, the CLL cells acquire new mutations that drive the conversion of the disease into a lymphoma. The disease course is aggressive and the prognosis poor. This group of patients exhibits abnormalities of chromosome 17, some of which affect the p53 TSG located at 17p13.

iii) Treatment

Untreated CLL patients survive for an overall average of 6 years following diagnosis. Because CLL patients are generally over 60 years of age, it is important to weigh the benefits of treatment against the significant side effects of chemotherapy. Infections due to chemotherapy-induced immunosuppression pose a real threat to elderly patients. Consequently, one of the most important management issues of CLL is to decide when to simply watch disease progression and when to medically intervene.

When CLL requires treatment, chemotherapy using a combination of drugs like fludarabine and cyclophosphamide is the preferred first line of attack. In addition, lenalidomide, a drug developed for the treatment of myeloma and therefore discussed in the next section, is being explored as a treatment for CLL. Clinicians are also turning to combinations of chemotherapy and immunotherapy. Anti-CD52 mAb, which binds to the CD52 marker expressed on most B and T lymphocytes, has been useful for CLL treatment. About 30% of CLL patients who do not benefit from chemotherapy will respond to anti-CD52 mAb therapy. Another subset of CLL patients responds to high doses of anti-CD20 mAb. IFNα, which works well for many CML patients, appears to be of minimal benefit to CLL patients. HSCT is rarely attempted as a treatment for CLL because most patients are at an age when they are less likely to survive the HSCT procedure. However, some younger CLL patients have enjoyed prolonged remission following HSCT.

> NOTE: Recent examinations of CLL cells have revealed that many of them are generated in the lymph nodes in a process driven by CD40-CD40L engagement mediated by Th cells. *In vitro*, a CLL cell that has been isolated from a lymph node and treated with anti-CD40 mAb to trigger CD40 signaling can resist apoptosis induced by agents that easily kill untreated CLL cells. Thus, agents that can block the CD40-CD40L signaling that occurs in the lymph node and promotes CLL cell proliferation and apoptotic resistance may be of future clinical benefit. The search for compounds and/or mAbs that can achieve this goal without being unacceptably toxic to normal cells is ongoing.

C. Myelomas

Normal plasma cells cannot divide and so die soon after doing their short-lived job of secreting antigen-specific antibody. In contrast, cancerous plasma cells divide uncontrollably and express huge quantities of antibodies or single Ig chains of unknown antigenic specificity. The Ig protein produced by the malignant plasma cell is called a *paraprotein* (as in "IgM paraprotein" or "para-IgM"). As a group, HCs involving transformed plasma cells are called **plasma cell dyscrasias**. Plasma cell dyscrasias are not considered leukemias because, although their paraprotein products may enter the blood, the cancerous plasma cells themselves are (at least initially) confined to the BM and do not enter the circulation. There is a broad spectrum of plasma cell dyscrasias whose severity ranges from the almost asymptomatic to life-threatening. At the mildest end of this spectrum is the colorfully named "monoclonal gammopathy of undetermined significance" (MGUS). About 70% of MGUS patients show no clinical symptoms. If

Plate 20-5
Myeloma

Bone marrow aspirate of a patient with multiple myeloma. A myeloma cell with its plasma cell-like morphology is indicated. *[Reproduced by permission of Doug Tkachuk, Lifelabs, Ontario, Toronto.]*

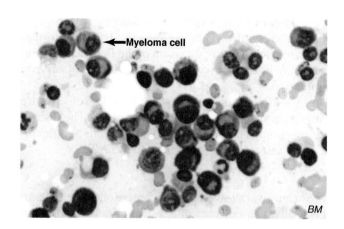

←Myeloma cell

BM

symptoms do occur, they take the form of mild anemia and usually disappear on their own with time. The most serious plasma cell dyscrasias are the **myelomas** (**Plate 20-5**). Myelomas can arise from pre-existing MGUS disease when a minimally transformed plasma cell acquires additional mutations that cause it to divide aggressively and synthesize high levels of a particular paraprotein. In its most advanced stages, large numbers of para-Ig-secreting tumor cells leave the BM and take up residence in multiple body sites, such that this disease is often referred to as *multiple myeloma*.

I. Clinical Features

Myelomas represent about 1% of all malignancies in North America, with over 80% of cases occurring in persons over 60 years of age. Myeloma incidence is slightly higher in men than women and occurs in twice as many blacks as whites. Exposure to ionizing radiation has been implicated in myelomagenesis. For example, epidemiological studies of individuals exposed to atomic radiation in Japan in 1945 demonstrated a 5-fold increase in myelomagenesis beginning about two decades later. Exposure to environmental carcinogens or certain chemicals also increases the rate of myeloma development.

Myeloma cells cause disease because their high numbers clog the BM and disrupt hematopoiesis such that the production of erythrocytes is decreased, resulting in anemia. Normal B cell production may also be impaired, leaving myeloma patients very vulnerable to infections. As well, the physical presence of high concentrations of para-Ig in the blood can cause coagulation and circulatory difficulties. Pulmonary and neurological effects may also be observed. In many myeloma cases, the transformed cells produce an abundance of Ig light chains in the absence of Ig heavy chains. These free Ig light chains can form aggregates with a glycoprotein in the urine, and deposition of these aggregates in the kidney can result in the failure of this organ.

Myeloma disease is classified according to the staging system shown in **Table 20-3**. These stages are defined by the number of malignant cells present; their pattern of distribution in the body; and blood levels of hemoglobin, calcium and paraproteins. In stage I myeloma disease, the myeloma clone remains partially reliant on BM stromal cell factors for its survival and so is confined to the BM. Serum hemoglobin and calcium levels are normal, the number of malignant cells is low and the production of Ig paraproteins is relatively low. Clinical symptoms are often mild and may go unnoticed. Stage II myeloma patients show subnormal hemoglobin, elevated serum calcium, moderate levels of malignant cells in the BM, and moderate levels of Ig paraprotein the blood. In patients with stage III myeloma disease, the number of malignant cells is high, and large masses may form in the BM. In addition, the myeloma cells accumulate oncogenic mutations that free them from their dependence on the BM stroma. Large numbers of malignant cells then migrate out of the BM and into the blood, infiltrating a variety of extramedullary tissues and disrupting their functions. Serum calcium is highly elevated and hemoglobin is severely decreased, frequently

TABLE 20-3 Stages of Myeloma	Stage I	Stage II	Stage III
Number of malignant cells	Low	Moderate	High
Location	Bone marrow	Bone marrow	Bone marrow, blood, extramedullary tissues
Serum Hb and Ca^{2+}*	Both normal	$\downarrow$ Hb $\uparrow$ Ca^{2+}	$\downarrow\downarrow$ Hb $\uparrow\uparrow$ Ca^{2+}
Ig paraprotein	Low	Moderate	High
Bone pain	No	No	Yes
Mean survival (untreated)	5 year	3 year	1 year

*Hb, hemoglobin; Ca^{2+}, calcium.

leading to renal complications and anemia. Levels of Ig paraproteins may be three times those in stage I patients.

One of the biochemical consequences of myelomagenesis is the deregulation of several cytokines and chemokines, the most prominent of which is IL-6. When overexpressed either by the myeloma cells themselves or by BM stromal cells, IL-6 appears to drive the development and maintenance of the cancer. When this cytokine is combined with IL-1, TNF and a chemokine called MIP-1α (all secreted by myeloma cells), bone metabolism is disrupted such that lesions develop in the bone structure. These lesions cause significant pain to many stage III patients. A protein called DKK1 also contributes to bone lesion development because DKK1 inhibits the intracellular signaling that supports bone formation. The overexpression of DKK1 by myeloma cells results in suppression of bone formation and stimulation of bone resorption. Additional cytokines implicated in myeloma disease progression include IL-2, IL-7, IL-11, LT and GM-CSF. The upregulation of these cytokines and their receptors tends to suppress Ig production by normal B cells and to inhibit helper T and NK cell functions, thereby increasing the susceptibility of these patients to infection.

II. Genetic Aberrations

The genetic aberrations leading to myelomagenesis are complex and have been much harder to identify than those associated with leukemias. There are some recurring chromosomal translocations associated with myelomas, but no one genetic change appears to be dominant or essential. Trisomy 3, 5, 6, 7, 9, 11, 15 or 19 has been reported in various myeloma clones. For unknown reasons, aberrations of chromosomes 6 and 9 are associated with the most favorable disease outcomes. In contrast, a deletion in the chromosomal region 13q14 that affects the Rb TSG is associated with a very poor prognosis, as is the so-called del(17p) deletion. Cases of stage III myeloma often show reciprocal translocations that affect the Igh locus (refer to **Box 20-2**). The genes involved in these translocations include the fibroblast growth factor receptor FGFR3 and the cell cycle regulator cyclin D1. More subtle genetic abnormalities include mutations of the c-Myc or p53 genes or activation of the Ras oncogene. These alterations are frequently reported in stage II and III myeloma patients. Other myeloma patients have mutations in the cyclin-dependent kinases associated with cell cycle control. Interestingly, regardless of the underlying genetic defect, almost all myeloma clones show upregulation of the anti-apoptotic genes Bcl-2 and Bcl-xL. Overexpression of these cell survival proteins, which is also seen in certain lymphomas, is thought to drive myeloma expansion.

III. Treatment

Myelomas are among the hardest HCs to treat. Chemotherapy is the first line approach for myeloma patients with progressing disease, but the 5-year survival rate is only about 40%. At later stages or in the most aggressive cases, *magnetic resonance imaging (MRI)* may be helpful in evaluating the number and location of myeloma masses. Localized irradiation can be effective for patients in whom these masses are isolated. As was true for CLL patients, the timing and choice of myeloma therapy are especially important for older individuals, for whom the side effects of medical intervention have to be balanced against quality of life.

The drug *thalidomide*, which was originally prescribed for the relief of nausea in pregnant women, became infamous in the 1960s for its devastating effects on fetuses. In the 21st century, this drug is finding a new use as a chemotherapeutic agent for myeloma. Thalidomide promotes the apoptosis of myeloma cells and blocks stromal cell secretion of IL-6 and the angiogenic factors that support myeloma growth. However, the side effects of thalidomide use are significant. Researchers have therefore devised several less toxic thalidomide analogs that are efficacious therapies not only for myeloma but for a range of HCs. The analog *lenalidomide* has the same direct and indirect killing effects on cancer cells as thalidomide but is a more potent immunoregulator. Accordingly, lenalidomide can also be used to treat inflammatory disorders.

A combination of lenalidomide and dexamethasone has been particularly helpful for patients with relapsed or refractory myeloma.

Because the tumor microenvironment is now known to play a key role in the survival and proliferation of myeloma cells, therapies have been designed to decrease the interactions between myeloma cells and the BM stromal cells that support them. In addition, attempts have been made to inhibit cytokine induction at affected local sites within the BM. For example, many myeloma patients have benefitted from treatment with a proteasome inhibitor called *bortezomib*. Bortezomib inhibits the normal degradation of proteins by the proteasome, including the degradation of an inhibitor of the transcription factor NF-κB that is required for IL-6 expression. If this inhibitor is not degraded, NF-κB cannot be activated, and the production of IL-6 that drives myeloma cell proliferation is blocked. Bortezomib also directly induces the apoptosis of myeloma cells and decreases angiogenesis.

Immunotherapeutic approaches to myeloma treatment include anti-CD20 mAb administration, but this is effective only for the 15–20% of patients whose myeloma cells express CD20. Other molecules targeted include growth factor receptors and adhesion molecules expressed by myeloma cells. An example of the latter class of mAbs is *elotuzumab*, which is directed against the glycoprotein adhesion molecule CS1. CS1 is highly expressed by the malignant cells in most myeloma patients but not by normal cells. Combinations of this mAb, or a mAb directed against IL-6 or the angiogenic factor VEGF, with either lenalidomide or bortezomib have been effective in clinical trials involving patients with relapsed or refractory myeloma. Additional mAbs directed against other cell surface molecules have been conjugated to microtubule-disrupting agents to form immunoconjugates. Some of these agents are yielding promising results in mouse models and early clinical trials. Although many of the molecules targeted by these mAbs are expressed by both myeloma cells and normal cells, the impact on the latter appears to be minimal such that these drugs are well tolerated by patients. Many other mAbs targeting various surface and internal molecules expressed by myeloma cells remain under pre-clinical investigation. In addition, researchers are starting to examine the possibility of exploiting potential T cell targets for myeloma treatment. For example, myeloma cells often overexpress the oncogenic TAA MUC1, which could in theory be used as the basis of a therapeutic antimyeloma vaccine.

Autologous HSCT is a realistic option for many myeloma patients, especially for those younger than 70 years of age. The mean survival of myeloma patients after autologous HSCT is about 5 years. Allogeneic HSCTs have also been performed in order to take advantage of an apparent GvC effect against myeloma cells. While this approach has allowed some patients to enjoy complete remission or reduced relapse of disease, others have suffered from significant morbidity or mortality due to GvHD. Attempts are now being made to avoid T cell-mediated GvHD by infusing patients with donor NK cells that are HLA-matched (to optimize engraftment) but KIR ligand-mismatched (to boost the GvC effect against myeloma cells).

D. Lymphomas

Lymphomas are solid cancers that initiate from the malignant transformation of a single lymphocyte. The affected lymphocyte is usually located in a lymph node but may be resident in another organized lymphoid tissue outside the BM such as the spleen or thymus. When the transformed lymphocyte is positioned in a diffuse lymphoid tissue such as the GALT, the lymphoma that develops is said to be *extranodal*. Lymphomas almost always depend on surrounding stromal cells for survival and growth factors as well as vital intercellular contacts, and so are generally restricted to sites within tissues. Thus, from its initiation site, a lymphoma tends to spread to additional secondary lymphoid tissues and eventually to non-lymphoid organs. Occasionally, a lymphoma cell undergoes additional mutations that allow it to survive and circulate in the blood; i.e., it becomes a leukemic cell. The disease may then be called a "leukemia/lymphoma."

The progression of any lymphoma can be described in four stages, as set out in **Table 20-4**. The identification of tumor sites and the staging of lymphomas have been

TABLE 20-4	Stages of Lymphoma
Stage	**Diagnostic Features**
I	One or more diseased lymph nodes are present in a single group of lymph nodes in one particular lymphoid tissue of the body.
II	Diseased lymph nodes are present in more than one group of lymph nodes, but all diseased nodes are contained either above or below the diaphragm. Tumor cells may also be present in a single organ near an affected node.
III	Diseased lymph nodes are present in two or more groups on both sides of the diaphragm. Tumor cells may also be present in the spleen and/or another organ near an affected node.
IV	Wide dissemination of tumor cells into multiple lymph nodes, bone marrow, liver and multiple organs.

made more accurate in recent years by the application of a diagnostic imaging technique called ^{18}F-FDG PET/CT. PET/CT stands for *positron emission tomography/ computerized tomography,* and ^{18}F-FDG is a radioactive form of glucose that emits positrons. Tumor cells use glucose at a much higher rate than normal cells, and so stand out in images of the body when the patient has been administered minute amounts of ^{18}F-FDG as a tracer and PET/CT imaging has been applied.

As introduced above, lymphomas are broadly classified into **Hodgkin lymphomas (HLs)** and **non-Hodgkin lymphomas (NHLs)**. According to current estimates of the American Cancer Society, HLs and NHLs will account for 6% and 47%, respectively, of all new cases of HC diagnosed in 2013. HLs and NHLs are distinguished by the architecture of the malignant mass, the morphology of its component cells, and the surface marker phenotype of these cells. In HL, only a tiny percentage of cells in the tumor mass are actually cancerous, and these cells are usually a peculiar B lineage-like cell type called *Reed–Sternberg* cells. The remainder of the tumor mass is made up of a so-called *reactive infiltrate* composed of non-transformed lymphocytes, fibroblasts and other cell types. In NHL, the solid mass consists almost entirely of transformed lymphocytes. The malignant cells are most often derived from a peripheral B cell but sometimes from a peripheral T cell. Occasionally, the cellular origin of the lymphoma cannot be clearly defined.

I. Hodgkin Lymphoma (HL)

In about 1900, researchers studying the enlarged lymph nodes of "Hodgkin's disease" patients found that each affected node contained a few large, multinucleated cells within an infiltrate of normal-looking cells. These unusual multinucleated cells became known as Reed–Sternberg (RS) cells in honor of their discoverers. RS cells were subsequently shown to be clonal in their growth, suggesting that they were transformed. Indeed, Hodgkin's disease was eventually demonstrated to be a true HC in which the RS cells were the tumor cells of the malignant mass. Hodgkin's disease was later renamed "Hodgkin lymphoma", and tissues showing the infiltration of an HL mass are said to be "HL-involved."

i) Clinical Features

HL is a rare disorder, representing less than 1% of all malignancies in North America. Unlike most solid NHCs, HL usually occurs in relatively young patients between the ages of 15 and 35 years. Males are affected slightly more often than females. Most patients initially present with a lump in the neck region due to one or more enlarged lymph nodes, although chest masses may appear upon radiography. Patients also experience a constellation of systemic complaints known as *B symptoms,* which are defined as unexplained rapid weight loss, fatigue, cyclic bouts of fever, and night sweats frequently accompanied by chest pain. *Pruritis* (intense itching) is present in about 20% of patients. HL patients are highly vulnerable to fungal and viral infections.

At least in the early stages of HL, RS cells do not establish just anywhere in the body. Instead, these cells appear to travel via the lymphatics in a sequential fashion from one lymph node to the next lymph node in the anatomical chain. It is not understood exactly how this unique pattern of HL spreading occurs, but it appears to depend on some type of close-range interaction either between cells or between cells and cytokines. Some scientists have speculated that cytokines accumulating in an affected lymph node spill down the efferent lymphatic, exiting that node to enter the next node in the chain. These cytokines may then alter conditions among the cells in the second node such that a reactive infiltrate develops. When an RS cell does break away from the original node and reaches the second node, conditions may then be ripe for its continued survival and proliferation. Histological studies have confirmed that lymph nodes near an HL-involved lymph node frequently contain reactive infiltrate but no RS cells. In the later stages of HL, however, the RS cells acquire additional mutations that allow them to become independent of the support of cytokines and the reactive infiltrate. The disease then spreads to additional lymph nodes in a less systematic fashion and invades non-lymphoid organs and tissues.

ii) Genetic Aberrations

It has been difficult to ascertain what types of mutations are associated with HL because of the challenges in obtaining adequate amounts of relatively pure populations of fresh RS cells for biochemical and molecular analyses. Although recurring chromosomal translocations are not common, the tumor cells of 25–75% of HL patients have cytogenetic anomalies such as trisomy 1, 2, 5, 12 or 21. Mutations of known TSGs and oncogenes have been found in only a tiny percentage of HL cases. There is some recent evidence that, in RS cells, abnormal hypermethylation of the DNA in promoters of various genes associated with cell cycle regulation and apoptosis may reduce the expression of these genes, driving RS cell survival and proliferation.

iii) Role of Cytokines in HL

Whatever the defect in the genome of an RS cell, the abnormality has a profound effect on the regulation of cytokines and their receptors that appears to be crucial for the development of HL disease. HL patients succumb readily to fungal and viral infections, indicating that Th1 and CTL responses are impaired. However, antibody responses to pathogens are functional. This reduction in cell-mediated immunity coupled with normal humoral immunity suggests that immune deviation to a Th2 response may be occurring during HL development.

A multitude of cytokines has been implicated in HL in various studies, but it is IL-13 that plays the most important role in HL pathogenesis. Only IL-13 is consistently and abundantly expressed by RS cells from different HL patients. The IL-13 receptor is also upregulated on RS cells, establishing a feedback loop in which the IL-13 produced by RS cells binds to IL-13R on the same (and other) RS cells. The persistent IL-13 signaling results in the constitutive activation of transcription factors that in turn activate the transcription of genes driving RS cell proliferation and survival. Some immunologists theorize that a minimally transformed precursor B cell which starts to overproduce IL-13 may respond to that IL-13 by becoming a full-fledged RS cell. IL-13 and other cytokines frequently elevated in HL may also contribute to the recruitment of the cells composing the reactive infiltrate, and this infiltrate may secrete still more cytokines and/or supply intercellular contacts that promote RS survival and proliferation.

iv) Viral Involvement?

The exact etiology of HL is unknown, but some scientists theorize that infection by EBV, a virus capable of transforming lymphocytes, may be a causative factor in some cases. Two EBV proteins can be found on the surfaces of RS cells in 40% of HL tumors, suggesting that these viral proteins might have a role in HL development. In addition, clinical studies have found that HL is 2- to 5-fold more frequent in patients who have had infectious mononucleosis, which is caused by EBV. However, 60% of HL tumors show no evidence of EBV proteins. Moreover, at least in North America, over 80% of the population has experienced an EBV infection at some point in their lives, but only a fraction of this population ever develops HL. Whether EBV is a contributing factor to HL therefore remains controversial.

v) Subtypes of HL

Two major subtypes of HL exist: *classical HL* (95% of cases) and *nodular lymphocyte predominant HL* (NLPHL; 5%). Classical HL is characterized by the presence of RS cells at a frequency of 0.1–1% (**Plate 20-6**). As well as by their multiple nuclei, RS cells are identified by their distinctive shape and their expression of the markers CD30 and CD15. In line with the hypothesis that RS cells are derived from early B cell precursors, the Ig genes are often rearranged in RS cells, but there is no evidence of somatic hypermutation. Rarely, there is some expression of Ig light chains. In about 40% of classical HL cases, the RS cells express CD20.

NLPHL is characterized by the presence of malignant *lymphocytic and histiocytic (L&H)* cells rather than RS cells. L&H cells are also called "popcorn cells" due to their unique appearance (**Plate 20-7**). L&H cells do not usually express CD30 or IL-13 but almost always express CD19 and CD20. Unlike the Ig genes in RS cells, the Ig genes in L&H cells show evidence of considerable somatic hypermutation. Ig light chains are more frequently expressed than in classical HL cases.

vi) Treatment of HL

The prognosis of HL patients was improved considerably in the early 1980s by the discovery that HL disease spread sequentially through the lymphatics to adjacent lymph nodes. Prophylactic irradiation of lymphatic channels in the immediate area of the affected node, a procedure called *extended field radiation*, became routine and increased the survival of many HL patients. Today, although some early stage HL patients may require only radiation treatment, chemotherapy is advised for individuals

Plate 20-6
Classical Hodgkin Lymphoma

Lymph node section from a patient with classical HL, indicating a binucleate RS cell with multiple red-staining nucleoli. [Reproduced by permission of Doug Tkachuk, Lifelabs, Ontario, Toronto.]

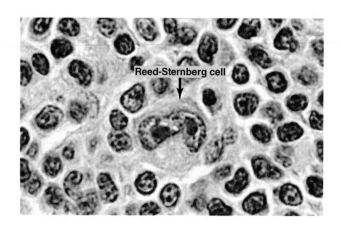

Plate 20-7
"Popcorn cells" in Nodular Lymphocyte Predominant Hodgkin Lymphoma

Lymph node section from a patient with non-classical HL. A popcorn cell is indicated. [Reproduced by permission of Doug Tkachuk, Lifelabs, Ontario, Toronto.]

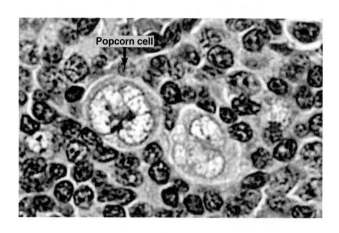

with more aggressive subtypes of the disease. Patients with advanced HL undergo multiple rounds of chemotherapy combined with aggressive local irradiation. These regimens, coupled with good clinical management to control infections, have enabled over 80% of both early and later stage HL patients to enjoy long-term disease-free survival.

Unfortunately, the increased doses of chemotherapy and radiation required to achieve these excellent results for advanced HL are associated with several serious side effects. The impact of these problems is made more devastating by the fact that HL patients are generally quite young (20–40 years of age). Sterility and hypothyroidism are not uncommon after extensive chemotherapy, and pulmonary and cardiac complications are frequent. However, the most sobering unintended consequence of aggressive HL treatment is a high rate of secondary tumorigenesis. AML, NHL and even some NHCs such as lung and breast cancers have appeared in HL patients within 5–15 years after initial treatment. Indeed, about 15% of conventionally treated HL patients develop a secondary cancer within 20 years.

The existence of HLs that resist aggressive chemotherapy and irradiation, coupled with the desire to eliminate secondary cancers and severe complications, drives the development of alternative therapies for HL. Patients now have several other options, some of which are still experimental in nature. Anti-CD20 mAb treatment has been helpful for HL cases featuring CD20+ RS cells. For patients suffering a second relapse of their disease, clinical trials are ongoing for treatment with an immunoconjugate called *brentuximab vedotin*. This compound contains an antimitotic agent (vedotin) coupled to a chimeric mAb (brentuximab) that is directed against the CD30 protein found on RS cells in most HL patients. The administration of neutralizing antibodies against IL-13, the growth factor postulated to drive RS cell proliferation and survival, is also being assessed as a novel therapy for HL. Autologous HSCT has been used to successfully treat some HL patients, and may be the treatment of choice after relapse. In addition, there has been some evidence for a GvC effect following allogeneic HSCT in some HL patients, in that relapse rates were decreased compared to HL patients treated with autologous HSCT. Allogeneic HSCT after reduced intensity conditioning is under continuing investigation, but is currently recommended only within a clinical trial setting for a very small subset of HL patients. It is not yet clear that the benefits of this approach outweigh its risks.

II. Non-Hodgkin Lymphoma (NHL)

The NHLs are a family of heterogeneous cancers. Unlike HL, the cancerous mass of an NHL is composed almost entirely of malignant lymphoid cells. In addition, NHLs lack unique tumor cell types such as RS cells or L&H cells. NHLs are the fifth most common cancer in the U.S., account for about 3–5% of all human cancers, and occur twice as often in men as in women. The median age of NHL patients is 66 years, considerably older than that for HL patients. Nevertheless, significant numbers of younger people between the ages of 30 and 40 develop NHL disease. Most NHLs are generally rare in children with the exception of *Burkitt lymphoma* (see later) in African children. Interestingly, the incidence of NHL in adults has doubled since the 1980s. Although a small percentage of this "epidemic" is due to a 10-fold increase in lymphomas in AIDS patients, an across-the-board elevation in incidence has been found for the general population. At least part of this increase has been attributed to a combination of factors, including (1) an expanding population of older individuals; (2) new imaging technologies that can detect previously unnoticed lymphomas; and (3) the reclassification of some HL malignancies under the NHL umbrella. Although specific genetic determinants promoting the development of most NHLs have yet to be conclusively identified, there is accumulating evidence that infection with certain pathogens can cause at least some NHLs. Prolonged exposure to various noxious substances has been cited as a possible cause of NHLs, but there is as yet no definitive proof of this link.

Based on U.S. National Institutes of Health data from 2005–2009, the median age at diagnosis of HL and NHL patients is 38 years and 66 years, respectively.

i) Clinical Features

NHL patients usually present with B symptoms (unexplained weight loss, fever, night sweats) accompanied by fatigue and pain in the bones, chest and abdomen.

Physical examination may reveal lymphadenopathy and/or hepatosplenomegaly. The lymphadenopathy may "wax and wane" over time, meaning that the lymph nodes swell and resolve repeatedly. Sometimes the NHL presentation is extranodal, in that the lymph nodes are not obviously affected and the suspicious lumps are present in other tissues such as the skin and GI tract. Most NHLs eventually go on to involve the BM.

Unlike HL, NHL does not usually spread sequentially from lymph node to lymph node. Rather, NHL cells may migrate from the initial affected lymph node and travel via the blood and lymphatics to scattered nodes in various regions of the body. It is postulated that the NHL tumor cells that have left the original lymph node are no longer reliant on a particular microenvironment or cytokine gradient, making local and distant lymph nodes equally suitable for invasion.

ii) Pathogen Involvement

Although the causes of most cases of NHL are unknown, infection by a pathogen has been linked to several types of NHLs. In some instances, an oncogenic virus transforms lymphocytes directly, leading to malignancy and lymphomagenesis. In other cases, chronic inflammation induced in response to a persistent pathogen is thought to interfere with normal lymphocyte proliferation and apoptosis. The affected cells may then accumulate mutations to DNA repair genes, oncogenes and/or TSGs that promote lymphomagenesis. This link to infection means that different NHLs are prevalent in different geographic areas, depending on the global distribution of the relevant pathogen. For example, the greatest numbers of T cell lymphoma cases occur in Japan, in the Caribbean islands and in the countries surrounding the Mediterranean Sea, locations where there is a high rate of infection by human T cell leukemia virus-1 (HTLV-1). HTLV-1 is an oncogenic retrovirus capable of inducing T cell malignancies (both leukemias and lymphomas) several decades after infection. Similarly, the incidence of gastric lymphoma is high in regions of the world (such as Italy) where *Helicobacter pylori* bacteria abound. Burkitt lymphoma, which is linked to EBV infection, is most often found in equatorial Africa where this virus is highly prevalent. Consistent with this pathogen's involvement in NHL initiation, SCID and AIDS patients infected with EBV have a dramatically enhanced risk of developing NHL compared to patients who escape EBV infection.

iii) Subtypes of NHL

NHLs occur in a wide range of subtypes that are classified based on a complex portfolio of morphological and immunophenotypic criteria, clinical characteristics, and genetic aberrations. Within a given subtype, there may be variations of the disease with features that substantially overlap those of another subtype or even those of a leukemia. In the majority of adult NHLs, the target cell is a peripheral B cell, but some are derived from a peripheral T cell. In children, B cell and T cell lymphomas occur at almost equal frequency. The clinical features and relative frequencies of the major B cell and T cell NHL subtypes are summarized in **Tables 20-5** and **20-6**, respectively. The distinguishing histological features of three NHLs are shown in **Plate 20-8**.

iv) Treatment of NHLs

The heterogeneity of NHL subtypes means that there is wide variability in their response to treatment. However, some common approaches are used at least initially for therapy of these malignancies.

Chemotherapy involving combinations of several anticancer drugs is the weapon of choice used to fight most NHLs, although irradiation of a localized mass can also be utilized. Several rounds of chemotherapy and irradiation may be required to achieve remission. Nevertheless, repeated relapses may occur if the lymphoma is particularly aggressive, like precursor B cell lymphoblastic leukemia/lymphoma (B-LBL) or follicular lymphoma (FL). If a lymphoma threatens to invade the CNS and brain, a procedure called *CNS prophylaxis* may be carried out. CNS prophylaxis relies on

TABLE 20-5   NHL Subtypes: B Cell Lymphomas

Abbreviation	Name	Frequency (% of All NHLs)*	Distinguishing Features
B-LBL	Precursor B cell lymphoblastic leukemia/lymphoma	<1% (A#) 2.5% (C)	Early[†] onset, very aggressive Tissue invasion by immature B cells Can have leukemia component Ig genes may be rearranged Good prognosis with vigorous therapy
MCL	Mantle cell lymphoma	7% (A)	Late[†] onset, often aggressive Invading B cells are from follicular mantle Often shows t(11;14)(q13;q32) [cyclin D1; Igh] Colon often shows invasion Improved prognosis with mAb therapy
B-CLL	B cell chronic lymphocytic lymphoma	7% (A)	Late onset, indolent[‡] Has leukemia component Good prognosis with therapy
FL	Follicular lymphoma	22% (A)	Late onset, indolent Invading B cells are from follicular center Often shows t(14;18)(q32;q21) [Igh; Bcl-2] Poor prognosis even with therapy
MALT-L	Mucosa-associated lymphoid tissue lymphoma	8% (A)	Late onset; usually indolent but can be aggressive Extranodal, usually in GI mucosa Some are associated with t(11;18)(q21;q21) [IAP-2[§]; MALT1] or t(1;14)(p22;q32) [Bcl-10; Igh] or t(14;18)(q32;q21) [Igh; MALT1] Others are associated with *H. pylori* infection Sometimes progresses to DLCL Good prognosis with therapy
DLCL	Diffuse large cell lymphoma	33% (A) 12% (C)	Early or late onset, aggressive Tumor cells are large and exhibit variable morphology Extranodal growth is common Often associated with t(3;14)(q26;q32) [Bcl-6; Igh] or t(3;22)(q26;q11) [Bcl-6; MN1] Fair prognosis with therapy
BL	Burkitt lymphoma	2% (A) 36% (C)	Usually early onset, very aggressive Often associated with t(8;14)(q24;q32) [c-Myc; Igh] Sometimes associated with t(2;8)(p11;q24) [Igk; c-Myc] or t(8:22)(q24;q32) [c-Myc; Igl] Good prognosis with vigorous therapy

*Frequencies of disease subtypes are approximate and incidence patterns may vary. Percentages do not add up to 100% due to the occurrence of unclassified NHLs.

#A, in adults; C, in children.

[†]Early onset, in children; late onset, in older adults.

[‡]Indolent, slow-growing low-grade lymphomas.

[§]Description of genes not in Box 20-2: IAP-2, apoptosis inhibitor; MALT1, scaffold protein required for intracellular signaling; Bcl-10, scaffold protein required for intracellular signaling; Igk, Ig kappa light chain; Igl, Ig lambda light chain; ACK, tyrosine kinase; nucleophosmin, nucleolar export protein.

intrathecal chemotherapy in which a drug that is slightly less toxic to the CNS than to proliferating lymphocytes is introduced just under the sheath covering the spinal cord. Localized irradiation of the CNS and brain may also be applied. CNS prophylaxis has proven especially effective in preventing the spread of BL and some forms of diffuse large cell lymphoma (DLCL). Just as is true for leukemias and myelomas, researchers are developing new classes of experimental drugs for NHL treatment that target anti-apoptotic proteins like Bcl-2 or the pathways that cancer cells manipulate to resist normal death signals.

If an NHL patient is younger than 60 years, lacks CNS involvement and is in a second remission, an autologous or allogeneic HSCT may be beneficial. For example,

TABLE 20-6	**NHL Subtypes: T Cell Lymphomas**		
Abbreviation	**Name**	**Frequency (% of All NHLs)[*]**	**Distinguishing Features**
T-LBL	Precursor T cell lymphoblastic leukemia/lymphoma	1.7% (A) 30% (C)	Usually early[†] onset, very aggressive Tissue invasion by immature T cells Can have leukemia component TCRB and TCRG may be rearranged Good prognosis with vigorous therapy
ATL	Adult T cell leukemia/ lymphoma	6% (A)	Late onset Indolent at first, then aggressive[‡] Tumor cells are mature T cells of heterogeneous morphology Often has leukemia component Often associated with HTLV-1 Fair prognosis with vigorous therapy
ALCL	Anaplastic large cell lymphoma	2% (A) 13% (C)	Usually early onset, aggressive Tumor cells are large mature T cells of distinctive morphology TCR genes are rearranged in most cases Often associated with t(2;5)(p23;q35) [ALK[§]; nucleophosmin] Good prognosis with therapy
AITL	Angioimmunoblastic T cell lymphoma	4% (A)	Late onset, very aggressive Tumor cells are mature T cells Some association with EBV infection Poor prognosis even with therapy
MF	Mycosis fungoides	5% (A)	Late onset, indolent Appearance and symptoms resemble fungal skin infection Tumor cells are mature T cells Sometimes associated with bacterial or HTLV-1 infections Good prognosis with early therapy

#A, in adults; C, in children.

*Frequencies of disease subtypes are approximate and incidence patterns may vary. Percentages do not add up to 100% due to the occurrence of unclassified NHLs.

†Early onset, in children; late onset, in older adults.

‡Indolent, slow-growing low grade lymphomas.

§Description of genes not in Box 20-2: IAP-2, apoptosis inhibitor; MALT1, scaffold protein required for intracellular signaling; Bcl-10, scaffold protein required for intracellular signaling; Igk, Ig kappa light chain; Igl, Ig lambda light chain; ACK, tyrosine kinase; nucleophosmin, nucleolar export protein.

autologous and allogeneic HSCTs have been successful in prolonging the lives of some younger FL, DLCL and adult T cell leukemia/lymphoma (ATL) patients. HSCT can be a particularly attractive option for young patients with very aggressive cancers like mantle cell lymphoma (MCL). If left untreated, aggressive NHLs are associated with such a short life expectancy that the risks associated with the disease usually outweigh the dangers of GvHD. Moreover, a putative GvC effect has been observed in some patients with FL who were treated with allogeneic HSCT plus a separate infusion of donor lymphocytes. If HLA matching between the patient and an allogeneic donor can be optimized to minimize GvHD and maximize GvC, the patient may experience an overall benefit.

Immunotherapy (often in combination with chemotherapy) is now the standard of care for several NHLs. For example, some older MCL or FL patients suffering a relapse have achieved a partial remission after administration of anti-CD20 mAb or IFNα. Anti-CD20 and anti-CD52 mAbs have also been employed with some success for cases of B cell chronic lymphocytic lymphoma (B-CLL). Some DLCL patients for whom chemotherapy has not worked show long-lasting responses to IFNα treatment, and about 40% of patients (usually those with small tumors) benefit to some degree from treatment with anti-CD20 mAb. Patients with mycosis fungoides (MF) also often find that IFNα or anti-CD52 mAb is helpful. As well, about half of the malignant cells

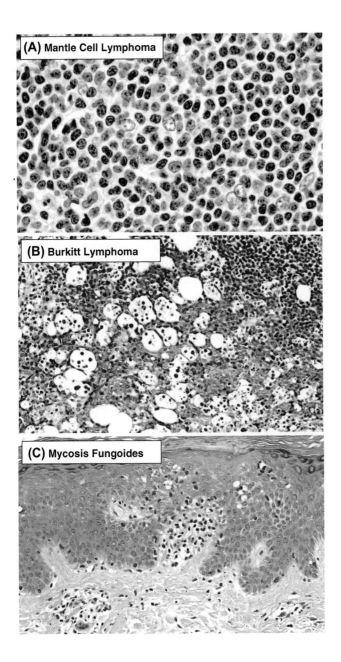

(A) Mantle Cell Lymphoma

(B) Burkitt Lymphoma

(C) Mycosis Fungoides

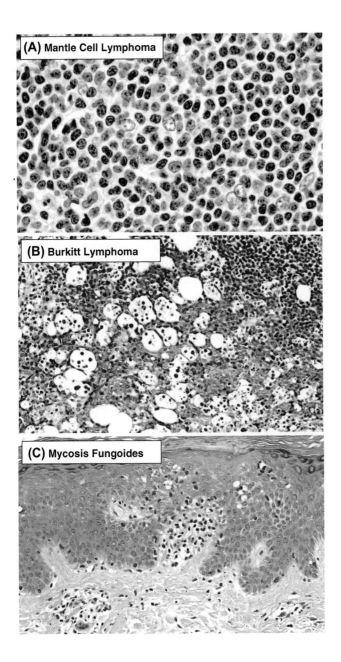

■- **Plate 20-8**
***Examples of Subtypes of
Non-Hodgkin Lymphoma***

(A) Lymph node section from a
patient with mantle cell lymphoma.
The tumor cells are irregularly
shaped B cells of small to medium
size. **(B)** Lymph node section from
a patient with Burkitt lymphoma.
The "starry sky" appearance is
due to the presence of vacuolated
macrophages that have phagocytized tumor cell debris. **(C)** Skin
section from a patient with mycosis
fungoides, showing aggressive
T cell infiltration of the epidermis.
*[Reproduced by permission of Doug
Tkachuk, Lifelabs, Ontario, Toronto.]*

in MF patients express IL-2R so that an immunotoxin created by fusing the receptor-
binding domain of IL-2 to diphtheria toxin has been investigated as a potential therapy. This agent, called denileukin diftitox, binds to IL-2R-expressing cells (including
the cancerous T cells of MF) and kills them by shutting down their protein synthesis.
Some success with this approach has been achieved in clinical trials. Similarly, studies
of mAbs targeting CD4 and CD30 as treatments for MF are ongoing. Combination
immunotherapies are now also under examination, as described in this chapter's Focus
on Relevant Research.

If a pathogen is associated with the onset of a lymphoma, treatment with an antipathogen agent may be effective. For example, about 80% of mucosa-associated lymphoid tissue lymphoma (MALT-L) cases are associated with *H. pylori* infection, and
about 70% of patients in this group (those that lack translocations) respond to antibiotics. As the bacteria are killed, the pathogen antigen putatively driving the lymphomagenesis disappears, and the tumor slowly resolves. Similarly, because ATL is often
associated with HTLV-1 infection, survival rates of ATL patients have been greatly
improved by the use of a combination of IFNα and the antiretroviral drug zidovudine
(azidothymidine; AZT).

"Anti-CD47 Antibody Synergizes with Rituximab to Promote Phagocytosis and Eradicate Non-Hodgkin Lymphoma" by Chao, M.P., Alizadeh, A.A., Tang, C., Myklebust, J.H., Varghese, B., Gill, S., Jan, M., Cha, A.C., Chan, C.K., Tan, B.T., Park, C.Y., Zhao, F., Kohrt, H.E., Malumbres, R., Briones, J., Gascoyne, R.D., Lossos, I.S., Levy, R., Weissman, I.L, & Majeti, R. (2010) *Cell* 142, 699–713.

Normal cells of the body have a limited life span such that the overall turnover rate is approximately 1 million cells per second! As described in Chapter 3, phagocytic cells play a key role in the removal of dead and dying cells. Phagocytes identify these spent cells by their expression of several molecules constituting a "find me and eat me" system [reviewed in Ravichandran (2011) *Immunity*, 35:445–455]. Researchers have discovered that tumor cells commonly express the surface molecule CD47, which engages a macrophage receptor called signal regulatory protein α (SIRPα). Engagement of this receptor sends a "don't eat me" signal to macrophages, permitting tumor cells to escape phagocytosis. Although mAb-based immunotherapy has become a powerful means of treating various leukemias and lymphomas, treatment failure is not uncommon, especially for NHL. In this research article, Chao *et al.* use a mouse model to explore a two-hit approach in which they combine rituximab, which binds to CD20 and opsonizes tumor cells for phagocytosis, with an anti-CD47 mAb that presumably blocks the "don't eat me" signal generated by the tumor cell (illustrated in the figure to the right). Immunodeficient mice were transplanted with tumor cells labeled with phosphorescent markers, and a sophisticated imaging system was used to track the fates of the labeled tumor cells in living mice. When the tumor-bearing mice were treated with the combination of anti-CD20 mAb and anti-CD47 mAb, complete eradication of the malignant

cells was observed. These results are very exciting because this same research group has recently reported that CD47 is expressed on virtually all types of cancer cells (*PNAS*, 2012, 109:6662–6667). Interestingly, although CD47 is also widely expressed on normal cells in mice, the animals do not appear to suffer undue toxicity from anti-CD47 mAb treatment. Thus, CD47 may constitute a "universal" cancer therapy target, and anti-CD47 mAb is now under investigation as a potential anticancer drug in humans.

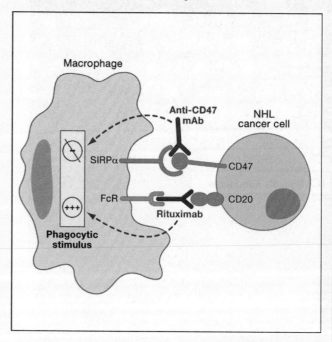

We have come to the end, not only of this chapter, but also of this book. Hopefully, we have given our readers a clear introduction to the principles of basic and clinical immunology and have perhaps inspired them to pursue further study in this field. Many intriguing problems in immunology remain to be resolved. More importantly, the growing links between the immune system and the nervous and endocrine systems, and between immune responses and cancer, mean that a solid understanding of immunology is more useful than ever before. Multidisciplinary approaches in both the laboratory and the clinic are the way of the future, and immunologists can expect to be valuable contributors to these endeavors.

Chapter 20 Take-Home Message

- Hematopoietic cancers (HCs) are malignancies of immune system cells.

- Unlike non-hematopoietic cancers, HCs are more commonly associated with gross chromosomal abnormalities such as recurring translocations.

- The three major types of HCs are leukemias, myelomas and lymphomas.

- Leukemias are "liquid tumors" in the blood and are derived from the transformation of either a hematopoietic precursor in the bone marrow or a mature hematopoietic cell in the blood. Leukemias can be lymphoid or myeloid, and acute or chronic.

- In myelomas, the transformed cell is a fully differentiated plasma cell. The proliferating malignant clone may be present either in dispersed form or as a solid mass in the bone marrow.

- In lymphomas, the transformed cell is a lymphocyte resident in a lymphoid tissue outside the bone marrow. The proliferating malignant clone creates a solid mass in this tissue.

- A lymphoma is classified as either a Hodgkin lymphoma (HL) or a non-Hodgkin lymphoma (NHL).

- In an HL, a reactive infiltrate of non-transformed cells is drawn to form a mass around a malignant clone of Reed–Sternberg cells. In an NHL, the entire cancerous mass develops from a transformed B or T lineage cell.

- Subtypes of HL and NHL are defined based on the architecture of the tumor, and the morphology, state of differentiation, surface marker expression and genetic aberrations of the transformed cells.

- Identification of the genetic aberration in an HC guides clinicians in optimizing treatment. Improvements in diagnosis and therapy, including the increasing use of molecularly targeted strategies, have led to increased patient survival over the past few decades.

Did You Get It? A Self-Test Quiz

Section A

1) Can you define these terms? leukemogenesis, myelomagenesis, lymphomagenesis, target cell, cancer stem cell, blast-like, relapse, remission, refractory disease, long-term survivor

2) What is the most common type of HC in North America?

3) Briefly distinguish between the three main types of HCs.

4) What are the hallmark symptoms of an HC?

5) What is a recurring chromosomal translocation?

6) Interpret the following: t(9;12)(q34;p13).

7) Compare the differentiation stage of the cells involved in high-grade versus low-grade HCs.

8) Briefly describe the principles of immunophenotyping, FISH and CGH. How are they helpful for HC diagnosis?

9) How is a response to chemotherapy or radiation treatment determined?

10) Give an example of how immunotherapy can treat HCs.

11) What is the GvC effect, what cells are responsible for it, and how is it mediated?

12) Outline how a cancer vaccine might benefit an HC patient.

Section B

1) Can you define these terms? hepatosplenomegaly, extramedullary, anemia, thrombocytopenia

2) Name the four major classes of leukemias and distinguish between them.

3) AML is dominated by what type of transformed cell?

4) Describe a recurring chromosomal translocation in AML.

5) Name a primary immunodeficiency associated with AML.

6) Name two genes whose disruption can lead to AML.

7) Outline two innovative approaches to treating AML.

8) What is WT1, and why is it of interest to oncologists?

9) Describe the three phases of CML.

10) What is the Philadelphia chromosome, and how does it arise?

11) What is Gleevec® and how does it work?

12) What approach is used to treat pediatric CML patients and why?

13) Name two peptides that have been tried as cancer vaccines for CML.

14) ALL is the most frequent form of cancer in what population?

15) ALL is dominated by what type of transformed cell?

16) Describe two recurring chromosomal translocations leading to ALL.

17) What non-translocation chromosomal anomaly is common in childhood ALL?

18) Describe two molecularly targeted agents being investigated for adult ALL treatment.

19) CLL is dominated by what type of transformed cell?

20) Describe two chromosomal anomalies associated with CLL.

Did You Get It? A Self-Test Quiz—Continued

21) What is atypical CLL?

22) Why is quality of life especially important to consider for CLL patients?

23) Describe two mAbs that have been useful for CLL therapy.

24) Why is HSCT used less often for CLL treatment than for ALL treatment?

Section C

1) Can you define these terms? paraprotein, plasma cell dyscrasia, MGUS, multiple myeloma

2) Why are plasma cell dyscrasias not generally considered leukemias?

3) How do normal plasma cells differ from myeloma cells?

4) Give three ways in which myeloma cells cause disease.

5) Outline the staging system used to classify myelomas.

6) What is the role of IL-6 in myeloma disease?

7) Describe three chromosomal anomalies associated with myelomagenesis.

8) How are thalidomide, lenalidomide, and bortezomib thought to be useful for myeloma treatment?

9) Name three immunotherapies used to treat myeloma patients.

10) How has HSCT been used to treat myeloma patients?

Section D

1) Can you define these terms? extranodal, Reed–Sternberg cell, L&H cell, reactive infiltrate, HL-involved tissue, B symptoms, ^{18}FDG-PET/CT

2) Why do lymphoma cells not usually migrate into the blood?

3) Distinguish between Hodgkin and non-Hodgkin lymphomas.

4) Outline the staging system used to classify lymphomas.

5) Describe the unique pattern of HL spreading in a patient.

6) Which cytokine is thought to be most important for HL development and why?

7) Distinguish between the two major subtypes of HL.

8) What is extended field radiation, and why does it improve the prognosis of HL patients?

9) What risk is associated with aggressive treatment for HL?

10) What is the incidence of NHL, and what age group is most often affected?

11) Give two reasons why the incidence of NHLs has increased over the past three decades.

12) How does the migration pattern of NHL cells differ from that of HL cells?

13) Give two examples of pathogens associated with NHL development.

14) What are the two most prevalent NHL subtypes in adults? In children?

15) What is CNS prophylaxis?

16) Describe how HSCT has been helpful for some NHL patients.

17) Give two examples of immunotherapeutics that have been helpful in NHL treatment.

18) Why are antibiotics effective in the treatment of some cases of MALT-L?

Can You Extrapolate? Some Conceptual Questions

1) In considering sources of stem cells for HSCT in the treatment of hematopoietic cancers, how might an autologous HSCT of cord blood saved since a person's birth

   a) Be advantageous relative to an allogeneic HSCT?

   b) Be advantageous relative to an autologous HSCT containing stem cells isolated from the patient's peripheral blood?

   c) Be disadvantageous relative to allogeneic HSCT?

2) Some scenarios related to leukemia diagnosis:

   a) Current statistics [U.S. National Institutes of Health (NIH) data for 2005–2009] show that the median age at diagnosis of three of the four main subtypes of leukemia (AML, CML, ALL or CLL) ranges from 62 to 72 years. The fourth subtype has a median age at diagnosis of 14 years. Which subtype would you expect to be the statistical outlier?

   b) A patient visits her doctor because of a vague feeling of fatigue and is sent for a routine blood analysis. The results show that, unexpectedly, this patient has leukemia. Further cytogenetic analysis reveals a chromosomal deletion at 13q14. Which leukemia does she have?

   c) A different patient visits his doctor with the same complaint and is also sent for a routine blood analysis. This patient also unexpectedly has leukemia, according to the blood results. However, cytogenetic analysis reveals the presence of the Philadelphia chromosome. Which leukemia does this patient have?

   d) One patient with Bloom syndrome, another patient with ataxia-telangiectasia, and another patient Li–Fraumeni syndrome are coping with their diseases but later begin to complain of malaise and fever. Which form of leukemia does their doctor suspect these patients may have developed?

Can You Extrapolate? Some Conceptual Questions—Continued

3) Since myelomas are cancers of antibody-secreting plasma cells, what might constitute a TSA in any given case of this disease? Would you expect this TSA to be a good candidate for the development of a cancer vaccine?

4) Lymphomas are highly dependent on intercellular contacts with surrounding stromal cells in order to survive, suggesting that a therapeutic approach which blocked such interactions might be effective. However, what disadvantage might there be to such an approach?

Would You Like To Read More?

Anguille, S., Van Tendeloo, V. F., & Berneman, Z. N. (2012). Leukemia-associated antigens and their relevance to the immunotherapy of acute myeloid leukemia. *Leukemia, 26*(10), 2186–2196.

Bacigalupo, A., Ballen, K., Rizzo, D., Giralt, S., Lazarus, H., Ho, V., et al. (2009). Defining the intensity of conditioning regimens: Working definitions. *Biology of Blood & Marrow Transplantation, 15*(12), 1628–1633.

Borchmann, P., Eichenauer, D. A., & Engert, A. (2012). State of the art in the treatment of Hodgkin lymphoma. *Nature Reviews Clinical Oncology, 9*(8), 450–459.

Eshaghian, S., & Berenson, J. R. (2012). Multiple myeloma: Improved outcomes with new therapeutic approaches. *Current Opinion in Supportive & Palliative Care, 6*(3), 330–336.

Gentile, M., Recchia, A. G., Mazzone, C., & Morabito, F. (2012). Emerging biological insights and novel treatment strategies in multiple myeloma. *Expert Opinion on Emerging Drugs, 17*(3), 407–438.

Graux, C. (2011). Biology of acute lymphoblastic leukemia (ALL): Clinical and therapeutic relevance. *Transfusion & Apheresis Science, 44*(2), 183–189.

Hayden, R. E., Pratt, G., Roberts, C., Drayson, M. T., & Bunce, C. M. (2012). Treatment of chronic lymphocytic leukemia requires targeting of the protective lymph node environment with novel therapeutic approaches. *Leukemia & Lymphoma, 53*(4), 537–549.

Kasner, M. T. (2010). Novel targets for treatment of adult acute lymphocytic leukemia. *Current Hematologic Malignancy Reports, 5*(4), 207–212.

Mahindra, A., Laubach, J., Raje, N., Munshi, N., Richardson, P. G., & Anderson, K. (2012). Latest advances and current challenges in the treatment of multiple myeloma. *Nature Reviews Clinical Oncology, 9*(3), 135–143.

Maloney, K. W., Giller, R., & Hunger, S. P. (2012). Recent advances in the understanding and treatment of pediatric leukemias. *Advances in Pediatrics, 59*(1), 329–358.

Norsworthy, K., Luznik, L., & Gojo, I. (2012). New treatment approaches in acute myeloid leukemia: Review of recent clinical studies. *Reviews on Recent Clinical Trials, 7*(3), 224–237.

Ringden, O., Karlsson, H., Olsson, R., Omazic, B., & Uhlin, M. (2009). The allogeneic graft-versus-cancer effect. *British Journal of Haematology, 147*(5), 614–633.

Rohon, P. (2012). Biological therapy and the immune system in patients with chronic myeloid leukemia. *International Journal of Hematology, 96*(1), 1–9.

Sawas, A., Diefenbach, C., & O'Connor, O. A. (2011). New therapeutic targets and drugs in non-Hodgkin's lymphoma. *Current Opinion in Hematology, 18*(4), 280–287.

van de Donk, N. W., Kamps, S., Mutis, T., & Lokhorst, H. M. (2012). Monoclonal antibody-based therapy as a new treatment strategy in multiple myeloma. *Leukemia, 26*(2), 199–213.

Velardi, A., Ruggeri, L., Mancusi, A., Aversa, F., & Christiansen, F. T. (2009). Natural killer cell allorecognition of missing self in allogeneic hematopoietic transplantation: A tool for immunotherapy of leukemia. *Current Opinion in Immunology, 21*(5), 525–530.

Selected Landmark Discoveries in Immunology

1798	Vaccination to prevent smallpox
1880s	Attenuated vaccines Phagocytic theory of immune defense Complement
1890s	Antibodies (antitoxins)
1900s	ABO blood groups Anaphylaxis Opsonization
1920s	Tuberculosis vaccine based on bacillus Calmette–Guérin (BCG) Bacterial toxin vaccine for diphtheria Delayed hypersensitivity
1930s	Histocompatibility antigens in mice Characterization of antibodies as the gamma globulin subset (immunoglobulins)
1940s	Adjuvant for vaccination (Freund's) Transplantation immunology Plasma cell production of antibodies
1950s	Lymphocytes Tolerance Protein structure and function of antibodies Clonal selection theory of antibody formation
1960s	Role of the thymus in immunity Hematopoietic stem cells Lymphokines (cytokines) B cell/T cell cooperation Distinct nature of helper T cells Primary sequence of an immunoglobulin molecule
1970s	Hypervariable regions of immunoglobulins Monoclonal antibody production by hybridomas Generation of antibody diversity by immunoglobulin gene rearrangement Link between immune responsiveness and histocompatibility genes Major histocompatibility complex (MHC) restriction of immune responses Distinct surface markers for Th and Tc cell subsets Processing and presentation of exogenous antigen Membrane attack complex in the complement cascade NK cells Drug-mediated immunosuppression (cyclosporine A)
1980s	Declaration of the worldwide eradication of smallpox Isolation of HIV from an AIDS patient Cloning of the T cell receptor genes Lymphocyte migration Role of coreceptors (CD4 and CD8) in T cell activation Processing and presentation of endogenous antigen Recognition of stress proteins by the immune system Th1/Th2 subsets of T helper cells γδ T cells Crystal structure of an MHC molecule

1990s	Thymic selection of T cells in establishing self tolerance
	Inhibition of NK cell inhibitory receptors by MHC class I
	Pattern recognition in innate immunity
	Mucosal immunity
2000s	Role of dendritic cells in T cell activation
	Role of PAMP/DAMP recognition by PRRs in activation of innate response
	NKT cells
	Th17 subset of T helper cells
	Role of regulatory T cells in controlling immune responses
	Balance of activatory/inhibitory receptors in controlling NK cell activation
	Crystal structure of TCR plus peptide–MHC complex
2010s	Role of chronic inflammation in autoimmunity, cancer, hypersensitivity, transplant rejection and chronic diseases
	Role of microbiota in regulation of innate and adaptive immunity
	Influence of tissue microenvironments on immune responses

Would You Like To Read More?

Calderon, M., Cardona, V., Demoly, P., & EAACI 100 Years of Immunotherapy Experts, Panel (2012). One hundred years of allergen immunotherapy. European Academy of Allergy and Clinical Immunology Celebration: Review of unanswered questions. *Allergy, 67*(4), 462–476.

Cohen, S. G. (2000). From immunity to autoimmune disease, a historic trail: Part III. *Allergy & Asthma Proceedings, 21*(3), 177–183.

Fitzhugh, D. J., & Lockey, R. F. (2011). Allergen immunotherapy: A history of the first 100 years. *Current Opinion in Allergy & Clinical Immunology, 11*(6), 554–559.

Imbach, P. (2012). Treatment of immune thrombocytopenia with intravenous immunoglobulin and insights for other diseases: A historical review. *Swiss Medical Weekly, 142* w13593.

Kaufmann, S. H. (2008). Immunology's foundation: The 100-year anniversary of the Nobel prize to Paul Ehrlich and Elie Metchnikoff. *Nature Immunology, 9*(7), 705–712.

Liston, A. (2011). Immunological tolerance 50 years after the Burnet Nobel Prize. *Immunology & Cell Biology, 89*(1), 14–15.

Nelson, H. S. (2011). Some highlights of the first century of immunotherapy. *Annals of Allergy, Asthma, & Immunology, 107*(5), 417–421.

Ochs, H. D., & Hitzig, W. H. (2012). History of primary immunodeficiency diseases. *Current Opinion in Allergy & Clinical Immunology, 12*(6), 577–587.

Passalacqua, G., & Bush, R. K. (2011). Specific immunotherapy, one century later. *Current Opinion in Allergy & Clinical Immunolog, 11*(6), 551–553.

Schwartz, R. H. (2012). Historical overview of immunological tolerance. *Cold Spring Harbor Perspectives in Biology, 4*(4) a006908.

Shahani, L., Singh, S., & Khardori, N. M. (2012). Immunotherapy in clinical medicine: Historical perspective and current status. *Medical Clinics of North America, 96*(3), 421–431.

Simpson, E., Scott, D., & Chandler, P. (1997). The male-specific histocompatibility antigen, H-Y: A history of transplantation, immune response genes, sex determination and expression cloning. *Annual Review of Immunology, 15*, 39–61.

Stiehm, E. R., & Johnston, R. B., Jr. (2005). A history of pediatric immunology. *Pediatric Research, 57*(3), 458–467.

Wagner, H. (2012). Innate immunity's path to the Nobel Prize 2011 and beyond. *European Journal of Immunology, 42*(5), 1089–1092.

Nobel Prizes Awarded for Work in Immunology

1901 **Emil von Behring**, for his discovery of serum antitoxins (antibodies) and serum therapy, and its application to the treatment of diphtheria.

1905 **Robert Koch**, for his investigations and discoveries in regard to tuberculosis. Koch developed tuberculin reactivity tests that later were important in the development of our current understanding of cellular immunity.

1908 **Elie Metchnikoff**, for his discovery of phagocytosis, and **Paul Ehrlich**, for his work on fundamental immunology.

1913 **Charles Robert Richet**, for his discovery of anaphylaxis.

1919 **Jules Bordet**, for his studies in regard to immunology, particularly complement-mediated lysis.

1930 **Karl Landsteiner**, for his discovery of the human blood groups.

1951 **Max Theiler**, for his development of vaccines against yellow fever.

1957 **Daniel Bovet**, for his development of antihistamines in the treatment of allergy.

1960 **Frank MacFarlane Burnet** and **Peter Brian Medawar**, for the discovery of acquired immunological tolerance.

1972 **Gerald Maurice Edelman** and **Rodney Robert Porter**, for their discoveries concerning the chemical structure of antibodies.

1977 **Rosalyn Yalow**, for the development of radioimmunoassays for peptide hormones.

1980 **Jean Dausset**, **George Davis Snell**, and **Baruj Benacerraf**, for the discoveries of the histocompatibility antigens on human and animal cells and their role in tissue and blood transplantation rejection (Dausset and Snell), and for work on the genetic control of immune responses (Benacerraf).

1984 **Georges J. F. Köhler** and **César Milstein**, for their development of cell hybridization as a technique to produce monoclonal antibodies, and **Niels K. Jerne**, for his many fundamental contributions to theoretical immunology.

1987 **Susumu Tonegawa**, for his work on the immunoglobulin genes and the mechanism by which antibody diversity is generated.

1990 **Joseph E. Murray** and **E. Donnall Thomas**, for their work on organ and bone marrow transplantation.

1996 **Peter C. Doherty** and **Rolf M. Zinkernagel**, for their discovery of the MHC restriction of T cell responses.

1997 **Stanley B. Prusiner**, for his discovery of prions, a new biological principle of infection.

2008 **Françoise Barré-Sinoussi** and **Luc Montagnier**, for their discovery of human immunodeficiency virus.

2011 **Ralph M. Steinman**, for his discovery of the dendritic cell and its role in adaptive immunity, and **Bruce A. Beutler** and **Jules A. Hoffmann**, for their discoveries concerning the activation of innate immunity.

Would You Like To Read More?

Barie, P. S. (2011). (Another) Nobel Prize in Physiology or Medicine awarded for work in inflammation and immunity. *Surgical Infections*, *12*(5), 337–338.

Ciechanover, A. J., & Sznajder, J. I. (2011). Innate and adaptive immunity: The 2011 Nobel Prize in Physiology or Medicine. *American Journal of Respiratory & Critical Care Medicine*, *184*(11), i–ii.

Gura, T. (1996). Nobel Prizes: Unraveling immune-cell mysteries. *Science*, *274*(5286), 345.

Masood, E., & Weiss, U. (1996). Nobel goes to T-cell pioneers whose work 'changed face of immunology'. *Nature*, *383*(6600), 465.

Nobel Prize to immunology. (2011). *Nature Reviews Immunology*, *11*(11), 714.

Pai, M. K., & Manjunatha, S. (2012). Immunology and Nobel Prize: A love story. *Indian Journal of Physiology & Pharmacology*, *56*(1), 1–6.

Zetterstrom, R. (2009). The 1908 Nobel Prize—Discovery of the basic principles of innate and acquired immunity. *Acta Paediatrica*, *98*(6), 1066–1069.

Comparative Immunology

All multicellular forms of life have a mechanism to distinguish self from non-self, and the means to preserve that self in the face of pathogen attack or competition for nutrients. These mechanisms and means were shaped by the various evolutionary and developmental pressures experienced by each organism. The elements of innate and specific immunity possessed by various phyla of the Animal Kingdom from the protozoans through the birds and mammals are summarized in **Figure C-1, Parts 1 and 2**. For reference, an evolutionary tree illustrating the hierarchical relationship among groups of phyla is shown in **Figure C-2**.

The most basic of multicellular organisms, such as the sponges of the phylum Porifera, grow in colonies of relatively undifferentiated cells. The functions of food gathering, waste product disposal and host defense are all carried out by the same type of cell. Nevertheless, these organisms can detect the encroachment of cells from another colony and kill the invaders. With the evolution in the lower invertebrates of circulatory systems and multiple body layers, the functions of host nutrition and host defense began to be carried out by separate cell types. Cells in the circulation of lower invertebrates and pre-vertebrates are specialized for detecting infected cells via the expression of a small number of pattern recognition molecules (PRMs) of relatively broad specificity. The primitive phagocytes bearing these PRMs are tasked with the disposal of non-self entities. Lectins, antimicrobial molecules, and at least some complement or complement-like components are also present in lower invertebrates and pre-vertebrates.

The range of PRMs expands as one proceeds higher up the pre-vertebrate and higher invertebrate evolutionary branches. For example, TLRs are found in all organisms from worms to mammals. Similarly, TLR ligands, such as flagellin, dsRNA and CpG motifs in DNA, are conserved from early fish and amphibians to mammals. These PAMPs are recognized by TLRs on more sophisticated phagocytes, which patrol the body and engulf invaders and produce large arrays of antimicrobial peptides and proteins. Many early protosomes and deuterosomes have molecules resembling complement components of the lectin pathway, but these proteins function only as opsonins. Some higher invertebrates possess a pathway called the *prophenoloxidase-activating (ProPO) system* which mediates cytotoxicity by coating and paralyzing a pathogen in the pigment melanin. Vertebrates do not have the ProPO system, making this defense mechanism unique to higher invertebrates.

No true lymphocytes, antibodies or MHC molecules are present in either the lower or higher invertebrates or pre-vertebrates. Thus, it seems that adaptive immunity is not needed for the survival of these species. These organisms have a limited habitat range, relatively short life spans (short reproductive cycle) and large reproductive capacities. The immune repertoire is limited, as must be true when the genes encoding defense molecules do not undergo somatic recombination and are "hard-wired" in the germline. However, this repertoire is sufficient for survival because a limited habitat means that the range of pathogens encountered is generally narrow. A short life span means the total number of pathogens encountered is relatively low, and a large reproductive capacity means that the huge numbers of offspring produced offset the loss of substantial numbers of them to pathogens.

Vertebrate species show enhanced anatomical complexity accompanied by increased mobility, such that these animals very often wander over great distances. In addition, vertebrates have longer life spans that correlate with an increased time to reach reproductive

PRM Recognition	Mammals	Birds	Reptiles	Amphibians	Bony fish	Cartilaginous fish	Jawless fish	Protochordates	Echinoderms	Higher invertebrates	Lower invertebrates	Protozoa
PRMs	+	+	+	+	+	+	+	+	+	+	+	+
Anti-microbials	+	+	+	+	+	+	+	+	+	+	+	+
Cytokines	+	+	+	+	+	+	+	+	+	+	+*	+*
Chemokines	+	+	+	+	+	+	+	+	?	+	−	−

ProPO System												
ProPO	−	−	−	−	−	−	−	−	−	+	−	−

Complement System												
CR	+	+	+	+*	+*	+*	+*	+*	+*	−	−	−
C3	+	+	+	+	+	+	+*	+*	+*	−	−	−
MBL/MASP	+	+	+?	+	+*	+*	+*	+*	+*	−	−	−
Terminal C'	+	+	+	+	+	+						
Factor I	+	+	+	+	+	+	−	−	−	−	−	−
Factor H	+	+	+	+	+	−	−	−	−	−	−	−
Lectin pathway	+	+	+	+	+	+	+	+	−	−	−	−
Classical pathway	+	+	+?	+	+	+	−	−?	−	−	−	−
Alternative pathway	+	+	+	+	+	+	−	−	−	−	−	−

Lymphoid Tissue												
Lymphocytes	T, B	T, B	T, B	T, B	T, B	T*, B*	T/B	+*	+*	+*	+*	−
GALT	+	+	+	+	+	+	+	−	−	−	−	−
Spleen and thymus	+	+	+	+	+	+	−	−	−	−	−	−
Bone marrow	+	+	+	+*	−	−	−	−	−	−	−	−
Lymph nodes (+ GC)	+	+	−	−	−	−	−	−	−	−	−	−
MALT/SALT	+	+	−	−	−	−	−	−	−	−	−	−

Fig. C-1, Part 1
Elements of Innate and Specific Immunity through Evolution

Key: *, like; ?, not definitively proven; ±, limited; C′, complement components; GC, germinal center; IgN, non-mammalian Ab isotype; •, non-MHC histocompatibility molecules exist; F, fast; S, slow; VS, very slow.

maturity. As a result, vertebrates frequently encounter a large number and wide variety of pathogens prior to successful reproduction. Vertebrates also produce many fewer offspring than either invertebrates or pre-vertebrates, such that severe losses to pathogens could threaten the species as a whole. These evolutionary pressures are thought to have promoted the development of adaptive immunity in vertebrates.

Although no conventional antibody can be detected in the lowest vertebrates (such as the jawless fish, Agnatha), primitive GALT is present. Cartilaginous fish like the sharks have hinged jaws, making them better predators able to take advantage of a broader range of nutritional opportunities. However, with such a diet comes an increased chance of internal injury and/or infection. Even with innate immunity in place, a strictly germline-encoded repertoire of non-self recognition molecules would not be sufficiently diverse to counter all the pathogens such vertebrates meet. Cartilaginous

	Mammals	Birds	Reptiles	Amphibians	Bony fish	Cartilaginous fish	Jawless fish	Protochordates	Echinoderms	Higher invertebrates	Lower invertebrates	Protozoa
Ig Responses												
RAG	+	+	+?	+	+	+	−	−	−	−	−	−
Tdt	+	+	+	+	+	+	−	−	−	−	−	−
Isotype switching	+	+	+	+	−	−	−	−	−	−	−	−
Somatic hypermutation	+	+	+	+	+	+	−	−	−	−	−	−
Affinity maturation	+	+	+?	+/−	+/−	−	−	−	−	−	−	−
Memory	+	+	+/−	+/−	+/−	−	−	−	−	−	−	−
Ig Isotypes												
IgM	+	+	+	+	+	+	−	−	−	−	−	−
IgD	+	+	+	+	+	+	−	−	−	−	−	−
IgG	+	−	−	−	−	−	−	−	−	−	−	−
IgA	+	+	−	−	−	−	−	−	−	−	−	−
IgE	+	−	−	−	−	−	−	−	−	−	−	−
IgN	−	+	+	+	+	+	−	−	−	−	−	−
MHC												
MHC class I	+	+	+	+	+	+	−•	−•	−	−	−	−
MHC class II	+	+	+	+	+	+	−	−	−	−	−	−
MHC class III	+	+*	+?	+	−	−	−	−	−	−	−	−
MHC class Ib	+	+	−	+	+	+	−	−	−	−	−	−
β2-microglobulin	+	+	+	+	+	+	−	−	−	+*	+*	−
Allograft rejection	F	F	F	F	F	S	VS	+*	+*	+*	+*	+*
TCR												
TCR genes	+	+	+?	+	+	+	−	−	−	−	−	−
Td responses	+	+	+	+	+	−	−	−	−	−	−	−
Ti responses	+	+	?	+	+	+	−	−	−	−	−	−

Fig. C-1, Part 2
Elements of Innate and Specific Immunity through Evolution

Key: *, like; ?, not definitively proven; ±, limited; C′, complement components; GC, germinal center; IgN, non-mammalian Ab isotype; •, non-MHC histocompatibility molecules exist; F, fast; S, slow; VS, very slow.

fish are thus the first organisms in which there exists a mechanism to somatically diversify immune system genes and expand the immune repertoire. A distinct thymus and spleen and true lymphocytes expressing forms of Ig and TCR molecules are present, with IgM, IgD and non-mammalian antibody isotypes being produced. The terminal complement components and MAC-mediated lysis are also first seen in the cartilaginous fish.

Vertebrates like amphibians that move from the sea to the land require additional host defense mechanisms to cope with the new environment. These animals have the limbs necessary to move on land and sophisticated vascular systems containing multiple types of circulating cells. Skins designed to shield the exposed animal from the sun's harmful rays are present, providing a physical barrier against pathogens. Bone marrow-like tissue serves as a source of distinct T and B cells, and lymphoid tissues

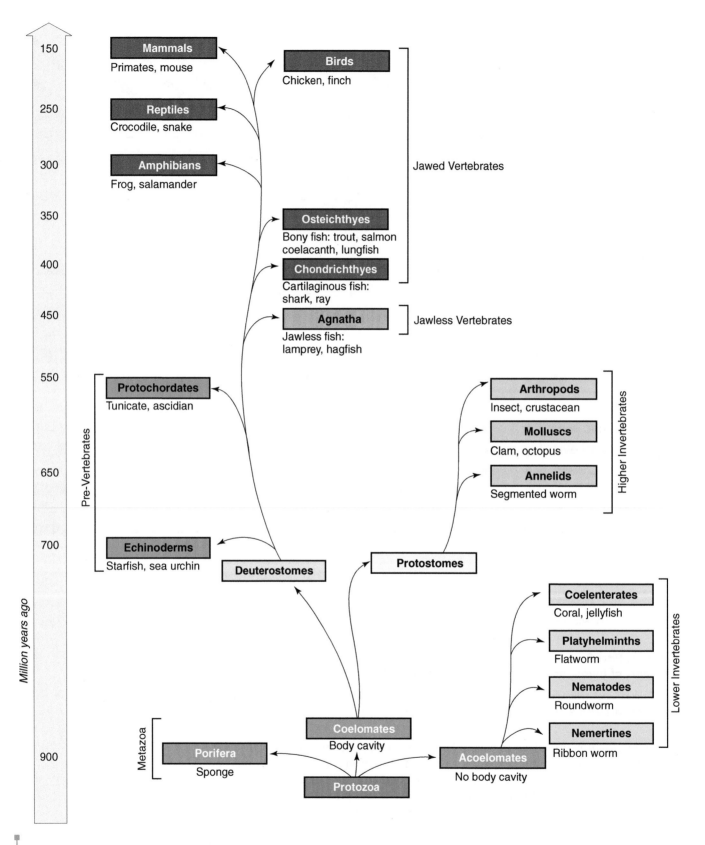

Fig. C-2
Evolutionary Tree of Kingdom Animalia

The major phyla and the approximate times when animals classified in these phyla first appeared are shown, starting with the protozoa at 900 million years ago. Phyla are grouped into the metazoa, lower invertebrates, higher invertebrates, pre-vertebrates and vertebrates as indicated. Time scale is approximate.

of increased complexity and wider distribution are present. In the reptiles, there are advanced lymphoid tissues plus eggs in which the young develop in a self-contained aqueous system enclosed by multiple membranes. This innovation frees these cold-blooded animals from having to return to the water to reproduce, and increases barrier protection for the offspring. However, the cold-bloodedness of the amphibians and reptiles affects their immune responses, causing measurable seasonal variations in the proliferative capacity of lymphocytes and the production of cytokines. The warm-bloodedness of birds and mammals removes this variation and allows these animals to forage and hunt at night when cold-blooded vertebrates are less active. However, with a permanently warm body comes an environment favoring pathogen growth. To counter the onslaught, birds and mammals have highly differentiated and structured lymphoid tissues, complete with distinct germinal centers and lymph nodes. Thus, if an innate response alone is unable to eliminate a threat, an adaptive immune response can be efficiently triggered within a local lymphoid tissue. This arrangement ensures that optimally coordinated cell-mediated and humoral responses against a pathogen are delivered in an effective but controlled way that causes the least amount of collateral damage to the host.

Would You Like To Read More?

Brown, G. D. (2010). How fungi have shaped our understanding of mammalian immunology. *Cell Host and Microbe*, *7*(1), 9–11.

Butler, J. E., Sinkora, M., Wertz, N., Holtmeier, W., & Lemke, C. D. (2006). Development of the neonatal B and T cell repertoire in swine: Implications for comparative and veterinary immunology. *Veterinary Research*, *37*(3), 417–441.

Cooper, E. L. (2003). Comparative immunology. *Current Pharmaceutical Design*, *9*(2), 119–131.

Das, S., Hirano, M., McCallister, C., Tako, R., & Nikolaidis, N. (2011). Comparative genomics and evolution of immunoglobulin-encoding loci in tetrapods. *Advances in Immunology*, *111*, 143–178.

Flajnik, M. F. (2002). Comparative analyses of immunoglobulin genes: Surprises and portents. *Nature Reviews Immunology*, *2*(9), 688–698.

Lillie, B. N., Brooks, A. S., Keirstead, N. D., & Hayes, M. A. (2005). Comparative genetics and innate immune functions of collagenous lectins in animals. *Veterinary Immunology & Immunopathology*, *108*(1–2), 97–110.

Robert, J., & Cohen, N. (2011). The genus Xenopus as a multispecies model for evolutionary and comparative immunobiology of the 21st century. *Developmental & Comparative Immunology*, *35*(9), 916–923.

Stuart, L. M., & Ezekowitz, R. A. (2008). Phagocytosis and comparative innate immunity: Learning on the fly. *Nature Reviews Immunology*, *8*(2), 131–141.

Sun, Y., Wei, Z., Li, N., & Zhao, Y. (2013). A comparative overview of immunoglobulin genes and the generation of their diversity in tetrapods. *Developmental & Comparative Immunology*, *39*(1–2), 103–109.

Zhu, L. Y., Nie, L., Zhu, G., Xiang, L. X., & Shao, J. Z. (2013). Advances in research of fish immune-relevant genes: A comparative overview of innate and adaptive immunity in teleosts. *Developmental & Comparative Immunology*, *39*(1–2), 39–62.

Selected CD Markers

CD Number*	Other Names†	Cellular/Tissue Expression	Functions
CD1a	R4	APCs#, cortical thymocytes	Structurally similar to MHC class I Presents lipid/glycolipid antigens to T cells
CD1b	R1	APCs, cortical thymocytes	Structurally similar to MHC class I Presents lipid/glycolipid antigens to T cells
CD1c	R7	APCs, cortical thymocytes, B cell subsets	Structurally similar to MHC class I Presents lipid/glycolipid antigens to T cells, especially $\gamma\delta$ T cells
CD1d	R3	B cell subsets, DCs, iIELs, cortical thymocytes	Structurally similar to MHC class I Presents glycolipid antigens to NKT cells
CD1e	R2	APCs, cortical thymocytes	Structurally similar to MHC class I Presents lipid/glycolipid antigens to T cells
CD2	LFA-2	Most T cells, thymocytes, NK cells, B cell subsets	Ig superfamily glycoprotein Adhesion molecule binding to LFA-3 (CD58); also binds to CD59 and CD15 Minor T cell costimulator
CD3c	CD3γ	T lineage cells	Ig superfamily glycoprotein Part of CD3 complex Required for cell surface expression and signal transduction of TCR
CD3d	CD3δ	T lineage cells	Ig superfamily glycoprotein Part of CD3 complex Required for cell surface expression and signal transduction of TCR
CD3e	CD3ϵ	T lineage cells	Non-glycosylated Ig superfamily protein Part of CD3 complex Required for cell surface expression and signal transduction of TCR
CD4	OKT4, T4, Leu-3	Subsets of thymocytes and Th cells; some DCs, monocytes and macrophages	Ig superfamily protein Coreceptor binding to MHC class II Required for thymocyte development and Th effector differentiation Binds to HIV gp120 protein
CD5	Leu-1, Ly-1	T lineage cells, neonatal B cells, B cell subsets	Transmembrane glycoprotein Influences T–B cell interaction CD5+ B cells are implicated in autoimmunity CD5+ T cells are seen in T-ALL
CD7	Leu-9, gp40	T lineage cells, NK cells, some myeloid progenitors	Ig superfamily glycoprotein Minor T cell costimulator CD7+ T cells are seen in T-ALL
CD8a	Leu-2, T8, Ly-2	Subsets of thymocytes and Tc cells, CTLs, some $\gamma\delta$ T cells, some NK cells, iIELs	Ig superfamily protein Coreceptor subunit binding to MHC class I Required for thymocyte development and CTL differentiation

(Continued)

—*Continued*

CD Number	Other Names	Cellular/Tissue Expression	Functions
CD8b	Lyt-3, Ly-3	Thymocyte and Tc cell subsets, CTLs	Ig superfamily protein Coreceptor subunit binding to MHC class I Required for thymocyte development and CTL differentiation
CD9	P24, MRP-1	Pre-B cells, granulocytes, endothelial cells, epithelial cells, activated T cells	Transmembrane glycoprotein Adhesion, migration Platelet activation and aggregation
CD10	CALLA, NEP	Pre-B cells, GC B cells, malignant B cells, fibroblasts	Membrane zinc metalloproteinase family Regulates B cell growth
CD11a	LFA-1 α chain, integrin αL	All leukocytes	CAM glycoprotein Complexes with CD18 to form LFA-1 Binds to ICAM1, ICAM2, ICAM3, ICAM4 on endothelial cells Adhesion and signal transduction during inflammation Minor T cell costimulator
CD11b	Mac-1 CR3 α chain integrin αM	Granulocytes, monocytes, macrophages, NK cells, subsets of T and B cells	CAM glycoprotein Complexes with CD18 to form CR3 Binds to iC3b, ICAM1, CD23 Adhesion, chemotaxis and signal transduction during inflammation
CD11c	P150 CR4 α chain	Granulocytes, monocytes, macrophages, NK cells, DCs, subsets of T and B cells	CAM glycoprotein Complexes with CD18 to form CR4 Binds to iC3b, LPS, ICAM-1, fibrinogen Adhesion and signal transduction during inflammation
CD14	LPS receptor	Monocytes, macrophages, LCs, granulocytes	Membrane protein Binds to TLR2, TLR3 or TLR4 plus LBP to transduce LPS signaling Mediates inflammation and endotoxic shock
CD15	Lewis X	Granulocytes, monocytes, RS cells	Pentasaccharide linked to protein or lipid Binds to CD62 (selectin proteins) to mediate granulocyte adhesion and activation
CD16a	FcγRIIIA	NK cells, macrophages, activated monocytes, mast cells	Ig superfamily glycoprotein Associates with CD3ζ chain or FcεRIγ chain to form low affinity IgG Fc receptor Mediates phagocytosis and ADCC
CD16b	FcγRIIIB	Neutrophils, activated eosinophils	Ig superfamily protein Binds to IgG and mediates phagocytosis or ADCC Traps immune complexes
CD18	Integrin β2 β-chain	All leukocytes	Transmembrane glycoprotein Associates with CD11a, b or c to form LFA-1, CR3 or CR4, respectively Binds to ICAMs, complement components Mediates cell adhesion to matrix proteins
CD19	B4	All B lineage cells except plasma cells, FDCs, malignant B cells	Ig superfamily protein Promotes BCR signal transduction
CD20	B1, Ly-44	All B lineage cells, including plasma cells, T cell subsets	Non-glycosylated protein Promotes B cell activation and proliferation Therapeutic target for lymphomas
CD21	CR2	Mature B cells, FDCs, DCs, epithelial cells, T cell subsets	Glycosylated RCA protein Promotes B cell signaling and activation Complement receptor for C3d Receptor for Epstein–Barr virus

—Continued

CD Number	Other Names	Cellular/Tissue Expression	Functions
CD22	BL-CAM, Siglec-2	Mature B cells, not plasma cells	Sialoadhesin superfamily protein Binds to sialic acids Adhesion and signal transduction Therapeutic target for lymphoma treatment
CD23	FcεRII, BLAST-2	Mature B cells, monocytes, eosinophils, FDCs, TECs, iIELs	Transmembrane glycoprotein Low affinity IgE Fc receptor Triggers NO and cytokine production by monocytes
CD25	Tac, p55, IL-2Rα	Activated T and B cells, Treg cells, thymocytes, DC subsets	Transmembrane glycoprotein Associates with IL-2Rβ (CD122) and IL-2Rγ (CD132) to form the high affinity IL-2 receptor
CD27	T14, TNFRSF7	T cells, mTECs, B cell subsets, NK cells	TNFR superfamily glycoprotein Binds to CD70 to deliver costimulatory signaling
CD28	T44	Activated T and B cells, thymocytes, NK cells	Transmembrane protein Binds to B7-1 (CD80) and B7-2 (CD86) on APCs Mediates T cell costimulation and stimulates T cell survival, proliferation, production of IL-2 and other cytokines TCR engagement in the absence of CD28 signaling can lead to anergy
CD29	VLA β chain, integrin β 1	Most leukocytes	Transmembrane protein Associates with CD49a–f to form VLA molecules Mediates cellular adhesion to VCAM-1, MAdCAM-1 and matrix proteins Critical for leukocyte migration
CD30	TNFRSF8	Activated B, T, and NK cells, monocytes, RS cells	TNFR superfamily glycoprotein May regulate development of thymocytes and proliferation of activated lymphocytes
CD31	PECAM-1	Monocytes, granulocytes, T and B cell subsets, endothelial cells	CAM glycoprotein Adhesion, activation, migration
CD32	FcγRII	B cells, monocytes, macrophages, granulocytes, endothelial cells	Transmembrane glycoprotein Binds to aggregated IgG complexes Promotes B cell development and activation Mediates phagocytosis and ADCC
CD33	My9 Siglec-3	All myeloid lineage cells	Sialoadhesin superfamily protein Binds to sialic acids May inhibit signal transduction
CD34	My10 Mucosialin	Hematopoietic precursors, BM stromal cells	Transmembrane glycoprotein Marker for human HSCs Mediates adhesion between hematopoietic precursors and BM stromal cells or matrix
CD35	CR1	Most blood cells (not platelets)	RCA glycoprotein Complement receptor for C3b and C4b bound to immune complexes Mediates opsonized phagocytosis, immune complex clearance, adhesion
CD36	GPIV	Platelets, monocytes, macrophages, erythroid precursors, endothelial cells	Scavenger receptor Binds to collagen, long-chain fatty acids and other ligands Mediates phagocytosis, platelet adhesion and aggregation
CD38	T10	Developing T and B cells, GC B cells, some hematopoietic progenitors	Adhesion glycoprotein Some hydrolase activity

(Continued)

—Continued

CD Number	Other Names	Cellular/Tissue Expression	Functions
CD40	TNFRSF5	Mature B cells, some pre-B cells, monocytes, DCs, T cell subsets, fibroblasts, keratinocytes	TNFR superfamily glycoprotein Binds to CD40L (CD154) on Th cells to costimulate B cell activation, GC formation and isotype switching Important for DC licensing
CD43	Leukosialin Sialophorin	Most leukocytes	Membrane glycoprotein Mediates adhesion by binding to ICAM-1 and selectins
CD44	Pgp-1, H-CAM	Leukocytes, erythrocytes, endothelial and epithelial cells (not platelets)	Heterogeneous family of glycoproteins Bind to hyaluronic acid, collagen, fibronectin Mediate cell–cell and cell–matrix adhesion and signaling Important for lymphocyte homing, recirculation, aggregation
CD45	B220	All leukocytes, hematopoietic progenitors	Heterogeneous family of glycoproteins Tyrosine phosphatase activity is important for signal transduction
CD46	MCP	Most leukocytes, endothelial and epithelial cells	RCA glycoprotein Controls complement activation by promoting cleavage of C3b and C4b
CD47	IAP	Most hematopoietic cells, epithelial cells, endothelial cells, fibroblasts, many types of tumor cells	Transmembrane glycoprotein Binds to SIRPα expressed on macrophages to deliver signaling suppressing phagocytosis
CD49a	VLA-1 α chain, α1 integrin	Activated T cells, monocytes, endothelial cells, melanoma cells	Integrin glycoprotein Binds to CD29 to form VLA-1 Mediates adhesion to collagen, laminin Upregulated during inflammation
CD49b	VLA-2 α chain, α2 integrin	Platelets, megakaryocytes, monocytes, epithelial and endothelial cells, NK cell subsets, activated T cells	Integrin glycoprotein Binds to CD29 to form VLA-2 Mediates cell and platelet adhesion to collagen, laminin, fibronectin, E-cadherin Promotes wound healing
CD49c	VLA-3 α chain, α3 integrin	B cells, monocytes	Integrin glycoprotein Binds to CD29 to form VLA-3 Mediates adhesion to collagen, laminin, fibronectin, thrombospondin
CD49d	VLA-4 α chain, α4 integrin	Most leukocytes (not platelets or neutrophils)	Integrin glycoprotein Binds to CD29 to form VLA-4 Mediates adhesion to VCAM-1, MAdCAM-1, fibronectin, thrombospondin Important for inflammation, HSC migration
CD49e	VLA-5 α chain, α5 integrin	Monocytes, DCs, NK cells	Integrin glycoprotein Binds to CD29 to form VLA-5 Mediates adhesion to fibronectin, fibrinogen
CD49f	VLA-6 α chain, α6 integrin	Platelets, megakaryocytes, monocytes, T cells, thymocytes, endothelial cells, epithelial cells	Integrin glycoprotein Binds to CD29 to form VLA-6 Mediates adhesion to laminin, other ligands Mediates interaction between epithelial cells and basement membrane during wound healing Involved in tumor cell metastasis
CD50	ICAM-3	Most leukocytes	Ig superfamily glycoprotein Binds to LFA-1 (CD11a/CD18) Contributes to APC adhesion, minor T cell costimulation

—*Continued*

CD Number	Other Names	Cellular/Tissue Expression	Functions
CD52	CAMPATH-1	Lymphocytes, monocytes, eosinophils, mast cells, sperm	Small membrane protein Physiological function unknown Anti-CD52 mAbs are used therapeutically to treat certain leukemias and lymphomas and to reduce GvHD
CD54	ICAM-1	Activated T cells, B cells, monocytes, endothelial cells, epithelial cells	Ig superfamily glycoprotein Binds to LFA-1, CR3, fibrinogen Signaling, adhesion and extravasation in inflammatory and adaptive immune responses
CD55	DAF	Most hematopoietic cells, many non-hematopoietic cells	RCA protein Binds C3b and C4b to inhibit C3 convertase formation, blocking complement activation Binds C3bBb and C4b2a to accelerate decay of C3 convertases
CD56	NCAM, Leu-19	Neural tissue, NK cells, NKT cells, T cell subsets	CAM glycoprotein Adhesion Signaling involved in migration, proliferation, apoptosis, differentiation, survival
CD58	LFA-3	Leukocytes, erythrocytes, endothelial and epithelial cells, fibroblasts	Ig superfamily protein Binds to CD2 Mediates adhesion between Th cells and APCs, Tc cells and target cells, thymocytes and TECs Minor T cell costimulation
CD59	MIRL, protectin	Most hematopoietic cells	RCA glycoprotein Binds to C8 and/or C9 and inhibits final steps of MAC formation
CD62E	E-selectin, LECAM-2	Activated endothelial cells, platelets, megakaryocytes	Adhesion glycoprotein Required for leukocyte extravasation and activation during inflammation Binds to glycoproteins and glycolipids, including CD162
CD62L	L-selectin, LECAM-1	Neutrophils, monocytes, NK cells, memory T cells, thymocytes, B cell subsets	Adhesion glycoprotein Required for leukocyte extravasation and activation during inflammation Mediates lymphocyte adherence to HEVs in lymph nodes Binds to glycoproteins and glycolipids, including CD162
CD62P	P-selectin, PADGEM	Megakaryocytes, activated platelets and endothelial cells	Adhesion glycoprotein Required for leukocyte extravasation and activation during inflammation Binds to glycoproteins and glycolipids, including CD162
CD63	LAMP-3, LIMP	Activated platelets, monocytes, macrophages, granulocytes, fibroblasts	Transmembrane protein Binds to VLA-3, VLA-4 to regulate cell migration
CD64	FcγRI	Monocytes, macrophages, neutrophils, DC subsets, activated granulocytes	Ig superfamily protein High affinity IgG Fc receptor Mediates phagocytosis, ADCC Mediates transfer of IgG across the placenta
CD66c	NCA	Neutrophils, epithelial cells, colon carcinoma cells	Ig superfamily protein Neutrophil activation, adhesion
CD66e	CEA	Adult colonic epithelial cells, colon carcinoma cells	Ig superfamily protein Adhesion

(Continued)

—*Continued*

CD Number	Other Names	Cellular/Tissue Expression	Functions
CD70	TNFSF7	Activated lymphocytes	TNF superfamily protein Binds to CD27 to deliver costimulatory signaling
CD71	TFR, TfR, T9	Proliferating cells, erythroid precursors	Integral membrane protein Transferrin receptor functioning in iron uptake
CD74	Invariant chain, Ii	Cells expressing MHC class II	Membrane glycoprotein Critical for MHC class II stabilization and antigen presentation Surface function not known
CD79a	Igα	All B cells and plasma cells	Ig superfamily glycoprotein with ITAMs Associates with Igβ (CD79b) and participates in the BCR complex to carry out signal transduction Essential for BCR function, B cell development
CD79b	Igβ	All B cells except plasma cells	Ig superfamily glycoprotein with ITAMs Associates with Igα (CD79a) and participates in the BCR complex to carry out signal transduction Essential for BCR function, B cell development
CD80	B7-1	Professional APCs, activated T cells	Ig superfamily glycoprotein Binds to CD28 to costimulate naïve T cells Binds to CTLA-4 (CD152) to inhibit T cell activation Highly induced on stimulated APCs
CD85a	ILT-5	Monocytes, granulocytes, DCs, T cell subsets	Leukocyte Ig-like receptor superfamily NK inhibitory receptor
CD85b	ILT-8	Monocytes, DCs, T cell subsets, B cells, NK cells	Leukocyte Ig-like receptor superfamily NK activatory receptor
CD85c	LIR-8	Monocytes, DCs, T cell subsets, B cells, NK cells	Leukocyte Ig-like receptor superfamily NK activatory receptor Activates NK cytotoxicity
CD85d	ILT-4	Monocytes, DCs, NK cells	Leukocyte Ig-like receptor superfamily NK inhibitory receptor
CD85e	ILT-6	Monocytes, DCs, T cell subsets, B cells, NK cells	Leukocyte Ig-like receptor superfamily NK activatory receptor
CD85f	LIT11	Monocytes, DCs, T cell subsets, B cells, NK cells	Leukocyte Ig-like receptor superfamily NK activatory receptor
CD85g	ILT-7	Monocytes, pDCs, T cell subsets, B cells, NK cells	Leukocyte Ig-like receptor superfamily NK activatory receptor
CD85h	LIR-7	Monocytes, granulocytes, DCs, T cell subsets, B cells, NK cells	Leukocyte Ig-like receptor superfamily NK activatory receptor
CD85i	LIR-6	Monocytes, DCs, T cell subsets, NK cells	Leukocyte Ig-like receptor superfamily NK activatory receptor
CD85j	ILT-2	Monocytes, DCs, subsets of T, B and NK cells	Leukocyte Ig-like receptor superfamily NK inhibitory receptor
CD85k	ILT-3	Monocytes, granulocytes, macrophages, DCs, NK cells	Leukocyte Ig-like receptor superfamily NK inhibitory receptor
CD85l	ILT-9	Monocytes, granulocytes, DCs, T cell subsets, B cells, NK cells	Leukocyte Ig-like receptor superfamily Effect on NK cytotoxicity unclear
CD85m	ILT-10	Monocytes, macrophages, DCs, T cell subsets, B cells	Leukocyte Ig-like receptor superfamily Effect on NK cytotoxicity unclear
CD86	B7-2	Professional APCs, endothelial cells	Ig superfamily protein Binds to CD28 to costimulate naïve T cells Binds to CTLA-4 (CD152) to inhibit T cell activation

—Continued

CD Number	Other Names	Cellular/Tissue Expression	Functions
CD88	C5aR	Most leukocytes, epithelial and endothelial cells, hepatocytes	Transmembrane protein Binds to anaphylatoxin C5a Stimulates granule release, chemotaxis, ROI and RNI production during inflammation
CD89	FcαR	Neutrophils, monocytes, macrophages, activated eosinophils	Ig superfamily glycoprotein IgA Fc receptor Stimulates phagocytosis, respiratory burst, degranulation, ADCC, release of inflammatory mediators and cytokines
CD90	Thy-1	Thymocytes, HEV endothelial cells, HSCs	Ig superfamily glycoprotein Adhesion, signaling
CD91	LRP	Monocytes, macrophages, hepatocytes	Scavenger receptor Binds to HSPs Mediates clearance of necrotic cells
CD94	Kp43	NK cells, NKT cells, some γδ T cells, some αβ CD8+ T cells	Glycoprotein with C-type lectin domain Associates with NKG2A (CD159a) and NKG2C (CD159c) to form NK inhibitory receptors Associates with NKG2D (CD314) to form an NK activatory receptor
CD95	Fas	Activated T and B cells, neutrophils, monocytes, fibroblasts	TNF superfamily glycoprotein Binds to FasL (CD178) to initiate apoptosis Important for peripheral tolerance and lymphocyte homeostasis
CD101	IGSF2	Monocytes, granulocytes, DCs, LCs, activated T cells	Ig superfamily glycoprotein Regulates T cell activation and proliferation by inhibiting IL-2R expression and IL-2 secretion
CD102	ICAM-2	Most leukocytes, endothelial cells	Ig superfamily glycoprotein Binds to LFA-1 (CD11a/CD18) to mediate adhesion and signal transduction important for T cell interactions, lymphocyte recirculation, NK cell migration
CD106	VCAM-1	Endothelial cells, DCs, FDCs, stromal cells	Ig superfamily glycoprotein Binds to VLA-4 (CD49d/CD29) to mediate leukocyte extravasation during inflammatory and adaptive immune responses
CD114	G-CSFR	Myeloid progenitors, granulocytes, monocytes, platelets, endothelial cells	Growth factor receptor Binds to G-CSF and regulates granulocyte differentiation and proliferation Stimulates mobilization of HSCs from BM into blood
CD115	M-CSFR	Mature and progenitor myeloid cells	Growth factor receptor Binds to M-CSF and regulates monocyte and macrophage differentiation and proliferation Promotes adhesion to BM stroma
CD116	GM-CSFR	Macrophages, neutrophils, eosinophils, DCs, myeloid and erythroid progenitors	Growth factor receptor subunit Associates with βc (CD131) to bind GM-CSF with high affinity Transduces signals promoting differentiation, proliferation and activation of myeloid and erythroid cells
CD117	c-kit, SCFR	HSCs and other progenitors, mast cells	Growth factor receptor Binds to SCF to mediate signal transduction promoting differentiation, especially HSCs

(Continued)

—*Continued*

CD Number	Other Names	Cellular/Tissue Expression	Functions
CD119	IFNγRα	Ubiquitous except RBCs	Cytokine receptor subunit Associates with IFNγR β subunit to form complex transducing IFNγ signaling Triggers numerous antiviral, antiproliferative and immunomodulatory effects during innate and adaptive responses
CD120a	TNFRI, p55	T and B cells; upregulated on most other cell types	TNFR superfamily protein Binds to TNF and LT Mediates signaling promoting inflammation, fever, shock, tumor necrosis, cell proliferation, differentiation and apoptosis
CD120b	TNFRII, p80	Most hematopoietic cells, especially myeloid cells, some non-hematopoietic cells	TNFR superfamily protein Binds to TNF and LT Promotes mainly cell survival and differentiation but sometimes necrosis or apoptosis
CD121a	IL-1RI	Almost ubiquitous	Ig superfamily glycoprotein Binds to IL-1 and mediates its pro-inflammatory activities
CD121b	IL-1RII	B cells, monocytes, macrophages, T cell subsets, keratinocytes	Ig superfamily glycoprotein Binds to IL-1 but is an inhibitory decoy receptor
CD122	IL-2Rβ	Constitutive low levels on T, B, NK cells and monocytes; upregulated by activation	Cytokine receptor subunit Associates with γc (IL-2Rγ; CD132) to form low affinity IL-2 receptor Associates with IL-2Rα (CD25) and CD132 to form high affinity IL-2 receptor Associates with IL-15Rα (CD215) and CD132 to form high affinity IL-15 receptor Binding to IL-2 promotes lymphocyte proliferation, differentiation and regulation contributing to peripheral tolerance Binding to IL-15 promotes NK cell proliferation
CD123	IL-3Rα	HSCs and other progenitors, NK cells, mast cells, endothelial cells, myeloid and B cell subsets	Cytokine receptor subunit Associates with CD131 to form IL-3 receptor Mediates signaling influencing growth and differentiation
CD124	IL-4R	T and B cells, hematopoietic precursors, fibroblasts, endothelial cells, epithelial cells	Cytokine receptor subunit Associates with CD132 to form IL-4 receptor IL-4 binding promotes growth of B and T cells Required for Th2 differentiation, IgE production and allergic inflammation Associates with IL-13Rα (CD213) to form IL-13 receptor
CD125	IL-5Rα	Eosinophils, basophils, activated B cells, mast cells	Cytokine receptor subunit Binds with CD131 to form IL-5 receptor Promotes eosinophil generation and activation
CD126	IL-6Rα	Mature T cells, B cell subsets, plasma cells, monocytes, granulocytes, epithelial cells, fibroblasts	Cytokine receptor subunit Associates with gp130 (CD130) to form IL-6 receptor Stimulates acute phase response in liver, regulates hematopoiesis Mediates growth signals for myelomas
CD127	IL-7Rα	Pro- and pre-B cells, thymocytes, T cell subsets	Cytokine receptor subunit Associates with CD132 to form IL-7 receptor Promotes T and B lymphopoiesis, γδ T cell development and survival (especially in mice)

—*Continued*

CD Number	Other Names	Cellular/Tissue Expression	Functions
CD129	IL-9Rα	Mast cells, macrophages, erythroid and myeloid precursors, activated granulocytes, thymocytes	Cytokine receptor Associates with CD132 to form IL-9 receptor Promotes growth of mast cells and erythroid and myeloid progenitors
CD130	gp130	Almost ubiquitous	Cytokine receptor subunit Common signaling chain for receptors binding IL-6, IL-11 or IL-27 (among others) Does not bind to cytokines itself
CD131	Common β chain, βc	HSCs and other progenitors, monocytic lineage cells	Cytokine receptor subunit Common signaling chain for receptors binding IL-3, IL-5 or GM-CSF Does not bind to cytokines itself Upregulated in response to inflammatory cytokines
CD132	Common γ chain, γc	T, B and NK cells; monocyte lineage cells; granulocytes; DCs	Cytokine receptor subunit Common signaling chain for receptors binding IL-2, -4, -7, -9, -15 or -21 Does not bind to cytokines itself
CD134	OX40, TNFRSF4	Activated T cells, regulatory T cells, hematopoietic precursors	TNFR superfamily protein Binds to OX40L (CD252) to promote T cell interaction with APCs Mediates adhesion of T cells to endothelium, T cell proliferation, differentiation, apoptosis
CD135	Flt3	HSCs, myeloid and early B cell precursors, thymocyte subsets	Ig superfamily protein Promotes proliferation of hematopoietic precursors
CD137	4-1BB, TNFRSF9	Activated T and B cells, monocytes, FDCs, epithelial cells	TNFR superfamily protein Binds to 4-1BB ligand (TNF superfamily protein) Additional costimulation for T cell activation, survival, proliferation and differentiation Supports survival and maturation of DCs
CD152	CTLA-4	Activated T cells, Treg cells, activated B cells	Ig superfamily protein Structurally similar to CD28 High avidity receptor for B7–1 (CD80) and B7–2 (CD86) Downregulates T cell activation, promotes Treg generation, contributes to maintenance of peripheral tolerance and lymphocyte homeostasis
CD153	CD30L, TNFSF8	Activated T cells, macrophages, neutrophils, eosinophils, B cells, RS cells	TNF superfamily glycoprotein Ligand for CD30 Precise function is unclear but may stimulate T–B cell interaction, proliferation and apoptosis
CD154	CD40L, TNFSF5	Activated T cells (CD4+), mast and NK cells, granulocytes, monocytes, activated platelets	TNF superfamily glycoprotein Ligand for CD40 Provides major costimulatory signal for B cell activation and survival signal to GC B cells Required for isotype switching, DC licensing
CD158a	KIR2DL1	Most NK cells, T cell subsets	Ig superfamily glycoprotein Binds to classical MHC class I molecules Inhibits NK cytotoxicity
CD158b1	KIR2DL2	Most NK cells, T cell subsets	Ig superfamily glycoprotein Binds to classical MHC class I molecules Inhibits NK cytotoxicity
CD158b2	KIR2DL3	Most NK cells, T cell subsets	Ig superfamily glycoprotein Binds to classical MHC class I molecules Inhibits NK cytotoxicity

(*Continued*)

—Continued

CD Number	Other Names	Cellular/Tissue Expression	Functions
CD158c	KIR3DP1	Most NK cells, T cell subsets	Ig superfamily glycoprotein Function unclear
CD158d	KIR2DL4	Most NK cells, T cell subsets	Ig superfamily glycoprotein Binds to HLA-G Activates NK cytotoxicity
CD158e1	KIR3DL1	Most NK cells, T cell subsets	Ig superfamily glycoprotein Binds to classical MHC class I molecules Inhibits NK cytotoxicity
CD158e2	KIR3DS1	Most NK cells, T cell subsets	Ig superfamily glycoprotein Binds to classical MHC class I molecules Activates NK cytotoxicity
CD158f	KIR2DL5A	Most NK cells, T cell subsets	Ig superfamily glycoprotein Binds to classical MHC class I molecules Inhibits NK cytotoxicity
CD158g	KIR2DS5	Most NK cells, T cell subsets	Ig superfamily glycoprotein Activates NK cytotoxicity
CD158h	KIR2DS1	Most NK cells, T cell subsets	Ig superfamily glycoprotein Binds to HLA-C Activates NK cytotoxicity
CD158j	KIR2DS2	Most NK cells, T cell subsets	Ig superfamily glycoprotein Binds to HLA-C Activates NK cytotoxicity
CD158k	KIR3DL2	Most NK cells, T cell subsets	Ig superfamily glycoprotein Binds to classical MHC class I molecules Inhibits NK cytotoxicity
CD158z	KIR3DL3	Most NK cells, T cell subsets	Ig superfamily glycoprotein Inhibits NK cytotoxicity
CD159a	NKG2A	NK cells, subset of CD8+ T cells	Glycoprotein with C-type lectin domain Associates with CD94 to form NK inhibitory receptor binding to HLA-E Inhibits NK cytotoxicity
CD159c	NKG2C	NK, NKT and $\gamma\delta$ T cells	Glycoprotein with C-type lectin domain Associates with CD94 to form NK activatory receptor binding to HLA-E Activates NK cytotoxicity May also activate NKT and $\gamma\delta$ T cells
CD161	NKR-P1A	Most NK cells, NKT cells, memory T cells, thymocytes	Glycoprotein with C-type lectin domain Costimulation of NK and NKT cell activation Stimulates thymocyte proliferation
CD172a	SIRPα	Monocytes, DCs, granulocytes, tissue stem cells	Ig superfamily glycoprotein Binds to CD47 to deliver a signal inhibiting phagocytosis Mediates adhesion
CD178	FasL, CD95L, TNFSF6	Activated T cells, NK cells, DCs, testicular cells, tumor cells	TNF superfamily glycoprotein Ligand for Fas (CD95) Induces apoptosis Involved in maintenance of peripheral tolerance
CD179a	V pre-β	Pro-B, pre-B cells	Protein homologous to Ig V region Associates with lambda 5 chain (CD179b) to form surrogate light chain required for pre-BCR expression and early B cell development

—*Continued*

CD Number	Other Names	Cellular/Tissue Expression	Functions
CD179b	Lambda 5	Pro-B, pre-B cells	Protein homologous to Ig lambda C region Associates with V pre-β chain (CD179a) to form surrogate light chain required for pre-BCR expression and early B cell development
CD181	CXCR1 IL-8RA IL-8R1	Granulocytes, some T cells, monocytes, mast cells, endothelial cells	Chemokine receptor Binds to IL-8 (only) and mediates signaling for neutrophil activation and chemotaxis Transduces signals for chemokine-mediated angiogenesis
CD182	CXCR2 IL-8RB IL-8R2	Granulocytes, some T cells, monocytes, mast cells, endothelial cells	Chemokine receptor Binds to IL-8 plus two related growth factors Mediates IL-8 signaling promoting neutrophil activation and chemotaxis Transduces signals for chemokine-mediated angiogenesis, hematopoiesis, and possibly metastasis
CD183	CXCR3	Activated T cells, Th effectors, NK cells, B cells proliferating endothelial cells	Chemokine receptor Binds to chemokines CXCL9, CXCL10, CXCL11 Facilitates T cell homing to the lung
CD184	CXCR4	T and B cell subsets, monocytes, HSCs, endothelial cells, breast cancer cells	Chemokine receptor Binds primarily to chemokine SDF-1 (CXCL12) Directs T cells to T cell-rich zones of secondary lymphoid tissues Coreceptor for HIV infection
CD185	CXCR5	T and B cell subsets, activated T cells, neurons	Chemokine receptor Binds primarily to chemokine CXCL13 Facilitates cell migration Directs B cells to B cell-rich zones of secondary lymphoid tissues and is required for GC formation
CD186	CXCR6	T, B and NK cell subsets	Chemokine receptor Binds primarily to chemokine CXCL16 Directs T cell migration and homing
CD191	CCR1	Lymphocytes, monocytes, macrophages, DCs	Chemokine receptor Binds to chemokines CCL3, -5, -7, -8, -14, -15, -23 Facilitates leukocyte migration and early events in inflammation
CD192	CCR2	Macrophages, monocytes, activated T cells, DCs, basophils, endothelial cells	Chemokine receptor Binds to chemokines CCL2, -7, -8, -12, -13, -16 Required for normal trafficking of APCs
CD193	CCR3 Eotaxin receptor	Eosinophils, DCs, activated T cells, Th2 effectors, basophils, airway epithelial cells	Chemokine receptor Binds to chemokines eotaxin (CCL11) as well as CCL3, -5, -7, -8, -13, -14, -15 Required for eosinophil trafficking Facilitates allergic reactions
CD194	CCR4	T cells, thymocytes, DCs	Chemokine receptor Binds to chemokines CCL17, -22 Facilitates T cell chemotaxis and homing to skin
CD195	CCR5	Lymphocytes, monocytes, macrophages, DCs	Chemokine receptor Binds to chemokines CCL3, -4, -5 Facilitates leukocyte chemotaxis Coreceptor for HIV infection

(Continued)

—Continued

CD Number	Other Names	Cellular/Tissue Expression	Functions
CD196	CCR6	Memory T and B cells, NK cells, LCs	Chemokine receptor Binds to chemokine CCL20 Facilitates cell migration Promotes Th17 cell differentiation
CD197	CCR7	Naïve T and B cells, mature DCs, Th1 effectors, some memory T cells	Chemokine receptor Binds to chemokines CCL21, -19 Directs lymphocyte homing to splenic PALS, lymph nodes Directs activated B cells to move to T cell zones of secondary lymphoid tissues
CD205	DEC-205	DCs, TECs, BM stromal cells	Glycoprotein with C-type lectin domain Binds to ligands of apoptotic or necrotic cells Facilitates endocytosis and antigen presentation
CD206	Mannose receptor	Macrophages, monocytes, DC subsets	Glycoprotein with C-type lectin domain Binds to mannose-bearing ligands Facilitates endocytosis and antigen presentation
CD209	DC-SIGN	DC subsets	Glycoprotein with C-type lectin domain Binds to mannose-bearing ligands, including those expressed by many viruses Facilitates endocytosis and antigen presentation
CD210a	IL-10R	T, B and NK cells; monocytes; macrophages; activated neutrophils	Cytokine receptor Binds to IL-10 Suppresses immune responses and inflammation
CD212	IL-12Rβ	Activated T and NK cells, some B cells and DCs	Cytokine receptor subunit Binds to IL-12Rα chain to form IL-12 receptor Promotes Th1 differentiation and IFNγ production Also binds to IL-23
CD213a1	IL-13Rα1	Vascular endothelial cells, monocytes, mast cells, B cells, RS cells	Cytokine receptor subunit Binds to IL-13 with low affinity Binds to the IL-4R chain (CD124) to form the high affinity IL-13 receptor Inhibits Th1 cytokine production but does not promote Th2 differentiation Promotes isotype switching to IgE
CD213a2	IL-13Rα2	B cells, monocytes, epithelial cells	Cytokine receptor subunit Binds to IL-13 but has no cytoplasmic domain and so does not mediate signaling Regulates signaling by IL-13R and IL-4R
CD215	IL-15Rα	Activated monocytes, T cell subsets, NK cells	Cytokine receptor subunit Associates with IL-2Rβ (CD122) and IL-2Rγ (CD132) to form the high affinity IL-15 receptor Promotes cell survival, differentiation, proliferation
CD217	IL-17R	B, T and NK cells; monocytes; macrophages; granulocytes; epithelial cells	Cytokine receptor Binds to IL-17 Promotes inflammation
CD220	Insulin receptor	Leukocytes, fibroblasts, epithelial cells, endothelial cells	Transmembrane receptor Binds to insulin Regulates glucose metabolism
CD221	IGF-IR	Leukocytes, many non-hematopoietic cells	Transmembrane receptor Binds to insulin-like growth factor-1 Regulates cell proliferation and differentiation

—Continued

CD Number	Other Names	Cellular/Tissue Expression	Functions
CD223	LAG-3	Activated T and NK cells	Ig superfamily protein resembling CD4 Binds to MHC class II Disrupts proliferation and homeostasis of T cells Blocks differentiation of macrophages and DCs
CD230	PrPc	Brain cells, DCs, T and B cells, monocytes	Membrane glycoprotein Function of normal PrPc protein is unknown PrPres is abnormal form that forms cytotoxic aggregates PrPres is infective agent in spongiform encephalopathies
CD247	CD3ζ	All T lineage cells, some NK cells	Ig superfamily protein Part of CD3 complex Required for cell surface expression and signal transduction of TCR
CD252	OX40L TNFSF4	Activated B cells, macrophages, DCs, endothelial cells, mast cells	TNF superfamily membrane protein Ligand for OX40 (CD134) Mediates adhesion of T cells to APCs Minor costimulation
CD253	TRAIL TNFSF10	Activated T cells, NK cells, B cells	TNF superfamily membrane protein Binds to DR4, DR5 Induces apoptosis
CD254	RANKL OPGL TNFSF11	Activated T cells, osteoblasts, stromal cells	TNF superfamily membrane protein Binds to RANK (CD265) Promotes T–B and T–DC interaction Regulates bone development
CD256	APRIL TNFSF13	Leukocytes, some non-hematopoietic cells	TNF superfamily membrane protein Binds to TACI (CD267) and BMCA (CD269) to promote B cell proliferation
CD257	BAFF TNFSF13b BLyS	Myeloid cells, activated monocytes, DCs	TNF superfamily membrane protein Can be cleaved to give secreted cytokine form Binds to TACI (CD267), BMCA (CD269) and BAFF-R (CD268) Essential for T1 to T2 transition during B cell development
CD261	TRAIL-R1 DR4 TNFRSF10a	Activated T cells, some tumor cells	TNFR superfamily glycoprotein Binds to TRAIL (CD253) to induce apoptosis
CD265	RANK TNFRSF11	DCs, osteoclasts	TNFR superfamily glycoprotein Binds to RANKL (CD254) Regulates cellular activation Promotes T cell–DC interaction
CD267	TACI TNFRSF13b	B cells, myeloma cells	TNFR superfamily glycoprotein Binds to BAFF (CD257) and APRIL (CD256) Inhibits B cell proliferation
CD268	BAFF-R TNFRSF13c	B cells, T cell subsets	TNFR superfamily glycoprotein Binds to BAFF (CD257) Promotes B cell survival, maturation Supports T cell activation
CD269	BCMA TNFRSF17	B cells, plasma cells	TNFR superfamily glycoprotein Binds to BAFF (CD257) and APRIL (CD256) Promotes plasma cell survival
CD273	PDL2 B7DC	DCs, activated monocytes and macrophages	Ig superfamily glycoprotein (B7 family) Binds to PD-1 (CD279) to inhibit T cell costimulation and activation

(Continued)

—*Continued*

CD Number	Other Names	Cellular/Tissue Expression	Functions
CD274	PDL1 B7H1	T, B and NK cells; DCs; macrophages; epithelial cells	Ig superfamily glycoprotein (B7 family) Binds to PD-1 (CD279) to inhibit T cell costimulation and activation
CD275	ICOSL B7H2	Activated monocytes, macrophages, DC subsets, B cells	Ig superfamily glycoprotein (B7 family) Binds to ICOS (CD278) to promote T cell costimulation leading to proliferation, cytokine secretion, Th2 responses, Td antibody responses Upregulated by inflammatory cytokines Required for T–B cooperation, GC formation, isotype switching, IgE and IgG1 production
CD278	ICOS	Activated T cells, Th2 cells, thymocyte subsets	Ig superfamily glycoprotein related to CD28 Binds to ICOSL (CD275) on APCs Induced on activated T cells Contributes to T cell costimulation for proliferation, cytokine secretion, Th2 responses, Td antibody responses Required for T–B cooperation, GC formation, isotype switching, IgE and IgG1 production
CD279	PD-1	Activated T and B cells, thymocyte subsets	Ig superfamily glycoprotein Binds to PDL1 (CD274) and PDL2 (CD273) to inhibit T cell costimulation and activation Important role in T cell tolerance
CD281	TLR1	Monocytes, macrophages DCs, keratinocytes	TLR glycoprotein Binds to lipoproteins Coreceptor of TLR2 (CD282)
CD282	TLR2	Monocytes, neutrophils, granulocytes, macrophages, DCs, epithelial cells, keratinocytes	TLR glycoprotein Cooperates with TLR1 or TLR6 to bind to lipoprotein and lipoglycan in bacterial cell walls Triggers innate antibacterial response Stimulates respiratory burst and production of NO and IL-12 by macrophages
CD283	TLR3	DCs, epithelial cells, fibroblasts	TLR glycoprotein Binds to double-stranded viral RNA Triggers innate antiviral response (IFN production)
CD284	TLR4	Myeloid cells, endothelial cells, B cells, granulocytes	TLR glycoprotein Binds to LPS of gram-negative bacterial cell walls Triggers innate antibacterial response Induces phagocytosis, inflammatory cytokines
CD286	TLR6	Monocytes, macrophages, DCs, granulocytes, epithelial cells	TLR glycoprotein Binds to lipoproteins and glycans Coreceptor of TLR2 (CD282)
CD288	TLR8	Monocytes, macrophages, DCs, neurons	TLR glycoprotein Binds to single-stranded viral RNA Triggers innate antiviral response Promotes brain development and hematopoiesis
CD289	TLR9	B cells, monocytes, pDCs	TLR glycoprotein Binds to CpG motifs in DNA Triggers innate antibacterial and antiviral responses
CD290	TLR10	B cells, pDCs	TLR glycoprotein Putative coreceptor of TLR2 (CD282)
CD314	NKG2D	NK cells, $\gamma\delta$ T cells, some CD8+ $\alpha\beta$ T cells	Glycoprotein with C-type lectin domain Associates with DAP10 to form NK activatory receptor Binds to stress antigens MICA, MICB Activates NK cytotoxicity and costimulates some T cells

—Continued

CD Number	Other Names	Cellular/Tissue Expression	Functions
CD335	NKp46 NCR1	Resting and activated NK cells	Ig superfamily glycoprotein Associates with CD3ζ or FcεRIγ to form NK activatory receptor Binds to non-MHC ligands, including viral hemagglutinins Activates NK cytotoxicity, cytokine production
CD336	NKp44 NCR2	Activated NK cells, some activated γδ T cells	Ig superfamily glycoprotein Associates with DAP12 to form NK activatory receptor Binds to a non-MHC ligand Increases efficiency of NK cytotoxicity, cytokine production
CD337	NKp30 NCR3	Resting and activated NK cells	Ig superfamily glycoprotein Associates with CD3ζ to form NK activatory receptor Binds to non-MHC ligands, including heparan sulfate proteoglycans Activates NK cytotoxicity, cytokine production Increases efficiency of tumor cell killing by activated NK cells
CD339	Jagged-1	BM stromal cells, TECs, endothelial cells, keratinocytes, cells of non-hematopoietic tissues	Transmembrane protein Binds to Notch 1, 2 and 3 to determine cell fate decisions during hematopoiesis
CD360	IL-21R	B cells, some T cells, NK cells, monocytes and DCs	Cytokine receptor Binds to IL-21 to promote T and NK cell activation, proliferation and development Induces B cell apoptosis

*CD numbers are for human proteins, and most are the same in the mouse. Expression and function of a given protein may vary between human and mouse.

†Additional alternative names may exist.

#Major abbreviations: ADCC, antibody-dependent cell-mediated cytotoxicity; ALL, acute lymphoblastic leukemia; APC, antigen-presenting cells; APRIL, a proliferation-inducing ligand; BAFF, B cell activating factor; BCR, B cell receptor; BM, bone marrow; BMCA, B cell maturation; CAM, cellular adhesion molecule; CR, complement receptor; CTLA-4, cytotoxic T lymphocyte antigen-4; DAF, decay accelerating factor; DC, dendritic cells; DR, death receptor; FDC, follicular dendritic cells; GC, germinal center; G-CSF, granulocyte colony stimulating factor; GM-CSF, granulocyte–macrophage colony stimulating factor; GvHD, graft-versus-host disease; HEV, high endothelial venules; HSC, hematopoietic stem cell; HSP, heat shock protein; ICAM, intercellular adhesion molecule; ICOS, inducible costimulator; ICOSL, inducible costimulator ligand; IFN, interferon; Ig, immunoglobulin; IGF, insulin-like growth factor; iIEL, intestinal intraepithelial lymphocyte; IL, interleukin; LFA, leukocyte function associated-1; ITAM, immunoreceptor tyrosine-based activation motif; LBP, LPS binding protein; LC, Langerhans cell; LPS, lipopolysaccharide; LT, lymphotoxin; mAb, monoclonal antibody; MAC, membrane attack complex; MAdCAM, mucosal addressin cellular adhesion molecule; M-CSF, macrophage colony stimulating factor; MCP, monocyte chemotactic protein; MICA/B, MHC class I-related gene A; MIRL, membrane inhibitor of reactive lysis; NKG, natural killer gene; NO, nitric oxide; OPGL, osteoprotegerin ligand; PALS, periarteriolar lymphoid sheath; PD, programmed death; pDC, plasmacytoid DC; RANK, receptor activator for NF-κB; PrP^c, prion protein, cellular; PrP^{res}, prion protein resistant; PRR, pattern recognition receptor; RBC, red blood cells; RCA, regulator of complement activation; RNI, reactive nitrogen intermediate; ROI, reactive oxygen intermediate; RS, Reed–Sternberg cells of Hodgkin's lymphoma; SCF, stem cell factor; SDF, stromal cell-derived factor; SIRPα, signal regulatory protein alpha; TACI, transmembrane activator and CAML (calcium-modulating cyclophilin ligand) interactor; TEC, thymic epithelial cell; TLR, Toll-like receptor; TNF, tumor necrosis factor; TNFR, tumor necrosis factor receptor; TRAIL, tumor necrosis factor-related apoptosis-inducing ligand; VCAM, vascular cellular adhesion molecule; VLA, very late antigen.

Cytokines, Chemokines and Receptors

| TABLE E-1 | Major Human Cytokines and Cytokine Receptors |

Abbreviation	Name of Cytokine	Function of Cytokine	Producers of Cytokine	Cytokine Receptor	Cells/Tissues Expressing Cytokine Receptor
Interferons					
IFNα*	Interferon alpha Type 1 IFN	Induces antiviral state ↓ Cell proliferation ↑ NK cell and CTL functions Influences isotype switching	Virus-infected host cells Activated macrophages, monocytes; some activated T cells	IFNα/βR (type 1 IFN receptor)	Virtually all cells
IFNβ	Interferon beta Type 1 IFN	Induces antiviral state ↓ Cell proliferation ↑ NK cell and CTL functions Influences isotype switching	Virus-infected host cells Fibroblasts	IFNα/βR	Virtually all cells
IFNγ	Interferon gamma Type 2 IFN	Induces antiviral state ↓ Cell proliferation ↑ NK cell and CTL functions Influences isotype switching and apoptosis ↑ APC production of IL-12 and Th1 cell differentiation ↓ IL-4 production and Th2 differentiation	Activated macrophages, NK cells, Th1 cells, CTLs	IFNγR (type 2 IFN receptor)	Virtually all cells; not erythrocytes
Interleukins					
IL-1	Interleukin-1	Pro-inflammatory ↑ Acute phase response Induces fever and wasting Mediates endotoxic shock	Macrophages, neutrophils, keratinocytes, epithelial cells, endothelial cells	IL-1R	Most cell types
IL-2	Interleukin-2	Th1 cytokine ↑ T and B cell activation, proliferation, differentiation ↑ NK cell proliferation and production of TNF, IFNγ Required for T cell peripheral tolerance and homeostasis Required for Treg cell differentiation	Activated T cells	IL-2R	Activated T, B and NK cells

(Continued)

Interleukins

Abbreviation	Name of Cytokine	Function of Cytokine	Producers of Cytokine	Cytokine Receptor	Cells/Tissues Expressing Cytokine Receptor
IL-3	Interleukin-3	Primarily a mast cell and basophil growth factor ↑ T cell production of IL-10, IL-13 Promotes antiparasite response	Activated T cells and mast cells	IL-3R	Early hematopoietic cells, most myeloid lineages, some B cells
IL-4	Interleukin-4	Th2 cytokine Required for Th2 cell differentiation ↓ Macrophage and IFNγ functions ↑ B cell proliferation, differentiation, isotype switching ↑ Mast cell proliferation	Activated T cells, basophils, mast cells, NKT cells	IL-4R	Hematopoietic cells
IL-5	Interleukin-5	Th2 cytokine ↑ Eosinophil chemotaxis and activation ↑ Mast cell histamine release	Activated Th2 cells, mast cells, NK cells, B cells, eosinophils	IL-5R	Eosinophils, mast cells, basophils
IL-6	Interleukin-6	Pro-inflammatory ↑ Acute phase response Induces fever ↑ Neutrophil microbicidal functions ↑ B cell terminal differentiation ↑ Th17 cell differentiation	Activated phagocytes, fibroblasts, endothelial cells, some activated T cells	IL-6R	Hepatocytes, monocytes, neutrophils, activated B cells, mature T cells
IL-7	Interleukin-7	Promotes lymphopoiesis ↑ Development of αβ T cells, γδ T cells, B cells ↑ Generation and maintenance of memory T cells	Primarily BM and thymic stromal cells	IL-7R	T, B, NK and NKT precursors
IL-8	Interleukin-8	CXC chemokine ↑ Neutrophil chemotaxis ↑ Neutrophil degranulation and microbicidal functions	All cell types encountering TNF, IL-1 or bacterial endotoxin	CXCR1 CXCR2	Neutrophils, NK cells, T cells, basophils
IL-9	Interleukin-9	Promotes erythroid, myeloid and neuronal precursor differentiation ↑ Mast cell proliferation and differentiation ↑ Antihelminth worm defense (with IL-4) ↑ Mucus production	Activated Th2 cells, memory CD4+ T cells	IL-9R	Many hematopoietic cell types
IL-10	Interleukin-10	Anti-inflammatory, immunosuppressive ↓ Activation of macrophages, neutrophils, mast cells, eosinophils ↓ Th1 cytokine production ↓ APC function	Activated macrophages, monocytes, Th2 cells, B cells, eosinophils, mast cells	IL-10R	Most hematopoietic cell types

Abbreviation	Name of Cytokine	Function of Cytokine	Producers of Cytokine	Cytokine Receptor	Cells/Tissues Expressing Cytokine Receptor
Interleukins					
IL-11	Interleukin-11	Promotes erythroid, myeloid and megakaryocyte precursor proliferation ↑ T and B cell and neutrophil proliferation ↓ Macrophage functions ↑ Fibroblast growth and collagen deposition	BM stromal cells, osteoblasts; cells in brain, joints, testes	IL-11R	Many hematopoietic and non-hematopoietic cell types
IL-12	Interleukin-12	Required for Th1 cell differentiation ↑ Production of IFNγ by macrophages, activated Th1 cells, NK cells ↑ DC and macrophage cytokine secretion ↑ CTL and NK cytotoxicity ↑ Memory T cell differentiation into Th1 cells Influences isotype switching	Activated macrophages, DCs; neutrophils, monocytes, B cells	IL-12R	Activated T and NK cells, B cells, DCs
IL-13	Interleukin-13	Th2 cytokine ↑ Th2 cell production of IL-4, IL-5, IL-10 Does not induce Th2 differentiation ↓ Macrophage cytokine secretion ↑ B cell proliferation and switching to IgE ↑ Anti-nematode defense ↑ Mucus production	Activated T cells, mast cells, basophils	IL-13R	Monocytes, macrophages, B cells, endothelial cells
IL-15	Interleukin-15	Required for NK cell development, proliferation and production of TNF, IFNγ ↑ γδT cell development ↑ T cell activation, proliferation, differentiation, homing and adhesion ↑ Memory CD8+ T cell survival ↑ Mast cell proliferation	Activated APCs	IL-15R (T, B and NK cells) IL-15RX (mast cells)	Lymphoid precursors; mature T, B and NK cells; mast cells
IL-17	Interleukin-17	Family of six closely related cytokines, IL-17A–F, that are structurally unique Antifungal and antibacterial protection ↑ Pro-inflammatory cytokine and chemokine production ↓ Th1 cell differentiation Mobilizes neutrophils in allergic and autoimmune responses Activates osteoclasts for bone resorption ↑ Hematopoietic growth factor synthesis	Th17 cells	IL-17R (binds to IL-17A and IL-17F) IL-17RB (binds to IL-17B, E)	Widely expressed, including peripheral T and B cells, some non-hematopoietic cells

(Continued)

Abbreviation	Name of Cytokine	Function of Cytokine	Producers of Cytokine	Cytokine Receptor	Cells/Tissues Expressing Cytokine Receptor
Interleukins					
IL-18	Interleukin-18	Synergizes with IL-12 functions during later stages of Th1 response ↑ Th1 cell proliferation, production of IFNγ, IL-2R ↑ NK cytotoxicity and production of IFNγ, TNF	Widely expressed	IL-18R	Virtually all cells
IL-21	Interleukin-21	Belongs to the IL-2 family ↑ Expansion of activated GC B cells ↑ Isotype switching, ↑ Plasma cell generation Contributes to Th17 cell differentiation Represses iTreg cell differentiation	Activated T cells (especially fTh cells and Th17 cells), NKT cells	IL-21R	Widely expressed
IL-22	Interleukin-22	Belongs to IL-10 family ↑ Wound healing and skin inflammation ↑ Cutaneous and mucosal immunity against pathogens ↑ Keratinocyte proliferation and antimicrobial peptide production	Activated Th17 cells and Th22 cells	IL-22R	Widely expressed
IL-23	Interleukin-23	Promotes expansion of Th17 cells ↑ Memory CD4+ T cell proliferation and differentiation into Th1 cells ↑ IFNγ production by DCs and memory Th1 effectors	Activated APCs	IL-23R	Memory CD4+ T cells, Th17 cells, DCs, NK cells
IL-25	Interleukin-25	Also known as IL-17E ↑ IL-4, IL-5, IL-13 production ↑ Memory Th2 responses ↑ Eosinophil expansion Involved in gut immunity and inflammation ↓ Th17 cell differentiation	Activated Th2 cells and mast cells	IL-17RB	Widely expressed
IL-27	Interleukin-27	Required for early stages of Th1 response ↑ Th1 and NK cell activation and production of IFNγ ↑ Pro-inflammatory cytokine production by mast cells, monocytes ↑ Activity of anti-tumor CTLs and NK cells ↑ Th17 cell differentiation	Activated APCs	IL-27R	Naïve CD4+ T cells, NK cells
IL-33	Interleukin-33	Belongs to IL-1 family ↑ Th2 cytokine production In excess, induces mucosal pathology	HEVs, fibroblasts, mast cells, many non-hematopoietic cells	IL-1RL1	Th2 cells, mast cells, eosinophils, basophils

Abbreviation	Name of Cytokine	Function of Cytokine	Producers of Cytokine	Cytokine Receptor	Cells/Tissues Expressing Cytokine Receptor
TNF-Related Cytokines					
TNF	Tumor necrosis factor	Potent inflammatory, immunoregulatory, cytotoxic, antiviral, pro-coagulatory, and growth stimulatory effects ↑ Proliferation, activation, adhesion, extravasation, cytokine production of hematopoietic cells ↑ Acute phase response ↑ Macrophage and neutrophil microbicidal functions ↑ APC functions ↑ B cell proliferation, Ab production, GC formation ↑ Tumor cell apoptosis and hemorrhagic necrosis High concentration induces wasting, endotoxic shock, fibrosis, bone destruction	Many types of activated hematopoietic and non-hematopoietic cells	TNFRI TNFRII	Widely expressed (but not resting T and B cells)
LT	Lymphotoxin	Secreted molecule with TNF-like activities	Activated Th1, B and NK cells	TNFRI TNFRII	Widely expressed (but not resting T and B cells)
BAFF	B cell activating factor	Secreted molecule essential for survival of transitional B cells during B cell development	Myeloid lineage cells	BAFF-R TACI BCMA	B lineage cells B lineage cells, some T cells B lineage cells, plasma cells
Transforming Growth Factor					
TGFβ	Transforming growth factor-β	Anti-inflammatory, immunosuppressive Chemoattractant for T cells, monocytes, neutrophils ↓ Activation, homing and effector functions of macrophages, DCs, T and B cells, CTLs, NK cells ↑ Angiogenesis and extracellular matrix protein production ↑ Th17 cell differentiation ↑ Treg cell differentiation	Most activated hematopoietic cells; some non-hemato-poietic cells	TGFβR	Widely expressed

(Continued)

Hematopoietic Growth Factors

Abbreviation	Name of Cytokine	Function of Cytokine	Producers of Cytokine	Cytokine Receptor	Cells/Tissues Expressing Cytokine Receptor
SCF	Stem cell factor	Promotes HSC survival, self-renewal and differentiation into hematopoietic progenitors ↑ Proliferation of lymphoid and myeloid precursors	Stromal cells in fetal liver, BM, thymus	c-kit	HSCs in BM, cells in CNS and gut
GM-CSF	Granulo-monocyte colony stimulating factor	Promotes generation and differentiation of monocyte and granulocyte precursors	BM stromal cells, activated T cells, endothelial cells, macrophages	GM-CSFR	Myeloid lineage precursors
G-CSF	Granulocyte colony stimulating factor	Acts on monocyte/granulocyte precursors to generate granulocytes ↑ Steady state and emergency production of neutrophils	BM stromal cells, activated T cells, endothelial cells, fibroblasts, macrophages	G-CSFR	Monocyte-granulocyte precursors
M-CSF	Monocyte colony stimulating factor	Acts on monocyte/granulocyte precursors to generate monocytes and macrophages ↑ Generation of bone-resorbing cells	BM stromal cells, endothelial cells, fibroblasts, macrophages	M-CSFR	Monocyte-granulocyte precursors
PDGF	Platelet-derived growth factor	Promotes division of undifferentiated cells of mesenchymal derivation, including endothelial cells, glial cells, smooth muscle cells ↑ Embryogenesis ↑ Angiogenesis ↑ Mesenchymal cell migration	Activated platelets, smooth muscle cells, endothelial cells, activated macrophages	PDGFR	Widely expressed
EPO	Erythropoietin	↑ Erythropoiesis ↑ Wound healing ↑ Neuronal injury response in brain	Kidney fibroblasts, liver sinusoidal cells	EPOR	Erythrocyte precursors (not mature erythrocytes) BM stromal cells Cells of CNS and peripheral nervous system

*Major abbreviations: BM, bone marrow; CNS, central nervous system; fTh, follicular T helper cell; HSC, hematopoietic stem cell; iTreg, induced T regulatory cell; R, receptor.

TABLE E-2   Major Human Chemokines and Chemokine Receptors

Chemokine Systematic Name	Chemokine Common Name	Chemokine Receptor[†]	Cells/Tissues Expressing Chemokine Receptor
CCL Chemokines			
CCL1	I-309[*]	CCR8	Act T, NK, Th2
CCL2	MCP-1	CCR4	Act T, Bas, Th2
CCL3	MIP-1α	CCR4	Act T, Bas, Th2
CCL4	MIP-1β	CCR5	Mac, Act T, Th1
CCL5	RANTES	CCR1	Mac, Act T, Neu, Bas
CCL8	MCP-2	CCR5	Mon, T, NK, Mast, Eo, Bas
CCL11	Eotaxin	CCR3	Eo, Act T, Bas, Th2
CCL15	MIP-5	CCR1	Neu, Mon, T, B
CCL17	TARC	CCR4	Thymocytes
CCL19	ELC	CCR7	DC, Act B, Memory T
CCL21	SLC	CCR7	Mature DC; naïve T and B
CCL23	MIP-3	CCR1	Mon, Neu, T
CCL26	MIP-4α	CCR3	Eo, Bas
CCL27	CTACK	CCR10	Memory T
CCL28	MEC	CCR3	Mucosal T and B
CXCL Chemokines			
CXCL2	MIP-2	CXCR2	Most leukocytes, HSC
CXCL4	PF4	CXCR3B	Neu, fibroblasts
CXCL5	ENA-78	CXCR2	Neu
CXCL6	GCP-2	CXCR2	Neu, NK
CXCL7	NAP-2	CXCR2	Neu, NK
CXCL8	IL-8	CXCR1	Neu, NK
CXCL10	IP-10	CXCR3	Act T, NK, Th1
CXCL9	MIG	CXCR3	T
CXCL12	SDF-1	CXCR4	T
CXCL13	BCA-1	CXCR5	B
CXCL14	BRAK	Unknown	Mon, DC, Act NK
CXCL16	SCYB16	CXCR6	T, NKT
CX3CL Chemokines			
CX3CL	Fractalkine	CX3CR1	T, Mon, Activated endothelial cells
XCL Chemokines			
XCL1	Lymphotactin	XCR1	T
XCL2	Lymphotactin	XCR1	T

[*]Major abbreviations: Act T, activated T cells; B, B cells; Bas, basophils; BCA-1, B cell attracting chemokine 1; BRAK, breast and kidney-expressed chemokine; CTACK, cutaneous T cell-attracting chemokine; DC, dendritic cells; ELC, EBV-induced molecule-1 ligand chemokine; ENA78, epithelial neutrophil-activating protein 78; Eo, eosinophils; GCP-2, granulocyte chemotactic protein 2; HSC, hematopoietic stem cell; IL-8, interleukin 8; IP-10, interferon-inducible protein 10; Mac, macrophages; MCP, monocyte chemotactic protein; MEC, mucosae-associated epithelial chemokine; MIG, monokine induced by interferon gamma; MIP, macrophage inflammatory protein; Mon, monocytes; NAP-2, neutrophil activating peptide-2; Neu, neutrophils; NK, natural killer cells; PF4, platelet factor 4; RANTES, regulated on activation normal T cell expressed and secreted; SCY, small cytokine; SDF-1, stromal cell-derived factor 1; SLC, secondary lymphoid tissue chemokine; T, T cells; Th1, T helper cell type 1; Th2, T helper cell type 2.

[†]Many chemokines bind to more than one chemokine receptor.

Appendix F

Selected Immunological Techniques

I. Experimental Techniques Using Antibodies

Scientists interested in dissecting a biological system often take advantage of the properties of the antigen–antibody bond because this interaction is highly specific. Where standard techniques may not be able to distinguish between very closely related molecules, specific antibodies for distinct epitopes on those molecules can do so with ease. Identification and purification of a single component from a complex mixture become a ready possibility because an antibody can detect one antigenic molecule among 10^8 other molecules. In addition, the antigen–antibody interaction is reversible and does not alter the antigen. For these reasons, techniques employing antibodies are used to purify, characterize and quantitate antigens, and to pinpoint their expression in cells or tissues. In the following sections, we provide illustrations and brief descriptions of the basic principles of several experimental techniques that use antibodies.

Two broad categories of techniques use antibodies to detect antigen. One category encompasses assays based on the detection of large immune complexes containing antibody and antigen trapped in a network. Examples of these techniques are the *precipitin reaction*, *agglutination*, and *complement fixation*. The other category includes techniques based on the formation of individual antigen–antibody pairs, in which detection relies on a "tag" chemically introduced onto either the antigen or the antibody molecule. Such tags are usually radioactive, enzymatic or fluorescent and give rise to easily detectable assay signals. These assays tend to be more sensitive than those based on immune complex formation. Techniques of this type include *radioimmunoassay (RIA)*, *enzyme-linked immunosorbent assay (ELISA)*, *immunofluorescence* and *flow cytometry*. A third category of antibody-based techniques involves the use of antibodies to isolate and characterize antigens, and includes *immunoprecipitation*, *affinity chromatography* and *Western blotting* (also known as *immunoblotting*). All of these techniques are described and illustrated in the following sections.

i) Techniques Based on Immune Complex Formation

Like all chemical reactions, the kinetics of antigen–antibody binding are driven by relative concentration. When complementary antibody and antigen (i.e., a binding pair) are mixed in a fluid in approximately equal amounts, non-covalent bonds are rapidly formed between individual molecules, resulting in a small *soluble* complex. However, because of bivalency, a single antibody molecule may use one antigen-combining site to bind to its epitope on one molecule of antigen molecule and the other antigen-combining site to bind the identical epitope on a second antigen molecule. Each of these two antigen molecules may possess additional epitopes for additional antibody binding, so that different antibody molecules mutually binding to this antigen are said to be *cross-linked*. Further cross-connections between additional antigen and antibody molecules result in the formation of an *immune complex* or *lattice*. As more and more antigen and antibody molecules become cross-linked, they form lattices large enough to precipitate out of solution and to become visible. These are the properties that were first exploited to examine antigen–antibody interactions, in the form of the *precipitin* and *immunodiffusion* assays. The collection of precipitated immune

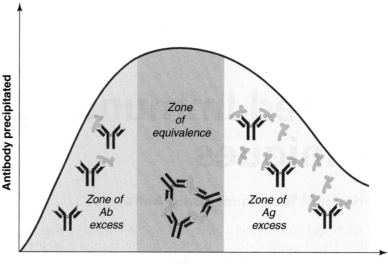

Fig. F-1
Precipitin Curve

To generate the precipitin curve, varying concentrations of antigen are added to a fixed concentration of antibody, and the amount of antibody precipitated is measured in each case. Optimal immune complex formation occurs in the zone of equivalence. Ab, antibody; Ag, antigen

complexes by centrifugation also provided the first means of isolating and purifying antigens.

As illustrated in the *precipitin curve* shown in **Figure F-1**, when increasing amounts of soluble antigen are added to a fixed amount of soluble complementary antibody, the amount of lattice or immune complex precipitated increases up to a broad peak called the *zone of equivalence* and then slowly declines again. In the zone of equivalence, the number of antigenic epitopes and antibody-combining sites is approximately equal, and the binding is optimal. Enough antigenic epitopes are present such that each individual antibody Fab site binds to a *different* antigen molecule, but the antigen concentration is still low enough that each antigen molecule is shared between (on average) two antibodies. Under these conditions, cross-linking readily occurs and immune complex precipitation is maximal. Where the amount of antigen added is very low and an excess of antibody over antigen is present, on average only one Fab site of an antibody is in use, meaning that cross-linking and immune complex formation are minimal. Free antibody remains in the supernatant after antigen–antibody complexes are removed by centrifugation. This area of the precipitin curve is known as the *zone of antibody excess*. When the antigen concentration increases past the zone of equivalence, a point is reached at which every available Fab site is bound to a separate antigen molecule: no antigen molecule is shared between two antibodies, and a large immune complex cannot form. Free antigen remains in the supernatant after antigen–antibody complexes are removed by centrifugation, giving rise to the area of the precipitin curve known as the *zone of antigen excess*. **Figures F-2, F-3, F-4 and F-5** illustrate several experimental techniques based on immune complex formation, including precipitin ring formation, various forms of immunodiffusion and immunoelectrophoresis, agglutination, and complement fixation.

ii) Techniques Based on "Tagging" Antigen–Antibody Pairs

Techniques based on immune complex formation exploit the fact that antigen–antibody immune complexes are relatively large in size and thus readily make themselves "visible" in assays. In contrast, unitary antibody–antigen pairs by definition are not found in extensive complexes. Thus, they cannot be made visible for quantitation

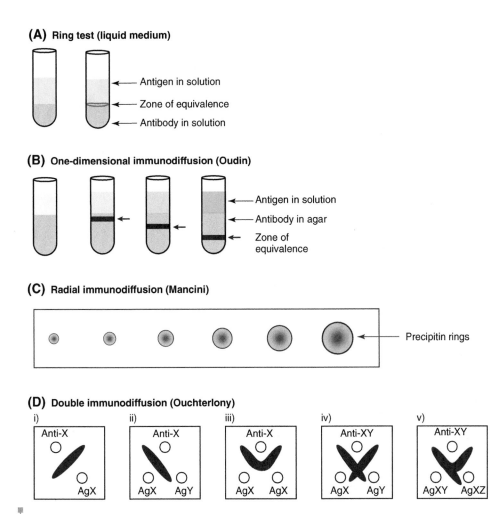

(A) Ring test (liquid medium)

Antigen in solution
Zone of equivalence
Antibody in solution

(B) One-dimensional immunodiffusion (Oudin)

Antigen in solution
Antibody in agar
Zone of equivalence

(C) Radial immunodiffusion (Mancini)

Precipitin rings

(D) Double immunodiffusion (Ouchterlony)

i)   Anti-X   AgX
ii)   Anti-X   AgX   AgY
iii) Anti-X   AgX   AgX
iv)   Anti-XY   AgX   AgY
v)   Anti-XY   AgXY AgXZ

Fig. F-2
Assays Based on the Precipitin Reaction and Immunodiffusion

(**A**) Ring test. Left-hand tube, no complementary antigen and antibody are present. Right-hand tube, complementary antigen and antibody cause a precipitin line to form at the solution interface (zone of equivalence). (**B**) One-dimensional immunodiffusion (Oudin) assay. With antibody-impregnated agar as the medium, increasing concentrations of antigen (left to right) form a series of descending precipitin lines after diffusion into the agar. (**C**) Radial immunodiffusion (Mancini) assay. Various amounts of antigen are placed in wells in antibody-impregnated agar. The greater the antigen concentration, the greater the diameter of the precipitin ring formed. (**D**) Double immunodiffusion (Ouchterlony) assay. Antibody and antigen samples diffuse toward each other from wells made in agar. A precipitin line will form wherever a zone of equivalence exists between a complementary antibody and antigen (i). In panels ii–v, a single antiserum is used to give information about the relationship between test antigens. The resulting patterns show the following relationships between test antigen X (AgX) and test antigen Y (AgY): (ii) non-identity (no epitopes shared between AgX and AgY), (iii) identity (all epitopes shared), (iv) non-identity, and (v) partial identity (some epitopes shared).

unless the pair is somehow labeled with an easily detected *tag*. Techniques that make use of detectable tags to track antigens or antibodies of interest have greatly increased the scope of antibody-based assays. Tags are generally radioisotopes, enzymes, or fluorochromes that are covalently bound to either the antigen or the antibody. In addition, one partner of the antigen–antibody pair must be immobilized in some way so that any tag that is not part of an antigen–antibody pair can be removed from the assay system by washing. As well as for detection, tag assays are frequently used for the quantitation of an antigen or antibody because the amount of tag detected is proportional to the number of antigen–antibody pairs in the sample. Tag assays can either be *direct* or *indirect*.

(A) Immunoelectrophoresis

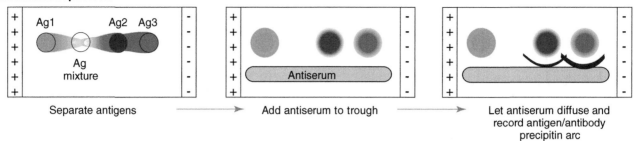

Separate antigens → Add antiserum to trough → Let antiserum diffuse and record antigen/antibody precipitin arc

(B) Rocket electrophoresis

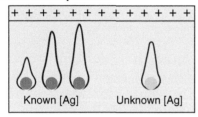

Electrophorese antigens on gel impregnated with antibody

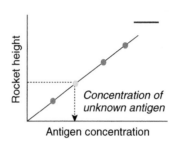

Antigen concentration

Concentration of unknown antigen

(C) Two-dimensional immunoelectrophoresis

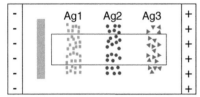

Electrophorese antigen mixture and cut out strip from gel

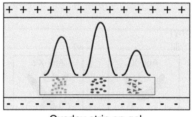

Overlay strip on gel impregnated with Ab

Fig. F-3
Types of Immunoelectrophoresis

(**A**) Immunoelectrophoresis. An antigen mixture is placed in a central well in a gel plate. An electrical current is applied to the plate that causes the antigens to separate to different positions in the gel on the basis of their charge. Antiserum is added to a well that runs parallel to the axis of antigen separation, and time is allowed for the antibodies to diffuse into the gel. Any interactions between antigens and antibodies are detected based on the precipitin reaction, as indicated. (**B**) Rocket electrophoresis. Electrophoresis of an antigen sample in an antibody-impregnated gel results in a precipitin line whose height is proportional to the antigen concentration (left panel). Using known antigen concentrations, a standard curve can be constructed to determine the sample concentration (right panel). (**C**) Two-dimensional immunoelectrophoresis. If a sample contains a mixture of antigens, they can first be separated in a gel by immunoelectrophoresis and then subjected to analysis by rocket electrophoresis.

a) Direct Tag Assays

Direct tag assays refer to single-step procedures in which a tagged antigen (or antibody) is used to detect the presence of its untagged antibody (or antigen) binding partner. The tagged antigen or antibody is incubated with the test sample to allow antigen–antibody pairs to form. Unbound tagged molecules are removed by washing, and the remaining tagged molecules are quantitated by measuring the amount of tag present relative to a standard curve. In many cases, the absolute number of antigen–antibody pairs present is not as important as the relative amount of tag (pairs) present compared to controls.

b) Indirect Tag Assays

When the antibody or antigen of interest is not available in pure form or is chemically difficult to tag, antigen–antibody binding can be detected by tagging a third component that binds to the unlabeled antigen–antibody pair of interest. Three reagents commonly used as the third component in indirect assays are secondary antibodies,

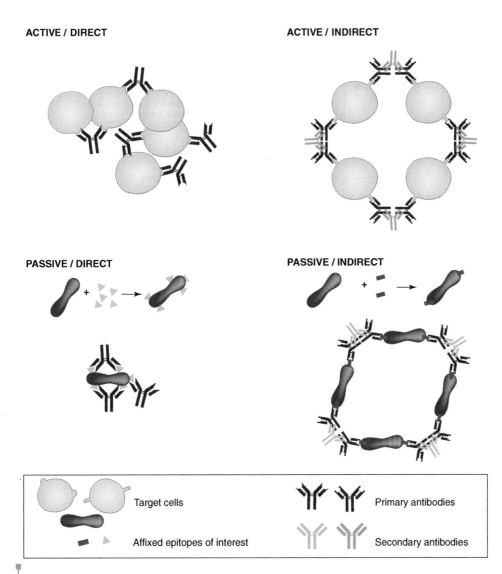

ACTIVE / DIRECT

ACTIVE / INDIRECT

PASSIVE / DIRECT

PASSIVE / INDIRECT

Target cells

Primary antibodies

Affixed epitopes of interest

Secondary antibodies

Fig. F-4
Types of Agglutination

Agglutination assays are used to test for the presence of an antibody of interest in a serum sample. In agglutination, the binding of specific antibody mediates the visible clumping of target cells (shown) or test particles (not shown). The epitope recognized by the antibody may occur naturally on the target cells (active) or may need to be affixed to target cells artificially (passive). The specific antibody itself may be able to agglutinate the target cells (direct). If not, a secondary antibody that recognizes and binds to the Fc region of the primary antibody must be used to mediate agglutination (indirect). In the examples shown, the target cells used in the passive agglutination assays (both direct and indirect) are red blood cells (RBCs), such that these assays are also referred to as "hemagglutination" assays.

Staphylococcus aureus Protein A or Protein G, and the biotin–avidin system (**Fig. F-6**). Secondary antibodies are usually prepared by immunizing another species with the primary antibody protein (used as the antigen).

c) Common Tag Assays for Antigen Detection

Several types of assays use tagged antigens or antibodies either directly or indirectly to detect antigens of interest in soluble, cellular or tissue samples. *Binder–ligand* assays include the radioimmunoassay (RIA) and the enzyme-linked immunosorbent assay (ELISA), in which the tag is a radioisotope in the former and a chromogenic substrate in the latter. How a binder–ligand assay can be used to detect antigen is illustrated for ELISA in **Figure F-7**. *Immunofluorescence* assays allow scientists to examine the

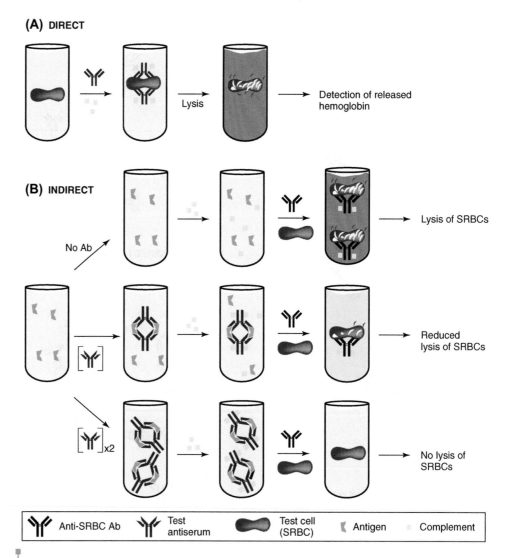

(A) DIRECT

Lysis → Detection of released hemoglobin

(B) INDIRECT

No Ab → Lysis of SRBCs

$[\text{Y}]$ → Reduced lysis of SRBCs

$[\text{Y}]_{x2}$ → No lysis of SRBCs

| Anti-SRBC Ab | Test antiserum | Test cell (SRBC) | Antigen | Complement |

Fig. F-5
Assays Based on Complement Fixation

Classical complement-mediated lysis of cells can be used to evaluate the presence of specific antibody. In these examples, the complement-mediated lysis of sheep red blood cells (SRBCs) used as test cells results in the release of hemoglobin that can be measured and directly related to the concentration of specific antibody present. Test cells other than SRBCs can also be used, in which case evaluation of lysis can be measured by the uptake of dyes or by the release of an internalized radioisotope. (**A**) Direct assay; the antigenic epitope of interest occurs on the test cell. (**B**) Indirect assay; the antigenic epitope of interest is soluble. After time is allowed for any antibody present to bind to the antigen, form complexes and consume ("fix") complement, the amount of complement remaining unfixed is measured by adding anti-SRBC antibodies and SRBCs. The amount of hemoglobin released is inversely proportional to the amount of antibody–antigen binding occurring in the first step of the assay. To simplify the illustration, the antigen–antibody complexes present since the first step of the indirect assay are not shown in the final tube, as they play no direct role at this point.

presence of an antigen *in situ* in a blood or tissue sample. In this case, the tag is a fluorochrome that emits light of a specific color after excitation by UV radiation (**Fig. F-8** and **Plate F-1**). Fluorochrome tag assays are also used for *flow cytometric analysis*, in which living cells bind antibodies tagged with particular fluorochromes. The cells can then be tracked and separated according to the tagged antibodies they have bound (**Plate F-2**). When this approach is used with multiple tagged antibodies to establish which cell surface proteins are expressed by a specific cell population, the technique

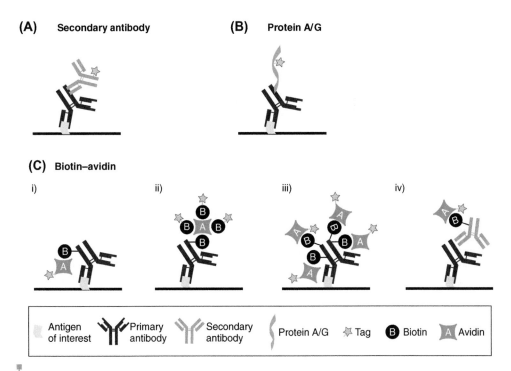

(A) Secondary antibody

(B) Protein A/G

(C) Biotin–avidin

i)

ii)

iii)

iv)

| Antigen of interest | Primary antibody | Secondary antibody | Protein A/G | Tag | B Biotin | A Avidin |

Fig. F-6
Third Components in Antigen–Antibody Assays

(A) A tagged secondary antibody is specific for the Fc region of the primary antibody. **(B)** Tagged Protein A or Protein G purified from the bacterium *Staphylococcus aureus* binds to the Fc region of primary IgG antibodies. **(C)** *Biotin* is a small molecule that binds with extremely high affinity to an egg white glycoprotein called *avidin*. Avidin contains four biotin-binding sites. A primary antibody can be conjugated with biotin before use in an assay so that the presence of antigen–antibody pairs can be later detected using tagged avidin (i). The assay sensitivity can be increased by using the avidin molecule as a multivalent bridge between the biotinylated antibody and tagged biotin (ii), or by ensuring that the biotinylated antibody has undergone biotin conjugation at multiple sites (iii). If desired, biotin/avidin-based approaches can also be applied using a secondary antibody that is biotinylated (iv).

is called *immunophenotyping* and generates a particular collection of marker proteins associated with that cell population.

iii) Techniques for the Isolation and Characterization of Antigens

Antigen–antibody interaction is frequently used to purify antigens and characterize their physical properties. The specificity of antibodies often makes extensive preliminary purification of the antigen unnecessary. The technique of *immunoprecipitation* by specific antibody can be used to isolate antigens present at low concentrations in complex mixtures of proteins. Specific antibody is added to the protein mixture (such as a cell extract), and the antigen–antibody complex is caused to precipitate out of solution by adding an insoluble agent to which the complexes will bind (**Fig. F-9**). Antibodies can also be used to isolate antigens using *affinity chromatography*. The protein mixture is passed over a column of inert beads to which specific antibody has been covalently fixed. The antigen is retained on the column, whereas other proteins are washed through (**Fig. F-10**). The *Western blot* uses the specificity of antigen–antibody binding in conjunction with gel electrophoresis to detect very small quantities of a protein of interest in a complex mixture. The proteins are first separated by size via electrophoresis and transferred to an inert membrane. The membrane is then incubated with a solution containing tagged specific antibody to identify the position of the antigen of interest (**Fig. F-11**).

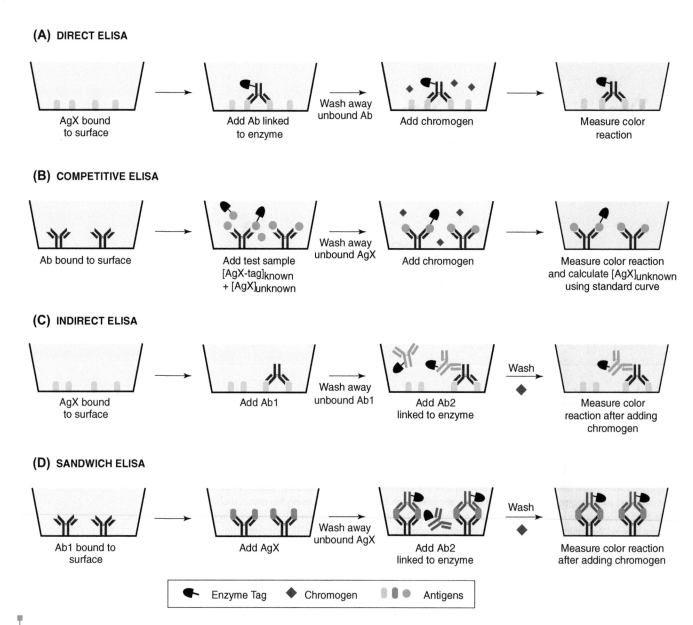

(A) DIRECT ELISA

AgX bound to surface → Add Ab linked to enzyme → *Wash away unbound Ab* → Add chromogen → Measure color reaction

(B) COMPETITIVE ELISA

Ab bound to surface → Add test sample $[AgX\text{-}tag]_{known}$ + $[AgX]_{unknown}$ → *Wash away unbound AgX* → Add chromogen → Measure color reaction and calculate $[AgX]_{unknown}$ using standard curve

(C) INDIRECT ELISA

AgX bound to surface → Add Ab1 → *Wash away unbound Ab1* → Add Ab2 linked to enzyme → *Wash* → Measure color reaction after adding chromogen

(D) SANDWICH ELISA

Ab1 bound to surface → Add AgX → *Wash away unbound AgX* → Add Ab2 linked to enzyme → *Wash* → Measure color reaction after adding chromogen

Enzyme Tag ◆ Chromogen ❙❙● Antigens

Fig. F-7
ELISAs as Examples of Binder–Ligand Assays

In all binder–ligand assays, one component of interest, either the antigen or antibody, is immobilized so that antigen–antibody pairs formed during the course of the assay are not lost during subsequent washing steps. If a radioisotope is used as the tag, the assay is an RIA (not shown). In an ELISA, the tag is an enzyme that acts on a chromogenic substrate to yield a detectable colored product. In the direct ELISA shown in **(A)**, antigen X of unknown concentration is coated onto the surface of a microtiter well, and tagged antibody is added to detect it. The concentration of antigen X is then calculated using a standard curve. In a competitive ELISA **(B)**, antibody specific for antigen X is coated onto the surface of a microtiter well. The presence of AgX in the test sample leads to a proportional displacement of tagged AgX from the immobilized antibody, decreasing the intensity of the final color reaction proportionally. For the indirect ELISA shown in **(C)**, a secondary antibody has been used as the third component; however, Protein A, Protein G or the biotin–avidin system could also be used for this step. The sandwich ELISA **(D)** is a variation of the indirect ELISA in which the secondary antibody is also specific for the antigen of interest. Note that in panels **(C)** and **(D)** a final washing step removes unbound secondary antibody prior to the addition of chromogen, so that any subsequent color reaction is due to specific binding of the enzyme-tagged antibody.

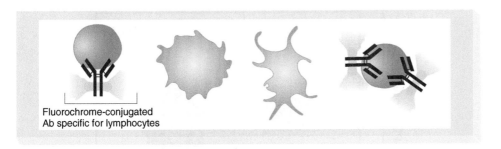

Fig. F-8
Immunofluorescence

In this schematic example, a blood cell sample (containing a spectrum of leukocytes) is tested for the presence of lymphocytes using direct immunofluorescence. The sample is fixed on a microscope slide and incubated with fluorochrome-conjugated antibody that binds to an antigen present only on lymphocytes. After excess unbound antibody is washed away, the slide is viewed under a fluorescence microscope. Fluorescing sites indicate the presence and location of lymphocytes on the slide.

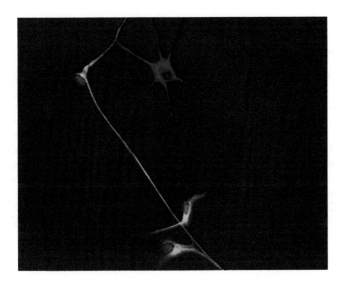

Plate F-1
Immunofluorescent Examination of Human Neural Cells

In this example, a neural filament protein (GFAP) is bound by an antibody tagged with a fluorochrome that emits red light. A different filament protein (nestin) is bound by an antibody tagged with a fluorochrome that emits green light. The yellow color associated with one of the neurons indicates that it expresses both GFAP and nestin. The nuclei of all the neurons are stained blued with Hoechst dye, a non-proteinaceous molecule that binds specifically to DNA. *[Reproduced by permission of Radha Chaddah, Neurobiology Research Group, University of Toronto.]*

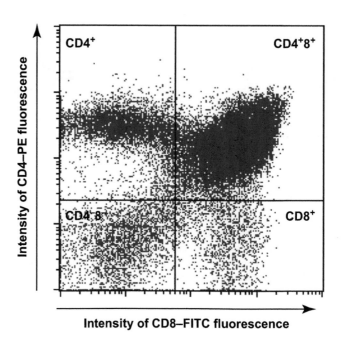

Plate F-2
Flow Cytometry

In this example, a mixed population of T lineage cells has been tested for levels of expression of CD4 and CD8. The T cells have been incubated with two antibodies: anti-CD4 antibody tagged with the fluorochrome PE and anti-CD8 antibody tagged with the fluorochrome FITC. The cells are then fed into a *flow cytometer* that separates the cells on the basis of the wavelength and intensity of the fluorescence emanating from the antibodies that have bound to each cell. These data indicate which antibodies (if any) each cell has bound and quantitates the amount of antibody bound by each cell (which is a measure of the level of expression of CD4 or CD8 by that cell). The results are presented as a quadrant graph. Mature T cells express either CD4 (upper left quadrant) or CD8 (lower right). Very early T cell precursors express neither CD4 nor CD8 (lower left), while slightly more advanced precursors express both CD4 and CD8 (upper right). [*Reproduced by permission of Juan-Carlos Zúñiga-Pflücker, Department of Immunology, University of Toronto and Sunnybrook Research Institute, Toronto.*]

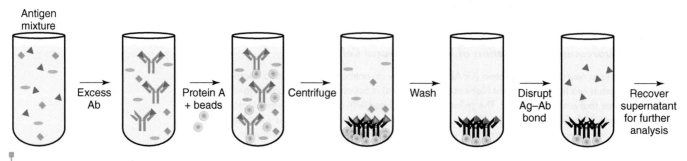

Fig. F-9
Immunoprecipitation

A protein mixture is incubated with specific antibody. Any antigen–antibody complexes that form are precipitated from solution by the addition of Protein A-coated beads that bind to the antibodies and collect at the bottom of the tube under the force of centrifugation. After washing, the desired antigen is released from the antibody-bound beads using altered pH and/or high salt concentration.

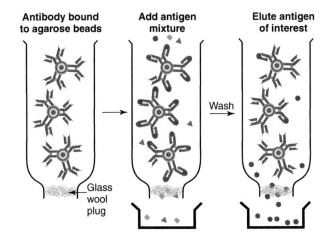

Fig. F-10
Affinity Chromatography

Agarose beads bearing immobilized specific antibody are placed into a column with a semi-permeable plug at the bottom. A solution containing antigen is passed slowly through the column, allowing the binding of specific antigen to the immobilized antibody. Unbound entities pass through the plug, and any molecule that binds to the beads non-specifically is removed by extensive washing. A solution with the appropriate pH and salt concentration to disrupt antigen–antibody binding is then passed through the column to elute (wash off) the antigen of interest.

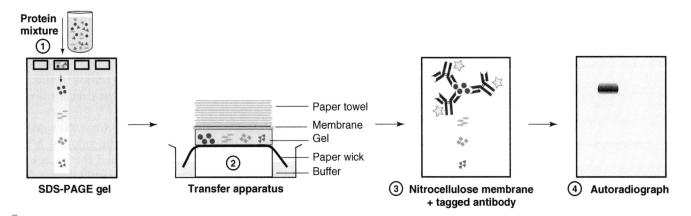

Fig. F-11
Western Blotting

Proteins are separated by charge using gel electrophoresis (**1**) and transferred onto an inert membrane. Originally, it was the capillary action of buffer drawn up through a filter paper wick toward absorbent paper towel that mediated the transfer of proteins onto the membrane (**2**). These days, electric current is used to speed the process. The antigenic protein of interest is then detected on the membrane by the binding of tagged antibody (**3**). If the tag is a radioisotope, its presence is detected by subjecting the membrane to autoradiography (**4**). If the tag is an enzyme (not shown), the membrane is exposed to the appropriate chromogenic substrate. A colored band develops where the protein is bound to the membrane.

II. The Polymerase Chain Reaction

In the early 1980s, a technique called the polymerase chain reaction (PCR) was developed that allows researchers to examine the DNA of cells that are available only in very limited numbers. PCR is capable of amplifying a single starting DNA molecule into more than a billion near-identical copies in just a few hours. The beauty of PCR is that absolutely any piece of DNA can be amplified as long as one knows the sequences of short stretches of DNA flanking the gene fragment of interest. First, oligonucleotides complementary to these short flanking sequences are synthesized, providing "PCR primers" that will be used to initiate replication of the desired stretch of DNA. A sample of the DNA of interest is then heated to denature the double-stranded structure into two single strands, each of which can serve as a template for copying. The separated DNA strands are then combined with an excess of the PCR primers, which bind to their complementary sequences on the template DNA strands. The templates are now considered to be "primed" and ready to be copied. Incubation of the primed template DNA strands with the four deoxynucleotide precursors (A, C, G, T) and a heat-resistant DNA polymerase called Taq results in the extension of the primers one nucleotide at a time, all the way toward the end of each DNA template. When the melting-annealing procedure is repeated for multiple "cycles," double-stranded amplification of the "target" DNA, the small stretch of DNA of interest between the primers, is readily achieved. **Table F-1** shows the exponential increase in DNA target copies produced during a 32-cycle PCR.

The combination of PCR with nucleotide sequencing revolutionized many types of molecular genetic analyses. Immunologists wasted no time in capitalizing on PCR to study the genes that encode the Ig, TCR, and MHC molecules, among others. PCR allows easy detection of Ig rearrangements, because if the gene segment of interest is not rearranged in a certain cell, the primers designed to amplify the target area of interest will be too far apart and the PCR reaction will fail. However, if rearrangement has occurred, the primers will bind at sites much closer together, permitting PCR to proceed. This technique has been used in various diagnostic applications. For example, certain follicular B cell lymphomas are caused by translocation of part of the immunoglobulin J_H locus on chromosome 14 to the locus for the Bcl-2 oncogene on chromosome 18. PCR permits this translocation to be detected with great precision in affected B cells (one cell in 10^6), facilitating either primary identification of the cancer or post-treatment monitoring.

Similar applications of PCR contributed to advances in other areas. New MHC alleles can be amplified and sequenced without ever isolating the original MHC gene from genomic DNA. Use of this methodology has streamlined the process of tissue typing prospective organ donors and recipients prior to transplantation (see Section III). Scientists have also taken full advantage of the fact that DNA (in the absence of nucleases) is stable for remarkably long periods. A wealth of otherwise inaccessible information has been gathered by performing PCR on cervical carcinoma biopsies embedded in paraffin for over 40 years, on 4000-year-old Egyptian mummies, and even on 18-million-year-old plant fragments preserved in shale fossil

TABLE F-1	The Power of PCR Amplification
Cycle Number	**Number of Copies of the Original DNA**
3	2
10	256
15	8192
20	262,144
25	8,388,608
32	1,073,741,824

beds. Forensic analysts can obtain useful information from miniscule amounts of DNA left at a crime scene. Researchers continue to innovate and refine techniques using PCR, developing ever more sensitive assay tools.

III. HLA Typing Techniques

i) Complement-Dependent Cytotoxicity

Historically, *complement-dependent cytotoxicity (CDC)* was the major means by which HLA typing was carried out. In this technique, blood leukocytes from the individual being typed are mixed in an array of microwells with defined collections of anti-HLA antibodies (**Fig. F-12**). The anti-HLA antibodies used are obtained from individuals exposed to allogeneic cells via a previous transplant, transfusion or pregnancy. Also added to the wells are complement and a visible dye (such as trypan blue) that is excluded from viable cells. In any given well, if the cells express HLA alleles recognized by the added anti-HLA antibodies, the cells are perforated by classical complement activation such that the dye can enter the cells and turn them blue. HLA mismatching of a prospective donor and recipient is thus revealed by different patterns of reactivity across the same array of antibody collections. CDC assays are quite labor-intensive and have been largely replaced by PCR-based methods.

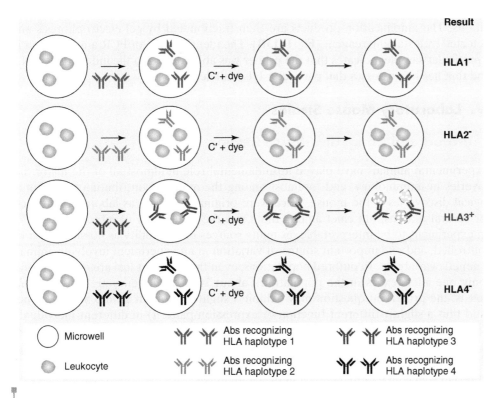

Fig. F-12
HLA Typing by CDC

In this example, leukocytes from an individual being typed are divided into four separate microwells. A preparation of purified antibody recognizing one of four different HLA haplotypes (HLA1, HLA2, HLA3 or HLA4) is added to the wells, as indicated. Rabbit complement (C') and trypan blue dye are then added to each microwell. Where the antibody preparation fails to bind to the leukocytes, complement is not activated; the cells remain intact and exclude the blue dye. Where the antibody preparation binds to the leukocytes, the cells are permeabilized by complement activation and take up the blue dye. Here we see that only the addition of antibody recognizing HLA haplotype 3 results in blue cells, meaning that the individual expresses HLA haplotype 3 but not haplotypes 1, 2 or 4. In an actual tissue typing scenario, a large number of microwells and anti-HLA antibodies would be used, such that the assay would detect the expression of a wide range of HLA alleles.

ii) PCR-based Methods of HLA Typing

As introduced in Chapter 17, PCR-SSOP and PCR-SSP are two PCR-based approaches commonly used for HLA typing of prospective tissue transplant donors and recipients. We provide here illustrations of the fundamentals of these techniques.

a) PCR Followed by Sequence-Specific Oligonucleotide Probing (PCR-SSOP)

Generic HLA primers are mixed with DNA samples extracted from cells of multiple prospective transplant donors, allowing for PCR amplification of all HLA alleles. The samples containing the amplified HLA sequences are split into multiple aliquots and dot-blotted onto inert membranes (**Fig. F-13A**). The membranes are hybridized to specific single-stranded oligonucleotide probes that represent diagnostic sequences from specific HLA alleles; these probes are labeled with a radioactive or non-radioactive tag. Detection of the complementary binding of a particular probe to an individual's DNA confirms that he/she expresses that HLA allele. Reverse PCR-SSOP can also be performed, in which the DNA samples are maintained in solution and mixed with specific oligonucleotide probes that have been fixed to an inert support matrix such as synthetic microbeads.

b) PCR Using Specific Primers (PCR-SSP)

In this *direct amplicon* method, aliquots of an individual's DNA are separately mixed with a series of HLA allele-specific oligonucleotide primers under PCR reaction conditions. The amplification products are then fractionated by gel electrophoresis and detected by a staining reagent (**Fig. F-13B**). The identification of PCR amplification in a particular sample indicates that the primer was able to bind to the individual's DNA and that he/she possesses that particular HLA allele.

IV. Laboratory Mouse Strains

i) Inbred and Congenic Mouse Strains

Experimental animals have played a fundamental role in almost all of the major discoveries in immunology, and foremost among the animals contributing to immunological discoveries is the mouse. Mice were originally chosen as laboratory subjects for several reasons, but chief among them was the availability of inbred strains. For an experiment to be interpretable, as many sources of variability as possible must be controlled, and one important source of variation in an experiment involving animals is genetic variability. In outbred populations, even though most loci are *monomorphic* (only one allele exists in the population), about 10% of all genes are *polymorphic*; that is, the protein in question may exhibit a slightly different amino acid sequence (and thus a slightly different function or expression pattern) in different individuals. Because of this variability, one cannot be sure that the results of experiments carried out in outbred populations are due to the experimental variable or to the influence of differing genetic factors. It is therefore desirable to conduct experiments using *inbred* strains in which all members are genetically identical.

The development of the inbred mouse as the premier laboratory animal for immunological experiments has an interesting history. For centuries, Chinese and Japanese hobbyists kept and bred mice that had attractive coat colors or other unusual characteristics such as crooked tails, obesity or dwarfism. In the 19th century, "mouse fancying," as it came to be called, spread to England and became quite popular. From England, it spread to America, and some American pet shop owners found that much of their business came from the sale of such mice. To maintain the desired characteristic, the pet shop owners bred siblings together (*sibling mating*), often for many consecutive generations. This took advantage of the fact that, after about 20 generations of sibling mating, heterozygosity is slowly lost and the progeny become essentially genetically identical; that is, they are homozygous at all loci. Thus, genetic variability can be eliminated as a factor influencing experimental

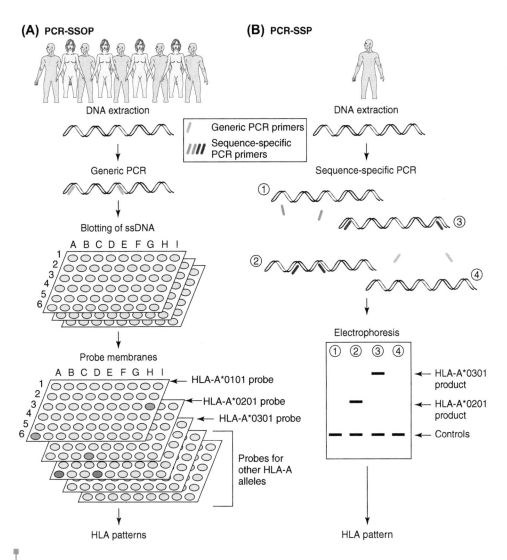

(A) PCR-SSOP

DNA extraction

Generic PCR primers

Sequence-specific PCR primers

Generic PCR

Blotting of ssDNA

A B C D E F G H I

Probe membranes

A B C D E F G H I

HLA-A*0101 probe

HLA-A*0201 probe

HLA-A*0301 probe

Probes for other HLA-A alleles

HLA patterns

(B) PCR-SSP

DNA extraction

Sequence-specific PCR

Electrophoresis

① ② ③ ④

HLA-A*0301 product

HLA-A*0201 product

Controls

HLA pattern

Fig. F-13
HLA Typing by PCR

(A) PCR-SSOP is generally used for the screening of large numbers of samples. In the example shown, DNA from each individual is extracted and hybridized separately to generic primers that allow amplification of all HLA alleles. The ssDNA PCR products are replicatively dot-blotted onto membranes, and each membrane is hybridized to a tagged probe specific for a different HLA allele. The hybridization pattern at a given coordinate on multiple membranes yields the HLA haplotype for a given individual (54 individuals in the example shown). For example, we see here that individual A6 expresses HLA-A*0101 and HLA-A*0301 but not HLA-A*0201. In contrast, individual D6 expresses HLA-A*0201 and HLA-A*0301 but not HLA-A*0101. **(B)** In PCR-SSP, an individual's DNA sample is incubated with various PCR primers that are complementary to allele-specific sequences within an HLA gene such as HLA-A. In the example shown, the individual's samples have been mixed (separately) in a PCR reaction mixture with PCR primers specific for HLA-A*0101, *0201, *0301, or *0401, as indicated by sample numbers 1, 2, 3 and 4, respectively. After time is allowed for hybridization and PCR amplification, each sample is subjected to gel electrophoresis to detect any PCR amplification product produced. The presence of a band on the gel that corresponds to the known size of the DNA fragment amplified by a given allele-specific primer indicates that the individual expresses the corresponding HLA-A allele. Here we see that samples 2 and 3 from the individual being typed give rise to bands on the gel of the sizes expected for HLA-A*0201 and HLA-A*0301, respectively. In contrast, no PCR products result from the use of primers specific for HLA-A*0101 or HLA-A*0401, indicating that the individual does not express these alleles. [*Adapted from Welsh, K. & Bunce, M. (2001) New methods in tissue typing. In Thiru, S. & Waldmann, H., eds.* Pathology and Immunology of Transplantation and Rejection. *Oxford: Blackwell Scientific Ltd..*]

results obtained using these animals. However, the health and reproductive capacity of inbred mice often decreases as genetic variability decreases, an effect known as *inbreeding depression*. Inbreeding depression results from the fact that all individuals harbor recessive, deleterious genetic defects that are masked by the normal, dominant allele. As inbreeding progresses, some of the disadvantageous alleles become homozygous, leading to reduced fitness, infertility, or death. Thus, many inbred lines die out before complete inbreeding is achieved.

Another type of mouse strain important in immunological research is the *congenic* mouse. Two congenic strains are identical at every genetic locus except one. The genotypes of outbred, inbred and congenic mice are compared schematically in **Figure F-14, panel A**. The creation of congenic mice via selective breeding between two different inbred strains is shown schematically in **Figure F-14, panel B** and described in its legend.

The use of inbred and congenic mouse strains has permitted scientists to dissect important aspects of the major histocompatibility complex, tumor rejection, organ transplantation, and autoimmunity, among other immunological phenomena. Such mice have also been used for similar investigations in other fields when genetic variability would be a confounding factor.

ii) Gene Targeted Mouse Strains

Many important discoveries in immunology have come about by investigating the molecular basis of diseases of the immune system that arise randomly in humans and mice. Sometimes it has been possible to uncover the role of a particular gene product in immune system development or function by observing what happens when a mutation renders that gene non-functional. However, waiting for a random mutation to arise in every gene of interest is not practical. In the 1980s, several research groups independently developed a technique that allowed them to deliberately mutate and inactivate a single gene in a mouse embryo and establish a mouse strain with the desired genetic deficiency. When such a "null" mutation causes the function of the gene to be totally lost, the animals are said to be **knockout mice** with respect to that gene. When the mutation is manipulated to take effect only during a certain developmental stage, or occur only in a certain tissue, the animals are said to be "conditional" knockout mice. For example, CD19 is a gene expressed almost exclusively in B cells. A mouse can be created in which the enzyme mediating the deletion of a gene of interest is placed under the control of the CD19 promoter. Thus, the gene of interest will be efficiently deleted only when and where the CD19 promoter is activated. As a result, the gene of interest will be lost only in the B cells of the mutant mouse but not in its other leukocytes or non-hematopoietic tissues. When a targeted mutation alters the function of the gene instead of obliterating it, the animals are said to be "knock-in" mutants. For example, a mutated version of the cytochrome c protein was created in which the electron transport function of this molecule was retained, but its pro-apoptotic function was lost. The creation of knock-in mutants expressing the altered cytochrome c protein allowed researchers to study *in vivo* the effects on cellular apoptosis of the loss of this one function of cytochrome c. The many manipulated mouse strains created by these methods have been and continue to be important for many fields of biological study, but are particularly well suited to the study of immunology because many immunologically important genes can be disrupted without killing the mouse.

The ability to create gene knockout and knock-in mice was made possible by the discovery of pluripotent *embryonic stem (ES) cells*. These cells differ from HSCs in that ES cells have the potential to give rise to not only cells of the blood and immune system but to *any* cell type found in the body, including germ cells. In addition, mouse ES cells can be cultured indefinitely *in vitro* under certain growth conditions, and plasmid DNA can readily be introduced into them. **Figure F-15** illustrates how targeted disruption of mouse ES cells is used to create a knockout mouse in which the gene of interest is rendered non-functional in every tissue of the animal's body. In contrast **Box F-1** describes how *RNA interference* techniques offer an alternative means of achieving targeted disruption of gene expression without the use of knockout mice.

(A) Types of Mouse Populations

(B) Breeding Congenic Mouse Strains

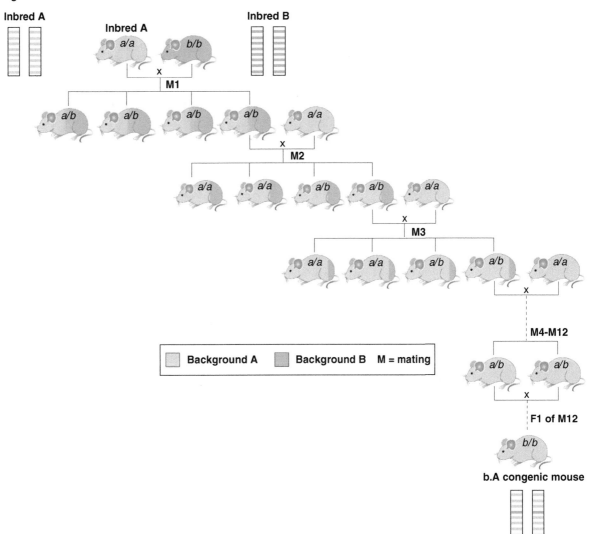

Background A   Background B   M = mating

Fig. F-14
Inbred and Congenic Mouse Strains

(**A**) Schematic comparison of the genotypes of outbred, inbred and congenic mice at genetic loci encoding polymorphic proteins, with different colors representing different alleles. At such loci, an outbred mouse is often heterozygous, and individuals within the population are often allogeneic. In contrast, an inbred mouse is homozygous at all genetic loci, and individuals within the population are syngeneic at all loci. A congenic mouse strain is genetically identical to an established inbred population with the exception of a single genetic locus at which these animals have been especially bred to be allogeneic. (**B**) Breeding process to obtain a hypothetical congenic mouse strain from two inbred mouse strains of genotype A (red) and genotype B (blue). B has a particular trait (b) that is determined by a single genetic locus that can be selected for, and that a researcher would like to have expressed in strain A mice; that is, he/she wants the B trait (b) on the A genetic background. First, A and B mice are crossed to produce AB progeny that are heterozygous at all loci: 50% of their genes are from A, and 50% are from B. The AB progeny with the desired B trait (b) are then bred back, or *backcrossed*, to strain A mice, and the progeny are again examined for the b characteristic. Those progeny with the b trait are identified as AB (now 75% A and 25% B genes) and are backcrossed again to strain A mice, and the AB progeny (87.5% A, 12.5% B genes) are again selected based on the presence of the desired b trait. Backcrossing is repeated for many generations, until the resulting mice are identical to strain A in every way except for the selected b trait. These congenic mice can then be used to study the effect of the B-derived gene, b, in the A genetic background.

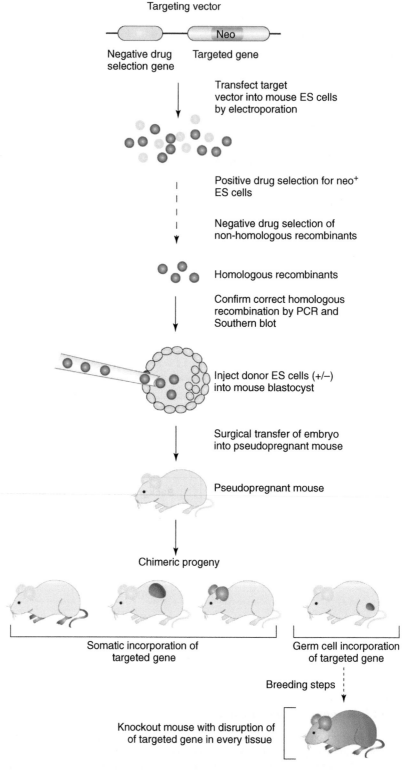

Targeting vector

Neo

Negative drug
selection gene

Targeted gene

Transfect target
vector into mouse ES cells
by electroporation

Positive drug selection for neo+
ES cells

Negative drug selection of
non-homologous recombinants

Homologous recombinants

Confirm correct homologous
recombination by PCR and
Southern blot

Inject donor ES cells (+/−)
into mouse blastocyst

Surgical transfer of embryo
into pseudopregnant mouse

Pseudopregnant mouse

Chimeric progeny

Somatic incorporation of
targeted gene

Germ cell incorporation
of targeted gene

Breeding steps

Knockout mouse with disruption of
of targeted gene in every tissue

Fig. F-15
Generation of Knockout Mice

To create ES cells bearing a targeted gene disruption, one needs to first build a targeting vector containing
the genomic clone of the gene of interest. Two "arms of homology," which are sequences from the genomic
region surrounding the gene of interest, are also included in the construct to facilitate insertion of the vector
into the desired part of the ES cell genome. The copy of the gene in the targeting vector is then disrupted
by inserting (into a part of the gene essential to its function) a selectable drug resistance marker such as the
neomycin resistance gene (*neo*). The targeting vector plasmid is introduced into the ES cells by electroporation, a method of transfection that uses electric current to make small pores in the ES cell membrane through

Box F-1 RNA Interference

Knockout mice are extremely important tools for determining the consequences of the loss of gene function, but generating them can take many months, if not years. In 1998, researchers discovered that cells can produce small double-stranded RNA molecules able to block the translation of mRNAs into proteins, effectively silencing the gene from which the mRNA was derived. This phenomenon was termed "post-transcriptional gene silencing," and the RNA molecules responsible were dubbed "micro RNAs" (miRNAs). Within a cell, the two strands of an miRNA molecule can separate, and the strand complementary in sequence to an mRNA binds to it and induces its degradation such that it is never translated. Such "RNA interference" plays a role in the regulation of gene expression and in protecting the cell against viruses. Scientists quickly learned to generate synthetic versions of miRNAs, which they called "small interfering RNAs" (siRNAs). Custom-made siRNAs can effectively inhibit the expression of a gene of interest, allowing very rapid determination (at least in isolated cells) of the consequences of the loss of expression of a particular gene. However, the blocking effect of siRNA is transient due to its rapid degradation by cellular nucleases.

A related technique uses "short hairpin RNA" (shRNA), which is a sequence of single-stranded RNA that can fold on itself to create a double-stranded RNA structure in the form of a hairpin. To deliberately "knock down" the expression of a gene of interest, an shRNA sequence is engineered into a viral vector that can integrate into the genome of the host cell such that the cell expresses a steady supply of single-stranded shRNA. As the shRNA is synthesized, it folds into its hairpin. However, this structure is quickly processed by the cell to generate a double-stranded siRNA in which one strand is complementary to an mRNA of interest and can bind to it, blocking its translation. Although shRNA is subject to rapid degradation by the same cellular nucleases that attack siRNA, because shRNA is continually generated by the cell into which it has been introduced, the silencing of the gene of interest is consistent and long-term. This vector-based approach has a downside, however, because if the shRNA vector integrates into the wrong place in the host cell genome, a deleterious mutation can result.

Such is the utility and importance of RNA interference techniques that Andrew Fire and Craig C. Mello shared the 2006 Nobel Prize in Physiology or Medicine for their pioneering work in this field.

which exogenous DNA can enter. Some of the cells will integrate the plasmid DNA into their genomes randomly (outside the sequence of the gene of interest), but a small proportion will incorporate the plasmid using *homologous recombination* between the arms of homology in the plasmid and the gene sequences of interest in the ES cell genome. The end result is the targeted disruption of one allele of the endogenous gene of interest by the insertion of the *neo* gene. Neomycin-resistant clones are then isolated by positive drug selection, in which neomycin treatment kills all cells that have failed to incorporate *neo*. Among the neomycin-resistant clones will be both cells that have integrated *neo* randomly and a small number of cells in which *neo* has been integrated into the gene of interest, disrupting it. These latter cells are correctly gene-targeted cells or *homologous recombinants*. Homologous recombinants can then be distinguished from random integrants by Southern blot or PCR screening. Sometimes these two groups of neomycin-resistant cells are distinguished by including a gene in the targeting plasmid *outside* the arms of homology that confers sensitivity to a second drug. This drug selection gene is then deleted from the ES cell only if homologous recombination occurs. Thus, if the targeting plasmid is randomly integrated, the second drug selection marker is retained in the ES cell, and the cell dies in the presence of the second drug (negative selection). However, homologous recombination of the targeting vector into the gene of interest deletes the second drug selection marker, rendering correctly gene-targeted ES cells resistant to the second drug. The gene-targeted ES cells are cultured to establish "knockout ES cell lines" mutated at one allele of a single gene. These cell lines can then be used to create "knockout mice" in the following way. The knockout ES cells are injected into a wild-type blastocyst (a 32-cell stage embryo) or cocultured with wild-type morulae (8- to 16-cell stage embryos). The targeted ES cells become incorporated into the embryo as it develops. These "blended" embryos are implanted into pseudo-pregnant female mice, which have been hormonally treated so as to create body conditions resembling pregnancy. The progeny of these implanted females will be chimeric in that some of their tissues will be derived from the original blastocyst or morula, but others will be derived from the injected knockout ES cells and will contain the disrupted allele. If a chimeric mouse happens to have germ cells derived from the ES cells, it can be bred to wild-type mice to generate progeny heterozygous for a targeted disruption in the endogenous gene. These heterozygotes are then interbred to produce F1 progeny, one-quarter of which are expected to be homozygous for the targeted mutation. The existence of the genetic disruption is confirmed by analysis of the DNA of the knockout mice. These null mutant mice will be deficient for the gene of interest in all their cells and tissues, and the effect of the loss of function of a single gene can thus be analyzed in a whole animal.

Would You Like To Read More?

Bontadini, A. (2012). HLA techniques: Typing and antibody detection in the laboratory of immunogenetics. *Methods (Duluth), 56*(4), 471–476.

Dickson, C. (2008). Protein techniques: Immunoprecipitation, *in vitro* kinase assays, and Western blotting. *Methods in Molecular Biology, 461*, 735–744.

Fall, B. I., & Niessner, R. (2009). Detection of known allergen-specific IgE antibodies by immunological methods. *Methods in Molecular Biology, 509*, 107–122.

Ghafouri-Fard, S., & Ghafouri-Fard, S. (2012). siRNA and cancer immunotherapy. *Immunotherapy, 4*(9), 907–917.

Howell, W. M., Carter, V., & Clark, B. (2010). The HLA system: Immunobiology, HLA typing, antibody screening and crossmatching techniques. *Journal of Clinical Pathology, 63*(5), 387–390.

Malou, N., & Raoult, D. (2011). Immuno-PCR: A promising ultrasensitive diagnostic method to detect antigens and antibodies. *Trends in Microbiology, 19*(6), 295–302.

Mayhew, T. M. (2011). Mapping the distributions and quantifying the labelling intensities of cell compartments by immunoelectron microscopy: Progress towards a coherent set of methods. *Journal of Anatomy, 219*(6), 647–660.

Morrison, H. W., & Downs, C. A. (2011). Immunological methods for nursing research: From cells to systems. *Biological Research for Nursing, 13*(3), 227–234.

Reid, M. E., & Denomme, G. A. (2011). DNA-based methods in the immunohematology reference laboratory. *Transfusion & Apheresis Science, 44*(1), 65–72.

Robbins, M., Judge, A., & MacLachlan, I. (2009). siRNA and innate immunity. *Oligonucleotides, 19*(2), 89–102.

Wasserman, J., Maddox, J., Racz, M., & Petronic-Rosic, V. (2009). Update on immunohistochemical methods relevant to dermatopathology. *Archives of Pathology & Laboratory Medicine, 133*(7), 1053–1061.

Wee, L. J., Lim, S. J., Ng, L. F., & Tong, J. C. (2012). Immunoinformatics: How in silico methods are re-shaping the investigation of peptide immune specificity. *Frontiers in Bioscience, 4*, 311–319.

Whitehead, K. A., Dahlman, J. E., Langer, R. S., & Anderson, D. G. (2011). Silencing or stimulation? siRNA delivery and the immune system. *Annual Review of Chemical & Biomolecular Engineering, 2*, 77–96.

Yan, Q. (2010). Immunoinformatics and systems biology methods for personalized medicine. *Methods in Molecular Biology, 662*, 203–220.

ABO antigens Family of glycoprotein antigens expressed on the surfaces of RBCs that define the ABO blood group system. Differences in ABO blood types can cause **hyperacute graft rejection** and severe blood **transfusion reactions**.

acquired immunodeficiency Loss of an immune system component due to an external factor such as a nutritional imbalance or a pathogen.

activation-induced cell death (AICD) Apoptotic death of lymphocytes subjected to prolonged antigenic stimulation. A means of eliminating effector cells after they are no longer needed.

active immunization Administration of an immunogen to an individual whose own body is then responsible for activating the lymphocytes required to provide defense against future attacks.

acute cellular rejection (ACR) A form of **acute graft rejection** in which allogeneic graft cells are destroyed by effector functions of recipient leukocytes.

acute disease A disease in which symptoms appear rapidly but remain for only a short time.

acute graft rejection Graft rejection occurring within days or weeks of a transplant. Can be cell-mediated (**acute cellular rejection**) or antibody-mediated (**acute humoral rejection**).

acute humoral rejection (AHR) A form of **acute graft rejection** due to production of antibodies directed against allo-MHC in the graft.

acute phase proteins Early inflammatory proteins made by hepatocytes. Acute phase proteins bind to cell wall components of microbes and activate complement. See also **inflammation**.

adaptive response An immune response mediated by uniquely specific recognition of a non-self entity by lymphocytes whose activation leads to elimination of the entity and the production of specific memory lymphocytes. Because these memory lymphocytes forestall disease in subsequent attacks by the same pathogen, the host immune system has "adapted" to cope with the entity.

addressins Cellular adhesion molecules expressed on **high endothelial venules (HEVs)** that mediate lymphocyte extravasation at particular sites in the body.

ADEPT =Antibody-directed enzyme/pro-drug therapy. Therapy employing an **immunoconjugate** in which the **monoclonal antibody** is linked to an enzyme capable of converting an inert pro-drug into an active cytotoxic drug. Used for anticancer therapy.

adjuvant A substance that, when mixed with an isolated antigen, increases its immunogenicity. Adjuvants provoke local inflammation, drawing immune system cells to the site and triggering maturation of DCs.

adoptive transfer The transfer of mature lymphocytes from one individual to another, most often by intravenous or peritoneal infusion.

adverse event (vaccine) Detrimental side effect associated with administration of a **vaccine**.

afferent lymphatic vessel A vessel that conveys **lymph** into a **lymph node**.

affinity A measure of the strength of the association established at a single point of binding between a receptor and its ligand. For **antibody**, the strength of the non-covalent association between a single antigen-binding site on the antibody and a single epitope on an **antigen.**

affinity maturation Positive selection of developing B cells with BCRs that have undergone **somatic hypermutation** resulting in increased **affinity** for specific antigen. Memory cells with this increased affinity for antigen are generated.

agglutination Aggregation of cells to form a visible particle.

AIDS =Acquired immunodeficiency syndrome. Failure of adaptive immunity due to T cell destruction caused by HIV infection.

alleles Two slightly different sequences of the same gene. Proteins produced from alleles have the same function.

allelic exclusion Cessation of somatic recombination in a second allele of an Ig locus due to productive recombination in the first allele.

allergen An antigen that is innocuous in most individuals but provokes **type I hypersensitivity** in some.

allergy Clinical manifestation of **type I hypersensitivity**. Mediated by mast cells armed with allergen-specific IgE.

alloantibodies Antibodies distinguishing between the different forms of a protein encoded by different alleles of a given gene. Often refers to antibodies specific for MHC molecules.

allogeneic Having different alleles at one or more loci in the genome compared with another individual of the same species.

allograft Tissue transplanted between allogeneic members of the same species.

allorecognition Recognition of allelic differences expressed by cells of one individual by the lymphocytes of another individual. Most often refers to recognition of MHC-encoded differences.

alternative complement activation Activation of the complement cascade initiated by direct binding of complement component C3b to a stabilizing ligand on a microbe. Involves cleavage and activation of Factors B and D and properdin.

anakinra Synthetic form of the natural antagonist IL-1Ra that binds to IL-1R and blocks its function, reducing inflammation.

anaphylatoxins Complement component cleavage products that have pro-inflammatory and chemoattractant effects; includes C3a, C4a and C5a.

anaphylaxis Systemic **type I hypersensitivity** response that may be fatal due to a catastrophic drop in blood pressure induced by the release of large quantities of inflammatory mediators.

anatomical barrier Body structure supplying non-specific, non-induced innate defense, such as intact skin.

anergy State of lymphocyte non-responsiveness to specific antigen. Induced by an encounter of the lymphocyte with cognate antigen under less than optimal conditions, such as in the absence of **costimulation**.

angiogenesis Process by which new blood vessels are formed.

antibody Secreted **immunoglobulin** that is produced by B lineage **plasma cells** and binds to specific antigens. Able to recognize antigens that are either soluble, or fixed in a tissue or on a cell surface.

antibody-dependent cell-mediated cytotoxicity (ADCC) An immune effector function that occurs when entities too large to be internalized are coated with antibodies which then bind to the **FcRs** of lytic cells such as NK cells or eosinophils, triggering **degranulation** and destruction of the entity.

antigen presentation Cell surface display of peptides in association with either MHC class I or II molecules allows T cell recognition of antigenic peptides on the surface of target cells or APCs, respectively.

antigen Entity (such as an element of an infectious pathogen, cancer or inert injurious material, or of a self tissue) that can bind to the antigen receptor of a T or B cell. This binding does not necessarily lead to lymphocyte activation.

antigen processing Degradation of a protein into peptides suitable for binding in the peptide binding grooves of either MHC class I or II molecules.

antigen receptor A cell surface receptor that recognizes an antigen.

antigenic determinant See **epitope**.

antigenic drift Subtle modification of pathogen antigens through random point mutations. Usually involves surface proteins that would normally be the target of neutralizing antibodies.

antigenic shift Dramatic modification of viral antigens due to reassortment of genomic segments of two different strains of a virus (that simultaneously infect the same individual) to generate progeny virions with new combinations of genome segments and thus new proteins.

antigen-presenting cell (APC) Cell expressing MHC class II and thus capable of presenting peptides to CD4⁺ Th cells. APCs are usually DCs, B cells or macrophages.

antiserum The clear liquid (serum) fraction of clotted blood containing antibodies produced when an individual or animal is exposed to a foreign substance or infectious agent.

antitoxins Antibodies made against bacterial **exotoxins** and **endotoxins**.

antiviral state A metabolic state of viral resistance induced in a cell by exposure to interferons released by neighboring virus-infected cells.

apoptosis The controlled death of a cell mediated by intracellular proteases that cause the orderly breakdown of the cell nucleus and its DNA. Death occurs without the release of internal contents and without triggering inflammation.

apoptotic body Small membrane-bound structure in which a cell that is dying by **apoptosis** packages its remaining organelles to prevent leakage of harmful enzymes or other contents.

Arthus reaction Localized **type III hypersensitivity** characterized by redness and swelling in the site of immune complex deposition.

atopy Signs and symptoms of **type I hypersensitivity** reactions.

autoantibody Antibody that recognizes a self epitope.

autoantigen Self antigen inducing an **autoreactive** response.

autoimmune disease Pathophysiological state in which the host's own tissues are damaged as a result of **autoimmunity**.

autoimmunity The response of an individual's immune system against self tissues. May cause **autoimmune disease** if unrestrained by mechanisms of **peripheral tolerance**.

autoinflammatory syndromes Inherited diseases in which patients suffer from recurrent severe local **inflammation** and prolonged periodic fevers in the absence of any obvious pathogenic cause. Caused by inappropriate activation of innate leukocytes mediating inflammatory responses.

autologous graft Transplantation of tissue from one part of an individual's body to another part of his or her body.

autophagy Engulfment process by which entities in the cytoplasm of a cell can be sequestered in endosomal compartments and digested, or delivered into the **exogenous antigen processing pathway**.

autoreactive Status of lymphocytes that recognize and are activated by self antigen are autoreactive.

avidity A measure of the total strength of all the associations established between a multivalent receptor and its multivalent ligand.

β2-microglobulin (β2m) Invariant chain of MHC class I proteins; encoded by a gene outside the **MHC**.

β-selection Process in which the TCRβ chain expressed by a DN3 **thymocyte** becomes part of a **pre-TCR** that allows the thymocyte to receive a survival/proliferation signal and become committed to the αβ T lineage. Required for a cell to pass the **pre-TCR checkpoint**, the first major checkpoint in T cell development.

B-1 cells   A subset of B cells that resides primarily in the peritoneal and pleural cavities and produces **natural antibodies**. Considered part of the **innate response** because they do not differentiate into **memory cells**.

B7 ligands   B7-1 (CD80) and B7-2 (CD86) are ligands for the costimulatory molecule CD28 and the negative regulator of T cell activation CTLA-4.

bacillus Calmette–Guérin   Strain of *Mycobacterium bovis* used as the BCG vaccine against tuberculosis.

basement membrane   A layer composed of collagen and other proteins and sugars that supports an overlying layer of epithelial or endothelial cells.

basophil   Circulating granulocyte with an irregularly shaped nucleus. Contains granules that stain a dark blue color with basic dyes. Basophil granules contain heparin and vasoactive amines, as well as many enzymes capable of promoting inflammation.

B cell receptor (BCR) complex   Antigen–receptor complex of B lineage cells. Composed of a membrane-bound Ig (mIg) monomer plus the **Igα/Igβ** complex required for **intracellular signaling**.

B cell receptor blockade   **Anergization** of a B cell due to very large amounts of antigen that persistently occupy the BCRs without cross-linking them.

benign tumor   A mass formed by abnormally dividing cells that are well differentiated and well organized. Resembles the normal tissue from which it originated and is securely encapsulated and relatively slow growing. Causes death only by indirect means.

B lymphocyte (B cell)   A **leukocyte** that matures in a **primary lymphoid tissue** (**bone marrow**), becomes activated in **secondary lymphoid tissues**, and mediates **adaptive immunity** by differentiating into antibody-producing **plasma cells**.

bone marrow (BM)   The **primary lymphoid tissue** (in the adult) in which all lymphocytes and other hematopoietic cells arise.

bone marrow transplant (BMT)   Replacement of a patient's dysfunctional immune system by **transplantation** of healthy whole **bone marrow** from a donor.

booster   A second or subsequent immunization (**vaccination**) to stimulate **memory cell** production.

brachytherapy   Implantation of a metal "seed" containing a radioisotope to mediate radiation therapy internally.

bronchi-associated lymphoid tissue (BALT)   Lymphoid patches and cells in the mucosae of the trachea and lungs. See also **mucosa-associated lymphoid tissue (MALT)**.

cachexia   Wasting of the body due to uncontrolled cellular catabolism. Cachexia in cancer patients is induced by their high levels of TNF.

cancer   See **malignant tumor**.

cancer immunoediting   Involvement of antitumor immune responses in exerting selective pressure to shape the nature of a cancer. Occurs in three phases: elimination, equilibrium and escape.

cancer stem cells (CSCs)   Rare cells within a tumor that are capable of establishing a new tumor if transplanted into a suitable recipient.

cancer-testis antigens   Proteins that are normally expressed solely in spermatogonia and spermatocytes but become **tumor-associated antigens (TAAs)** when transformation causes them to be expressed on other cell types.

carcinogen   Any substance or agent that significantly increases the incidence of **malignant tumors** by mutating or deregulating **oncogenes**, **tumor suppressor genes** or **DNA repair genes**.

carcinogenesis   Multistep process by which malignant transformation of a cell culminates in cell growth deregulation and the formation of a **malignant tumor**. Steps are initiation, promotion, progression and malignant conversion.

CD marker   =Cluster of differentiation marker. A cell surface protein identified by the binding to that protein of a cluster of different antibodies. Each CD marker is given a unique numeric designation. Also known as a *CD molecule*.

CD1 molecules   Unconventional **antigen presentation** molecules that present non-peptide antigens (lipid or glycolipid antigens) to subsets of αβ and γδ T cells and NKT cells.

CD3 complex   Family of five **ITAM**-containing accessory chains necessary for **TCR** signaling and insertion of the TCR complex in the membrane. See also **TCR complex**.

cell-mediated immunity   Originally, adaptive immune responses mediated by effector actions of cytotoxic T lymphocytes. Now includes effector actions of NK and NKT cells

central memory T cells (Tcm)   Class of memory T cells that act as a long-term central reservoir of memory T cells in the lymph nodes and other secondary lymphoid tissues.

central tolerance   Mechanism that eliminates most **autoreactive** T and B cells during lymphocyte development. Established by **negative selection** processes in the thymus and bone marrow.

centroblasts   Rapidly proliferating, antigen-activated B cells that fill the **dark zone** of a **germinal center (GC)** and undergo **somatic hypermutation**.

centrocytes   Smaller, non-dividing cells arising from centroblasts that migrate to the **light zone** of the **Germinal center (GC)** and undergo **affinity maturation** and **isotype switching**. Give rise to **plasma cells** and memory cells.

chemokines   Cytokines that induce leukocyte migration. See also **chemotaxis**.

chemotaxis   The directed movement of cells along the concentration gradient of a **chemotactic factor**. Serves to draw neutrophils, lymphocytes, and other leukocytes from the circulation into an injured or infected tissue.

chronic disease A disease in which symptoms are experienced on an ongoing or recurring basis. See also **persistent infection**.

chronic graft rejection (CGR) Immune response against an **allograft** that occurs several months after transplantation and leads to loss of allograft function.

classical complement activation Initiated by the Fc region of an antigen-bound antibody binding to C1, followed by recruitment and cleavage of C4, C2, C3 and assembly of the terminal complement components. See also **complement**.

clathrin-mediated endocytosis Process in which binding of a soluble macromolecule to its complementary cell surface receptor induces cellular internalization via polymerization of clathrin, a protein component of the microtubule network.

clinical trials A series of controlled tests of a drug, vaccine or treatment in human volunteers that is used to determine if its safety and efficacy warrant licensing and use in the general population. Generally involves four phases from I to IV.

CLIP =Class II invariant chain peptide. A small fragment of **invariant chain** that sits in the MHC class II groove and prevents the premature binding of peptides.

clonal deletion The induction of **apoptosis** in B or T lymphocytes that have bound their **cognate** antigen.

clonal exhaustion A process by which an activated lymphocyte divides so quickly in the face of persistent antigen that its progeny burn out before memory cells have been generated.

clonal selection The activation of only those clones of lymphocytes bearing receptors specific for a given antigen.

cocooning (vaccination) **Herd immunity** established by vaccination within a single household.

codominance Expression of a given gene from both the maternal and paternal chromosomes.

cognate antigen An antigen known to be recognized by a given lymphocyte antigen receptor because it was used for the original activation of that lymphocyte.

cold chain (vaccination) Constant refrigeration of a labile vaccine from the point of manufacture and shipping to the point of use in the field.

collectins Soluble **pattern recognition molecules (PRMs)** that mediate pathogen clearance by **opsonization**, **agglutination** or the **lectin pathway of complement activation**. Includes **mannose-binding lectin (MBL)**.

colony-stimulating factor (CSF) Growth factor promoting the proliferation of granulocytes.

combination vaccine Single **vaccine** containing immunogens from several different pathogens.

combined immunodeficiency **Primary immunodeficiency** in which both T and B cell responses are compromised. NK and NKT cell functions may also be affected.

commensal organisms Beneficial microbes that normally inhabit the surfaces of skin and body tracts. Also called **microbiota** or **microflora**.

common lymphoid progenitor (CLP) Early descendant of **multipotent progenitors (MPPs)** that gives rise to B lineage cells.

common mucosal immune system (CMIS) A system of extended mucosal defense established when lymphocytes activated in one mucosal **inductive site** migrate to a large number of **effector sites** in various mucosal tissues.

common myeloid progenitor (CMP) Early descendant of **MPPs** that give rise to **myeloid cells**.

complement System of over 30 soluble and membrane-bound proteins that act through a tightly regulated cascade of pro-protein cleavage and activation to mediate cell lysis through assembly of the **membrane attack complex (MAC)** in a target cell membrane. Intermediates in the complement cascade play a variety of roles in antigen clearance.

complementarity-determining region (CDR) See **hypervariable region**.

conformational determinant A protein **epitope** in which the contributing amino acids are located far apart in the linear sequence but which become juxtaposed when the protein is folded in its native shape.

conjugate A structure resulting from the physical joining of two other structures. Two cells may form a conjugate, such as Th and B cells during the provision of B–T cooperation, or CTLs and target cells during cytolysis. Two molecules may also form a conjugate, such as the covalent joining of a small inorganic molecule to a protein carrier to induce an immune response.

conjugate vaccine A vaccine based on the covalent linkage of a pathogen carbohydrate epitope to a protein carrier such that it induces a B cell response to **Td antigens** rather than **Ti antigens.**

constant domain A domain of an Ig or TCR chain that is encoded by the corresponding **constant exon**. The constant domains have very little amino acid variability.

constant exon Exon encoding a **constant domain** of either an Ig or a TCR protein. A C exon is spliced at the mRNA level to a rearranged variable (V) exon to produce a transcript of a complete Ig or TCR gene.

constant region The relatively invariant C-terminal portion of an Ig or TCR molecule. Comprises the constant domains of all the polypeptides involved.

coreceptor Protein that enhances the binding of a primary receptor to a ligand. The T cell coreceptors CD4 and CD8 bind to non-polymorphic sites on MHC class II and I, respectively, that are outside the peptide-binding groove. This binding stabilizes the contact between the peptide-MHC complex (pMHC) and the TCR, and also recruits intracellular signaling enzymes.

cortex The outer layer of an organ such as the **thymus**.

cortical thymic epithelial cells (cTECs) See **thymic epithelial cells (TECs)**.

costimulation The second signal required for completion of lymphocyte activation and prevention of **anergy**. Supplied by engagement of CD28 by B7 (T cells), and of CD40 by CD40L (B cells).

coverage (vaccination) See **efficacy (vaccine)**.

cross-matching Analysis of a recipient's blood to ensure that he/she does not possess any pre-formed antibodies that could attack a graft from a particular donor. A positive cross-match occurs when an individual's blood contains pre-formed alloantibodies to one or more HLA molecules expressed on an **allogeneic** donor organ, indicating that the transplant should not proceed.

cross-presentation Transfers peptides from the **exogenous antigen processing pathway** into the **endogenous antigen processing pathway**.

cross-reactivity Recognition by a lymphocyte or antibody of an antigen other than the **cognate antigen**. Cross-reactivity results either when the same epitope is found on two different antigens or when two epitopes on separate antigens are similar.

C-type lectin receptors (CLRs) Pattern recognition receptors that act as **lectins** by binding to bacterial carbohydrates.

cutaneous immune responses Immune responses mediated by leukocytes in the **SALT** that can respond to antigen attacking the skin without necessarily involving a lymph node.

cytokine withdrawal Induction of activated T cell death due to insufficient levels of cytokines, particularly IL-2.

cytokines Low molecular weight, soluble proteins that bind specific cell surface receptors whose engagement leads to intracellular signaling, triggering the activation, proliferation, differentiation, effector action, or death of the cell. Synthesized by leukocytes and some non-hematopoietic cells under tight regulatory controls.

cytoplasmic domain Portion of a transmembrane protein that is in the cytoplasm of a cell and transduces the signal initiated by engagement of the **extracellular domain** into the cell's interior.

cytotoxic cytokines Cytokines that directly induce the death of cells. For example, TNF and LT.

cytotoxic T cell (Tc) One of two major subsets of naïve T lymphocytes.

cytotoxic T lymphocytes (CTLs) **Effector** progeny of an activated Tc cell. CTLs recognize and destroy target cells displaying foreign peptide complexed to MHC class I. Target cell killing occurs via **cytotoxic cytokine** secretion, **Fas** ligation, or **perforin/granzyme-mediated cytotoxicity**.

damage-associated molecular patterns (DAMPs) Molecules released by stressed or dying cells that bind to **pattern recognition molecules (PRMs)** and induce inflammation.

dark zone Region of a **germinal center** where **somatic hypermutation** occurs.

DC licensing See **licensing**.

degranulation Extracellular release by a **granulocyte** of the destructive contents of its granules.

delayed-type hypersensitivity (DTH) Immunopathological damage occurring 24–72 hours after exposure of a sensitized individual to an antigen. Cell-mediated rather than antibody-mediated. See also **type IV hypersensitivity**.

delivery vehicle Inert, non-toxic structure designed to protect **vaccine** antigens from nuclease- or protease-mediated degradation. May also act as an **adjuvant** or increase antigen display. Includes liposomes, ISCOMs, virosomes and virus-like particles.

dendritic cells (DCs) Irregularly shaped phagocytic leukocytes with finger-like processes resembling neuronal dendrites (except plasmacytoid subset). DC subsets arise from both the myeloid and lymphoid lineages and include conventional and plasmacytoid DCs.

dermis Lower layer of skin beneath the **epidermis** and **basement membrane**. Contains lymphatics and blood vessels.

desmosomes Junctions between keratinocytes that hold a skin layer together.

direct allorecognition A transplant recipient's T cells recognize peptide/allo-MHC epitopes presented on the surfaces of **allogeneic** donor cells in the graft. See also **allorecognition**.

DN thymocytes =Double negative **thymocytes**. Thymocytes that express neither CD4 nor CD8 molecules. Includes DN1–4 subsets.

DNA repair genes Genes encoding elements of intracellular pathways that correct mutations in DNA to maintain genomic stability.

dome region Region of mucosal lymphoid follicle overlying a **germinal center**.

DP thymocytes =Double positive **thymocytes**. Thymocytes that express both CD4 and CD8 molecules.

draining lymph node Nearest lymph node that receives the lymph, antigens, lymphocytes and APCs emanating from a particular tissue.

early phase reaction In the **effector** phase of **type I hypersensitivity**, the rapid onset of clinical symptoms induced by pre-formed inflammatory mediators immediately released via **mast cell** degranulation.

effector cells The differentiated progeny of an activated leukocyte that act to eliminate a non-self entity. Includes **plasma cells**, **Th effector cells** and **CTLs**.

effector functions The actions taken by effector cells and antibodies to eliminate foreign entities. Includes cytokine secretion, cytotoxicity, and antibody-mediated clearance.

effector memory T cells (Tem) Class of memory T cells that is short-lived and circulates through non-lymphoid tissues where pathogens are likely to attack a second time. These cells can also enter sites of inflammation and infection.

effector site Remote mucosal location where lymphocytes activated in a mucosal **inductive site** differentiate and exert effector actions.

effector stage (hypersensitivity) The excessive, abnormal secondary response to an antigen (sensitizing agent) that results in inflammatory tissue damage. See **type I–IV hypersensitivities**.

efferent lymphatic vessel A vessel that conveys **lymph** away from a **lymph node**.

efficacy (vaccine) The ability of a **vaccine** to effectively protect individuals from disease. Expressed as the percentage of individuals vaccinated that develop immune responses to the pathogen. Also called "coverage."

ELISA =Enzyme-linked immunosorbent assay. Binder–ligand assay in which the antibody or antigen used is linked to an enzyme. The presence of binder–ligand pairs can be determined by adding a chromogenic substrate whose enzymatic conversion causes a detectable color change.

endocytic processing Use of an intracellular system of membrane-bound endosomes and endolysosomes containing hydrolytic enzymes and other substances to digest internalized materials. Responsible for **exogenous antigen processing and presentation**.

endogenous antigen An antigenic protein that originates within a cell in the host, as in a protein synthesized in a cell infected by a virus or intracellular bacterium.

endogenous antigen processing and presentation Mechanism by which **endogenous antigens** in the cytosol are degraded into peptides via **proteasomes** and complexed to MHC class I in the endoplasmic reticulum (ER). The peptide–MHC class I complex is then displayed on the cell surface. This pathway operates in almost all nucleated cell types.

endolysosome Intracellular membrane-bound structure formed by the fusion of a late endosome and a lysosome. Degrades internalized entities.

endosome Intracellular membrane-bound vesicle.

endotoxic shock A sometimes fatal collapse of circulatory and metabolic systems induced by overwhelming amounts of cytokines (particularly IL-1 and TNF) released in response to bacterial **endotoxins**.

endotoxin Toxin released from the cell walls of damaged **gram-negative bacteria**; lipid portion of **lipopolysaccharide (LPS)**.

eosinophils Connective tissue granulocytes with granules that stain reddish with acidic dyes. The granules contain highly basic proteins and enzymes effective in the killing of larger parasites. Eosinophils also play a role in **allergy**.

epidermis Top layers of skin that contain mainly **keratinocytes** plus elements of **SALT** but no blood vessels.

epitope The small region of a macromolecule that specifically binds to the **antigen receptor** of a B or T lymphocyte. B cell epitopes can be composed of almost any structure. T cell epitopes are a complex of antigenic peptide bound to either MHC class I or II.

epitope spreading An immune response against one **epitope** causes tissue destruction that exposes previously hidden epitopes, activating additional lymphocyte clones.

ER stress Cellular stress resulting when the endoplasmic reticulum (ER) machinery of a host cell is overheated due to excessive synthesis of large quantities of proteins, such as occurs during viral infection. Can trigger **apoptosis**.

erythrocytes Red blood cells (RBCs); carry oxygen to the tissues.

exocytosis Process by which the membrane of an exocytic vesicle fuses with the plasma membrane, everting the contents into the extracellular fluid.

exogenous antigen An antigenic protein that originates outside the cells of the host, as in a bacterial toxin.

exogenous antigen processing and presentation Mechanism by which **exogenous antigens** are internalized into an APC, degraded by **endocytic processing**, and complexed to MHC class II in an endolysosomal vesicle. The peptide–MHC class II complex is then displayed on the cell surface. This pathway operates almost exclusively in APCs.

experimental tolerance Lack of an immune response to a foreign antigen induced by treatment of a mature animal with either a non-immunogenic form of the antigen and using the usual dose and route of immunization, or an immunogenic form of the antigen and using a non-immunogenic dose or route of administration.

extracellular bacteria Bacteria that do not have to enter host cells to reproduce.

extracellular domain Portion of a transmembrane protein that is on the exterior of a cell and binds to a specific ligand.

extracellular pathogen A pathogen that does not enter host cells but reproduces in the interstitial fluid, blood or lumens of the respiratory, urogenital and gastrointestinal tracts.

extramedullary tissues Tissues and organs that are neither **lymphoid tissues** nor **bone marrow**.

extrathymic T cell development Development and maturation of T cells in tissues outside the **thymus**.

extravasation Exit of leukocytes from the blood circulation into the tissues in response to inflammatory signals.

Fab fragment/region =Fragment, antigen binding. Originally, the N-terminal portion of an Ig molecule left after digestion with papain. The Fab region contains the two antigen-binding sites of the antibody.

familial cancer Cancer that develops with increased frequency in genetically related individuals.

Fas Transmembrane death receptor (CD95) that induces cellular apoptosis upon engagement by Fas ligand (FasL).

Fc fragment/region =Fragment, crystallizable. Originally, the C-terminal portion of Ig molecule left after digestion with papain; crystallizes at low temperature. The Fc region contains the **constant region** of the antibody.

Fc receptor (FcR) Leukocyte receptor that binds to the **Fc region** of a specific antibody isotype. Engagement can trigger **clathrin-mediated endocytosis**, **phagocytosis**, **ADCC**, degranulation or cytokine release. Members of the FcR family are structurally diverse.

follicle-associated epithelium (FAE) Epithelium lying directly over single or aggregated lymphoid follicles. FAE is specialized for carrying out **transcytosis** due to the presence of **M cells**.

follicular dendritic cells (FDCs) Distinct lineage of DCs found in B cell-rich areas of lymphoid organs. FDCs do not internalize antigen and do not function as APCs but rather trap antigen–antibody complexes on their cell surfaces and display them for extended periods.

framework regions Relatively invariant parts of the V domain of an Ig or TCR chain that are outside the **hypervariable regions**.

fungus Eukaryotic organism that can exist outside a host but will invade and colonize if conditions permit. May be single-celled or multicellular.

gamma–delta (γδ) T cells T lymphocytes bearing γδ TCRs. Considered cells of innate immunity.

gamma–delta (γδ) TCRs Heterodimer of TCRγ and δ chains plus the **CD3 complex**. Rather than pMHC, γδ TCRs recognize antigens such as stress proteins and **heat shock proteins**. See also **TCR complex**.

gene segment A short, germline sequence of DNA from either the variable (V), diversity (D), or joining (J) families that randomly joins via **V(D)J recombination** with one or two other gene segments to complete a **V exon** in either the Ig or TCR loci.

genetic escape mutant Pathogen that evades neutralizing antibodies and antigen-specific CTLs due to antigenic variation arising from mutations that occur during replication.

germinal center (GC) Aggregations of rapidly proliferating B cells and differentiating memory B and plasma cells that develops in a **secondary lymphoid follicles**. Site of **isotype switching**, **somatic hypermutation** and **affinity maturation**.

graft rejection Attack by recipient immune system on transplanted donor tissue (a graft). See also **hyperacute, acute** and **chronic graft rejection (CGR)**. Also known as **transplant rejection.**

graft-versus-cancer (GvC) effect In BMT/HSCT for cancer treatment, destruction of residual cancer cells in a recipient's solid tumor by T cells from an **allogeneic** donor.

graft-versus-host disease (GvHD) Attack by immunocompetent cells in transplanted donated tissue on recipient tissues due to **MHC** or **MiHA** mismatches.

graft-versus-leukemia (GvL) effect In BMT/HSCT for cancer treatment, destruction of residual recipient leukemia cells by T cells from an **allogeneic** donor.

gram-negative bacteria Bacteria that have thin cell walls containing LPS.

gram-positive bacteria Bacteria that have thick cell walls containing peptidoglycan and lipoteichoic and teichoic acids.

granulocytes Myeloid leukocytes that harbor large intracellular granules containing microbe-destroying hydrolytic enzymes. Include **neutrophils, basophils** and **eosinophils**.

granuloma Structure formed by a group of **hyperactivated macrophages** that fuse together to wall off a persistent pathogen from the rest of the body. Also contains CD4$^+$ and CD8$^+$ T cells. Formation depends on TNF produced by activated Th1 effectors.

gut-associated lymphoid tissue (GALT) The Peyer's patches, appendix and diffuse collections of immune system cells in the linings of the small and large intestine. See also **mucosa-associated lymphoid tissue (MALT)**.

H-2 complex Murine **major histocompatibility complex (MHC)**.

haplotype (MHC) The set of MHC alleles contained on a single chromosome of an individual.

heat shock proteins (HSPs) Proteins whose expression is upregulated in cells subjected to environmental stresses such as heat or inflammation.

helper T cell (Th) One of two major subsets of T lymphocytes.

hematopoiesis The generation of hematopoietic cells from **HSCs** in the bone marrow or fetal liver.

hematopoietic cells Red and white blood cells (erythrocytes and leukocytes).

hematopoietic stem cell (HSC) Pluripotent hematopoietic precursor that either self-renews or differentiates into lymphoid, myeloid or mast cell precursors.

hematopoietic stem cell transplant (HSCT) Replacement of a damaged immune system by transplantation of isolated healthy HSCs from a donor.

herd immunity Protection of non-immune individuals in a population from a given pathogen due to effective vaccination of the majority of the population.

high endothelial venules (HEVs) Specialized post-capillary venules in most secondary lymphoid tissues that allow lymphocyte extravasation from the blood into these sites.

hinge region Site in the Ig monomer where the **Fab region** joins the **Fc region**.

histocompatibility Ability of a recipient to accept a tissue graft from another individual.

HIV = Human immunodeficiency virus. HIV infection causes AIDS.

HLA complex =Human leukocyte antigen complex. Human **major histocompatibility complex (MHC)**.

HLA typing Identification of HLA alleles expressed on an individual's cells. Used to determine the degree of HLA mismatching between a donor and recipient in a transplant situation.

Hodgkin's lymphoma **Lymphoma** in which the tumor mass is made up of a reactive infiltrate of non-transformed lymphocytes, macrophages and fibroblasts plus scattered, malignant Reed–Sternberg cells.

homeostasis The natural state of physiological balance of all organs, tissues and cells within a living organism.

homing receptors Receptors expressed by lymphocytes that direct lymphocyte trafficking by binding to specific **addressins** expressed on particular tissues at particular times.

humanized antibodies Engineered antibodies in which the V domains of non-human antibodies that recognize a human antigen are combined genetically with the C domains of human antibodies.

humoral immunity Adaptive immune responses mediated by B cells that differentiate into plasma cells producing antibodies.

hybridoma An immortalized cell secreting large amounts of pure **monoclonal antibody** of a single known specificity; created by fusion of a **myeloma** cell with an activated B cell producing antibody of known specificity.

hygiene hypothesis The theory that excessive zeal in preventing exposure to pathogens in infancy leads to a lack of activation of the immature immune system. The resulting bias toward Th2 responses may predispose an individual to hypersensitivity and/or autoimmunity.

hyperactivated macrophage See **macrophage**.

hyperacute graft rejection (HAR) Extremely rapid destruction of a graft almost immediately after transplantation. Mediated by classical **complement** activation initiated by pre-formed antibodies recognizing **allogeneic** epitopes on the graft vasculature. A form of **type II hypersensitivity**.

hypersensitivity See **immune hypersensitivity**.

hypervariable region Region of extreme amino acid variability in the V domain of an Ig or TCR chain. The hypervariable regions largely form the antigen-binding site. Also known as *complementarity-determining regions (CDRs)*.

hypodermis Fatty layer of the skin below the **dermis**.

Ig domain Protein motif of 70–110 amino acids with intrachain disulfide bond. Folds back on itself to form a characteristic Ig barrel (or Ig fold) structure. Mediates inter- and intramolecular interactions.

Ig superfamily Group of proteins in which each contains one or more **Ig domains**.

Igα/Igβ complex Accessory heterodimer within the **BCR complex**. Required to transduce intracellular signaling initiated by mIg engagement by antigen.

immediate hypersensitivity See **type I hypersensitivity**.

immune blockade Artificial interference with T cell negative regulatory molecules like CTLA-4 and PD-1 to liberate protective T cell responses, or with T cell costimulatory molecules like CD28 and CD40 to shut down harmful T cell responses.

immune complex Lattice-like structure composed of interlinked antigen–antibody complexes. Large immune complexes are insoluble and can become trapped in vessel walls or narrow body channels, provoking inflammation and **type III hypersensitivity**.

immune deviation Conversion of an adaptive immune response that is harmful to a less harmful response via a switch between Th1 and Th2 responses. A form of peripheral T cell control.

immune hypersensitivity Excessive immune reactivity to a generally innocuous antigen that results in inflammation and/or tissue damage. Includes **type I–IV hypersensitivities**.

immune privilege Property of certain anatomical sites in which immune responses are actively or passively suppressed.

immune regulation (tolerance) The absence of an immune response to a given antigen due to the regulation of the activities of effector leukocytes.

immune response A coordinated action by numerous cellular and soluble components in a network of tissues and circulating systems that combats pathogens, injury by inert materials, and cancers.

immune system A system of cells, tissues and their soluble products that recognizes, attacks and destroys internal entities that threaten to endanger the health of an individual.

immunity The ability to rid the body successfully of a foreign entity.

immunoconjugate Chimeric protein in which a whole **mAb** (or a structural derivative of that mAb) is linked to a cytokine, radioisotope or toxin either chemically or at the DNA level.

immunodeficiency A failure in some aspect of immune system function. See also **primary immunodeficiency** and **secondary immunodeficiency**.

immunodominant epitope The **epitope** against which the majority of antibodies is raised, or to which the majority of T cells responds.

immunogen An antigen that can induce lymphocyte activation.

immunoglobulin (Ig) Antigen-binding protein expressed by B lineage cells. An Ig monomer is composed of two identical light and two identical heavy chains (H_2L_2). In its plasma membrane-bound form, an Ig is the antigen-binding component of the **BCR complex**. In its secreted or secretory form, an Ig is an antibody.

immunological memory During the **primary adaptive response** to a pathogen, each antigen-specific lymphocyte clone that is activated generates many memory lymphocytes of identical specificity and greater affinity. When the same pathogen attacks the body a second time, it is eliminated more rapidly and efficiently by the **secondary response** mediated by the activation of these memory lymphocytes.

immunological synapse The contact zone between a T cell and an APC.

immunology The study of the cells and tissues that mediate **immunity** and the investigation of the genes and proteins underlying their function.

immunopathic damage Collateral damage to tissues caused by an immune response.

immunosuppressive molecules Molecules, such as the cytokines IL-10 and TGFβ or certain drugs, that reduce or eliminate immune responses.

immunosurveillance Concept that the immune system monitors the body for pathogens and tumor cells and destroys them.

immunotherapy The manipulation of the immune system to prevent, mitigate or cure disease.

indirect allorecognition Peptides derived from **allogeneic** proteins of the graft are presented to recipient T cells by recipient APCs. See also **allorecognition**.

induced fit Antigen influences the conformation of the antigen binding site of an antibody such that it better accommodates the antigen.

inducible nitric oxide synthetase (iNOS) Enzyme induced mainly in phagocytes by the presence of microbial products or pro-inflammatory cytokines. Generates nitric oxide, which is toxic to endocytosed pathogens.

inductive site A local area of the **mucosae** where antigen is encountered and a primary mucosal response initiates.

infection The attachment and entry of a pathogen into a host's body such that the pathogen successfully reproduces.

inflammasome Cytoplasmic multimeric complex that activates caspase enzymes required for the processing of certain pro-inflammatory **interleukins** into their active forms.

inflammation A local response at a site of infection or injury initiated by an influx of innate leukocytes that fight infections using broadly specific recognition mechanisms. Clinically, inflammation is characterized by heat and pain as well as swelling and redness. Also called *inflammatory response*.

innate response Non-specific and broadly specific mechanisms that deter entry or promote elimination of foreign entities. These include (1) physical, chemical and molecular barriers that exclude antigens in a totally non-specific way and (2) soluble and membrane-bound that recognize a limited number of molecular patterns that are common to a wide variety of pathogens or produced by host cells under stress.

intercellular adhesion molecule (ICAM) Molecules facilitating cell–cell and cell–matrix adhesion as well as extravasation. See also **homing receptors** and **addressins**.

interferons (IFNs) A major family of **cytokines**.

interfollicular region Tissue between **lymphoid follicles** positioned in a group. Contains mature T cells surrounding **HEVs**.

interleukins (ILs) A major family of **cytokines**.

intestinal follicles Lymphoid follicles located in the intestinal **lamina propria**, either grouped in **Peyer's patches** or in the appendix, or scattered singly. Composed of a **germinal center** containing B cells and **FDCs** topped by a **dome** containing **APCs** and T cells.

intracellular bacteria Bacteria that must enter host cells to reproduce.

intracellular pathogen Pathogen that spends a significant portion of its life cycle within a host cell; reproduction is usually within the host cell.

intracellular signaling The binding of a ligand to its receptor initiates a series of interactions between various proteins which culminate in the activation of transcription factors that enter the nucleus and alter the transcription patterns of genes controlling cellular proliferation, differentiation and effector functions.

intraepithelial lymphocytes (IELs) αβ and γδ T cells dispersed among the epithelial cells lining a body tract.

intraepithelial pocket Pocket within the **FAE** created by invagination of the basolateral surface of an **M cell**.

invariant chain (Ii) Transmembrane protein that binds to the peptide-binding groove of a newly synthesized **MHC class II molecules** and chaperones it out of the rER into the endocytic system. Upon cleavage, Ii gives rise to **CLIP**.

invasive pathogen Pathogen that is highly adept at entering and spreading within the body.

isograft Graft between two genetically identical individuals.

isotype switching Mechanism by which a B cell producing an Ig heavy chain of one isotype can then switch to producing Ig of the same variable region but a different constant region. See also **switch recombination**.

isotypes Classes of Igs defined on the basis of the amino acid sequences of their **constant regions**. Include IgM, IgD, IgA, IgG and IgE heavy chain isotypes and Igκ and Igλ light chain isotypes.

ITAM =Immunoreceptor tyrosine-based activation motif. Activatory sequence in receptor tails that facilitates recruitment of kinases promoting intracellular signaling.

ITIM =Immunoreceptor tyrosine-based inhibition motif. Inhibitory sequence in receptor tails that facilitates recruitment of phosphatases downregulating intracellular signaling.

iTreg cell CD4$^+$Foxp3$^+$ regulatory T cells that block conventional T cell activation via IL-10 and TGF secretion. Generated in the periphery from Th0 cells that have interacted with a mature DC.

IV-IG = Intravenous immunoglobulin. A pooled plasma preparation from thousands of healthy blood donors that is administered intravenously to an antibody-deficient patient as a form of immunoglobulin replacement therapy.

J chain =Joining chain. Small polypeptide that binds to the tail pieces of α and μ Ig heavy chains; stabilizes polymeric IgA or IgM.

junctional diversity Variation in the amino acid sequences of Igs and TCRs that arises during **V(D)J recombination**. Due to imprecise joining of gene segments as well as deletion and/or addition of nucleotides to the joint.

keratinocytes Epidermal squamous epithelial cells that produce keratin. Stratified layers are held together as units by **desmosomes**.

killed vaccine A vaccine in which the whole pathogen of interest is killed or inactivated by treatment with gamma irradiation or a chemical agent.

knockout mouse (KO) Mutant mouse strain in which a single gene in the DNA of a mouse embryo is deliberately deleted or rendered defective by genetic engineering techniques.

Kupffer cells Macrophages in the liver.

lamina propria Layer of loose connective tissue between the basolateral surface of the mucosal epithelium and the underlying muscle layer.

Langerhans cells (LCs) Epidermal DCs.

late phase reaction In the effector phase of **type I hypersensitivity**, leukocytes (especially eosinophils) that entered the site of allergen penetration in the **early phase reaction** release additional cytokines, enzymes and inflammatory mediators that do further damage.

latency An extended period during which a pathogen is present in the body but is non-infectious and does not cause clinical symptoms.

lectin A protein that binds to particular carbohydrate moieties on membrane glycoproteins or glycolipids on cell surfaces.

lectin-mediated complement activation Complement activation initiated by the binding of **mannose-binding lectin (MBL)** to pathogen monosaccharides. Involves the MASP complex and C4, C2 and C3. See also **complement**.

leukapheresis Removal of leukocytes from a donor's blood, followed by return of the remaining blood products to the donor.

leukemia "Liquid" malignancy of hematopoietic cells. Manifested as greatly increased cell numbers in the blood and bone marrow.

leukocytes White blood cells, including lymphocytes, granulocytes, monocytes, macrophages, NK and NKT cells.

licensing (DC) Th-induced upregulation of **dendritic cell (DC)** costimulatory molecules that allow the DC to mediate full activation of an antigen-stimulated Tc cell in the absence of further Th cell involvement.

light zone Region of the **germinal center** where **isotype switching** and **affinity maturation** occur.

linked recognition The requirement that B and T cell **epitopes** be physically linked to induce an efficient humoral response.

lipopolysaccharide (LPS) Component of **Gram-negative** bacterial cell walls that generates **endotoxin** and induces **endotoxic shock**.

low zone tolerance **Experimental tolerance** induced by administration of very small doses of immunogen over an extended period.

lymph Nutrient-rich interstitial fluid that bathes all cells in the body. Lymph filters slowly through the tissues, collects antigen, and eventually enters the **lymphatic system**.

lymph nodes Bean-shaped, encapsulated **secondary lymphoid tissues** located in clusters along the length of the lymphatic system. Contain the concentrations of the T and B lymphocytes and DCs required for primary adaptive immune responses.

lymphatic system A network of lymphatic capillaries, vessels and trunks through which lymph and lymphocytes travel. The lymphatic trunks connect with certain veins of the blood circulation.

lymphoblasts Daughter cells immediately produced by the division of an activated naïve lymphocyte. Lymphoblasts rapidly proliferate and differentiate into effector lymphocytes.

lymphocyte Small round leukocytes with a large nucleus and little cytoplasm. Concentrated in the **secondary lymphoid tissues** while in the resting state. Two major types exist, T cells and B cells, which are responsible for adaptive immune responses.

lymphocyte recirculation Continual migration of lymphocytes from the tissues back into the blood via the **lymphatic system**, followed by return to the tissues via **extravasation** from the blood circulation.

lymphoid cells T and B lymphocytes, **NK cells** and **NKT cells**.

lymphoid follicles Organized spherical aggregates of lymphocytes.

lymphoid organs Encapsulated groups of lymphoid follicles.

lymphoid patches Groups of lymphoid follicles that are not encapsulated.

lymphoid tissues Specialized anatomical regions containing lymphocytes, such as the spleen, tonsils and lymph nodes.

lymphoma Solid malignancy of **lymphoid cells** arising in a structured or diffuse secondary lymphoid tissue rather than in the blood or bone marrow. Classified as **Hodgkin's** or **non-Hodgkin's**.

lymphopoiesis The process of **HSC** differentiation into **lymphoid cells**.

lysozyme Protease that digests particular bacterial cell wall components. Found in lysosomes, neutrophil granules, tears and body secretions.

M cells =Membranous or microfold cells. Large epithelial cells with an **intraepithelial pocket**. **Transcytose** antigens from a body tract lumen across the epithelial layer. Most M cells reside in the **dome** overlying **intestinal follicles**.

macroautophagy **Autophagy** involving preliminary formation of an isolation membrane in the cytoplasm that circularizes around a large cytosolic entity to form an autophagosome. Fusion of the autophagosome with a lysosome forms an autophagolysosome in which the entity is degraded.

macrophage Powerful **phagocyte** that also secretes a large array of proteases, cytokines and growth factors and can act as an **APC**. In the presence of high levels of IFNγ, an activated macrophage becomes "hyperactivated" and acquires enhanced antipathogen activities and the capacity to kill tumor cells.

macropinocytosis Internalization of extracellular fluid containing soluble macromolecules. Droplets are engulfed and form **macropinosomes** that undergo **endocytic processing**.

macropinosome Membrane-bound vesicle formed by plasma cell invagination around a droplet of extracellular fluid. See also **macropinocytosis**.

major histocompatibility complex (MHC) Region of the genome containing genes encoding the chains of the **MHC class I, class II** and **class III proteins**. MHC class I and class II proteins combine with antigenic peptides and display them on the surface of host cells for recognition by T cells. The MHC class III genes encode various proteins important in complement activation, inflammation and stress responses.

malignant tumor =Cancer. A mass formed by abnormally dividing cells. Appears disorganized, is rarely encapsulated, and may undergo **metastasis**. Directly lethal to the host unless removed or killed.

mannose-binding lectin (MBL) A serum **collectin** that specifically binds to distinctive mannose structures on microbial pathogens in the blood. Engagement of MBL triggers **lectin-mediated complement activation**.

mast cell progenitor (MCP) Early descendant of **MPPs** that gives rise to mast cells.

mast cells Leukocytes with granules containing preformed mediators such as histamine. Mast cell degranulation is important for **inflammation** and **allergy**.

maternal-fetal tolerance Tolerance of a mammalian mother for her fetuses despite their expression of **allogeneic** paternal MHC.

medulla Inner region of an organ such as the **thymus**.

medullary thymic epithelial cells (mTECs) See **thymic epithelial cells**.

megakaryocytes Multinucleate **myeloid** leukocytes giving rise to platelets.

membrane attack complex (MAC) Pore-shaped structure assembled in the membrane of a pathogen or target cell as a consequence of **complement** activation. Facilitates osmotic imbalance and lysis.

membrane-bound Ig (mIg) Cell surface form of Ig molecule. **BCR** antigen-binding moiety.

memory cells Lymphocytes generated during a **primary immune response** that remain in a quiescent state until fully activated by a subsequent exposure to specific antigen (a **secondary immune response**).

metastases Secondary tumors established by **metastasis** of a **primary tumor** to sites in a different organ or tissue.

metastasis Process by which malignant cells break away from a **primary tumor** and spread via the blood to secondary sites.

MHC class I proteins Cell surface proteins composed of a polymorphic MHC class I α chain non-covalently associated with the invariant **β2-microglobulin** (β2m) chain. Expressed by most nucleated cells. Present peptide antigens to CD8+ T cells.

MHC class Ib and IIb proteins Non-classical MHC proteins with restricted polymorphism. Most are not involved in antigen presentation.

MHC class II proteins Cell surface proteins composed of a polymorphic MHC class II α chain and a polymorphic MHC class II β chain. Expressed on APCs. Present peptide antigens to CD4+ T cells.

MHC class III proteins Proteins encoded by the MHC class III loci. Include certain complement proteins, heat shock proteins, TNF and LT.

MHC-like proteins Non-polymorphic proteins encoded outside the MHC (e.g., CD1). Structurally and functionally similar to MHC proteins.

MHC molecules Proteins encoded by genes within the **major histocompatibility complex**. See also **MHC class I, Ib, IIb, II,** and **III proteins** above.

MHC restriction The principle that a T cell recognizing a given antigenic peptide presented by a particular MHC molecule will not recognize the same peptide if presented by a different MHC molecule.

MICA, MICB =MHC class I chain related A and B. Human stress ligands upregulated in response to heat, infection or transformation.

microbiota See **commensal organisms**.

microbiota-associated molecular patterns (MAMPs) Common molecular patterns borne by **microbiota**. May engage certain PRRs in the same way DAMPs or PAMPs engage other PRRs, enhancing inflammation.

microflora See **commensal organisms**.

microglia Macrophages in the brain.

MIICs =MHC class II compartments. Late endosomal compartments that are part of the **exogenous antigen processing** pathway.

minor H peptides   Peptides derived from donor **minor histocompatibility antigens**.

minor histocompatibility antigens (MiHA)   Proteins that exist in a small number of different allelic forms in a population. In a transplant situation, peptides derived from **allogeneic** MiHA of the donor are recognized as nonself by the recipient's T cells. Generally invoke slower, weaker **graft rejection** than MHC incompatibilities.

minor histocompatibility loci   Genes encoding **minor histocompatibility antigens**.

mitogen   Molecule that non-specifically stimulates cells to initiate mitosis.

molecular mimicry   Concept that a pathogen **epitope** may resemble a self epitope closely enough to activate an **autoreactive** lymphocyte, if the appropriate **cytokine** microenvironment is present.

monoclonal antibody (mAb)   Antibodies of a single, known specificity produced by a single B cell clone. See also **hybridoma**.

monocyte   Myeloid cell in the blood that enters the tissues and matures into a macrophage.

monomorphic gene   A gene that exists in a single allelic form within a species so that all individuals share the same nucleotide sequence at that locus.

mucosa-associated lymphoid tissue (MALT)   Collections of APCs and lymphocytes in the mucosae. Includes the **gut-associated lymphoid tissue (GALT)**, the **bronchi-associated lymphoid tissue (BALT)** and the **nasopharynx-associated lymphoid tissue (NALT)**.

mucosae   Mucosal epithelial layers that cover the luminal surfaces of body passages such as the gastrointestinal, respiratory and urogenital tracts. Singular is "mucosa."

mucosal immune responses   Immune responses mediated by immune cells in the **MALT** that can respond to antigen attacking the **mucosae** without necessarily involving a **lymph node**

lymph node.   Viscous fluid that coats the surface of cells of the mucosae. Contains secretory antibodies and antimicrobial molecules.

multipotent progenitor (MPP)   Descendant of **HSCs** that gives rise to **NK/T**, **CLP** and **MCP** precursors.

myeloablative conditioning   Complete elimination of a patient's hematopoietic cells in the bone marrow using chemotherapy and irradiation, leading to eventual depletion of immune system cells from the peripheral blood and all secondary lymphoid tissues.

myeloid cells   Cells that develop from common myeloid progenitors (CMPs). Include erythrocytes, **neutrophils**, **monocyte/macrophages**, **eosinophils**, **basophils** and **megakaryocytes**.

myeloma   **Plasma cell** tumor that secretes large quantities of an Ig protein of (usually) unknown specificity. When tumors are present in multiple body sites, the disease is referred to as *multiple myeloma*.

myelopoiesis   The process of **HSC** differentiation into myeloid cells.

naïve lymphocytes   Resting mature B and T cells that have not interacted with specific antigen; also known as *virgin* or *unprimed*.

naked DNA vaccine   A vaccine based on an isolated DNA plasmid (no vector) encoding the vaccine antigen.

nasopharynx-associated lymphoid tissue (NALT)   Mucosal lymphoid elements in the tonsils and upper respiratory epithelium. See also **mucosa-associated lymphoid tissue (MALT)**.

natural antibodies   Serum IgM antibodies that are produced by B-1 cells and generated without the need for exogenous antigen. Often recognize fungal components.

natural cytotoxicity   **Perforin/granzyme-mediated cytotoxicity** of target cells carried out by **NK cells**.

natural killer (NK) cells   **Lymphoid** lineage cells that recognize non-self entities with broad specificity. Activated when a target cell expresses ligands that bind to **NK activatory receptors** but lacks sufficient MHC class I to adequately engage **NK inhibitory receptors**. Sentinels of innate immunity.

natural killer T (NKT) cells   **Lymphoid** lineage cells that express a semi-invariant **TCR** recognizing glycolipid or lipid antigens presented on CD1d. Activated NKT cells quickly secrete cytokines that affect DCs, and NK, T and B cells. Sentinels of innate immunity.

necrosis   Sudden, uncontrolled cell death due to **infection** or trauma. The cell spills its contents into the surrounding milieu, releasing DAMPs that trigger **inflammation**.

negative selection   A **central tolerance** process that removes **autoreactive** cells from the developing B or T lymphocyte pool destined for the **peripheral tissues**. Based on high **affinity/avidity** of antigen receptors for self antigens.

neo-antigen   An antigen formed when a small, chemically reactive molecule binds to a self protein. Can precipitate contact hypersensitivity, a form of **type IV hypersensitivity**.

neonatal immunity   Immunity in the newborn due to maternal circulating antibodies passed on to the fetus via the placenta, or maternal secretory antibodies consumed by the newborn in breast milk. A form of **passive immunization**.

neonatal tolerance   Phenomenon that **tolerance** to an antigen is established more easily in neonatal than mature animals. Due to functional immaturity and low numbers of neonatal T and B cells, DCs, macrophages and FDCs, and altered lymphocyte recirculation.

neoplasm   See **tumor**.

neutralization   Ability of an antibody to bind to an antigen and physically prevent it from binding to and harming a host cell.

neutrophil extracellular trap (NET)   Microbe-trapping network formed from chromatin, histones and granule proteases released from a dying neutrophil.

neutrophils Most common **leukocytes**. Function as both **granulocytes** and **phagocytes**. Enter tissues from the circulation immediately in great numbers in response to injury or pathogen attack. Neutrally staining cytoplasmic granules.

NK activatory receptors Receptors whose engagement induces **natural cytotoxicity** and **cytotoxic cytokine** secretion by NK cells if not counteracted sufficiently by inhibitory receptor engagement.

NK inhibitory receptors Receptors whose engagement by self MHC class I molecules on a potential target counteracts the effects of **NK activatory receptor** engagement, preventing target cell destruction.

NK/T precursor **MPP**-derived precursor that can develop into T, NKT or NK cells but not B cells.

N nucleotides =Non-templated nucleotides. Nucleotides that are added randomly by **TdT** onto the ends of two antigen receptor gene segments undergoing **V(D)J recombination**.

NOD proteins =Nucleotide-binding oligomerization domain proteins. Member of the NLR family of cytoplasmic **PRMs** that detect products of intracellular pathogens. Engagement induces **inflammasome** formation and inflammatory cytokine production.

NOD-like receptors (NLRs) Cytoplasmic PRMs whose engagement induces **inflammasome** formation and inflammatory cytokine production.

non-Hodgkin's lymphoma Heterogeneous group of **lymphomas** in which the solid tumor mass consists almost entirely of malignant lymphocytes.

non-homologous end joining (NHEJ) Pathway of DNA repair in which double-stranded DNA breaks are joined in the absence of a homologous sequence to guide repair. Certain enzymes involved are also active in **somatic recombination**.

non-myeloablative conditioning Partial elimination of a patient's hematopoietic cells in the bone marrow using chemotherapy and irradiation, leading to eventual reductions of immune system cells in the peripheral blood and secondary lymphoid tissues. Less toxic than myeloablative conditioning.

non-responder Individual who fails to mount an immune response to a foreign protein that provokes a strong response in other individuals.

non-selection In **central tolerance**, the apoptotic death of CD4$^+$CD8$^+$ **thymocytes** expressing TCRs with little or no affinity for self MHC. Also called "neglect."

nTreg cells CD4$^+$Foxp3$^+$ regulatory T cells that block conventional T cell activation via intercellular contacts rather than immunosuppressive cytokines. Generated in the thymus from DP **thymocytes**.

nuclear transcription factors Proteins that reside in or translocate into the nucleus, bind to gene promoters, and initiate or regulate mRNA transcription.

oncogene A gene whose deregulation is directly associated with **carcinogenesis**. Oncogenes often encode positive regulators of cell growth.

opportunistic pathogen A pathogen that does not cause disease unless offered an unexpected opportunity by a failure in host defense.

opsonin A host protein that coats a foreign entity such that it binds more easily to phagocyte receptors, enhancing **phagocytosis**.

oral tolerance Experimental **tolerance** induced by feeding of an **immunogen**.

osteoclasts Macrophages in bone.

paracortex In an organ such as the **thymus**, the layer of cells lying just under the cortex.

parasite A pathogen that depends on a host organism for both habitat and nutrition at some point in its life cycle. Includes protozoans, helminth worms and ectoparasites.

parenteral Administration of a substance by a non-oral route (injection).

passaging Repeated forced replication of a microorganism in cell cultures in a laboratory. A method of producing a live attenuated **vaccine**.

passive immunization Transfer of **antibodies** to a non-immune recipient to provide immediate protection against a particular pathogen.

pathogen Organism that causes **disease** in its host as it attempts to reproduce. Includes **extracellular bacteria, intracellular bacteria, viruses, parasites, fungi** and **prions**.

pathogen-associated molecular patterns (PAMPs) Structural patterns present in components or products common to a wide variety of microbes (but not host cells). Ligands for **pattern recognition molecules (PRMs)**.

pattern recognition molecules (PRMs) Proteins recognizing **PAMPs**. Soluble PRMs include the **collectins (MBL)**, **acute phase proteins** and **NOD proteins**. Membrane-bound PRMs are **pattern recognition receptors (PRRs)**.

pattern recognition receptors (PRRs) Widely distributed membrane-bound **PRMs** fixed in either the plasma membrane of a cell or in the membranes of its endocytic vesicles. Includes **Toll-like receptors (TLRs)** and scavenger receptors. Engagement of PRRs induces pro-inflammatory cytokines.

peptide vaccine A **vaccine** that uses a small antigenic peptide for immunization.

perforin/granzyme-mediated cytotoxicity Mechanism of apoptotic target cell destruction triggered when CTLs or NK cells degranulate to release granzymes (proteases) and perforin (pore-forming protein).

periarteriolar lymphoid sheath (PALS) A cylindrical lymphoid tissue surrounding each splenic arteriole. Populated by mature T cells, some B cells, plasma cells, macrophages and DCs.

peripheral tissues   Tissues and organs other than the bone marrow and thymus; also referred to as *the periphery*.

peripheral tolerance   Functional silencing or deletion of **autoreactive** peripheral lymphocytes that escaped elimination by **central tolerance**.

persistent infection   An infection in which the pathogen remains in the body for a prolonged period. May be **latent** or cause **chronic disease**.

phagocyte   Cell capable of carrying out **phagocytosis**.

phagocytosis   Process by which a **phagocyte** captures particulate entities by membrane-mediated engulfment.

phagolysosome   Vesicle formed by fusion of a **phagosome** with a lysosome during **phagocytosis**.

phagosome   Intracellular vesicle in which a captured entity is first sequestered during **phagocytosis**.

physiological barriers   Body elements supplying non-specific, non-induced innate defense, such as the low pH of stomach acid and hydrolytic enzymes in body secretions.

plasma   Fluid component of blood.

plasma cell dyscrasias   Hematopoietic cancers of plasma cells.

plasma cells   Terminally differentiated B cells that secrete **antibody**. May be short-lived (no isotype switching or somatic hypermutation) or long-lived (undergo isotype switching and somatic hypermutation).

plasmablasts   Proliferating progeny of an activated B cell. Become plasma cells.

plasmapheresis   A process by which an individual's blood is withdrawn and passed through a machine designed to remove antibody proteins. The machine then returns the treated blood to the patient.

platelet-activating factor (PAF)   A lipid inflammatory mediator that activates platelets.

P nucleotides   If two gene segments undergoing **V(D)J recombination** are nicked elsewhere than at their precise ends, a recessed strand end and an overhang are generated. The nucleotides added to fill the gaps on both strands are considered P nucleotides.

pMHC   The peptide-MHC complex formed by the covalent binding of peptide in the peptide-binding groove of an MHC molecule.

polyclonal antiserum   Antiserum that contains antibodies produced by many different B cell clones responding to different epitopes of an antigen.

polygenicity   The existence of multiple genetic loci encoding polypeptides with identical function.

poly-Ig receptor (pIgR)   =Polymeric immunoglobulin receptor. Receptor positioned on the basolateral surface of mucosal epithelial cells. Binds to the **J chain** in secreted polymeric Ig and facilitates **transcytosis** of the Ig into external secretions. Cleavage of pIgR leaves **secretory component** attached to the Ig molecule.

polymorphism   Existence of different alleles of a gene within a population.

polymorphonuclear leukocytes   Original name for neutrophils, referring to their irregularly shaped, multi-lobed nuclei.

positive selection   A **central tolerance** process that promotes the survival and maturation of developing B or T lymphocytes with the potential to bind to non-self antigen upon release into the periphery.

post-transplant lymphoproliferative disease (PTLD)   A complication of transplantation in which recipient B cells that are infected with EBV proliferate uncontrollably due to the immunosuppressive treatment necessary to prevent **graft rejection**.

preactivation   The state of a **provirus** that remains untranscribed in the host cell genome due to the absence of host cell stimulation.

pre-BCR   Complex composed of **surrogate light chain (SLC)**, a candidate μ chain, and the **Igα/Igβ** complex. The pre-BCR is inserted transiently in the membrane of a developing B cell to test the functionality of a particular heavy chain VDJ combination.

pre-clinical trials   A series of controlled tests of a drug, vaccine or treatment that is carried out in cell cultures or experimental animals and used to determine if the treatment should progress to **clinical trials** in humans.

pre-T alpha chain   Invariant TCRα-like chain expressed only in DN **thymocytes**. Used to test the functionality of candidate TCRβ chains. Not required for γδ T cell development. See also **pre-TCR**.

pre-TCR   Transient complex composed of a candidate TCRβ chain plus the **pTα** chain plus the CD3 chains. Used to test the functionality of a particular VDJ rearrangement in the TCRβ gene.

pre-TCR checkpoint   First major checkpoint in T cell development, dependent on **β-selection** and formation of a **pre-TCR** that allows a **thymocyte** to receive a survival/proliferation signal and become committed to the αβ T lineage.

primary follicles   Spherical aggregates of resting mature B cells, macrophages, and FDCs within B cell-rich regions of **secondary lymphoid tissues** such as spleen, lymph nodes and Peyer's patches.

primary immune response   The adaptive immune response mounted upon a first exposure to a non-self entity. The primary response is slower and weaker than **secondary** (or subsequent) **immune responses**.

primary immunodeficiency (PID)   Failure of a component of the immune system due to an inborn genetic mutation.

primary lymphoid tissue   Lymphoid tissues (**bone marrow** and **thymus**) where lymphocytes are generated and mature.

primary tumor   Original tumor mass established by the first transformed cell.

priming   First encounter of a naïve lymphocyte with specific antigen. Leads to a **primary immune response**.

prion Infectious protein of abnormal conformation. Prions spread by altering the conformation of their normal protein counterparts in infected brain, causing spongiform encephalopathies.

prophylactic vaccination See **vaccination (prophylactic)**.

proteasome Cytoplasmic organelle containing multiple proteases that digest proteins into peptides. Integral component of the **endogenous antigen processing and presentation pathway**. "Standard" proteasomes carry out housekeeping degradation of self proteins in all cells, while "immunoproteasomes" are induced in cells exposed to inflammatory cytokines and play a role in digestion of foreign proteins for **antigen presentation**.

protective epitopes In **vaccination**, **epitopes** of a pathogen that induce an immune response preventing subsequent infection by that pathogen.

protein tyrosine kinase (PTK) Enzymes that, when activated, phosphorylate the tyrosine residues of substrate proteins.

provirus Viral DNA that has been integrated into the host cell genome.

pus Cream-colored substance at a site of injury or infection. Accumulation of leukocytes that have died fighting infection.

R5 viruses HIV strains that bind to CCR5 and infect macrophages as well as CD4⁺ T cells.

RAG recombinases =Recombination activation gene recombinases. RAG-1 and RAG-2 mediate **V(D)J recombination**.

reactive nitrogen intermediates (RNIs) Nitrogen-derived free radicals that kill microbes.

reactive oxygen intermediates (ROIs) Oxygen-derived free radicals that kill microbes.

receptor editing Secondary rearrangements of Ig loci carried out during the maturation of an autoreactive B cell in an effort to alter its antigenic specificity so that it becomes non-autoreactive.

receptor for advanced glycation end products (RAGE) Pattern recognition receptor that binds to a range of extracellular **DAMPs** and triggers signaling that amplifies the innate response.

recombinant vector vaccine A vaccine in which the DNA encoding the vaccine antigen is incorporated into a vector that enters host cells and promotes translation of the vaccine antigen directly within them.

recurring chromosomal translocation A chromosomal translocation that appears in many different patients. Most often observed in hematopoietic cancers.

reduced intensity conditioning Partial elimination of a patient's hematopoietic cells in the bone marrow using chemotherapy and irradiation. More toxic than **non-myeloablative conditioning** but less toxic than **myeloablative conditioning**.

regulators of complement activation (RCA) Soluble and membrane proteins that bind to C4 and C3 products and interfere with **complement** activation.

regulatory T cells T cells that inhibit the responses of other immune system cells, including Th and Tc cells, by intercellular contact and/or immunosuppressive cytokine secretion. Includes **nTreg, iTreg, Tr1** and **Th3** cells.

repertoire Pool of antigenic specificities represented in the total population of T or B lymphocytes.

reservoir Non-human species or environmental niche in which a **pathogen** that normally infects humans can survive.

respiratory burst Significant increase in oxygen utilization by the NADPH oxidases that generate **reactive oxygen intermediates (ROIs)**. Observed during degranulation.

retinoic acid inducible gene-1-like receptors (RLRs) Cytosolic **PRMs** that recognize intracellular viral RNAs in the cytoplasm of the cell.

Rh disease Destruction of the erythrocytes of a fetus during the pregnancy of an Rh⁻ mother carrying her second (or subsequent) Rh⁺ fetus. Also known as *hemolytic disease of the newborn* or *erythroblastosis fetalis*.

rheumatoid factor Collection of autoantibodies directed against IgG molecules.

RSS =Recombination signal sequence. The 12-RSS and the 23-RSS flank germline V, D and J gene segments and align them for **V(D)J recombination**.

scavenger receptors (SRs) Pattern recognition receptors that bind to a wide variety of lipid-related ligands from pathogens and damaged or dying host cells.

SCID =Severe combined immunodeficiency disease. Family of PIDs characterized by a lack of T and B cell functions.

SC-IG = Subcutaneous immunoglobulin. A pooled plasma preparation from thousands of healthy blood donors that is administered subcutaneously to an antibody-deficient patient as a form of immunoglobulin replacement therapy.

secondary follicles Once the **primary follicles** are infiltrated by activated T and B cells, they become secondary follicles that form **germinal centers** and foster terminal differentiation of activated B cells into **memory** B and **plasma cells**.

secondary immune response A secondary response is mounted by antigen-specific **memory** lymphocytes activated by a subsequent exposure to a given non-self entity. Faster and stronger than the **primary immune response**.

secondary immunodeficiency A non-genetic, acquired failure in the immune system. Often caused by severe infection, immunosuppressive drugs, cancer therapy, trauma, or malnutrition. HIV infection is also a major cause and if untreated usually leads to development of **acquired immunodeficiency syndrome (AIDS)**.

secondary lymphoid tissues Peripheral lymphoid tissues inhabited by mature lymphocytes. Includes the **spleen**, **lymph nodes**, **mucosa-associated lymphoid tissues** and **skin-associated lymphoid tissues**. Adaptive immune responses are initiated in these sites.

secreted antibody (sIg) Soluble form of Ig serving as circulating antibody in the blood.

secretory antibody (SIg) Secreted antibodies containing **secretory component**. Present in body secretions such as tears and mucus.

secretory component (SC) Protein fragment of the **poly-Ig receptor** that remains associated with the Ig after cleavage of the receptor during Ig **transcytosis**.

sensitization (hypersensitivity) An abnormal primary response to an antigen (sensitizing agent) such that, in a subsequent exposure, the individual mounts an excessive or abnormal secondary response that causes disease rather than immunity. See also **effector stage (hypersensitivity)** and **immune hypersensitivity**.

serotype One strain or subtype of a species of microorganism with members being defined by the sharing of common antibody-inducing antigens.

siderophores Small iron-chelating molecules produced by microbes attempting to grow under conditions of low iron.

single nucleotide polymorphism (SNP) A difference between **alleles** that occurs at one nucleotide position in the relevant gene.

skin-associated lymphoid tissue (SALT) The diffuse collections of $\alpha\beta$ and $\gamma\delta$ T cells and DCs (**Langerhans cells**) in the **epidermis**, and $\alpha\beta$ T cells, fibroblasts, DCs, macrophages and lymphatic vessels in the **dermis**.

somatic hypermutation Introduction of random point mutations at an unusually high frequency into the V exons of Ig genes. Increases V region variability in Ig proteins.

somatic recombination Site-specific recombination of pre-existing V, D and J **gene segments** in the Ig and TCR loci to generate unique **variable (V) exons**. Also known as **V(D)J recombination**.

SP thymocytes =Single positive **thymocytes**. Thymocytes that express either CD4 or CD8 but not both.

spleen An organ in the abdomen containing **secondary lymphoid tissue**. Traps blood-borne antigens.

sporadic cancer A **malignant tumor** in humans that is caused by transformation of a somatic cell of a tissue and is not inherited.

structural isoforms Proteins that are encoded by the same gene but have slight differences in amino acid structure and function due to differential processing that occurs after gene transcription, such as membrane-bound Ig versus secreted Ig made by a given B cell.

subunit vaccine A vaccine based on an isolated pathogen component such as a viral protein or bacterial polysaccharide.

supermolecular activation cluster (SMAC) Three concentric rings of molecules at the interface between an activated T cell and a DC. The inner *central SMAC* (cSMAC) contains aggregated TCRs and costimulatory molecules. The middle *peripheral SMAC* (pSMAC) contains signaling adaptors, integrins and adhesion molecules. The outer *distal SMAC* (dSMAC) contains actin-based cytoskeletal structures and large excluded proteins.

surrogate light chain (SLC) Two polypeptides called V_{preB} and $\lambda 5$ assemble non-covalently to form the SLC. The SLC participates in the **pre-BCR**.

switch recombination Mechanism of Ig **isotype switching**. The switch region of the C_H exon originally in a VDJ-C gene pairs with the switch region of a downstream C_H exon such that the intervening C_H exons are excised. The VDJ exon is then joined to the new C_H exon. Cytokines influence which C_H exon is selected.

syngeneic Individuals are syngeneic at a given genetic locus if they have the same alleles at that locus.

T4 count Number of CD4$^+$ T cells in a cubic millimeter of blood. A normal T4 count is 800 CD4$^+$ T cells/mm^3, whereas late-stage AIDS patients often have a T4 count of less than 200 CD4$^+$ T cells/mm^3.

tailpiece Short C-terminal domain in the heavy chains of **secreted antibody**.

TAP =Transporter of antigen processing. A complex positioned in the rER membrane that transfers peptides from the cytosol into the rER lumen for loading onto MHC class I.

tapasin Protein in the ER that links TAP and the MHC class I α chain.

target cell (cancer) The first cell that undergoes a tumorigenic mutation during the initiation step of **carcinogenesis**. Also called *cell of origin*.

target cell (cytolysis) An altered self cell, such as an infected cell, a cancer cell or a graft cell, that is destroyed by cytotoxicity or cytokine secretion mediated by CTLs or NK cells.

Tc cells Cytotoxic T cells that generally express the CD8 coreceptor and recognize non-self peptide presented on MHC class I. Upon activation, Tc cells differentiate into CTL effectors that kill **target cells** by **perforin/granzyme-mediated cytotoxicity** or by secretion of **cytotoxic cytokines**.

T cell help Provision of cytokines and/or costimulatory contacts by **Th cells** to **B cells** and **Tc cells** to support their activation.

T cell receptor (TCR) Antigen receptor expressed by T cells. Composed of an α and β chain, or a γ and δ chain.

TCR $\alpha\beta$ checkpoint Second major checkpoint in T cell development. **Positive selection** of **DP thymocytes** expressing a fully functional $\alpha\beta$TCR promotes the survival of thymocyte clones recognizing self-MHC alleles with moderate **avidity**, while **negative selection** deletes high avidity clones that are potentially **autoreactive**.

TCR complex   Complete antigen receptor of T lineage cells. Contains αβ or γδ TCR plus five CD3 chains.

T-dependent (Td) antigens   Antigens that bind to BCRs and initiate B cell activation but cannot induce B cell differentiation or Ig production without direct contact between the B cell and an activated Th cell. Td antigens are proteins containing both B and T cell epitopes.

TdT   =Terminal dideoxy transferase. Randomly adds non-templated **N nucleotides** onto the ends of nicked DNA strands in VD and DJ joints during **somatic recombination**.

terminal complement components   Complement components C5–9. Required for **MAC** formation.

tertiary lymphoid tissue   Site of inflammation that experiences an influx of leukocytes and comes to transiently resemble a secondary lymphoid tissue. Postcapillary venules in the site may take on the characteristics of **high endothelial venules (HEVs)**.

Th cells   Helper T cells that generally express the CD4 coreceptor and recognize non-self peptide presented on MHC class II molecules expressed by an APC. Effector Th cells, including the **Th1**, **Th2** and **Th17** subsets, are generated that secrete different panels of cytokines.

Th1 effector cells   Secrete IFNγ and IL-2. Generally combat intracellular bacteria and viruses, and induce isotype switching in humans to IgG1 and IgG3.

Th17 effector cells   Secrete IL-17 and IL-6. Generally mediate immune defense against certain bacteria and fungi, particularly at mucosal surfaces. Excessive normal function may contribute to autoimmune diseases and hypersensitivities.

Th2 effector cells   Secrete IL-4, IL-5 and IL-10. Generally act against extracellular bacteria and parasites, and induce isotype switching to IgA, IgE and IgG4.

therapeutic vaccination   See **vaccination (therapeutic)**.

thymic epithelial cells (TECs)   Stromal epithelial cells in the cortex (cTECs) or medulla (mTECs) of the **thymus**. Involved in positive and negative thymocyte selection.

thymic involution   After puberty, the replacement of the lymphoid components of the thymus with fatty connective tissue.

thymic selection   Processes of **positive selection** and **negative selection** that occur in the **thymus** and determine the specificities comprising the mature T cell **repertoire**.

thymocytes   T cell precursors developing in the thymus.

thymus   A small bilobed organ located above the heart, consisting of the medulla, the cortex, and the subcapsule. **Primary lymphoid tissue** for T cell maturation.

T-independent (Ti) antigens   Antigens that can stimulate B cells in the absence of T cell help but induce only very limited isotype switching, somatic hypermutation and memory B cell generation. Usually large polymeric proteins or carbohydrates with repetitive elements.

titer   Concentration of antigen-specific antibodies in an **antiserum**.

T lymphocyte (T cell)   A leukocyte that matures in a **primary lymphoid tissue** (**thymus**), becomes activated in **secondary lymphoid tissues**, and mediates **adaptive immunity** by differentiating into either **Th** or **Tc** cells.

tolerance   The absence of an immune response to a given antigen due to the prevention of lymphocyte activation. See also **experimental tolerance, central tolerance,** and **peripheral tolerance**.

tolerogen   An experimental foreign antigen that is recognized by a T or B lymphocyte but **anergizes** rather than activates these cells.

tolerogenic DCs   A DC that **anergizes** rather than activates naïve T cells. Tolerogenic DCs may express costimulatory molecules but cannot deliver T cell activation signal 2.

Toll-like receptors (TLR)   Membrane-bound **pattern recognition receptors (PRRs)** that bind to pathogen products such as **lipopolysaccharide (LPS)**.

tonsil   A network of cells that supports **lymphoid follicles** and **interfollicular regions** that are part of the **NALT**. The nasopharyngeal tonsil is called "the adenoids."

toxin   Pathogen-derived molecule that damages or kills host cells. See also **exotoxin** and **endotoxin**.

toxoid   Chemically inactivated pathogen **toxin** that retains its immunogenicity but not its toxicity.

Tr1 cells   CD4$^+$Foxp3$^-$ **regulatory T cells** that secrete IL-10 plus low amounts of TGFβ. Most are generated from naïve T cells that interact with a **tolerogenic DC**.

transcytosis   Transport of a molecule from one surface of a cell to its opposite surface, via an intracellular transport vesicle.

transfusion reaction   Destruction of transfused blood cells by pre-formed recipient antibodies specific for donor blood group antigens.

transgenic organism   Organism carrying one or more foreign genes in its DNA.

transmembrane domain   Portion of a transmembrane protein that spans a cell's plasma membrane and links the **extracellular domain** with the **cytoplasmic domain**.

transmigration   Process by which a leukocyte moves across an endothelial layer into the tissue below during **extravasation**.

transplant rejection   Attack by recipient immune system on transplanted donor tissue (a graft). See also **hyperacute, acute** and **chronic graft rejection**.

Treg cells   CD4$^+$CD25$^+$Foxp3$^+$ **regulatory T cells** that **anergize** other T cells non-specifically via intercellular contacts. May be **nTreg** or **iTreg** cells.

tumor   An abnormal tissue mass. May be **benign** or **malignant**. Also called *neoplasm*.

tumor-associated antigen (TAA)   Structurally normal protein or carbohydrate expressed in a tumor at a concentration, location or time that is abnormal relative to its status in the healthy, fully differentiated cells. TAAs are encoded by normal cellular genes that are dysregulated.

tumor-associated macrophages (TAMs) Macrophages found within a tumor.

tumor-infiltrating lymphocytes (TILs) CTLs, NK or NKT cells found within a tumor.

tumor regression Disappearance of a **malignant tumor**. May be induced by anticancer treatment or occur spontaneously.

tumor rejection Regression of a tumor induced by an obvious immune response.

tumor-specific antigen (TSA) A macromolecule that is unique to a tumor and not produced by any type of normal cell. TSAs are encoded by mutated cellular genes or by viral oncogenes.

tumor suppressor gene (TSG) Gene encoding a protein whose absence promotes **carcinogenesis**. Often encode negative regulators of cell growth or survival.

type I hypersensitivity **Hypersensitivity** arising from the synthesis of IgE antibodies directed against an antigen (sensitizing agent). These antibodies arm mast cells via FcεR binding. See also **allergy** and **anaphylaxis**.

type II hypersensitivity **Hypersensitivity** arising from direct antibody-mediated cytotoxicity. Target cells may be mobile (blood cells) or fixed as part of a solid tissue. Antibodies may be IgM or IgG. If autoantibodies are involved, the reaction may be a component of an **autoimmune disease**.

type III hypersensitivity Immune complex-mediated **hypersensitivity**. A soluble antigen forms large insoluble immune complexes with IgM or IgG in the circulation. Deposition of these complexes in narrow body channels triggers damaging inflammation. If autoantibodies are involved, the reaction may contribute to **autoimmune disease**.

type IV hypersensitivity Cell-mediated **hypersensitivity** in which Th cells activate Tc cells and macrophages that then damage or destroy host cells.

unprimed lymphocytes See **naïve lymphocytes**. Also known as *virgin lymphocytes*.

V(D)J recombination Site-specific recombination of pre-existing V, D and J **gene segments** in the Ig and TCR loci to generate unique **variable (V) exons**. Also known as **somatic recombination.**

vaccination (prophylactic) Administration of a non-pathogenic form of a pathogen or its components (the **vaccine**) prior to natural exposure in order to prime an adaptive response and generate pathogen-specific memory T and B cells. A natural exposure to the pathogen then triggers a secondary, rather than primary, response and protects against overt disease.

vaccination (therapeutic) Administration of a non-pathogenic form of a pathogen or its components (the **vaccine**) to induce an adaptive response that will cure or mitigate established disease, rather than prevent it.

vaccine A modified, non-pathogenic form of a natural **immunogen**. May be a killed, inactivated or **attenuated** form of the pathogen, or composed of pathogen proteins, DNA or other molecules.

vaccine-associated paralytic polio (VAPP) Poliomyelitis in an individual vaccinated with an oral, live attenuated poliovirus vaccine in which one of the attenuated viruses reverts to pathogenicity.

variable (V) domain The domain of an Ig or TCR chain that is encoded by the corresponding **variable (V) exon**. The variable domains have a high degree of amino acid variability.

variable (V) exon Exon encoding the **variable domain** of an Ig or TCR protein. V exons in the Igk, Igl, TCRA and TCRD loci are randomly assembled from V and J **gene segments**, while the V exons in the Igh, TCRB and TCRG loci contain V, D and J segments.

variable (V) region The highly variable N-terminal portion of an Ig or TCR molecule. Composed of the **variable domains** of all the polypeptides involved. Responsible for antigen recognition.

vasodilation An expansion in the diameter of local blood vessels. A characteristic feature of **inflammation**.

virgin lymphocytes See **naïve lymphocytes**. Also known as *unprimed lymphocytes*.

virulence Ability of a **pathogen** to invade host tissues and cause **disease**.

virus Submicroscopic acellular particle consisting of a protein coat surrounding an RNA or DNA genome. Must enter host cells to replicate.

X4 viruses HIV strains that bind to CXCR4 and infect CD4+ T cells but not macrophages.

xenograft Tissue transplanted between members of two different species. Also called a *xenotransplant*.

Index

Note: Page numbers with "f" denote figures; "t" tables; "b" boxes.

A

Abnormalities
  in APCs, 524
  in B cells, 524
  in cytokine production, 524
  in regulatory t cells, 523–524
Acquired immunodeficiency syndrome
    (AIDS), 17–18, 372, 377, 401, 403t
  impact of, 401–403
  infections and neoplasms in, 410t
Activation-induced cell death
    (AICD), 220–221
Activation-induced cytidine deaminase
    (AID), 126, 392
Acute cellular rejection (ACR), 465
Acute humoral rejection (AHR), 466
Acute lymphoblastic leukemia
    (ALL), 566–567
  clinical features, 567
  genetic aberrations, 567–568
  treatment, 568
Acute myeloid leukemia (AML), 561
  clinical features, 562
  genetic aberrations, 562–563
  treatment, 563–564
Acute rheumatic fever (ARF), 530
ADA. *See* Adenosine deaminase
ADCC. *See* Antibody-dependent cell-
    mediated cytotoxicity
Addressins, 52
Adenosine (Ado), 386b
Adenosine deaminase (ADA), 384
Adenylate kinase-2 (AK2), 385
ADEPT. *See* Antibody-directed enzyme/
    pro-drug therapy
Adjuvant system 04 (AS04), 347
Adjuvants, 173b
  alum, 346–347
  lipid-or oil-based, 347
  PRR ligands, 347–348
Ado. *See* Adenosine
Adult T cell leukemia/lymphoma
    (ATL), 579–580
Affinity, 106
  chromatography, 631f
  maturation, 106, 126f, 127
AFP. *See* Alpha-fetoprotein
Agglutination, 625f
AHR. *See* Acute humoral rejection
AICD. *See* Activation-induced cell death
AID. *See* Activation-induced cytidine
    deaminase
AIDS. *See* Acquired immunodeficiency
    syndrome
AIRE. *See* Autoimmune regulator

AK2. *See* Adenylate kinase-2
ALL. *See* Acute lymphoblastic leukemia
Alleles, 152–153
Allergen(s), 487–488, 488t
  airborne, 493
  allergen-specific immunotherapy,
    501–503, 503b
  antigen into, 496–497
  breaching mucosal barrier, 490f
  hypoallergenic, 503b
  peptides, 489
Allergenicity, 497b
Allergome, 496b
Allergy, 18
allo-MHC. *See* Allogeneic MHC molecule
Alloantibodies, 459
Alloantibody analysis
  alloantibody screening, 470–471
  cross-matching test, 471
Allogeneic, 153, 457–458
  HSCT, 560–561
Allogeneic MHC molecule
    (allo-MHC), 459
Allografts, 457–458
Alloimmune
  hemolytic anemias, 504–506
  thrombocytopenia, 506
Allorecognition, 459
Allorecognition, direct, 460
Alpha-fetoprotein (AFP), 430
Alum, 346–347
Alymphocytosis. *See RAG SCID* disease
AML. *See* Acute myeloid leukemia
Amplification, 67
Anakinra, 543–544
Anaphylatoxins, 71
Anaphylaxis, 369, 369b, 488, 495
Anatomical barriers, 9
Ancestral haplotypes, 154
Anergization, 229
Ankylosing spondylitis (AS), 157–158,
    532–533
Anthrax, 351
Anthrax vaccine adsorbed (AVA), 351
Anti-HLA antibodies, 633
Anti-IgE immunotherapy, 501
Anti-inflammatory drugs, 542
Antibody, 6, 11–12
  natural, 325
  synthesis mechanism, 129–130
Antibody diversity
  immunoglobulin chain pairing, 103–104
  junctional diversity, 102–103
  multiplicity and combinatorial
    joining, 102

by somatic recombination, 102
  total diversity estimates, 104
Antibody-dependent cell-mediated cyto-
    toxicity (ADCC), 131–133, 133f
Antibody-directed enzyme/pro-drug
    therapy (ADEPT), 447
Antibody–antigen interaction
  affinity, 106
  avidity, 107
  CDR regions, 106f
  cross-reactivity, 107–108, 107f
  intermolecular forces, 105
  structural requirements, 105
Antigen, 7
  PAMP *vs.* 8f
  presentation, 161
    MHC class Ib molecules, 176
    non-peptide antigen presentation,
      176–177
    pathway types, 161–162
    T cells, 162
  processing, 161
    cross-presentation, 161–162
    endogenous processing pathway,
      161–162, 162f
    exogenous processing pathway,
      161–162, 162f
  receptors, 7
  sampling
    FAE, 274–275
    GALT DCs, 275
Antigen-binding sites, 182, 192
Antigen-presenting cells (APCs), 12,
    162–163
  abnormalities in, 524
  comparison of professional, 167t
  dendritic cells, 163–168
  non-professional, 162–163, 163b
Antigen–antibody interaction, 627
Antigenic, 228
  drift, 312
  epitope, 105
  mimicry, 358b
  shift, 312, 313f
Antigenic determinant.
    See Antigenic—epitope
Antigens, protective, 339
Antihistamines, 501
Antipolysaccharide antibody deficiency
    (APAD), 392
Antiretroviral drugs
  chemokine receptor inhibitors, 416
  fusion inhibitors, 416
  ISTIs, 416
  NNRTIs, 415

Printed and bound by CPI Group (UK) Ltd, Croydon, CR0 4YY

03/10/2024

01040318-0019